Advances in Food Analysis

Advances in Food Analysis

Topical Collection Editors

Alessandra Gentili
Chiara Fanali

MDPI • Basel • Beijing • Wuhan • Barcelona • Belgrade

MDPI

Topical Collection Editors
Alessandra Gentili   Chiara Fanali
University of Rome   Università Campus Bio-Medico
Italy   of Rome
                                      Italy

Editorial Office
MDPI
St. Alban-Anlage 66
4052 Basel, Switzerland

This is a reprint of articles from the Topical Collection published online in the open access journal *Molecules* (ISSN 1420-3049) from 2018 to 2019 (available at: https://www.mdpi.com/journal/molecules/special_issues/food_analysis).

For citation purposes, cite each article independently as indicated on the article page online and as indicated below:

LastName, A.A.; LastName, B.B.; LastName, C.C. Article Title. *Journal Name* **Year**, *Article Number, Page Range.*

ISBN 978-3-03921-742-7 (Pbk)
ISBN 978-3-03921-743-4 (PDF)

Contents

About the Topical Collection Editors

Alessandra Gentili is Associate Professor of Analytical Chemistry at Sapienza Università di Roma, where she received her Master's degree, Magna cum Laude, in Industrial Chemistry and her PhD degree in Chemical Sciences. She is also Director of Sapienza's Research Centre HYDRO-ECO, which comprises four departments from the Faculties of Science and Engineering. Her research activity essentially concerns the study of original analytical methodologies aimed at solving problems in different areas of Chemistry, namely Clinical, Food, and Environmental Chemistry. The themes of her research include the development of original extraction procedures based on last-generation sorbent materials or neoteric solvents. The results of her research have been published over 96 publications, including 86 papers in international peer-reviewed journals and 10 chapters in international books. She is a member of the Editorial Board of *Molecules* (Section: Analytical Chemistry), *Current Analytical Chemistry*, and *Journal of Chromatography A* (Advisory Editorial Board).

Chiara Fanali is Associate Professor of Analytical Chemistry at Università Campus Bio-Medico of Rome (Italy), ResearchUunit of Food Science and Nutrition. In 2008, she received her PhD degree in Biochemical Studies of Proteome at Catholic University of Rome (Italy). Since February 2010 she has carried out her research at Università Campus Bio-Medico of Rome. Her research interests mainly concern the application of modern and innovative analytical techniques to the analysis and characterization of food bioactive compounds as well as peptides and proteins in biological fluids. The techniques employed in this research include high-performance liquid chromatography (HPLC) and nanoliquid chromatography coupled to such mass spectrometers as ion trap (IT), single quadrupole, and high-resolution linear ion trap Orbitrap and time of flight (TOF). Chiara Fanali is co-author of more than 80 publications in international journals (ISI indexed), 5 book chapters as well as proceedings in journals and contributions in national and international symposia. As of October 2019, there are 90 documents in Scopus database corresponding to 1728 citations and an h-index of 24.

Preface to "Advances in Food Analysis"

The interest in innovative and advanced analytical techniques has been growing in recent years due to the renewed necessity for analyzing complex matrices like foods. Knowing foods means being able to elucidate their constituent composition as well as to control contamination and preserve them from adulteration. Every single food is a very complex matrix whose chemical nature differs greatly with regard to constituents (amino acids, polysaccharides, proteins, lipids, nucleic acids, sterols, etc.) and concentrations, which can range from the micromole to femtomole scale. Besides the importance of nutrient characterization, there is deep interest in the definition of food nutraceutical properties. Another aspect of fundamental importance is the identification and quantification of residues resulting from different processes such as cultivation, fermentation, release from packaging, etc., in order to ensure high standards in quality assurance and process control. For all these reasons, analytical chemistry related to food analysis is a rapidly growing research area. Constant efforts have been devoted to developing more sensitive, fast, and cost-effective analytical methods to guarantee the safety, quality, and traceability of foods in compliance with legislation and consumer demands. Sample preparation is the first critical step of analysis, and innovative extraction techniques such as supercritical fluid extraction (SFE), microwave-assisted extraction (MAE), subcritical water extraction (SWE), QuEChERS (quick, easy, cheap, effective, rugged, and safe) methodology, ultrasound-assisted extraction have also been applied to the extraction of food constituents. Physical techniques employing powerful instrumentation—including spectroscopy, chromatography and electrophoresis, biochemical analysis, and sensory analysis techniques—have replaced the old methods used at the beginning of the 20th century. The advantages and drawbacks of each approach are always taken into consideration. This Topical Collection provides readers with a good overview of the current status and exciting developments in this field. It includes papers focused on modern analytical instrumentation, new methods and their application to food science, as well as works on quality control and safety, nutritional value, processing effects, storage, bioactivity, and so forth. We would like to thank all contributors and colleagues who chose to publish their works here as well as the reviewers who dedicated their time, effort, and expertise in evaluating the submissions and assuring the high quality of the published work. We would also like to thank the publisher, MDPI, and the editorial staff of the journal for their constant and professional support as well as for their invitation to edit this Special Issue.

Alessandra Gentili, Chiara Fanali
Topical Collection Editors

molecules
MDPI

Article

Transfer of a Multiclass Method for over 60 Antibiotics in Food from High Resolution to Low Resolution Mass Spectrometry

Danilo Giusepponi [1]**, Fabiola Paoletti** [1]**, Carolina Barola** [1]**, Simone Moretti** [1]**, Giorgio Saluti** [1]**,
Federica Ianni** [2]**, Roccaldo Sardella** [2]◉ **and Roberta Galarini** [1,*]

1   Istituto Zooprofilattico Sperimentale dell'Umbria e delle Marche "Togo Rosati", 06126 Perugia, Italy
2   Department of Pharmaceutical Sciences, University of Perugia, 06123 Perugia, Italy
*   Correspondence: r.galarini@izsum.it; Tel.: +39-075-343-272

Academic Editors: Alessandra Gentili and Chiara Fanali
Received: 21 July 2019; Accepted: 10 August 2019; Published: 13 August 2019

check for updates

Abstract: A multiclass method has been developed to screen and confirm a wide range of anti-microbial residues in muscle and milk, and validated using liquid-chromatography coupled to (low-resolution, LR) tandem mass spectrometry (LC-QqQ). Over sixty antibiotics, belonging to ten distinct families, were included in the method scope. The development process was rapidly concluded as a result of two previously implemented methods. This consisted of identical sample treatments, followed by liquid chromatography, and coupled with high-resolution (HR) mass spectrometry (LC-Q-Orbitrap). The validation study was performed in the range between 10–1500 $\mu g \cdot kg^{-1}$ for muscles and 2–333 $\mu g \cdot kg^{-1}$ for milk. The main performance characteristics were estimated and, then, compared to those previously obtained with HR technique. The validity of the method transfer was ascertained also through inter-laboratory studies.

Keywords: antibiotics; liquid chromatography mass spectrometry; milk; muscle; validation

---

1. Introduction

Antibiotics are widely used in livestock breeding to treat several diseases that appear in all the food producing animal species. To guarantee public health protection, the European Union requires member states to implement yearly monitoring plans to control the presence of antibiotic residues in food. Therefore, surveillance should be aimed particularly at controlling compliance with the maximum residue limits (MRLs), fixed in Table 1 of the Annex of Regulation (EC) No 37/2010 [1]. For several antibiotics, MRLs have been set in various matrices, such as eggs, fat, honey, kidney, liver, milk, and muscles and still, today, new MRLs are being fixed. In the early 2000s, the liquid chromatography coupled to tandem mass spectrometry technique (LC-QqQ) became essential in the routine analysis of single class of veterinary drug residues in food. Indeed triple quadrupole mass spectrometry analyzers were able to assure both greater sensitivity and selectivity than the traditional LC detectors, based on UV-Vis and fluorescence spectroscopy. In addition, for some important classes, such as aminoglycosides or avermectins, the need of a derivation step could be avoided. In the last ten years, the improvement of LC-QqQ systems allowed the realization of a further step in drug residue analysis, introducing procedures that are able to determine simultaneously more than one drug class [2–4]. As consequence, a remarkable effort has been made to progressively replace single-class with multiclass protocols, since this is a cost-effective way to improve the current residue control programs, thereby ensuring the determination of a wide number of compounds, with only few methods. Reviewing the main relevant published papers, some research groups recurred (Table 1). Among the control laboratories, the Official Food Control Authority of Zurich (Zurich, Switzerland), the RIKILT (Wageningen, Netherlands), the

Molecules **2019**, *24*, 2935

European Union Reference Laboratory for Antimicrobial Residues in Food (EURL, Fougères, France), the National Institute for Agrarian and Veterinary Research (INIAV, Vila do Conde, Portugal), the Istituto Zooprofilattico Sperimentale dell'Umbria e delle Marche (IZSUM, Perugia, Italy), the Canadian Food Inspection Agency (Calgary, Canada), the Residue Analysis Laboratory of Laboratório Nacional Agropecuário (LANAGRO, Porto Alegre, Brazil), and the US Department of Agriculture (USDA, Wyndmoor, PA, USA) are mentioning.

Molecules **2019**, *24*, 2935

Table 1. Overview of multiclass methods for the determination of veterinary drug residues in tissues and milk.

N° of Veterinary Drugs	Matrix	Equipment	Reference	Laboratory/Centre [a]
1	18	Milk	LC-QqQ	Aguilera-Luiz et al. 2008 [5]	Almeria University (Spain)
2	39	Chicken muscle	LC-QqQ	Chico et al. 2008 [6]	Barcelona University (Spain)
3	>100	Muscle	LC-TOF	Kaufmann et al. 2008 [7]	OFCA-Zurich (Switerland)
4	>100	Milk	LC-TOF	Stolker et al. 2008 [8]	RIKILT (The Netherlands)
5	Ca 100	Meat and other food	LC-TOF	Peters et al. 2009 [9]	RIKILT (The Netherlands)
6	Ca 26	Animal tissues	LC-QqQ	Stubbings et al. 2009 [10]	FERA (UK)
7	58	Milk	LC-QqQ	Gaugain-Juhel et al. 2009 [11]	EURL (France)
8	21	Milk	LC-QqQ	Martínez-Vidal et al. 2010 [12]	Almeria University (Spain)
9	30	Milk	LC-Orbitrap, LC-Q-TOF, LC-QqQ	Romero-González et al. 2011 [5]	Almeria University (Spain)
10	>100	Meat and other food	LC-Orbitrap	Kaufmann et al. 2011 [13]	OFCA-Zurich (Switzerland)
11	>60	Meat	LC-LTQ-Orbitrap	Hurtaud-Pessel et al. 2011 [14]	EURL (France)
12	59	Milk and honey	LC-Q-TOF	Wang et al. 2012 [15]	CFIA-Calgary (Canada)
13	21	Meat	LC-QqQ	Bittencourt et al. 2012 [16]	LANAGRO (Brazil)
14	24	Milk and liver	LC-QqQ	Martins et al. 2014 [17]	LANAGRO (Brazil)
15	>100	Milk	LC-Q-Orbitrap	Kaufmann et al. 2014 [18]	OFCA-Zurich (Switzerland)
16	39	Liver	LC-QqQ	Freitas et al. 2015 [19]	INIAV (Portugal)
17	23	Liver	LC-QqQ	Martins et al. 2015 [20]	LANAGRO (Brazil)
18	>100	Milk	LC-Q-Orbitrap	Wang et al. 2015 [21]	CFIA-Calgary (Canada)
19	>100	Various food	LC-Q-TOF	Dasenaki et al. 2015 [22]	University of Athens (Greece)
20	76	Bovine muscle	LC-QqQ	Dasenaki et al. 2016 [23]	University of Athens (Greece)
21	62	Animal muscle	LC-Q-Orbitrap	Moretti et al. 2016 [24]	IZSUM (Italy)
22	62	Milk	LC-Q-Orbitrap	Moretti et al. 2016 [25]	IZSUM (Italy)
23	>120	Animal tissues	LC-QqQ/LC-Q-TOF	Anumol et al. 2017 [26]	USDA (USA)
24	174	Bovine tissues	LC-QqQ	Lehotay et al. 2018 [27]	USDA (USA)
25	44	Salmon	LC-Q-TOF	Gaspar et al. 2019 [28]	INIAV (Portugal)

[a] OFCA = Official Food Control Authority; CFIA:Canada Food Inspection Agency; FERA: The Food and Environment Research Agency; LANAGRO: Laboratório Nacional Agropecuário; INIAV: Instituto Nacional de Investigação Agrária e Veterinária; USDA: United States Department of Agriculture; IZSUM: Istituto Zooprofilattico Sperimentale dell'Umbria e delle Marche.

The universities of Barcelona (Spain), Almeria (Spain), and Athens (Greece) have been the most active in this analytical field. LC-QqQ techniques are the most consolidated and most common multiclass procedures for veterinary drugs. These techniques have been mainly developed using this type of equipment [5,6,10–12,16,17,19,20,23,27]. In 2008–2009, the Official Food Control Authority of Zurich and the Dutch RIKILT Institute proposed, for the first time, the application of high-resolution (HR) mass spectrometry, based on time-of-flight (TOF) technology [7–9]. About three years later, the same Laboratory of Zurich, and the research group of Almeria University developed multiclass procedures for veterinary drugs, respectively, in meat, and milk, using LC-Orbitrap technique, a new MS analyzer, that was commercialized in 2005 [13,29]. Later, the introduction of benchtop hybrid high-resolution mass spectrometers (mainly, Q-TOF and Q-Orbitrap) produced further advantages in terms of selectivity and accuracy and, accordingly, these kinds of equipment has been more commonly applied (Table 1) [14,15,18,21,22,24–26,28,29].

Based on all the above, multiclass methods are no longer innovative procedures, and there is interest in their wide diffusion. The possibility of easy implementation and sustainable daily management, independent from the available LC-MS equipment. The aim of this work was to discuss the transfer of previously developed multiclass methods for more than sixty antibiotics in meat and milk from an LC-Q-Orbitrap platform to an LC-QqQ one [24,25]. The performance characteristics of the new LC-QqQ methods were estimated by means of full validation studies carried out according to European Commission Decision 2002/657/EC [30]. Finally, a comparison between the two techniques was carried out in the light of their cost-effectiveness in routine analysis of veterinary drug residues.

2. Results and Discussion

2.1. Optimization of LC-MS/MS Conditions

The choice of analytes has been carried out using the most administered antibiotics in farm. Only the classes of aminoglycosides and colistins were excluded, as their high polarity hampers the chromatographic retention, based on the reversed-phase mechanism (C18 column). On the other hand, the addition of ion-pairing agents on the mobile phase produced remarkable ion suppression, with detrimental effects on all the other analytes [24]. The chromatographic conditions were optimized starting from the parameters set for the LC-Q-Orbitrap methods. In order to profitably increase analyte retention, the percentage of methanol (eluent B) was reduced from 5% down to 2% (by volume). According to a typical reversed-phase mechanism, this change allowed us to obtain retention times of about 0.5 min higher than the initial tested conditions (Figure S1).

The MS conditions were established without the infusion of the individual solutions of analytes, but by setting the transitions on the basis of the ion fragments previously studied [24]. As shown in Table 2, apart few exceptions, such as some beta-lactams ($[M + Na]^+$), sulfanilamide ($[M + H - NH_3]^+$), spiramycin, neospiramycin, cefquinome, tildipirosin, tilmicosin, tulathromycin marker, and tulathromycin ($[M + 2H]^{++}$), the selected precursor ion species were generally the protonated molecular ions ($[M + H]^+$). For macrolides, it is not uncommon for the choice of bi-charged ions to be used as a precursor, due to their favorable abundance among the formed charged species [31]. The sample preparation was exactly the same as that previously optimized by Moretti et al. [24,25]; however, two internal standards (ISs) were replaced, in order to either, decrease costs (metacycline instead of tetracycline-d6), or to improve the MS response (ceftiofur-d3 instead of cefadroxil-d4). In this context, the ISs were not used for quantification purposes, but only to perform the internal quality control by checking the success of the analytical operations, during the routine application of the procedure as well as to monitor the run-to-run differences in the retention times [8]. For this purpose, at the beginning of sample treatment, IS were added at 10 µg·kg^{-1} and, before the release of the results, the presence (S/N > 3) of all eight compounds must be verified. The analyte quantification was achieved by matrix-matched curves (external standardization), which corrected the concentration for the relevant recovery factor [32]. The LC-QqQ chromatograms of a blank muscle, and of the same spiked at 10 µg·kg^{-1}, are reported in Figures 1 and 2, respectively. Eight representative analytes are shown, starting from the polar metabolite of florfenicol (florfenicol amine, RT = 3.4 min) to the last eluting compound (rifaximin, RT = 20.7 min). The analogous chromatograms are shown also for milk (Figures 3 and 4).

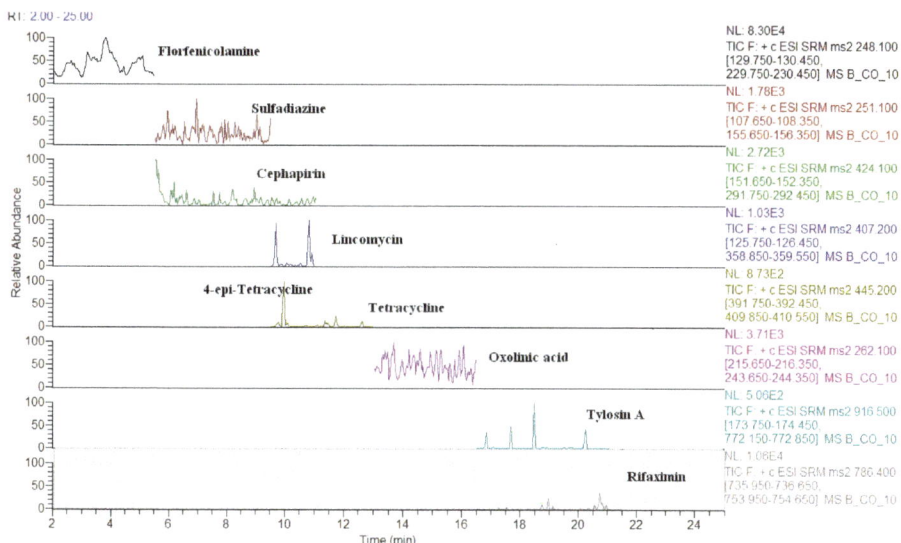

Figure 1. LC-QqQ chromatograms of a blank bovine muscle.

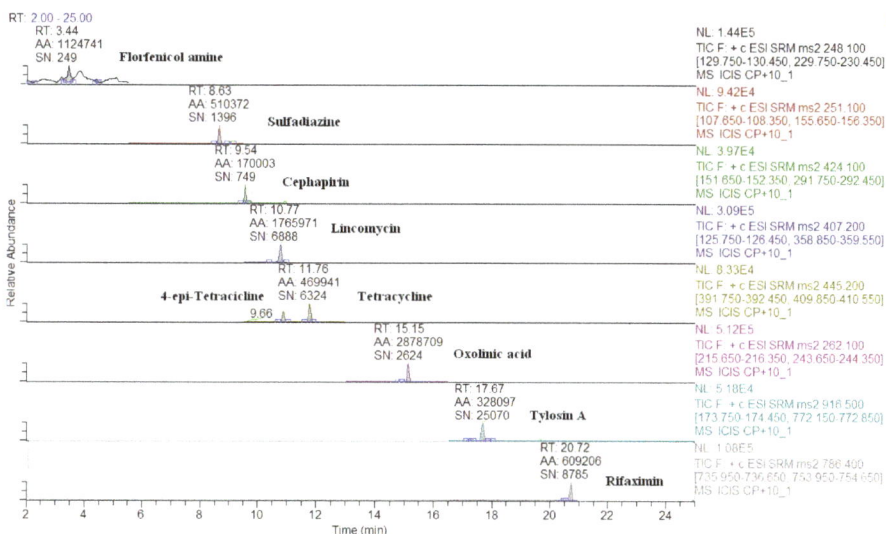

Figure 2. LC-QqQ chromatograms of a spiked bovine muscle (10 µg·kg^{-1}).

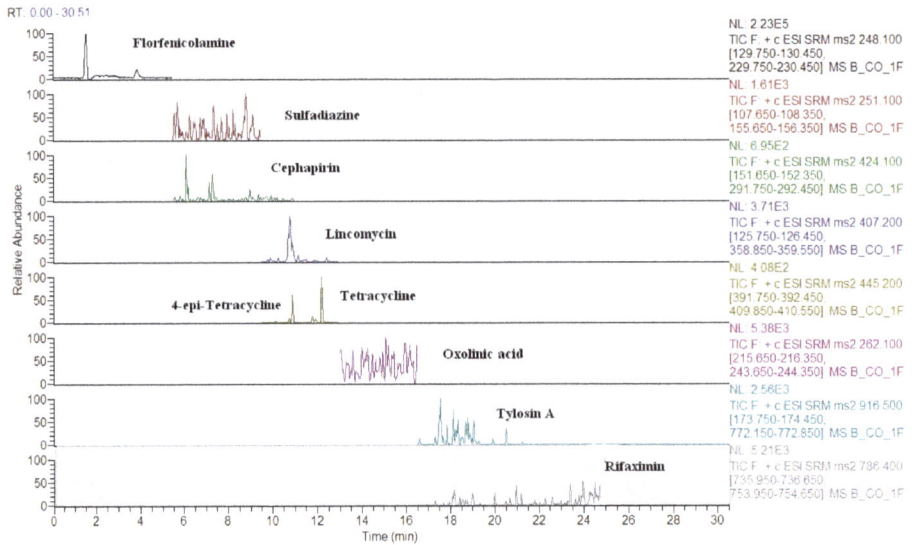

Figure 3. LC-QqQ chromatograms of a blank bovine milk.

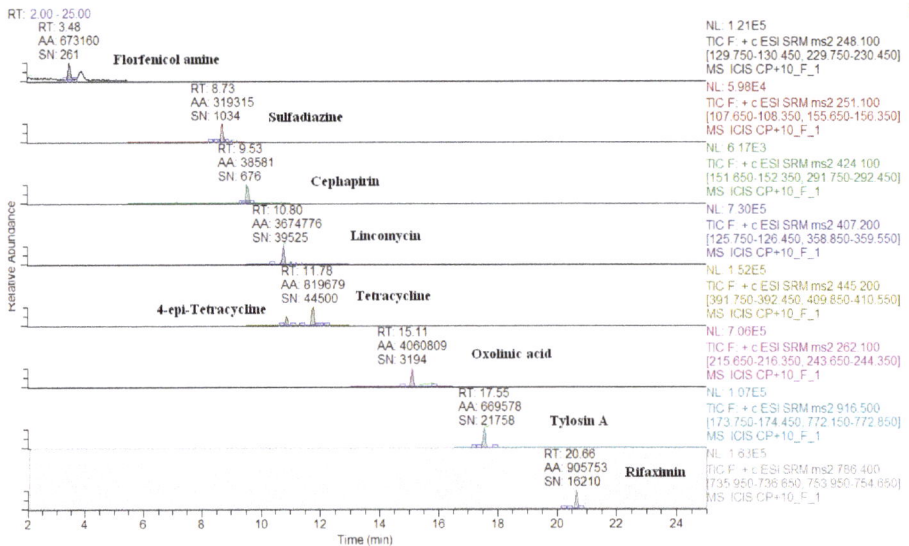

Figure 4. LC-QqQ chromatograms of a spiked bovine milk (10 μg·kg^{-1}).

2.2. Method Validation

Selectivity requirements are reported in Commission Decision 2002/657/EC [30]. The ion ratio of the two selected transitions (Table 2), and their relative retention times (<2.5%), were checked to confirm analyte identification. Linearity in the matrix was evaluated with five-points matrix-matched curves: 2, 10, 33, 100, and 150 μg·kg^{-1}. Therefore, levels higher than 150 μg·kg^{-1} had to be tested, and the final extract was diluted ten-fold or more, as reported in Tables S1 and S2. The linearity data are summarized in Table S3. For several analytes, the first calibration point (2 μg·kg^{-1}) had to be discarded, due to the scarce response. In other more critical cases (e.g., cefacetrile in meat/muscle, tildipirosin and tulathromycin markers in milk) additional points have been removed. Since Commission Decision 657/2002/EC [30] does not furnish precise criteria for evaluating linearity, the "Guidance document on analytical quality control and validation procedures for pesticide residues analysis in food and feed" was followed [33]. The percentage deviation of the back-calculated concentrations ($C_{measured}$) from the true concentrations (C_{true}) was calculated (1):

$$Deviation\ (\%) = \frac{(C_{measured} - C_{true})}{C_{true}} \cdot 100 \tag{1}$$

Table 2. Summary of the selected reactions transitions (SRM) monitored for the sixty-four targeted analytes.

N°	Analyte	Retention Time (min)	Adduct (m/z)	Precursor Ion (m/z)	Product Ions (m/z)	Collision Energy (eV)
1	Sulfaguanidine	2.85	$[M + H]^+$	215.1	92.0	15
					156.0	20
2	Florfenicolamine	3.20	$[M + H]^+$	248.1	230.1	10
					130.1	30
3	Sulfanilamide	3.30	$[M + H - NH_3]^+$	156.0	92.0	12
					108.1	10
	Sulfanilamide-13C6	3.30	$[M + H - NH_3]^+$	162.0	98.1	13
					114.1	13
4	Desacetylcephapyrin	6.80	$[M + H]^+$	382.1	152.0	30
					226.0	20
5	Amoxicillin	8.30	$[M + H]^+$	366.1	349.1	10
					114.0	20
6	Sulfadiazine	8.50	$[M + H]^+$	251.1	108.0	26
					156.0	15
7	Sulfathiazole	9.20	$[M + H]^+$	256.0	92.1	28
					156.0	15
8	Cephapyrin	9.45	$[M + H]^+$	424.1	292.1	20
					152.0	30
9	Sulfapyridine	9.50	$[M + H]^+$	250.1	108.0	26
					156.0	17
10	Tildipirosin	9.90	$[M+2H]^{++}$	367.7	281.2	20
					98.1	18
11	Sulfamerazine	9.90	$[M + H]^+$	265.1	108.0	27
					156.0	17
12	Cefquinome	10.00	$[M + 2H]^{++}$	265.1	134.2	20
					199.1	20
13	Cefacetrile	10.15	$[M + Na]^+$	362.0	258.0	10
					302.0	10
14	Cefalonium	10.50	$[M + H]^+$	459.1	337.0	10
					152.0	20
15	Lincomycin	10.50	$[M + H]^+$	407.2	126.1	30
					359.2	10
16	Tulathromycin marker	10.60	$[M + 2H]^{++}$	289.0	158.3	17
					420.5	17
17	Thiamphenicol	10.60	$[M + H]^+$	356.0	308.0	20
					229.0	20
18	Epitetracycline	10.60	$[M + H]^+$	445.2	410.2	20
					392.1	30
19	Trimethoprim	10.70	$[M + H]^+$	291.1	261.1	30
					230.1	30

Table 2. *Cont.*

N°	Analyte	Retention Time (min)	Adduct (*m/z*)	Precursor Ion (*m/z*)	Product Ions (*m/z*)	Collision Energy (eV)
20	Marbofloxacin	10.80	[M + H]$^+$	363.1	276.1	14
					320.1	14
21	Sulfamethazine	11.10	[M + H]$^+$	279.1	92.1	31
					124.1	28
	Sulfamethazine-^{13}C6	11.10	[M + H]$^+$	285.1	186.1	17
22	Epioxytetracycline	11.35	[M + H]$^+$	461.2	426.1	20
					337.1	30
23	Norfloxacin[a]	11.50	[M + H]$^+$	320.1	231.2	39
					282.1	29
24	Tetracycline	11.50	[M + H]$^+$	445.2	410.2	20
					392.1	30
25	Cefalexin	11.70	[M + H]$^+$	348.1	158.0	10
					174.1	20
26	Oxytetracycline	11.80	[M + H]$^+$	461.2	426.1	20
					337.1	30
27	Ciprofloxacin	11.80	[M + H]$^+$	332.1	245.1	23
					288.1	17
28	Enrofloxacin	11.90	[M + H]$^+$	360.2	245.0	26
					316.1	19
	Enrofloxacin -d5	11.90	[M + H]$^+$	365.2	321.4	18
29	Tulathromycin	11.90	[M + 2H]$^{++}$	404.0	158.1	20
					116.1	20
30	Danofloxacin	11.95	[M + H]$^+$	358.2	283.1	24
					340.1	22
31	Cefazolin	12.00	[M + H]$^+$	455.0	323.1	10
					156.0	20
32	Sulfamethoxazole	12.10	[M + H]$^+$	254.1	108.1	28
					156.0	16
33	Difloxacin	12.30	[M + H]$^+$	400.1	299.1	28
					356.1	18
34	Ampicillin	12.30	[M + H]$^+$	350.1	106.1	20
					160.0	20
35	Sulfamonomethoxine	12.30	[M + H]$^+$	281.1	108.1	28
					156.0	16
36	Florfenicol	12.40	[M + H]$^+$	358.0	241.0	20
					340.0	10
	Florfenicol -d3	12.40	[M + H]$^+$	361.0	241.0	16
37	Cefoperazone	12.60	[M + H]$^+$	646.1	530.3	10
					143.1	30
38	Sarafloxacin	12.60	[M + H]$^+$	386.1	342.1	18
					299.1	26
39	Epichlortetracycline	12.85	[M + H]$^+$	479.1	444.1	20
					154.0	30
40	Neospiramycin	13.40	[M + 2H]$^{++}$	350.2	160.1	10
					174.1	20
41	Chlortetracycline	13.80	[M + H]$^+$	479.1	441.1	20
					154.0	30
42	Spiramycin	14.05	[M + 2H]$^{++}$	422.3	702.4	10
					174.1	20
	Spiramycin -d3	14.05	[M + 2H]$^{++}$	423.8	174.0	20
43	Sulfadimethoxine	14.40	[M + H]$^+$	311.1	108.1	29
					156.0	21
44	Sulfaquinoxaline	14.80	[M + H]$^+$	301.1	92.1	30
					156.0	21

Table 2. *Cont.*

N°	Analyte	Retention Time (min)	Adduct (*m/z*)	Precursor Ion (*m/z*)	Product Ions (*m/z*)	Collision Energy (eV)
45	Oxolinic Acid	15.00	[M + H]⁺	262.1	216.0 / 244.0	29 / 18
46	Ceftiofur	15.10	[M + H]⁺	524.0	241.0 / 126.0	20 / 30
	Ceftiofur-d3	15.10	[M + H]⁺	527.0	244.1	15
	Metacycline	15.15	[M + H]⁺	443.1	426.2	16
47	Gamithromycin	15.40	[M + H]⁺	777.5	158.1 / 619.7	39 / 32
48	Tilmicosin	15.70	[M + 2H]⁺⁺	435.3	695.5 / 174.1	20 / 30
49	Doxycycline	15.50	[M + H]⁺	445.2	428.1 / 321.1	10 / 35
50	Nalidixic Acidᵃ	16.80	[M + H]⁺	233.1	159.0 / 187.0	30 / 25
51	Tiamulin	17.10	[M + H]⁺	494.3	192.1 / 119.0	20 / 30
52	Penicillin G	17.15	[M + Na]⁺	357.1	198.1 / 182.0	20 / 20
	Penicillin G-d7	17.15	[M + Na]⁺	364.0	205.2	13
53	Flumequine	17.20	[M + H]⁺	262.1	202.0 / 244.0	33 / 19
54	Tylosina A	17.40	[M + H]⁺	916.5	174.1 / 772.5	36 / 28
55	Erythromycin	17.60	[M + H]⁺	734.5	576.4 / 158.1	20 / 30
56	3-O-Acetyltylosin	17.75	[M + H]⁺	958.5	174.0 / 772.5	36 / 28
57	Oxacillin	18.20	[M + Na]⁺	424.1	265.1 / 182.0	20 / 20
58	Penicillin V	18.20	[M + Na]⁺	373.1	182.0 / 214.0	20 / 20
59	Cloxacillin	18.50	[M + Na]⁺	458.1	299.0 / 182.0	20 / 20
60	Valnemulin	19.10	[M + H]⁺	565.4	263.1 / 72.1	20 / 30
61	Dicloxacillin	19.20	[M + Na]⁺	492.0	333.0 / 182.0	20 / 20
62	Nafcillin	19.30	[M + H]⁺	415.1	171.1 / 199.1	34 / 13
63	Tilvalosin	19.45	[M + H]⁺	1042.6	174.0 / 814.5	39 / 30
64	Rifaximin	20.60	[M + H]⁺	786.4	754.3 / 736.3	20 / 30

ᵃ Acquired only in milk.

For each calibration point, it was verified that its value was not more than ± 20%. As an example, the calibration data of six analytes, belonging to different antibiotic families, are reported in Table S4 and Figures S2 and S3.

The overall recoveries and precisions data are listed in Table 3. For meat, seven validation levels were performed in the range 10–1500 µg·kg⁻¹, whereas, in milk, five levels were investigated (10–333 µg·kg⁻¹), with an additional concentration at 2 µg·kg⁻¹ to check amoxicillin, ampicillin, and penicillin G accuracy at $\frac{1}{2}$ MRL, as required by the Commission Decision 2002/657/EC [30]. Moreover, in milk, two additional molecules were tested with respect to the original group of 62 compounds, that is, the two quinolones, nalidixic acid, and norfloxacin.

Table 3. Precision, recovery and matrix effect (muscle and milk).

Analyte [a,b]	Muscle				Milk			
	$CV_{r,pooled}$ (%)	$CV_{Rw, pooled}$ (%)	Rec (%)	ME [c] (%)	$CV_{r,pooled}$ (%)	$CV_{Rw, pooled}$ (%)	Rec (%)	ME [c] (%)
Sulfaguanidine	5.8	10	83	−14	5.0	14	61	**253**
Florfenicol Amine	3.2	6.1	85	**−22**	2.5	3.8	94	−13
Sulfanilamide	6.6	8.9	74	**−35**	5.8	12	66	**−21**
Desacetylcephapirin	7.6	7.3	76	−1	5.9	6.0	91	**−21**
Amoxicillin	4.5	6.3	64	−1	5.8	5.8	89	−10
Sulfadiazine	5.1	8.2	86	−18	4.8	8.6	74	−16
Sulfathiazole	5.2	8.0	83	−6	5.2	8.3	71	−15
Cephapirin	6.9	7.8	75	1	12	12	94	−5
Sulfapyridine	4.6	7.2	85	13	5.8	9.5	67	−13
Tildipirosin	6.2	10	72	−8	6.0	20	87	5
Cefquinome	6.1	7.2	97	−4	11	19	78	−16
Sulfamerazine	3.8	5.9	89	11	6.7	8.7	70	−13
Cefacetrile	12	13	80	**−26**	22	27	92	−5
Cefalonium	6.7	7.3	78	3	8.7	8.7	93	−6
Lincomycin	4.3	7.2	83	−12	2.3	4.4	92	**30**
Epitetracycline	6.2	9.2	66	**22**	4.7	7.6	96	5
Trimethoprim	3.4	7.8	90	−3	2.7	4.9	94	4
Thiamphenicol	10	11	87	−13	13	13	92	−13
Tulathromycin marker	5.6	8.6	81	−13	8.7	26	81	−18
Marbofloxacin	6.1	7.0	87	**−27**	4.3	4.9	96	−11
Sulfamethazine	4.1	7.2	85	−1	7.8	14	68	−13
Epioxytetracycline	8.8	13	62	**40**	13	15	90	−3
Norfloxacin	-	-	-	-	7.0	8.0	94	3
Tetracycline	6.3	9.4	71	**27**	4.5	5.0	91	**40**
Cefalexin	6.2	8.4	64	1	6.2	6.7	90	−7
Oxytetracycline	6.7	8.1	63	10	4.4	5.2	90	1
Ciprofloxacin	7.5	8.4	81	**−25**	6.1	7.5	95	−10
Enrofloxacin	4.7	6.6	93	−16	4.1	5.9	99	−1
Tulathromycin	8.3	16	69	10	7.5	13	94	5
Danofloxacin	5.7	7.2	90	−17	4.2	5.0	97	3
Cefazolin	8.4	9.0	83	−8	14	14	94	−11
Sulfamethoxazole	7.7	9.7	88	−16	6.7	6.7	84	**−24**
Difloxacin	5.0	7.4	93	−15	3.3	4.2	96	−13
Ampicillin	6.6	8.0	68	−2	9.4	10	88	−10
Sulfamonomethoxine	4.6	7.4	86	0	6.4	9.9	78	−19
Florfenicol	9.8	16	90	−17	13	13	94	**−28**
Cefoperazone	6.8	8.4	85	−13	16	18	103	**−25**
Sarafloxacin	5.8	7.4	86	**−26**	4.8	6.6	99	**−22**
Epichlorotetracycline	9.3	13	71	**65**	7.4	11	100	**22**
Neospiramycin	7.8	16	67	12	7.3	12	86	13
Chlortetracycline	5.8	7.7	69	**47**	5.7	7.2	92	**48**
Spiramycin	8.4	17	74	20	4.5	7.6	91	**21**
Sulfadimethoxine	4.5	8.1	88	−12	3.9	3.9	90	**−29**
Sulfaquinoxaline	6.1	7.7	86	**−26**	4.6	4.8	91	**−38**
Oxolinic Acid	4.9	6.7	97	4	2.6	6.6	96	−3
Ceftiofur	7.5	9.7	72	**−21**	6.7	7.1	95	**−24**
Gamithromycin	5.6	7.1	94	**55**	3.2	4.8	98	**26**
Tilmicosin	6.9	10	88	**49**	3.2	4.4	95	**38**
Doxycycline	7.0	9.5	69	1	6.6	7.4	95	10
Nalidixic Acid	-	-	-	-	3.6	6.0	94	−3
Penicillin G	7.9	9.0	85	**−21**	12	13	92	**37**
Tiamulin	9.3	13	88	**−21**	2.5	4.1	98	−5
Flumequine	4.3	7.3	95	−3	2.9	4.5	95	**52**
Tylosin A	7.7	15	85	**29**	2.5	4.8	96	**54**
Erythromycin A	4.9	8.6	89	**−27**	4.7	12	67	**−21**
3-O-Acetyltylosin	9.3	16	88	**42**	3.7	5.0	95	**53**
Oxacillin	6.0	11	83	**−29**	8.0	8.5	100	1
Penicillin V	7.9	11	85	**−28**	7.5	7.7	99	6

Table 3. *Cont.*

	Muscle				Milk			
Analyte [a,b]	$CV_{r,pooled}$ (%)	$CV_{Rw, pooled}$ (%)	Rec (%)	ME [c] (%)	$CV_{r,pooled}$ (%)	$CV_{Rw, pooled}$ (%)	Rec (%)	ME [c] (%)
Cloxacillin	8.2	11	84	**−27**	9.8	12	110	**−25**
Valnemulin	19	31	75	**−26**	2.7	5.4	108	**55**
Dicloxacillin	6.7	10	81	**−25**	13	13	99	−10
Nafcillin	5.0	8.2	84	−7	5.0	5.4	99	−19
Tylvalosin	9.0	19	93	**50**	5.4	6.9	101	**90**
Rifaximin	9.0	12	91	**33**	4.8	9.6	98	7

[a] Nalidixic acid and norfloxacin were not included in muscle method since these antibiotics were introduced later; [b] For valnemulin (muscle), cefacetrile (milk) and cefquinome (milk) the method can be used only for screening purposes (inadequate accuracy); [c] Values of matrix effect (ME) in bold are considered significant (>|20|%).

In this work, the classical validation scheme (0.5 MRL, MRL and 1.5 MRL), described in Commission Decision 2002/657/EC, was not applied. Paragraph 3.1.3. of the same decision allows the introduction of alternative models [30] and, since the validation studies of multiclass procedures have to consider dozens of MRLs, which can vary also in function of the animal species, the adoption of progressive validation levels, that are equal for all the analytes, was fully justified. On the other hand, when the Commission Decision was issued (2002), the development of multiclass procedures for the control of veterinary drug residues was not initiated. The spiking ranges were chosen, considering, on the one side, the reachable concentrations, and, on the other side, the relevant MRLs [1] in all the couplings analyte/matrix. Preliminary experiments demonstrated that, at levels lower than 10 µg·kg^{-1}, the precision of several amphenicols, macrolides, beta-lactams, and tetracyclines became unsatisfactory.

The recovery factors (Rec) were established by comparing the peak area of each compound in the spiked samples against the peak area in the matrix matched standards, by which the antibiotics were added immediately prior to LC injection. The data in Table 3 summarize the average recoveries obtained in the whole validation range. In muscles, all recoveries were higher than 70%, except the majority of tetracyclines (62–69%), three beta-lactams (64–68%), and two macrolides (neospiramycin, 67% and tulathromycin, 69%). Analogously, in milk, recoveries higher than 70% were generally obtained, except for the more polar sulfonamides (61–70%) and one macrolide (erythromycin A, 67%). Since the quantification was performed with an external standardization, the raw results were always corrected for the relevant recovery factor, in order to correct the systematic error [32]. With regard to precision, the coefficients of variations (CVs) were calculated at each validation level both, repeatedly, and in intra-lab reproducibility conditions (CV_r, and CV_{wR}, respectively), by applying ANOVA. Moreover the CV_{wR} (and CV_r) were pooled to obtain an overall precision index, namely $CV_{wR,pooled}$ (2),

$$CV_{wR,pooled} = \sqrt{\frac{(n_1 - 1)CV^2_{wR1} + (n_2 - 1)CV^2_{wR2} + ...(n_n - 1)CV^2_{wRn}}{(n_1 - 1) + (n_2 - 1) + ...}} \qquad (2)$$

where $CV_{wR1}, CV_{wR2} ... CV_{wRn}$ were the coefficients of variation at the increasing levels 1, 2 ... n; n_1, $n_2 ... n$ were the number of replicates at each level [34]. For a certain analyte, the $CV_{wR,pooled}$ give a single estimate of precision, which can be applied to calculate decision limits (CCα) and detection capabilities (CCβ) at whatever MRL value. Therefore, the $CV_{wR,pooled}$ can also be used to obtain CCα and CCβ, where new MRLs were fixed in Regulation 37/2010 [1], maximizing the cost-effectiveness of the validation study. The decision limit and detection capability were calculated as follows (equations 3 and 4):

$$CC\alpha_{MRL} = MRL + 1.64 \cdot CV_{wR,pooled} \cdot MRL \qquad (3)$$

$$CC\beta_{MRL} = CC\alpha_{MRL} + 1.64 \cdot CV_{wR,pooled} \cdot CC\alpha_{MRL} \qquad (4)$$

In Tables S4 and S5, the MRLs of the 64 antibiotics, in various food-producing animal species, and in bovine milk, together with the relevant CCαs and CCβs, are listed. The MRLs are those reported in the last consolidated text of Regulation 37/2010 [1,35]. In muscles, valnemulin demonstrated high imprecision ($CV_{wR,pooled}$ = 31%) and, therefore, for these compounds, the developed procedure could be only used for screening purposes (Table 3). For all the other antibiotics, $CV_{wR,pooled}$ was always lower than 20%. On the other hand, in milk, cefacetrile and tulathromycin marker demonstrated insufficient precision ($CV_{wR,pooled}$ > 22%). Matrix effects (ME%) listed in Table 3 were calculated as follows (equation 5),

$$ME(\%) = \frac{b_{MM}}{b_S} \times 100 \qquad (5)$$

where $b_{MM,}$ and b_S were the slopes of matrix matched curves, and solvent standard curves, i.e., curves prepared in ammonium acetate 0.2 M, respectively. In the whole, although the sample purification was scarce, the matrix effects (suppression or enhancement) were limited and very few compounds demonstrated ME (%) higher than |50|%. This was probably because the long chromatographic run (30.5 min) allowed the distribution of matrix-interfering compounds and analytes, from preventing excessive bunching [26].

Since the signal-to-noise approach (S/N) is rather subjective [36], "operative" (fit for purpose) LOD and LOQ were fixed by examining the precision at each validation level (Table S7). All the compounds were detectable at the first concentration, i.e., 2 µg·kg^{-1} for amoxicillin, ampicillin, and penicillin G in milk and 10 µg·kg^{-1} for all the others. The only exception was cefacetrile, which is a scarcely ionizable molecule and, therefore, detectable and quantifiable from 33 µg·kg^{-1} in muscle (CV_r = 13%, CV_{wR} = 14%, recovery = 76%) and only detectable in milk (insufficient precision). A satisfactory accuracy for both amoxicillin and ampicillin was obtained at 2 µg·kg^{-1}, whereas penicillin G demonstrated unsatisfactory precision at this level (CV_r = 23%, CV_{wR} = 26%) and, therefore, this beta-lactam could be quantified only starting from 10 µg·kg^{-1} (CV_r = 7.5%, CV_{wR} = 11%). Therefore, penicillin G, a fundamental drug for the treatment of sub-clinical mastitis in **lactating cows**, could not be quantified in milk at $\frac{1}{2}$ MRL (2 µg·kg^{-1}) and the method was suitable only for screening purpose [30]. In Figure S4 the LC-QqQ chromatograms of these three compounds are shown in a blank milk sample and in the same sample spiked at 2 µg·kg^{-1}.

2.3. Comparison of LC-QqQ and LC-Q-Orbitrap Methods

Comparing the recovery factors of the methods developed in meat, similar results were always obtained (differences < 15%), except for cefquinome, which demonstrated a higher recovery when LC-QqQ was applied (+ 26%: 97% LR vs. 71% HR) and valnemulin which, nevertheless, was not accurately quantifiable with the LR procedure, as discussed above. The recovery differences are visualized in Figure S5. Examining the data of the more polar cephalosporins (desacetylcephapirin, cephapirin, cefacetrile and cefalonium), cefquinome recovery appeared suspect (over-estimated) using QqQ technique. Since the sample preparation protocol was the same, this difference should be attributed to an instrumental technique. Interestingly, cefquinome co-eluted with sulfamerazine and, in addition, accidentally their precursor ion had the same nominal mass charge ratio i.e., *m/z* 265 (Table 2). In our laboratory, at the beginning of the development of multiclass methods for antibiotics in food (2014), two separate matrix-matched curves were prepared, one for beta-lactams and another for all other analytes [24]. This measure was precautionary in order to avoid the possible negative effects of methanol, contained in the intermediate solution of antibiotics other than beta-lactams (see Section 2.2) on beta-lactam stability, as described in the literature [37,38]. In saying that, the artefacts in the quantification of cefquinome and/or sulfamerazine could occur if "hidden" transitions were shared between these two compounds. This means that a transition, monitored only for one of the two analytes (Table 2), was shared by the other one, too. In order to verify this hypothesis, two individual solutions (50 ng·mL^{-1}) were separately injected, by simultaneously monitoring the four relevant SRMs (Figure S6). It was evident that sulfamerazine shared a "hidden" transition with cefquinome (*m/z* 265 >

m/z 199—left side of Figure S6), but also viceversa (*m/z* 265 > *m/z* 156—right side of Figure S6). For sulfamerazine the fragment ion species at *m/z* 199 was formed by the rearrangement ion, by losing H_2SO_2 from the protonated molecule [39]. On the other hand, for cefquinome, the ion *m/z* 156 derived by the cleavage of C-C bond between oxime and the carbonyl group of C^7 amide ($[C_5H_6N_3OS]^+$) of the beta-lactam ring [40]. However, from a quantitative point of view, sulfamerazine could notably affect the peak area of cefquinome, but the contrary was much less, since sulfamerazine responded more significantly than cefquinome (about 4.9×10^6 vs. 1.1×10^6). Accordingly, when sulfamerazine was co-present, cefquinome concentration was significantly over-estimated. Later, observing that, in the adopted experimental conditions, beta-lactams were not deteriorated by methanol, the validation study in milk was performed by preparing only one matrix-matched calibration curve with all the analytes, including beta-lactams. In summary, although the development of LR procedures has been very fast and effective, due to previously-studied conditions [24,25], for one analyte, i.e., cefquinome, the choice of the precursor ion should be re-evaluated.

In Figure S7 the differences between LR and HR recovery factors in milk are shown. Tulathromycin marker and cefquinome had better recovery rates using HR detection, whereas valnemulin, cloxacillin, and rifaximin, showed better results using the LR system. As reported in our previous paper [25], LC-Q-Orbitrap suffers from "post interface ion suppression", which consists of instrument saturation when intense matrix-related compounds are present [41]. This phenomenon was more pronounced for the last eluting compounds, such as valnemulin, cloxacillin, and rifaximin, which explains the observed data in milk extracts. These have more interfering substances with respect to muscle [25]. Comparing the precision data (intra-laboratory reproducibility, $CV_{wR,pooled}$, see Tables S8 and S9), remarkable differences ($\geq 15\%$) between the two techniques were observed only in milk and, again, tulathromycin marker and cefquinome revealed the worse performances when determined by LC-QqQ. $C_{wR,pooled}$ of tylosin, cloxacillin, and tylvalosin were about 10% higher when analyzed by LC-Q-Orbitrap (Figure S9).

According to accreditation rules [42], since 2014 our laboratory participated in Proficiency Test Schemes in meat and milk, by applying the LC-Q-Orbitrap methods in these products. Some of the stored test materials were then re-tested, by applying the new developed procedures. The results, together with the consensus values assigned by the Organizers, are listed in Table 4. Examining the acceptability ranges, satisfactory z-scores would have been always obtained, except in the case of amoxicillin in milk (sample code: MI1532-A1). This latter result was explainable with the well-known instability of penicillin antibiotics [43,44].

Table 4. Participation in Proficiency Test Schemes: comparison between methods.

Method		LC-Q-Orbitrap		LC-QqQ		
Sample Code/Year	Matrix	Analyte	Found Concentration (µg·kg⁻¹)	Found Concentration (µg·kg⁻¹)	Consensus Value (µg·kg⁻¹)	Acceptability Range (µg·kg⁻¹)
MI1432-A1/2014 [a]	Milk	Sulfamethazine	144	96	103	57–150
MI1432-A2/2014 [a]	Milk	Amoxicillin	5	ND	Not assigned	-
M1435-A1/2014 [a]	Pig muscle	Sulfamethazine	88	75	69	36–102
	Pig muscle	Sulfadimethoxine	32	23	27	12–42
	Turkey muscle	Ciprofloxacin	5.5	5	5.6	3.2–8.1
M1433-A2/2014 [a]	Turkey muscle	Enrofloxacin	173	152	160	92–227
MI1532-A1/2015 [a]	Milk	Amoxicillin	16	ND	14	5.6–22
MI1532-A2/2015 [a]	Milk	Sulfamethazine	165	131	134	75–191
MI1623-A1/2016 [a]	Milk	Flumequine	91	111	88	47–129
MI1623-A2/2016 [a]	Milk	Oxytetracycline	93	55	91	49–132
MI1715-A2/2017 [a]	Milk	Danofloxacin	91	80	74	39–109
484 (material C)/2018 [b]	Bovine Muscle	Marbofloxacin	178	193	170	100–240
484 (material C)/2018 [b]	Bovine muscle	Oxytetracycline	89	79	106	59–152
334/2019 [b]	Bovine muscle	Ciprofloxacin	10	10	NA[c]	-
334/2019 [b]	Bovine muscle	Enrofloxacin	82	84	NA[c]	-
544/2019 [b]	Bovine muscle	Tylosin A	54	87	NA[c]	-

[a] Test Veritas, Padova, Italy; [b] RIKILT, Wageningen, Netherlands; [c] The final Report is not yet available.

In summary, the main advantages and disadvantages of the two techniques are undoubtedly, the QqQ analyser, which only involves only a kind of acquisition, i.e., SRM mode. On the contrary, Q-Orbitrap forces more complex experiments, also because the detection of analytes at trace levels is complicated by the "post interface ion suppression" phenomenon. On this subject, we demonstrated that, in certain chromatographic regions, for milk it is not possible to reach the required limits using full-scan experiments, since the massive presence of interfering substances can drastically worsen the sensitivity [25]. With regard to the sample throughput, the sample preparation protocol is identical and, therefore, there is no great difference. Moreover, it must be highlighted that some performances of LR system are due to the obsolescence of the available equipment. For example, most likely, a more recent LR platform could reach comparable LODs with HR, i.e., lower than 10 µg·kg^{-1}. All that said, the LC-QqQ technique is more suitable for routine laboratories, considering its user-friendliness and lesser cost (about three time lesser than LC-Q-Orbitrap).

3. Experimental

3.1. Chemical and Reagents

Acetonitrile (ACN) and methanol (LC-MS grade) were from Carlo Erba Reagents (Milan, Italy). Formic acid (50%) and *N,N'*-dimethylformamide (DMF) were purchased from Sigma-Aldrich (St. Louis, MO, USA). EDTA sodium salt dehydrate and ammonium acetate were provided by Sigma-Aldrich. Ultra-pure deionized water was generated by a Milli-Q purification apparatus (Millipore, Bedford, MA, USA). Amoxicillin, ampicillin, cloxacillin, dicloxacillin, nafcillin, oxacillin, penicillin G (benzylpenicillin), penicillin G-d7, penicillin V (phenoxymethylpenicillin), cefalonium, cefoperazone, cefquinome, ceftiofur, cephalexin, ciprofloxacin, danofloxacin, difloxacin, enrofloxacin, flumequine, marbofloxacin, nalidixic acid, norfloxacin, oxolinic acid, sarafloxacin, erithromycin A, spiramycin I, tylosin A, tilmicosin, sulfadiazine, sulfaguanidine, sulfadimethoxine, sulfamerazine, sulfamethazine (sulfadimidine), sulfamethoxazole, sulfanilamide, sulfapyridine, sulfaquinoxaline, sulfathiazole, trimethoprim, chlortetracycline, doxycycline, metacycline, oxytetracycline, tetracycline, florfenicol, florfenicol amine, thiamphenicol, lincomycin, rifaximin, tiamulin and valnemulin were obtained from Sigma-Aldrich. Sulfamonomethoxine was purchased from Dr. Ehrenstorfer (Augsburg, Germany); cefazolin, cefacetrile, ceftiofur-d3, cephapirin, desacetylcephapirin, florfenicol-d3, 3-*O*-acetyltylosin, gamithromycin, neospiramycin, spiramycin I-d3, tildipirosin, tulathromycin, tylvalosin, 4-epi-chlortetracycline, 4-epi-tetracycline, 4-epi-oxytetracycline, and tulathromycin marker (CP-60,300) were purchased from TRC Inc. (Toronto, Canada); sulfamethazine-[13]C6, sulfanilamide-[13]C6 and enrofloxacin-d5 were obtained from WITEGA (Berlin, Germany).

3.2. Standard Solutions

Individual stock standard solutions of 100 µg·mL^{-1} were prepared with methanol (amphenicols, lincosamides, macrolides, pleuromutilins, sulphonamides, tetracyclines and trimethoprim). The solubilization and storage conditions were previously studied [24]. Beta lactams were solubilized in H$_2$O/ACN 75/25 (*v/v*), except ceftiofur (DMF). Quinolones in MeOH/H$_2$O 80/20 (*v/v*), except ciprofloxacin, nalidixic acid, norfloxacin, enrofloxacin and oxolinic acid (DMF). Rifaximin was prepared in MeOH/H$_2$O 50/50 (*v/v*). The stock solutions were stored at −20 °C with variable storage times: From 1 month (cefquinome) to 24 months (sulphonamides). Working solutions (10, 1 and 0.1 µg·mL^{-1}) were prepared from the relevant stock solutions diluting with H$_2$O/ACN 75/25 (*v/v*) for beta-lactams and with methanol for all the other antibiotics. The solutions of internal standards were prepared according to their native compound or class.

3.3. LC-MS-MS Conditions

LC-MS/MS measurements were performed by a Surveyor LC pump, coupled with a triple quadrupole mass spectrometer (TSQ Quantum Ultra, Thermo Fisher, San Jose, CA, USA), an electrospray source included, and operating in a positive ionization mode. Separation was achieved on a Poroshell 120 EC-C18 column (3.0 × 100 mm; 2.7 μm particle diameter), which was connected to a guard column Poroshell (2.1 × 5 mm), both from Agilent Technologies (Santa Clara, CA, USA). The flow rate used was 0.25 mL·min^{-1} and the column temperature set at 30 °C. Mobile phase A was an aqueous solution 0.1% (*v/v*) formic acid and eluent B methanol. The gradient profile started at 2% eluent B for 1 min and increased linearly up to 95% B in 19.5 min; this condition was maintained for 5 min before returning to initial condition in 1 min (2% B) and held for 4 min to equilibrate the column. The sample temperature was kept at 16 °C and the injection volume 10 μL. For MS detection, the parameters were the follows: Capillary temperature 300 °C, vaporizer 320 °C, spray voltage 3 kV, and a resolution setting of Q1 and Q3 *m/z* 0.7. Sheath gas and auxiliary gas (nitrogen) pressures were set at 35, and 15 arbitrary units, respectively. Collision gas (argon) pressure: 1.5 mtorr. The collision energies, that were associated with each transition, are listed in Table 2.

3.4. Sample Preparation

The sample preparation was described elsewhere [24,25]. Briefly, (1.50 ± 0.01) g of minced muscle or milk were weighed in a 50 mL Falcon tube. The sample was spiked with: (i) 15 μL of the solution of the two internal standards (ISs) of beta-lactams at 1 μg·mL^{-1} (ceftiofur-d3 and penicillin G-d7); (ii) 15 μL of the solution of the other six ISs (sulfanilamide-^{13}C6, sulfamethazine-^{13}C6, enrofloxacin-d5, florfenicol-d3, spiramycin-d3 and metacycline) at 1 μg mL^{-1}. For muscles, one hundred microliters of 0.1 M of EDTA was then added, and the sample was extracted with 3 mL of a mixture of acetonitrile/water 80/20 (*v/v*). Milk was extracted with one milliliter of 0.1 M of EDTA and 3 mL of acetonitrile. A second extraction with 3 mL of pure acetonitrile was performed for both matrices. After centrifugation, the reunited extracts were evaporated and solubilized in 1.5 mL of ammonium acetate 0.2 M. Ten μL was injected in the LC system.

3.5. Method Validation

The validation was carried out following the Commission Decision 2002/657/EC [30]. To test the selectivity, blank muscle samples, belonging to the main animal species (bovine, swine and poultry) and bovine milk samples from different origins, were analyzed. The linearity in the matrix has been evaluated in the range 2–150 μg·kg^{-1} (2, 10, 33, 100 and 150 μg·kg^{-1}). The matrix-matched solutions were prepared by adding the analytes immediately prior to LC injection. The curves were constructed, including at zero (blank), and by plotting the average peak area of analyte (three injections for each concentration point) against its concentration. An unweighted linear regression model was applied to the calibration data. The precision (repeatability and within-laboratory reproducibility), recovery (trueness), decision limit (CCα), and detection capability (CCβ) were studied following the experimental plans, described in Table S1 (muscle) and Table S2 (milk) of the Supplementary Material. A blank bovine muscle or milk was spiked at the beginning of the extraction procedure with the appropriate standard solutions. Four replicates (*n* = 4) at each level were carried out on the same day, along with the relevant matrix-matched calibration curve. Each series was repeated on three different days, and at varying times, operator, and calibration status of the LC-MS system. The spiking levels in muscles were; 10, 33, 100, 150, 333, 1000, and 1500 μg·kg^{-1} and 10, 33, 100, 150, and 333 μg kg^{-1} in milk. In milk, penicillin G, amoxicillin, and ampicillin were also tested at 2 μg·kg^{-1}. The Limits of Detection (LODs) and Quantification (LOQs) were estimated on the basis of the observed accuracy (recovery and precision) at the first validation concentration or, if necessary, at the second one.

4. Conclusions

The validity of the transfer, from LC-Q-Orbitrap to LC-QqQ of two multiclass methods for veterinary drugs in food, has been demonstrated. Using the LR platform, valnemulin in muscle, cefacetrile, and tulathromycin marker in milk did not reach acceptable precision. In addition, examining LC-QqQ results, a series of accidental events (co-elution, selected precursor ions, fragmentation pathway etc) produced an over-estimation of cefquinome in meat, due to the co-presence of sulfamerazine. If neglected, this phenomenon could give false positive results. Since the quantification process of the LC-Q-Orbitrap system was based on a different principle (peak area of the precursor ion measured with MS accuracy < 5 ppm), this drawback was not observed. However, in milk, LC-Q-Orbitrap achieved worse recoveries for some of the more non-polar analytes, eluting in the chromatographic zone in which the interfering substances were more abundant. The "post interface ion suppression", which is a specific phenomenon of Orbitrap mass analysers could explain this latter evidence. A further consideration concerns the chromatographic separation, which is fundamental when highly selective detectors are applied. A good separation of peaks, avoiding the overlapping and bunching of both analytes and endogenous substances, minimizes the risk of false positive results and reduces matrix effects. Finally, the described method transfer has been successfully performed on an obsolete LC-QqQ platform (fifteen year-old), encouraging the implementation of multiclass strategy also in routine laboratories with limited instrumental resources.

Supplementary Materials: The following are available online, Figure S1: LC-QqQ chromatograms of the three more polar analytes (sulfanilamide, sulfaguanidine and florfenicolamine) with different % MeOH at the beginning of the gradient: 2% (left) and 5% (right), Figure S2: Matrix-matched calibration curves of danofloxacin, doxycycline, oxacillin, spiramycin, sulfamethazine and tiamulin in meat, Figure S3: Matrix-matched calibration curves of of danofloxacin, doxycycline, oxacillin, spiramycin, sulfamethazine and tiamulin in milk, Figure S4: LC-QqQ chromatograms of a blank (a) and a milk sample spiked at 2 μg kg^{-1} (b) of the three penicillins with MRL at 4 μg kg^{-1}, Figure S5: Recovery differences between the LC-QqQ and LC-Q-Orbitrap methods in meat. Positive values (%) indicate better recovery of LC-QqQ procedure and vice versa, Figure S6: LC-QqQ chromatograms of individual solutions (50 ng mL^{-1}) of sulfamerazine (left) and cefquinome (right), Figure S7: Recovery differences between the LC-QqQ and LC-Q-Orbitrap methods in bovine milk. Positive values (%) indicate better recovery of LC-QqQ procedure and vice versa, Figure S8: Precision differences between the LC-QqQ and LC-Q-Orbitrap methods in meat. Positive values (%) indicate better precision of LC-QqQ procedure and vice versa, Figure S9: Precision differences between the LC-QqQ and LC-Q-Orbitrap methods in bovine milk. Positive values (%) indicate better precision of LC-QqQ procedure and vice versa, Table S1: Validation plan in bovine muscle, Table S2: Validation plan in milk, Table S3: Linearity studies in matrix (matrix-matched curves), Table S4: Calibration data (matrix-matched curves) of six representative analytes, Table S5: MRL (μg kg^{-1}), decision limits (μg kg^{-1}) and detection capabilities (μg kg^{-1}) of the 64 tested antibiotics in the main food-producing species, Table S6: MRL (μg kg^{-1}), decision limits (μg kg^{-1}) and detection capabilities (μg kg^{-1}) of the 64 antibiotics in aquaculture, rabbits, horses, Table S7: Estimated LODs and LOQs based on the observed accuracy at the first and at the second validation level.

Author Contributions: Conceptualization, D.G., S.M., G.S. and F.P.; methodology, G.S. and S.M.; validation, D.G., F.P. and C.B.; investigation, D.G., F.P., S.M, G.S. and C.B.; data curation, R.S. and F.I.; writing—original draft preparation, R.G.; writing—review and editing, R.G., R.S. and F.I.; supervision and funding acquisition, R.G.

Funding: This research was supported by Ministero della Salute, Ricerca Corrente IZSUM 002 2015

Conflicts of Interest: The authors declare no conflict of interest.

References

1. European Communities Commission Regulation (EU) No 37/2010 of 22 December 2009 on pharmacologically active substances and their classification regarding maximum residue limits in foodstuffs of animal origin. *Off. J. Eur. Communities* **2010**, *L15*, 1–72. Available online: https://eur-lex.europa.eu/legal-content/EN/TXT/?uri=CELEX%3A32010R0037 (accessed on 30 May 2019).

2. Masiá, A.; Suarez-Varela, M.M.; Llopis-Gonzalez, A.; Picó, Y. Determination of pesticides and veterinary drug residues in food by liquid chromatography-mass spectrometry: A review. *Anal. Chim. Acta* **2016**, *936*, 40–61. [CrossRef] [PubMed]

3. Mainero Rocca, L.; Gentili, A.; Pérez-Fernández, V.; Tomai, P. Veterinary drugs residues: A review of the latest analytical research on sample preparation and LC-MS based methods. *Food Addit. Contam. - Part A Chem. Anal. Control. Expo. Risk Assess.* **2017**, *34*, 766–784. [CrossRef] [PubMed]

4. Rossi, R.; Saluti, G.; Moretti, S.; Diamanti, I.; Giusepponi, D.; Galarini, R. Multiclass methods for the analysis of antibiotic residues in milk by liquid chromatography coupled to mass spectrometry: A review. *Food Addit. Contam. - Part A Chem. Anal. Control. Expo. Risk Assess.* **2018**, *35*, 241–257. [CrossRef] [PubMed]

5. Aguilera-Luiz, M.M.; Vidal, J.L.M.; Romero-González, R.; Frenich, A.G. Multi-residue determination of veterinary drugs in milk by ultra-high-pressure liquid chromatography-tandem mass spectrometry. *J. Chromatogr. A* **2008**, *1205*, 10–16. [CrossRef] [PubMed]
6. Chico, J.; Rúbies, A.; Centrich, F.; Companyó, R.; Prat, M.D.; Granados, M. High-throughput multiclass method for antibiotic residue analysis by liquid chromatography-tandem mass spectrometry. *J. Chromatogr. A* **2008**, *1213*, 189–199. [CrossRef] [PubMed]
7. Kaufmann, A.; Butcher, P.; Maden, K.; Widmer, M. Quantitative multiresidue method for about 100 veterinary drugs in different meat matrices by sub 2-microm particulate high-performance liquid chromatography coupled to time of flight mass spectrometry. *J. Chromatogr. A* **2008**, *1194*, 66–79. [CrossRef] [PubMed]
8. Stolker, A.A.M.; Rutgers, P.; Oosterink, E.; Lasaroms, J.J.P.; Peters, R.J.B.; Van Rhijn, J.A.; Nielen, M.W.F. Comprehensive screening and quantification of veterinary drugs in milk using UPLC-ToF-MS. *Anal. Bioanal. Chem.* **2008**, *391*, 2309–2322. [CrossRef]
9. Peters, R.J.B.B.; Bolck, Y.J.C.C.; Rutgers, P.; Stolker, A.A.M.M.; Nielen, M.W.F.F. Multi-residue screening of veterinary drugs in egg, fish and meat using high-resolution liquid chromatography accurate mass time-of-flight mass spectrometry. *J. Chromatogr. A* **2009**, *1216*, 8206–8216. [CrossRef]
10. Stubbings, G.; Bigwood, T. The development and validation of a multiclass liquid chromatography tandem mass spectrometry (LC-MS/MS) procedure for the determination of veterinary drug residues in animal tissue using a QuEChERS (QUick, Easy, CHeap, Effective, Rugged and Safe) approac. *Anal. Chim. Acta* **2009**, *637*, 68–78. [CrossRef]
11. Gaugain-Juhel, M.; Delépine, B.; Gautier, S.; Fourmond, M.P.; Gaudin, V.; Hurtaud-Pessel, D.; Verdon, E.; Sanders, P. Validation of a liquid chromatography-tandem mass spectrometry screening method to monitor 58 antibiotics in milk: A qualitative approach. *Food Addit. Contam. - Part A Chem. Anal. Control. Expo. Risk Assess.* **2009**, *26*, 1459–1471. [CrossRef] [PubMed]
12. Martínez Vidal, J.L.; Frenich, A.G.; Aguilera-Luiz, M.M.; Romero-González, R. Development of fast screening methods for the analysis of veterinary drug residues in milk by liquid chromatography-triple quadrupole mass spectrometry. *Analytical and Bioanalytical Chemistry.* **2010**, *397*, 2777–2790. [CrossRef] [PubMed]
13. Kaufmann, A.; Butcher, P.; Maden, K.; Walker, S.; Widmer, M. Development of an improved high resolution mass spectrometry based multi-residue method for veterinary drugs in various food matrices. *Anal. Chim. Acta* **2011**, *700*, 86–94. [CrossRef] [PubMed]
14. Hurtaud-Pessel, D.; Jagadeshwar-Reddy, T.; Verdon, E. Development of a new screening method for the detection of antibiotic residues in muscle tissues using liquid chromatography and high resolution mass spectrometry with a LC-LTQ-Orbitrap instrument. *Food Addit. Contam. - Part A Chem. Anal. Control. Expo. Risk Assess.* **2011**, *28*, 1340–1351. [CrossRef] [PubMed]
15. Wang, J.; Leung, D. The challenges of developing a generic extraction procedure to analyze multi-class veterinary drug residues in milk and honey using ultra-high pressure liquid chromatography quadrupole time-of-flight mass spectrometry. *Drug Test. Anal.* **2012**, *1*, 103–111. [CrossRef] [PubMed]
16. Bittencourt, M.S.; Martins, M.T.; de Albuquerque, F.G.S.; Barreto, F.; Hoff, R. High-throughput multiclass screening method for antibiotic residue analysis in meat using liquid chromatography-tandem mass spectrometry: A novel minimum sample preparation procedure. *Food Addit. Contam. - Part A Chem. Anal. Control. Expo. Risk Assess.* **2012**, *29*, 508–516. [CrossRef] [PubMed]
17. Martins, M.T.; Melo, J.; Barreto, F.; Barcellos Hoff, R.; Jank, L.; Soares Bittencourt, M.; Bazzan Arsand, J.; Scherman Schapoval, E.E. A simple, fast and cheap non-SPE screening method for antibacterial residue analysis in milk and liver using liquid chromatography-tandem mass spectrometry. *Talanta* **2014**, *129*, 374–383. [CrossRef] [PubMed]
18. Kaufmann, A.; Butcher, P.; Maden, K.; Walker, S.; Widmer, M. Multi-residue quantification of veterinary drugs in milk with a novel extraction and cleanup technique: Salting out supported liquid extraction (SOSLE). *Anal. Chim. Acta* **2014**, *820*, 56–68. [CrossRef]
19. Freitas, A.; Barbosa, J.; Ramos, F. Multidetection of antibiotics in liver tissue by ultra-high-pressure-liquid-chromatography-tandem mass spectrometry. *J. Chromatogr. B Anal. Technol. Biomed. Life Sci.* **2015**, *976-977*, 49–54. [CrossRef]
20. Martins, M.T.; Barreto, F.; Hoff, R.B.; Jank, L.; Arsand, J.B.; Feijó, T.C.; Schapoval, E.E.S. Determination of quinolones and fluoroquinolones, tetracyclines and sulfonamides in bovine, swine and poultry liver using LC-MS/MS. *Food Addit. Contam. - Part A Chem. Anal. Control. Expo. Risk Assess.* **2015**, *32*, 333–341. [CrossRef]

21. Wang, J.; Leung, D.; Chow, W.; Chang, J.; Wong, J.W. Development and Validation of a Multiclass Method for Analysis of Veterinary Drug Residues in Milk Using Ultrahigh Performance Liquid Chromatography Electrospray Ionization Quadrupole Orbitrap Mass Spectrometry. *J. Agric. Food Chem.* **2015**, *63*, 9175–9187. [CrossRef] [PubMed]

22. Dasenaki, M.E.; Thomaidis, N.S. Multi-residue determination of 115 veterinary drugs and pharmaceutical residues in milk powder, butter, fish tissue and eggs using liquid chromatography-tandem mass spectrometry. *Anal. Chim. Acta* **2015**, *880*, 103–121. [CrossRef] [PubMed]

23. Dasenaki, M.E.; Michali, C.S.; Thomaidis, N.S. Analysis of 76 veterinary pharmaceuticals from 13 classes including aminoglycosides in bovine muscle by hydrophilic interaction liquid chromatography–tandem mass spectrometry. *J. Chromatogr. A* **2016**, *1452*, 67–80. [CrossRef] [PubMed]

24. Moretti, S.; Dusi, G.; Giusepponi, D.; Pellicciotti, S.; Rossi, R.; Saluti, G.; Cruciani, G.; Galarini, R. Screening and confirmatory method for multiclass determination of 62 antibiotics in meat. *J. Chromatogr. A* **2016**, *1429*, 175–188. [CrossRef] [PubMed]

25. Moretti, S.; Cruciani, G.; Romanelli, S.; Rossi, R.; Saluti, G.; Galarini, R. Multiclass method for the determination of 62 antibiotics in milk. *J. Mass Spectrom.* **2016**, *51*, 792–804. [CrossRef] [PubMed]

26. Anumol, T.; Lehotay, S.J.; Stevens, J.; Zweigenbaum, J. Comparison of veterinary drug residue results in animal tissues by ultrahigh-performance liquid chromatography coupled to triple quadrupole or quadrupole–time-of-flight tandem mass spectrometry after different sample preparation methods, including use of. *Anal. Bioanal. Chem.* **2017**, *409*, 2639–2653. [CrossRef] [PubMed]

27. Lehotay, S.J.; Lightfield, A.R. Simultaneous analysis of aminoglycosides with many other classes of drug residues in bovine tissues by ultrahigh-performance liquid chromatography–tandem mass spectrometry using an ion-pairing reagent added to final extracts. *Anal. Bioanal. Chem.* **2018**, *2018. 410*, 1095–1109. [CrossRef]

28. Gaspar, A.F.; Santos, L.; Rosa, J.; Leston, S.; Barbosa, J.; Vila Pouca, A.S.; Freitas, A.; Ramos, F. Development and validation of a multi-residue and multi-class screening method of 44 antibiotics in salmon (Salmo salar) using ultra-high-performance liquid chromatography/time-of-flight mass spectrometry: Application to farmed salmon. *J. Chromatogr. B Anal. Technol. Biomed. Life Sci.* **2019**, *1118–1119*, 78–84. [CrossRef]

29. Romero-González, R.; Aguilera-Luiz, M.M.; Plaza-Bolaños, P.; Frenich, A.G.; Vidal, J.L.M. Food contaminant analysis at high resolution mass spectrometry: Application for the determination of veterinary drugs in milk. *J. Chromatogr. A* **2011**, *1218*, 9353–9365. [CrossRef]

30. Commission Decision (2002/657/EC) of 12 August 2002 implementing Council Directive 96/23/EC concerning the performance of analytical methods and the interpretation of results. *Off. J. Eur. Communities* **2002**, *L221*, 8–36.

31. Dickson, L.C. Performance characterization of a quantitative liquid chromatography-tandem mass spectrometric method for 12 macrolide and lincosamide antibiotics in salmon, shrimp and tilapia. *J. Chromatogr. B Anal. Technol. Biomed. Life Sci.* **2014**, *967*, 203–210. [CrossRef] [PubMed]

32. SANCO/2004/2726-rev 4-December 2008 Guidelines for the Implementation of Decision 2002/657/EC. European Commission Health & Consumer Protection Directorate-General. 2008. Available online: https://ec.europa.eu/food/sites/food/files/safety/docs/cs_vet-med-residues_cons_2004-2726rev4_en.pdf (accessed on 30 May 2019).

33. SANTE/11813/2017. Guidance document on analytical quality control and validation procedures for pesticide residues analysis in food and feed. In European Commission Health & Consumer Protection Directorate-General. 2017. Available online: https://ec.europa.eu/food/sites/food/files/plant/docs/pesticides_mrl_guidelines_wrkdoc_2017-11813.pdfwebsite (accessed on 30 May 2019).

34. Barvick, V.J.; Wllison, S.R.L. Part (d): Protocol for uncertainty evaluation from validation data. In *VAM Project 3.2.1. Development and Harmonisation of Measurement Uncertainty Principles*; LGC: Teddington, UK, 2000.

35. European Communities Commission Regulation (EU) No 37/2010 of 22 December 2009 on pharmacologically active substances and their classification regarding maximum residue limits in foodstuffs of animal origin. Consolidated test. 3 March 2019. Available online: https://eur-lex.europa.eu/collection/eu-law/consleg.html (accessed on 30 May 2019).

36. Galarini, R.; Moretti, S.; Saluti, G. Quality Assurance and Validation General Considerations and Trends. In *Chromatographic Analysis of the Environment Mass Spectrometry Based Approaches, Fourth Edition*; Leo, M.L., Nollet, D.A.L., Eds.; CRC Press: Boca Raton, FL, USA, 2017; ISBN 9781315316208.

37. Grujic, S.; Vasiljevic, T.; Lausevic, M.; Ast, T. Study on the formation of an amoxicillin adduct with methanol using electrospray ion trap tandem mass spectrometry. *Rapid Commun. Mass Spectrom.* **2008**, *22*, 67–74. [CrossRef] [PubMed]

38. Mastovska, K.; Lightfield, A.R. Streamlining methodology for the multiresidue analysis of beta-lactam antibiotics in bovine kidney using liquid chromatography-tandem mass spectrometry. *J. Chromatogr. A* **2008**, *1202*, 118–123. [CrossRef] [PubMed]

39. Asteggiante, L.G.; Nunez, A.; Lehotay, S.J.; Lightfield, A.R. Structural characterization of product ions by electrospray ionization and quadrupole time-of-flight mass spectrometry to support regulatory analysis of veterinary drug residues in foods. *Rapid Commun. Mass Spectrom.* **2014**, *28*, 1061–1081. [CrossRef] [PubMed]

40. Niessen, W.M.A.; Correa, R.A.C. *Interpretation of MS-MS Mass Spectra of Drugs and Pesticides*; John Wilewy and Sons: Hoboken, NJ, USA, 2017.

41. Kaufmann, A.; Walker, S. Extension of the Q Orbitrap intrascan dynamic range by using a dedicated customized scan. *Rapid. Commun. Mass Spectrom.* **2016**, *30*, 1087–1095. [CrossRef] [PubMed]

42. *General requirements for the competence of testing and calibration laboratories*, Second ed.; International Standard Organization: Geneva, Switzerland, 2005.

43. Berendsen, B.J.A.A.; Elbers, I.J.W.W.; Stolker, A.A.M.M. Determination of the stability of antibiotics in matrix and reference solutions using a straightforward procedure applying mass spectrometric detection. *Food Addit. Contam. - Part A Chem. Anal. Control. Expo. Risk Assess.* **2011**, *28*, 1657–1666. [CrossRef]

44. Gaugain, M.; Chotard, M.P.; Verdon, E. Stability study for 53 antibiotics in solution and in fortified biological matrixes by LC/MS/MS. *J. AOAC Int.* **2013**, *96*, 471–480. [CrossRef]

Sample Availability: Not available.

molecules

MDPI

Review

Ion Mobility Spectrometry in Food Analysis: Principles, Current Applications and Future Trends

Maykel Hernández-Mesa [1,2,3,*], David Ropartz [2], Ana M. García-Campaña [1], Hélène Rogniaux [2], Gaud Dervilly-Pinel [3] and Bruno Le Bizec [3]

1   Department of Analytical Chemistry, Faculty of Sciences, University of Granada, Campus Fuentenueva s/n, E-18071 Granada, Spain
2   INRA, UR1268 Biopolymers Interactions Assemblies, F-44316 Nantes, France
3   Laboratoire d'Etude des Résidus et Contaminants dans les Aliments (LABERCA), Oniris, INRA UMR 1329, Route de Gachet-CS 50707, F-44307 Nantes CEDEX 3, France
*   Correspondence: maykelhm@ugr.es

Academic Editors: Alessandra Gentili and Chiara Fanali
Received: 24 June 2019; Accepted: 22 July 2019; Published: 25 July 2019

check for updates

Abstract: In the last decade, ion mobility spectrometry (IMS) has reemerged as an analytical separation technique, especially due to the commercialization of ion mobility mass spectrometers. Its applicability has been extended beyond classical applications such as the determination of chemical warfare agents and nowadays it is widely used for the characterization of biomolecules (e.g., proteins, glycans, lipids, etc.) and, more recently, of small molecules (e.g., metabolites, xenobiotics, etc.). Following this trend, the interest in this technique is growing among researchers from different fields including food science. Several advantages are attributed to IMS when integrated in traditional liquid chromatography (LC) and gas chromatography (GC) mass spectrometry (MS) workflows: (1) it improves method selectivity by providing an additional separation dimension that allows the separation of isobaric and isomeric compounds; (2) it increases method sensitivity by isolating the compounds of interest from background noise; (3) and it provides complementary information to mass spectra and retention time, the so-called collision cross section (CCS), so compounds can be identified with more confidence, either in targeted or non-targeted approaches. In this context, the number of applications focused on food analysis has increased exponentially in the last few years. This review provides an overview of the current status of IMS technology and its applicability in different areas of food analysis (i.e., food composition, process control, authentication, adulteration and safety).

Keywords: food quality; IMS; food composition; food process control; food authentication; food adulteration; food safety

---

1. Introduction

In the current context of food trade globalization and due to the recognized impact of the diet on human health, food analysis has become more important than ever. Food analysis has today gone beyond the traditional analysis of the major components of food and is more complex and broader. In addition to the nutritional value of foodstuffs (i.e., carbohydrates, proteins, lipids, vitamins, minerals and water), food analysis has been focused on food safety for a long time, mainly in the determination of residues of pesticides and veterinary drugs. Nowadays, food safety analysis encompasses a wide variety of compounds including natural contaminants (e.g., toxins) or anthropogenic contaminants such as persistent organic pollutants (POPs) [1]. Understanding the interactions of food with the environment and consequences (i.e., large-scale production, organic production, environmental contamination, etc. and including the control of food processes, packaging, etc.), as well as its effects on consumers (e.g., investigation of bioactive compounds), is also gaining great importance. Other issues such as food authentication (i.e., quality, origin, etc.), adulteration and fraud detection have also acquired great relevance in the field of food chemistry in the last few years [2].

Food analysis involves the determination of a wide range of compounds with different chemical nature. Chromatographic techniques (i.e., liquid chromatography (LC) and gas chromatography (GC)) coupled to mass spectrometry (MS) are the gold standard for this purpose because they allow the analysis of molecules with different polarity and volatility as well as provide mass spectra for compound identification. However, LC–MS and GC–MS methods still face several challenges related to the complexity of food matrices, the presence of compounds at different concentration levels (from pg/kg and pg/L to mg/kg and mg/L levels), and the existence of isobars and isomers that are not separated in the chromatographic dimension and cannot be distinguished by MS.

Consequently, the development of more advanced analytical strategies is required for the analysis of food composition, including nutritive and bioactive components, as well as to guarantee food safety and avoid food fraud. Within this framework, ion mobility spectrometry (IMS) has been recently introduced in the food chemistry field in order to improve LC–MS and GC–MS workflows. Despite the fact that IMS can be used alone as analytical tool, it can also be coupled to other analytical separation techniques such as LC, GC, capillary electrophoresis (CE), or supercritical fluid chromatography (SFC) as well as to MS, enhancing their performance characteristics in terms of sensitivity, peak capacity, and compound identification [3,4]. Its coupling with front-end separation techniques and with MS has emerged as a useful approach to extend the current boundaries of analytical methods in food science, and it can be anticipated that it will rapidly be growing in this field.

The implementation of IMS within the food analytical field is quite new [5–8], and is still barely known by many researchers in this scientific area. Despite IMS fundamentals have been developed since the beginning of the 20th century [9], it has not been until the recent commercialization of hyphenated ion mobility-mass spectrometry (IM–MS) instruments when this technique has really caught the attention of researchers from multiple fields, including food science. As a result, the number of publications on IMS applications in food analysis has rapidly increased over the last few years (Figure 1). Within this context, this review provides a general overview of IMS principles and presents the current state of the art of this technology for food analysis purposes.

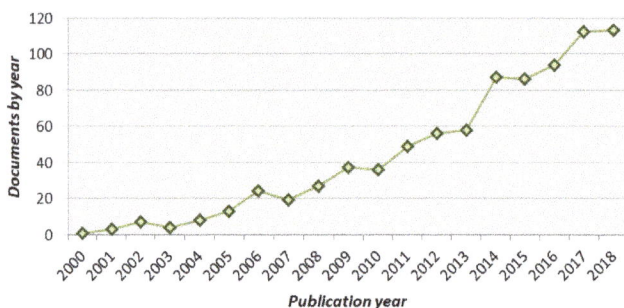

Figure 1. Search of literature related to ion mobility spectrometry (IMS) in food analysis from 2000 to 2018 on Scopus database. The terms "ion mobility spectrometry" and "food" have been included in the search topic.

2. Overview of Ion Mobility Spectrometry (IMS) Technique and Potential

IMS is an electrophoretic separation technique in which ionized compounds are separated in a neutral gas phase at atmospheric or near to atmospheric pressure. Therefore, separation takes places under an electric field (E) and is the result of the difference in mobility (K) of ions in the drift cell. K refers to the time (t_d) required by the ions to traverse the length (l) of the mobility cell and is related to the electric field according to Equation (1). In this equation, v_d represents the steady-state net ion/gas relative velocity [10].

$$K = \frac{1}{t_d \, E} = \frac{v_d}{E} \tag{1}$$

This physical property depends on several experimental conditions such as temperature (T), pressure (p) and gas number density (N). The reduced mobility (K_0), which refers to standard conditions (Equation (2); $N_0 = 2.687 \times 10^{25}$ m^{-3}, $p_0 = 760$ Torr, $T_0 = 273.16$ K), is typically reported instead of K in order to allow comparison between studies independently of the experimental conditions.

$$K_0 = K \frac{N}{N_0} = K \frac{p}{p_0} \frac{T_0}{T} \tag{2}$$

Although the 'momentum transfer collision integral' (Ω), commonly referred to as the collision cross section (CCS), is reported as the response resulting from ion mobility measurements [11], t_d or K are actually the variables that are measured when performing IMS experiments. Nevertheless, both parameters can be easily correlated according to Mason–Schamp equation (Equation (3)) when the separation occurs at low electric fields [12].

$$K = \frac{3}{16} \frac{ze}{N} \left(\frac{2\pi}{\mu k_B T} \right)^{1/2} \frac{1}{\Omega} \tag{3}$$

where z and e represent the absolute charge of the ion and the elementary charge, respectively; μ encompasses the reduced mass of the ion–neutral drift gas pair (i.e., $\mu = mM/(m + M)$; m and M are the ion and gas-particle masses, respectively); and k_B is the Boltzmann constant.

CCS represents the averaged momentum transfer impact area of the ion and is a molecular parameter related to ions size, shape and charge state. Therefore, CCS is widely used for structural elucidation since it provides knowledge about the three dimensional (3D) conformation of ions in the gas phase [13,14]. The correlation existing between this parameter and the mass-to-charge ratio (*m/z*) is not negligible. However, it also gives additional information to retention index (e.g., retention time, electrophoretic mobility, migration time, etc.), mass spectra, fragmentation and isotopic patterns, etc., for peak annotation in analytical workflows, especially in omics approaches (e.g., metabolomics, lipidomics) and, ultimately, for compound identification [15,16].

2.1. IMS Instrumentation

Nowadays, there is a wide variety of stand-alone IMS and IM–MS instruments on the market, but they are based on different technologies that offer different advantages [4,14,17]. They can be classified in time-dispersive, space-dispersive and trapping (i.e., ion confinement and release) technologies [18]. Drift tube ion mobility spectrometry (DTIMS) and travelling tube ion mobility spectrometry (TWIMS) are two different types of time-dispersive forms. For example, DTIMS is currently commercialized by Agilent (i.e., 6560 Ion Mobility LC/quadrupole-time of flight (Q–ToF)), Excellims Corporation (i.e., high-performance ion mobility spectrometry (HPIMS) systems including MA3100 and RA4100 HPIMS–MS instruments), Gesellschaft für Analytische Sensorsysteme mbH (G.A.S.) (i.e., GC–IMS systems) and TOFWerk (i.e., IMS-ToF), whereas TWIMS is available from Waters Corporation (i.e., Synapt G2-Si and Vion IMS QTof). High-field asymmetric waveform ion mobility spectrometry (FAIMS) and differential ion mobility spectrometry (DIMS or DMS) belong to space-dispersive techniques. Both FAIMS and DMS are based on the same principles of operation and mainly differ in the geometry of the cell. FAIMS cells are curved whereas DMS cells present a planar geometry and, consequently, they can lead to different analytical properties [19]. FAIMS systems are currently available from Owlstone Medical (i.e., ultraFAIMS) and Thermo Fisher Scientific (i.e., Thermo Scientific FAIMS Pro interface), whereas DMS is commercialized by SCIEX (i.e., SelexION). Until now, trapped ion mobility spectrometry (TIMS), which represents a type of trapping technology, is the only IMS system of its class that is currently commercially available (i.e., timsTOF from Bruker Daltonics).

Regarding CCS-related measurements, they can only be carried out using IMS instruments that operate at low electric fields (e.g., DTIMS) because the reduced mobility is independent of the electric field under this condition. K_0 becomes dependent of the reduced field strength (E/N) at high electric fields (E/N > 4–10 Townsends (Td); 1 Td = 10^{-21} V m^2) and Equation (3) is no longer applicable [10]. In this sense, only primary DTIMS methods can be applied to obtain CCS values directly, whereas secondary DTIMS, TWIMS and TIMS methods require system calibration using reference compounds of known CCS. Only those compounds characterized in terms of CCS by primary DTIMS methods can be used as calibrants [20]. The same results should be obtained for the CCS of a specific ion independently of the IMS platform employed if secondary methods operate under the same experimental conditions (i.e., temperature, drift gas, E/N, etc.) as primary methods. Since this is not always possible, CCS is a method-dependent value [20]. Specific annotation ($^{IMS\ form}CCS_{drift\ gas}$, IMS form: DT for DTIMS, TW for TWIMS, TIMS for TIMS; and drift gas: N_2, CO_2, He, etc.) is currently accepted to indicate the type of IMS technology and drift gas used for CCS measurements [11].

In DTIMS and TWIMS methods, ions travel through the drift cell against the buffer gas describing a similar path, so they are separated according to their mobility in the drift cell (Figure 2). Compact molecules collide less frequently with the molecules of drift gas (normally N_2 or He, but also CO_2), so they present a higher mobility (i.e., a smaller CCS) and cross the cell faster than elongated molecules.

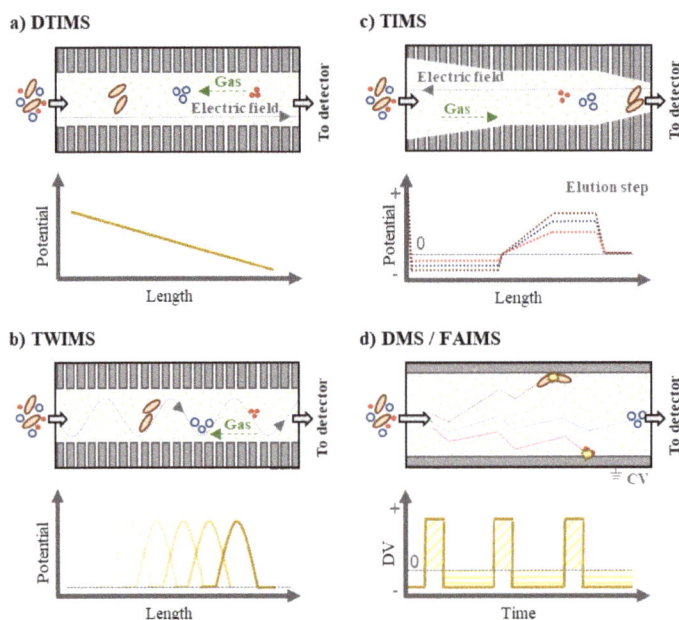

Figure 2. Schematic representation of commercially available IMS forms.

DTIMS is the former and simplest form of IMS, and DTIMS cells consist of a series of piled electrodes that generate a weak uniform electric field. In general, E/N varies between 1 to 15 Td [20]. DTIMS works as a primary method when stepped field experiments are carried out. In this case, the arrival time (t_A) is measured at multiple fixed drift voltages, and each t_A value represents the sum of drift time (t_d) of ions and the time elapsing between the moment when ions exit the drift tube and their detection (t_0). Subsequently, the reduced mobility can be directly obtained by applying Equation (4) and the CCS is further calculated by Mason–Schamp equation, Equation (3) [21].

$$t_A = t_d + t_0 = \left(\frac{l^2}{K_0} \times \frac{T_0 P}{T P_0}\right) \times \frac{1}{\Delta V} = slope \times \frac{1}{\Delta V} + t_0 \qquad (4)$$

In general, the term 'arrival time distribution' (ATD) should be used because IMS measurements are based on a population of ions rather than on a single ion and they do not reach one point of the instrument at the same time. ATD gives information about how homogeneous is the population of ions, and it should be reported in addition to t_A [20]. Furthermore, it must be noted that stepped field methods are limited to the K_0/CCS characterization of molecules because the analysis of complex samples is not feasible with them. Consequently, DTIMS usually works in single-field mode in which a single linear voltage is applied along the drift tube. In this condition, DTIMS operates as a secondary method and CCS calibration is required to measure this molecular descriptor [22].

Unlike in DTIMS, a dynamic electric field is applied in TWIMS to separate the ions [17]. TWIMS systems consist of a stacked ring ion guide where a pulsed differential voltage is applied to each electrode. As a result, a wave of electric potential travels along the drift tube and propels the ions axially. Ions are subjected to a varying voltage with a maximum E/N of 160 Td [10], and their separation depends on the speed and magnitude of the voltage wave. Voltage waves move faster than the ions, so they roll over the wave and require a succession of them to reach the exit of the mobility cell. In addition, in order to prevent the ions being pushed towards the drift tube wall, radio-frequency (RF) voltages of opposite phases are periodically applied to adjacent electrodes causing the radial confinement of ions [4,23].

On the other hand, TIMS instruments consist of three regions of electrodes: entrance funnel, TIMS tunnel, and exit funnel (Figure 2). Unlike in DTIMS and TWIMS where an electric field is applied to push the ions through the cell, in TIMS systems, an electric field (i.e., 45–150 Td) is applied to trap the ions and they are only dragged through the drift tube towards the detector by the gas [24]. Initially, ions are accumulated for a fixed period of time and directed to the entrance funnel. Subsequently, they are released into the TIMS funnel traversing an axial electric field gradient (EFG). When the velocity of the buffer gas (v_g) is equaled by the opposite steady state drift velocity of the ion (v_d), ions reach a stationary state and are trapped in a moving column of gas. Finally, the EFG is decreased and ions are released towards the exit funnel. Large ions with lower K are eluted before more compact ions since the dependence of the drift time on K_0 is opposite to DTIMS [25].

Finally, FAIMS and DMS instruments do not separate ions in function of their mobility in a neutral gas as in DTIMS, TWIMS or TIMS, but rather by the ratio of low-field to high-field mobilities [17]. Consequently, CCS values cannot be obtained by these systems. In FAIMS/DMS systems, a time-dependent electric field is applied between two parallel electrodes. Low and high electric fields of opposite polarities are simultaneously applied (i.e., high electric fields > 30 Td) [26], taking into account that the product of the voltage (commonly referred as dispersive voltage, DV) and time for each condition must be the same as shown in Figure 2 [19]. Low electric fields are applied for longer periods than high electric fields. Moreover, a compensation voltage (CV) is applied to one of the electrodes with the aim of avoiding that ions collide against them. As a result, the trajectory of ions is altered ensuring that they migrate to the exit of the cell. In this sense, FAIMS and DMS act as selective instruments because the CV is an analyte-dependent parameter and they are not able to scan the complete CV range during the transition of ions through the cell. Unlike in other IMS forms where all the ions are normally detected, in FAIMS/DMS technology, only those analytes related to the selected CVs reach the detector.

More detailed information about the operation and physical principles of the different IMS forms can be found in specialized literature [17,19,25,27,28]. Furthermore, a guide about how IMS experiments must be reported, including CCS measurements, has been recently published [20]. It is recommended to follow this guide for the communication of IMS-related results because it will probably set the basis for future guidelines and standards in ion mobility.

2.2. Collision Cross Section (CCS) for Structural Elucidation

As previously mentioned, IMS also gives access to the CCS characteristic which, in certain cases, provides additional information to mass spectra and retention index. It can be highly valuable for a higher confidence in the determination of residues and contaminants in food safety or for achieving a more complete fingerprint of food products. In the last few years, several CCS databases have been built in an attempt to use this characteristic as an identification parameter [11,15]. In general, there is still much controversy about the added value provided by this parameter in comparison to mass spectra. It cannot be denied that CCS and *m/z* are not fully orthogonal parameters since a close correlation exists between them, as observed in Figure 3B. Nevertheless, slight differences existing between the CCS of molecules with similar or equal *m/z* (i.e., isobars and isomers) can be enough to distinguish them (differences of at least 1.5–1.8% (2.0–6.5 Å^2) in apparent CCS values for accurate CCS determination [29]).

Regarding the application of CCS databases, an error threshold of ±2% is currently accepted for CCS measurements in comparison to CCS values reported by them (Equation (5)). However, the threshold of ±2% could potentially be reduced, which will give more confidence to the results [30]. Even so, this threshold for CCS measurements can still result in being more effective than isotopic pattern or fragmentation criteria to reduce the number of false positive results in automated screening workflows and avoid the requirement of post manual verification or confirmation analysis [31].

$$\% \text{ CCS error} = \frac{\left(\text{CCS}^{\text{measured}} - \text{CCS}^{\text{database}}\right)}{\text{CCS}^{\text{database}}} \times 100 \tag{5}$$

2.3. IMS Hyphenation

IMS separations typically take place in the millisecond range, so they can be easily carried out after traditional chromatographic or electrophoretic separations (i.e., LC, GC, SFC, and CE), which occur in the second range. The selection of the chromatographic/electrophoretic technique is obviously influenced by the nature of analytes (e.g., SFC for polar compounds, CE for polar and ionic compounds, GC for volatiles). Moreover, MS separations last microseconds so IM-MS hyphenation is also possible. IMS is normally coupled to time-of-flight (ToF)-MS due to its high acquisition rate. Indeed, as indicated in Section 2.1, several IMS systems combined with ToF-MS technology are already available on the market as integrated instruments.

GC was one of the first analytical techniques to be coupled to IMS, which has been mainly used as a detector in GC–IMS configurations [3,32]. GC–IMS hyphenation has been widely applied to the analysis of volatile compounds in food samples because stand-alone IMS systems generally provide low resolution. Both analytical techniques present a high degree of orthogonality, so selectivity and peak capacity are improved. However, LC–IM–MS platforms are currently becoming very popular and are used for numerous applications [4]. In this context, the integration of IMS in LC–MS workflows introduces a third separation dimension that improves peak capacity and allows the separation of isobars and isomers [33]. The implementation of IMS in LC–MS workflows increase peak capacity at least 2 or 3-fold [34,35], although this improvement ultimately depends on IMS resolution and the analytical application. Moreover, chromatographic peaks are extracted from background noise which provides cleaned-up chromatograms and mass spectra, leading to an improvement of the limits of detection (LODs) and sensitivity. Mass spectral data can also be interpreted more easily. Undoubtedly, these characteristics have contributed to the implementation of IMS technology in the food science field where samples present a high complexity and some compounds need to be determined at low concentration levels (i.e., ppb and ppt ranges).

From a technical point of view, limited dynamic range and ion loss have been traditionally attributed to LC–IM–MS in comparison to LC–MS methods [36]. However, several technological advances have been accomplished in the last years and, for example, some recent studies indicate that hyphenation with an ion mobility device does not affect the linear dynamic range [37]. Furthermore, it is important to remark that each IMS form can improve the performance of LC–MS methods in one way or another, but they are not exempt of limitations [4,14]. Therefore, the choice of the IMS technology will depend on the purpose of the intended application. In general, FAIMS and DMS provide higher orthogonality to MS than DTIMS and TWIMS but they act as signal filters according to the CV selected. Consequently, information about the sample is lost because the number of detected compounds is limited.

Finally, independent of whether stand-alone IMS, IM–MS or LC–IM–MS platforms are used, samples need to be ionized prior to entering into the mobility cell. Nowadays, electrospray ionization (ESI) is the gold standard for IMS analyses, although ^{63}Ni radioactive ionization sources are widely used in portable IMS instruments. Other ionization sources such as photoionization, corona discharge (CD) and pulse glow discharge (PGD) have also been used in food applications, but to a lesser extent [8]. Sample ionization is a process to be controlled because, as well as ion transportation and storage, it influences the conformation of ions and, consequently, their mobility. In addition, different adducts and ions with different charges states and sites (e.g., protomers) can be detected for the same compound [20], and their formation during the ionization can be altered by experimental conditions (i.e., ionization source and related parameters, solvents composition, etc.). In the context of food analysis, specifically in food safety, the identification of protomers is of high relevance since it could justify why some compounds lead to non-compliance with confirmation criteria based on ion fragmentation ratios and are not detected in screening analyses [38,39]. In addition, if it is applicable, a different t_d, K or CCS could potentially be obtained for each ion which provides more information for peak annotation. But it also hinders data treatment due to the number of species or 'features' that are detected. In this sense, data treatment generally remains as the main bottleneck of hyphenated LC–IM–MS systems, and a higher development is still required to successfully integrate IMS data in current LC–MS workflows.

3. Applications of IMS in Food Analysis

The application of IMS in food science is still in its early years, especially those approaches involving IM–MS or LC–IM–MS hyphenation. Despite IMS, and specifically IM-MS, has found its main application in omics approaches (i.e., proteomics and metabolomics, including lipidomics and glycomics) [40–42], this technique has barely been exploited for the specific analysis of composition and nutritional value of food, and their health benefits and risks (i.e., foodomics). Nevertheless, it is expected that its use will be extended rapidly to foodomics where a wide range of food metabolites need to be characterized for understanding their effects on human health. For example, the effects of coffee consumption on lipids profile (i.e., 853 lipid species from 14 lipid classes) have been investigated by DMS-triple quadrupole (QqQ)/linear ion trap (LIT)–MS [43]. This lipidomics approach has suggested that coffee intake alters glycerophospholipid metabolism and supports previous studies about the health benefits of drinking coffee.

The interest for this technology is growing and an increasing number of IMS-based methods have already been developed for food analysis, especially for food safety and food authentication purposes. A selection of the most recent applications of IMS in food science, including food composition, process, authentication, adulteration, and safety, is presented below. This section intends to provide a general vision of the potential of this technique for food analysis rather than being a comprehensive summary of all related articles that have been reported in the last years.

In general, IMS applications in food science cover from the determination of a predetermined number of compounds (i.e., targeted analysis) to the analysis of a large non-predetermined set of molecules (i.e., non-targeted analysis). As will be shown in the following sections, IMS has mainly been applied to targeted or semi-targeted analysis until now. However, it is expected that it will be widely used in non-targeted approaches, such as in food fingerprinting, where a large number of compounds are detected and the performance characteristics of analytical methods need to be improved (i.e., requirement of higher resolving power, tools to support molecular identification, simplification of mass spectra, etc.). On the other hand, current IMS applications in food analysis can be classified in three types: 1) those using stand-alone IMS instruments in which their intrinsic low selectivity is not a limitation, 2) approaches where IMS is coupled to a chromatography technique, usually GC, and mainly acts as a detector, and 3) applications based on LC–IM–MS workflows (alternatively

GC–IM–MS) in which the potential of IMS is fully exploited [i.e., extra separation dimension that improves selectivity and sensitivity (separation of isomers and isobars, and isolation of compounds of interest form background noise), compound selection based on their CVs (in FAIMS and DMS), CCS as an additional molecular descriptor (except in FAIMS and DMS)].

3.1. Food Composition

From a chemical composition point of view, food consists of a wide variety of compounds that can be divided in macronutrients (i.e., proteins, carbohydrates, lipids and vitamins) and micronutrients (e.g., polyphenols, etc.). Nowadays, IMS applications in food composition cover a wide variety of substances (e.g., lipids, peptides, phenolic compounds, terpenes, etc.) [44–47], including allergens [48], and samples (e.g., flours, olive oil, mushrooms, etc.) [48–50]. Due to the complexity of food matrices, stand-alone IMS has rarely been investigated in food composition studies, although it has found a wider application in food process analysis where few and specific compounds are determined, as will be shown in Section 3.2. As an example, stand-alone IMS has been used for the analysis of seven alcohol sweeteners in chewing gum, identifying the presence of sorbitol [51]. Other peaks were also observed in the ion mobility spectrum which were attributed to gum base components. However, this fact highlights the requirement of IM–MS for proper identification. Furthermore, more analytical information is obtained when IMS is combined with other analytical techniques such as LC or GC. Chromatographic and IMS separations are not correlated [44,46] (Figure 3), so their coupling improves peak capacity and enables the detection of a larger number of compounds. This advantage is widely used for food fingerprinting at low cost [50,52,53], without applying MS which is an expensive technology. In this case, if complete food characterization is required, standards of those substances that are potentially present in the sample must also be characterized in terms of retention time and mobility, drift time, or CCS. Consequently, peaks resulting from sample analysis can be tentatively identified (MS is mandatory for identity confirmation).

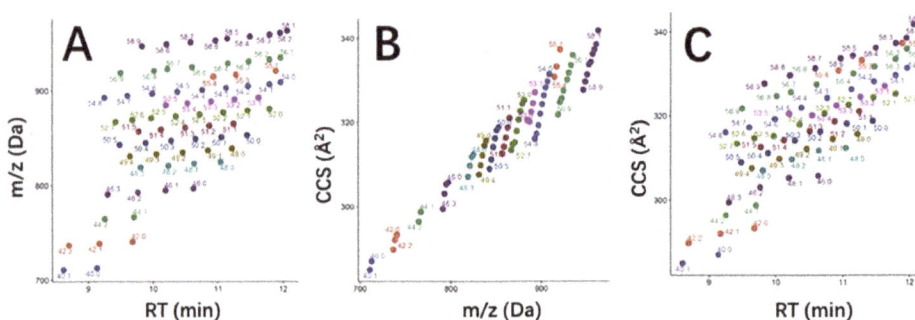

Figure 3. Selected triacylglycerides found in milk and comparison of the measured orthogonal parameters: (**A**) *m/z* vs. retention time (RT), (**B**) collision cross section (CCS) vs. *m/z*, and (**C**) CCS vs. RT. Figure reprinted with permission from [44]. Copyright (2018) American Chemical Society.

In this context, the volatile organic compounds (VOCs) fraction of olive oils has been widely studied by GC–IMS, and headspace (HS) analysis has been usually carried out for this purpose [49,52,53]. In general, IMS offers high sensitivity allowing the detection of olive oil volatiles in the range of 0.01–0.05 ppm, which is usually below the odor threshold of many compounds in extra virgin olive oils (EVOO) [53]. Several aldehydes (e.g., hexanal, 2-hexenal), ketones (e.g., 1-penten-3-ol), alcohols (e.g., 1-pentanol, 1-hexanol, etc.) and esters (e.g., hexylacetate), which are related to desirable attributes of olive oils, are present in their VOCs fraction. Their targeted analysis by HS–GC–DTIMS combined with chemometrics has been shown to be an effective strategy to characterize olive oils and distinguish them according to their quality (i.e., extra virgin, virgin and lampante) [49]. As alternative to HS analysis, olive oils samples can be submitted to laser desorption–GC–IMS analysis for the simultaneous characterization of VOCs fraction and detection of semi- and non-volatile compounds (e.g., (*E*,*Z*)-2,4-dodecadiene, 1-dodecene, (*E*)-2-hexenal, phenol, ß-pinene, benzaldehyde, acetic acid, limonene, 1-hexanol, nonanal, 1-heptanol, and octanal [54]. In comparison to HS, a higher number of signals were detected when laser desorption was applied to the analysis of oil samples. Despite the analysis of certain markers providing useful information about olive oils quality, statistical models for olive oils classification are improved when oil fingerprinting (i.e., semi-targeted or non-targeted analysis) is carried out [52,53]. This is because food fingerprinting gives a more complete picture of samples composition than the analysis of specific compounds. In the case of using GC–IMS methods, this implies that samples are differentiated according to biomarkers not identified, but characterized analytically. In addition to olive oils, HS–GC–DTIMS has also been applied to characterize the volatile fingerprint of fresh and dried *Tricholoma matsutake* Singer samples [50] and green tea aromas [55].

Unlike GC–IMS, LC–IMS coupling has not been evaluated so far for food composition analysis according to the literature. Nevertheless, the determination of phenolic acids in seedling roots by high-performance liquid chromatography (HPLC)–ESI–DTIMS has been recently reported [56], and this application can be extended to food composition analysis. Several compounds co-eluted in the chromatographic dimension (i.e., vanillic acid and caffeic acid; ferulic acid and sinapic acid; benzoic acid and salicylic acid), but they were separated in the ion mobility dimension. It demonstrated the potential of IMS for improving analytical separations at low analysis times (< 30 ms).

A new trend is to integrate IMS in LC–MS workflows for achieving the separation of isomers and isobars that are not resolved in the chromatographic dimension. For example, milk oligosaccharide isomers lacto-N-hexaose (LNH) and lacto-N-neo-hexaose (LNnH) are difficult to separate by LC and tandem MS provides similar fragmentation patterns in both positive and negative mode. However, both carbohydrates are baseline resolved by IMS under negative ionization conditions [57]. In this case, the statement 'IMS separates ions rather than molecules' is clearly exemplified. Deprotonated forms of LNH and LNnH were separated because they presented different CCS, but it was not the case for their sodium adducts. Therefore, ionization conditions have a special relevance for IMS results and, if there are not ionization constraints (i.e., ion suppression, low ionization for targeted analytes), they must be investigated when performing IMS experiments.

In general, IMS is considered as a third separation dimension in LC–MS workflows, but it can also act as a fourth dimension if a two-dimensional (2D) chromatographic separation is performed prior to IM–MS analysis. Food matrices are highly complex and one single LC or GC separation can be insufficient to provide a satisfactory degree of resolving power for their characterization. Advanced approaches such as LC × LC and GC × GC have been developed to enhance peak capacity which is increased even more when IMS is integrated in LC × LC–MS workflows, as shown in the analysis of phenolic compounds in chestnut (i.e., ellagitannins and gallotannins), grape seed (i.e., procyanidins), rooibos tea and red wine (i.e., flavonoid and non-flavonid phenolics) [58]. TWIMS improved the practical peak capacity from 7- to 17-fold (depending on the sample) in comparison to a 2D–hydrophilic interaction liquid chromatography (HILIC)-reverse phase (RP)–LC–ToF–MS method. Regarding the orthogonality degree of separation methods, lower orthogonality was found for a HILIC–TWIMS combination (52%), whereas the orthogonality existing between RP-LC and TWIMS was surprisingly higher (73%) than the orthogonality resulted from HILIC–RP–LC combination (67%). 2D–LC–IMS–MS approaches clearly reveal a higher complexity of food samples and improve the probabilities of distinguishing between isobaric and isomeric compounds [59]. For example, 2D–ultra-high performance liquid chromatography (UHPLC)–TWIMS–ToF–MS has been applied to the characterization of ginsenosides in white and red ginsengs which are diet supplements, and from the 201 compounds identified, 10 pairs of co-eluting isobaric ginsenosides were resolved only by IMS [60]. On the other hand, 2D–GC–IMS–MS methods have not been reported for the analysis of food samples, but the potential of this strategy has already been shown for the analysis of volatiles in medical herbs [61]. Consequently, similar approaches can be applied to the determination of the VOCs fraction of foodstuff.

Finally, IMS provides structural information of the ionized compounds (i.e., CCS), giving a more complete overview of food composition [44,62]. In the context of food composition analysis, $^{DT}CCSN_2$ libraries have been developed for the characterization of phenolic acids in wine [46] and lipids in bovine milk [44]. In this last case, lipid identification rates were increased when CCS, in addition to retention time and accurate *m/z*, was considered as molecular descriptor. Machine-learning algorithms were developed for the classification of 429 lipids according to their family class (i.e., triacylglycerides, diacylglycerides, phosphatidylcholines, and sphingomyelins) and carbon number. In general, a satisfactory classification rate (84.01%) was achieved when retention time and *m/z* were selected as analytical parameters. Classification accuracy was increased to 91.78% when CCS was included in the model, mainly due to the separation of isomeric species. This model was further improved to include the unsaturation level and, as a proof of concept, finally applied to the classification of 2087 bovine milk lipidomics data. As a result, 429 lipids, which were previously identified, were accurately classified whereas 179 unknown lipids were also annotated confidently. The identification of steviol glycosides represents another example of the applicability of CCS values for food characterization [62]. Steviol glycosides encompass a wide range of isomeric compounds and related molecules that can co-elute or result in similar fragment ions, impacting identification certainty. In order to improve the characterization of food products containing steviol sweeteners, a library of $^{TW}CCSN_2$ values was developed and applied to the fingerprinting of 55 food commodities. $^{TW}CCSN_2$ values contributed to the identification of several isomeric pairs such as rebaudioside E and rebaudioside A ([M − H]⁻ *m/z* 965.4230, CCS = 289.2 and 298.8 Å^2) and stevioside and rebaudioside B ([M − H]⁻ *m/z* 803.3701, CCS = 269.6 and 261.2 Å^2), without requiring ion fragmentation for identification, which cannot provide relevant information at low analyte concentrations.

3.2. Food Process Control

Food process control requires rapid response analytical tools, mainly intended as the analysis of many volatile compounds, in order to monitor industrial processes in near-real time and make rapid decisions if needed. In this sense, VOCs fraction can be related to storage conditions and production process, being also an indicator of the shelf life of food products. Consequently, volatiles are usually analyzed as part of food process and quality control. Stand-alone IMS represents a good option for this purpose because it allows the quick detection of volatile compounds (~20 s) at low cost, which is also attractive for industrial companies.

IMS has been widely applied to control the freshness of food products with special attention to the determination of biogenic amines [7,8], which are usually associated with fermentation processes and food degradation. Stand-alone IMS devices, which typically consist of DTIMS systems, are commercialized with different ionization sources. Therefore, the instrument to be used depends on the nature of analytes and their ionization characteristics. This is not a disadvantage for the analysis of biogenic amines by IMS because similar LODs can be reached (i.e., 0.1–1.2 ppm expressed as vapor in air for trimethylamine, putrescine and cadaverine), independent of the ionization source (i.e., photo-ionization, corona discharge, and radioactive ion sources) [63]. Stand-alone IMS methods have been reported for the determination of histamine in tuna fish [64], trimethylamine in seafood [65], and the simultaneous analysis of histamine, putrescine, cadaverine, tyramine in canned fish [66]. Stand-alone IMS applications have also been developed to detect the presence of off-flavors and contaminants generated during food processing and storage such as 2,4,6-trichloroanisole [67], and furfural and hydroxymethylfurfural toxicants [68].

Due to the low selectivity of stand-alone IMS instruments, the combination of HS–GC with IMS is recommended for the simultaneous analysis of several volatile compounds in liquid and solid samples in food quality control [5]. Despite chromatographic separations requiring several minutes, HS–GC–IMS can be less time-consuming and more environmentally friendly than other analytical and physico-chemical approaches currently used in food process control [69]. HS–GC–IMS methods have been applied to study lipid oxidation processes experienced by roasted peanuts [70] and EVOO [69] during storage.

In addition to the selective analysis of specific volatile substances, the profiling of VOCs fraction by IMS approaches can provide more complete information about the freshness and storage conditions of food products such as fish [71] and eggs [72]. VOCs profiling by GC–IMS usually requires chemometrics for data treatment in order to classify the samples, although the majority of compounds remain unidentified. It does not represent a limitation for sample classification, but compound identification is essential to understand fermentation and other decay processes. For example, up to 35 potential markers related to freshness were detected in the VOC fraction of eggs, but only a few compounds (i.e., butyl acetate, heptanal, dimethyl disulfide, dimethyl trisulfide and 1-butanol) were identified (Figure 4) [72]. Nevertheless, eggs were correctly classified according to their freshness (from 0 to 5 days at room temperature) by a principal component analysis (PCA) model. A supervised orthogonal partial least square discriminant analysis (OPLS–DA) model was finally developed to maximize the separation of both groups (i.e., fresh eggs vs. non-fresh eggs).

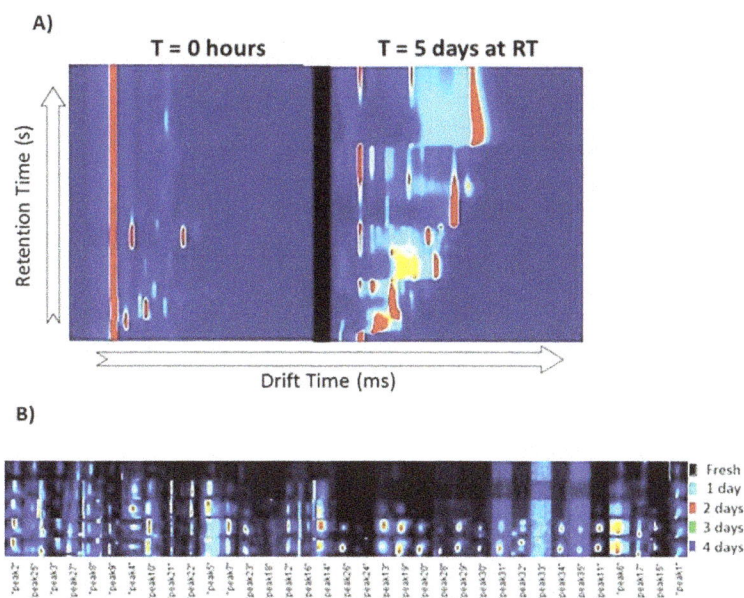

Figure 4. (**A**) Ion mobility spectrum of an egg product at T = 0 h (**left**) and after 5 days at room temperature (**right**). The red line identifies the reaction ion peak (RIP) position. (**B**) Global overview of the spots or 'features' identified in one egg product at different time points (from 0 to 4 days). Figure reprinted with permission from [72]. Copyright (2019) Elsevier.

Volatile compounds are not only related to the shelf life of food products and their degradation. The monitoring of VOCs fraction can also be decisive in food process and quality. For example, beer fermentation is conditioned by the presence of diacetyl and 2,3-pentadione whose concentrations must be reduced below the human odor threshold. In comparison to traditional methods, the use of GC–IMS for their determination allows decreasing the analysis time from 3 h to 10 min [5]. GC-IMS has also been applied to the characterization of VOCs fraction during the fermentation of lychee beverages [73] and wine [74]. One of the advantages of GC–IMS over other analytical tools is that it enables online monitoring and control of bioprocesses [74]. However, GC separations still require several minutes, which is a disadvantage of real-time decisions. By contrast, IM–MS analyses take a few milliseconds allowing the direct monitoring of VOCs evolution, which involves time- and cost-saving on sample preparation as well as avoids process interruption. Corona discharge–DTIMS–ToF–MS has been proposed for the on-line monitoring of VOCs formation in coffee roasting processes [75]. More than 150 VOCs were observed during the roasting of Brazilian *Coffea arabica* beans, and several alkyl pyrazines, fatty acids and other organic acids were identified. Some of them were isomers and isobars, so MS was not enough to identify them. Nevertheless, the integration of DTIMS in the analytical workflow allowed their separation according to their $^{DT}CCSN_2$ (e.g., unsaturated fatty acids presented smaller $^{DT}CCSN_2$), facilitating their identification.

3.3. Food Authentication

Chemical fingerprinting of food products and chemometrics are widely used for food authentication and ultimately to identify food fraud [76]. Within this framework, metabolomics fingerprinting is a promising strategy for food authentication that combines both approaches. In this sense, it is very efficient and has already shown clear benefits over traditional methods [77]. At the moment, only one study has been reported about the applicability of LC–IM–MS in metabolomics fingerprinting for food authentication [78]. The metabolic fingerprint of 42 red wines from the Republic of Macedonia was obtained by LC–ESI–DTIMS–ToF–MS, and the detected features were characterized in terms of retention time, accurate mass and $^{DT}CCSN_2$ under positive and negative ionization conditions. $^{DT}CCSN_2$ values added an extra-identification point for the putative identification of several phenolic compounds and other grape reaction products. After data treatment by PCA, Vranec wines were clearly distinguished from other varieties such as Cabernet Sauvignon or Merlot wines.

In general, IMS-based methods for food authentication have mainly been developed for the fingerprinting of volatile and semi-volatile compounds, usually involving GC–IMS hyphenation. GC–IMS methods in combination with chemometrics have been successfully applied to the authentication assessment of oils [79,80], meat products [81,82], wines [83] and honey [84]. For example, EVOO are 'designation of origin' (DO) products of high value, so their authentication is required not only to detect any fraud related to oil quality but also to identify mislabeling regarding origin. Based on the analysis of the VOCs by HS–GC–DTIMS, high discrimination rates (i.e., 98% and 92% by PCA-linear discriminant analysis (LDA) and PCA-k-nearest neighbors (kNN), respectively) have been reached to effectively discriminate between EVOO from Italy and Spain [80]. A similar HS–GC–DTIMS approach has been followed to authenticate Iberian hams and discard mislabeling [81]. The monitoring of specific non-identified markers by HD–GC–DTIMS and the application of orthogonal projections to latent structures discriminant analysis allowed to distinguish hams with a discrimination rate of 100%. As a third example, a HS–GC–DTIMS method has also been applied to identify the botanical source of different types of honey [84], since faulty declaration of the botanical source constitutes one of the main frauds on honey products. VOC profiling by HS–GC–DTIMS provided a discrimination rate (>98.6% according to PCA-linear-discriminant analysis (LDA)) as high as proton nuclear magnetic resonance (^{1}H-NMR) spectroscopy, which has been widely investigated for honey authentication in the last years.

Despite the efficiency showed by GC–IMS for the authentication of food products, the information resulting from GC–IMS analyses is very limited since MS is not included in the analytical workflow and markers cannot be identified. Only those compounds whose standards are available in the laboratory can be identified according to their retention and drift time, whereas the majority of compounds in the sample remain unknown. Nevertheless, discrimination of food products of different geographical origin, nature, quality, etc. can be achieved by GC–IMS–chemometrics methods based on unidentified markers. These strategies are usually recommended as screening approaches since molecular identities are generally required for confirmation.

3.4. Food Adulteration

The detection of food adulteration represents a particular case of food authentication in which substances are intentionally added to food products to decrease and mask their quality or valuable ingredients are removed in order to obtain higher economic profit. Consequently, similar IMS-based approaches as those shown in Section 3.3 have been proposed for the identification of this type of food fraud. HS–GC–IMS in combination with chemometrics has been evaluated for detecting the adulteration of winter honey (derived from *Schefflera actinophylla* (Endl.) *Harms* and wild *Eurya* spp. (Theaceae)) with cheaper *Sapium* honey [85], canola oil with other vegetable oils (i.e., sunflower, soybean, and peanut oils) [86], and crude palm oil with process byproducts of lower quality (i.e., palm fiber oil and sludge palm oil) [87]. As previously mentioned, GC–IMS methods should mainly be used as screening approaches because MS results essential for confirmation. However, screening methods are required for rapid analysis at lower cost and are of special interest for carrying out in situ analyses. Thus, only a few suspicious samples are sent to the laboratory for confirmation. In this context, stand-alone in combination with chemometrics has been investigated for the rapid detection of EVOO [88], sesame oil [89], and flaxseed oil [90] adulteration with vegetable oils of lower quality. Therefore, and despite the lack of selectivity of IMS, these applications demonstrate its efficiency to initially discriminate between allegedly adulterated samples and samples that are in compliance with their quality and labeling.

3.5. Chemical Food Safety

Chemical food safety covers the determination of a wide range of residues and contaminants (i.e., pesticides, veterinary drugs, toxins, environmental contaminants, etc.) in feed and food-related matrices; being one of the food science fields where IMS has found a wider application. Within this framework, the analysis of pesticides in a great variety of vegetables (i.e., apples, tomatoes, cucumbers, etc.), juices, oils, animal feed and water samples has been the topic most investigated [8]. IMS has been shown to be a solution for answering current society's concerns such as the presence of glyphosate in drinking water (as well as in other food products). The determination of glyphosate is quite difficult due to its ionic character, low volatility and low molecular weight. However, the LOD achieved by IMS for drinking water analysis (i.e., 10 μg/L) is comparable to those reported by HPLC and GC–MS (i.e., 0.02–50 μg/L) or ion chromatography (i.e., 15.4 μg/L) methods [91]. Despite of the potential of IMS for food safety applications, other food contaminants and residues rather than pesticides have been scarcely studied by IMS [8].

The analysis of residues and contaminants in food products faces the same issues as other areas of food analysis (i.e., high complexity of food matrices requiring high selective analytical techniques, presence of isomers and isobars, and requirement of detecting compounds at very low concentration levels) but, in this particular case, it must also comply several regulations according to the compounds analyzed. From a legal perspective, the employment of MS is practically mandatory in order to reach enough identification points (IP), so IM–MS related methods are of major interest for chemical food safety applications. However, IMS-based methods do not follow current guidelines and regulations concerning analytical methods intended to the determination of pesticides (e.g., SANTE/11813/2017 [92]), veterinary drugs (e.g., Regulation 2002/657/EC [93] or mycotoxins (e.g., SANTE/12089/2016 [94]). Indeed, because this particular technique has not been considered in the corresponding documents and no IP is allocated to CCS determination yet, this impairs its use as official identification criteria. For instance, scientists currently working is the revision of decision 2002/657/EC are debating the value of this new parameter as a criterion for identification and, if necessary, the value of the IP awarded [(acceded on 12 June 2019), [95]].

Under this context, stand-alone IMS and GC–IMS methods have been widely proposed for chemical food safety analyses [8], but they are outside the current frame of food law enforcement. Moreover, despite the use of stand-alone IMS has been shown to be an effective strategy for the rapid and in situ detection of residues and contaminants [96], its selectivity is limited and samples must normally be submitted to selective sample treatments prior analysis. In this sense, molecularly imprinted solid-phase extraction (MIPSE) and immunoaffinity chromatography (IAC) have been commonly employed for the selective extraction of the analytes of interest and to avoid matrix background.

For example, IAC has been applied to the analysis of fungicides in strawberry juices and wines [97] and mycotoxins (i.e., aflatoxins B1 and B2) in pistachios [98], whereas MISPE has been use to study the uptake and translocation of a neonicotinoid pesticide (i.e., imidacloprid) in chili and tomato plants [99]. Regarding GC–IMS methods, IMS has been useful to separate pesticides that were not resolved in the chromatographic dimension [100,101], although in general it merely acts as detector [32].

In chemical food safety, tandem mass spectrometry (MS/MS) and high-resolution mass spectrometry (HRMS) are usually employed with the aim of achieving the unequivocal identification of compounds. The combination of IMS and HRMS or MS/MS has commonly been applied to clean up mass spectra by removing matrix interferences and signal background. As a result, the task of mass spectra interpretation is reduced and compounds at lower concentrations are detected more easily. FAIMS and DMS are ion filters and this characteristic has been exploited to reduce the signal background in the determination of the mycotoxin zearalenone and its metabolites in cornmeal [102]. LODs were consequently improved up to 25 and 42.5-fold when applying FAIMS–ESI–MS instead of ESI–MS/MS or ESI–MS, respectively. Despite the characteristic of ion filtering is usually assigned to FAIMS/DMS technology, DTIMS and TWIMS are also able to isolate analyte signals from chemical background, improving concentration sensitivity [30,103]. In the context of chemical food safety, signal to noise ratio (S/N) of LC–ESI–ToF–MS methods intended to monitoring steroids and their metabolites in livestock were increased between 2 and 7 fold by the integration of TWIMS in the analytical workflow (Figure 5) [30].

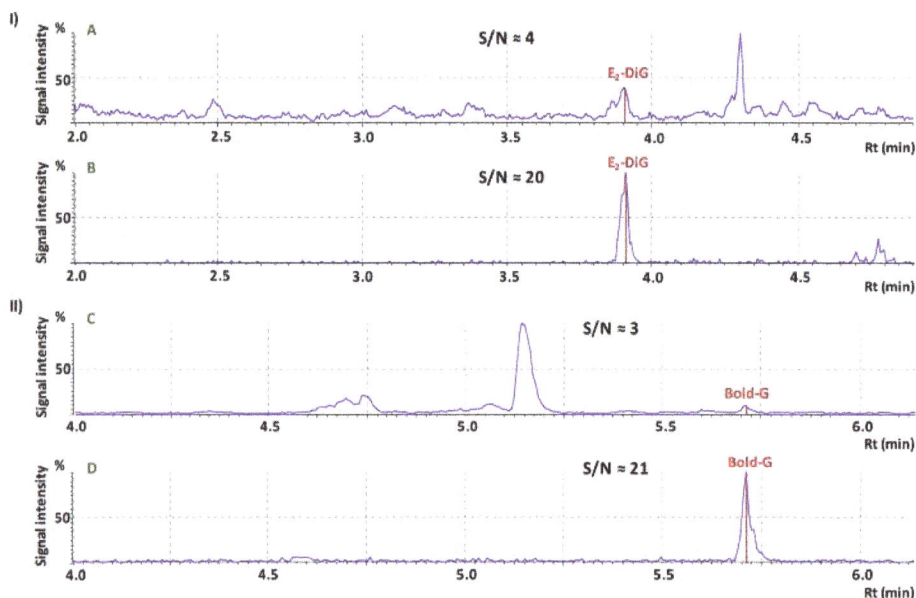

Figure 5. Extracted ion chromatograms (EICs) resulted from the analysis of: I) estradiol diglucuronide (E$_2$-DiG; 2 µg mL^{-1}; [M + Na]$^+$), and II) boldenone glucuronide (Bold-G; 0.2 µg mL^{-1}; [M − H]$^-$) in adult bovine urine samples. The following filters were applied for signal processing of related total ion chromatograms: A) *m/z* 647, B) *m/z* 647 and drift time range between 11.3 and 11.7 ms, C) *m/z* 461, D) *m/z* 461 and drift time range between 4.9 and 5.2 ms. Figure adapted from [30], which is licensed under CC BY-NC-ND 4.0 (changes: example III has been removed).

On the other hand, IMS–ESI–MS methods provide rapid throughput as observed in the separation of isomeric perfluoroalkyl substances (PFAS) by DMS–trap quadrupole (LTQ)–MS [104]. Nevertheless, food samples are very complex and matrix compounds can cause ion suppression when applying

ESI ionization. This issue has a great impact on signal sensitivity so it must be avoided for achieving robust and confident analyses. Thus, samples must normally be submitted to LC (or GC, SFC, CE) separation after specific sample treatments and prior to IM–MS analysis in order to overcome this effect. For example, FAIMS coupled to LTQ–Orbitrap has been shown to be a useful tool to separate paralytic shellfish toxins epimeric pairs [105], but their analysis in shellfish tissue extracts required a previous HILIC separation due to severe ionization suppression from matrix components. In general, 3D–LC–IM–MS separation is usually recommended when IMS is applied within the framework of food safety.

In this sense, LC and IMS are complementary separation techniques because they are based on different separation principles. Saxitoxins, which are a class of marine neurotoxins present in shellfish, comprise a group of isomers that are not differentiated by MS. Some diastereomers can be separated by HILIC but it does not distinguish between non-sulphated saxitoxins analogues. These molecules can be separated by TWIMS which makes HILIC–TWIMS–MS the best solution for the analysis of these group of substances in one single run [106]. Other examples of isobars and isomers separation by IMS include veterinary drugs such as ractopamine/isoxuprine ([M + H]$^+$; *m/z* 302.1751, ^{TW}CCSN$_2$ = 171.9 Å^2 and *m/z* 302.2025, ^{TW}CCSN$_2$ = 173.6 Å^2, respectively) by TWIMS [107] and isomeric environmental contaminants such as 2,2′,4,5′,6-pentachlorobiphenyl (PCB-103) and 3,3′,4,4′,5-pentachlorobiphenyl (PCB-126) ([M − Cl + O]$^-$; *m/z* 304.9095, ^{DT}CCSN$_2$ = 160.7 and 164.4 Å^2, respectively) by DTIMS [108]. Furthermore, if the separation of isobars and isomers cannot be achieved under standard conditions in the drift cell (i.e., only containing the drift gas), organic solvents (e.g., acetonitrile, methanol, isopropanol, etc.) can be added into the buffer gas in DMS and FAIMS systems in order to improve ion mobility separation [109]. This approach has been followed for the separation of the neurotoxin β-*N*-methylamino-L-alanine (BMAA) and its isomer β-amino-N-methylalanine (BAMA) in mussel tissues [110]. Both isomers were only resolved by HILIC–DMS–QqQ/LIT–MS when 0.35% acetonitrile was added in the DMS carrier gas. DMS also improved the low sensitivity traditionally observed for the analysis of BMAA by LC–MS methods.

Finally, as previously mentioned, IMS also provides additional analytical information to support molecular assignment. For example, drift time has been used to identify 100 pesticides in different vegetables and fruits by LC–TWIMS–ToF–MS [111]. This analytical property is not influenced by the matrix but drift times are instrument-dependent, so these data cannot be extrapolated to other IMS platforms. For this reason, the use of the CCS as the information provided by IMS has been extended. Consequently, CCS databases can be created and applied to support the determination of compounds, reducing the number of false negative/positives found in classical LC–MS workflows [112], and increasing the detection rates at low concentration levels of residues [39]. In the current context of Regulation 2002/657/EC revision, it has emerged the proposal to include the CCS as IP for the identification of veterinary drugs in food [95]. Scientists and experts are currently debating the value of grating IPs to the CCS and, if so, to determine its contribution in the 4 and 5 IPs required for the confirmation of substances with maximum residues limits (MRLs) and non-authorized substances, respectively. However, several concerns have been raised such as the lack of CCS databases and the tolerance accepted for CCS measurements. While, until a few years ago, CCS had been barely used as verification parameter in screening approaches due to the lack of databases, this situation has now recently changed and several CCS databases for pesticides [112,113], veterinary drugs [114–116], mycotoxins [117] and other contaminants [108], have been reported in the last five years.

Finally, recent applications of LC–ESI–TWIMS–ToF–MS methods, which include the CCS as signal filter in data processing, report the analysis of mycotoxins in cereals [117] and pesticides in green tea powder, fresh garlic, leek, fresh herb chives and rye [112], fish feed [31], and in 20 food products (e.g., strawberries, honey, chia seed, etc.) that belong to six different commodities according to SANTE/11813/2017 [39]. Other LC–TWIMS–MS applications in the context of food safety encompass the identification and structural characterization of residues and contaminants, either for evaluating exposure [118] or for studying their biotransformation [119].

4. Current Perspectives of Ion Mobility Spectrometry

Significant progress has been made in IMS technology in the last years, especially in IM–MS hyphenation. This fact has been directly reflected in the number of IMS-related applications that are currently published in various scientific fields including food science, and the number of related papers will keep growing due to technological developments in this analytical technique. Advances are expected in three different senses: IMS coupling with MS technologies offering higher resolution than ToF-MS (i.e., Fourier transform ion cyclotron resonance (FTICR) and Orbitrap–MS), improvement in IMS resolution, and implementation of the CCS as a parameter to support molecular assignment. In this section, only the last two topics will be discussed in more detail because the first is more related to improvements in MS acquisition rates than to IMS evolution. Several prototypes based on DTIMS/TWIMS–FTICR/Orbitrap–MS hyphenation have already been proposed; so further developments are expected [41]. These developments will certainly open up new possibilities for enhanced food fingerprinting, compounds identification and, ultimately, the discovery of substances, including contaminants, present in food but not detected until now.

4.1. Improvement in Peak Resolution

The resolving power (R_p) of IMS is typically expressed in the scale of CCS for IMS-platform comparison, although it excludes FAIMS/DMS systems. R_p is calculated according to Equation (6) where ΔCCS represents the full width of the peak at half its maximum height (FWHM).

$$R_p = \frac{CCS}{\Delta CCS} \tag{6}$$

As a consequence of the R_p of IMS, the integration of this analytical technique in LC–MS workflows increase their peak capacity. Thus, it can be expected that LC-IM-MS will quickly replace LC–MS methods in many applications, including food analysis. Current IMS instrumentation provide a R_p up to 300, so molecules presenting CCS differences (Equation (7)) as small as 0.5% can be separated [120]. However, commercial systems are not able to routinely offer this type of performance. TIMS is the only capable to provide a R_p higher than 200 [41], which is considered the lower limit of ultra-high resolution (UHR)IMS [23]. R_p of DTIMS systems is around 100 whereas TWIMS instruments currently reach a maximum R_p of 40 [14,23]. Therefore, the development of UHRIMS and its implementation in routine analysis obviously represent some of the main challenges of IMS, but it will also contribute to extend the use of this technique.

$$\Delta CCS_{A,B}(\%) = \frac{CCS_B - CCS_A}{\text{average } CCS_{A,B}} \times 100 \tag{7}$$

As a result of the improvements on R_p expected in UHRIMS technology, IMS separations will be able to provide similar peak capacities as LC at lower analysis times, which will be revolutionary from an analytical point of view and transformative for several applications [121]. In general, limitations on the applied field involve that R_p can only be increased by extending the drift path, although it presents some practical constraints [17,23]. In this sense, two IMS approaches (i.e., cyclic-TWIMS and structure for lossless ion manipulations (SLIM)–TWIMS) have successfully overcome these drawbacks and, consequently, are gaining attention due to their enhanced R_p. In addition, a third alternative was previously explored, namely the high resolution ion cyclotron mobility spectrometry (cyclic-DTIMS) [122], which is a circular 180.88 cm-length drift tube consisting of four quarter-circle drift tubes with ion funnels in between to re-focus the ions. Ions undergo multiple passes through the drift path in order to improve their separation. Successive improved versions of this technology have been shown to reach a R_p in excess of 1000 (as a consequence of 100 transits or cycles and involving a drift length of over 180 m) [123]. Nevertheless, no further research has been reported about cyclic-DTIMS, probably due to the long measurement times required and the significant loss of ions at long drift times of more than a second [23].

In cyclic-TWIMS, the drift cell is arranged in a circle configuration with pre- and post-store cells for ejecting the ions in and out [124,125]. As a result, ions can be submitted to several passes enhancing their separation. The resolution of multiple passes is given by Equation (8) where A is the single pass resolution (~65, 98 cm pass length), n represents the number of passes, and z is the ion charge state.

$$R_P = A(nz)^{1/2} \tag{8}$$

One single separation can be enough to obtain a general overview of samples and select the ATD regions of interest that will be submitted to multiple passes. The selection of ATD regions is required in order to avoid that the fastest ions trap the slowest ones, causing a 'wrapping effect'. Due to its enhanced R_p, cyclic-TWIMS provides new insights of ions conformation not previously described, as observed for the three protomeric species of the veterinary drug danofloxacin (Figure 6) [126]. As discussed above, it is relevant to characterize protomers in the food safety context because they do not fragment in the same way and results could be non-compliant with confirmation criteria based on ion fragmentation ratios. This technology was recently launched to the market by Waters Corporation during the last American Society for Mass Spectrometry (ASMS) meeting (2019).

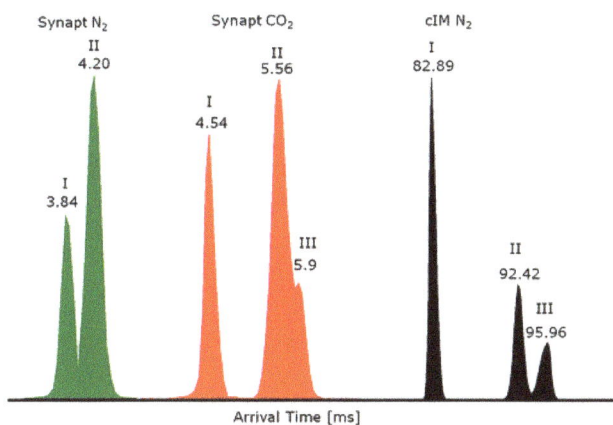

Figure 6. Danofloxacin IMS protomer separation using the Synapt (N_2 and CO_2 IMS gas) and cyclic-IMS (N_2 IMS gas, 5 passes) systems. Figure reprinted with permission from [126]. Copyright (2019) Wiley.

SLIM–TWIMS technology is not yet commercially available, but it is also catching the attention due to its high R_p resulted from its long drift path (~337, 540 m; 40 passes, 13.5-m path) [121]. As occurs with cyclic-IMS, SLIM–IMS can be based on DTIMS or TWIMS separations. However, TWIMS technology is usually applied because it allows longer path lengths than DTIMS, which presents voltage limitations [23]. SLIM–TWIMS systems consist of two planar surfaces fabricated using printed circuit boards and containing an array of electrodes. Ions travel through a serpentine path and are confined due to the application of DC and RF potentials [127]. SLIM–TWIMS offers a longer drift path than cyclic-TWIMS and, as a consequence, a larger range of mobilities can be monitored simultaneously. Although SLIM–TWIMS approaches have not been directly applied to food analysis, it has been shown to improved separations of lipid and peptide isomers [128] as well as glycans that differ in CCS by as little as 0.2% [129], which can have potential applicability in food composition analysis and food fingerprinting.

4.2. Implementation of CCS in Current Analytical Workflows

The CCS parameter is closely correlated to *m/z*, but its use as molecular descriptor in traditional analytical workflows has been shown to be effective in improving data processing by reducing false positive/negative results [112,113]. Despite its potential to support compound identification, the lack of CCS libraries is considered to be the major drawback for reaching this goal and implementing the CCS parameter in current analytical workflows [8]. In an effort to overcome this issue, several CCS databases have been reported in the last years [15], although they are not enough considering the number of molecules that remain uncharacterized. In order to extend the knowledge provided by CCS databases created experimentally, other strategies such as computational modeling and, more recently, machine-learning based prediction have also been investigated [130]. Machine-learning approaches are very effective for CCS prediction and are becoming very popular for generating CCS values of molecules on a large scale because their application only takes a few seconds/minutes whereas computational modeling is computationally intensive. For example, a machine-learning model based on artificial neural networks (ANNs) has been developed for the CCS prediction of small molecules selecting 205 compounds for model training, verification and blind test sets (ratio 68:16:16) [131]. This model was subsequently applied to the analysis of ten pesticides in spinach samples, and deviations between the observed and predicted CCS values for the protonated ions of pesticides were smaller than 5.3%.

CCS databases are typically based on one-platform measurements and, taking into account that CCSs are conditional values, the implementation of current libraries in other IMS platforms already introduces uncertainty into CCS measurements [20]. Theoretically, CCS values are platform-independent but they are influenced by several experimental parameters (i.e., E/N, temperature, and buffer gas) as indicated above. Specifically, the nature of buffer gas has a great influence on the CCS of molecules [132], and CCS values measured in different drift gases cannot be directly compared, as shown for a wide variety of pesticides whose CCSs have been obtained in He, CO_2, N_2O and SF_6 by DTIMS [133]. Within this framework, it may actually be more appropriate to develop standardized CCS libraries based on multiple platform measurements. Inter-platform studies should also consider different IMS forms (i.e., DTIMS, TWIMS, TIMS) to establish if standardized CCSs can be implemented in all of them or whether, on the contrary, ^{DT}CCS, ^{TW}CCS and ^{TIMS}CCS values must be reported and used. In general, similar CCS values are provided by these three IMS technologies when using the same buffer gas as observed for the sodium adduct of 25-hydroxyvitamin D3 [134]. However, another recent study about the CCS of 35 pharmaceuticals, 64 pesticides, and 25 metabolites of pesticides has shown that, although the majority of ions present differences lower than 1.1% between ^{DT}CCS and ^{TW}CCS values, both values cannot be compared in all cases [135]. Deviations up to 6.2% between ^{DT}CCS and ^{TW}CCS values were observed for several ions.

In the same vein, there is still a knowledge gap in the use of CCS databases and the variability associated with CCS measurements. Recent studies have shown high repeatability over time [30,112], and negligible impact of sample matrix on CCS values [39,113]. Despite this, the use of this parameter for specific analytical applications regulated by guidelines requires a comprehensive assessment of reproducibility across different laboratories and instrument types. Inter-laboratory reproducibility has barely been evaluated, and only few studies have tackled this issue [22,112,136]. This information is highly relevant to establish confidence intervals for CCS measurements and thresholds for their comparison with CCS values in databases. An inter-laboratory study involving four DTIMS-platforms have recently reported absolute bias of 0.54% to the standardized stepped field ^{DT}CCSN$_2$ values on the reference system [22]. Under this context, the current threshold of 2% accepted for CCS measurements seems to be too wide and could be reduced. As a result, the number of possibilities for peak assignment will be decreased when using the CCS for compound identification. Based on current evidence, this threshold could potentially be reduced to at least 1–1.5% [30,112,113]. Looking ahead to the implementation of the CCS parameter to support compound identification, a score system based on deviations from standardized CCS values should also be developed. Scoring systems based on mass spectra are typically used for the putative identification of molecules, so a similar approach could be suitable for the integration of CCS values in data processing workflows.

In addition to the standardization of CCS libraries and establishment of thresholds for CCS measurements, there is a third issue that requires further development. CCS calibration must be carried out in DTIMS (in single-field mode), TWIMS and TIMS systems before CCS measurements, and must be performed under the same operational conditions applied to obtain the CCS of analytes. Until now, there is neither consensus on the application of standardized calibration protocols nor on the application for this purpose of primary standards, which also serve for preparing reference materials [20]. A protocol has been reported for TWIMS calibration and operation [137], but different calibration procedures and calibrants are applied in other IMS systems. Under this context, 'Agilent tunemix calibration standard' (i.e., mixture of hexakis-fluoropropoxyphosphazines) [22], poly-DL-alanine [107], and 'major mix IMS/ToF calibration kit' from Waters (i.e., mixture of small molecules such as caffeine or sulfadimethoxine, and including poly-DL-alanine) [116] are currently widely used as CCS calibrants. Other alternative CCS calibrants such as the dextran have also been investigated for obtaining the CCS of carbohydrates [138]. Therefore, more studies about CCS calibration are still needed since, in the case of some classes of compounds, the accuracy of CCS measurements depends on the chemical nature of the calibrant [139]. Within this framework, both standardized calibration procedures and primary standards of different chemical class seem to be crucial in order to increase confidence in CCS comparison and measurements. The development of reference materials and standardized calibration protocols will contribute to the implementation of the CCS, but it calls for a great effort from IMS community including IMS suppliers.

5. Conclusions and Perspectives in Food Analysis

The implementation of IMS in food science is quite new but its applicability is growing exponentially. Nowadays, there is a wide range of methods intended for food composition, process control, authentication, adulteration and safety analysis, covering a great variety of molecules (i.e., macronutrients such as proteins, lipids, carbohydrates, vitamins; micronutrients such as polyphenols; process and biodegradation products such as biogenic amines; residues such as pesticides and veterinary drugs; or contaminants such as toxins and environmental pollutants), and involving different IMS forms (i.e., DTIMS, TWIMS, TIMS, FAIMS and DMS).

Stand-alone IMS instruments are portable and provide a quick response (< 30 ms), so they have been shown to be very efficient for in situ analysis and real-monitoring, such as in food process control where rapid decisions have to be made. It can be expected that miniaturized and portable IMS instruments will be implemented for food analysis as already occurs in the analysis of chemical warfare agents and drugs in airports, courts, etc. Potential applications are in situ monitoring of ripening processes of crops, depletion of residues after veterinary treatments in farms, etc.

In many applications, stand-alone IMS approaches offer low selectivity, so they are limited to the determination of specific compounds and usually require exhaustive sample treatments prior analysis due to the complexity of food matrices. In order to improve selectivity, IMS is commonly coupled to a chromatographic technique, especially GC if MS detection is not used. GC–IMS methods are very popular for the analysis of VOC fraction of food. This approach is widely used in food authentication where HS analysis is typically carried out. HS–GC–IMS in combination with chemometrics is broadly applied for food fingerprinting and product discrimination according to their quality, origin, etc. However, this strategy barely exploits the potential of the IMS technique (since it usually acts as simple detector) and is limited by the low number of compounds that are identified (i.e., only substances with standards available in the laboratory are identified based on their retention and drift times). The development of GC–IM–MS strategies will give more knowledge about VOCs composition, overcoming the current boundaries of GC–IMS approaches for food characterization and authentication. Consequently, for example, food fermentation and decay processes will be better understood since mass spectra will be obtained.

The recent commercialization of hyphenated IM–MS instruments, usually as part of LC–IM–MS platforms, has been the reason why IMS is becoming very popular in food science. IMS is extending the current boundaries of LC–IM–MS methods by introducing an extra separation dimension that allows the separation of isobars and isomers (not always separated in the chromatographic dimension and undistinguished by MS) and the isolation of analytes of interest from chemical background (so improving S/N and enhancing sensitivity). Although the correlation existing between IMS and MS has been a topic of discussion for a long time, the applications included in this review show that a higher number of compounds are detected by LC–IMS–MS in comparison to traditional LC–MS workflows, which gives a more complete picture for food characterization. Moreover, the integration of IMS provides cleaned-up chromatograms and mass spectra facilitating data interpretation.

In addition, IMS gives additional information to retention index and mass spectra of molecules (i.e., K_0, drift time, CV and/or CCS) which can ultimately be used for compound identification. Within this framework, the CCS of molecules has acquired great relevance. Despite this parameter is correlated to m/z, it has been shown to be useful to distinguish compounds with different chemical nature, but also to differentiate among close chemical molecules. From a regulatory point of view, the application of the CCS parameter for the identification of residues and contaminants in food is currently under discussion. Nowadays, the development of open-access CCS databases (based on inter-laboratory measurements for proposing normalized reference values), the requirement of standardized calibration procedures and calibrants for CCS measurements, and the establishment of thresholds for CCS measurements, are viewed as the main challenges to tackle for the implementation of CCS in food analysis. Based on current knowledge and trends, CCS will definitely be included in current analytical workflows to support food characterization and, probably, legally accepted as complementary information to confirm the presence or absence of residues and contaminants in food products.

Technological developments experienced by IMS, and especially by IM–MS hyphenation, are going to have a great impact on the implementation of this technique in the food science field. Enhancements in R_p, currently lead by cyclic-TWIMS and SLIM–TWIMS technologies, will contribute to overcome the challenges arise from the complexity of food matrices (i.e., high number of compounds with different chemical nature at different concentration levels). Under this context, they will enable the discovery of unknown food components with bioactive properties which are normally at low concentration levels and are typically masked by isobaric and isomeric compounds or by food major components. In addition, more complete fingerprints of food products will be achieved, providing more detailed information for food authentication and detection of food adulteration.

Finally, the integration of IMS in current LC–MS (or GC–MS, CE–MS, SFC–MS) workflows involves certain technical challenges for operators already working in food analysis and for whom this technique is generally unknown. The addition of a third dimension also involves more complex data. This issue hinders data interpretation, especially in the case of food fingerprinting where large datasets are generated. Therefore, the development of simple benchtop LC–IM–MS platforms and user-friendly software for IMS operation and data treatment will be crucial to the success of the implementation of IMS in food analysis.

Author Contributions: M.H.-M. conceptualized and wrote the manuscript. A.M.G.-C., G.D.-P. and B.L.B. provided scientific knowledge about food science and its related challenges. D.R., H.R., G.D.-P., and B.L.B. contributed to develop the ideas about the potential of IMS for food analysis. D.R., H.R., A.M.G.-C., G.D.-P. and B.L.B. reviewed and edited the manuscript. H.R. and A.M.G.-C. provided scientific and administrative support to M.H.-M.'s fellowship project. G.D.-P. and B.L.B. initiated and supervised the project. All authors edited, read and approved the final manuscript.

Funding: M.H.-M. was granted a postdoctoral fellowship (University Research Plan, Program "Perfeccionamiento de doctores en el extranjero 2017") by the University of Granada (Spain). This research received no other external funding.

Conflicts of Interest: The authors declare no conflict of interest.

References

1. Lehotay, S.J.; Chen, Y. Hits and misses in research trends to monitor contaminants in foods. *Anal. Bioanal. Chem.* **2018**, *410*, 5331–5351. [CrossRef] [PubMed]
2. Medina, S.; Pereira, J.A.; Silva, P.; Perestrelo, R.; Câmara, J.S. Food fingerprints—A valuable tool to monitor food authenticity and safety. *Food Chem.* **2019**, *278*, 144–162. [CrossRef] [PubMed]
3. Zheng, X.; Wojcik, R.; Zhang, X.; Ibrahim, Y.M.; Burnum-Johnson, K.E.; Orton, D.J.; Monroe, M.E.; Smith, R.D.; Baker, E.S. Coupling front-end separations, ion mobility spectrometry, and mass spectrometry for enhanced multidimensional biological and environmental analyses. *Annu. Rev. Anal. Chem.* **2017**, *10*, 71–92. [CrossRef] [PubMed]
4. D'Atri, V.; Causon, T.; Hernandez-Alba, O.; Mutabazi, A.; Veuthey, J.-L.; Cianferani, S.; Guillarme, D. Adding a new separation dimension to MS and LC-MS: What is the utility of ion mobility spectrometry? *J. Sep. Sci.* **2018**, *41*, 20–67. [CrossRef] [PubMed]
5. Vautz, W.; Zimmermann, D.; Hartmann, M.; Baumbach, J.I.; Nolte, J.; Jung, J. Ion mobility spectrometry for food quality and safety. *Food Addit. Contam.* **2006**, *23*, 1064–1073. [CrossRef] [PubMed]
6. Arce, L.; Valcárcel, M. The role of ion mobility spectrometry to support the food protected designation of origin. In *Comprehensive Analytical Chemistry—Food Protected Designation of Origin: Methodologies and Applications*; de la Guardia, M., González, A., Eds.; Elsevier: Amsterdam, The Netherlands, 2013; Volume 60, pp. 221–249.
7. Karpas, Z. Applications of ion mobility spectrometry (IMS) in the field of foodomics. *Food Res. Int.* **2013**, *54*, 1146–1151. [CrossRef]
8. Hernández-Mesa, M.; Escorrou, A.; Monteau, F.; Le Bizec, B.; Dervilly-Pinel, G. Current applications and perspectives of ion mobility spectrometry to answer chemical food safety issues. *TrAC Trends Anal. Chem.* **2017**, *94*, 39–53. [CrossRef]
9. Uetrecht, C.; Rose, R.J.; van Duijn, E.; Lorenzen, K.; Heck, A.J.R. Ion mobility mass spectrometry of proteins and protein assemblies. *Chem. Soc. Rev.* **2010**, *39*, 1633–1655. [CrossRef]
10. Gabelica, V.; Marklund, E. Fundamentals of ion mobility spectrometry. *Curr. Opin. Chem. Biol.* **2018**, *42*, 51–59. [CrossRef]
11. May, J.C.; Morris, C.B.; McLean, J.A. Ion mobility collision cross section compendium. *Anal. Chem.* **2017**, *89*, 1032–1044. [CrossRef]
12. Revercomb, H.E.; Mason, E.A. Theory of plasma chromatography/gaseous electrophoresis—A review. *Anal. Chem.* **1975**, *47*, 970–983. [CrossRef]
13. Creaser, C.S.; Griffiths, J.R.; Bramwell, C.J.; Noreen, S.; Hill, C.A.; Thomas, C.L.P. Ion mobility spectrometry: A review. Part 1. Structural analysis by mobility measurement. *Analyst* **2004**, *129*, 984–994. [CrossRef]
14. Lanucara, F.; Holman, S.W.; Gray, C.J.; Eyers, C.E. The power of ion mobility-mass spectrometry for structural characterization and the study of conformational dynamics. *Nat. Chem.* **2014**, *6*, 281–294. [CrossRef] [PubMed]
15. Picache, J.A.; Rose, B.S.; Balinski, A.; Leaptrot, K.L.; Sherrod, S.D.; May, J.C.; McLean, J.A. Collision cross section compendium to annotate and predict multi-omic compound identities. *Chem. Sci.* **2019**, *10*, 983–993. [CrossRef] [PubMed]
16. Mairinger, T.; Causon, T.J.; Hann, S. The potential of ion mobility-mass spectrometry for non-targeted metabolomics. *Curr. Opin. Chem. Biol.* **2018**, *42*, 9–15. [CrossRef] [PubMed]
17. Ewing, M.A.; Glover, M.S.; Clemmer, D.E. Hybrid ion mobility and mass spectrometry as a separation tool. *J. Chromatogr. A* **2016**, *1439*, 3–25. [CrossRef] [PubMed]
18. May, J.C.; McLean, J.A. Ion mobility-mass spectrometry: Time-dispersive instrumentation. *Anal. Chem.* **2015**, *87*, 1422–1438. [CrossRef] [PubMed]
19. Schneider, B.B.; Nazarov, E.G.; Londry, F.; Vouros, P.; Covey, T.R. Differential mobility spectrometry/mass spectrometry history, theory, design optimization, simulations, and applications. *Mass Spectrom. Rev.* **2016**, *35*, 687–737. [CrossRef] [PubMed]
20. Gabelica, V.; Shvartsburg, A.A.; Afonso, C.; Barran, P.; Benesch, J.L.P.; Bleiholder, C.; Bowers, M.T.; Bilbao, A.; Bush, M.F.; Campbell, J.L.; et al. Recommendations for reporting ion mobility mass spectrometry measurements. *Mass Spectrom. Rev.* **2019**, *38*, 291–320. [CrossRef]

21. Nichols, C.M.; May, J.C.; Sherrod, S.D.; McLean, J.A. Automated flow injection method for the high precision determination of drift tube ion mobility collision cross sections. *Analyst* **2018**, *143*, 1556–1559. [CrossRef]

22. Stow, S.M.; Causon, T.J.; Zheng, X.; Kurulugama, R.T.; Mairinger, T.; May, J.C.; Rennie, E.E.; Baker, E.S.; Smith, R.D.; McLean, J.A.; et al. An interlaboratory evaluation of drift tube ion mobility-mass spectrometry collision cross section measurements. *Anal. Chem.* **2017**, *89*, 9048–9055. [CrossRef] [PubMed]

23. Kirk, A.T.; Bohnhorst, A.; Raddatz, C.-R.; Allers, M.; Zimmermann, S. Ultra-high-resolution ion mobility spectrometry—Current instrumentation, limitations, and future developments. *Anal. Bioanal. Chem.* **2019**, in press. [CrossRef] [PubMed]

24. Ridgeway, M.E.; Lubeck, M.; Jordens, J.; Mann, M.; Park, M.A. Trapped ion mobility spectrometry: A short review. *Int. J. Mass Spectrom.* **2018**, *425*, 22–35. [CrossRef]

25. Michelmann, K.; Silveira, J.A.; Ridgeway, M.E.; Park, M.A. Fundamentals of trapped ion mobility spectrometry. *J. Am. Soc. Mass Spectrom.* **2015**, *26*, 14–24. [CrossRef]

26. Cumeras, R.; Figueras, E.; Davis, C.E.; Baumbach, J.I.; Gràcia, I. Review on ion mobility spectrometry. Part 1: Current instrumentation. *Analyst* **2015**, *140*, 1376–1390. [CrossRef] [PubMed]

27. Eiceman, G.A.; Karpas, Z. *Ion. Mobility Spectrometry*, 2nd ed.; CRC Press, Taylor & Francis: Boca Raton, FL, USA, 2005.

28. Shvartsburg, A.A.; Smith, R.D. Fundamentals of traveling wave ion mobility spectrometry. *Anal. Chem.* **2008**, *80*, 9689–9699. [CrossRef]

29. Causon, T.J.; Hann, S. Theoretical evaluation of peak capacity improvements by use of liquid chromatography combined with drift tube ion mobility-mass spectrometry. *J. Chromatogr. A* **2015**, *1416*, 47–56. [CrossRef] [PubMed]

30. Hernández-Mesa, M.; Monteau, F.; Le Bizec, B.; Dervilly-Pinel, G. Potential of ion mobility-mass spectrometry for both targeted and non-targeted analysis of phase II steroid metabolites in urine. *Anal. Chim. Acta X* **2019**, *1*, 100006. [CrossRef]

31. Regueiro, J.; Negreira, N.; Hannisdal, R.; Berntssen, M.H.G. Targeted approach for qualitative screening of pesticides in salmon feed by liquid chromatography coupled to traveling-wave ion mobility/quadrupole time-of-flight mass spectrometry. *Food Control* **2017**, *78*, 116–125. [CrossRef]

32. Kanu, A.B.; Hill, H.H., Jr. Ion mobility spectrometry detection for gas chromatography. *J. Chromatogr. A* **2008**, *1177*, 12–27. [CrossRef]

33. Lapthorn, C.; Pullen, F.; Chowdhry, B.Z. Ion mobility spectrometry-mass spectrometry (IMS-MS) of small molecules: Separating and assigning structures to ions. *Mass Spectrom. Rev.* **2013**, *32*, 43–71. [CrossRef] [PubMed]

34. Haynes, S.E.; Polasky, D.A.; Dixit, S.M.; Majmudar, J.D.; Neeson, K.; Ruotolo, B.T.; Martin, B.R. Variable-velocity traveling-wave ion mobility separation enhancing peak capacity for data-independent acquisition proteomics. *Anal. Chem.* **2017**, *89*, 5669–5672. [CrossRef] [PubMed]

35. Arthur, K.L.; Turner, M.A.; Reynolds, J.C.; Creaser, C.S. Increasing peak capacity in nontargeted omics applications by combining full scan field asymmetric waveform ion mobility spectrometry with liquid chromatography-mass spectrometry. *Anal. Chem.* **2017**, *89*, 3452–3459. [CrossRef] [PubMed]

36. Hill, H.H.; Simpson, G. Capabilities and limitations of ion mobility spectrometry for field screening applications. *Field Anal. Chem. Technol.* **1997**, *1*, 119–134. [CrossRef]

37. Kaufmann, A.; Walker, S. Comparison of linear intrascan and interscan dynamic ranges of Orbitrap and ion-mobility time-of-flight mass spectrometers. *Rapid Commun. Mass Spectrom.* **2017**, *31*, 1915–1926. [CrossRef] [PubMed]

38. Kaufmann, A.; Butcher, P.; Maden, K.; Widmer, M.; Giles, K.; Uría, D. Are liquid chromatography/electrospray tandem quadrupole fragmentation rations unequivocal confirmation criteria? *Rapid Commun. Mass Spectrom.* **2009**, *23*, 958–998. [CrossRef] [PubMed]

39. Bauer, A.; Kuballa, J.; Rohn, S.; Jantzen, E.; Luetjohann, J. Evaluation and validation of an ion mobility quadrupole time-of-flight mass spectrometry pesticide screening approach. *J. Sep. Sci.* **2018**, *41*, 2178–2187. [CrossRef]

40. Tu, J.; Zhou, Z.; Li, T.; Zhu, Z.-J. The emerging role of ion mobility-mass spectrometry in lipidomics to facilitate lipid separation and identification. *TrAC Trends Anal. Chem.* **2019**, *116*, 332–339. [CrossRef]

41. Chouinard, C.D.; Nagy, G.; Smith, R.D.; Baker, E.S. Ion mobility-mass spectrometry in metabolomic, lipidomic, and proteomic analyses. *Compr. Anal. Chem.* **2019**, *83*, 123–159. [CrossRef]

42. Manz, C.; Pagel, K. Glycan analysis by ion mobility-mass spectrometry and gas-phase spectroscopy. *Curr. Opin. Chem. Biol.* **2018**, *42*, 16–24. [CrossRef]

43. Kuang, A.; Erlund, I.; Herder, C.; Westerhuis, J.A.; Tuomilehto, J.; Cornelis, M.C. Lipidomic response to coffee consumption. *Nutrients* **2018**, *10*, 1851. [CrossRef] [PubMed]

44. Blaženović, I.; Shen, T.; Mehta, S.S.; Kind, T.; Ji, J.; Piparo, M.; Cacciola, F.; Mondello, L.; Fiehn, O. Increasing compound identification rates in untargeted lipidomics research with liquid chromatography drift time-ion mobility mass spectrometry. *Anal. Chem.* **2018**, *90*, 10758–10764. [CrossRef] [PubMed]

45. López-Morales, C.A.; Vázquez-Leyva, S.; Vallejo-Castillo, L.; Carballo-Uicab, G.; Muñoz-García, L.; Herbet-Pucheta, J.E.; Zepeda-Vallejo, L.G.; Velasco-Velázquez, M.; Pavón, L.; Pérez-Tapia, S.M.; et al. Determination of peptide profile consistency and safety of collagen hydrolysates as quality attributes. *J. Food Sci.* **2019**, *430*–439. [CrossRef] [PubMed]

46. Causon, T.J.; Došen, M.; Reznicek, G.; Hann, S. Workflow development for the analysis of phenolic compounds in wine using liquid chromatography combined with drift-tube ion mobility-mass spectrometry. *LC-GC N. Am.* **2016**, *34*, 854–867.

47. Rodríguez-Maecker, R.; Vyhmeister, E.; Meisen, S.; Martinez Rosales, A.; Kuklya, A.; Telgheder, U. Identification of terpenes and essential oils by means of static headspace gas chromatography-ion mobility spectrometry. *Anal. Bioanal. Chem.* **2017**, *409*, 6595–6603. [CrossRef] [PubMed]

48. Alves, T.O.; D'Almeida, C.T.S.; Victorio, V.C.M.; Souza, G.H.M.F.; Cameron, L.C.; Ferreira, M.S.L. Immunogenic and allergenic profile of wheat flours from different technological qualities revealed by ion mobility mass spectrometry. *J. Food Compos. Anal.* **2018**, *73*, 67–75. [CrossRef]

49. Garrido-Delgado, R.; Dobao-Prieto, M.M.; Arce, L.; Valcárcel, M. Determination of volatile compounds by GC-IMS to assign the quality of virgin olive oil. *Food Chem.* **2015**, *187*, 572–579. [CrossRef]

50. Guo, Y.; Chen, D.; Dong, Y.; Ju, H.; Wu, C.; Lin, S. Characteristic volatiles fingerprints and changes of volatile compounds in fresh and dried *Tricholoma matsutake* Singer by HS-GC-IMS and HS-SPME-GC-MS. *J. Chromatogr. B* **2018**, *1099*, 46–55. [CrossRef]

51. Browne, C.A.; Forbes, T.P.; Sisco, E. Detection and identification of sugar alcohol sweeteners by ion mobility spectrometry. *Anal. Methods* **2016**, *8*, 5611–5618. [CrossRef]

52. Contreras, M.D.M.; Jurado-Campos, N.; Arce, L.; Arroyo-Manzanares, N. A robustness study of calibration models for olive oil classification: Targeted and non-targeted fingerprint approaches based on GC-IMS. *Food Chem.* **2019**, *288*, 315–324. [CrossRef]

53. Gerhardt, N.; Schwolow, S.; Rohn, S.; Pérez-Cacho, P.R.; Galán-Soldevilla, H.; Arce, L.; Weller, P. Quality assessment of olive oils based on temperature-ramped HS-GC-IMS and sensory evaluation: Comparison of different processing approaches by LDA, kNN, and SVM. *Food Chem.* **2019**, *278*, 720–728. [CrossRef] [PubMed]

54. Liedtke, S.; Seifert, L.; Ahlmann, N.; Hariharan, C.; Franzke, J.; Vautz, W. Coupling laser desorption with gas chromatography and ion mobility spectrometry for improved olive oil characterization. *Food Chem.* **2018**, *255*, 323–331. [CrossRef] [PubMed]

55. Li, J.; Yuan, H.; Yao, Y.; Hua, J.; Yang, Y.; Dong, C.; Deng, Y.; Wang, J.; Li, H.; Jiang, Y.; et al. Rapid volatiles fingerprinting by dopant-assisted positive photoionization ion mobility spectrometry for discrimination and characterization of Green Tea aromas. *Talanta* **2019**, *191*, 39–45. [CrossRef] [PubMed]

56. Lu, Y.; Guo, J.; Yu, J.; Guo, J.; Jia, X.; Liu, W.; Tian, P. Two-dimensional analysis of phenolic acids in seedling roots by high performance liquid chromatography electrospray ionization-ion mobility spectrometry. *Anal. Methods* **2019**, *11*, 610–617. [CrossRef]

57. Struwe, W.B.; Baldauf, C.; Hofmann, J.; Rudd, P.M.; Pagel, K. Ion mobility separation of deprotonated oligosaccharide isomers—Evidence for gas-phase charge migration. *Chem. Commun.* **2016**, 12353–12356. [CrossRef] [PubMed]

58. Venter, P.; Muller, M.; Vestner, J.; Stander, M.A.; Tredoux, A.G.J.; Pasch, H.; De Villiers, A. Comprehensive three-dimensional LC × LC × ion mobility spectrometry separation combined with high-resolution MS for the analysis of complex samples. *Anal. Chem.* **2018**, *90*, 11643–11650. [CrossRef] [PubMed]

59. Stephan, S.; Jakob, C.; Hippler, J.; Schmitz, O.J. A novel four-dimensional analytical approach for analysis of complex samples. *Anal. Bioanal. Chem.* **2016**, *408*, 3751–3759. [CrossRef]

60. Zhang, H.; Jiang, J.M.; Zheng, D.; Yuan, M.; Wang, Z.Y.; Zhang, H.M.; Zheng, C.W.; Xiao, L.B.; Xu, H.X. A multidimensional analytical approach based on time-decoupled online comprehensive two-dimensional liquid chromatography coupled with ion mobility quadrupole time-of-flight mass spectrometry for the analysis of ginsenosides from white and red ginsengs. *J. Pharm. Biomed. Anal.* **2019**, *163*, 24–33. [CrossRef]

61. Lipok, C.; Hippler, J.; Schmitz, O.J. A four dimensional separation method based on continuous heart-cutting gas chromatography with ion mobility and high resolution mass spectrometry. *J. Chromatogr. A* **2018**, *1536*, 50–57. [CrossRef]

62. McCullagh, M.; Douce, D.; Hoeck, V.; Goscinny, S. Exploring the complexity of steviol glycosides analysis using ion mobility mass spectrometry. *Anal. Chem.* **2018**, *90*, 4585–4595. [CrossRef]

63. Karpas, Z.; Guamán, A.V.; Pardo, A.; Marco, S. Comparison of the performance of three ion mobility spectrometers for measurement of biogenic amines. *Anal. Chim. Acta* **2013**, *758*, 122–129. [CrossRef] [PubMed]

64. Cohen, G.; Rudnik, D.D.; Laloush, M.; Yakir, D.; Karpas, Z. A novel method for determination of histamine in tuna fish by ion mobility spectrometry. *Food Anal. Methods* **2015**, *8*, 2376–2382. [CrossRef]

65. Cheng, S.; Li, H.; Jiang, D.; Chen, C.; Zhang, T.; Li, Y.; Wang, H.; Zhou, Q.; Li, H.; Tan, M. Sensitive detection of trimethylamine based on dopant-assisted positive photoionization ion mobility spectrometry. *Talanta* **2017**, *162*, 398–402. [CrossRef] [PubMed]

66. Parchami, R.; Kamalabadi, M.; Alizadeh, N. Determination of biogenic amines in canned fish samples using head-space solid phase microextraction based on nanostructured polypyrrole fiber coupled to modified ionization región ion mobility spectrometry. *J. Chromatogr. A* **2017**, *1481*, 37–43. [CrossRef] [PubMed]

67. Zarpas, Z.; Guamán, A.V.; Calvo, D.; Pardo, A.; Marco, S. The potential of ion mobility spectrometry (IMS) for detection of 2,4,6-trichloroanisole (2,4,6-TCA) in wine. *Talanta* **2012**, *93*, 200–205. [CrossRef]

68. Kamalabadi, M.; Ghaemi, E.; Mohammadi, A.; Alizadeh, N. Determination of furfural and hydroxymethylfurfural from baby formula using headspace microextraction base don nanostrutured polypyrrole fiber coupled with ion mobility spectrometry. *Food Chem.* **2015**, *181*, 72–77. [CrossRef]

69. Garrido-Delgado, R.; Dobao-Prieto, M.M.; Arce, L.; Aguilar, J.; Cumplido, J.L.; Valcárcel, M. Ion mobility spectrometry versus classical physico-chemical analysis for assessing the shelf life of extra virgin olive oil according to container type and storage conditions. *J. Agric. Food Chem.* **2015**, *63*, 2179–2188. [CrossRef]

70. Tzschoppe, M.; Haase, H.; Höhnisch, M.; Jaros, D.; Rohm, H. Using ion mobility spectrometry for screening the autoxidation of peanuts. *Food Control.* **2016**, *64*, 17–21. [CrossRef]

71. Raatikainen, O.; Reinikainen, V.; Minkkinen, P.; Ritvanen, T.; Muje, P.; Pursiainen, J.; Hiltunen, T.; Hyvönen, P.; Von Wright, A.; Reinikainen, S.-P. Multivariate modelling of fish freshness index based on ion mobility spectrometry measurements. *Anal. Chim. Acta* **2005**, *544*, 128–134. [CrossRef]

72. Cavanna, D.; Zanardi, S.; Dall'Asta, C.; Suman, M. Ion mobility spectrometry coupled to gas chromatography: A rapid tool to assess eggs freshness. *Food Chem.* **2019**, *271*, 691–696. [CrossRef]

73. Tang, Z.-S.; Zeng, X.-A.; Brennan, M.A.; Han, Z.; Niu, D.; Huo, Y. Characterization of aroma profile and characteristic aromas during lychee wine fermentation. *J. Food Process. Preserv.* **2019**, in press. [CrossRef]

74. Halbfeld, C.; Ebert, B.E.; Blank, L.M. Multi-capillary column-ion mobility spectrometry of volatile metabolites emitted by *Saccharomyces Cerevisiae*. *Metabolites* **2014**, *4*, 751–774. [CrossRef] [PubMed]

75. Gloess, A.N.; Yeretzian, C.; Knochenmuss, R.; Groessl, M. On-line analysis of coffee roasting with ion mobility spectrometry-mass spectrometry (IMS-MS). *Int. J. Mass Spectrom.* **2018**, *424*, 49–57. [CrossRef]

76. Danezis, G.P.; Tsagkaris, A.S.; Camin, F.; Brusic, V.; Georgiou, C.A. Food authentication: Techniques, trends & emerging approaches. *TrAC Trends Anal. Chem.* **2016**, *85*, 123–132. [CrossRef]

77. Cubero-Leon, E.; Peñalver, R.; Maquet, A. Review on metabolomics for food authentication. *Food Res. Int.* **2014**, *60*, 95–107. [CrossRef]

78. Causon, T.J.; Ivanova-Petropulos, V.; Petrusheva, D.; Bogeva, E.; Hann, S. Fingerprinting of traditionally produced red wines using liquid chromatography combined with drift tube ion mobility-mass spectrometry. *Anal. Chim. Acta* **2019**, *1052*, 179–189. [CrossRef] [PubMed]

79. Garrido-Delgado, R.; Mercader-Trejo, F.; Sielemann, S.; de Bruyn, W.; Arce, L.; Valcárcel, M. Direct classification of olive oils by using two types of ion mobility spectrometers. *Anal. Chim. Acta* **2011**, *696*, 108–115. [CrossRef]

80. Gerhardt, N.; Birkenmeier, M.; Sanders, D.; Rohn, S.; Weller, P. Resolution-optimized headspace gas chromatography-ion mobility spectrometry (HS-GC-IMS) for non-targeted olive oil profiling. *Anal. Bianal. Chem.* **2017**, *409*, 3933–3942. [CrossRef] [PubMed]

81. Arroyo-Manzanares, N.; Martín-Gómez, A.; Jurado-Campos, N.; Garrido-Delgado, R.; Arce, C.; Arce, L. Target vs spectral fingerprint data analysis of Iberian ham samples for avoiding labelling fraud using headspace– gas chromatography—ion mobility spectrometry. *Food Chem.* **2018**, *246*, 65–73. [CrossRef] [PubMed]

82. Martín-Gómez, A.; Arroyo-Manzanares, N.; Rodríguez-Estévez, V.; Arce, L. Use of a non-destructive sampling method for characterization of Iberian cured ham breed and feeding regime using GC-IMS. *Meat Sci.* **2019**, 146–154. [CrossRef]

83. Garrido-Delgado, R.; Arce, L.; Guamán, A.V.; Pardo, A.; Marco, S.; Valcárcel, M. Direct coupling of a gas-liquid separator to an ion mobility spectrometer for the classification of different white wines using chemometrics tools. *Talanta* **2011**, *84*, 471–479. [CrossRef] [PubMed]

84. Gerhardt, N.; Birkenmeier, M.; Schwolow, S.; Rohn, S.; Weller, P. Volatile-compound fingerprinting by headspace-gas-chromatography ion-mobility spectrometry (HS-GC-IMS) as a benchtop alternative to ^{1}H NMR profiling for assessment of the authenticity of honey. *Anal. Chem.* **2018**, *90*, 1777–1785. [CrossRef] [PubMed]

85. Wang, X.; Yang, S.; He, J.; Chen, L.; Zhang, J.; Jin, Y.; Zhou, J.; Zhang, Y. A green triple-locked strategy based on volatile-compound imaging, chemometrics, and markers to discriminate winter honey and *sapium* honey using headspace gas chromatography-ion mobility spectrometry. *Food Res. Int.* **2019**, *119*, 960–967. [CrossRef] [PubMed]

86. Chen, T.; Chen, X.; Lu, D.; Chen, B. Detection of adulteration in canola oil by using GC-IMS and chemometric analysis. *Int. J. Anal. Chem.* **2018**, *2018*, 3160265. [CrossRef] [PubMed]

87. Othman, A.; Goggin, K.A.; Tahir, N.I.; Brodrick, E.; Singh, R.; Sambanthamurthi, R.; Parveez, K.A.; Davies, A.N.; Murad, A.J.; Muhammad, N.H.; et al. Use of headspace-gas chromatography-ion mobility spectrometry to detect volatile fingerprints of palm fibre oil and sludge palm oil in samples of crude palm oil. *BMC Res. Notes* **2019**, *12*, 229. [CrossRef] [PubMed]

88. Garrido-Delgado, R.; Muñoz-Pérez, E.; Arce, L. Detection of adulteration in extra virgin olive oils by using UV-IMS and chemometric analysis. *Food Control.* **2018**, *85*, 292–299. [CrossRef]

89. Zhang, L.; Shuai, Q.; Li, P.; Zhang, Q.; Ma, F.; Zhang, W.; Ding, X. Ion mobility spectrometry fingerprints: A rapid detection technology for adulteration of sesame oil. *Food Chem.* **2016**, *192*, 60–66. [CrossRef]

90. Shuai, Q.; Zhang, L.; Li, P.; Zhang, Q.; Wang, X.; Ding, X.; Zhang, W. Rapid adulteration detection for flaxseed oil using ion mobility spectrometry and chemometric methods. *Anal. Methods* **2014**, *6*, 9575–9580. [CrossRef]

91. Khademi, S.M.S.; Telgheder, U.; Valadbeigi, Y.; Ilbeigi, V.; Tabrizchi, M. Direct detection of glyphosate in drinking water using corona-discharge ion mobility spectrometry: A theoretical and experimental study. *Int. J. Mass Spectrom.* **2019**, *442*, 29–34. [CrossRef]

92. European Commission. *Guidance Document on Analytical Quality Control and Method Validation Procedures for Pesticides Residues Analysis in Food and Feed*; SANTE/11813/2017; European Commission: Brussels, Belgium, 2017.

93. European Commission. Commission Decision (EEC) 2002/657/EC. *Off. J. Eur. Commun.* **2002**, *L221*, 8.

94. European Commission. *Guidance Document on Identification of Mycotoxins in Food and Feed*; SANTE/12089 /2016; European Commission: Brussels, Belgium, 2016.

95. Wageningen University & Research. Available online: https://www.wur.nl/upload_mm/e/6/e/0d3c53a2-28b8-4e4d-a436-b84cf471f20e_20181015-9%20Revision%202002-657%20confirmation%20criteria.pdf (accessed on 24 July 2019).

96. Weickhardt, C.; Kaiser, N.; Borsdorf, H. Ion mobility spectrometry of laser desorbed pesticides from fruit surfaces. *Int. J. Ion. Mobil. Spectrom.* **2012**, *15*, 55–62. [CrossRef]

97. Armenta, S.; de la Guardia, M.; Abad-Fuentes, A.; Abad-Somovilla, A.; Esteve-Turrillas, F.A. Off-line coupling of multidimensional immunoaffinity chromatography and ion mobility spectrometry: A promising partnership. *J. Chromatogr. A* **2015**, *1426*, 110–117. [CrossRef] [PubMed]

98. Sheibani, A.; Tabrizchi, M.; Ghaziaskar, H.S. Determination of aflatoxins B1 and B2 using ion mobility spectrometry. *Talanta* **2008**, *75*, 233–238. [CrossRef] [PubMed]

99. Aria, A.A.; Sorribes-Soriano, A.; Jafari, M.T.; Nourbakhsh, F.; Esteve-Turrilas, F.A.; Armenta, S.; Herrero-Martínez, J.M.; de la Guardia, M. Uptake and translocation monitoring of imidacloprid to chili and tomato plants by molecularly imprinting extraction—Ion mobility spectrometry. *Microchem. J.* **2019**, *144*, 195–202. [CrossRef]

100. Saraji, M.; Jafari, M.T.; Mossaddegh, M. Carbon nanotubes@silicon dioxide nanohybrids coating for solid-phase microextraction of organophosphorus pesticides followed by gas chromatography-corona discharge ion mobility spectrometric detection. *J. Chromatogr. A* **2016**, 30–39. [CrossRef] [PubMed]

101. Kermani, M.; Jafari, M.; Saraji, M. Porous magnetized carbon sheet nanocomposites for dispersive solid-phase microextraction of organophosphorus pesticides prior to analysis by gas chromatography-ion mobility spectrometry. *Microchim. Acta* **2019**, *186*, 88. [CrossRef] [PubMed]

102. McCooeye, M.; Kolakowski, B.; Boison, J.; Mester, Z. Evaluation of high-field asymmetric waveform ion mobility spectrometry mass spectrometry for the analysis of the mycotoxin zearalenone. *Anal. Chim. Acta* **2008**, *627*, 112–116. [CrossRef] [PubMed]

103. Xu, Z.; Li, J.; Chen, A.; Ma, X.; Yang, S. A new retrospective, multi-evidence veterinary drug screening method using drift tube ion mobility mass spectrometry. *Rapid Commun. Mass Spectrom.* **2018**, *32*, 1141–1148. [CrossRef]

104. Ahmed, E.; Kabir, K.M.M.; Wang, H.; Xiao, D.; Fletcher, J.; Donald, W.A. Rapid separation of isomeric perfluoroalkyl substances by high-resolution differential ion mobility mass spectrometry. *Anal. Chim. Acta* **2019**, *1058*, 127–135. [CrossRef]

105. Beach, D.G.; Melanson, J.E.; Purves, R.W. Analyis of paralytic shellfish toxins using high-field asymmetric waveform ion mobility spectrometry with liquid chromatography-mass spectrometry. *Anal. Bioanal. Chem.* **2015**, *407*, 2473–2484. [CrossRef]

106. Poyer, S.; Loutelier-Bourhis, C.; Coadou, G.; Mondeguer, F.; Enche, J.; Bossée, A.; Hess, P.; Afonso, C. Identification and separation of saxitoxins using hydrophilic interaction liquid chromatography coupled to travelling wave ion mobility-mass spectrometry. *J. Mass Spectrom.* **2015**, *50*, 175–181. [CrossRef] [PubMed]

107. Beucher, L.; Dervilly-Pinel, G.; Prévost, S.; Monteau, F.; Le Bizec, B. Determination of a large set of β-adrenergic agonists in animal matrices based on ion mobility and mass separations. *Anal. Chem.* **2015**, *87*, 9234–9242. [CrossRef] [PubMed]

108. Zheng, X.; Dupuis, K.T.; Aly, N.A.; Zhou, Y.; Smith, F.B.; Tang, K.; Smith, R.D.; Baker, E.S. Utilizing ion mobility spectrometry and mass spectrometry for the analysis of polyciyclic aromatic hydrocarbons, polychlorinated biphenyls, polybrominated diphenyl ethers and their metabolites. *Anal. Chim. Acta* **2018**, *1037*, 265–273. [CrossRef] [PubMed]

109. Varesio, E.; Le Blanc, J.C.Y.; Hopfgartner, G. Real-time 2D separation by LC x differential ion mobility hyphenated to mass spectrometry. *Anal. Bioanal. Chem.* **2012**, *402*, 2555–2564. [CrossRef] [PubMed]

110. Beach, D.G.; Kerrin, E.S.; Quilliam, M.A. Selective quantitation of the neurotoxin BMAA by use of hydrophilic-interaction liquid chromatography-differential mobility spectrometry-tandem mass spectrometry (HILIC-DMS-MS/MS). *Anal. Bioanal. Chem.* **2015**, *407*, 8397–8409. [CrossRef] [PubMed]

111. Goscinny, S.; Joly, L.; De Pauw, E.; Hanot, V.; Eppe, G. Travelling-wave ion mobility time-of-flight mass spectrometry as an alternative strategy for screening of multi-class pesticides in fruits and vegetables. *J. Chromatogr. A* **2015**, *1405*, 85–93. [CrossRef] [PubMed]

112. Goscinny, S.; McCullagh, M.; Far, J.; De Pauw, E.; Eppe, G. Towards the use of ion mobility mass spectrometry derived collision cross section as a screening approach for unambiguous identification of targeted pesticides in food. *Rapid Commun. Mass Spectrom.* **2019**, *33*, 34–48. [CrossRef]

113. Regueiro, J.; Negreira, N.; Berntssen, M.H.G. Ion-mobility-derived collision cross section as an additional identification point for multiresidue screening of pesticides in fish feed. *Anal. Chem.* **2016**, *88*, 11169–11177. [CrossRef]

114. Hines, K.M.; Ross, D.H.; Davidson, K.L.; Bush, M.F.; Xu, L. Large-scale structural characterization of drug and drug-like compounds by high-throughput ion mobility-mass spectrometry. *Anal. Chem.* **2017**, *89*, 9023–9030. [CrossRef]

115. Hernández-Mesa, M.; Le Bizec, B.; Monteau, F.; García-Campaña, A.M.; Dervilly-Pinel, G. Collision Cross Section (CCS) database: An additional measure to characterize steroids. *Anal. Chem.* **2018**, *90*, 4616–4625. [CrossRef]

116. Tejada-Casado, C.; Hernández-Mesa, M.; Monteau, F.; Lara, F.J.; del Olmo-Iruela, M.; García-Campaña, A.M.; Le Bizec, B.; Dervilly-Pinel, G. Collision cross section (CCS) as a complementary parameter to characterize human and veterinary drugs. *Anal. Chim. Acta* **2018**, *1043*, 52–63. [CrossRef] [PubMed]

117. Righetti, L.; Bergmann, A.; Galaverna, G.; Rolfsson, O.; Paglia, G.; Dall'Asta, C. Ion mobility-derived collision cross section database: Application to mycotoxin analysis. *Anal. Chim. Acta* **2018**, *1014*, 50–57. [CrossRef] [PubMed]

118. Righetti, L.; Fenclova, M.; Dellafiora, L.; Hajslova, J.; Stranska-Zachariasova, M.; Dall'Asta, C. High resolution-ion mobility mass spectrometry as an additional powerful tool for structural characterization of mycotoxin metabolites. *Food Chem.* **2018**, *245*, 768–771. [CrossRef] [PubMed]

119. Bauer, A.; Luetjohann, J.; Hanschen, F.S.; Schreiner, M.; Kuballa, J.; Jantzen, E.; Rohn, S. Identification and characterization of pesticide metabolites in *Brassica* species by liquid chromatography travelling wave ion mobility quadrupole time-of-flight mass spectrometry (UPLC-TWIMS-QTOF-MS). *Food Chem.* **2018**, *244*, 292–303. [CrossRef] [PubMed]

120. Dodds, J.N.; May, J.C.; McLean, J.A. Correlating resolving power, resolution, and collision cross section: Unifying cross-platform assessment of separation efficiency in ion mobility spectrometry. *Anal. Chem.* **2019**, *89*, 12176–12184. [CrossRef] [PubMed]

121. Deng, L.; Webb, I.K.; Garimella, S.V.B.; Hamid, A.M.; Zheng, X.; Norheim, R.V.; Prost, S.A.; Anderson, G.A.; Sandoval, J.A.; Baker, E.S.; et al. Serpentine ultralong path with extended routing (SUPER) high resolution traveling wave ion mobility-MS using structures for lossless ion manipulations. *Anal. Chem.* **2017**, *89*, 4628–4634. [CrossRef] [PubMed]

122. Merenbloom, S.I.; Glaskin, R.S.; Henson, Z.B.; Clemmer, D.E. High-resolution ion cyclotron mobility spectrometry. *Anal. Chem.* **2009**, *81*, 1482–1487. [CrossRef] [PubMed]

123. Glaskin, R.S.; Ewing, M.A.; Clemmer, D.E. Ion trapping for ion mobility spectrometry measurements in a cyclical drift tube. *Anal. Chem.* **2013**, *85*, 7003–7008. [CrossRef]

124. Giles, K.; Wildgoose, J.; Pringle, S.; Garside, J.; Carney, P.; Nixon, P.; Langridge, D. Design and utility of a multi-pass cyclic ion mobility separator. In *Annual Conference Proceedings*; ASMS: Baltimore, MD, USA, 2014.

125. Ujma, J.; Ropartz, D.; Giles, K.; Richardson, K.; Langridge, D.; Wildgoose, J.; Green, M. Cyclic ion mobility mass spectrometry distinguishes anomers and open-ring forms of pentasaccharides. *J. Am. Soc. Mass Spectrom.* **2019**, *30*, 1028–1037. [CrossRef]

126. McCullagh, M.; Giles, K.; Richardson, K.; Stead, S.; Palmer, M. Investigations into the performance of travelling wave enabled conventional and cyclic ion mobility systems to characterize protomers of fluoroquinolone antibiotic residues. *Rapid Commun. Mass Spectrom.* **2019**, *33*, 11–21. [CrossRef]

127. Hamid, A.M.; Ibrahim, Y.M.; Garimella, S.V.B.; Webb, I.K.; Deng, L.; Chen, T.-C.; Anderson, G.A.; Prost, S.A.; Norheim, R.V.; Tolmachev, A.V.; et al. Characterization of traveling wave ion mobility separations in structures for lossless ion manipulations. *Anal. Chem.* **2015**, *87*, 11301–11308. [CrossRef] [PubMed]

128. Deng, L.; Ibrahim, Y.M.; Baker, E.S.; Aly, N.A.; Hamid, A.M.; Zhang, X.; Zheng, X.; Garimella, S.V.B.; Webb, I.K.; Prost, S.A.; et al. Ion mobility separations of isomers based upon long path length structures for lossless ion manipulations combined with mass spectrometry. *ChemistrySelect* **2016**, *1*, 2396–2399. [CrossRef] [PubMed]

129. Faleh, A.B.; Warnke, S.; Rizzo, T.R. Combining ultrahigh-resolution ion-mobility spectrometry with cryogenic infrared spectroscopy for the analysis of glycan mixtures. *Anal. Chem.* **2019**, *91*, 4876–4882. [CrossRef] [PubMed]

130. Zhou, Z.; Tu, J.; Zhu, Z.-J. Advancing the large-scale CCS database for metabolomics and lipidomics at the machine-learning era. *Curr. Opin. Chem. Biol.* **2018**, *42*, 34–41. [CrossRef] [PubMed]

131. Bijlsma, L.; Bade, R.; Celma, A.; Mullin, L.; Cleland, G.; Stead, S.; Hernandez, F.; Sancho, J.V. Prediction of collision cross-section values for small molecules: Application to pesticide residue analysis. *Anal. Chem.* **2017**, *89*, 6583–6589. [CrossRef] [PubMed]

132. Morris, C.B.; May, J.C.; Leaptrot, K.L.; McLean, J.A. Evaluating separation selectivity and collision cross section measurement reproducibility in helium, nitrogen, argon, and carbon dioxide drift gases for drift tube ion mobility-mass spectrometry. *J. Am. Soc. Mass Spectrom.* **2019**, *30*, 1059–1068. [CrossRef] [PubMed]

133. Kurulugama, R.T.; Darland, E.; Kuhlmann, F.; Stafford, G.; Fjeldsted, J. Evaluation of drift gas selection in complex sample analysis using a high performance drift tube ion mobility-QTOF mass spectrometer. *Analyst* **2015**, *140*, 6834–6844. [CrossRef]

134. Oranzi, N.R.; Kemperman, R.H.J.; Wei, M.S.; Petkovska, V.I.; Granato, S.W.; Rochon, B.; Kaszycki, J.; La Rotta, A.; Jeanne Dit Fouque, K.; Fernandez-Lima, F.; et al. Measuring the integrity of gas-phase conformers of sodiated 25-hydroxyvitamin d3 by drift tube, traveling wave, trapped, and high-field asymmetric ion mobility. *Anal. Chem.* **2019**, *91*, 4092–4099. [CrossRef]

135. Hinnenkamp, V.; Klein, J.; Mecklmann, S.W.; Balsaa, P.; Schmidt, T.C.; Schmitz, O.J. Comparison of CCS values determined by traveling wave ion mobility mass spectrometry and drift tube ion mobility mass spectrometry. *Anal. Chem.* **2018**, *90*, 12042–12050. [CrossRef]

136. Paglia, G.; Williams, J.P.; Menikarachchi, L.; Thompson, J.W.; Tyldesley-Worster, R.; Halldórson, S.; Rolfsson, O.; Moseley, A.; Grant, D.; Langridge, J.; et al. Ion mobility derived collision cross sections to support metabolomics applications. *Anal. Chem.* **2014**, *86*, 3985–3993. [CrossRef]

137. Paglia, G.; Astarita, G. Metabolomics and lipidomics using traveling-wave ion mobility mass spectrometry. *Nat. Protoc.* **2017**, *12*, 797–813. [CrossRef] [PubMed]

138. Hofmann, J.; Struwe, W.B.; Scarff, C.A.; Scrivens, J.H.; Harvey, D.J.; Pagel, K. Estimating collision cross sections of negatively charged n-glycans using traveling wave ion mobility-mass spectrometry. *Anal. Chem.* **2014**, *86*, 10789–10795. [CrossRef] [PubMed]

139. Hines, K.M.; May, J.C.; McLean, J.A.; Xu, L. Evaluation of collision cross section calibrants for structural analysis of lipids by travelling wave ion mobility-mass spectrometry. *Anal. Chem.* **2016**, *88*, 7329–7336. [CrossRef] [PubMed]

molecules

MDPI

Article

Discrimination of Natural Mature Acacia Honey Based on Multi-Physicochemical Parameters Combined with Chemometric Analysis

Tianchen Ma [1], Haoan Zhao [2], Caiyun Liu [3], Min Zhu [2], Hui Gao [1], Ni Cheng [1,3] and Wei Cao [1,3,*]

[1] School of Food Science and Engineering, Northwest University, Xi'an 710069, China
[2] School of Chemical Engineering, Northwest University, Xi'an 710069, China
[3] Bee Product Research Center of Shaanxi Province, Xi'an 710065, China
* Correspondence: caowei@nwu.edu.cn; Tel./Fax: +86-29-8830-2213

Received: 29 May 2019; Accepted: 18 July 2019; Published: 23 July 2019

check for updates

Abstract: Honey maturity is an important factor in evaluating the quality of honey. We established a method for the identification of natural mature acacia honey with eighteen physicochemical parameters combined with chemometric analysis. The analysis of variance showed significant differences between mature and immature acacia honey in physicochemical parameters. The principal component analysis explained 82.64% of the variance among samples, and indicated that total phenolic content, total protein content, and total sugar (glucose, fructose, sucrose) were the major variables. The cluster analysis and orthogonal partial least squares-discriminant analysis demonstrated that samples were grouped in relation to the maturity coinciding with the results of the principal component analysis. Meanwhile, the 35 test samples were classified with 100% accuracy with the method of multi-physicochemical parameters combined with chemometric analysis. All the results presented above proved the possibility of identifying mature acacia honey and immature acacia honey according to the chemometric analysis based on the multi-physicochemical parameters.

Keywords: natural mature honey; immature honey; chemometric analysis; multi-physicochemical parameters

1. Introduction

Acacia honey is the natural sweet substance produced by honeybees, which collect nectar from the flowers of *Robinia pesudoacacia* (Figure 1), transform and combine it with specific substances of their own, store it, and leave it in the honeycomb to ripen and mature [1]. In the process of maturation, honey properties and chemical compositions are also changed as a result of biotransformation, biodegradation, and bioaccumulation, which have a great influence on the quality and authenticity of natural honey [2–5]. Foraging bees collect nectar through their proboscis, and place it in the proventriculus (honey stomach) [4,6–9]. Meanwhile, proteins and salivary enzymes from the hypopharyngeal glands of bees begin breaking down sugars from the nectar [7,10–13]. Hydrogen peroxide and gluconic acid, formed by the degradation of glucose oxidase, are partly able to suppress bacterial growth and are responsible for increasing the acidity of honey [14,15]. As nectar collection is completed, the hive bees continually digest nectar and hydrolyze sucrose into glucose and fructose by using bee digestive enzymes [2,7,8,16–19]. Glucose and fructose indicate about 75% of the sugars found in honey, which play important roles in honey quality control and authenticity. The ratio between fructose and glucose, as well as their concentration, are commonly used for predicting honey crystallization and are a beneficial index for the classification of monofloral honeys [20,21]. Another major transformation that occurs during the maturity process is moisture evaporation. The hive bees store digested nectar in the honeycomb cells and transfer it from one cell to another. Hive bees flutter their wings continuously to circulate air and to evaporate moisture from the honey to about 18% and eventually cover the cells with wax to seal them [2,7,22]. Reduction of moisture content below 18% is deemed to be a secure level for retarding yeast activity, decreasing the rate of fermentation and avoiding the appearance of undesirable flavor [23–27]. According to all the above, honey with less than 18% moisture, a sugar concentration above the saturation point, and sealed honeycomb cells may be considered as natural mature honey.

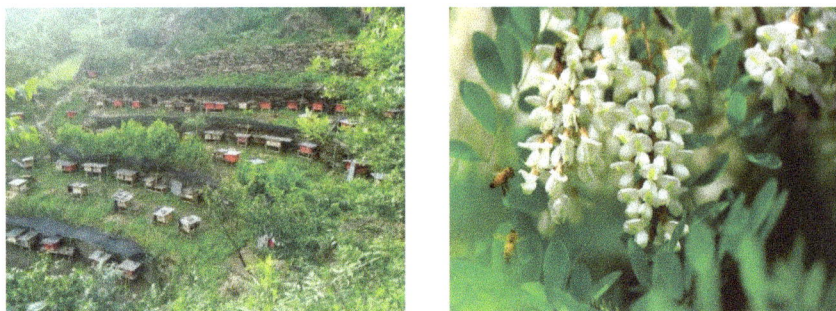

Figure 1. Photograph of mature honey demonstration basement and acacia tree flower.

Because of the influence of components change during the honey maturation process, natural mature honey has its own characteristics, which differ from immature honey. Biologically speaking, immature honey lacks many of the positive properties of natural mature honey. Natural mature honey is a highly complex product with around 200 different substances, which cannot be artificially emulated [28,29]. Physicochemical parameters can provide useful and complete information for the composition and properties of honey and are effective to assess the honey quality and authenticity [30–33]. Chemometric methods can lessen the complexity of large data sets and offer better explication and construction of data sets, as well as identify the natural clustering pattern and group variables based on similarities between samples [34–36]. Combining the great deal of data acquired from physicochemical parameters with chemometrics may be an excellent measure for discrimination of natural mature honey.

The purpose of our research is to identify natural mature honey by establishing the method of multi-physicochemical parameters combined with chemometric analysis. Our study stands out as the first report comparing mature honey with immature honey using principal component analysis (PCA), cluster analysis (CA), and orthogonal partial least squares-discriminant analysis (OPLS-DA). Mature honey was differentiated by evaluating the similarities and discriminant features of honey samples qualitatively, a method that is expected to be applicable to quality control and authenticity identification of natural mature honey.

2. Materials and Methods

2.1. Honey Sample

A total of 85 acacia (*Robinia pseudoacacia*) honey samples were collected from several geographical areas of Shaanxi, China (Table 1) and were kept at 4 °C prior to analysis. The botanical origin of the samples was confirmed by the method of Lutier and Vassiere [37]. All honey samples were collected from a specific mature honey demonstration basement (Figure 1). The natural mature acacia honey (A1–A29) was collected from honey that was capped in the hive and brewed by bees for 7 to 10 days. Immature acacia honey (B30–B85) was collected by hive bees that brewed honey for one to three days.

In addition, all honey samples were extracted using a honey extractor and filtered to remove beeswax and other debris.

Table 1. Characterization of the analyzed acacia honey samples.

Samples	Type of Honey	Botanical Source	Production Region	Predominant Pollen (%)
A1–A29	Monofloral	*Robinia pseudoacacia*	Yan'an, Shaanxi	88.38 ± 2.56
B30–B45	Monofloral	*Robinia pseudoacacia*	Yan'an, Shaanxi	87.21 ± 1.78
B46–B55	Monofloral	*Robinia pseudoacacia*	Chunhua, Shaanxi	86.31 ± 3.01
B56–B60	Monofloral	*Robinia pseudoacacia*	Luochuan, Shaanxi	82.61 ± 1.89
B61–B65	Monofloral	*Robinia pseudoacacia*	Fufeng, Shaanxi	85.77 ± 3.21
B66–B70	Monofloral	*Robinia pseudoacacia*	Qianyang, Shaanxi	83.60 ± 2.18
B71–B75	Monofloral	*Robinia pseudoacacia*	Longxian, Shaanxi	84.34 ± 1.63
B76–B80	Monofloral	*Robinia pseudoacacia*	Yongshou, Shaanxi	88.63 ± 3.84
B81–B85	Monofloral	*Robinia pseudoacacia*	Tongchuan, Shaanxi	86.67 ± 1.67

The honey samples numbered A16–A29, B37–B45, B46–B53, B56–B58, B61–B63, B66–B68, B71–B73, B76–B78, and B81–B83 were randomly selected for the calibration set, and the remaining 35 samples were used as test samples to verify the accuracy of this method.

2.2. Pollen Analysis

The botanical origin of the samples was determined using the method of Lutier and Vassiere [37]. For floral identification, honey samples (5 g) were thoroughly mixed with distilled water (5 mL), and centrifuged at 3000 rpm for 10 min, to separate the pollens. Samples of separated pollen grains were spread with the help of a brush on a slide containing a drop of lactophenol. The slides were examined microscopically at 45× magnification, using a bright-field microscope (Olympus, Tokyo, Japan). According to the different volumes, contours, grooves, holes, and other characteristics of pollen morphology, as well as pictures of different varieties, pollen varieties were identified. A total of 40 horizons and a certain number of pollen grains were observed (the total number of pollen should be more than 100 grains).

$$Pollen\ Content\ (\%) = \frac{a\ certain\ pollen\ number\ of\ 40\ horizons}{a\ total\ pollen\ number\ of\ 40\ horizons} \times 100\%$$

2.3. Physicochemical Properties

The methods used for the quantitative analysis of physicochemical properties were determined mainly according to the Association of Official Analytical Chemists (AOAC) [38]. The details of the methods used in this study are summarized in Supplementary Material.

2.4. HPLC Conditions

The contents of fructose, glucose, and sucrose were determined by high-performance liquid chromatography (HPLC) and a refractive index detector (Shodex R1-201H, Shanghai, China). The column was a Waters carbohydrate high performance (4.6 × 250 mm, 4 µm; Waters). Honey samples (5 g) were thoroughly mixed with ultrapure water (60 mL), and the total volume of the mixture was adjusted to 100 mL with acetonitrile. The mobile phase was 78% acetonitrile and 22% ultrapure water (*v/v*), using an isocratic method. The solutions were filtered through a 0.45 µm membrane filter prior to use. The column was operated at 35 °C, the detector pool temperature was 35 °C, the flow-rate was 1.0 mL min^{-1}, and the injection volume was 15 µL.

2.5. Data Analysis

The multivariate statistical analysis was analyzed by SIMCA (Version 14.1, Umetrics, Umeå, Sweden). The main chemometrics methods used were principal component analysis (PCA), cluster analysis (CA), and orthogonal partial least squares-discriminant analysis (OPLS-DA). PCA simplifies multiple indexes into a small number of comprehensive indexes and uses as many variables as possible to reflect the information of the original variables [39]. CA is the aggregation of samples according to the similarity degree of quality characteristics, and the most similar priority polymerization [40,41]. OPLS-DA is a regression modeling method from multiple dependent variables to multiple independent variables [42]. CA and PCA were used to analyze the comprehensive change of honey, with mean-centered, UV scaled, and log-transformed data before building the PCA model. OPLS-DA was carried out to discriminate features with mean-centered, Pareto scaled and log-transformed data, and the validation of the model was tested using seven-fold internal cross-validation and permutation tests for 200 times. Significant differences were determined using Mann–Whitney U tests, and $p < 0.05$ was considered to be statistically significant. In order to avoid the influence of moisture on other variables, we preprocessed the variables and adopted the dry-weight value of each variable, that is, normalized the moisture content of the variables, and used the ratio of the actual variable value to the content of moisture as the dry weight value of the corresponding variables.

3. Results and Discussion

3.1. Pollen Analysis

Table 1 shows the floral origin of acacia honeys determined by microscopy pollen analysis. The data indicate that all the honey samples were monofloral. *Robinia pseudoacacia* pollen was detected in all samples more than 80.00%.

3.2. Physicochemical Parameters Analysis

Table 2 shows the mean values of the physicochemical parameters of natural mature acacia honey (NMH) and immature acacia honey (IMH) from different regions of the Shaanxi Province. Comparing the physicochemical parameters of NMH and IMH, we found significant differences between NMH and IMH in the mean values of moisture, sugars, protein content, total phenolic, and proline (Figure 2).

Moisture is an important standard for evaluating honey quality, as it can determine the shelf life of honey and its ability to resist fermentation deterioration [43]. The European and Codex standards established a limit of 20% in the case of honey, which can be kept for long periods of time without becoming spoiled [2,9,27]. The moisture of NMH ranged from 15.63% to 16.61%, and IMH was between 21.14% and 26.61%, exceeding the limit, making it difficult to store, producing acids and alcohols more easily, and seriously affecting the quality of honey [5,20].

Honey is a saturated solution of sugars, which accounts for about 70–75% of soluble sugar. Fructose and glucose account for the largest proportion of honey composition, but a small quantity of sucrose was also discovered [25,30]. They are the building blocks of more complex sugars such as disaccharides and maltose [33]. The content was within the limits of European and Codex standards of 65% minimum for glucose and fructose, where sucrose content should be not more than 5%. The sugar concentration (fructose and glucose) of NMH was approximately from 74.01% to 84.06%, which confirmed that samples were genuine honeys [30,32,44].

The total protein content of NMH was between 510.49 mg/kg and 622.29 mg/kg and that of IMH was lower (369.79–538.35 mg/kg). It is universally known that honey contains a trace amount of protein, usually formed by bees, which constantly swallow honey and flutter their wings [31]. The variability in the protein content of different maturity of types honey may be related to its brewing time and brewing degree [45].

During the ripening process, enzymes are proteins that help to speed up many chemical reactions in living organisms, and to convert certain substances into different products [46], especially diastase, invertase, and glucose oxidase. Owing to the high enzyme activity, NMH showed a richer nutritional value than IMH.

Table 2. Physicochemical property of samples (dry-weight basis).

Samples	L*	a*	b*	Conductivity (µS/cm)	pH	Free Acid (meq/kg Dry Matter)	Lacton (meq/kg Dry Matter)	Acid Value (meq/kg Dry Matter)	HMF (mg/kg Dry Matter)
A1–A29	47.60 ± 2.82 [a]	101.99 ± 4.30 [a]	23.49 ± 6.72	136.25 ± 3.38 [g]	3.17 ± 0.06 [c]	27.17 ± 1.72 [c]	5.98 ± 0.99 [a,b]	33.15 ± 1.58 [d]	N.D
B30–B45	70.99 ± 1.53 [b,c]	124.24 ± 2.29 [a]	8.79 ± 2.73 [a]	108.66 ± 5.58 [c,d]	2.88 ± 0.08 [a,b]	21.21 ± 1.09 [a,b]	4.94 ± 1.71 [a]	26.15 ± 2.38 [a,b]	N.D
B46–B55	70.71 ± 0.71 [b,c]	125.64 ± 1.71 [b]	3.62 ± 0.08 [a]	103.70 ± 1.58 [c]	2.89 ± 0.07 [a,b]	20.27 ± 0.74 [a]	3.78 ± 0.57 [a]	24.05 ± 0.94 [a]	N.D
B56–B60	72.95 ± 0.86 [d]	123.25 ± 2.41 [b]	5.92 ± 0.35 [a]	113.88 ± 0.67 [d,e]	2.94 ± 0.05 [a,b]	22.77 ± 0.25 [b]	4.76 ± 0.49 [a]	27.54 ± 0.25 [b]	N.D
B61–B65	72.25 ± 1.07 [c]	129.64 ± 1.88 [b]	4.01 ± 0.29 [a]	110.71 ± 3.93 [c,d]	2.89 ± 0.04 [a,b]	23.04 ± 0.86 [b]	4.86 ± 0.49 [a]	27.91 ± 1.31 [b,c]	N.D
B66–B70	71.45 ± 0.55 [b,c]	125.50 ± 1.65 [b]	3.75 ± 0.13 [a]	84.61 ± 2.25 [a]	3.09 ± 0.03 [c]	23.83 ± 0.49 [b]	8.79 ± 0.55 [c]	31.62 ± 1.04 [d]	N.D
B71–B75	69.48 ± 0.44 [b,c]	116.85 ± 0.58 [b]	6.29 ± 0.21 [a]	117.68 ± 4.80 [f]	2.83 ± 0.02 [a]	22.94 ± 0.83 [b]	7.98 ± 0.35 [b,c]	30.92 ± 0.84 [c,d]	N.D
B76–B80	71.65 ± 0.16 [b,c]	130.71 ± 1.11 [a]	2.56 ± 0.17 [a]	93.43 ± 2.15 [b]	2.99 ± 0.02 [b]	23.48 ± 0.43 [b]	7.35 ± 0.63 [b,c]	30.83 ± 0.98 [c,d]	N.D
B81–B85	68.34 ± 0.14 [b]	120.92 ± 1.44 [b]	3.73 ± 0.21 [a]	105.69 ± 1.09 [c]	2.86 ± 0.02 [a,b]	23.25 ± 0.59 [b]	7.75 ± 0.19 [b,c]	31.00 ± 0.44 [c,d]	N.D

Samples	Glucose (g/100 g Dry Matter)	Fructose (g/100 g Dry Matter)	Sucrose (g/100 g Dry Matter)	Total Sugar (g/100 g Dry Matter)	Total Phenolic (mg/kg Dry Matter)	Total Protein (mg/kg Dry Matter)	Amylase Activity (° Gothe)	Proline (mg/kg Dry Matter)	Glucose Oxidase (U/g Dry Matter)
A1–A29	26.04 ± 1.22 [b,c]	37.39 ± 1.12 [c]	1.97 ± 0.26 [d]	65.40 ± 1.71 [c]	126.59 ± 7.85 [c]	454.09 ± 11.48 [f]	39.15 ± 2.44 [f]	343.35 ± 11.42 [f]	1.26 ± 0.21 [a]
B30–B45	25.20 ± 0.36 [b]	34.07 ± 0.50 [b]	1.17 ± 0.17 [a,b]	60.44 ± 0.58 [b]	89.06 ± 1.21 [b]	361.14 ± 10.13 [c,d]	33.86 ± 1.65 [d,e]	238.21 ± 11.65 [d]	2.36 ± 0.18 [b]
B46–B55	25.18 ± 0.24 [b]	33.93 ± 0.73 [b]	1.15 ± 0.11 [a,b]	60.26 ± 0.66 [b]	87.81 ± 2.46 [b]	388.65 ± 2.28 [d,e]	35.50 ± 1.23 [e]	232.51 ± 1.54 [d]	2.29 ± 0.30 [b]
B56–B60	25.61 ± 0.45 [b,c]	34.07 ± 0.43 [b]	1.43 ± 0.03 [b,c]	61.11 ± 0.10 [b]	88.71 ± 1.12 [b]	329.02 ± 3.11 [a,b,c]	33.34 ± 1.03 [d,e]	266.43 ± 13.60 [e]	2.53 ± 0.16 [b]
B61–B65	24.67 ± 0.53 [b]	34.11 ± 0.87 [b]	1.14 ± 0.15 [a,b]	59.93 ± 0.59 [b]	79.35 ± 6.51 [a,b]	300.09 ± 11.81 [a]	29.62 ± 1.60 [b,c]	183.34 ± 0.41 [b]	3.25 ± 0.61 [c]
B66–B70	27.28 ± 1.14 [c]	35.61 ± 0.53 [b]	1.41 ± 0.15 [b,c]	64.29 ± 1.15 [c]	74.78 ± 4.65 [a]	416.62 ± 8.67 [e]	27.61 ± 1.38 [b]	133.24 ± 3.41 [a]	1.96 ± 0.58 [b]
B71–B75	24.89 ± 0.59 [b]	34.31 ± 0.83 [b]	0.89 ± 0.11 [a]	60.09 ± 1.48 [b]	89.39 ± 2.16 [c]	319.56 ± 12.93 [a,b]	31.41 ± 0.68 [c,d]	213.51 ± 2.637 [c]	2.37 ± 0.86 [b]
B76–B80	24.88 ± 0.31 [b]	34.32 ± 0.13 [b]	1.68 ± 0.46 [c,d]	60.88 ± 0.68 [b]	74.97 ± 0.89 [a]	350.63 ± 12.13 [b,c,d]	21.40 ± 0.59 [a]	179.95 ± 0.32 [b]	1.22 ± 0.21 [a]
B81–B85	22.46 ± 1.86 [a]	30.52 ± 1.79 [a]	1.54 ± 0.07 [b,c]	54.52 ± 1.67 [a]	82.92 ± 2.77 [a,b]	366.10 ± 8.53 [c,d]	21.82 ± 1.03 [a]	199.24 ± 1.97 [b,c]	0.66 ± 0.15 [a]

A1–A29: mature honey; B30–B85: immature honey; N.D: not detected. Results presented in the table are expressed as the mean values ± standard deviation (SD). Different lower case letters correspond to significant differences at $p < 0.05$.

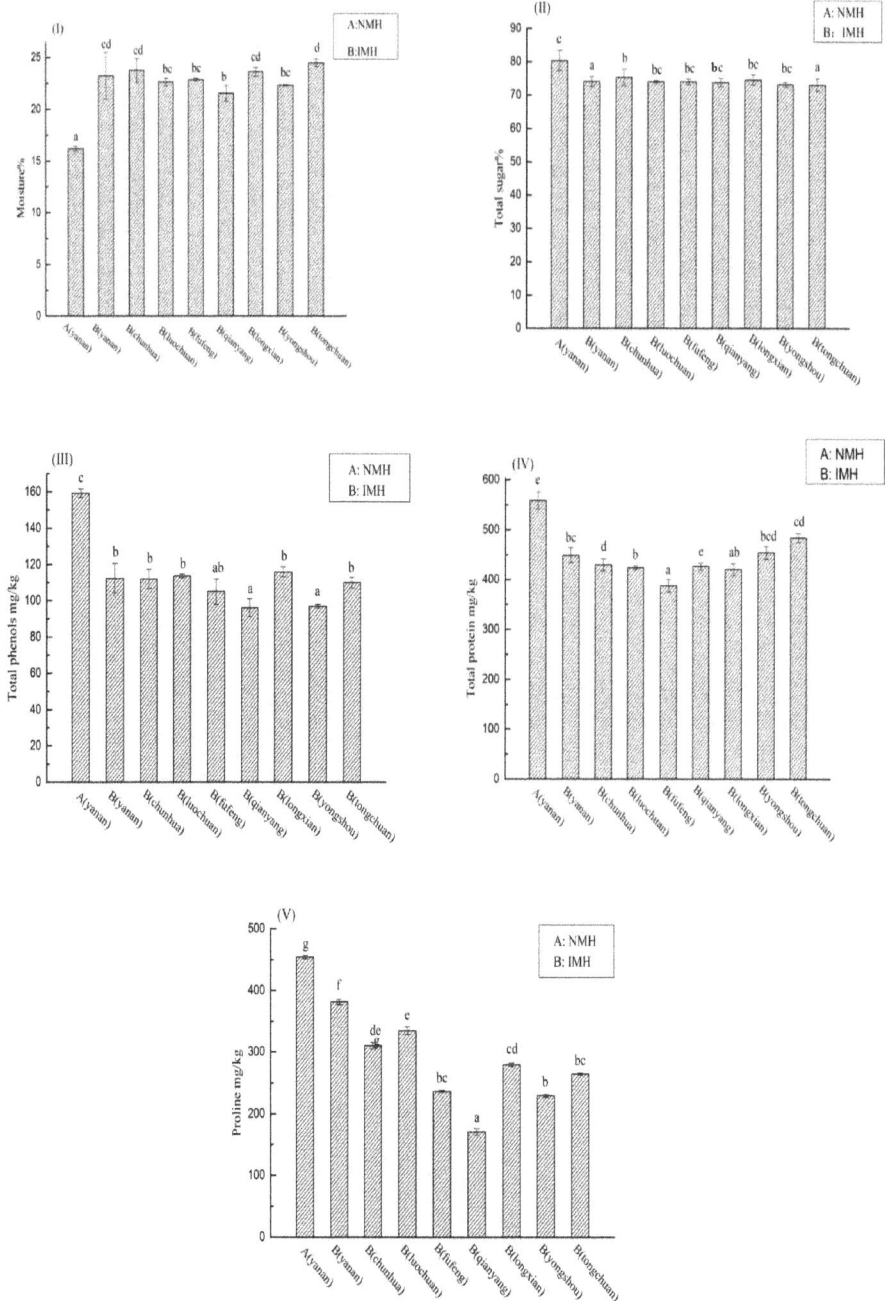

Figure 2. Significant physicochemical parameters of natural mature acacia honey (NMH) and immature acacia honey (IMH). Different lower case letters correspond to significant differences at $p < 0.05$. (I) represents moisture; (II) represents total sugar content; (III) represents total phenols content; (IV) represents total protein content; (V) represents proline content.

The electrical conductivity (EC) is based on the acid contents and ash of honey [27,31], and the free acidity is the content of free acids, which is determined by the equivalence point titration [47]. Furthermore, the pH is correlative with the stability and the shelf life of honey. The pH values of honey usually range from 3.5 to 5.5 [30,33]. The results of statistical analysis were not significantly different in electrical conductivity, pH, and free acidity. 5-hydroxymethylfurfural (HMF) is considered to be a useful indicator for heat treatment and long-term storage of honey [36]. An excessive amount of HMF has been considered evidence of overheating, characterized by a darkening of color and a loss of freshness of honey [23]. In this study, HMF was found to be below the detection limit, indicating that all samples were fresh and not overheating.

3.3. Principal Component Analysis (PCA)

PCA is a statistical method that simplifies several related indicators into a few comprehensive indicators [48]. PCA was combined with the physicochemical parameters in Table 2, in order to avoid the influence of moisture on other physicochemical parameters. The moisture was eliminated when the model was established, and the data of other parameters were normalized and used on a dry-weight basis. The principal component analysis results of the 17 indicators for 50 samples are shown in Table 3. There were three principal components, which are in agreement with the results in Table 3.

Table 3. Variance contribution rate and composition load matrix. PC, principal component.

Quality Index	PC1	PC2	PC3
L*	−0.963	0.029	0.153
a*	−0.870	0.014	0.175
b*	0.905	0.069	−0.134
Conductivity (μS/cm)	0.847	0.142	−0.333
pH	0.852	−0.142	0.218
Free acid (meq/kg)	0.829	−0.314	−0.034
Lacton (meq/kg)	0.106	−0.875	0.240
Acid value (meq/kg)	0.676	−0.651	0.088
Glucose %	0.425	0.195	0.785
Fructose %	0.817	0.164	0.338
Sucrose %	0.777	−0.252	−0.113
Total sugar %	0.814	0.154	0.533
Total phenolic content (mg/kg)	0.936	0.146	−0.160
Total Protein (mg/kg)	0.798	0.073	0.045
Amylase activity (°Gothe)	0.654	0.692	0.008
Proline (mg/kg)	0.872	0.294	−0.301
Glucose oxidase (U/g)	−0.538	0.516	0.234
Eigenvalues	10.208	2.343	1.497
Contribution rate %	60.045	13.785	8.806
Cumulative contribution%	60.045	73.830	82.636

We standardized data to ensure that all the elements had an equal influence over the results. The eigenvalues and the percentage variance, explained by principal components, are shown in Table 3. Three components, eigenvalues > 1, were extracted and used to examine the dataset. The first three components accounted for 82.64% of the total variance. Principal component 1 (PC1) expressed 60.05% of the variance, and the next principal components explained 13.79% and 8.81% of the variance, respectively. The loadings of each compound on the principal component analysis explicitly showed that the grouping of the different maturity honey was mainly influenced by certain compounds. PC1 and PC2 of all the samples explained 73.83% of the total variance at length. PC1 was directly relevant to L*, total phenolic content, and proline, and the dominant variables were mainly affected by protein, total phenol, and sucrose in PC2 (Figure 3B: loading). Moreover, honey samples were properly classified, NMH was classified into one category, and IMH was separated into another category

(Figure 3A: scores of sample). In summary, mature honey samples are significantly different from immature honey, which is in agreement with the result of the physicochemical parameters analysis.

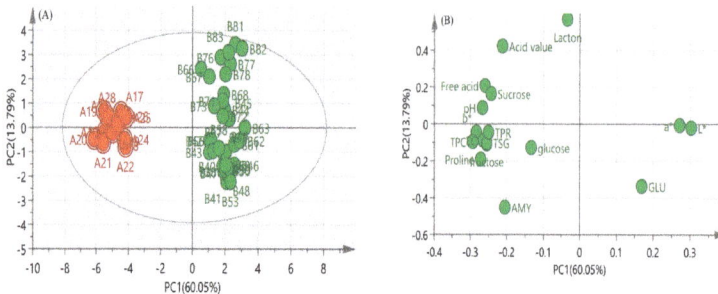

Figure 3. The principal component analysis (PCA) score plots (**A**) and PCA loading plots (**B**) of acacia honey samples. GLU: glucose oxidase; TSG: total sugar; AMY: amylase activity; EC: conductivity; TPC: total phenolic content; TPR: total protein content (for interpretation of the references to color in this figure legend, the reader is referred to the Web version of this article).

3.4. Cluster Analysis

Cluster analysis is a kind of statistical method used to classify the characteristics of multi-indicators and multi-objects [34]. It gradually aggregates according to the similarity of sample quality characteristics. The greatest degree of similarity is aggregated, and multiple varieties are integrated according to the comprehensive nature of the categories [41]. In this study, based on the physicochemical parameters, 50 samples were subjected to the Ward method. The cluster analysis of the pedigree chart is shown in Figure 4. Acacia honey samples were separated into two categories. The first category contained NMH and the second group, namely the remaining samples, contained IMH. The distance between these two categories was more than 100.

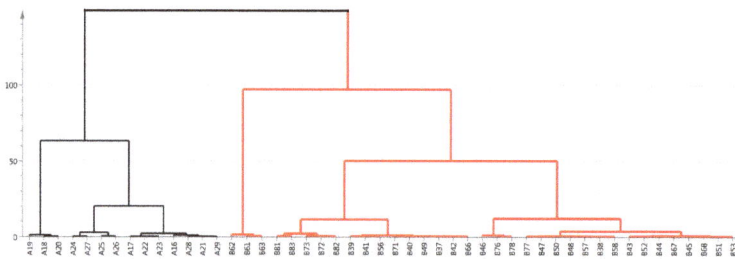

Figure 4. Results of hierarchical cluster analysis of samples.

According to the CA and the PCA, all honey samples were divided into two types, which were basically consistent with the analysis of the parameters mentioned above: color, electrical conductivity, pH, free acid, lactone, total acid, glucose, fructose, sucrose, total sugar content, total phenolic content, amylase activity, HMF, total protein content, glucose oxidase, and other basic physicochemical parameters.

3.5. Orthogonal Partial Least Squares-Discriminant Analysis

Orthogonal partial least squares-discriminant analysis (OPLS-DA) was developed to distinguish between NMH and IMH, and score plots of these models were applied for separating samples. The cross-validation method was used to verify the model. A total of three principal components were selected. The fitted model's index R^2X (cum) was 0.928, indicating that the four principal components explained 92.80% of the X variables, and the index of fitting the dependent variable R^2Y (cum) was 0.978, indicating that the four principal components interpreted 97.80% of the Y variable. The model prediction index Q^2 (cum) was 0.971, explaining that the model had a predictive power of 97.10% for mature honey and immature honey and that this model was stable and reliable. Permutation was conducted 200 times in order to further test the predictability of the OPLS-DA model. To show the predictability, all R^2 and Q^2 (Figure 5B) were > 0 and <−0.5, respectively.

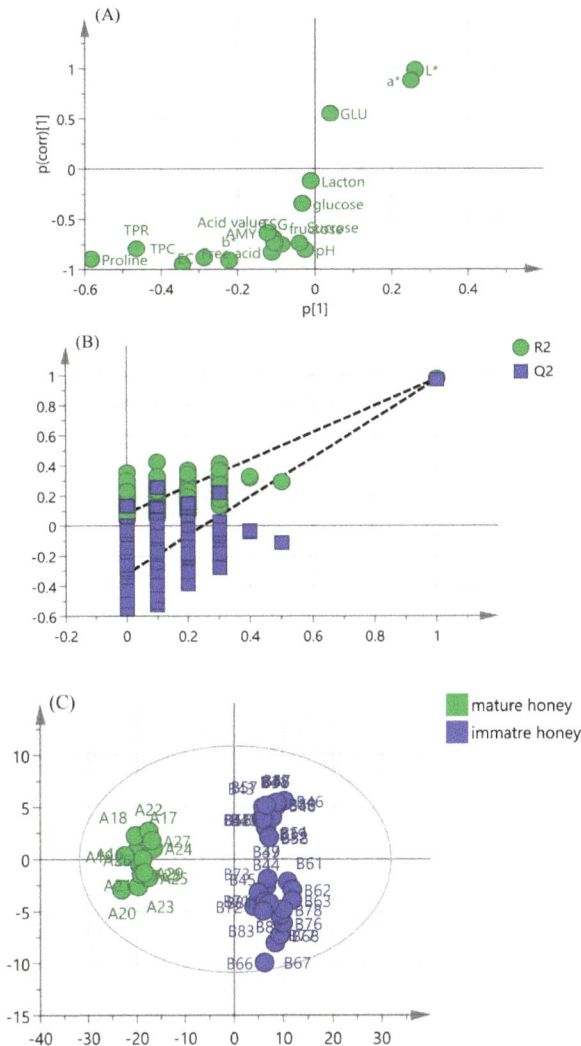

Figure 5. The orthogonal partial least squares-discriminant analysis (OPLS-DA) S-plots (**A**), validation plot (**B**), and score plots (**C**) of honey samples. P [1] is the loading vector of covariance in the first principal component. P (corr) [1] is loading vector of correlation in the first principal component. Variables with $|p| \geq 0.05$ and $|p$ (corr)$| \geq 0.5$ are considered statistically significant. R^2 is the fitted model's index; Q^2 is the model prediction index. GLU: glucose oxidase; TSG: total sugar; AMY: amylase activity; EC: conductivity; TPC: total phenolic content; TPR: total protein content (for interpretation of the references to color in this figure legend, the reader is referred to the Web version of this article).

An S-plot was adopted to visualize the influence of sensitive indices on the mature and immature honey (Figure 5A). Discriminant markers are located in the upper right and lower left corners of the S-plot, with higher absolute p [1] and p (corr) values to select the potential markers [49]. In this model, the sensitive markers were proline, total protein content, EC, L*, a*, b*, and total phenolic content. The variable importance to projection (VIP) values of these indicators are more important than 1. Figure 5 demonstrates a direct separation between NMH and IMH.

3.6. Test Samples Analysis

Thirty-five samples were tested. The results of the PCA score plots (Supplementary Figure S1) show that NMH samples were classified into one category and IMH samples into another. We took distance as a dependent variable and the test samples as independent variables for cluster analysis. Supplementary Figure S2 shows that 35 test samples were properly classified. The accuracy of both the method and cross-verification is 100%.

The score plots of the OPLS-DA models were developed for separating test samples. The R^2X of test honey supervised models was 0.901, R^2Y was 0.985, and Q^2 calculated from seven-fold cross-validation was 0.978. All R^2 and Q^2 were > 0 and <−0.5, respectively, showing a predictability. The classification results are shown in Supplementary Figure S3A,B. Thirty-five test samples were correctly classified on the basis of their maturity. The overall correct classification rate of the original and cross-validation methods is 100%.

Both the 50 honey samples and the 35 test samples achieved appropriate classification results according to the physicochemical parameters combined with chemometric methods. Therefore, physicochemical parameters combined with chemometric methods could be used in the classification of NMH and IMH.

4. Conclusions

In conclusion, the current results clearly show that there are significant differences in physicochemical parameters between natural mature acacia honey and immature acacia honey. Principal component analysis showed that total phenolic content, total protein content, and total sugar (glucose, fructose, sucrose) were the main parameters affected seriously by honey maturity. The cluster analysis and orthogonal partial least squares-discriminant analysis showed that samples were grouped in relation to the maturity and the overall correct classification rate reached 100%. The approach of multi-physicochemical parameters combined with chemometrics is effective in discriminating natural acacia mature honey. However, to confirm the applicability of the other monofloral honeys, this approach needs to be validated. We look forward to the development of this as a promising method for authenticity identification of natural mature honey in any quality control of businesses and government agencies.

Supplementary Materials: The following are available online at http://www.mdpi.com/1420-3049/24/14/2674/s1, Figure S1: PCA score plots, Figure S2.: Results of hierarchical cluster analysis of test sample, Figure S3: The OPLS-DA score plots (A) and OPLS-DA validation plot (B).

Author Contributions: T.M. and W.C. designed the study and interpreted the results. C.L. and M.Z. collected test data. T. M., W.C., H.G., and N.C. drafted the manuscript.

Funding: This work was financially supported by the National Natural Science Foundation of China (no.31272510 and 31871876) and the Science Foundation of Forestry Department of Shaanxi Province (no.2017-01).

Acknowledgments: We are grateful to Liming Wu and Xiaofeng Xue from the Institute of Apicultural Research, Chinese Academy of Agricultural Sciences for their assistance, and also thank the beekeepers for providing honey samples.

Conflicts of Interest: The authors declare no conflict of interest

References

1. Council, E.U. Council Directive 2001/110/EC of 20 December 2001 relating to honey Official. *J. Eur. Communities L* **2002**, *10*, 47–52.
2. Ball, D.W. The chemical composition of honey. *J. Chem. Educ.* **2007**, *84*, 1647–1650. [CrossRef]
3. Chua, L.S.; Abdul-Rahaman, N.L.; Sarmidi, M.R.; Aziz, R. Multi-elemental composition and physical properties of honey samples from malaysia. *Food Chem.* **2012**, *135*, 880–887. [CrossRef]
4. Verma, L.R.; Partap, U. Foraging behaviour of apis cerana on cauliflower and cabbage and its impact on seed production. *J. Apic. Res.* **1994**, *33*, 231–236. [CrossRef]
5. Di Rosa, A.R.; Leone, F.; Cheli, F.; Chiofalo, V. Novel approach for the characterisation of Sicilian honeys based on the correlation of physicochemical parameters and artificial senses. *Ital. J. Anim. Sci.* **2018**, *18*, 389–397. [CrossRef]
6. Bailey, L. The action of the proventriculus of the honeybee (Apis Mellifera L.). *Bee World* **1952**, *32*, 92. [CrossRef]
7. Graham, J.M. The hive and the honey bee. *Bull. ESA* **1992**, *10*, 62.
8. Blatt, J.; Roces, F. The control of the proventriculus in the honeybee (Apis mellifera carnica L.) ii. Feedback mechanisms. *J. Insect Physiol.* **2002**, *48*, 683–691. [CrossRef]
9. Pita-Calvo, C.; Guerra-Rodriguez, M.E.; Vazquez, M. A review of the analytical methods used in the quality control of honey. *J. Agric. Food Chem.* **2017**, *65*, 690–703. [CrossRef]
10. Crailsheim, K. Intestinal transport of sugars in the honeybee (Apis mellifera L.). *J. Insect Physiol.* **1988**, *34*, 840–845. [CrossRef]
11. Almamary, M.; Almeeri, A.; Alhabori, M. Antioxidant activities and total phenolics of different types of honey. *Nutr. Res.* **2002**, *22*, 1041–1047. [CrossRef]
12. Cocker, L. The enzymic production of acid in honey. *J. Sci. Food Agric.* **2010**, *2*, 411–414. [CrossRef]
13. Oroian, M.; Ropciuc, S. Honey authentication based on physicochemical parameters and phenolic compounds. *Comput. Electron. Agric.* **2017**, *138*, 148–156. [CrossRef]
14. White, J.W.; Subers, M.H.; Schepartz, A.I. The identification of inhibine, the antibacterial factor in honey, as hydrogen peroxide and its origin in a honey glucose-oxidase system. *Biochim. Biophys. Acta* **1963**, *73*, 57–70. [CrossRef]
15. Zhu, W.; Hu, F.; Li, Y.; Zhan, Y. The antibacterial mechanism and the affected factors of honey. *Nat. Prod. Res. Dev.* **2004**, *16*, 372–374.
16. Free, J.B. Biology and behaviour of the honey bee Apis florea, and possibilities for beekeeping. *Bee World* **1981**, *62*, 46–59. [CrossRef]
17. Crane, E. Honey from honeybees and other insects. *Ethol. Ecol. Evol.* **1991**, *3*, 100–105. [CrossRef]
18. Peng, Y.S.; Marston, J.M. Filtering mechanism of the honey bee proventriculus. *Physiol. Entomol.* **2010**, *11*, 433–439. [CrossRef]
19. Bentabol Manzanares, A.; García, Z.H.; Galdón, B.R.; Rodríguez, E.R.; Romero, C.D. Differentiation of blossom and honeydew honeys using multivariate analysis on the physicochemical parameters and sugar composition. *Food Chem.* **2011**, *126*, 664–672. [CrossRef]
20. Silva, P.M.D.; Gauche, C.; Gonzaga, L.V. Honey chemical composition, stability and authenticity. *Food Chem.* **2016**, *196*, 309–323. [CrossRef]
21. Wu, L.; Du, B.; Vander Heyden, Y.; Chen, L.; Zhao, L.; Wang, M.; Xue, X. Recent advancements in detecting sugar-based adulterants in honey—A challenge. *TrAC Trends Anal. Chem.* **2017**, *86*, 25–38. [CrossRef]
22. Bogdanov, S.; Jurendic, T.; Sieber, R.; Gallmann, P. Honey for nutrition and health: A review. *J. Am. Coll. Nutr.* **2009**, *27*, 677–689. [CrossRef]
23. Subramanian, R.; Umesh Hebbar, H.; Rastogi, N.K. Processing of honey: A review. *Int. J. Food Prop.* **2007**, *10*, 127–143. [CrossRef]
24. Alvarez-Suarez, J.M.; Tulipani, S.; Romandini, S.; Bertoli, E.; Battino, M. Contribution of honey in nutrition and human health: A review. *Nutr. Metab.* **2010**, *3*, 15–23. [CrossRef]
25. Doner, L.W. The sugars of honey—A review. *J. Sci. Food Agric.* **2010**, *28*, 443–456. [CrossRef]
26. Gupta, R.K.; Reybroeck, W.; Veen, J.W.V.; Gupta, A. Beekeeping for poverty alleviation and livelihood security: II Quality control of honey and bee products. In *Beekeeping for Poverty Alleviation and Livelihood Security*; Springer: Berlin, Germany, 2014.

27. Pita-Calvo, C.; Vázquez, M. Differences between honeydew and blossom honeys: A review. *Trends Food Sci. Technol.* **2017**, *59*, 79–87. [CrossRef]

28. Grout, R.A. American Bee Journal Editor, M.G. Dadant. *Am. Bee J.* **2018**, *158*, 741–747.

29. Jibril, F.I.; Hilmi, A.B.M.; Manivannan, L. Isolation and characterization of polyphenols in natural honey for the treatment of human diseases. *Bull. Nat. Res. Cent.* **2019**, *43*, 4–5. [CrossRef]

30. Silvano, M.F.; Varela, M.S.; Palacio, M.A.; Ruffinengo, S.; Yamul, D.K. Physicochemical parameters and sensory properties of honeys from buenos aires region. *Food Chem.* **2014**, *152*, 500–507. [CrossRef]

31. Sohaimy, S.A.E.; Masry, S.H.D.; Shehata, M.G. Physicochemical characteristics of honey from different origins. *Ann. Agric. Sci.* **2015**, *60*, 279–287. [CrossRef]

32. Siddiqui, A.J.; Musharraf, S.G.; Choudhary, M.I.; Rahman, A. Application of analytical methods in authentication and adulteration of honey. *Food Chem.* **2017**, *217*, 687–698. [CrossRef] [PubMed]

33. Popek, S.; Halagarda, M.; Kursa, K. A new model to identify botanical origin of polish honeys based on the physicochemical parameters and chemometric analysis. *LWT-Food Sci. Technol.* **2017**, *77*, 482–487. [CrossRef]

34. Yücel, Y.; Demir, C. Principal component analysis and cluster analysis for the characterisation of marbles by capillary electrophoresis. *Talanta* **2004**, *63*, 451–459. [CrossRef] [PubMed]

35. Yücel, Y.; Sultanoğlu, P. Characterization of hatay honeys according to their multi-element analysis using ICP-OES combined with chemometrics. *Food Chem.* **2013**, *140*, 231–237. [CrossRef] [PubMed]

36. Zhao, J.; Du, X.; Cheng, N.; Chen, L.; Xue, X.; Wu, L.; Cao, W. Identification of monofloral honeys using HPLC-ECD and chemometrics. *Food Chem.* **2016**, *194*, 167–174. [CrossRef] [PubMed]

37. Lutier, P.M.; Vaissière, B.E. An improved method for pollen analysis of honey. *Rev. Palaeobot. Palynol.* **1993**, *78*, 129–144. [CrossRef]

38. Association of Official Analytical Chemists (AOAC). Available online: www.aoac.org/aoac_prod_imis/AOAC/Publications/Official_Methods_of_Analysis/AOAC_Member/Pubs/OMA/AOAC_Official_Methods_of_Analysis.aspx (accessed on 21 July 2019).

39. Geladi, P.; Isaksson, H.; Lindqvist, L.; Wold, S.; Esbensen, K. Principal component analysis of multivariate images. *Chemom. Intell. Lab.* **1989**, *5*, 209–220. [CrossRef]

40. Everitt, B. *Cluster Analysis*; Quality & Quantity; Kluwer Academic Publishers: Norwell, MA, USA, 1980; Volume 14, pp. 75–100.

41. Everitt, B. *The Cambridge Dictionary of Statistics*; Cambridge University Press: Cambridge, UK, 2002.

42. Boccard, J.; Rutledge, D.N. A consensus orthogonal partial least squares discriminant analysis (OPLD-DA) strategy for multiblock omics data fusion. *Analytica Chimica Acta* **2013**, *769*, 30–39. [CrossRef]

43. Ouchemoukh, S.; Louaileche, H.; Schweitzer, P. Physicochemical characteristics and pollen spectrum of some algerian honeys. *Food Control* **2007**, *18*, 52–58. [CrossRef]

44. Lazarević, K.B.; Andrić, F.; Trifković, J.; Tešić, Ž.; Milojković-Opsenica, D. Characterisation of serbian unifloral honeys according to their physicochemical parameters. *Food Chem.* **2012**, *132*, 2060–2064. [CrossRef]

45. Azeredo, L.D.C.; Azeredo, M.A.A.; Souza, S.R.D.; Dutra, V.M.L. Protein contents and physicochemical properties in honey samples of apis mellifera of different floral origins. *Food Chem.* **2003**, *80*, 249–254. [CrossRef]

46. Weirich, G.F.; Collins, A.M.; Williams, V.P. Antioxidant enzymes in the honey bee, Apis mellifera. *Apidologie* **2002**, *33*, 3–14. [CrossRef]

47. Alqarni, A.S.; Owayss, A.A.; Mahmoud, A.A. Physicochemical characteristics, total phenols and pigments of national and international honeys in saudi arabia. *Arab. J. Chem.* **2016**, *9*, 114–120. [CrossRef]

48. Ares, G.; Jach, A. Applications of principal component analysis (PCA) in food science and technology. In *Mathematical and Statistical Methods in Food Science and Technology*; John Wiley & Sons, Ltd.: Hoboken, NJ, USA, 2014.

49. Zhao, H.; Cheng, N.; Zhang, Y.; Sun, Z.; Zhou, W.; Wang, Y.; Cao, W. The effects of different thermal treatments on amino acid contents and chemometric-based identification of overheated honey. *LWT-Food Sci. Technol.* **2018**, *96*, 133–139. [CrossRef]

molecules

MDPI

Article

Use of ¹H NMR to Detect the Percentage of Pure Fruit Juices in Blends

Lucia Marchetti [1,2], **Federica Pellati** [1,*], **Stefania Benvenuti** [1] **and Davide Bertelli** [1]

1   Department of Life Sciences, University of Modena and Reggio Emilia, Via G. Campi 103, 41125 Modena, Italy

2   Doctorate School in Clinical and Experimental Medicine (CEM), University of Modena and Reggio Emilia, 41125 Modena, Italy

*   Correspondence: federica.pellati@unimore.it; Tel.: +39-059-205-8565

Received: 24 June 2019; Accepted: 16 July 2019; Published: 17 July 2019

check for updates

Abstract: The consumption of high-nutritional-value juice blends is increasing worldwide and, considering the large market volume, fraud and adulteration represent an ongoing problem. Therefore, advanced anti-fraud tools are needed. This study aims to verify the potential of ¹H NMR combined with partial least squares regression (PLS) to determine the relative percentage of pure fruit juices in commercial blends. Apple, orange, pineapple, and pomegranate juices were selected to set up an experimental plan and then mixed in different proportions according to a central composite design (CCD). NOESY (nuclear Overhauser enhancement spectroscopy) experiments that suppress the water signal were used. Considering the high complexity of the spectra, it was necessary to pretreat and then analyze by chemometric tools the large amount of information contained in the raw data. PLS analysis was performed using venetian-blind internal cross-validation, and the model was established using different chemometric indicators (RMSEC, RMSECV, RMSEP, R^2_{CAL}, R^2_{CV}, R^2_{PRED}). PLS produced the best model, using five factors explaining 94.51 and 88.62% of the total variance in X and Y, respectively. The present work shows the feasibility and advantages of using ¹H NMR spectral data in combination with multivariate analysis to develop and optimize calibration models potentially useful for detecting fruit juice adulteration.

Keywords: fruit juice; blends; adulteration; ¹H NMR; PLS; chemometrics

---

1. Introduction

Fruit juice consumption is increasing worldwide since fruit-based products are promoted as healthy foods and, in addition, producers try to be more and more innovative by developing the market segment of blends and mixed juices of high nutritional value. According to the European Fruit Juice Association (AIJN), consumption of 38.5 million L of fruit juice was registered worldwide in 2015 and, in the same year, the consumption estimated per head of population was about 19 L in the European Union, 26 L in North America, 6 L in South America and 2 L in Pacific Asia [1]. Considering the large market volume, economic frauds and adulteration in this sector represent an ongoing problem and have often been reported; hence, there is a need for advanced and suitable anti-fraud tools [2,3]. Tackling the total cost of analysis to assess fruit juice authenticity can be very expensive; mostly the aim is to detect water dilution, the addition of inexpensive juice blends to higher-value fruit juice, or the addition of pure beet sugar. Currently, the most common techniques applied to reveal fraud in this field involve whole-food profiling or the search for a number of compounds (targeted analysis). The main disadvantages of these analyses are the high cost and time-consuming nature, the fact that adulterants at concentrations lower than 10% are difficult to detect, and the fact that, in most cases, only one type of adulteration can be unmasked at a time [4].

For this reason, advanced analytical methods are highly recommended to prevent fraudulent practices and to protect the rights of producers as well as those of consumers, with particular attention to safety issues. Since many factors contribute to the variation in juice composition, e.g., fruit geographical origin and climate, maturity degree, technological processes, and storage conditions, a comprehensive method based on the fingerprinting approach is recommended, in which a large number of unknown metabolites is included in the analysis and contributes to the results. For this purpose, the possibility of using nuclear magnetic resonance (NMR) spectroscopy for monitoring fruit juice and other fruit-derived products is well known and has been demonstrated to be an efficient tool in beverage authentication, since the spectral data cover a wide range of compounds [5]. NMR has many advantages over the most common separation methods based on GC and HPLC, which may be cheaper but they seem to be more appropriate for target analysis [6]. Indeed, NMR can detect many different compounds in one sample run, it is non-destructive, stable over time, and it requires only a limited sample preparation [6]. Various applications of NMR spectroscopy are now available for beverage quality control, as reported for wines, spirits, and juices [7,8]. ^{1}H NMR spectroscopy has shown great potential for determining the country of origin of green tea samples [9]. Moreover, the same technique has been demonstrated to be very accurate in the determination of the origin of fruit, and it can be used to examine the source of the raw material used in the preparation of juices [10]. NMR spectroscopy has also been recently applied to alcoholic beverages for authentication purposes, taking into account their high prices and the high risk of fraud by adulteration or deliberate mislabeling. The potential of NMR spectroscopy has also been evaluated to verify the composition of beer and to correlate it to the brewing site and to the date of production, as well as for quality control [11]. In particular, the use of NMR spectroscopy has become an indispensable tool for authenticity studies on fruit juice and has now reached the commercial level. The NMR technique is able to evaluate simultaneously, from a single dataset, a multitude of parameters related to the quality and authenticity of juice, providing targeted and non-targeted multi-marker analysis. In addition, the spectral results can be compared with databases of reference juice [10,12,13]. However, to the best of our knowledge, NMR application for the qualitative and quantitative determination of the composition in juice blends has not been previously described in the literature.

The application of ^{1}H NMR to complex matrices, as in the case of fruit juice, which normally contains a large amount of natural or added sugars and organic acids, and lower but significant amounts of other organic substances (phenolics, terpenes, amino acids, etc.), may result in complex spectra, with a lot of crowded signals. This condition often makes it difficult to define the assignments of substances and to correctly proceed with the integration without the use of deconvolution. Deconvolution is an algorithm-based process that allows us to resolve or decompose a set of overlapping peaks into their separate additive components. In these cases, a non-targeted approach based on a sample fingerprint is advantageous. In particular, the use of the entire ^{1}H NMR spectrum profile without any kind of targeted measurement was evaluated in this work. Fingerprinting techniques require the use of adequate statistical methods to extract information from raw data. Chemometrics can reveal latent correlations in the data and can be useful for both qualitative and quantitative purposes.

Partial least squares (PLS) modeling is a powerful multivariate statistical tool that has often been applied to spectral analysis. PLS is related to other multivariate calibration methods, such as classical least squares (CLS), inverse least squares (ILS), and principal component regression (PCR) methods [14]. The main scope of PLS is to eliminate multicollinearity in the set of explanatory variables X of a regression model, reducing the dimension of the set in such a way that the resulting subset of descriptive variables is optimal for predicting the dependent variable Y. The values of Y typically represent the analyte concentrations (or any sample properties) [14]. For the quantitative approach, PLS is the most frequently used multivariate statistical method. Hence, this study aims to verify the potential of proton NMR (^{1}H NMR) combined with PLS to construct models for the determination of the relative percentage of pure fruit juices in a blend.

Molecules **2019**, 24, 2592

2. Results and Discussion

The use of ^{1}H NMR coupled with PLS was proposed here to develop and optimize multivariate calibration models to determine the relative concentration of four pure juices in blends. The PLS model was established by using different chemometric indicators (root mean square error of calibration (RMSEC); root mean square error of cross-validation (RMSECV); root mean square error of prediction (RMSEP); coefficient of determination for calibration (R^2_{CAL}), cross-validation (R^2_{CV}) and prediction (R^2_{PRED}).

Figure 1 shows the NMR spectra of the four pure juices considered, and Figure 2 shows the typical spectrum of a mixture containing equal percentages of apple, orange, pineapple, and pomegranate juice. As expected, the spectra present wide regions with overlapping phenomena, which makes it difficult to proceed with the peak integration. The typical ^{1}H NMR spectrum of a juice shows three defined regions. In the first, ranging from 0.5 to 3.0 ppm, protons of organic acids (citric and malic) and amino acids (alanine, valine, and proline) are present. The second region (3.0–6.0 ppm) is typical for carbohydrates, with sucrose, α-glucose, β-glucose, and fructose being the most abundant [7]. The last region, ranging from 6.0 to 8.5 ppm, shows phenolic metabolites and aromatic protons. Moreover, each juice presents the peak assigned to the methyl group of ethanol at 1.17 ppm [6,15]. Spectra of pure juice have similar shapes in the aliphatic region and limited quantitative differences, except for the two peaks at 5.40 and 4.22 ppm related to sucrose, which is absent in pomegranate. The region ranging from 6 to 10 ppm is meanwhile more typical; here, aromatic and phenolic compounds are normally present. In this case, the signals are lower in intensity with respect to the aliphatics. As regards the mixtures, the enlargement of the aromatic region in Figure 2 appears even more complex, due to the fact that the weak aromatic signals of each single juice are more diluted in the blend and the number of signals is increased, owing to the simultaneous presence of many compounds. All these considerations led us to choose an untargeted approach combined with the consolidated PLS chemometric method [16].

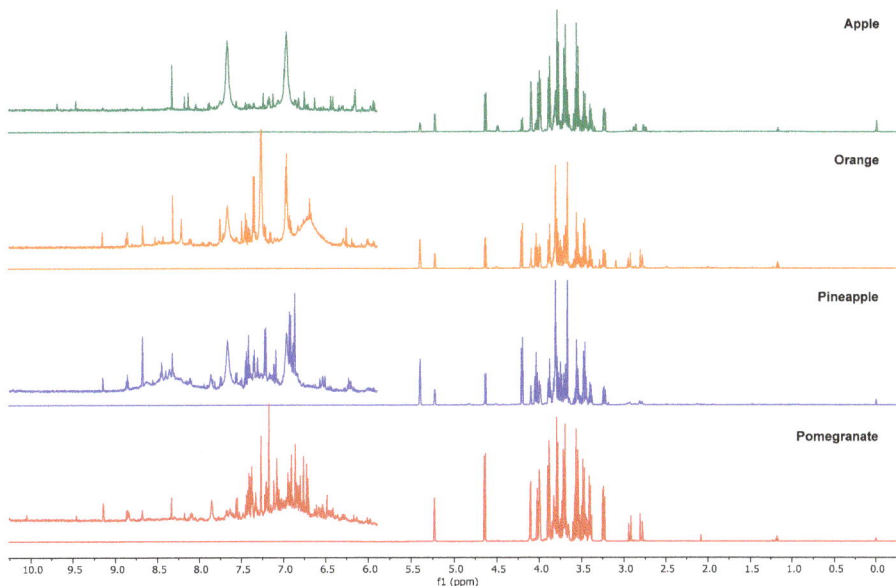

Figure 1. 1D NOESY spectra of four pure juices, with the 6–10 ppm regions enlarged.

Molecules **2019**, *24*, 2592

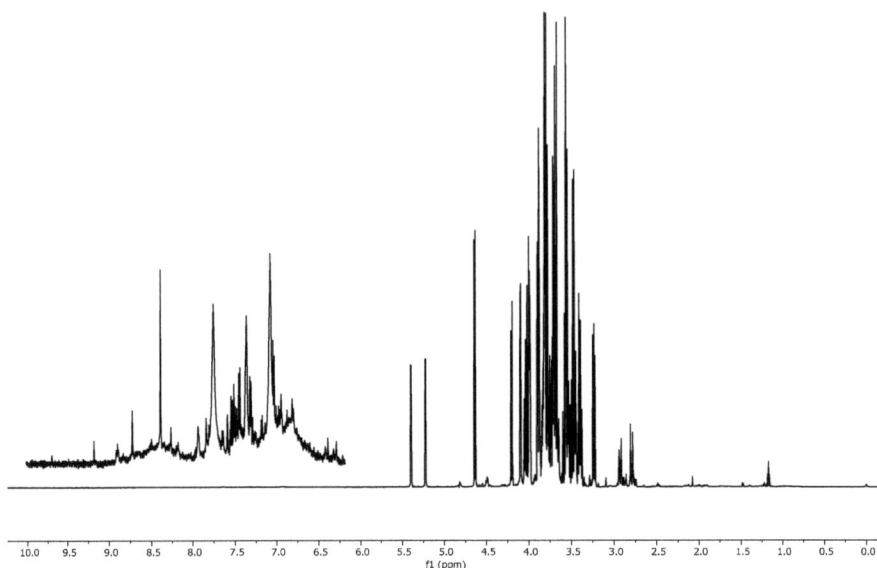

Figure 2. Typical 1D NOESY spectrum of mixture containing equal percentages of pure apple, orange, pineapple, and pomegranate juice.

The number of latent variables to be included in the PLS model was selected in such a way that the RMSEC, RMSECV (obtained from calibration and internal cross-validation, respectively) and RESEP (obtained from external validation test set) were reduced to the lowest values, ensuring at the same time the highest possible predictive capacity. Other statistical parameters to be considered are R^2 and the amount of explained variance. The best model was built by using five factors, explaining 94.51% and 88.62% of total variance in X and Y, respectively. All the calculated RMSEs (except RMSECV for pineapple and pomegranate) are lower than 10 and each R^2 value is acceptable, especially for prediction values. All the chemometric indicators are reported in Table 1.

Table 1. PLS model results of four fruit juice samples.

Model	Number of Factors: 5 Variance: Y = 88.62%; X = 94.51%			
	RMSEC [a]	RMSECV [b]	RMSEP [c]	R^2 [d]
Apple	6.869	8.732	2.324	$R^2_{CAL} = 0.912$ $R^2_{CV} = 0.899$ $R^2_{PRED} = 0.987$
Orange	6.333	9.435	4.435	$R^2_{CAL} = 0.914$ $R^2_{CV} = 0.882$ $R^2_{PRED} = 0.950$
Pineapple	8.634	12.631	5.438	$R^2_{CAL} = 0.885$ $R^2_{CV} = 0.821$ $R^2_{PRED} = 0.946$
Pomegranate	7.182	10.511	7.092	$R^2_{CAL} = 0.950$ $R^2_{CV} = 0.860$ $R^2_{PRED} = 0.929$

[a] RMSEC: root mean square error of calibration; [b] RMSECV: root mean square error of cross-validation; [c] RMSEP: root mean square error of prediction. [d] R^2: coefficient of determination for calibration (CAL), cross-validation (CV) and prediction (P).

In Figure 3 the Hotelling's T^2 vs. Q residuals graphic is shown. The presence of potential spectral outliers was verified by applying a 95% confidence interval, any outlier is present. Four samples with high Q residuals and three with high T^2 scores are present, but none of them was excluded from the model. All the test set samples (shown in red) are in the 95% confidence area, demonstrating the high predictive capacity of the model.

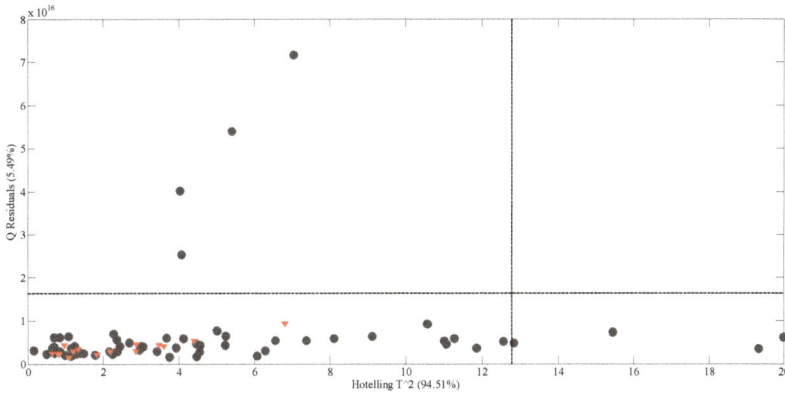

Figure 3. Q residuals versus Hotelling's T^2 plot for the PLS model of 60 fruit juice samples (•) and test set samples (▼).

In Figures 4 and 5 the loadings for the five extracted latent variables and the VIP (variable importance in the projection) scores, respectively, are shown for each juice. A VIP score is a measure of a variable importance in the model. It summarizes the contribution of a variable to the model. The VIP score is calculated as a weighted sum of the squared correlations between the PLS components and the original variables. As regards loadings, highly correlated variables have similar weights in the loading vectors. Thus, a different pattern of variables is significant for each extracted factor. In this case, for all the latent variables any significance arises for the aromatic signals, even though this spectral region (6–8 ppm) seems to be more characteristic for each juice in comparison with the aliphatic areas (3–4 ppm), which appear quite similar (Figure 1). Considering VIP scores, apple, pineapple, and pomegranate share the same pattern of variables, while in the case of orange juice the most important signals are those located at low frequencies, corresponding to acid compounds. In any case, and also for orange juice, the significance of the aromatic and phenolic region is limited. Some different pre-treatment procedures were attempted to increase the significance of the aromatic spectral region, but no substantial improvement was achieved.

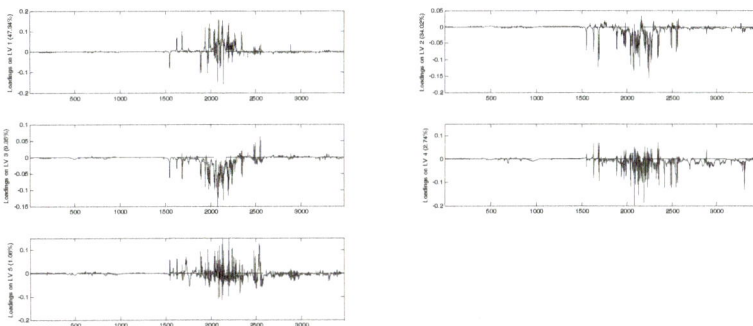

Figure 4. Loadings for the five extracted latent variables.

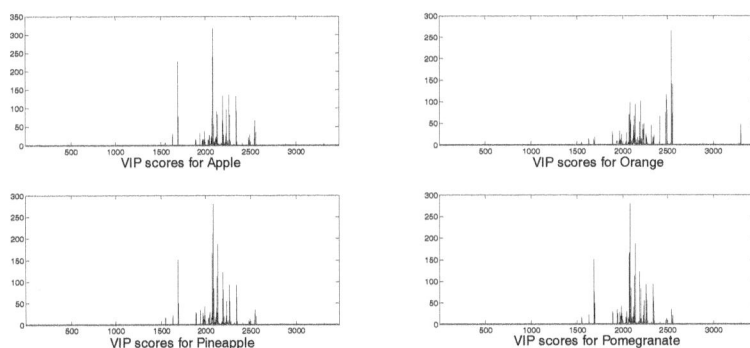

Figure 5. VIP scores obtained from the selected model for the four juices.

The regression vectors for each juice are shown in Figure 6. The four plots are different from each other, confirming the same tendency of VIP scores, thus in the definition of regression models different variables play a significant role for each juice.

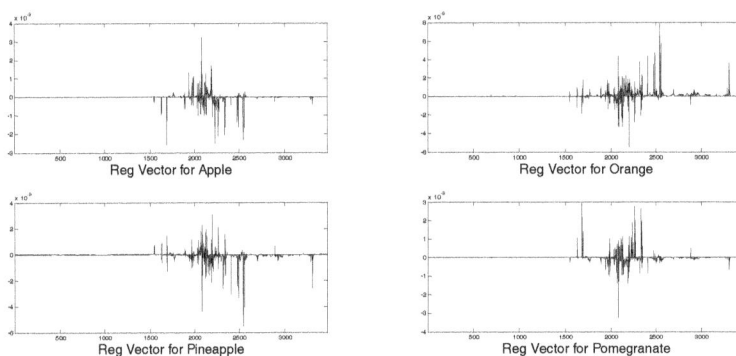

Figure 6. Regression vectors for the four juices considered in the model.

Figure 7 shows the scatter plot of measured values vs. predicted values in mixture samples for calibration and test sets (regression data are reported in Table 1). The regression models are similar in quality and predictive capacity. The best-performing model is apple, while the worst is pomegranate (Figure 7: top left and bottom right, respectively).

The method discussed here is suitable and sufficiently effective for the quantification of the considered mixtures. To validate this approach, the next step will be the evaluation of the predictive capacity of mixtures prepared with juices of other brands and also of mixtures in which one or more components are different from those covered by the model. Moreover, this method showed its effectiveness in a wide range of component percentages (6.25–100%), thus proving useful for both high and low concentrations. As a consequence, it should also be able to detect small differences from the composition declared on the labels of the commercial products.

The results presented herein show that the NMR analysis coupled with chemometrics provides adequate results in a comparable time with respect to other analytical approaches. The main advantages of the method proposed include the reduced sample preparation and lack of extraction and purification steps. In addition, this technique is effective at reducing the amounts of reagents and solvents, making it more competitive and environmentally sustainable than the most common separative techniques.

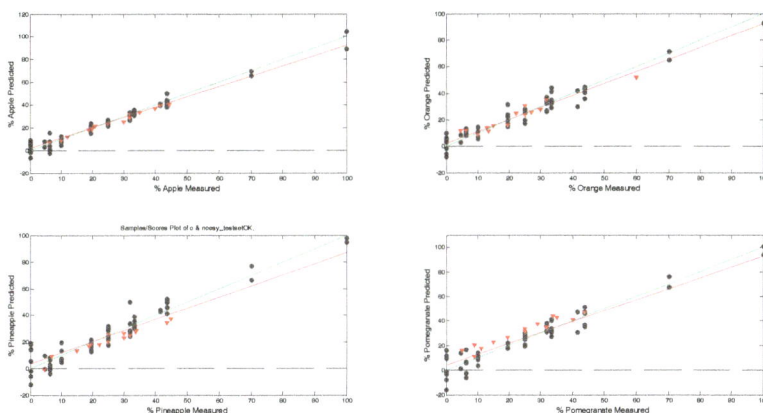

Figure 7. Correlation between the prediction and measured values of juice samples, showing calibration (•) and test set (▼) samples.

3. Materials and Methods

3.1. Samples

To verify the feasibility of ^{1}H NMR combined with chemometrics to determine the percentage of pure fruit juices in a mixture, the following experimental plan was designed. To minimize the system complexity in this step of the research, four fruit juices (apple, orange, pineapple, and pomegranate) from a single brand were purchased from a local marketplace. These samples represent some of the most consumed fruit juices, often combined for the preparation of more attractive mixtures for consumers.

To obtain the highest significant distribution of mixtures to be analyzed, they were prepared in accordance with an experimental plan, obtained by means of a central composite design (CCD), with $\alpha = 1.4826$, based on a 2^4 full factorial design, plus eight axial points, plus two replicates in the center of the domain and four pure juice samples [17]. The ratios provided by the model were transformed into a percentage composition to get the final experimental plan and then were used as dependent variables for the subsequent PLS analysis (Table 2).

Table 2. Percentage composition of juice samples.

Sample	% Apple	% Orange	% Pineapple	% Pomegranate
1	100	0	0	0
2	0	100	0	0
3	0	0	100	0
4	0	0	0	100
5	43.75	6.25	43.75	6.25
6	25	25	25	25
7	6.25	6.25	43.75	43.75
8	43.75	6.25	6.25	43.75
9	25	25	25	25
10	19.5	19.5	19.5	41.5
11	10	10	10	70
12	33.33	33.33	0	33.33
13	31.82	4.54	31.82	31.82
14	33.33	33.33	33.33	0
15	6.25	43.75	43.75	6.25
16	0	33.33	33.33	33.33
17	10	70	10	10
18	19.51	41.46	19.51	19.51

Table 2. *Cont.*

Sample	% Apple	% Orange	% Pineapple	% Pomegranate
19	19.51	19.51	41.46	19.51
20	10	10	70	10
21	25	25	25	25
22	31.82	31.82	4.54	31.82
23	6.25	43.75	6.25	43.75
24	4.54	31.82	31.82	31.82
25	33.33	0	33.33	33.33
26	43.75	43.75	6.25	6.25
27	31.82	31.82	31.82	4.54
28	41.46	19.51	19.51	19.51
29	25	25	25	25
30	70	10	10	10

Analyses were performed in duplicate, making sure to repeat the analysis on different days. In this way, the 60 experiments were conducted in a randomized order. The intra- and inter-day variability, previously evaluated on one sample, were analyzed 10 times on the same and on different days, giving 1.4 and 4.0 CV%, respectively. In addition, an external validation test set was constructed including five randomly selected CCD replicates (newly prepared and analyzed) and 10 extra mixtures outside the CCD, with random percentages of the four juices (Table 3).

Table 3. Percentage composition of test set samples.

Sample	% Apple	% Orange	% Pineapple	% Pomegranate
TS1	20	15	30	35
TS2	44.44	22.22	22.22	11.11
TS3	40	13	7	40
TS4	18.92	13.51	33.78	33.78
TS5	20	60	15	5
TS6	30	30	25	15
TS7	12	60	19	9
TS8	35	10	30	25
TS9	21	25	45	9
TS10	10	27	34	29
7 *	6.25	6.25	43.75	43.75
9 *	25	25	25	25
13 *	31.82	4.54	31.82	31.82
22 *	31.82	31.82	4.54	31.82
28 *	41.46	19.51	19.51	19.51

* CCD samples randomly selected.

For the preparation of samples, each juice was first centrifuged at 5000 rpm for 5 min to remove all the material in suspension. Subsequently, for NMR analysis, different juices were mixed together in the ratios indicated by the CCD.

For organic acids, amino acids, and each compound whose ionization changes depending on pH, the chemical shift varies accordingly. These spectral variations may affect or invalidate the results of the multivariate analysis, whose basic principle is that the corresponding signals in different samples must have the same chemical shift. To minimize shifts, the pH of the samples with a buffered solution was standardized. To 9 mL of each mixture, 1 mL of 1 M KH_2PO_4 was added and the pH was measured; eventually, it was adjusted to 3.1–3.2 by using small amounts of 1 M HCl. An aliquot of 630 µL of this buffered mixture were then added to 70 µL of 13 mM 3-(trimethylsilyl)propionic-2,2,3,3-d_4 acid sodium salt (TSP) and 0.1% sodium azide in 99.9% deuterium oxide (D_2O). TSP was employed for internal referencing of ^1H chemical shifts, while sodium azide as the preserving agent. The final solution

(700 µL) was transferred into a WILMAD® NMR tube, 5 mm, Ultra-Imperial grade, L 7 in., 528-PP purchased from Sigma-Aldrich (Milan, Italy). All chemicals and solvents were of analytical grade and they were purchased from Sigma-Aldrich.

3.2. NMR Spectroscopy

One-dimensional ^{1}H NMR spectra of juice mixtures were acquired with a Bruker FT-NMR Avance III HD 600 MHz spectrometer (Ettlingen, Germany), and all the NMR experiments were performed at 300 °K. After 10 min of thermal equilibration inside the probe, the solvent (D_2O) was locked to assure the maximum sensitivity, the probe was manually tuned and matched, and the 90° pulse was calibrated; subsequently, the power level for pre-saturation was calculated, and finally the receiver gain was automatically set. 1D NOESY experiments were acquired by using the Bruker sequence "noesygppr1d" in order to suppress the water signal. The pre-saturation frequency was adjusted for each sample before acquisition. The acquisition parameters were as follows: time domain (number of data points), 64 K; dummy scans, 0; number of scans, 32; acquisition time, 3.90 s; delay time, 8 s; spectral width, 14 ppm (8403.4 Hz), fid resolution, 0.2564 Hz; digitization mode, baseopt, total acquisition time, 6 min 22 s.

3.3. Spectra Pretreatment

The application of the ^{1}H NMR technique to juice samples generates complex and crowded spectra, which need to be previously pre-treated and then analyzed by chemometric tools to handle the high amount of information contained in the raw data. Assuming that peak intensities are directly proportional to the concentration of compounds, each spectral point was used as an absolute intensity value, without performing any type of signal integration. This approach allows for overcoming issues related to the understanding and integration of overlapped signals. First, ^{1}H NMR spectra were phased and calibrated using the TSP signal for chemical shifts referencing, and the baseline was adjusted. All spectra processing was performed using TopSpin 3.5 software package (Bruker Biospin GmbH Rheinstetten). Each spectrum generates a 64 K data point file. In order to perform the spectra alignment, the files from each sample were exported and collected in a dataset consisting in 64 K spectral variables for 75 samples (60 calibration samples plus 15 test set samples). The alignment was performed by using Icoshift 1.0 toolbox for MATLAB® (Mathworks Inc., Natick, MA, USA) to reduce the lack of homogeneity in chemical shifts that principally occurs for pH-dependent signals [18]. Afterwards, in order to reduce the number of data points and manipulate datasets more easily, non-significant spectral regions were deleted to obtain a new dataset with 34,740 data points and, finally, the resolution was reduced by selecting one out of every 10 points, thus obtaining 3474 data points for each spectrum.

3.4. Statistical Analysis

With a view to performing the statistical analysis, the dataset used for the alignment and resolution reduction (75 samples) was again split into two datasets, i.e., one containing the 60 samples for calibration and the second containing the 15 samples for the external validation test set. All multivariate data analyses and calculations were performed by using the software PLS_Toolbox 5.2.2 (Eigenvector Research Inc., Manson, WA, USA) for MATLAB®. The PLS method is based on the SIMPLS algorithm. All data were mean-centered and scaled by applying the Pareto scaling method, which is useful when spectral noise is expected to be proportional to the standard deviation square root of variables. Pareto scaling can reduce the relative importance of large values, keeping the data structure partially intact [19].

Partial least squares regression (PLS) is an extension of the multiple linear regression model that does not impose the restrictions employed by discriminant analysis, principal components regression, or canonical correlation. In PLS, prediction functions are represented by factors extracted from the Y'XX'Y matrix, and it is probably the least restrictive of the multiple linear regression extensions. PLS analysis was performed using a venetian-blind internal cross-validation method with the number of data splitting set to 7. The number of factors (latent variables) in the models and the model performance were assessed by using the root mean square error of calibration (RMSEC) and the root mean square error of cross-validation (REMSCV). The presence of potential spectral outliers was verified by applying a 95% confidence interval to the Q residuals and Hotelling's T^2 scores (Figure 3).

The final regression model was assessed by the coefficients of determination for calibration, cross-validation and test sets (R^2_{CAL}, R^2_{CV}), the root mean square error of calibration and the root mean square error of cross-validation (RMSEC, RMSECV). The prediction ability of the model was evaluated by external validation set (test set), through the coefficient of determination for the test set (R^2_{PRED}) and the root mean square error of prediction (RMSEP).

4. Conclusions

The present work proposes the use of ^{1}H NMR applied to the development and optimization of multivariate calibration models to determine the relative concentration of four juices in blends. The PLS model was established by using different chemometric indicators (RMSEC, RMSECV, RMSEP, R^2_{CAL}, R^2_{CV} and R^2_{PRED}). The results demonstrate the feasibility and several advantages of using ^{1}H NMR spectral data in combination with multivariate data analysis to build a model capable of detecting adulterations in fruit juice.

Author Contributions: Conceptualization, D.B.; methodology, D.B. and L.M.; software, D.B. and L.M.; validation, D.B. and L.M.; formal analysis, L.M.; investigation, L.M.; resources, D.B.; data curation, F.P.; writing—original draft preparation, L.M.; writing—review and editing, D.B. and F.P.; visualization, L.M.; supervision, S.B.; project administration, D.B.

Funding: This research received no external funding.

Acknowledgments: The authors want to express their thanks to the C.I.G.S. staff (Centro Interdipartimentale Grandi Strumenti, Modena, Italy) for assistance during the experimental work and to Fondazione Cassa di Risparmio di Modena for the purchase of a Bruker FT-NMR Avance III HD 600 MHz spectrometer.

Conflicts of Interest: The authors declare no conflict of interest. The funders had no role in the design of the study; in the collection, analyses, or interpretation of data; in the writing of the manuscript, or in the decision to publish the results.

References

1. Market Report 2016. Available online: https://aijn.eu/en/publications/market-reports-1/market-report-2016 (accessed on 9 May 2019).
2. Bat, K.B.; Eler, K.; Mazej, D.; Vodopivec, B.M.; Mulič, I.; Kump, P.; Ogrinc, N. Isotopic and elemental characterisation of Slovenian apple juice according to geographical origin: Preliminary results. *Food Chem.* **2016**, *203*, 86–94.
3. Kamiloglu, S. Authenticity and traceability in beverages. *Food Chem.* **2019**, *277*, 12–24. [CrossRef] [PubMed]
4. GAO/RCED-96-18-Fruit Juice Adulteration: Detection Is Difficult, and Enhanced Efforts Would Be Costly -Content Details-GAOREPORTS-RCED-96-18. Available online: https://www.govinfo.gov/app/details/GAOREPORTS-RCED-96-18 (accessed on 9 May 2019).
5. Cuny, M.; Vigneau, E.; Le Gall, G.; Colquhoun, I.; Lees, M.; Rutledge, D.N. Fruit juice authentication by 1H NMR spectroscopy in combination with different chemometrics tools. *Anal. Bioanal. Chem.* **2008**, *390*, 419–427. [CrossRef] [PubMed]
6. Salazar, M.O.; Pisano, P.L.; Sierra, M.G.; Furlán, R.L. NMR and multivariate data analysis to assess traceability of argentine citrus. *Microchem. J.* **2018**, *141*, 264–270. [CrossRef]
7. Cusano, E.; Simonato, B.; Consonni, R. Fermentation process of apple juice investigated by NMR spectroscopy. *LWT* **2018**, *96*, 147–151. [CrossRef]
8. Jamin, E.; Thomas, F. SNIF-NMR Applications in an Economic Context: Fraud Detection in Food Products. In *Modern Magnetic Resonance*; Springer Science and Business Media LLC: Berlin/Heidelberg, Germany, 2018; pp. 1405–1416.
9. Le Gall, G.; Colquhoun, I.J.; Defernez, M. Metabolite Profiling Using 1H NMR Spectroscopy for Quality Assessment of Green Tea, Camellia sinensis (L.). *J. Agric. Food Chem.* **2004**, *52*, 692–700. [CrossRef] [PubMed]

10. Spraul, M.; Schütz, B.; Rinke, P.; Koswig, S.; Humpfer, E.; Schäfer, H.; Mörtter, M.; Fang, F.; Marx, U.C.; Minoja, A. NMR-Based Multi Parametric Quality Control of Fruit Juices: SGF Profiling. *Nutrients* **2009**, *1*, 148–155. [CrossRef] [PubMed]

11. Almeida, C.; Duarte, I.F.; Barros, A.; Rodrigues, J.E.; Spraul, M.; Gil, A.M. Composition of Beer by1H NMR Spectroscopy: Effects of Brewing Site and Date of Production. *J. Agric. Food Chem.* **2006**, *54*, 700–706. [CrossRef] [PubMed]

12. Minoja, A.P.; Napoli, C. NMR screening in the quality control of food and nutraceuticals. *Food Res. Int.* **2014**, *63*, 126–131. [CrossRef]

13. Mixture Analysis by NMR as Applied to Fruit Juice Quality Control-Spraul-2009-Magnetic Resonance in Chemistry -Wiley Online Library. Available online: https://onlinelibrary.wiley.com/doi/abs/10.1002/mrc.2528 (accessed on 9 May 2019).

14. Haaland, D.M.; Thomas, E.V. Partial least-squares methods for spectral analyses. 1. Relation to other quantitative calibration methods and the extraction of qualitative information. *Anal. Chem.* **1988**, *60*, 1193–1202. [CrossRef]

15. Villa-Ruano, N.; Pérez-Hernández, N.; Zepeda-Vallejo, L.G.; Quiroz-Acosta, T.; Mendieta-Moctezuma, A.; Montoya-García, C.O.; García-Nava, M.L.; Martínez, E.B. 1 H-NMR Based Metabolomics Profiling of Citrus Juices Produced in Veracruz, México. *Chem. Biodivers.* **2019**, *16*, e1800479. [CrossRef] [PubMed]

16. Ebrahimi, P.; Viereck, N.; Bro, R.; Engelsen, S.B. Chemometric Analysis of NMR Spectra. In *Modern Magnetic Resonance*; Springer Science and Business Media LLC: Berlin/Heidelberg, Germany, 2018; pp. 1649–1668.

17. Lundstedt, T.; Seifert, E.; Abramo, L.; Thelin, B.; Nyström, Å.; Pettersen, J.; Bergman, R. Experimental design and optimization. *Chemom. Intell. Lab. Syst.* **1998**, *42*, 3–40. [CrossRef]

18. Savorani, F.; Tomasi, G.; Engelsen, S.B. icoshift: A versatile tool for the rapid alignment of 1D NMR spectra. *J. Magn. Reson.* **2010**, *202*, 190–202. [CrossRef] [PubMed]

19. Winning, H.; Roldán-Marín, E.; Dragsted, L.O.; Viereck, N.; Poulsen, M.; Sanchez-Moreno, C.; Cano, M.P.; Engelsen, S.B. An exploratory NMR nutri-metabonomic investigation reveals dimethyl sulfone as a dietary biomarker for onion intake. *Analyst* **2009**, *134*, 2344–2351. [CrossRef] [PubMed]

Sample Availability: Fruit juice samples are available from the authors.

molecules

MDPI

Article

Extraction of Carotenoids and Fat-Soluble Vitamins from *Tetradesmus Obliquus* Microalgae: An Optimized Approach by Using Supercritical CO_2

Laura Chronopoulou [1], Chiara Dal Bosco [1], Fabrizio Di Caprio [1], Letizia Prosini [1], Alessandra Gentili [1,2], Francesca Pagnanelli [1] and Cleofe Palocci [1,2,*]

[1]   Chemistry Department, University of Rome La Sapienza, 00185 Rome, Italy
[2]   CIABC, University of Rome La Sapienza, 00185 Rome, Italy
*   Correspondence: cleofe.palocci@uniroma1.it

Academic Editors: Alessandra Gentili and Chiara Fanali
Received: 1 July 2019; Accepted: 15 July 2019; Published: 16 July 2019

check for updates

Abstract: In recent years, great attention has been focused on rapid, selective, and environmentally friendly extraction methods to recover pigments and antioxidants from microalgae. Among these, supercritical fluid extraction (SFE) represents one of the most important alternatives to traditional extraction methods carried out with the use of organic solvents. In this study, the influence of parameters such as pressure, temperature, and the addition of a polar co-solvent in the SFE yields of carotenoids and fat-soluble vitamins from *T. obliquus* biomass were evaluated. The highest extraction of alpha-tocopherol, gamma-tocopherol, and retinol was achieved at a pressure of 30 MPa and a temperature of 40 °C. It was observed that overall, the extraction yield increased considerably when a preliminary step of sample pre-treatment, based on a matrix solid phase dispersion, was applied using diatomaceous earth as a dispersing agent. The use of ethanol as a co-solvent, under certain conditions of pressure and temperature, resulted in selectively increasing the yields of only some compounds. In particular, a remarkable selectivity was observed if the extraction was carried out in the presence of ethanol at 10 MPa and 40 °C: under these conditions, it was possible to isolate menaquinone-7, a homologous of vitamin K2, which, otherwise, cannot not recovered by using traditional extraction procedures.

Keywords: microalgae; *Scenedesmus*; supercritical fluid extraction; carotenoids; fat-soluble vitamins; antioxidants

1. Introduction

Aquatic species are promising sources of products for the fine chemicals industry, and this has aroused a growing interest toward such organisms for several applications such as the production of biofuels, the extraction of food additives or active ingredients for cosmetic formulations [1–3]. In particular, algae represent an attractive source for the extraction of vitamin K, carotenoids, and other fat-soluble vitamins.

Vitamin K is a family of structurally similar chemical compounds including phylloquinone (vitamin K1), which occurs in green plants, and menaquinones (vitamin K2 vitamers), which are predominantly of microbial origin [4,5]. Besides acting as a cofactor for the enzyme γ-glutamylcarboxylase, recent research has shown that vitamin K can protect against intracellular oxidative stress and cognitive decline [6–9]. Regarding the vitamin K content in common macroalgae, extremely variable concentrations of phylloquinone have been observed [10,11], however, it was not detected in *P. tricornutum* [11], while its concentration reached 750 μg/100 g in *Sargassum muticum* (commonly known as Japanese wireweed), which is a significantly higher value than that observed in terrestrial plants [10]. To the best of our knowledge, no information on the distribution of menaquinones has so far been reported.

Carotenoids are tetraterpenoids with numerous biological functions synthesized by plants, algae, fungi, and bacteria. They are essential for photosynthesis and, in general, for life in the presence of oxygen. Due to their chemical structure, based on a long chain of conjugated double bonds, these micronutrients are highly lipophilic, variously colored, and exhibit antioxidant properties [12]. Due to changes in lifestyle and the rising health consciousness of the average population, the demand for nutrient-rich supplements with health benefits has risen substantially. Carotenoids also have various medicinal properties and are widely used as preventatives against diseases such as cancer, diabetes, and cataracts [13]. Carotenoids are also currently used in food supplements, cosmetics, and pharmaceuticals [14].

Hydrocarbon carotenoids are referred to as carotenes, while their oxygenated derivatives are known as xanthophylls. Within the latter group, the oxygen atom may be present in the form of hydroxyl groups (as in the case of lutein) or as keto groups (as in the case of canthaxanthin) or as a combination of both (as in astaxanthin) [15]; other oxygenated functional groups occurring in xanthophylls are the epoxy group and carboxylic group. Moreover, all carotenoids having a polyene chain with 11 carbon atoms, and at least one unsubstituted β-ionone ring contributes to the synthesis of vitamin A by means of their conversion into retinol. Increasingly restrictive legislation regarding the origin of food preservatives (e.g., antioxidants and antimicrobials), coupled with the growing demand for natural compounds, have renewed the interest in biomass as a potential source for such compounds rather than resorting to chemical synthesis [16,17]. The best candidates for carotenoid biosynthesis are microalgae because they have many useful features from an industrial point of view: a high surface-to-volume ratio; high growth rates; different metabolic pathways; environmental adaptability; and the simplicity of screening and genetic manipulation [10]. Typical maximum growth rates of microalgae are between 0.5 and 2 d^{-1} that correspond to the duplication times of a few hours, which is much higher than the growth rates of terrestrial plants. Although microalgae are photosynthetic microorganisms, they can also grow in mixotrophic and heterotrophic conditions [18]. Microalgae generally have an optimal pH between 7–9 and an optimal temperature between 25–30 °C. Some carotenoids are synthesized by microalgae as primary metabolites to protect the photosystems from photodamage caused by excessive light exposition and to enlarge the harvesting spectrum of light (e.g., lutein and fucoxanthin). Others carotenoids such as astaxanthin and β-carotene are accumulated as secondary metabolites under stress conditions (e.g., nutrients starvation, osmotic shock, high temperature) to protect the cells from oxidative stresses [14,19]. Among the various species of microalgae, *Tetradesmus obliquus* (generally known as *Scenedesmus obliquus*), an ubiquitous microorganism in lakes and freshwater rivers, is one of the most studied strains for large scale industrial applications, thanks to its ability to grow well in the non-optimal conditions typical of industrial outdoor plants [20]. *T obliquus* can be cultivated with biomass productivities until 2 g/L d in photoautotrophic conditions [21] and until 6 g/L d in heterotrophic conditions [18]. This microalga appears to be a promising source of carotenoid-rich extracts [22]. However, to date, *T. obliquus* is still not exploited for the industrial production of carotenoids or vitamins [19]. One relevant limitation comes from the absence of extraction processes that are sufficiently selective and efficient [23]. Among the most widely used organic solvents for carotenoid extraction, there is hexane (particularly effective for carotenes), ethanol (advantageous for xanthophylls), and acetone. However, the use of organic solvents for extraction is not the preferred way to produce healthy foods due to the toxicity of solvent residues in food as well as the issues of environmental pollution. The development of an efficient extraction technique for the isolation of pharmaceutical compounds from natural resources is necessary.

Extraction using supercritical fluids (SFE) is currently considered an important green alternative to traditional methods. The properties of supercritical fluids (SFs) can be considered as intermediate between the ones of liquids and gases. Similarly to gases, SFs are highly compressible, but have high densities comparable to those of liquids. The combination of some of the properties of liquids with those of gases provides supercritical fluids with some very interesting features. For example, supercritical fluids can effuse through solid materials like a gas, but can also act like a liquid and dissolve substances. In SFE, the organic phase used in typical solid–liquid extractions (SLE) is substituted by a supercritical fluid. The manipulation of both the temperature and pressure of the fluid can solubilize the substance of interest in a complex matrix and selectively extract it. Compared to SLE, SFE is indeed simpler, faster, and more efficient but without consuming large quantities of organic solvents, which are both expensive and potentially harmful. Other immediate advantages of SFE compared to traditional extraction techniques are process flexibility due to the continuous modulation of the solvent power/selectivity of the supercritical fluid and the elimination of polluting organic solvents, which also prevents expensive post-processing of the extracts for solvent removal. CO_2 is the most commonly used supercritical fluid thanks to its non-toxicity, chemical inertia, low cost, and most importantly, low critical values. Its low critical temperature (below 32 °C) makes CO_2 ideal for the extraction of thermolabile compounds. For these reasons, the use of CO_2 as an extraction solvent has been successfully reported in the literature for the isolation of many compounds from various sources [24–28]. For example, supercritical CO_2 has been tested for the extraction of carotenoids and triglycerides from microalgae such as *Hematococcus pluvialis*, *Scenedesmus* sp., and *Chlorella* sp. in different previous works, by mainly using ethanol as the co-solvent, as reported in a recent review [29]. However, there is scarce information about the co-extraction of vitamins and carotenoids from microalgae. In this study, the possibility of extracting carotenoids and fat-soluble vitamins from *T. obliquus* by means of CO_2 in the supercritical phase was evaluated. The effect of several parameters such as the CO_2 physical variables, the addition of co-solvents (methanol and limonene), and an inert dispersing phase on the recovery of different carotenoids and fat-soluble vitamins was investigated. The supercritical fluid extraction was also compared in terms of yield and selectivity with conventional extraction methods.

2. Results and Discussion

2.1. Extraction via Matrix Solid-Phase Dispersion

According to the methodology reported in the literature [30,31], the HPLC-MS analysis of the extracts obtained by the matrix solid-phase dispersion (MSPD) showed that the most abundant carotenoids and fat-soluble vitamins in the algal biomass were lutein and α-tocopherol, respectively. The MSPD extraction, applied individually in accordance with that described in Section 3.4, allowed 10 different compounds to be isolated and identified via HPLC-MS.

2.2. Supercritical CO_2 (SCCO$_2$) Extraction

2.2.1. Evaluation of the Optimal Extraction Time

A series of preliminary SFE was performed on samples of *T. obliquus* at constant pressure (P) and temperature (T) values and by varying the extraction time between 1 and 4 h in order to determine its optimal value. In particular, by using a P_{CO2} of 30 MPa and a T_{CO2} of 50 °C, the percentage of extracted carotenoids was evaluated as a function of the incubation time. The 90-min extraction time was selected as the most appropriate one since, at higher incubation times, no increase in the amount of extracted material was recorded. Therefore, all subsequent extractions in the static approach were carried out for 90 min. For operations in the dynamic mode, an initial static extraction time of 90 min was used, followed by a dynamic extraction step for 10 min. In both cases, at the end of the process, CO_2 was withdrawn from the cell and the extract was recovered in 2 mL of ethanol.

2.2.2. Addition of Diatomaceous Earth as a Dispersing Phase

Before extraction, the algal biomass was mixed with diatomaceous earth in a 1:10 *w/w* proportion, as reported in the experimental section. This sample preparation was selected with the aim of obtaining a better yield and extraction reproducibility (data not shown). Indeed, mixing with diatomaceous earth can help to break the cell walls and membranes of the microalgae, exposing the cellular content to the action of the extracting fluid as well as increasing the contact area between the sample and the solvent.

2.3. Evaluation of the Influence of Pressure and Temperature on the Recovery of Carotenoids from T. Obliquus

2.3.1. Extraction Conditions and Extraction Variables

For all of the SFE and MSPD extraction procedures, a fixed quantity of diatomaceous earth/microalga mixture (equal to 0.200 g of sample, of which 0.01818 g was *T. obliquus* biomass) was used. Each experiment was repeated at least twice. Three different CO_2 pressure values were selected during the extractions (25, 30, and 35 MPa, respectively) and for each pressure, the extractions were performed at three different temperatures (40, 50, and 60 °C). The pressure and temperature values were selected on the basis of the literature data and taking into account the thermolability of the carotenoids. In general, as expected, it was observed that as the pressure of the supercritical CO_2 increases at constant temperature, the extraction yields increase as a consequence of an increase in the solvent power. In contrast, as the temperature increases at constant pressure, the solvent power of CO_2 decreases, and therefore carotenoid extraction yields are reduced.

Moreover, we investigated the effect of modifiers such as methanol (5% *v/v*) and limonene (5% *v/v*), on the composition of the extracts. As can be seen in Figures 1 and 2, the extractions carried out with the addition of limonene did not show a qualitative or quantitative improvement in the composition of the extract, except for the best extraction of phytofluene (a non-polar compound structurally very similar to limonene), while as expected, the addition of methanol allowed a better recovery of all the more polar carotenoids.

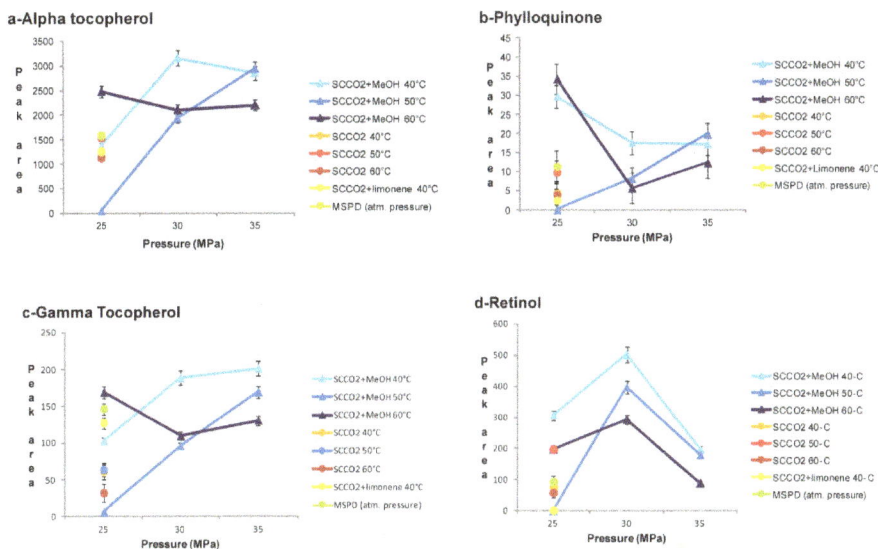

Figure 1. Extraction yields (average peak area ± standard error) of (**a**) alpha tocopherol, (**b**) phylloquinone, (**c**) gamma tocopherol, and (**d**) retinol obtained with SFE and MSPD.

2.3.2. SCCO$_2$ Extraction of Fat-Soluble Vitamins and Carotenoids from T. Obliquus in Comparison with MSPD

The supercritical fluid extraction of micronutrients from algal biomass in comparison with the solid/liquid extraction methodology is reported in Figures 1 and 2. We optimized the extraction conditions in SCCO$_2$ for the following compounds: α-tocopherol, canthaxantin, γ-tocopherol, lutein, phylloquinone, phytofluene, retinol, and menaquinone-7, whose structures are shown in Figures 3 and 4. In Figures 5 and 6, the mass spectra of each extracted compound are reported.

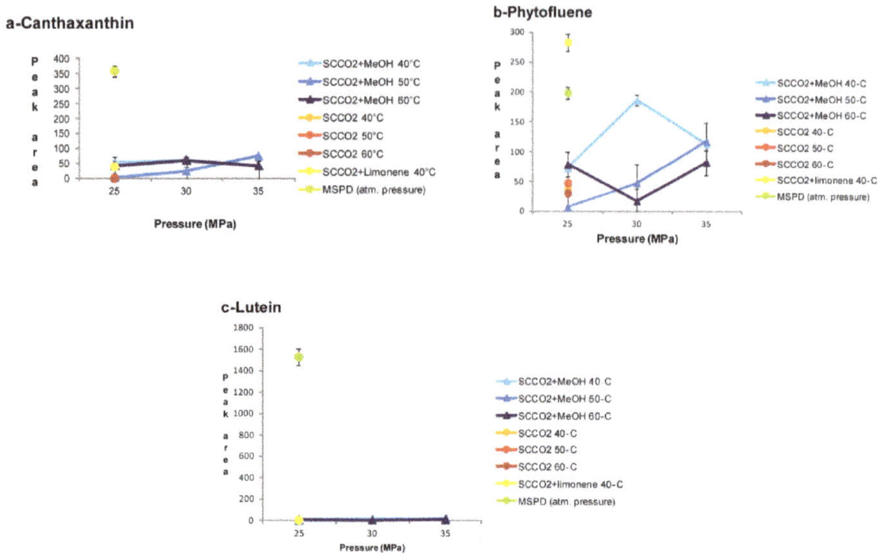

Figure 2. Extraction yields (average peak area ± standard error) of (**a**) canthaxanthin, (**b**) phytofluene, and (**c**) lutein obtained with SFE and MSPD.

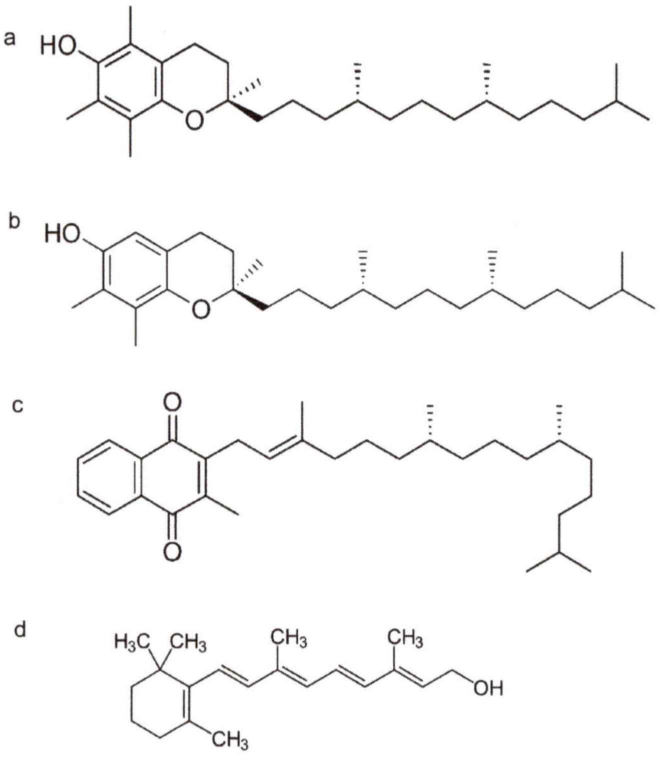

Figure 3. Chemical structures of: (**a**) alpha tocopherol, (**b**) gamma-tocopherol, (**c**) phylloquinone, and (**d**) retinol.

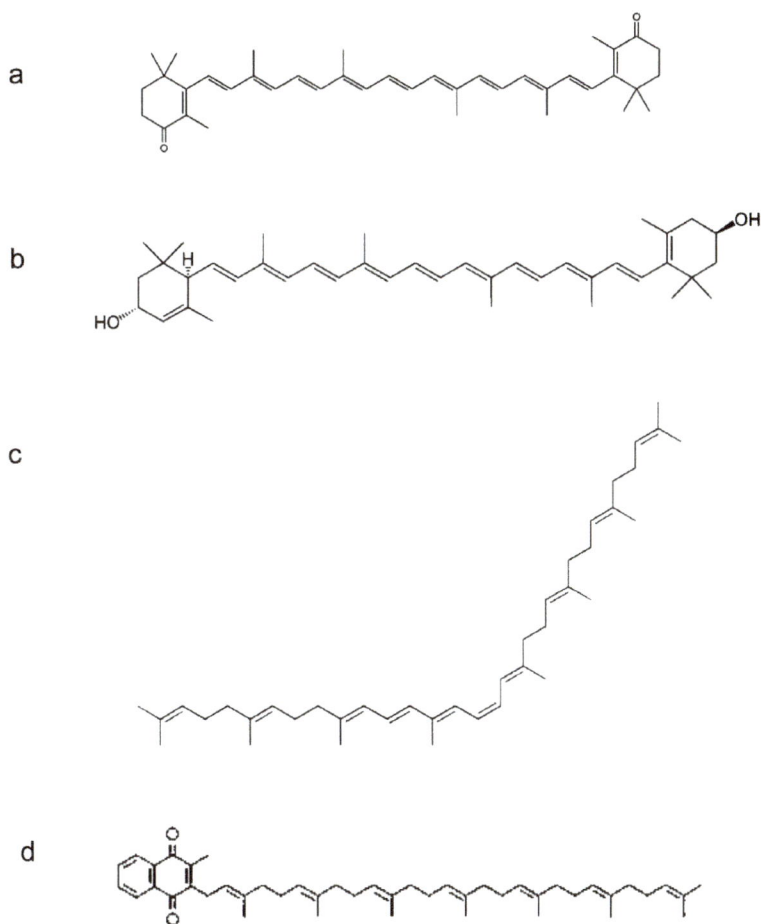

Figure 4. Chemical structures of: (**a**) canthaxanthin, (**b**) lutein, (**c**) phytofluene, and (**d**) menaquinone-7.

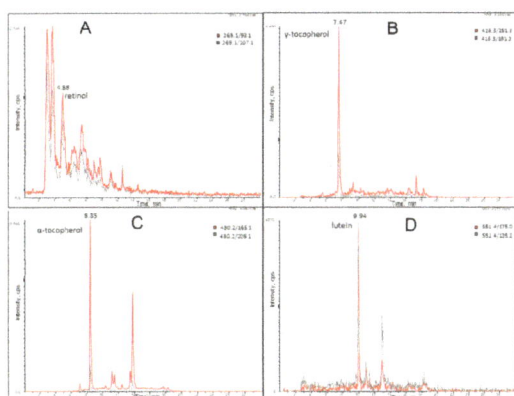

Figure 5. LC-MRM profile of retinol (**A**), gamma-tocopherol (**B**), alpha tocopherol (**C**) and lutein (**D**) extracted from a microalga sample by SFE. Extraction conditions: P_{CO2} = 35 MPa, T_{CO2} = 40 °C, 5% MeOH.

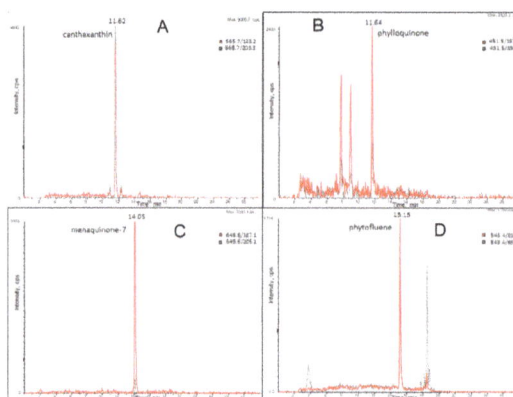

Figure 6. LC-MRM profile of canthaxanthin (**A**), phylloquinone (**B**), menaquinone-7 (**C**), and phytofluene (**D**) extracted from a microalga sample by SFE. Extraction conditions: A, B and D: P_{CO2} = 35 MPa, T_{CO2} = 40 °C, 5% MeOH. C: P_{CO2} = 10 MPa, T_{CO2} = 40 °C.

As reported in Figure 1a at P_{CO2} 30 MPa and T_{CO2} 40 °C with the addition of MeOH, the concentration of α-tocopherol detected in the extract was 61.8 μg/mL. The extraction with organic solvents allowed a lower recovery of α-tocopherol, with a concentration of 27.7 μg/mL. The molecular structure of α-tocopherol contains an oxygen atom that makes the molecule more polar than the carotenoids: this explains its higher SF extraction when adding MeOH as a cosolvent for CO_2.

Regarding alpha tocopherol extraction from different types of algae, a strong variability of its content has been reported in the literature and often very close taxa have very different alpha tocopherol contents. Such results point out the importance of growth conditions for obtaining higher quantities of this compound [32]. On this basis, it is very difficult to find useful extraction data to compare the efficiency of the extraction technique used.

Canthaxanthin is co-extracted with $SCCO_2$ in the same experimental conditions (Figure 2a). The extraction yields, however, were not quantitatively comparable with those obtained by MSPD. It is most likely that the chemical structure of canthaxanthin, which contains two carbonyl groups, makes it very poorly soluble in supercritical CO_2 (even at high density and in the presence of methanol). Indeed, the recovery of canthaxanthin in CO_2-only extraction was below the limit of detection, while better results were obtained by increasing the polarity of the solvent phase with the addition of MeOH.

Phylloquinone contains two carbonyl groups together with the presence of a long hydrocarbon chain in its molecular structure. This feature can be responsible for a greater solubility in supercritical CO_2 than in organic solvent mixtures. In the case of phylloquinone, the best extraction conditions were obtained at CO_2 pressure values of 25 MPa and T = 60 °C in the presence of MeOH (Figure 1b).

γ-tocopherol, one of the eight vitamers of vitamin E, showed a better extraction profile with $SCCO_2$, and the addition of MeOH at a temperature of 40 °C and pressure of both 35 and 30 MPa, produced quantitatively higher extraction yields than those obtained with the MSPD technique (Figure 1c). α-tocopherol and γ-tocopherol are structurally similar molecules, and despite being extracted under the same experimental conditions, their recovery was different due to the different relative quantities contained in the microalgae.

Retinol, or vitamin A, as a metabolite of provitamin A carotenoids, could be formed during the extraction procedure by increasing the extraction temperature. In fact, in almost all the $SCCO_2$ extraction conditions tested, a larger recovery of this vitamin was observed when compared to the MSPD technique (Figure 1d).

Phytofluene, a colorless carotenoid precursor, has a structure with 40 carbon atoms and five conjugated double bonds. It showed a better recovery profile with the MSPD technique compared to SFE, although the addition of limonene allowed a better extraction of this carotenoid with the SFE technique (Figure 2b). This may be due to its remarkably apolar structure, while further degradation occurs at temperatures above 40 °C and at higher CO_2 pressures.

Lutein, known as E161b in the European codification of food additives, was extracted in more significant amounts with the MSPD technique than with $SCCO_2$ (Figure 2c). Lutein contains two hydroxyl groups within the molecule and is a very polar compound. It is partially recovered in SFE extractions with the addition of MeOH, but in lower relative yields.

Finally, extractions on the microalgae carried out in $SCCO_2$ at a pressure of 10 MPa and T = 40 °C allowed the selective extraction of menaquinone-7 (Figure 6), which was not detected in the MSPD extractions. $SCCO_2$ extractions at higher pressures and temperatures did not show the presence of menaquinone-7, confirming the hypothesis that it could be chemically degraded at high CO_2 pressure or temperature values.

Menaquinone-7, like the other menaquinones, has a bacterial origin. It is not usually synthesized from algae, although in the literature, its presence has been hypothesized in the microalgae of the genus *Scenedesmus* [33]. Moreover, the microalgae used in this study were grown in a non-sterile environment, therefore they may have contained the products of a unique system formed by the microbiota in symbiosis with microalgae [34]. It has been proven that several microalgae species cannot survive without such associated bacteria because these latter furnish essential vitamins (such as vitamin B_{12}) to the microalgae [35].

Our results demonstrate that SFE with $SCCO_2$ is a green method for the extraction of high purity thermolabile compounds such as carotenoids. However, the yield of polar carotenoids, as reported in the literature, is often low [36]. By optimizing some key parameters (use of entrainers), it is possible to improve the solubility of more polar analytes in $SCCO_2$. Table 1 reports a comparison of the obtained SFE extraction yields based on the peak areas obtained from the mass spectra with those obtained with MSPD. By varying the SFE conditions, we were able to obtain comparable and sometimes higher extraction yields than MSPD for alpha and gamma tocopherol, canthaxanthin, phylloquinone, phytofluene, retinol, and menaquinone-7.

Table 1. The SFE extraction yields of alpha tocopherol, phylloquinone, gamma-tocopherol, retinol, canthaxanthin, phytofluene, and lutein using different experimental conditions. The values reported are expressed as percentages of MSPD extraction yields ± standard error.

Compound	SCCO$_2$ 25 MPa			SCCO$_2$ + limonene	SCCO$_2$ + MeOH 25 MPa			SCCO$_2$ + MeOH 30 MPa			SCCO$_2$ + MeOH 35 MPa		
	40 °C	50 °C	60 °C		40 °C	50 °C	60 °C	40 °C	50 °C	60 °C	40 °C	50 °C	60 °C
Alpha-tocopherol	(77.71 ± 4.95)%	(97.3 ± 3.59)%	(71.12 ± 3.91)%	(78.73 ± 3.28)%	(86.25 ± 5.16)%	(2.88 ± 2.76)%	(156.95 ± 5.31)%	(200.15 ± 4.64)%	(122.39 ± 4.61)%	(133.44 ± 5.45)%	(180.73 ± 5.33)%	(185.79 ± 3.78)%	(139.39 ± 3.74)%
Phylloquinone	(38.83 ± 4.61)%	(86.50 ± 4.31)%	(34.77 ± 4.31)%	(21.45 ± 4.15)%	(262.14 ± 5.16)%	0	(302.74 ± 5.89)%	(154.46 ± 5.72)%	(71.67 ± 6.11)%	(50.31 ± 6.73)%	(151.81 ± 5.63)%	(174.76 ± 6.28)%	(108.56 ± 6.45)%
Gamma-tocopherol	(41.42 ± 3.76)%	(43.26 ± 3.73)%	(21.77 ± 5.71)%	(86.73 ± 3.82)%	(69.74 ± 3.16)%	(4.09 ± 3.16)%	(115.46 ± 3.81)%	(128.90 ± 3.45)%	(65.30 ± 3.18)%	(74.72 ± 3.62)%	(137.43 ± 3.92)%	(115.52 ± 3.75)%	(88.98 ± 3.49)%
Retinol	(81.44 ± 3.18)%	(212.56 ± 3.47)%	(62.03 ± 3.94)%	0	(332.72 ± 2.58)%	0	(213.75 ± 5.76)%	(543.00 ± 5.73)%	(430.97 ± 8.56)%	(317.32 ± 4.61)%	(213.00 ± 3.94)%	(192.93 ± 6.31)%	(94.78 ± 4.72)%
Canthaxanthin	0	0	0	(8.09 ± 2.98)%	(11.02 ± 3.11)%	(0.56 ± 0.44)%	(9.16 ± 3.26)%	(12.99 ± 3.45)%	(4.99 ± 2.31)%	(12.99 ± 4.23)%	(9.25 ± 3.62)%	(16.14 ± 3.15)%	(9.25 ± 3.52)%
Phytofluene	(18.24 ± 5.64)%	(23.50 ± 4.25)%	(15.21 ± 2.58)%	(142.65 ± 7.51)%	(36.43 ± 4.39)%	(94.04 ± 3.98)%	(56.80 ± 4.58)%	(3.34 ± 2.13)%	(23.50 ± 8.57)%	(58.92 ± 8.63)%	(39.46 ± 8.91)%	(8.44 ± 7.32)%	(41.13 ± 8.69)%
Lutein	0	0	0	0	(0.74 ± 0.53)%	0	(0.70 ± 0.45)%	(1.03 ± 0.74)%	(0.29 ± 0.25)%	(0.22 ± 0.20)%	(0.67 ± 0.64)%	(1.25 ± 0.50)%	(0.40 ± 0.40)%

3. Materials and Methods

3.1. Biomass Production

A strain of the microalgae *Tetradesmus obliquus* was maintained in the laboratory under phototrophic conditions as previously described [36]. *T. obliquus* is generally known as *Scenedesmus obliquus*, but it has been recently reclassified by Wynne and Hallan [37].

The biomass used for the extraction tests was produced by diluting 1 to 10 (*v/v*) microalgae from the maintenance flasks in two column photobioreactors (ø = 9 cm, h = 65 cm) with the cultivation medium. The initial biomass concentration was 0.05 g/L. The cultivation medium used was a "tap water based medium", which is a cultivation medium obtained by adding $NaNO_3$ and K_2HPO_4 to the local tap water; its exact chemical composition has been described in a previous work [38,39]. The photobioreactors were maintained under 24 h/24 constant illumination at 100 μmol m^{-2}s^{-1}, by means of cool-white florescent lamps and constant air feeding (0.5 L/min) at a room temperature of 27 ± 3 °C. After 15 days of cultivation, the produced microalgae biomass was harvested by centrifugation at 1370× *g* for 5 min and then the obtained pellet was freeze dried.

3.2. Chemicals and Solvents

Methanol, ethanol, 2-propanol, hexane, and acetone (HPLC grade) were purchased from Sigma (St. Louis, MO, USA). Dichloromethane and acetonitrile (analytical grade) were obtained from Chromasolv (Barcelona, Spain). Diatomaceous earth SPE-ED MATRIX 38 was purchased from Applied Separations (Allentown, PA, USA). Syringe-like polypropylene tubes (i.d. 26 mm, 75 mL capacity) and polyethylene frits were obtained from Alltech (Deerfield, IL, USA).

The following standards were purchased from Aldrich-Fluka-Sigma Chemical (St. Louis, MO, USA): retinol, ergocalciferol, δ-tocopherol, β-tocopherol, γ-tocopherol, cholecalciferol, α-tocopherol, menaquinone-4, menaquinone-7, phylloquinone, all-trans-lutein, all trans-zeaxanthin, all-trans-β-cryptoxanthin, and all-trans-β-carotene. Standards of α-tocotrienol, β-tocotrienol, δ-tocotrienol, and γ-tocotrienol were bought from LGC Standards (Middlesex, U.K.). Standards of 15-cis-phytoene, all-trans-phytoene, all-trans-phytofluene, 13-cis-β-carotene, 9-cis-β-carotene, all-trans-ζ-carotene, all-trans-γ-carotene, all-trans-lycopene, and 5-cis-lycopene were purchased from CaroteNature GmbH (Ostermundigen, Switzerland). All chemicals had a purity grade of >97%.

3.3. Biomass Pretreatment

The biomass was freeze dried and stored at −20 °C. Before extraction, the biomass was manually ground into a fine powder. When diatomaceous earth was used, it was mixed with the ground biomass at a 1:10 *w/w* ratio (algae:diatomaceous earth).

3.4. MSPD Extraction

The results obtained by SFE were compared with those from the MSPD extraction. The last one was performed as follows: 200 mg of sample (biomass and diatomaceous earth 1:10 *w/w*) was ground with a pestle into a ceramic mortar until an evenly colored powder was obtained. Subsequently, this powder was used for filling a syringe-like polypropylene tube previously prepared with a first layer of C18 sorbent (0.4 g). The resultant chromatographic bed was held among two polyethylene frits. Vacuum-assisted elution was conducted with 15 mL of methanol, 5 mL of 2-propanol, and 20 mL of hexane by collecting the analytes into a 50 mL falcon. Samples were then centrifuged at 6000 rpm for 10 min. The supernatant was poured out into a glass tube with a conical bottom (i.d. 2 cm) and evaporated up to dryness under a gentle flow of nitrogen in a water bath kept at 25 °C. Finally, the dry extract was dissolved in 2 mL of ethanol, sonicated for 2 min, and passed through a PTFE 0.45 μm filter. Forty microliters were injected into the chromatographic column for the LC-MS analysis.

3.5. Supercritical Fluid Extraction

Supercritical fluid extractions (SFE) were performed on a SFE 300 analytical extractor manufactured by Carlo Erba Instruments. A scheme of the extraction apparatus is shown in Figure 1. The extractions took place in a metal tubular reactor of 1 cm^3 where the samples were introduced. Cooled CO_2 was fed into the high-pressure reactor, pressurized at the desired target pressure by a syringe pump, and heated to the desired temperature with a system of recirculating air in the thermostated chamber where the reactor was located. In all of the experiments, a static extraction was followed by a 10-min dynamic extraction performed through the depressurization of $SCCO_2$ in 2 mL of ethanol.

For each extraction experiment, 200 mg of ground powder (containing biomass and diatomaceous earth 1:10 *w/w*) were placed in the extraction cell. The operating parameters were varied in the pressure range of 10–35 MPa and in the temperature range of 40–60 °C. A series of extractions was also performed in the presence of a modifier by adding 5% (*v/v*) methanol (T = 40, 50, and 60 °C; P = 25, 30, and 35 MPa) or limonene (T = 40 °C; P = 30 MPa) to the sample inside the extraction cell.

3.6. Mass Spectrometry Experiments

Analytes were detected by a 4000 Qtrap (AB SCIEX, Foster City, CA, USA.) mass spectrometer equipped with an atmospheric pressure chemical ionization (APCI) probe on a Turbo V source. A positive ionization mode was used, setting a needle current (NC) of 3 µA and a probe temperature of 450 °C. High-purity nitrogen was used as the curtain (40 psi) and collision (4 mTorr) gas, whereas air was the nebulizer (55 psi) and makeup (30 psi) gas. The preliminary calibration of Q1 and Q3 mass analyzers was conducted by infusing a polypropylene glycol solution at 10 µL/min. The unit mass resolution was established by maintaining a full width at half-maximum (fwhm) of approximately 0.7 ± 0.1 unit in each mass-resolving quadrupole. APCI–Q1–full scan spectra and product ion scan spectra of the analytes were acquired by working in flow injection analysis (1–10 ng injected, 1 mL/min flow rate).

3.7. Liquid Chromatography

Liquid chromatography (LC) was performed by a micro HPLC series 200 (PerkinElmer, Norwalk, CT, USA.) equipped with an autosampler, vacuum degasser, and column chiller. Analytes were separated on a ProntoSIL C30 column (4.6 mm × 250 mm, 3 µm) from Bischoff Chromatography (Leonberg, Germany), protected by a guard C30 column (4.0 mm × 10 mm, 5 µm), under non aqueous-reversed phase (NARP) conditions at 19 °C. The elution profile by using methanol (phase A) and 2-propanol/hexane (50:50, *v/v*; phase B) was as follows: 0–1 min, 0% B; 1–15 min, 0–75% B; 15–15.1 min, 75–99.5% B; and 15.1–30.1 min, 99.5% B. The mobile phase was entirely introduced into the MS detector at a flow rate of 1 mL/min. Phase B was also used to wash the autosampler injection device.

The separation and detection of MK-7 were confirmed by using a specific chromatographic method with increased efficiency in separating vitamin K homologues from interfering compounds. This method differs from the previous one for the use of two reversed-phase columns connected in series (SUPELCOSILTM C18, 4.6 mm × 50 mm, 5 mm, Supelco–Sigma–Aldrich, Bellefonte, PA, USA; and Alltima C18, 4.6 mm × 250 mm; 5 mm, Alltech, Deerfield, IL, USA).

4. Conclusions

In this study, the influence of parameters such as pressure, temperature, and the addition of a polar co-solvent on the SFE yields of carotenoids and fat-soluble vitamins from *T. obliquus* biomass was studied. The optimized extraction conditions revealed the possibility to substantially increase the yields of some compounds with respect to conventional solid–liquid extraction. In particular, by varying the SFE polarity, we were able to obtain comparable and sometimes higher extraction yields than MSPD for many low or medium-polar carotenoids and vitamins. We also obtained a remarkable selectivity (at 10 MPa and 40 °C) for the extraction of the compound menaquinone-7, whose extraction has been rarely achieved by using traditional procedures.

Author Contributions: Investigation, L.C., L.P., C.D.B., A.G., and F.D.C.; Writing—original draft preparation, L.C., L.P., A.G., and C.D.B.; Writing—review and editing, F.D.C., F.P., C.D.B., A.G., L.C., and C.P.; Supervision, C.P.

Funding: This research received no external funding.

Conflicts of Interest: The authors declare no conflicts of interest.

References

1. Sabu, S.; Bright-Singh, I.; Joseph, V. Molecular identification and comparative evaluation of tropical marine microalgae for biodiesel production. *Mar. Biotechnol.* **2017**, *19*, 328–344. [CrossRef] [PubMed]

2. Matsumoto, M.; Nojima, D.; Nonoyama, T.; Ikeda, K.; Maeda, Y.; Yoshino, T.; Tanaka, T. Outdoor cultivation of marine diatoms for year-round production of biofuels. *Mar. Drugs* **2017**, *15*, 94. [CrossRef] [PubMed]

3. Ahmed, A.B.; Adel, M.; Karimi, P.; Peidayesh, M. Pharmaceutical, cosmeceutical, and traditional applications of marine carbohydrates. *Adv. Food Nutr. Res.* **2014**, *73*, 197–220. [PubMed]

4. Nelsestuen, G.L.; Shah, A.M.; Harvey, S.B. Vitamin K-dependent proteins. *Vitam. Horm.* **2000**, *58*, 355–389. [PubMed]

5. Gentili, A.; Miccheli, A.; Tomai, P.; Baldassarre, M.E.; Curini, R.; Pérez-Fernández, V. Liquid chromatography–tandem mass spectrometry method for the determination of vitamin K homologues in human milk after overnight cold saponification. *J. Food Compos. Anal.* **2016**, *47*, 21–30. [CrossRef]

6. Diener, H.C.; Hart, R.G.; Koudstaal, P.J.; Lane, D.A.; Lip, G.Y.H. Atrial fibrillation and cognitive function: Jacc review topic of the week. *J. Am. Coll. Cardiol.* **2019**, *73*, 612–619. [CrossRef] [PubMed]

7. Brangier, A.; Ferland, G.; Rolland, Y.; Gautier, J.; Féart, C.; Annweiler, C. Vitamin K antagonists and cognitive decline in older adults: A 24-month follow-up. *Nutrients* **2018**, *10*, 666. [CrossRef]

8. Vos, M.; Esposito, G.; Edirisinghe, J.N.; Vilain, S.; Haddad, D.M.; Slabbaert, J.R.; Van Meensel, S.; Schaap, O.; De Strooper, B.; Meganathan, R.; et al. Vitamin K2 is a mitochondrial electron carrier that rescues pink1 deficiency. *Science* **2012**, *336*, 1306–1310. [CrossRef] [PubMed]

9. Gentili, A.; Cafolla, A.; Gasperi, T.; Bellante, S.; Caretti, F.; Curini, R.; Pérez-Fernández, V. Rapid, high performance method for the determination of vitamin K1, menaquinone-4 and vitamin K1 2, 3-epoxide in human serum and plasma using liquid chromatography-hybrid quadrupole linear ion trap mass spectrometry. *J. Chromatogr. A* **2014**, *1338*, 102–110. [CrossRef]

10. Roeck-Holtzhauer, Y.D.; Quere, I.; Claire, C. Vitamin analysis of five planktonic microalgae and one macroalga. *J. Appl. Phycol.* **1991**, *3*, 259–264. [CrossRef]

11. Tarento, T.D.C.; McClure, D.D.; Vasiljevski, E.; Schindeler, A.; Dehghani, F.; Kavanagh, J.M. Microalgae as a source of vitamin K1. *Algal Res.* **2018**, *36*, 77–87. [CrossRef]

12. Flores-Hidalgo, M.; Torres-Rivas, F.; Monzon-Bensojo, J.; Escobedo-Bretado, M.; Glossman-Mitnik, D.; Barraza-Jimenez, D. Electronic Structure of Carotenoids in Natural and Artificial Photosynthesis. In *Carotenoids*; Cvetkovic, D.J., Nikolic, G.S., Eds.; IntechOpen: London, UK, 2017; pp. 17–33. [CrossRef]

13. Mein, J.R.; Lian, F.; Wang, X.D. Biological activity of lycopene metabolites: Implications for cancer prevention. *Nutr. Rev.* **2008**, *66*, 667–683. [CrossRef] [PubMed]

14. Sathasivam, R.; Ki, J.S. A Review of the Biological Activities of Microalgal Carotenoids and Their Potential Use in Healthcare and Cosmetic Industries. *Mar. Drugs* **2018**, *16*, 26. [CrossRef] [PubMed]

15. Higuera-Ciapara, I.; Félix-Valenzuela, L.; Goycoolea, F.M. Astaxanthin: A review of its chemistry and applications. *Crit. Rev. Food Sci. Nutr.* **2006**, *46*, 185–196. [CrossRef] [PubMed]

16. Gordon, H.T.; Bauernfeind, J.C. Carotenoids as food colorants. *Crit. Rev. Food Sci. Nutr.* **1982**, *18*, 59–97. [CrossRef] [PubMed]

17. Macías-Sánchez, M.D.; Mantell Serrano, C.; Rodríguez Rodríguez, M.; Martínez de la Ossa, E.; Lubián, L.M.; Montero, O. Extraction of carotenoids and chlorophyll from microalgae with supercritical carbon dioxide and ethanol as cosolvent. *J. Sep. Sci.* **2008**, *31*, 1352–1362. [CrossRef] [PubMed]

18. Di Caprio, F.; Altimari, P.; Iaquaniello, G.; Toro, L.; Pagnanelli, F. Heterotrophic cultivation of T. obliquus under non-axenic conditions by uncoupled supply of nitrogen and glucose. *Biochem. Eng. J.* **2019**, *145*, 127–136. [CrossRef]

19. Ambati, R.R.; Gogisetty, D.; Aswathanarayana, R.G.; Ravi, S.; Bikkina, P.N.; Bo, L.; Yuepeng, S. Industrial potential of carotenoid pigments from microalgae: Current trends and future prospects. *Crit. Rev. Food Sci.* **2019**, *59*, 1880–1902. [CrossRef]

20. Wu, Y.H.; Hu, H.Y.; Yu, Y.; Zhang, T.Y.; Zhu, S.F.; Zhuang, L.L.; Zhang, X.; Lu, Y. Microalgal species for sustainable biomass/lipid production using wastewater as resource: A review. *Renew. Sust. Energ. Rev.* **2014**, *33*, 675–688. [CrossRef]

21. Breuer, G.; Lamers, P.P.; Martens, D.E.; Draaisma, R.B.; Wijffels, R.H. Effect of light intensity, pH, and temperature on triacylglycerol (TAG) accumulation induced by nitrogen starvation in *Scenedesmus obliquus*. *Bioresour. Technol.* **2013**, *143*, 1–9. [CrossRef]

22. Amaro, H.M.; Fernandes, F.; Valentão, P.; Andrade, P.B.; Sousa-Pinto, I.; Malcata, F.X.; Guedes, A.C. Effect of Solvent System on Extractability of Lipidic Components of Scenedesmusobliquus (M2-1) and Gloeothece sp. on Antioxidant Scavenging Capacity Thereof. *Mar. Drugs.* **2015**, *13*, 6453–6471. [CrossRef] [PubMed]

23. Postma, P.R.; 't Lam, G.P.; Barbosa, M.J.; Wijffels, R.H.; Eppink, M.H.M.; Olivieri, G. Microalgal Biorefinery for Bulk and High-Value Products: Product Extraction Within Cell Disintegration. In *Handbook of Electroporation*; Miklavcic, D., Ed.; Springer International Publishing: Berlin, Germany, 2016; pp. 1–20. [CrossRef]

24. De Andrade Lima, M.; Kestekoglou, I.; Charalampopoulos, D.; Chatzifragkou, A. Supercritical Fluid Extraction of Carotenoids from Vegetable Waste Matrices. *Molecules* **2019**, *24*, 466. [CrossRef] [PubMed]

25. Mo, Z.Z.; Lin, Z.X.; Su, Z.R.; Zheng, L.; Li, H.L.; Xie, J.H.; Xian, Y.F.; Yi, T.G.; Huang, S.Q.; Chen, J.P. *Angelica sinensis* Supercritical Fluid CO_2 Extract Attenuates D-Galactose-Induced Liver and Kidney Impairment in Mice by Suppressing Oxidative Stress and Inflammation. *J. Med. Food* **2018**, *21*, 887–898. [CrossRef] [PubMed]

26. Chronopoulou, L.; Agatone, A.; Palocci, C. Supercritical CO2 extraction of oleanolic acid from grape pomace. *Int. J. Food Sci. Tech.* **2013**, *48*, 1854–1860. [CrossRef]

27. Molino, A.; Larocca, V.; Di Sanzo, G.; Martino, M.; Casella, P.; Marino, T.; Karatza, D.; Musmarra, D. Extraction of Bioactive Compounds Using Supercritical Carbon Dioxide. *Molecules* **2019**, *24*, 782. [CrossRef] [PubMed]

28. Tyśkiewicz, K.; Dębczak, A.; Gieysztor, R.; Szymczak, T.; Rój, E. Determination of fat- and water-soluble vitamins by supercritical fluid chromatography: A review. *J. Sep. Sci.* **2018**, *41*, 336–350. [CrossRef] [PubMed]

29. Gallego, R.; Bueno, M.; Herrero, M. Sub- and supercritical fluid extraction of bioactive compounds from plants, food-by-products, seaweeds and microalgae: An update. *TrAC, Trends Anal. Chem.* **2019**, *116*, 198–213. [CrossRef]

30. Gentili, A.; Dal Bosco, C.; Fanali, S.; Fanali, C. Large-scale profiling of carotenoids by using non aqueous reversed phase liquid chromatography—Photodiode array detection—Triple quadrupole linear ion trap mass spectrometry: Application to some varieties of sweet pepper (*Capsicum annuum* L.). *J. Pharm. Biomed. Anal.* **2019**, *164*, 759–767. [CrossRef]

31. Gentili, A.; Caretti, F.; Ventura, S.; Pérez-Fernández, V.; Venditti, A.; Curini, R. Screening of carotenoids in tomato fruits by using liquid chromatography with diode array–linear ion trap mass spectrometry detection. *J. Agric. Food Chem.* **2015**, *63*, 7428–7439. [CrossRef]

32. Mudimu, O.; Koopmann, I.K.; Rybalka, N.; Friedl, T.; Schulz, R.; Bilger, W. Screening of microalgae and cyanobacteria strains for α-tocopherol content at different growth phases and the influence of nitrate reduction on α-tocopherol production. *J. Appl. Phycol.* **2017**, *29*, 2867–2875. [CrossRef]

33. Schwender, J.; Seemann, M.; Lichtenthaler, H.K.; Rohmer, M. Biosynthesis of isoprenoids (carotenoids, sterols, prenyl side-chains of chlorophylls and plastoquinone) via a novel pyruvate/glyceraldehyde 3-phosphate non-mevalonate pathway in the green alga Scenedesmus obliquus. *Biochem. J.* **1996**, *316*, 73–80. [CrossRef] [PubMed]

34. Lian, J.; Wijffels, R.H.; Smidt, H.; Sipkema, D. The effect of the algal microbiome on industrial production of microalgae. *Microb. Biotechnol.* **2018**, *11*, 806–818. [CrossRef] [PubMed]

35. Croft, M.T.; Lawrence, A.D.; Raux-Deery, E.; Warren, M.J.; Smith, A.G. Algae acquire vitamin B12 through a symbiotic relationship with bacteria. *Nature* **2005**, *438*, 90–93. [CrossRef] [PubMed]

36. Pour, H.S.R.; Tavakoli, O.; Sarrafzadeh, M.H. Experimental optimization of SC-CO2 extraction of carotenoids from *Dunaliella salina*. *J. Supercrit. Fluids* **2017**, *121*, 89–95. [CrossRef]

37. Di Caprio, F.; Scarponi, P.; Altimari, P.; Iaquaniello, G.; Pagnanelli, F. The influence of phenols extracted from olive mill wastewater on the heterotrophic and mixotrophic growth of *Scenedesmus* sp. *J. Chem. Technol. Biotechnol.* **2018**, *93*, 3619–3626. [CrossRef]

38. Wynne, M.J.; Hallan, J.K. Reinstatement of Tetradesmus, G.M. Smith (Sphaeropleales, Chlorophyta). *Feddes Repertorium* **2016**, *126*, 83–86. [CrossRef]
39. Di Caprio, F.; Altimari, P.; Pagnanelli, F. Integrated microalgae biomass production and olive mill wastewater biodegradation: Optimization of the wastewater supply strategy. *Chem. Eng. J.* **2018**, *349*, 539–546. [CrossRef]

Sample Availability: Samples of the compounds are not available from the authors.

molecules

MDPI

Article

Simultaneous Characterization and Quantification of Varied Ingredients from *Sojae semen praeparatum* in Fermentation Using UFLC–TripleTOF MS

Chuan Chai [ID], Xiaobing Cui, Chenxiao Shan, Sheng Yu, Xinzhi Wang and Hongmei Wen *

School of Pharmacy, Nanjing University of Chinese Medicine, Nanjing 210029, Jiangsu, China;
echo_0523@hotmail.com (C.C.); xiaobingcui@163.com (X.C.); thomastiger@163.com (C.S.);
yusheng1219@163.com (S.Y.); wxzatnj@sina.com (X.W.)
* Correspondence: njwenhm@126.com; Tel./Fax: +86-25-8581-1839

Academic Editors: Alessandra Gentili and Chiara Fanali
Received: 23 April 2019; Accepted: 14 May 2019; Published: 15 May 2019

check for
updates

Abstract: Systematic comparison of active ingredients in *Sojae semen praeparatum* (SSP) during fermentation was performed using ultra-fast liquid chromatography (UFLC)–TripleTOF MS and principal component analysis (PCA). By using this strategy, a total of 25 varied compounds from various biosynthetic groups were assigned and relatively quantified in the positive or negative ion mode, including two oligosaccharides, twelve isoflavones, eight fatty acids, N–(3–Indolylacetyl)–DL–aspartic acid, methylarginine, and sorbitol. Additionally, as the representative constituents, six targeted isoflavones were sought in a targeted manner and accurately quantified using extracted ion chromatograms (XIC) manager (AB SCIEX, Los Angeles, CA, USA) combined with MultiQuant software (AB SCIEX, Los Angeles, CA, USA). During the fermentation process, the relative contents of oligoses decreased gradually, while the fatty acids increased. Furthermore, the accurate contents of isoflavone glycosides decreased, while aglycones increased and reached a maximum in eight days, which indicated that the ingredients converted obviously and regularly throughout the SSP fermentation. In combination with the morphological changes, which meet the requirements of China Pharmacopoeia, this work suggested that eight days is the optimal time for fermentation of SSP from the aspects of morphology and content.

Keywords: *Sojae semen praeparatum* (SSP); fermentation; conversion; ultra-fast liquid chromatography (UFLC)–TripleTOF MS; principal component analysis (PCA)

1. Introduction

Fermentation is one of the major processes used in the production of food from soybeans and has played an important role in human life for centuries [1,2]. Many studies have reported the components that are converted and how bioactivities increased in soybean products during the fermentation process [3–7].

Sojae semen praeparatum (SSP), whose Chinese herbal name is dandouchi, which is a product of Chinese fermented preparation obtained from the ripe seed of soybean (*Glycine max* (L.) Merr.), has been used as an important component in traditional diets and as an effective traditional Chinese medicine (TCM) among the Chinese community worldwide. More people are expected to consume SSP if the fermentation process includes quality assessment and quality control. Other studies have focused on the active ingredients in SSP, such as isoflavone [8–11], peptides [12], biogenic amines [13,14], and volatile components [15], the physiological properties of SSP such as anti-oxidative activity [16], anti-proliferative activity [17], anti-α-glucosidase activity [18] and anti-hypertensive effects [19], and species and quantities of fermenting bacteria in SSP spontaneous fermentation, such as bacterial fermentation and fungus fermentation [20,21]. However, no systematic comparison has been conducted of the active ingredients among the raw materials and the SSP products collected at different fermentation stages.

Isoflavones were reported to be representative constituents affecting soybean due to their significant estrogen-like bioactivity [22,23], which increased after fermentation [24,25]. The main isoflavones found in soybean are daidzein, genistein and glycitein, which are present either in glycosidic or aglycone form, mainly with β–glycosides and some 6″–O–malonyl or 6″–O–acetylglucose [26,27]. Aglyca were reported to have more bioactivity compared to the corresponding glycosides [28]. Some studies suggested that bacterial or fungal β–glycosidases are attractive candidates for use in converting β–glycosides isoflavone to their aglycones, thus enhancing the nutritional value of soy products [29,30]. However, the composition and contents change trends of isoflavones contained in SSP during fermentation have not yet been reported.

Therefore, we aimed to characterize the conversion of ingredients associated with the SSP fermentation process using ultra–fast liquid chromatography-triple time of flight mass spectrometry (UFLC–TripleTOF MS) [31], and accurately quantify the major components that vary using extracted ion chromatograms (XIC) manager (AB SCIEX, Los Angeles, CA, USA) with standard injections, thereby providing some technological supports for the optimization and quality control of SSP fermentation.

2. Results and Discussion

2.1. Morphologic Changes

Morphological changes in soybean and SSP products during fermentation were shown in Figure 1. With the increase in fermentation time, black soybean was overgrown with white hyphae, which then changed to yellow in SSP fermented for six days (S6), turned yellow completely in SSP fermented for eight days (S8), and finally hardened. In accordance with the 2015 edition of China Pharmacopoeia [32], moisturized soybeans should be fermented with boiled Artemisiae annuae herba and Mori folium until "yellow cladding". The morphological changes in S8 were consistent with these requirements, so we speculated that eight days is the optimal time for the fermentation of SSP.

2.2. Qualitative Analysis and Principle Component Analysis (PCA)

Using UFLC–TripleTOF MS analysis, information on intact precursors and fragment ions were obtained from a single injection.

The base peak chromatograms (BPCs) of soybean and S8 using both positive and negative ion modes are shown in Figure 2. The BPCs of S8 were significantly different from those of soybeans. Compared to the BPC of soybean in positive ion mode (Figure 2A), the BPC of S8 showed much higher peak intensities at t_R = 10–20 min (Figure 2C). Much higher peak intensities occurred at t_R = 0–10 min of the BPC for S8 (Figure 2D) in negative ion mode compared to soybean (Figure 2B), which shows that the ingredients converted during the SSP fermentation.

Figure 3A shows that the score plot of soybean and SSP products in positive ion mode is separated into three significant clusters ($P < 0.05$) for the first and the second principal components (PCs). Here, the green cluster (S2 (SSP fermented for two days), S4 (SSP fermented for four days) and S6) and the red cluster (S8, S10 (SSP fermented for ten days) and S15 (SSP fermented for fifteen days)) are separated by the first PC, whereas the blue cluster (soybean and S0) and the green cluster (S2, S4 and S6) are separated by the second PC. The first and second PCs' values are both 14.5%.

Similarly, Figure 3C illustrates the score plot of soybean and SSP products in negative ion mode, which is separated into three significant clusters for the first and the second PCs. The blue cluster (soybean) and the red cluster (S6, S8, S10 and S15) are separated by the first PC, whereas the blue cluster (soybean) and the green cluster (S0, S2 and S4) are separated by the second PC. The first and second PCs' values are both 14.5%.

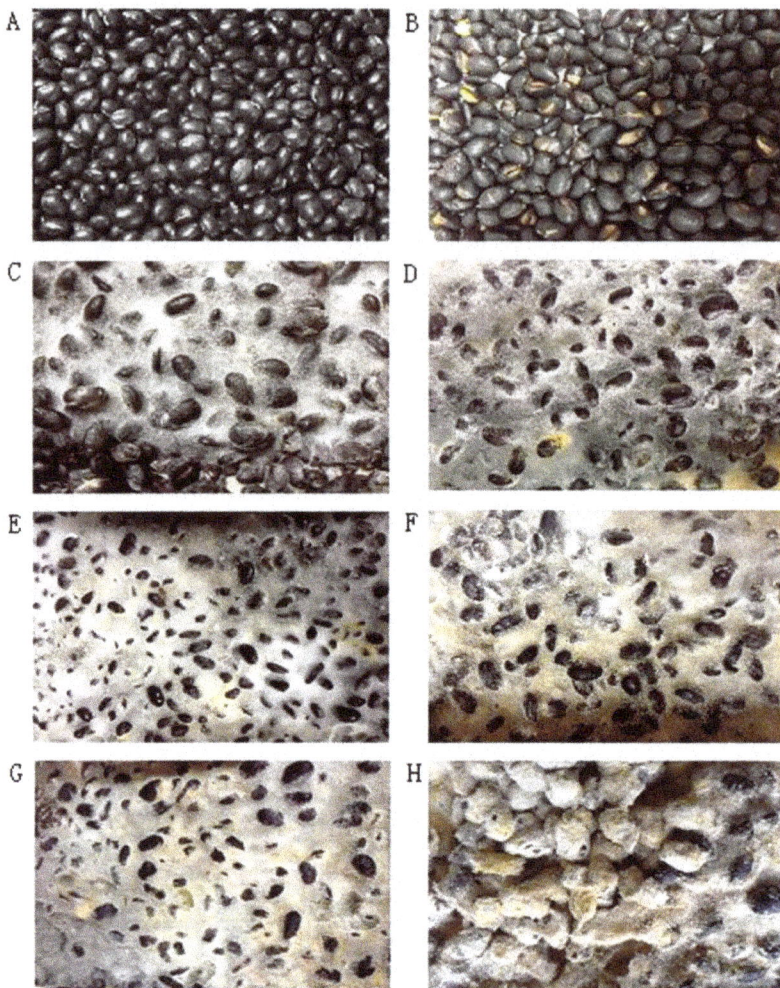

Figure 1. The morphology of soybean (**A**), S0 (**B**), S2 (**C**), S4 (**D**), S6 (**E**), S8 (**F**), S10 (**G**), S15 (**H**) days.

From the corresponding loadings plots (Figure 3B,D), a significant number of variables are located around the observations of the samples, indicating that SSP converted significantly throughout the fermentation process.

The ion species, retention times, molecular formulas, mean measured mass, mass accuracies and assigned identities of the significantly variables are shown in Table 1. $[M + H]^+$, $[M + Na]^+$, $[M + K]^+$, $[M + NH_4]^+$, $[M + H + CH_3OH]^+$, and $[M + H - H_2O]^+$ ion species were found in positive ion mode and $[M - H]^-$, $[M - H - H_2O]^-$, $[M + CH_3COO]^-$, and $[M + HCOO]^-$ were found in negative ion mode.

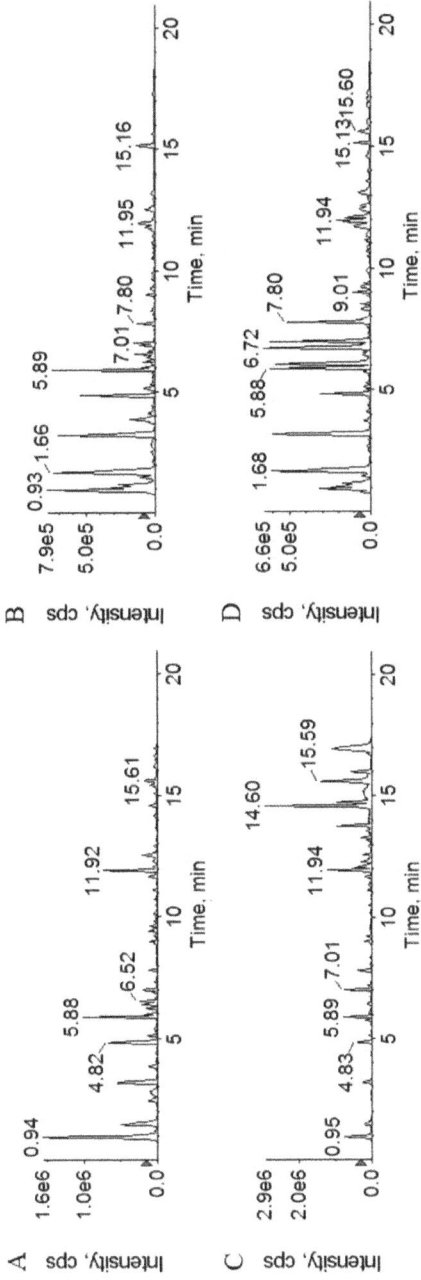

Figure 2. UFLC–TripleTOF MS base peak chromatograms (BPCs) of soybean in positive (**A**) and negative (**B**) ion modes and S8 in positive (**C**) and negative (**D**) ion modes.

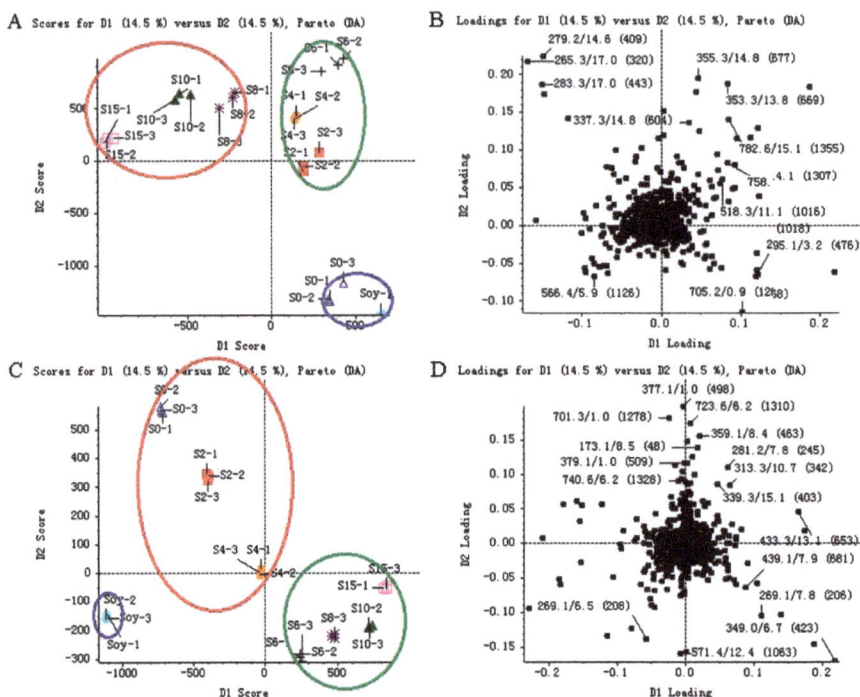

Figure 3. Score plots (**A,C**) and loading plots (**B,D**) of metabolites determined in soybean (Soy) and *Sojae semen praeparatum* products (S0, S2, S4, S6, S8, S10 and S15) by UFLC–TripleTOF MS in both positive (A and B) and negative (C and D) ion modes. The three clusters (green, blue and red) were used for the color coding of different groups separated by the first and second principal components.

We found 29 components of varied classes in positive ion mode, and 22 of them were inferred to be raffinose (**2**), stachyose (**3**), *N*–(3–Indolylacetyl)–DL–aspartic acid (**5**), daidzin (**6**), glycitin (**7**), genistin (**8**), 6″–*O*–malonyldaidzin (**9**), 6″–*O*–malonylglycitin (**10**), 6″–*O*–acetyldaidzin (**11**), 6″–*O*–acetylglycitin (**12**), 6″–*O*–malonylgenistin (**13**), daidzein (**14**), 6″–*O*–acetylgenistin (**15**), glycitein (**16**), methylarginine (**17**), genistein (**18**), dimorphecolic acid (**20**), α–linolenic acid (**22**), linoleic acid (**24**), oleic acid (**25**), palmitic acid (**27**) and stearic acid (**28**) with the help of Peakview® software (AB SCIEX, Los Angeles, CA, USA). We identified 23 of varied classes in negative ion mode, of which 10 were putatively identified as stachyose (**3**), sorbitol (**32**), *N*–(3–Indolylacetyl)–DL–aspartic acid (**5**), daidzin (**6**), genistin (**8**), 6″–*O*–acetyldaidzin (**11**), 6″–*O*–acetylgenistin (**15**), glycitein (**16**), gheddic acid (**42**) and nonadecanoic acid (**43**) by linking the masses of ions to structures. We found 7 assigned and 1 unassigned variable in both positive and negative ion modes.

All chemical structures, selected ion intensity trend plots, mass spectra, and mass spectral interpretation of putatively assigned identities were listed in Figure 4, Figures S1 and S2, and Table 2.

Table 1. Varied components putatively identified from soybean and *Sojae semen praeparatum* products in both positive and negative ion modes.

Ionization Mode	Compound No.	t_R [a] (min)	Molecular Formula	Mass (Da)	Ion Species	Mean Measured Mass (Da)	Mass Accuracy (ppm)	Assigned Identity	References
Positive	1	0.93	$C_{12}H_{18}N_6O_6$	342.1288	$[M+K]^+$	381.0907	−0.4	/[b]	– [c]
	2	0.93	$C_{18}H_{32}O_{16}$	504.1690	$[M+K]^+$	543.1474	−1.9	Raffinose	[33]
	3	0.95	$C_{24}H_{42}O_{21}$	666.2213	$[M+K]^+$	705.2045	−1.2	Stachyose	[33]
	4	1.44	$C_{13}H_{15}N_3O_5$	293.1012	$[M+NH_4]^+$	311.1328	−2.2	/[b]	– [c]
	5	3.86	$C_{14}H_{14}N_2O_5$	290.0903	$[M+H]^+$	291.0955	−4.6	N-(3-Indolylacetyl)-DL-aspartic acid	– [c]
	6	4.81	$C_{21}H_{20}O_9$	416.1107	$[M+H]^+$	417.1307	0.8	Daidzin	[37]
	7	5.14	$C_{22}H_{22}O_{10}$	446.1213	$[M+H]^+$	447.1429	−0.3	Glycitin	[37]
	8	5.89	$C_{21}H_{20}O_{10}$	432.1057	$[M+H]^+$	433.1266	0.9	Genistin	[37]
	9	6.03	$C_{24}H_{24}O_{12}$	502.1111	$[M+H]^+$	503.1335	−0.6	6″-O-malonyldaidzin	[37]
	10	6.04	$C_{25}H_{24}O_{13}$	532.1217	$[M+H]^+$	533.1451	−1.3	6″-O-malonylglycitin	[37]
	11	6.44	$C_{23}H_{22}O_{10}$	458.1213	$[M+H]^+$	459.1435	−0.2	6″-O-acetyldaidzin	[37]
	12	6.51	$C_{24}H_{24}O_{11}$	488.1319	$[M+H]^+$	489.1535	−1.0	6″-O-acetylglycitin	[37]
	13	6.57	$C_{24}H_{22}O_{13}$	518.1060	$[M+H]^+$	519.1285	−0.8	6″-O-malonylgenistin	[37]
	14	7.00	$C_{15}H_{10}O_4$	254.0579	$[M+H]^+$	255.0726	0.9	Daidzein	[37]
	15	7.05	$C_{23}H_{22}O_{11}$	474.1162	$[M+H]^+$	475.1379	−1.0	6″-O-acetylgenistin	[37]
	16	7.17	$C_{16}H_{12}O_5$	284.0685	$[M+H]^+$	285.0840	0.3	Glycitein	[37]
	17	7.69	$C_7H_{16}N_4O_2$	188.1268	$[M+H+H_2O]^+$	207.1452	−3.9	Methylarginine	– [c]
	18	7.80	$C_{15}H_{10}O_5$	270.0528	$[M+H]^+$	271.0680	1.0	Genistein	[37]
	19	11.94	$C_{21}H_{45}N_9O_6$	519.3493	$[M+H]^+$	520.3534	1.0	/[b]	– [c]
	20	13.29	$C_{18}H_{32}O_3$	296.2710	$[M+H]^+$	297.2517	−3.9	/[b]	– [c]
	21	13.77	$C_{23}H_{44}O_2$	352.3336	$[M+H]^+$	353.3297	0.1	Dimorphecolic acid	– [c]
	22	14.61	$C_{18}H_{30}O_2$	278.2246	$[M+H]^+$	279.2408	−0.4	α–Linolenic acid	[34]
	23	14.76	$C_{23}H_{46}O_2$	354.3492	$[M+H]^+$	355.2950	0.8	/[b]	– [c]
	24	15.62	$C_{18}H_{32}O_2$	280.2402	$[M+H+CH_3OH]^+$	281.2567	0.7	Linoleic acid	[34]
	25	15.70	$C_{18}H_{34}O_2$	282.2559	$[M+H+CH_3OH]^+$	315.2915	0.6	Oleic acid	[34]
	26	16.01	$C_{21}H_{40}O_2$	324.3023	$[M+H]^+$	357.3101	2.9	/[b]	– [c]
	27	16.67	$C_{16}H_{32}O_2$	256.2402	$[M+H]^+$	257.2565	−0.3	Palmitic acid	[34]
	28	17.04	$C_{18}H_{36}O_2$	284.2715	$[M+H]^+$	285.3050	−0.8	Stearic acid	[34]
Negative	29	0.92	$C_{12}H_{18}N_6O_6$	342.1288	$[M+HCOO]^-$	387.1623	1.7	/[b]	– [c]
	3	0.95	$C_{24}H_{42}O_{21}$	666.2213	$[M+HCOO]^-$	711.3077	1.3	Stachyose	[33]
	30	1.05	$C_{30}H_{32}N_6O_9$	620.2220	$[M−H−H_2O]^-$	601.2133	−1.1	/[b]	– [c]
	31	3.12	$C_{16}H_{18}N_6$	294.1582	$[M−H]^-$	293.1552	−1.4	/[b]	– [c]
	32	3.21	$C_6H_{14}O_6$	182.0790	$[M+CH_3COO]^-$	241.0915	0.8	Sorbitol	[35]
	5	3.85	$C_{14}H_{14}N_2O_5$	290.0903	$[M−H]^-$	289.1078	−1.3	N-(3-Indolylacetyl)-DL-aspartic acid	– [c]

Table 1. *Cont.*

Ionization Mode	Compound No.	tR [a] (min)	Molecular Formula	Mass (Da)	Ion Species	Mean Measured Mass (Da)	Mass Accuracy (ppm)	Assigned Identity	References
	6	4.81	$C_{21}H_{20}O_9$	416.1107	$[M + HCOO]^-$	461.1650	0.2	Daidzin	[37]
	33	5.64	$C_{18}H_{10}N_2O_6$	350.0528	$[M − H]^-$	349.0451	−2	/[b]	−[c]
	8	5.89	$C_{21}H_{20}O_{10}$	432.1057	$[M + HCOO]^-$	477.1629	−2.5	Genistin	[37]
	34	6.15	$C_{18}H_4N_6$	304.0497	$[M + CH_3COO]^-$	363.0635	−0.5	/[b]	−[c]
	11	6.42	$C_{23}H_{22}O_{10}$	458.1213	$[M + HCOO]^-$	503.1805	−2.3	6″–O–acetyldaidzin	[37]
	35	6.47	$C_{13}H_{10}O_6$	262.0466	$[M − H]^-$	261.0400	2.4	/[b]	−[c]
	36	6.68	$C_{19}H_6N_6O_3$	366.0490	$[M − H]^-$	365.0430	−1.3	/[b]	−[c]
	15	7.04	$C_{23}H_{22}O_{11}$	474.1162	$[M + HCOO]^-$	519.1784	2.3	6″–O–acetylgenistin	[37]
Negative	16	7.15	$C_{16}H_{12}O_5$	284.0685	$[M − H]^-$	283.0969	2.1	Glycitein	[37]
	37	7.91	$C_{18}H_{18}O_{10}$	394.0895	$[M + HCOO]^-$	439.0878	−2.0	/[b]	−[c]
	38	10.83	$C_{31}H_{50}N_2O_9$	594.3505	$[M − H]^-$	593.3490	−0.6	/[b]	−[c]
	39	11.29	$C_{19}H_{36}O_3$	312.2654	$[M − H − H_2O]^-$	293.2483	1.7	/[b]	−[c]
	40	11.66	$C_{32}H_{48}N_6O_5$	596.3675	$[M − H]^-$	595.3613	1.9	/[b]	−[c]
	41	12.06	$C_{17}H_{36}N_8O_5$	432.2798	$[M − H]^-$	431.2728	0.1	/[b]	−[c]
	42	12.54	$C_{34}H_{68}O_2$	508.3389	$[M − H]^-$	507.3340	0.1	Cheddic acid	[36]
	43	12.58	$C_{19}H_{38}O_2$	298.2861	$[M − H]^-$	297.2814	4.0	Nonadecanoic acid	−[c]
	44	13.57	$C_{18}H_{42}N_{10}O$	414.3543	$[M + HCOO]^-$	459.3520	−1.3	/[b]	−[c]
	45	15.55	$C_{12}H_{28}N_8O_6$	380.2121	$[M − H]^-$	379.2049	−2.6	/[b]	−[c]

[a] tR, retention time; [b] Not assigned; [c] No reference.

Table 2. Mass spectral interpretation of assigned compounds in soybean and *Sojae semen praeparatum* products.

Compound No.	Assigned Identity (Ion Mode)	MS/MS Fragments Ions
2	Raffinose (+)	$543.1474[M + K]^+$, $381.0905[M + K − glu + H_2O]^+$,
3	Stachyose (+)	$705.2045[M + K]^+$, $543.1471[M+K − glu]^+$,
5	N-(3-Indolylacetyl)-DL-aspartic acid (+)	$711.3077[M + HCOO]^-$, $665.3014[M − H]^-$, $485.2130[M − H − glu]^-$, $341.1514[M − H − 2glu + 2H_2O]^-$, $291.0955[M + H]^+$, $161.0645[C_5H_6O_2N + H]^+$, $139.0428[M + H − C_8H_6N − 2H_2O]^+$,
6	Daidzin (+)	$289.1078[M − H]^-$, $271.0968[M − H − H_2O]^-$, $245.1127[M − H − CO_2]^-$, $227.1008[M − H − CO_2 − H_2O]^-$, $439.1128[M + Na]^+$, $417.1307[M + H]^+$, $277.0551[M + Na − glu + H_2O]^+$, $255.0728[M + H − glu + H_2O]^+$,
7	Glycitin (+)	$461.1650[M + HCOO]^-$, $415.1556[M − H]^-$, $253.0825[M − H − glu + H_2O]^-$, $469.1235[M + Na]^+$, $447.1429[M + H]^+$, $307.0662[M + Na − glu + H_2O]^+$, $285.0839[M + H − glu + H_2O]^+$,
8	Genistin (+)	$433.1266[M + H]^+$, $271.0674[M + H − glu + H_2O]^+$, $243.0712[M + H − glu + H_2O − CO]^+$, $215.0752[M + H − glu + H_2O − 2CO]^+$, $153.0218[M + H − C_{13}H_{12}O_7]^+$, $477.1629[M + HCOO]^-$, $431.1529[M − H]^-$, $269.0795[M − H − glu + H_2O]^-$,
9	6″–O–malonyldaidzin (+)	$525.1146[M + Na]^+$, $503.1335[M + H]^+$, $481.1244[M + Na − CO_2]^+$, $439.1133[M + Na − malonyl − H_2O]^+$, $277.0549[M + Na − malonyl − glu]^+$, $255.0728[M + H − malonyl − glu]^+$,

Table 2. *Cont.*

Compound No.	Assigned Identity (Ion Mode)	MS/MS Fragments Ions
10	6″–O–malonylglycitin (+)	$533.1451[M + H]^+$, $285.0845[M + H - \text{malonyl} - \text{glu}]^+$,
11	6″–O–acetyldaidzin (+) (−)	$459.1435[M + H]^+$, $255.0726[M + H - \text{acetyl} - \text{glu}]^+$, $503.1805[M + HCOO]^-$, $457.1720[M - H]^-$, $253.0822[M - H - \text{acetyl} - \text{glu}]^-$,
12	6″–O–acetylglycitin (+)	$489.1535[M + H]^+$, $285.0839[M + H - \text{acetyl} - \text{glu}]^+$,
13	6″–O–malonylgenistin (+)	$541.1093[M + Na]^+$, $523.0995[M + Na - H_2O]^+$, $519.1285[M + H]^+$, $497.1167[M + Na - CO_2]^+$, $455.1096[M + Na - \text{malonyl} - H_2O]^+$, $293.0505[M + Na - \text{malonyl} - \text{glu}]^+$, $271.0678[M + H - \text{malonyl} - \text{glu}]^+$,
14	Daidzein (+)	$277.0550[M + Na]^+$, $255.0726[M + H]^+$, $237.0611[M + H - H_2O]^+$, $227.0762[M + H - CO]^+$, $199.0806[M + H - 2CO]^+$, $181.0695[M - 2CO - H_2O]^+$, $137.0273[M - C_8H_6O]^+$, $91.0582[M + H - H_2O - C_9H_6O_2]^+$,
15	6″–O–acetylgenistin (+) (−)	$475.1379[M + H]^+$, $271.0682[M + H - \text{acetyl} - \text{glu}]^+$, $519.1784[M + HCOO]^-$, $473.1689[M - H]^-$, $269.0795[M - H - \text{acetyl} - \text{glu}]^-$,
16	Glycitein (+) (−)	$307.0672[M + Na]^+$, $285.0840[M + H]^+$, $270.0599[M + H - CH_3]^+$, $242.0642[M + H - CH_3 - CO]^+$, $169.0614[M + H - C_8H_4O]^+$, $141.0740[M + H - C_9H_4O_2]^+$, $283.0969[M - H]^-$, $268.0714[M - H - CH_3]^-$, $240.0729[M - H - CH_3 - CO]^-$, $196.0776[M - C_2H_3O - OH]^-$,
17	Methylarginine (+)	$207.1452[M + H + H_2O]^+$, $189.1321[M + H]^+$, $161.1377[M + H - H_2O]^+$,
18	Genistein (+)	$271.0680[M + H]^+$, $253.0565[M + H - H_2O]^+$, $243.0716[M + H - CO]^+$, $215.0757[M + H - 2CO]^+$, $153.0223[M + H - C_8H_6O]^+$,
20	Dimorphecolic acid (+)	$297.2517[M + H]^+$, $279.2398[M + H - H_2O]^+$, $261.2287[M + H - 2H_2O]^+$, $233.2328[M + H - 2H_2O - CO]^+$, $109.1051[M + H - C_9H_{17}COOH - CH_4]^+$, $97.1051[M + H - C_9H_{17}COOH - C_2H_4]^+$, $81.0743[M + H - C_9H_{17}COOH - C_3H_8]^+$, $67.0590[M + H - C_9H_{17}COOH - C_4H_{10}]^+$,
22	α–Linolenic acid (+)	$297.2517[M + H + H_2O]^+$, $279.2408[M + H]^+$, $149.0276[M + H - C_6H_3COOH]^+$, $135.1206[M + H - C_7H_5COOH]^+$, $125.1000[M + H - C_8H_{13}COOH]^+$, $123.1200[M + H - C_8H_{15}COOH]^+$, $109.1045[M + H - C_9H_{17}COOH]^+$, $95.0892[M + H - C_{10}H_{19}COOH]^+$, $81.0738[M + H - C_{11}H_{21}COOH]^+$, $67.0588[M + H - C_{12}H_{23}COOH]^+$,
24	Linoleic acid (+)	$313.2827[M + H + CH_3OH]^+$, $281.2567[M + H]^+$, $263.2443[M + H - H_2O]^+$, $239.2433[M + H + CH_3OH - H_2O - C_4H_8]^+$, $221.2318[M + H - CH_3COOH]^+$, $147.1210[M + H + CH_3OH - C_{12}H_{22}]^+$ $133.1051[M + H + CH_3OH - C_{13}H_{24}]^+$, $109.1049[M + H - C_9H_{19}COOH]^+$, $95.0895[M + H - C_{10}H_{21}COOH]^+$, $71.0902[M + H - C_{12}H_{21}COOH]^+$, $57.0755[M + H - C_{13}H_{23}COOH]^+$,
25	Oleic acid (+)	$315.2915[M + H + CH_3OH]^+$, $283.2461[M + H]^+$, $271.5449[M + H + CH_3OH - CO_2]^+$, $267.0220[M + H - CH_4]^+$, $265.2757[M + H - H_2O]^+$, $187.1151[M + H + CH_3OH - C_9H_{20}]^+$, $171.0300[M + H - C_8H_{16}]^+$, $114.9656[M + H - C_9H_{15}COOH]^+$, $96.9540[M + H - C_{10}H_{21}COOH]^+$, $83.0875[M + H - C_{11}H_{23}COOH]^+$, $57.0750[M + H - C_{13}H_{25}COOH]^+$
27	Palmitic acid (+)	$257.2565[M + H]^+$, $201.1894[M + H - C_4H_8]^+$, $97.1025[M + H - C_{10}H_{22}]^+$, $71.0902[M + H - C_{10}H_{21}COOH]^+$, $57.0753[M + H - C_{11}H_{29}COOH]^+$,
28	Stearic acid (+)	$285.3050[M + H]^+$, $267.2656[M + H - H_2O]^+$, $126.9036[M + H - C_8H_{17}COOH]^+$, $83.0895[M + H - C_{10}H_{21}COOH - CH_4]^+$, $69.0739[M + H - C_{11}H_{23}COOH]^+$, $57.0742[M + H - C_{13}H_{27}COOH]^+$,
32	Sorbitol (−)	$241.0915[M + CH_3COO - H_2O]^-$, $223.0801[M + CH_3COO - H_2O]^-$, $181.0622[M - H]^-$, $149.0803[M - H - CH_3OH]^-$,
42	Gheddic acid (−)	$507.3340[M - H]^-$, $279.2671[M - H - C_{13}H_{27}COOH]^-$, $153.0148[M - H - C_{22}H_{45}COOH]^-$,
43	Nonadecanoic acid (−)	$297.2814[M - H]^-$, $279.2672[M - H - H_2O]^-$, $183.1614[M - H - C_5H_9COOH]^-$,

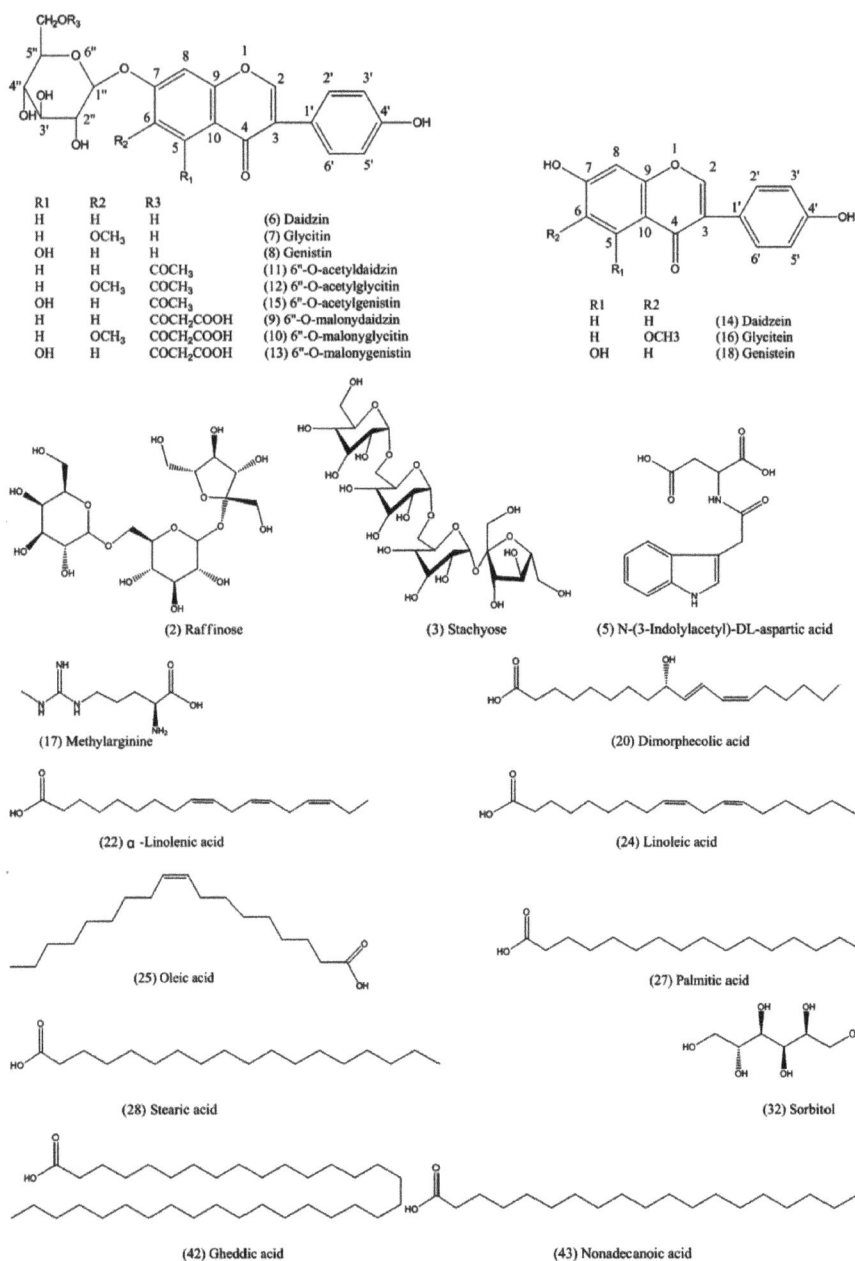

Figure 4. Chemical structures of assigned compounds.

Compounds **2** and **3** were inferred as raffinose and stachyose, respectively, for the loss of aglyca. Compounds **14**, **16** and **18** were putatively identified as isoflavone aglycones for their fragment ions at m/z 137, 153 and 169, respectively, after retro–Diels–Alder reaction. Compounds **6–8** were assigned as isoflavone glycosides for their glycones $[M + H - glu + H_2O]^+$ at m/z 255, 271 and 285 and $[M - H - glu + H_2O]^-$ at m/z 253, 269 and 283 after deglycosylation. Further dissociation of the glycones yielded a serial of fragments in agreement with the aglycones. The six isoflavone compounds were also confirmed by injecting a mix of standard solutions (Figure 6). Compounds **9**, **10** and **13** were confirmed as isoflavone glycoside malonates: compounds **11**, **12** and **15** were identified as isoflavone acetyl glycosides for their common glycones in comparison with glycosides. Compounds **20**, **22**, **24**, **25**, **27** and **28** were assumed to be a series of fatty acids for the homologous fragment ions at m/z (67, 81, 95, 109 and 123); (83, 97 and 111); and (57, 71 and 85), the difference between every pair of the fragment ions was 14 ($-CH_2-$). Twenty of the assigned compounds were previously reported in soybean [33–37]. However, N–(3–Indolylacetyl)–DL–aspartic acid, methylarginine, dimorphecolic acid, gheddic acid, and nonadecanoic acid have never been reported in soybean; sorbitol was only detected in germinating soybean seeds [35], and gheddic acid was identified in *Mori folium* [36]. Six constituents were presumed to be introduced from processing adjuvants or produced during the SSP fermentation process.

2.3. Relative Quantitative Analysis

In agreement with previous results [38–40], the relative contents of raffinose and stachyose decreased gradually during the entire fermentation process due to degradation by bacteria. As stachyose and raffinose cause indigestion and flatulence in animals after ingestion, the reduction of oligosaccharides is an indication that fermentation can promote the absorption of soybean nutrients. Isoflavone was inferred to be the principle difference among the products obtained from the SSP fermentation process due to its high proportions of varied components. With increasing fermentation time, the relative contents of isoflavone glycosides decreased while the isoflavone aglycones increased, reaching a maximum in S8. The isoflavone glycoside malonates decreased while the isoflavone acetyl glycosides increased to a maximum in S2 and then dropped. We assumed that the isoflavone glycoside malonates were transformed to acetyl glycosides in the early fermentation period due to their heat instability, and that all the isoflavone glycosides were converted to isoflavone aglycones. The increase in fatty acids showed that lipids could be degraded during fermentation. All the results provide some technological supports for the optimization and quality control of SSP fermentation. Given of the regular component conversion in SSP during fermentation, further identification of the unassigned varied ingredients present in SSP is needed.

2.4. Accurate Quantitative Analysis

XIC manager combined with MultiQuant software was used to automatically highlight all findings above a defined thresholds at an exactive mass of ± 0.02 Da and to quantitatively compare samples with a series of standard injections.

The validation values are summarized in Table 3. The calibration curves show satisfactory linearity. The correlation coefficient (r) ranged from 0.9840–0.9981 for all the isoflavones. Limit of detection (LOD) and limit of quantitation (LOQ) values were 0.1–50.0 and 2.0–250.0 ng/mL, respectively. The intra and inter-day precisions were less than 0.48% and 2.87%, respectively. The repeatability was within 2.53–4.82%. The recoveries were between 97.61 ± 3.73% and 104.84 ± 2.58% at different spiking concentration levels. The short-term stability analyzed at various periods was less than 4.10%. The above results demonstrate that the established method is accurate and reproducible for determining the six isoflavones in SSP.

Controlled by mix standard solution, the accurate quantitative results of six isoflavones in SSP were summarized in Figure 5. All six isoflavones were identified in soybean and SSP products. With increasing fermentation time, daidzin, glycitin and genistin decreased while daidzein, glycitein and genistein increased and raised to the top in S8 at 74.50, 13.52 and 47.42 mg/100 g dry weight respectively. Total glycoside and aglycone were also calculated as the sum of each individual isoflavone and presented in Figure 6. As the fermentation time increased, total glycoside contents decreased, while total aglycone contents increased significantly and rose to the top in the S8. The total glycoside content in S8 was less than half a percent of soybean's while the total aglycone content in S8 was 4.8 times higher than that in soybean, indicating that the ingredients converted regularly during the fermentation process.

Molecules **2019**, *24*, 1864

Table 3. Validation data of targeted analytes.

Isoflavone	RT [a] (min)	Regression Equation [b]	Linear Range (µg/mL)	r	LOD [c] (ng/mL)	LOQ [d] (ng/mL)	Precision (RSD % [e]) Intra Day	Inter Day	Repeatability (RSD, %)	Recovery (Mean [f] ± RSD %)	Stability (RSD % [a])
Daidzin	4.81	Y = 58544X + 384206	0.010–100.0	0.9979	5.0	10.0	1.14	4.24	3.98	103.83 ± 3.15	4.10
Glycitin	5.14	Y = 49463X + 421236	0.010–100.0	0.9981	5.0	10.0	0.57	2.87	3.27	104.84 ± 2.58	1.84
Genistin	5.89	Y = 60953X +286019	0.100–600.0	0.9980	10.0	50.0	0.48	4.48	4.41	97.76 ± 4.70	3.04
Daidzein	7.02	Y = 76972 X + 705619	0.100–100.0	0.9840	0.1	3.0	1.61	3.50	2.53	97.61 ± 3.73	3.34
Glycitein	7.17	Y = 52697 X + 643182	0.010–200.0	0.9973	0.1	2.0	1.42	2.61	4.19	103.71 ± 2.69	3.78
Genistein	7.80	Y = 78899 X + 342189	0.50–200.0	0.9973	50.0	250.0	2.35	3.99	4.82	99.67 ± 3.16	2.39

[a] RT, retention time; [b] Y, peak area; X, concentration (µg/mL); [c] LOD, Limit of detection (S/N = 3); [d] LOQ, Limit of quantification (S/N = 10). [e] Relative standard deviation (%) = (standard deviation / mean) × 100. (n = 3); [f] Mean extraction yield (%) = (detected amount − original amount)/spiked amount × 100. (n = 3).

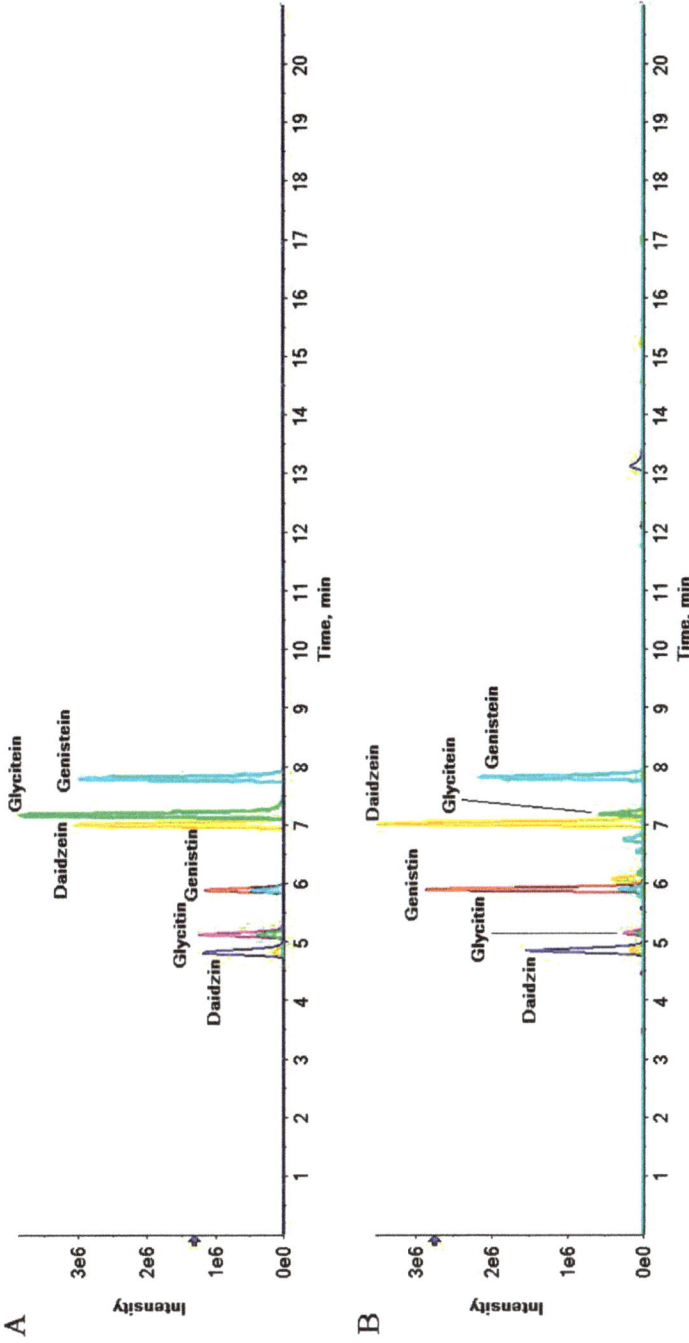

Figure 5. Representative extract ions chromatograms (XIC) of mix standard solution (**A**) and S8 (**B**) at *m/z* 417.118 ± 0.02 (daidzin), 447.129 ± 0.02 (glycitin), 433.113 ± 0.02 (genistin), 255.065 ± 0.02 (daidzein), 285.076 ± 0.02 (glycitein) and 271.060 ± 0.02 (genistein).

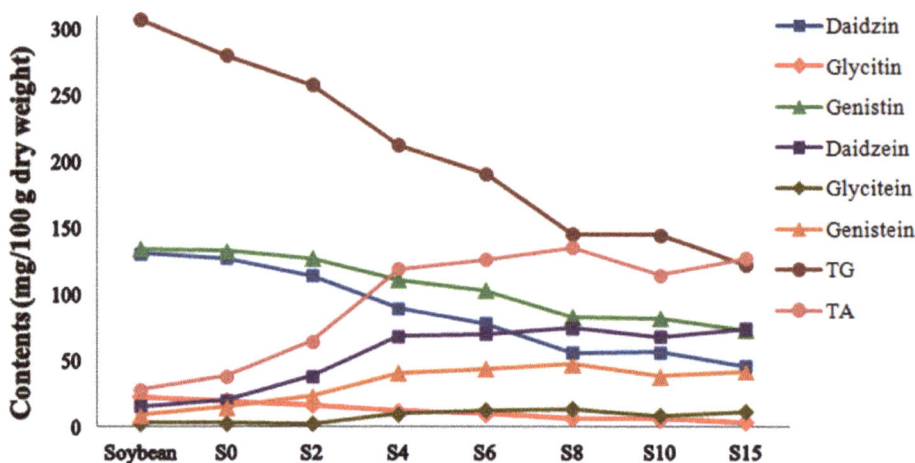

Figure 6. Content profile plots for six isoflavones, total glycoside (TG), and aglycone (TA) in SSP products collected at different fermentation stages.

Previous research has suggested that the increase in aglycone content and β–glucosidase activity during the fermentation of soybean show a similar trend [41]. As an attractive candidate to convert isoflavone glycosides to their aglycones, β–glucosidase reached its maximum activity on the eighth days of fermentation.

As aglycone possesses higher bioactivity and bioavailability compared to the β–glycosides isoflavone, the quantitative results of the representative constituents illustrated that eight days is the optimal time for the fermentation of SSP, which agrees with the morphologic changes (Figure 1). We reasoned that the bioactivities may be related to the variations in isoflavone content and β–glucosidase activity during SSP fermentation. Moreover, we created a perfect setup for SSP fermentation quality assessment and quality control. To determine the impact of bacterial or fungi on SSP fermentation, –further study is required.

3. Materials and Methods

3.1. Chemicals and Reagents

The reference standards of daidzin, glycitin, genistin, daidzein, glycitein and genistein were all purchased from Sigma–Aldrich (St. Louis, MO, USA)

Liquid chromatography (LC)/MS–grade acetonitrile, formic acid, methanol, and water were purchased from Merck Co. (Darmstadt, Germany).

Soybean, Artemisiae annuae herba and Mori folium used in the fermentation were purchased from YiFeng TCM shop (Nanjing, China) and authenticated by Associate Professor Jianwei Chen (Department of Pharmacy, Nanjing University of Chinese Medicine, Nanjing, Jiangsu, China).

3.2. SSP Fermentation

SSP was fermented in the laboratory and the preparation was performed as described in detail by the 2015 edition of China Pharmacopoeia as illustrated in Figure 7. The steps and parameters were as follows:

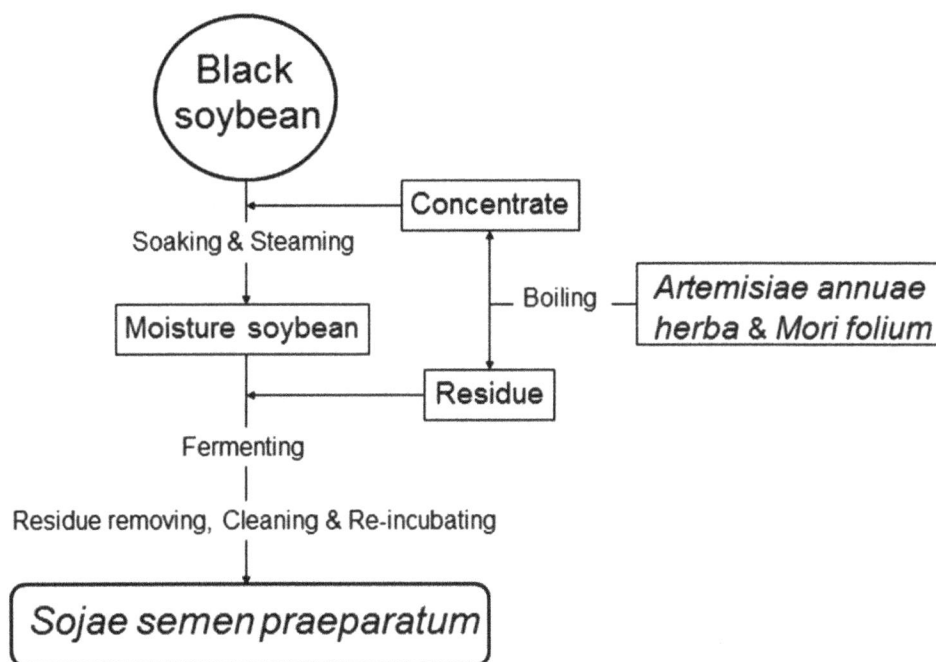

Figure 7. Flow diagram for fermentation of soybean to *Sojae semen praeparatum* (SSP).

Artemisiae annuae herba (100 g dry weight) and Mori folium (90 g dry weight) were washed and boiled with water (3600 mL) for 1 h in triplicate, and the decoction was concentrated to a relative density of 1.10–1.12 g/cm^3 concentrate. Black soybean (1000 g dry weight) was soaked in the concentrate overnight and steamed for 1.5 h while covered with wet cheesecloth. Aliquots (100 g wet weight) of the moisture soybean were placed on enamel trays covered with wet cheesecloth, which were covered with the residue of boiled Artemisiae annuae herba and Mori folium and incubated at 37 °C with 60–80% humidity. After fermenting for 0, 2, 4, 6, 8, 10, 15 days and removal of residues, wash cleaning, and re-incubating until a sweet smell drifting out, the SSP samples (S0, S2, S4, S6, S8, S10, S12, and S15) were then dried and pulverized into powder using an electric mill and sieved through 80 mesh sieves.

3.3. Sample Extraction

One gram of the powdered samples was accurately weighed and extracted with 25 mL of 75% methanol at 80 °C for 30 min using a Soxhlet extractor. This was followed by centrifugation for 15 min at 12000 rpm. The extraction supernatants were then diluted 10 times and filtered through a 0.45 μm filter unit.

3.4. Standard Solutions Preparation

Mix standard solutions were prepared by accurately weighing the standard substances and mixing them in 75% methanol. This standard mixture was filtered through a 0.45 μm filter unit.

3.5. LC-MS Spectrometric Conditions

An ultra-fast liquid chromatography system (Shimadzu Corporation UFLC XR; Kyoto, Japan) was connected to a triple time−of−flight mass spectrometer (TripleTOF 5600 system, AB SCIEX, Los Angeles, CA, USA) with an electrospray ionization source.

All samples were separated by an ACQUITY UPLC BEH C_{18} column (2.1 mm × 100 mm, 1.7 μm, Waters Corp., Milford, MA, USA). A binary solvent gradient consisting of solvent A (water with 0.2% formic acid) and solvent B (acetonitrile with 0.2% formic acid) was used. The flow rate was 300 μL/min. The total run time was 21 min with a gradient as follows: 0–3 min, 10–16% B; 3–7 min, 16–50% B; 7–12 min, 50–80% B; 12–15 min, 80–90% B; 15–17 min, 90–10% B; and 17–21 min, 10% B for column equilibration before the next run. An injection volume of 1 μL was used. The column temperature was 40 °C.

The samples were analyzed by acquiring full scan MS data in both positive and negative ion modes. The automatic data-dependent information product ion spectra (IDA−MS/MS) without any predefinition of the ions was checked. A calibrated delivery system was used to ensure the accuracy error of masses less than 1 ppm. The settings were nitrogen gas for nebulization at 55 psi, heater gas pressure at 55 psi, curtain gas at 35 psi, temperature of 500 °C, and ion spray voltage at 5500 V in positive ion mode, and −4500 V in negative ion mode. The acquisition of a survey tripleTOF MS spectrum was operated under high-resolution settings. The optimized declustering potential (DP) and collision energy (CE) were set at 80 eV and 15 eV in positive ion mode, and to −80 eV and −15 eV in negative ion mode, respectively. A sweeping collision energy setting at 35/−35 eV ± 15 eV was applied for collision-induced dissociation (CID).

3.6. Method Validation

The method was fully validated in accordance with guidelines on linearity, precision, recovery, detection limit, quantification limit, and stability. Calibration curves were generated by plotting peak area against the concentration of standard solutions. The intra-day precision was examined for six replicates of injections with the mixed standard solutions in one day, and the inter-precision was determined by injection in duplicates over three consecutive days. All the results are expressed using the relative standard deviation (RSD). The LOD and LOQ were calculated based on the peak−to−noise ratios of 3:1 and 10:1, respectively. The repeat, recovery, and stability were tested on the analytes in S8, and the repeatability was analyzed on six sample solutions from the same sample in parallel. The recovery was used to evaluate the accuracy at different spiking concentration levels (80%, 100%, and 120% as compared to the nominal concentration) of standard solutions. and the sample stability was tested by periodic analysis at room temperature for various periods (0, 2, 4, 8, 12, 16, 20, and 24 h).

3.7. Data Processing

TOF−MS data were collected using Analyst® version 1.6 software (AB SCIEX, Los Angeles, CA, USA) and processed using PeakView® version 1.2 software (AB SCIEX, Los Angeles, CA, USA) with the XIC Manager (AB SCIEX, Los Angeles, CA, USA) add-in and MultiQuant™ version 2.1 software (AB SCIEX, Los Angeles, CA, USA).

The PeakView® software contained a simple fragment ion predictor to help link the MS/MS spectrum to structures (saved as .mol files) and to provide insights into fragmentation mechanisms.

XIC Manager was used for targeted and non-targeted data processing, which consisted of a table for defining a list of masses or formulae to generate extracted ion chromatograms (XIC), and to review the identification of detected compounds. Our high confidence in results is based on retention times, accurate mass, isotopic pattern and MS/MS library searching.

The PCA was performed using MarkerView® software, where three repeated spectra for each sample were imported and analyzed with Pareto scaling. The T−value and corresponding P−value were calculated between each group and all the other eight groups. The program was linked back to the raw data so that differences could be directly visualized in spectra or chromatograms. The converted components were putatively identified by PeakView® software.

4. Conclusions

This was the first systematic comparison of active ingredients in the raw materials and processed products obtained during *Sojae semen praeparatum* (SSP) fermentation. Simultaneous characterization and quantification were performed using ultra-fast liquid chromatography (UFLC)−TripleTOF MS combined with XIC manager. The quantitative results verified that the components converted during the SSP fermentation, and we identified 45 components in positive ion mode and negative ion mode, in which 25 were putatively identified and a high proportion was isoflavone. *N*-(3-Indolylacetyl)-DL-aspartic acid, methylarginine, dimorphecolic acid, sorbitol, gheddic acid, and nonadecanoic acid were presumed to be introduced from processing adjuvants or produced during the fermentation process. The relative contents of raffinose and stachyose decreased gradually, while the fatty acids and isoflavone aglycones increased, which indicated that fermentation promotes the absorption of soybean nutrition and lipids degradation. The accurate quantitation of isoflavone, the representative constituents in soybean, revealed that fermentation for eight days produced a marked increase in the content of aglycone, the bioactive isoflavone, and a significant reduction in the content of β−glycosides isoflavone compared with unfermented soybean. This illustrated that eight days is the optimal time for the fermentation of SSP from the aspects of content, in agreement with the morphologic changes. We reasoned that the bioactivities of SSP might be related to isoflavone. Our study has provided some technological support for the optimization and quality control of SSP fermentation.

Supplementary Materials: The following are available online, Figure S1: Selected ion intensity trend plots of assigned identities., Figure S2: Mass spectrum of assigned compounds in soybean and *Semen sojae praeparatum* products.

Author Contributions: Conceptualization, C.C., X.C. and H.W.; data curation, C.C. and C.S.; formal analysis, C.C. and C.S.; funding acquisition, H.W.; methodology, C.C.; resources, S.Y. and H.W.; software, C.C. and X.W.; supervision, X.C. and S.Y.; validation, C.C.; writing—original draft, C.C.; writing—review & editing, H.W.

Funding: This work was supported by National Natural Science Foundation of China (51403104), Jiangsu Key Laboratory for Functional Substance of Chinese Medicine, the Natural Science Foundation of the Jiangsu Higher Education Institutions of China (08KJB360004), School Foundation of Nanjing University of Chinese Medicine, the People's Republic of China under Grant No. 12XZR24 and A Project Funded by the Priority Academic Program Development of Jiangsu Higher Education Institutions, Top−notch Academic Programs Project of Jiangsu Higher Education Institutions (TAPP−PPZY2015A070).

Conflicts of Interest: The authors declare no conflicts of interest.

References

1. Choi, M.-Y.; Chai, C.; Park, J.H.; Lim, J.; Lee, J.; Kwon, S.W. Effects of storage period and heat treatment on phenolic compound composition in dried citrus peels (chenpi) and discrimination of chenpi with different storage periods through targeted metabolomic study using hplc-dad analysis. *J. Pharm. Biomed. Anal.* **2011**, *54*, 638–645. [CrossRef]
2. Mortera, P.; Zuljan, F.A.; Magni, C.; Bortolato, S.A.; Alarcon, S.H. Multivariate analysis of organic acids in fermented food from reversed-phase high-performance liquid chromatography data. *Talanta* **2018**, *178*, 15–23. [CrossRef] [PubMed]
3. Chai, C.; Ju, H.K.; Kim, S.C.; Park, J.H.; Lim, J.; Kwon, S.W.; Lee, J. Determination of bioactive compounds in fermented soybean products using gc/ms and further investigation of correlation of their bioactivities. *J. Chromatogr. B* **2012**, *880*, 42–49. [CrossRef] [PubMed]
4. Zhu, H.; Bi, K.; Han, F.; Guan, J.; Zhang, X.; Mao, X.; Zhao, L.; Li, Q.; Hou, X.; Yin, R. Identification of the absorbed components and metabolites of zhi-zi-da-huang decoction in rat plasma by ultra-high performance liquid chromatography coupled with quadrupole-time-of-flight mass spectrometry. *J. Pharm. Biomed. Anal.* **2015**, *111*, 277–287. [CrossRef]
5. Zhang, P.; Zhang, P.; Xie, M.; An, F.; Qiu, B.; Wu, R. Metaproteomics of microbiota in naturally fermented soybean paste, da-jiang. *J. Food Sci.* **2018**, *83*, 1342–1349. [CrossRef] [PubMed]
6. Kim, M.S.; Kim, B.; Park, H.; Ji, Y.; Holzapfel, W.; Kim, D.Y.; Hyun, C.K. Long-term fermented soybean paste improves metabolic parameters associated with non-alcoholic fatty liver disease and insulin resistance in high-fat diet-induced obese mice. *Biochem. Biophys. Res. Commun.* **2018**, *495*, 1744–1751. [CrossRef] [PubMed]

7. Ali, M.W.; Shahzad, R.; Bilal, S.; Adhikari, B.; Kim, I.D.; Lee, J.D.; Lee, I.J.; Kim, B.O.; Shin, D.H. Comparison of antioxidants potential, metabolites, and nutritional profiles of korean fermented soybean (cheonggukjang) with bacillus subtilis kctc 13241. *J. Food Sci. Technol.* **2018**, *55*, 2871–2880. [CrossRef]

8. Qu, L.; Fan, G.; Peng, J.; Mi, H. Isolation of six isoflavones from semen sojae praeparatum by preparative HPLC. *Fitoterapia* **2007**, *78*, 200–204. [CrossRef] [PubMed]

9. Chai, C.; Bai, Y.T.; Wen, H.M.; Song, L.H.; Tu, J.Y.; Shan, C.X. Determination of 3 kinds of isoflavone aglycones in semen sojae praeparatum by uplc. *Chin. J. Ethnomed. Ethnopharm.* **2012**, *12*, 2.

10. Guo, H.; Zhang, Z.; Yao, Y.; Liu, J.; Chang, R.; Liu, Z.; Hao, H.; Huang, T.; Wen, J.; Zhou, T. A new strategy for statistical analysis-based fingerprint establishment: Application to quality assessment of semen sojae praeparatum. *Food Chem.* **2018**, *258*, 189–198. [CrossRef] [PubMed]

11. Qiu, F.; Shi, L.; Wang, S.; Wu, S.; Wang, M. Simultaneous high-performance liquid chromatography with diode array detection and time-of-flight mass spectrometric confirmation of the ten bioactive compounds in semen sojae preparatum. *J. Sep. Sci.* **2018**, *41*, 3360–3371. [CrossRef]

12. Zhang, J.-H.; Tatsumi, E.; Ding, C.-H.; Li, L.-T. Angiotensin i-converting enzyme inhibitory peptides in douchi, a chinese traditional fermented soybean product. *Food Chem.* **2006**, *98*, 551–557. [CrossRef]

13. Tsai, Y.-H.; Kung, H.-F.; Chang, S.-C.; Lee, T.-M.; Wei, C.-I. Histamine formation by histamine-forming bacteria in douchi, a chinese traditional fermented soybean product. *Food Chem.* **2007**, *103*, 1305–1311. [CrossRef]

14. Chai, C.; Cui, X.; Shan, C.; Yu, S.; Wen, H. Contents variation analysis of free amino acids, nucleosides and nucleobases in semen sojae praeparatum fermentation using uflc-qtrap ms. *Biomed. Chromatogr.* **2017**, *31*, 1–11. [CrossRef]

15. Chai, C.; Yu, S.; Cui, X.B.; Zhang, A.H.; Zhu, D.; Shan, C.X.; Wen, H.M. Analysis of volatile components in semen sojae praepatum with automatic static headspace and gas chromatography-mass spectrometry. *Food Res. Dev.* **2013**, *34*, 4.

16. Suo, H.; Feng, X.; Zhu, K.; Wang, C.; Zhao, X.; Kan, J. Shuidouchi (fermented soybean) fermented in different vessels attenuates hcl/ethanol-induced gastric mucosal injury. *Molecules* **2015**, *20*, 19748–19763. [CrossRef]

17. Qu, K.; Zhao, L.; Luo, X.; Zhang, C.; Hou, P.; Bi, K.; Chen, X. An LC-MS method for simultaneous determination of five iridoids from zhi-zi-chi decoction in rat brain microdialysates and tissue homogenates: Towards an in depth study for its antidepressive activity. *J. Chromatogr. B Anal. Technol. Biomed. Life Sci.* **2014**, *965*, 206–215. [CrossRef]

18. Chen, J.; Cheng, Y.-Q.; Yamaki, K.; Li, L.-T. Anti-α-glucosidase activity of chinese traditionally fermented soybean (douchi). *Food Chem.* **2007**, *103*, 1091–1096. [CrossRef]

19. McClean, S.; Beggs, L.B.; Welch, R.W. Antimicrobial activity of antihypertensive food-derived peptides and selected alanine analogues. *Food Chem.* **2014**, *146*, 443–447. [CrossRef]

20. Chen, L.; Liu, Q.; Sun, Y.; Wang, P.; Zhang, L.; Wang, W.J.C.P. Screening and enzymatic activity analysis of dominant fermentive bacteria of sojae semen praeparatum from different production places. *China Pharma.* **2017**, *28*, 4359–4361.

21. Li, G.; Long, K.; Su, M.S.; Liang, Y.H.; Yang, A.J.; Xie, X.M. Preliminary study on dynamic change of microbial flora in fermentation process to 'yellow cladding' of sojoe semen praeparatum. *Chin. J. Exp. Tradit. Med. Formulae* **2014**, *20*, 139–142.

22. Blay, M.; Espinel, A.E.; Delgado, M.A.; Baiges, I.; Bladé, C.; Arola, L.; Salvadó, J. Isoflavone effect on gene expression profile and biomarkers of inflammation. *J. Pharm. Biomed. Anal.* **2010**, *51*, 382–390. [CrossRef]

23. Kao, T.H.; Chen, B.H. Effects of different carriers on the production of isoflavone powder from soybean cake. *Molecules* **2007**, *12*, 917–931. [CrossRef]

24. Saha, J.; Biswas, A.; Chhetri, A.; Sarkar, P.K. Response surface optimisation of antioxidant extraction from kinema, a bacillus-fermented soybean food. *Food Chem.* **2011**, *129*, 507–513. [CrossRef]

25. Wang, L.; Yin, L.; Li, D.; Zou, L.; Saito, M.; Tatsumi, E.; Li, L. Influences of processing and nacl supplementation on isoflavone contents and composition during douchi manufacturing. *Food Chem.* **2007**, *101*, 1247–1253. [CrossRef]

26. Hurtado-Fernández, E.; Gómez-Romero, M.; Carrasco-Pancorbo, A.; Fernández-Gutiérrez, A. Application and potential of capillary electroseparation methods to determine antioxidant phenolic compounds from plant food material. *J. Pharm. Biomed. Anal.* **2010**, *53*, 1130–1160. [CrossRef]

27. Gaya, P.; Medina, M.; Sanchez-Jimenez, A.; Landete, J.M. Phytoestrogen metabolism by adult human gut microbiota. *Molecules* **2016**, *21*, 1034. [CrossRef]

28. Chung, I.-M.; Seo, S.-H.; Ahn, J.-K.; Kim, S.-H. Effect of processing, fermentation, and aging treatment to content and profile of phenolic compounds in soybean seed, soy curd and soy paste. *Food Chem.* **2011**, *127*, 960–967. [CrossRef]

29. Yang, S.; Wang, L.; Yan, Q.; Jiang, Z.; Li, L. Hydrolysis of soybean isoflavone glycosides by a thermostable β-glucosidase from paecilomyces thermophila. *Food Chem.* **2009**, *115*, 1247–1252. [CrossRef]

30. Kaya, M.; Ito, J.; Kotaka, A.; Matsumura, K.; Bando, H.; Sahara, H.; Ogino, C.; Shibasaki, S.; Kuroda, K.; Ueda, M.; et al. Isoflavone aglycones production from isoflavone glycosides by display of beta-glucosidase from aspergillus oryzae on yeast cell surface. *Appl. Microbiol. Biotechnol.* **2008**, *79*, 51–60. [CrossRef]

31. Zeng, X.; Su, W.; Zheng, Y.; Liu, H.; Li, P.; Zhang, W.; Liang, Y.; Bai, Y.; Peng, W.; Yao, H. UFLC-Q-TOF-MS/MS-based screening and identification of flavonoids and derived metabolites in human urine after oral administration of exocarpium citri grandis extract. *Molecules* **2018**, *23*, 895. [CrossRef] [PubMed]

32. The Pharmacopoeia Commission of the Ministry of Health of the People's Republic of China. *Pharmacopoeia of People's Republic of China; Part I*; Medical Science and Technology Press: Beijing, China, 2015; p. 328.

33. Huang, Y.; Zhang, H.; Ben, P.; Duan, Y.; Lu, M.; Li, Z.; Cui, Z. Characterization of a novel gh36 alpha-galactosidase from bacillus megaterium and its application in degradation of raffinose family oligosaccharides. *Int. J. Biol. Macromol.* **2018**, *108*, 98–104. [CrossRef]

34. Yu, F.X.; Chen, X.; Chen, Z.W.; Wei, X.J. Fatty acid analysis of edible oils. *Adv. Mater. Res.* **2014**, *962–965*, 1222–1225. [CrossRef]

35. Kuo, T.M.; Doehlert, D.C.; Crawford, C.G. Sugar metabolism in germinating soybean seeds. Evidence for the sorbitol pathway in soyabean axes. *Plant Physiol.* **1990**, *93*, 1514–1520. [CrossRef] [PubMed]

36. Wang, X.J.; Wang, Y.S.; Qiu, J. Study on chemical composition of folium mori. *Food Drug* **2007**, *30*, 1–3.

37. Li, S.; Li, S.; Liu, C.; Liu, C.; Zhang, Y. Extraction and isolation of potential anti-stroke compounds from flowers of pueraria lobata guided by in vitro pc12 cell model. *J. chromatogr. B Anal. Technol. Biomed. Life Sci.* **2017**, *1048*, 111–120. [CrossRef] [PubMed]

38. Yang, Y.J.; Yao, Y.S.; Qin, Y.C.; Qiu, J.; Li, J.G.; Li, J.; Gu, X. Investigation and analysis of main afn in soybean meal and fermented soybean meal. *Sci. Agric. Sin.* **2016**, *49*, 573–580.

39. Liu, Y.C.; Cao, Y.; Zhang, J.X.; Zhang, N.; Lin-Jie, B.I.; Liu, T.; Chen, J.J. The condition optimization of bacterial combinations for fermentation of bean and determination of the content of stachyose and raffinose. *Hubei Agric. Sci.* **2015**, *54*, 4007–4011.

40. Fu, T.; Qi, W.; Li, A.; Liu, J.; Liang, X.; Yun, T.; Wang, Y. Effects of different bacteria on fermentative degradation of stachyose and raffinose in soybean meal and detection technology. *J. Chin. Cereals Oils Assoc.* **2014**, *29*, 111–118.

41. Lee, I.H.; Chou, C.C. Distribution profiles of isoflavone isomers in black bean kojis prepared with various filamentous fungi. *J. Agric. Food Chem.* **2006**, *54*, 1309–1314. [CrossRef]

Sample Availability: Samples of the compounds are available from the authors.

molecules

MDPI

Article

Development of an Accelerated Solvent Extraction-Ultra-Performance Liquid Chromatography-Fluorescence Detection Method for Quantitative Analysis of Thiamphenicol, Florfenicol and Florfenicol Amine in Poultry Eggs

Bo Wang [1,2,†], Xing Xie [3,†], Xia Zhao [2,4], Kaizhou Xie [2,4,*], Zhixiang Diao [2,4], Genxi Zhang [2,4], Tao Zhang [2,4] and Guojun Dai [2,4]

1   College of Veterinary Medicine, Yangzhou University, Yangzhou 225009, China; yzwbo168@163.com
2   Joint International Research Laboratory of Agriculture & Agri-Product Safety, Yangzhou University, Yangzhou 225009, China; 18252711481@139.com (X.Z.); 18352764521@163.com (Z.D.); zgx1588@126.com (G.Z.); zhangt@yzu.edu.cn (T.Z.); gjdai@163.com (G.D.)
3   Key Laboratory of Veterinary Biological Engineering and Technology, Ministry of Agriculture, Institute of Veterinary Medicine, Jiangsu Academy of Agricultural Sciences, Nanjing 210014, China; yzxx1989@163.com
4   College of Animal Science and Technology, Yangzhou University, Yangzhou 225009, China
*   Correspondence: yzxkz168@163.com; Tel.: +86-13952750925
†   These authors contributed equally to this work.

Academic Editors: Alessandra Gentili and Chiara Fanali
Received: 13 April 2019; Accepted: 10 May 2019; Published: 13 May 2019

check for updates

Abstract: A simple, rapid and novel method for the detection of residues of thiamphenicol (TAP), florfenicol (FF) and its metabolite, florfenicol amine (FFA), in poultry eggs by ultra-performance liquid chromatography-fluorescence detection (UPLC-FLD) was developed. The samples were extracted with acetonitrile-ammonia (98:2, *v/v*) using accelerated solvent extraction (ASE) and purified by manual degreasing with acetonitrile-saturated n-hexane. The target compounds were separated on an ACQUITY UPLC® BEH C_{18} (2.1 mm × 100 mm, 1.7 μm) chromatographic column using a mobile phase composed of 0.005 mol/L NaH_2PO_4, 0.003 mol/L sodium lauryl sulfate and 0.05% trimethylamine, adjusted to pH 5.3 ± 0.1 by phosphoric acid and acetonitrile (64:36, *v/v*). The limits of detection (LODs) and limits of quantification (LOQs) of the three target compounds in poultry eggs were 1.8–4.9 μg/kg and 4.3–11.7 μg/kg, respectively. The recoveries of the three target compounds in poultry eggs were above 80.1% when the spiked concentrations of three phenicols were the LOQ, 0.5 maximum residue limit (MRL), 1.0 MRL and 2.0 MRL. The intraday relative standard deviations (RSDs) were less than 5.5%, and the interday RSDs were less than 6.6%. Finally, this new detection method was successfully applied to the quantitative analysis of TAP, FF and FFA in 150 commercial poultry eggs.

Keywords: poultry eggs; thiamphenicol; florfenicol; florfenicol amine; ASE; UPLC-FLD

1. Introduction

Poultry eggs contain high levels of essential amino acids and vitamins as well as various major and trace elements required by the human body, and as a result they have become an increasingly popular consumer product [1]. Consumer demand for poultry products has promoted the growth of the poultry industry, and intensive farming has also increased the morbidity and mortality of poultry. To control the occurrence of diseases and reduce mortality in poultry, antibiotics are widely used to prevent poultry diseases, increase feed conversion rates, and promote animal growth [2].

Thiamphenicol (TAP) and florfenicol (FF) are synthetic chloramphenicols (CAPs) used as broad-spectrum antibiotics, and they have chemical structures and efficacies similar to that of CAP as well as good therapeutic effects on various bacterial strains common in poultry. These compounds are widely used in actual production [3]. The only difference between TAP and CAP is the structure of the substituents on their benzene rings. CAP has a nitro group on the phenyl ring, and TAP has a methyl sulfone group. However, TAP is much less toxic than CAP, its blood toxicity effects are reversible, and it does not cause aplastic anemia. On the other hand, TAP can inhibit the formation of red blood cells, white blood cells and platelets; it has a strong immunosuppressive effect; and it has weaker antibacterial effects than CAP, which limit its practical use [4–6]. Many countries use these compounds as veterinary drugs, but they are banned from use in food animals. FF is a fluorinated analogue of TAP with a molecular structure similar to that of CAP, but it lacks the nitro group on the aromatic ring. Studies have shown that this substituent is the key molecular characteristic of CAP causing dose-independent irreversible aplastic anemia in the human body [7]. However, FF can theoretically cause serious adverse reactions similar to those caused by CAP. Therefore, FF can only be used for the treatment of animal diseases [8]. FF shows the most potent antibacterial activity among CAP drugs, and it has many advantages (wide spectrum of antibacterial activity, good oral absorption, wide distribution in the body, high bioavailability, good safety profile, etc.), making it a broad-spectrum antibiotic with great potential for practical applications [9]. At present, the drug is on the market in many countries, and it is widely used in animal husbandry and aquaculture for disease prevention. However, reproductive toxicity tests have shown that FF has certain embryotoxic effects, and as a result the detection of its residue in animal foods such as livestock, poultry, and aquatic products has attracted increasing attention [10].

Because both TAP and FF have toxic side effects, the EU [11], US [12] and China's Ministry of Agriculture [13] have established maximum residue limits (MRLs) for these compounds in poultry tissues (TAP: 50 µg/kg, FF: 100 µg/kg in muscle), and stipulated that the limit of FF residue in poultry tissues is based on the total amount of both the prototype drug (FF) and its metabolite, florfenicol amine (FFA), and that these drugs should not be detected in poultry eggs. Therefore, developing different detection methods to determine whether veterinary drug residues meet the legal requirements before the animal foods are marketed is of great importance. Establishing and improving the detection methods for TAP, FF and FFA residues in poultry eggs is also necessary.

At present, there are many reported methods for detecting TAP, FF and FFA residues, including methods based on high-performance liquid chromatography (HPLC) [14–16], gas chromatography (GC) [17,18], liquid chromatography-tandem mass spectrometry (LC-MS/MS) [19–22], and gas chromatography-tandem mass spectrometry (GC-MS) [23,24]. When using GC to analyze CAP drug residues, the target needs to be derivatized, making the analysis process cumbersome. The most widely used analytical method involves LC-MS, and although this method offers qualitative and quantitative (mass spectrometry) accuracy and high sensitivity, the instrument is expensive, and the detection cost is high. Meanwhile, fluorescence detection is commonly used for veterinary drug residues and environmental analysis because of its advantages of speed, ease of operation, and low cost of detection [14,25]. Therefore, developing a simple, fast and low-cost analytical method that meets the detection requirements is of great importance. Moreover, in the reported detection methods, the sample matrices used for analysis were generally animal tissues [15,26] or aquatic products [21,27], and there are few reports on detection methods for poultry eggs [14]. Xie et al. [14] established an HPLC-FLD method for the determination of TAP, FF and FFA residues in eggs and sample pretreatment using a liquid-liquid extraction method, ethyl acetate:acetonitrile:ammonium hydroxide (49:49:2, *v/v/v*) as an extractant, delipided in n-hexane. Based on previous research, a comprehensive method using ultra-performance liquid chromatography-fluorescence detection (UPLC-FLD) for the determination of TAP, FF and FFA residues in poultry eggs (hen eggs, duck eggs, goose eggs, pigeon eggs and quail eggs) is reported here. Compared with the previously studied HPLC-FLD method [14], the UPLC-FLD method has the advantages of fast analysis speed (detection time < 5 min), strong separation ability (recoveries were 80.1%–98.6%), high sensitivity, and low consumption of reagents. This study intends to use accelerated solvent extraction (ASE) as the sample pretreatment procedure to extract the target analytes from the samples, aiming to establish an ASE-UPLC-FLD method for the determination of TAP, FF and FFA residues in poultry eggs. This technique will provide a new, simple, inexpensive, highly efficient and rapid method for the detection of these analytes. In addition, the effects of ultrasonic extraction, vortex oscillation extraction, vortex oscillation + ultrasonic extraction and ASE extraction are compared in this study. Compared to other extraction methods, ASE was investigated as a novel alternative technology, which has the advantages of automation (saving time and human effort), consuming less reagents, higher recovery rate, and suitability for batch processing of samples.

2. Results and Discussion

2.1. Selection of the Chromatographic Column and Mobile Phase

The composition of the mobile phase and the type of chromatographic column have a substantial influence on the separation and peak shape of the analytes. Among the reported methods, the most commonly used columns for the detection of CAP drugs are C_{18} columns [28,29], produced by various manufacturers. Meanwhile, the ACQUITY UPLC® BEH C_{18} (2.1 mm × 100 mm, 1.7 μm) column offers outstanding chemical stability, a wide range of pH conditions (pH 1-12) and a wide range of mobile phases, which provides a versatile and reliable separation technique for method development. Therefore, this study used an ACQUITY UPLC® BEH C_{18} (2.1 mm × 100 mm, 1.7 μm) column as the analytical column. For the mobile phase, acetonitrile-water [16,30] mixtures are often used to determine CAP residues in LC-MS methods. FFA is a weakly basic substance that does not generally remain on the C_{18} column, and it elutes with the dead volume. A common solution to this is to add ammonium formate or ammonium acetate [27,31] to the mobile phase system to enhance the retention of FFA on the C_{18} column. When detecting CAPs with a fluorescence detector, to enhance the retention of FFA on a C_{18} column, an ion-pair reagent is usually added to the aqueous phase to react with the FFA to form a weakly bound ion pair, and the pH is adjusted with a buffer to keep the whole system weakly acidic and prevent dissociation of the ion pair [14,32]. Yang et al. [33] reported a liquid chromatography-fluorescence detection method for the determination of TAP, FF and FFA residues in aquatic products. Sodium heptane sulfonate was added to the mobile phase, and the target compounds were well separated. Commonly used ion-pairing reagents are sodium heptane sulfonate and sodium lauryl sulfate, and these ion-pairing reagents provide a good separation for the target compounds. Because it is inexpensive, sodium lauryl sulfate is used as an ion pairing reagent in this test. Sodium lauryl sulfate and FFA form weakly polar pairs, which are distributed on the surface of the hydrophobic stationary phase and then eluted by the mobile phase. The buffer system uses phosphate-phosphoric acid and triethylamine to improve peak shape and reduce peak tailing. In this study, the amounts of ion-pairing reagent (sodium lauryl sulfate), buffer system (phosphate-phosphoric acid), and triethylamine were optimized. The effects of 1, 3, 5, and 10 mM sodium lauryl sulfate were investigated. As the concentration increased, the retention time of FFA increased, and a concentration of 3 mM could ensure that FFA eluted first. The effects of 0, 3, 5, 10, and 20 mM NaH_2PO_4 are compared in Figure 1a. The retention time of FFA was slightly shorter with increasing NaH_2PO_4 concentration, but the use of salt impacted the instrument, and the chromatographic column was easily blocked; thus, considering the FFA retention time and the effect on the instrument, a concentration of 5 mM NaH_2PO_4 was selected. The pH impacted the response and retention time of FFA, as shown in Figure 1b. Under neutral conditions, the retention time of FFA was shorter, and decreasing the pH gradually increased the response of FFA (Figure 2) but increased the retention time. When the pH was 5.4, the response of FFA reached its highest value, and further reducing the pH had little effect on the response. However, the retention time was too high, increasing the overall detection time, so a pH of 5.3 ± 0.1 was selected. The amount of triethylamine was also investigated. It was found that 0.01% triethylamine could reduce peak trailing. However, as the amount of triethylamine was increased (0.03%, 0.05%, and 0.1%), the retention time of FFA increased. Based on all these factors, the concentration of triethylamine was set as 0.05%. To separate the targets from impurities, the ratio of the solvents (63:37, 64:36, 65:35, 66:34, *v/v*) in the mobile phase was optimized. When the mobile phase ratio was 64:36 (*v/v*), the targets and impurities were separated, and the peak shapes were good. In summary, the final mobile phase conditions were water (containing 5 mM NaH_2PO_4, 3 mM lauryl sodium sulfate, 0.05% triethylamine, adjusted to pH 5.3 ± 0.1) and acetonitrile in a 64:36 (*v/v*) ratio. According to the chemical nature of the ACQUITY UPLC® BEH C18 (2.1 mm × 100 mm, 1.7 μm) column and final mobile phase composition, the elution order and resolution of the target compound were analyzed; TAP was preferentially eluted, followed by FF and then FFA, and examining the fluorescence intensity of the target compound showed that the resolution was greatly improved.

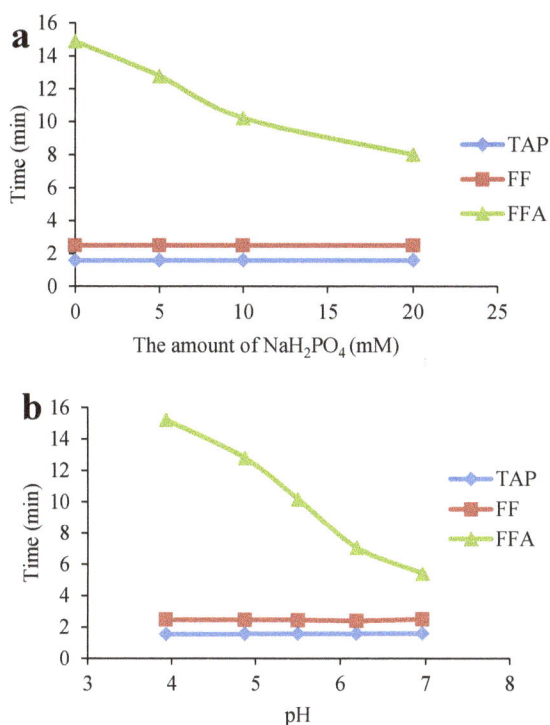

Figure 1. Effects of the amount of NaH_2PO_4 (**a**) and the mobile phase pH (**b**) on the retention times of the targets.

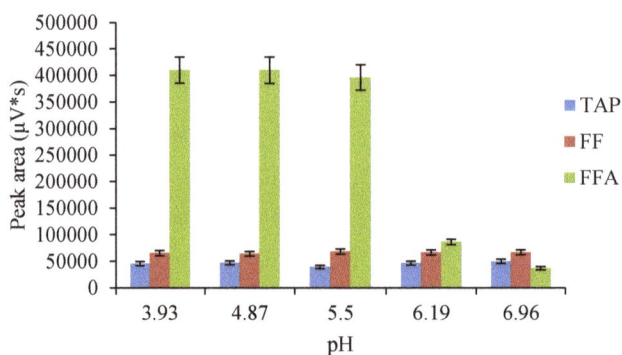

Figure 2. Effects of pH on the target responses.

2.2. Determination of the Detection Wavelength

Liquid chromatography-fluorescence detection methods are commonly used in veterinary drug residues in animal-derived foods, pesticide residues in agricultural products, and environmental analysis. LC-FLD can detect compounds containing fluorophores, which facilitates the development of simple and rapid methods for the determination of target compounds containing fluorophores. In the literature, when using ultraviolet detectors to identify CAP drugs, the most commonly used detection wavelengths include 220, 223, 224, 225, and 228 nm [15,16,33,34]; when detecting CAP drugs with a fluorescence detector, the excitation and emission wavelengths include 224 and 290 nm, 224 and 295 nm, and 225 and 290 nm, respectively [33,35,36]. However, the optimal detection wavelength for these targets may be different under different detection conditions because of variations in detection devices and experimental conditions. The optimal excitation and emission wavelengths of TAP obtained by fluorescence scanning were 229.8 nm and 285.3 nm, respectively; those of FF were 229.8 nm and 283.9 nm, respectively; and those of FFA were 227.9 nm and 283.9 nm. Therefore, based on the literature and the optimum excitation and emission wavelengths for each target, the optimal detection wavelength for TAP, FF and FFA under the conditions used in this study was simultaneously determined by scanning with a fluorescence detector. An excitation wavelength of 233 nm and an emission wavelength of 284 nm were ultimately selected to simultaneously measure TAP, FF and FFA. These wavelengths ensured both high response values of the targets and no interference from impurities.

2.3. Selection and Optimization of the Extraction Solvent and Extraction Method

In the reported literature, the extraction of CAP drugs is typically carried out by liquid-liquid extraction. The most commonly used extractants are acetonitrile, ethyl acetate and a mixture of acetonitrile and ethyl acetate (different proportions of ammonia-containing extractants will provide extracts containing FFA) [33,37,38]. There are few reports on the extraction of CAPs from matrices by automatic extraction equipment-ASE. Only Yang et al. [39] used ASE to extract CAP and FF from aquatic products, and they used static extraction with ethyl acetate at 100 °C for 5 min; their obtained recoveries of the samples were from 90.2% to 109%. Xiao et al. [40] used subcritical water as the extraction solvent to extract traces of CAP, TAP, FF and FFA from poultry tissues using pressurized liquid extractors operating at 150 °C and 100 bar (static extraction, two extraction cycles, 3 min each cycle), and the average recoveries of the four analytes from the samples were 86.8–101.5%. However, a method for simultaneously extracting TAP, FF and FFA from poultry eggs using a modern automatic extraction instrument-ASE instrument has not been reported. In this study, the effects of mixtures of acetonitrile and ammonia (98:2, *v/v*), ethyl acetate and ammonia (98:2, *v/v*) and acetonitrile and ammonia with ethyl acetate (49:49:2, *v/v/v*) were compared. The results showed that the above extractants can extract TAP, FF and FFA from the matrix and that the recoveries meet the detection requirements. However, acetonitrile offers better deproteinization, and the substance in eggs that causes the most interference is protein. To improve the detection results, acetonitrile containing ammonia (98:2, *v/v*) was selected as the extractant. Because FFA is a weakly alkaline compound with an amino group, it is more advantageous to extract FFA under alkaline conditions because greater similarity leads to better solubility. Under optimized ASE conditions, this study compared the effects of different ratios of extractants (acetonitrile: ammonia = 99:1, 98:2, 97:3, 96:4, and 95:5, *v/v*) on poultry egg recovery. Table 1 shows that as the proportion of ammonia increased, the recoveries of FFA also gradually increased; however, when the content of ammonia exceeded 2% of the total volume, the recoveries of FFA decreased, and excessive ammonia also reduced the recoveries of TAP and FF. In summary, the extraction outcome with acetonitrile:ammonia (98:2, *v/v*) was best, and the recoveries of all targets were above 91.0%.

Table 1. Effects of different ratios of extraction reagents (acetonitrile:ammonia) for ASE on the recoveries of 25 µg/kg TAP and 50 µg/kg FF and FFA from poultry eggs (%) ($n = 6$).

Matrix	Analyte	Extraction Reagents (acetonitrile:ammonia, *v/v*)				
		99:1	98:2	97:3	96:4	95:5
	TAP	93.4 ± 1.8	92.8 ± 1.1	89.4 ± 2.3	81.6 ± 2.2	73.9 ± 2.0
Hen eggs	FF	92.6 ± 2.7	93.2 ± 3.0	84.2 ± 2.2	74.3 ± 1.9	69.6 ± 2.0
	FFA	86.9 ± 2.1	92.4 ± 2.2	88.4 ± 3.1	78.8 ± 2.9	69.3 ± 1.9
	TAP	90.2 ± 2.3	91.3 ± 2.3	85.3 ± 2.4	80.8 ± 2.4	70.3 ± 2.2
Duck eggs	FF	93.0 ± 1.8	91.2 ± 2.5	84.2 ± 2.0	74.6 ± 1.9	68.7 ± 2.3
	FFA	85.0 ± 2.5	93.2 ± 1.8	84.3 ± 2.3	73.6 ± 1.9	61.5 ± 1.9
	TAP	90.7 ± 2.2	93.4 ± 2.6	86.7 ± 2.4	75.8 ± 2.7	67.9 ± 2.0
Goose eggs	FF	93.1 ± 1.8	91.3 ± 1.9	87.3 ± 2.3	71.4 ± 2.2	60.4 ± 2.4
	FFA	85.0 ± 1.8	92.3 ± 2.0	84.1 ± 1.8	73.0 ± 1.9	63.8 ± 2.2
	TAP	92.4 ± 2.6	92.6 ± 2.1	82.3 ± 3.1	75.5 ± 2.4	63.2 ± 2.3
Pigeon eggs	FF	92.3 ± 2.4	93.1 ± 2.0	83.8 ± 1.9	76.1 ± 3.0	61.0 ± 2.2
	FFA	88.2 ± 2.2	96.3 ± 2.1	86.5 ± 2.7	73.1 ± 2.1	64.8 ± 2.3
	TAP	91.3 ± 2.3	92.8 ± 2.3	87.3 ± 2.0	70.8 ± 1.9	65.7 ± 2.1
Quail eggs	FF	92.7 ± 2.2	94.2 ± 2.5	86.2 ± 2.5	74.5 ± 2.4	68.5 ± 2.5
	FFA	86.5 ± 1.8	93.0 ± 1.9	84.7 ± 2.5	76.2 ± 1.9	70.4 ± 1.8

In this study, under the conditions of acetonitrile:ammonia (98:2, *v/v*) as an extractant, the effects of ultrasonic extraction, vortex oscillation extraction, vortex oscillation + ultrasonic extraction and ASE extraction were compared. Table 2 shows that compared to those with other sample preparation processes, the recoveries with ASE extraction were best; in addition, the ASE method saves time and is suitable for batch processing of samples. The time required to prepare a sample with ASE (15 min) was half that required for vortexing + ultrasonic extraction (30 min). Moreover, using ASE to process samples avoided contact between the experimenter and the reagent, which is more consistent with the detection requirements of health and environmental protection. Therefore, this study ultimately selected ASE for sample extraction.

2.4. Optimization of the ASE Method

The sample extraction step is often considered to be the bottleneck of the entire analytical process. To simplify the pretreatment of the sample and improve the efficiency of sample preparation, a variety of pretreatment methods have been developed. Since the introduction of ASE in 1995, it has rapidly become an acceptable alternative to traditional extraction methods. ASE uses high-temperature and high-pressure conditions, which result in greatly improved extraction efficiency. This study explored the effects of various operating parameters (temperature, time, volume of solvent used and so on) on ASE performance. The effects of different temperatures (40, 60, 80, 100 and 120 °C) on the recoveries of the targets were studied. The recoveries of the targets gradually increased with increasing extraction temperature. When the temperature exceeded 80 °C, the recoveries decreased (Figure 3a). Studies have shown that increasing the time of static extraction provides the target sufficient time to diffuse into the extraction solvent, improving the efficiency of the extraction. However, this study compared different static extraction times (2, 3, 5, and 8 min) and found that for TAP and FF, static extraction for 5 min provides high recoveries (>90%), while the recovery of FFA decreases with increasing static extraction time (Figure 3b). As shown in Figure 3b, 3 min and 5 min had little effect on TAP and FF extraction, and a high recovery rate for FFA was observed when the static extraction time was 3 min. Therefore, to ensure that the recoveries of all the targets met the requirements of detection under these ASE conditions, 3 min was selected as the static extraction time. This study also explored the effect of the number of static extraction cycles and the volume of the extractant on the extraction outcome. Extraction of the target twice provided a better recovery than one extraction cycle. When the volume of the extractant was 40% by volume, the target could be efficiently extracted, so there was no need to increase the amount of extractant. In summary, the final ASE conditions were 80 °C, 1500 psi, 40% pool volume, static extraction for 3 min, and two static extraction cycles.

Table 2. Effects of different extraction methods on the recoveries of 25 µg/kg TAP and 50 µg/kg FF and FFA from poultry eggs (%) (*n* = 6).

Matrix	Analyte	Extraction Method			
		Ultrasonic	Vortex Oscillation	Vortex Oscillation + Ultrasonic	ASE
Hen egg	TAP	34.4 ± 2.5	78.6 ± 2.6	88.3 ± 2.2	92.8 ± 2.4
	FF	37.1 ± 2.3	81.5 ± 3.1	92.0 ± 2.7	96.0 ± 2.3
	FFA	54.4 ± 2.2	84.0 ± 2.9	91.9 ± 2.8	93.3 ± 2.0
Duck eggs	TAP	30.9 ± 2.0	70.0 ± 2.7	84.9 ± 2.2	90.2 ± 1.9
	FF	33.4 ± 2.5	79.5 ± 2.7	89.7 ± 2.4	92.5 ± 2.1
	FFA	49.0 ± 2.4	74.4 ± 3.2	86.0 ± 2.9	91.7 ± 2.2
Goose eggs	TAP	31.3 ± 2.9	71.9 ± 3.0	82.7 ± 2.6	93.3 ± 1.9
	FF	35.6 ± 2.7	72.0 ± 2.6	88.4 ± 3.0	92.4 ± 2.5
	FFA	40.5 ± 3.4	77.0 ± 2.8	85.0 ± 2.4	90.4 ± 2.1
Pigeon eggs	TAP	34.4 ± 3.0	69.9 ± 2.7	85.3 ± 2.9	86.9 ± 2.2
	FF	34.7 ± 2.6	70.0 ± 3.2	87.0 ± 2.6	89.0 ± 2.4
	FFA	40.7 ± 2.8	72.6 ± 2.9	80.3 ± 2.3	95.1 ± 1.6
Quail eggs	TAP	36.1 ± 2.9	72.1 ± 2.8	83.5 ± 2.7	90.1 ± 2.0
	FF	30.1 ± 3.0	73.4 ± 2.3	88.5 ± 3.3	92.6 ± 2.4
	FFA	45.5 ± 2.8	69.7 ± 3.1	83.7 ± 2.5	92.7 ± 2.7

2.5. Bioanalytical Method Validation

In the blank poultry eggs, TAP was spiked at a concentration from limit of quantification (LOQ)-250 µg/kg, and FF and FFA were added at LOQ-400 µg/kg. The peak area was correlated with the spiked concentration of the analyte, and the linearity was good. The linear equations, linear ranges and coefficients of determination of TAP, FF and FFA in poultry eggs are shown in Table 3.

The recoveries and precisions of TAP, FF and FFA in different blank poultry egg samples are shown in Tables 4 and 5, respectively. As shown in Tables 4 and 5, the recoveries of TAP, FF and FFA in poultry eggs were 80.1%–98.6%, the relative standard deviations (RSDs) were 1.2%–4.3%, the intraday RSDs were 1.2%–5.5%, and the interday RSDs were 1.8%–6.6%. According to the EU 2002/675/EC resolution and the FDA [41,42], the acceptable range of recoveries for multidrug residue testing procedures is 70–120%. The average recoveries of TAP, FF and FFA from different blank poultry egg samples were all above 80.0%, which are consistent with the EU's requirements for the recoveries of analytes. The limits of detection (LODs) and LOQs of TAP, FF and FFA in different blank poultry egg samples using the optimized pretreatment method and instrument analysis method are shown in Table 3. The LODs of TAP, FF and FFA were 3.3–3.4 µg/kg, 4.7–4.9 µg/kg and 1.8–1.9 µg/kg, respectively, and the LOQs were 9.7–9.9 µg/kg, 10.5–11.7 µg/kg and 4.3–4.8 µg/kg, respectively.

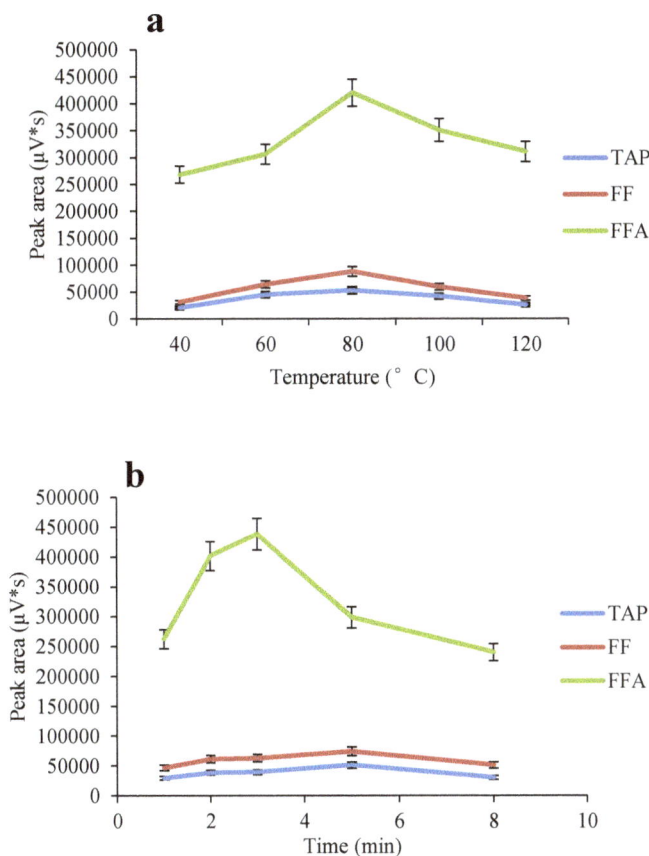

Figure 3. Effects of temperature (**a**) and time (**b**) on ASE extraction.

Table 3. The linear ranges, linear regression equations, determination coefficients, LODs and LOQs of TAP, FF and FFA from poultry eggs.

Matrix	Analyte	Linear Range (µg/kg)	Linear Regression Equation	Determination Coefficient (R^2)	LOD (µg/kg)	LOQ (µg/kg)
	TAP	9.7–250.0	y = 1030.2x + 295.52	0.9996	3.3	9.7
Hen eggs	FF	10.5–400.0	y = 737.43x + 714.06	0.9998	4.7	10.5
	FFA	4.3–400.0	y = 3020.3x + 164.01	0.9998	1.8	4.3
	TAP	9.9–250.0	y = 562.84x + 634.05	0.9997	3.4	9.9
Duck eggs	FF	11.7–400.0	y = 844.1x + 616.34	0.9997	4.9	11.7
	FFA	4.7–400.0	y = 4495.8x + 520.7	0.9996	1.9	4.7
	TAP	9.8–250.0	y = 618.73x + 139.95	0.9996	3.4	9.8
Goose eggs	FF	11.2–400.0	y = 713.68x + 738.78	0.9996	4.8	11.2
	FFA	4.7–400.0	y = 3081.1x − 258.32	0.9997	1.9	4.7
	TAP	9.9–250.0	y = 683.44x + 543.48	0.9998	3.4	9.9
Pigeon eggs	FF	11.2–400.0	y = 762.9x + 761.11	0.9993	4.8	11.2
	FFA	4.8–400.0	y = 4019.1x + 490.45	0.9998	1.9	4.8
	TAP	9.7–250.0	y = 667.44x + 483.91	0.9993	3.3	9.7
Quail eggs	FF	10.6–400.0	y = 753.98x − 193.97	0.9999	4.7	10.6
	FFA	4.6–400.0	y = 4824.6x − 229.06	0.9994	1.8	4.6

Table 4. Recoveries and precisions of TAP, FF and FFA from spiked blank hen eggs, duck eggs and goose eggs.

Matrix	Analyte	Spiking Level (µg/kg)	Recovery (%) (*n* = 6)	RSD (%) (*n* = 6)	Intraday RSD (%) (*n* = 6)	Interday RSD (%) (*n* = 18)
Hen eggs	TAP	9.7	85.6 ± 1.8	2.1	2.6	3.6
		25	90.5 ± 2.7	3.0	3.7	5.0
		50 $^\alpha$	92.7 ± 1.5	1.6	2.5	3.1
		100	91.5 ± 2.0	2.2	2.5	2.9
	FF	10.5	84.9 ± 3.4	4.0	4.7	5.1
		50	90.2 ± 2.5	2.8	3.8	4.8
		100 $^\alpha$	93.5 ± 2.6	2.8	4.0	4.3
		200	94.9 ± 2.7	2.8	2.3	3.4
	FFA	4.3	86.7 ± 3.6	4.2	3.9	5.3
		50	91.5 ± 1.8	2.0	3.1	3.6
		100 $^\alpha$	96.7 ± 3.5	3.6	3.2	2.8
		200	98.0 ± 1.8	1.8	2.4	2.7
Duck eggs	TAP	9.9	84.8 ± 2.0	2.4	3.9	4.2
		25	93.8 ± 1.8	1.9	3.6	4.6
		50 $^\alpha$	94.5 ± 2.4	2.5	1.9	2.6
		100	92.8 ± 1.1	1.2	2.4	2.4
	FF	11.7	85.1 ± 1.6	1.9	1.2	2.4
		50	89.4 ± 2.0	2.2	1.8	1.8
		100 $^\alpha$	94.8 ± 1.5	1.6	2.5	3.6
		200	94.5 ± 3.3	3.5	2.6	4.1
	FFA	4.7	87.5 ± 1.9	2.2	3.1	3.7
		50	96.5 ± 2.2	2.3	3.1	3.4
		100 $^\alpha$	96.9 ± 1.7	1.8	2.7	3.6
		200	96.1 ± 1.7	1.8	3.1	3.3
Goose eggs	TAP	9.8	85.5 ± 1.7	2.0	2.5	2.9
		25	93.3 ± 2.1	2.3	3.5	4.0
		50 $^\alpha$	93.1 ± 2.9	3.1	4.3	5.4
		100	94.5 ± 4.0	4.2	3.9	4.0
	FF	11.2	80.7 ± 3.5	4.3	5.2	6.2
		50	93.9 ± 2.2	2.3	2.7	3.1
		100 $^\alpha$	95.2 ± 1.5	1.6	2.2	3.0
		200	95.2 ± 2.5	2.6	3.1	3.4
	FFA	4.7	83.9 ± 2.8	3.3	3.5	4.1
		50	93.7 ± 2.5	2.7	3.2	3.8
		100 $^\alpha$	94.5 ± 2.0	2.1	3.2	3.5
		200	96.1 ± 2.6	2.7	3.3	4.4

Note: α. Maximum Residue Limits.

Under the optimized UPLC-FLD conditions, the retention times of TAP, FF and FFA from different poultry eggs were 1.50, 2.10 and 3.80 min, respectively. The peak shapes were good, and the blank samples had no interference peaks around these retention times. Taking hen egg samples as an example, the chromatograms of the standards, blank hen egg samples and blank hen egg samples spiked with standards are shown in Figures 4–6.

Table 5. Recoveries and precisions of TAP, FF and FFA from spiked blank pigeon eggs and quail eggs.

Matrix	Analyte	Spiking Level (µg/kg)	Recovery (%) (n = 6)	RSD (%) (n = 6)	Intraday RSD (%) (n = 6)	Interday RSD (%) (n = 18)
Pigeon eggs	TAP	9.9	80.1 ± 2.5	3.1	2.5	3.6
		25	93.6 ± 3.3	3.5	4.7	5.4
		50 $^\alpha$	92.3 ± 2.4	2.6	2.7	3.2
		100	94.4 ± 3.3	3.5	2.2	4.3
	FF	11.2	84.4 ± 2.2	2.6	3.2	3.3
		50	95.9 ± 2.5	2.6	4.4	4.2
		100 $^\alpha$	97.9 ± 4.1	4.2	3.3	5.0
		200	98.6 ± 2.8	2.8	3.1	3.1
	FFA	4.8	85.7 ± 2.0	2.3	3.8	4.1
		50	94.0 ± 2.3	2.4	2.9	3.2
		100 $^\alpha$	95.7 ± 3.2	3.3	4.2	5.4
		200	97.7 ± 1.8	1.8	2.3	3.7
Quail eggs	TAP	9.7	84.1 ± 3.0	3.6	3.9	4.0
		25	95.4 ± 2.0	2.1	3.3	3.9
		50 $^\alpha$	93.4 ± 2.8	3.0	3.9	4.4
		100	96.3 ± 2.2	2.3	3.7	5.4
	FF	10.6	86.5 ± 3.5	4.0	5.5	5.5
		50	94.9 ± 2.1	2.2	2.7	3.8
		100 $^\alpha$	95.9 ± 3.1	3.2	4.0	4.6
		200	96.7 ± 2.3	2.4	3.2	3.4
	FFA	4.6	87.5 ± 2.7	3.1	3.7	4.1
		50	96.0 ± 3.0	3.1	4.7	5.6
		100 $^\alpha$	95.5 ± 3.4	3.6	3.5	5.5
		200	96.2 ± 3.6	3.7	5.4	6.6

Note: α. Maximum Residue Limits.

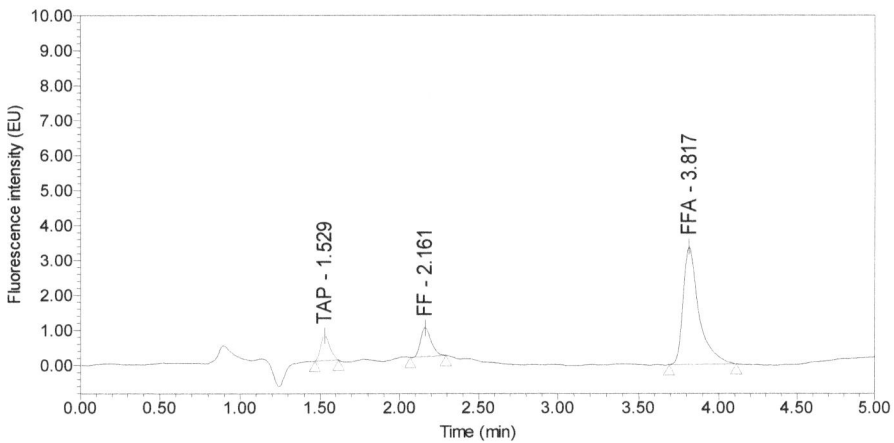

Figure 4. Chromatogram of the standards (25 µg/kg TAP and 50 µg/kg FF and FFA standards).

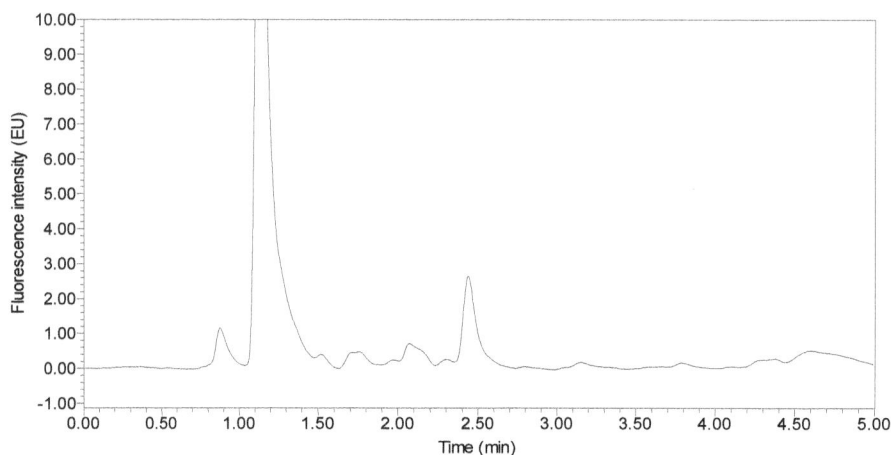

Figure 5. Chromatogram of the blank hen egg.

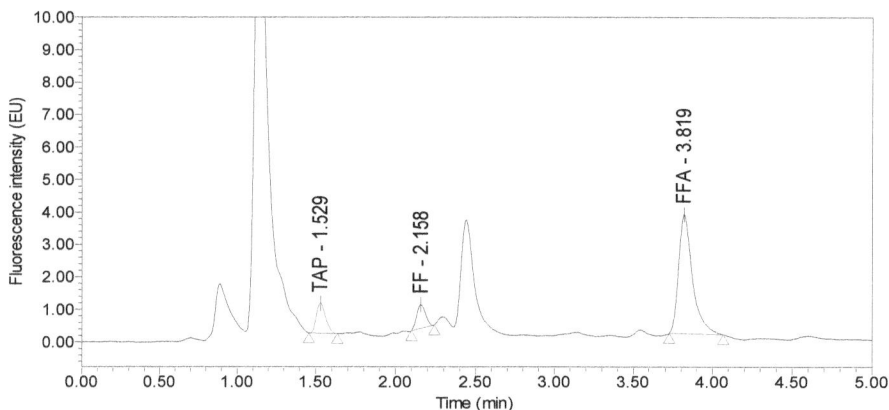

Figure 6. Chromatogram of the blank hen egg spiked with 25 µg/kg TAP and 50 µg/kg FF and FFA standards.

2.6. Real Sample Analysis

Our newly developed detection method was applied to the evaluation of real samples. A total of 150 commercial poultry egg samples (30 hen eggs, 30 duck eggs, 30 goose eggs, 30 pigeon eggs and 30 quail eggs) from a local supermarket were analyzed by the described method. Only hen eggs (34 and 20 µg/kg) and duck eggs (44 and 18 µg/kg) were found to contain FF and FFA residues, and none of the samples exceeded the MRL of 100 µg/kg (EU standard). From these data, we can evaluate the applicability and reliability of the newly developed method. Thus, this new UPLC-FLD method can be applied to the quantification of these drugs in poultry egg samples.

3. Materials and Methods

3.1. Chemicals and Reagents

TAP (99.0% purity), FF (99.0% purity) and FFA (99.8% purity) standards were obtained from Dr. Ehrenstorfer GmbH (Augsburg, Germany). HPLC-grade acetonitrile and triethylamine were purchased from EMD Millipore Company Inc. (Billerica, MA, USA) and Tedia Company Inc. (Fairfield,

OH, USA), respectively. Other reagents were of analytical grade and were supplied by Sinopharm Chemical Reagent Co. Ltd. (Shanghai, China).

3.2. Standard and Working Solutions

Stock solutions of TAP and FF at concentrations of 400.0 mg/L and of FFA at a concentration of 100.0 mg/L were prepared by dissolving TAP, FF, and FFA (initially dissolved in 1 mL of ultrapure water), respectively, in acetonitrile. Working standard solutions of TAP, FF, and FFA at different concentrations were prepared by diluting the stock solutions with acetonitrile-water (36:64, *v/v*). The stock solutions were stable for five months at −70 °C. Fresh working solutions were prepared by dilution of the stock solution before use.

3.3. UPLC-FLD Instrumentation and Conditions

A Waters ACQUITY UPLC System and a Waters fluorescence detector (Waters Corp., Milford, MA, USA) were used. The separation was achieved on an ACQUITY UPLC® BEH C_{18} (2.1 mm × 100 mm, 1.7 μm) chromatographic column. The column temperature was maintained at 30 °C. The injection volume was 10 μL. The analysis was carried out using acetonitrile and 0.005 mol/L NaH_2PO_4 solution containing 0.003 mol/L sodium dodecyl sulfate and 0.05% triethylamine with pH (FE20, METTLER TOLEDO, Shanghai, China) adjusted to 5.3 ± 0.1 with 85% phosphoric acid as the mobile phase (36:64, *v/v*) at a flow rate of 0.2 mL/min, and the excitation and emission wavelengths were 233 nm and 284 nm, respectively.

3.4. Sample Preparation

3.4.1. ASE Extraction

Homogeneous blank poultry egg samples (2.0 ± 0.02 g) were accurately weighed in a mortar and then ground with the appropriate amount of diatomite. The samples were then transferred into a 22 mL extraction tank, placed on the mechanical arm of the accelerated solvent extractor for extraction, and the extractant was acetonitrile:ammonia (98:2, *v/v*). The extraction method and parameters were as follows: extraction pressure of 1500 psi, extraction temperature of 80 °C, extraction time of 3 min, solvent flush of approximately 40% of the tank volume, and two cycles of static extraction. Finally, the extract was collected and left to stand.

3.4.2. Ultrasonic Extraction

Homogeneous blank poultry egg samples (2.0 ± 0.02 g) were accurately weighed into 50 mL polypropylene centrifuge tubes, and 1 mL of 30% acetonitrile solution was added. The solution was vortexed, and then 10 mL of acetonitrile:ammonia (98:2, *v/v*) was added. After ultrasonic extraction for 15 min via an ultrasonic cleaning machine (P300H Elma, Konstanz, Germany), the solution was centrifuged at 8000× *g* for 10 min in a desktop high-speed refrigerated centrifuge (5810R, Eppendorf, Hamburg, Germany), and the supernatant was transferred to a 50 mL propylene centrifuge tube. The extraction process was repeated a second time, and the two extracts were combined for further analysis.

3.4.3. Vortex Shock Extraction

Homogeneous blank poultry egg samples (2.0 ± 0.02 g) were accurately weighed into 50 mL polypropylene centrifuge tubes, and 1 mL of 30% acetonitrile solution was added. The solution was vortexed on a vortex oscillator (G560E, Scientific Industries Ltd., Bohemia, New York, NY, USA), 10 mL of acetonitrile:ammonia (98:2, *v/v*) was added, and the sample was vortexed for 2 min. The solution was centrifuged at 8000× *g* for 10 min and then transferred to 50 mL polypropylene centrifuge tubes. The extraction process was repeated a second time, and the two extracts were combined for further analysis.

3.4.4. Vortex Oscillating + Ultrasonic Extraction

Homogeneous blank poultry egg samples (2.0 ± 0.02 g) were accurately weighed into 50 mL polypropylene centrifuge tubes, and 1 mL of 30% acetonitrile solution was added. The samples were vortex mixed, 10 mL of acetonitrile:ammonia (98:2, *v/v*) was added, and they were vortex mixed for an additional 2 min. Then, they were subjected to ultrasonication for 15 min. After centrifugation at 8000× *g* for 10 min, the supernatant was transferred to a 50 mL polypropylene centrifuge tube. The extraction process was repeated a second time, and the two extracts were combined for further analysis.

3.4.5. Sample Purification

The collected extract was concentrated to near dryness in a centrifugal concentrator, and the residue was dissolved in 1 mL of acetonitrile and then defatted with acetonitrile-saturated n-hexane (hen, duck, and goose whole egg and yolk with 10 × 10 mL; egg white with 5 mL; and pigeon eggs and quail eggs with 8 × 8 mL). After vortexing for 1 min, the upper layer was left to rest for 5 min. The centrifuge tube was placed into the centrifugal concentrator to concentrate the lower liquid layer to dryness. The dried sample was reconstituted with 2.0 mL of the mobile phase, vortexed for 1 min, and centrifuged at 12,000× *g* for 15 min. The supernatant was passed through a 0.22 μm PVDF needle, and the filtrate was analyzed by UPLC-FLD.

3.5. Method Validation

This optimized method was validated according to the requirements defined by the EU and the FDA [41,42]. The selectivity of the method was estimated by preparing and analyzing 20 blank and spiked samples. The probable interferences from endogenous substances were assessed based on the chromatograms of blank and spiked poultry egg samples. The sensitivity of the method was assessed in terms of the LODs and LOQs. The LODs were defined by the concentration of each of the three analytes in the sample matrix that resulted in a signal-to-noise (S/N) ratio of 3:1. The LOQs were defined as the lowest concentration on the calibration curve of each of the three analytes giving a S/N ratio of 10:1.

The calibration curves were prepared based on the peak areas and the concentrations of the working solutions. A series of working standard solutions at concentrations of LOQ, 20.0, 50.0, 100.0, 150.0, 200.0, and 250.0 μg/kg for TAP; LOQ, 20.0, 50.0, 100.0, 200.0, 300.0, and 400.0 μg/kg for FF; and LOQ, 10.0, 20.0, 50.0, 100.0, 200.0, and 400.0 μg/kg for FFA were prepared by diluting the stock solutions with extract of blank sample matrix, and then these solutions were analyzed by the UPLC-FLD method.

The accuracy and precision of the method were evaluated by determining the recoveries of TAP, FF and FFA in poultry egg samples at concentrations of LOQ, 0.5 MRL, 1.0 MRL and 2.0 MRL. The recovery of the method was calculated by comparing the determined concentrations of samples to their theoretical concentrations.

4. Conclusions

In this study, an ASE method for the extraction of TAP, FF and FFA from poultry eggs was established. This extraction method has high extraction efficiencies and high recoveries (>80.0%), and the entire process is automated. This UPLC-FLD method for the quantitative determination of TAP, FF and FFA is accurate and sensitive. The parameters of this method were shown to meet the requirements of the Chinese Ministry of Agriculture, the EU and the FDA for the detection of veterinary drug residues.

Author Contributions: Conceptualization, K.X.; Data curation, B.W. and X.Z.; Formal analysis, G.Z., T.Z. and G.D. Funding acquisition, B.W., X.X. and K.X. Investigation, X.Z. Methodology, X.X. and K.X. Resources, Z.D. Software, Z.D. and G.D. Validation, Z.D., G.Z. and T.Z. Writing–original draft, B.W. and X.Z. Writing–review & editing, B.W. and X.X.

Funding: This research was financially supported by the China Agriculture Research System (CARS-41-G23), the Priority Academic Programme Development of Jiangsu Higher Education Institutions (PAPD), the National Natural and Science Foundation of China (31800161), Natural Sciences Foundation of Jiangsu Province (BK20180297), the Yangzhou University High-end Talent Support Programme and the Yangzhou University International Academic Exchange Foundation.

Conflicts of Interest: The authors declare no conflicts of interest.

References

1. Lesnierowski, G.; Stangierski, J. What's new in chicken egg research and technology for human health promotion? - a review. *Trends Food Sci. Technol.* **2018**, *71*, 46–51. [CrossRef]

2. Ashraf, A.; Rahman, F.A.; Abdullah, N. Poultry feed in Malaysia: an insight into the Halalan Toyyiban issues. In Proceedings of the 3rd International Halal Conference (INHAC 2016), Selangor, Malaysia, 21–22 November 2016; Hashim, N.M., Shariff, N.N.M., Mahamood, S.F., Harun, H.M.F., Shahruddin, M.S., Bhari, A., Eds.; Springer: Singapore, 2018; pp. 511–531.

3. Tao, X.; He, Z.; Cao, X.; Jiang, H.; Li, H. Approaches for the determination of florfenicol and thiamphenicol in pork using a chemiluminescent ELISA. *Anal. Methods* **2015**, *7*, 8386–8392. [CrossRef]

4. Festing, M.F.; Diamanti, P.; Turton, J.A. Strain differences in haematological response to chloramphenicol succinate in mice: implications for toxicological research. *Food Chem. Toxicol.* **2001**, *39*, 375–383. [CrossRef]

5. Hanekamp, J.C.; Bast, A. Antibiotics exposure and health risks: chloramphenicol. *Environ. Toxicol. Pharmacol.* **2015**, *39*, 213–220. [CrossRef]

6. Hu, D.; Han, Z.; Li, C.; Lv, L.; Cheng, Z.; Liu, S. Florfenicol induces more severe hemotoxicity and immunotoxicity than equal doses of chloramphenicol and thiamphenicol in Kunming mice. *Immunopharmacol. Immunotoxicol.* **2016**, *38*, 472–485. [CrossRef]

7. Sadeghi, A.S.; Mohsenzadeh, M.; Abnous, K.; Taghdisi, S.M.; Ramezani, M. Development and characterization of DNA aptamers against florfenicol: fabrication of a sensitive fluorescent aptasensor for specific detection of florfenicol in milk. *Talanta* **2018**, *182*, 193–201. [CrossRef]

8. Sams, R.A. Florfenicol: chemistry and metabolism of a novel-broad-spectrum antibiotic. *Tieraerztliche Umschau* **1995**, *50*, 703–707.

9. Cornejo, J.; Pokrant, E.; Riquelme, R.; Briceno, C.; Maddaleno, A.; Araya-Jordan, C.; Martin, B.S. Single-laboratory validation of an LC-MS/MS method for determining florfenicol (FF) and florfenicol amine (FFA) residues in chicken feathers and application to a residue-depletion study. *Food Addit. Contam. Part A* **2017**, *34*, 469–476. [CrossRef]

10. Yang, F.; Yang, F.; Kong, T.; Wang, G.; Bai, D.; Liu, B. Pharmacokinetics of florfenicol and its metabolite florfenicol amine in crucian carp (*Carassius auratus*) at three temperatures after one single intramuscular injection. *J. Vet. Pharmacol. Ther.* **2018**, *41*, 739–745. [CrossRef]

11. The European Medicines Agency. *Commission Regulation (EU) No. 37/2010 of 22 December 2009 on Pharmacologically Active Substances and Their Classification Regarding Maximum Residue Limits in Foodstuffs of Animal Origin*; The European Commission: Brussels, Belgium, 2010.

12. U.S. Food and Drug Administration. *CFR-Code of Federal Regulations Title 21 Part 556 Tolerances for Residue of New Animal Drugs in Food*; U.S. Food and Drug Administration: Rockville, MD, USA, 2014.

13. Ministry of Agriculture of the People 's Republic of China. *Maxium Residue Level of Veterinary Drugs in Food of Animal Origin. Notice No. 235*; Ministry of Agriculture: Beijing, China, 2002.

14. Xie, K.; Jia, L.; Yao, Y.; Xu, D.; Chen, S.; Xie, X.; Pei, Y.; Bao, W.; Dai, G.; Wang, J.; et al. Simultaneous determination of thiamphenicol, florfenicol and florfenicol amine in eggs by reversed-phase high-performance liquid chromatography with fluorescence detection. *J. Chromatogr. B* **2011**, *879*, 2351–2354. [CrossRef]

15. Huang, S.; Gan, N.; Liu, H.; Zhou, Y.; Chen, Y.; Cao, Y. Simultaneous and specific enrichment of several amphenicol antibiotics residues in food based on novel aptamer functionalized magnetic adsorbents using HPLC-DAD. *J. Chromatogr. B Analyt. Technol. Biomed. Life Sci.* **2017**, *1060*, 247–254. [CrossRef]

16. Yang, J.; Sun, G.; Qian, M.; Huang, L.; Ke, X.; Yang, B. Development of a high-performance liquid chromatography method for the determination of florfenicol in animal feedstuffs. *J. Chromatogr. B Analyt. Technol. Biomed. Life Sci.* **2017**, *1068–1069*, 9–14. [CrossRef]

17. Yang, Q.H.; Xiao-Hui, A.I.; Rong, L.I. Simultaneous determination of chloramphenicol, thiamphnicol, florfenicol and forfenicol-amine in aquatic products by gas chromatographic method with solid phase extraction. *Chin. J. Anal. Lab.* **2015**, *34*, 533–537.

18. Peng, S.W.; Xin, L.J.; Mei, S.S. Determination of 3 chloramphenicols medicine residuesin animal-origin Chinese medicinal materials pheretima and hirudoin the retailer by gas chromatography. *Chin. J. Vet. Drug* **2017**, *51*, 65–69.

19. Chen, D.; Yu, J.; Tao, Y.; Pan, Y.; Xie, S.; Huang, L.; Peng, D.; Wang, X.; Wang, Y.; Liu, Z.; et al. Qualitative screening of veterinary anti-microbial agents in tissues, milk, and eggs of food-producing animals using liquid chromatography coupled with tandem mass spectrometry. *J. Chromatogr. B. Analyt. Technol. Biomed. Life Sci.* **2016**, *1017–1018*, 82–88. [CrossRef]

20. Imran, M.; Habib, F.E.; Majeed, S.; Tawab, A.; Rauf, W.; Rahman, M.; Umer, M.; Iqbal, M. LC-MS/MS-based determination of chloramphenicol, thiamphenicol, florfenicol and florfenicol amine in poultry meat from the Punjab-Pakistan. *Food Addit. Contam. Part A Chem. Anal. Control. Expo. Risk Assess.* **2018**, *35*, 1530–1542. [CrossRef]

21. Marques, T.V.; Paschoal, J.A.R.; Barone, R.S.C.; Cyrino, J.E.P.; Rath, S. Depletion study and estimation of withdrawal periods for florfenicol and florfenicol amine in pacu (*Piaractus mesopotamicus*). *Aquac. Res.* **2018**, *49*, 111–119. [CrossRef]

22. Xie, X.; Wang, B.; Pang, M.; Zhao, X.; Xie, K.; Zhang, Y.; Wang, Y.; Guo, Y.; Liu, C.; Bu, X.; et al. Quantitative analysis of chloramphenicol, thiamphenicol, florfenicol and florfenicol amine in eggs via liquid chromatography-electrospray ionization tandem mass spectrometry. *Food Chem.* **2018**, *269*, 542–548. [CrossRef]

23. Shen, J.; Xia, X.; Jiang, H.; Li, C.; Li, J.; Li, X.; Ding, S. Determination of chloramphenicol, thiamphenicol, florfenicol, and florfenicol amine in poultry and porcine muscle and liver by gas chromatography-negative chemical ionization mass spectrometry. *J. Chromatogr. B. Analyt. Technol. Biomed. Life Sci.* **2009**, *877*, 1523–1529. [CrossRef]

24. Azzouz, A.; Ballesteros, E. Multiresidue method for the determination of pharmacologically active substances in egg and honey using a continuous solid-phase extraction system and gas chromatography-mass spectrometry. *Food Chem.* **2015**, *178*, 63–69. [CrossRef]

25. Amorello, D.; Barreca, S.; Gambacurta, S.; Gulotta, M.G.; Orecchio, S.; Pace, A. An analytical method for monitoring micro-traces of landfill leachate in groundwater using fluorescence excitation–emission matrix spectroscopy. *Anal. Methods* **2016**, *8*, 3475–3480. [CrossRef]

26. Fedeniuk, R.W.; Mizuno, M.; Neiser, C.; O'Byrne, C. Development of LC-MS/MS methodology for the detection/determination and confirmation of chloramphenicol, chloramphenicol 3-O-beta-d-glucuronide, florfenicol, florfenicol amine and thiamphenicol residues in bovine, equine and porcine liver. *J. Chromatogr. B. Analyt. Technol. Biomed. Life Sci.* **2015**, *991*, 68–78. [CrossRef]

27. Shin, D.; Kang, H.-S.; Jeong, J.; Kim, J.; Choe, W.J.; Lee, K.S.; Rhee, G.-S. Multi-residue determination of veterinary drugs in fishery products using liquid chromatography-tandem mass spectrometry. *Food Anal. Methods* **2018**, *11*, 1815–1831. [CrossRef]

28. Moretti, S.; Dusi, G.; Giusepponi, D.; Pellicciotti, S.; Rossi, R.; Saluti, G.; Cruciani, G.; Galarini, R. Screening and confirmatory method for multiclass determination of 62 antibiotics in meat. *J. Chromatogr. A* **2016**, *1429*, 175–188. [CrossRef]

29. Wang, K.; Lin, K.; Huang, X.; Chen, M. A simple and fast extraction method for the determination of multiclass antibiotics in eggs using LC-MS/MS. *J. Agric. Food Chem.* **2017**, *65*, 5064–5073. [CrossRef]

30. Orlando, E.A.; Costa Roque, A.G.; Losekann, M.E.; Simionato, A.V.C. UPLC-MS/MS determination of florfenicol and florfenicol amine antimicrobial residues in tilapia muscle. *J. Chromatogr. B Analyt. Technol. Biomed. Life Sci.* **2016**, *1035*, 8–15. [CrossRef]

31. Barreto, F.; Ribeiro, C.; Barcellos Hoff, R.; Costa, T.D. Determination of chloramphenicol, thiamphenicol, florfenicol and florfenicol amine in poultry, swine, bovine and fish by liquid chromatography-tandem mass spectrometry. *J. Chromatogr. A.* **2016**, *1449*, 48–53. [CrossRef]

32. Feng, J.B.; Huang, D.R.; Zhong, M.; Liu, P.; Dong, J.D. Pharmacokinetics of florfenicol and behaviour of its metabolite florfenicol amine in orange-spotted grouper (*Epinephelus coioides*) after oral administration. *J. Fish. Dis.* **2016**, *39*, 833–843. [CrossRef]

33. Yang, F.; Fang, Y.; Liu, Z.C. Determination of thiamphenicol, florfenicol and florfenicol amine residues in aquatic products by HPLC with fluorescence detection. *Chin. J. Vet. Drug* **2008**, *42*, 14–16.

34. Karami-Osboo, R.; Miri, R.; Javidnia, K.; Kobarfard, F. Simultaneous chloramphenicol and florfenicol determination by a validated DLLME-HPLC-UV method in pasteurized milk. *Iran. J. Pharm. Res.* **2016**, *15*, 361–368.

35. Zhai, Z.Y.; Xu, Y. Determination of thiamphenicol, florfenicol and florfenicol amine in lyophilized royal jelly powder by HPLC. *J. Prev. Med. Inf.* **2010**, *26*, 82–84.

36. Xie, K.Z.; Xu, D.; Chen, S.Q.; Xie, X.; Jia, L.F.; Huang, Y.P.; Guo, H.S.; Wang, J.Y.; Liu, Z.P. Simultaneous determination of residues of thiamphenicol forfenicol and forfenicol amine in chicken muscle by HPLC with fluorescence detection. *Chin. J. Anal. Lab.* **2011**, *30*, 31–35.

37. Alechaga, E.; Moyano, E.; Galceran, M.T. Ultra-high performance liquid chromatography-tandem mass spectrometry for the analysis of phenicol drugs and florfenicol-amine in foods. *Analyst* **2012**, *137*, 2486–2494. [CrossRef]

38. Lehotay, S.J.; Lightfield, A.R. Simultaneous analysis of aminoglycosides with many other classes of drug residues in bovine tissues by ultrahigh-performance liquid chromatography-tandem mass spectrometry using an ion-pairing reagent added to final extracts. *Anal. Bioanal. Chem.* **2018**, *410*, 1095–1109. [CrossRef]

39. Yang, H.S.; Meng, Y.Z.; Meiqin, W.G. Simultaneous UHPLC-MS/MS determination of chloramphenicol and florfenicol in aquatic products with accelerated solvent extraction. *Phy. Chem. Test. Chem.* **2012**, *48*, 1353–1356.

40. Xiao, Z.; Song, R.; Rao, Z.; Wei, S.; Jia, Z.; Suo, D.; Fan, X. Development of a subcritical water extraction approach for trace analysis of chloramphenicol, thiamphenicol, florfenicol, and florfenicol amine in poultry tissues. *J. Chromatogr. A* **2015**, *1418*, 29–35. [CrossRef] [PubMed]

41. US Department of Health and Human Services, Food and Drug Administration, Center for Drug Evaluation and Research, Center for Veterinary Medicine. *Guidance for Industry: Bioanalytical Method Validation*; US Department of Health and Human Services: Washington, DC, USA, 2001.

42. The European Communities. Commission decision 2002/657/EC of 12 august 2002 implementing council directive 96/23/EC concerning the performance of analytical methods and the interpretation of results. *Off. J. Eur. Commun.* **2002**, *221*, 8–36.

Sample Availability: Samples of the compounds are not available from the authors.

molecules

MDPI

Article

¹H-NMR Profiling and Carbon Isotope Discrimination as Tools for the Comparative Assessment of Walnut (*Juglans regia* L.) Cultivars with Various Geographical and Genetic Origins—A Preliminary Study

Raluca Popescu [1], Roxana Elena Ionete [1,*], Oana Romina Botoran [1], Diana Costinel [1],
Felicia Bucura [1], Elisabeta Irina Geana [1], Yazan Falah Jadee 'Alabedallat [2] and Mihai Botu [2,3]

[1] National Research and Development Institute for Cryogenics and Isotopic Technologies-ICSI Rm. Valcea,
    ICSI Analytics Group, 4 Uzinei Street, RO-240050 Râmnicu Vâlcea, Romania; raluca.popescu@icsi.ro (R.P.);
    oana.dinca@icsi.ro (O.R.B.); diana.costinel@icsi.ro (D.C.); felicia.bucura@icsi.ro (F.B.);
    irina.geana@icsi.ro (E.I.G.)
[2] University of Craiova, Faculty of Horticulture, Department of Horticulture and Food Science, 13 Al. I. Cuza
    Street, RO-200585 Craiova, Romania; yazan.fja@gmail.com (Y.F.J.'A.); btmihai2@yahoo.com (M.B.)
[3] University of Craiova, Fruit Growing Research Station (SCDP) Vâlcea, 464 Calea lui Traian Street,
    RO-240273 Râmnicu Vâlcea, Romania
*   Correspondence: roxana.ionete@icsi.ro; Tel.: +4-0250-732744; Fax: +4-0250-732746

Academic Editors: Alessandra Gentili and Chiara Fanali
Received: 12 March 2019; Accepted: 1 April 2019; Published: 8 April 2019

check for
updates

Abstract: The aim of the study was to investigate the differences between walnut genotypes of various geographical and genetic origins grown under the same or different environmental conditions. The biological material analyzed consisted in walnut kernels of 34 cultivars, nine advanced selections, and six hybrids harvested in 2015 and 2016, summing up to a total of 64 samples. The walnuts, walnut oil, and residue were characterized in respect to their chemical (proximate composition—fat, protein, nutritional value, fatty acids profile by ¹H-NMR) and carbon-13 isotopic composition. The data was used to statistically discriminate the cultivars according to composition, geographical area of origin, and year of harvest, comparing the Romanian cultivars, selections, and hybrids with the internationally available ones.

Keywords: δ^{13}C-IRMS; fatty acids composition; ¹H-NMR; walnut varieties

1. Introduction

Rich in omega-3 fats, proteins, and with a higher number of antioxidants than most other foods, walnuts are recognized as an important component of a healthy diet. The walnut kernels' composition contains between 52 and 77% fats, $12 \div 25\%$ protein, $5 \div 24\%$ carbohydrates, quinones, tannins, minerals, and vitamins (A, B1, B2, P, and C), that may significantly differ with the genotype, growth location, and environmental conditions [1]. Compared to other types of nuts, they are rich in terms of polyunsaturated fatty acids (PUFA), such as omega-6 and omega-3 [2], the major fatty acids found in walnut being linoleic (C18:2), oleic (C18:1), α-linolenic (C18:3), palmitic (C16:0), and stearic acids (C18:0) [3,4].

Consumers are becoming increasingly aware of food quality and authenticity of the nutritional composition and health-promoting components. In this regard, walnuts have generated considerable interest since it was confirmed by various survey studies [5–7] that their consumption can lower the cholesterol, improve the arterial function (by decreasing total and LDL-cholesterol, and increasing

HDL-cholesterol), reduce inflammation, and decrease the likelihood of diabetes and neurological diseases. The beneficial components for health found in walnuts are the polyunsaturated fatty acids, phytosterols, proteins, biogenic amines (melatonin and serotonin), dietary fiber, folate, tannins, polyphenols, and minerals (magnesium, potassium, calcium, etc.) [1,2].

Covering a wide and diverse expanse of land in Central Asia [8], the walnut has spread in China, India, Europe, and America, and is nowadays cultivated in almost all countries with temperate climates. The walnut belongs to the Juglandaceae family that contains six genera for the temperate region, with the most important two being *Juglans* and *Pterocarya*. The *Juglans* genus contains 12 species, from which the high feasibility in obtaining new cultivars and parent stock are the *Juglans regia* L. and *Juglans nigra* [9]. Persian walnut (*Juglans regia* L.) is mostly cultivated in Asia (with China as the top producer), the U.S., and Europe. Of the European countries, Romania is in the top three for the yield/hectare and in the top 10 for the production and export [10], cultivating different varieties chosen according to their high yield, disease resistance, or product quality.

Evaluating and certifying the identity/origin and quality of walnuts, and their byproduct, the oil, is a challenge for the food industry, regulatory bodies, and consumers. There are various developed methods to differentiate vegetable oils, including nuts oil, and the most recent are the nuclear magnetic resonance-based metabolomics [11,12] and stable isotopes investigation [13–15], along with chemometrics [16]. Generally, the studies on walnuts investigated the major and minor compounds, such as fatty acids, sterols, polyphenols, volatiles, and minerals [17,18], or assessed the oxidative stability, antioxidant, and antimicrobial activity [1,19,20] to determine the variations given by the location, environmental conditions, cultivars, and technological processing [21–23]. Although the main purpose of isotopic studies is to identify the botanical and geographical origins of different food products, only a few were applied to walnuts and walnut oil [24–26], but lacked reference to the variation given by the cultivar/genotype, geographical origin, or year of production (environmental conditions—temperature, rainfall, and light).

As a consequence, the aim of the present work was to investigate the differences between walnut genotypes of various geographical and genetic origins grown under the same or different environmental conditions. The analyzed biological material consisted in walnut kernels of 34 cultivars, nine advanced selections, and six hybrids from the breeding program, harvested in 2015 and 2016. The walnuts, walnut oil, and residue were characterized in terms of chemical proximate composition (fats and protein content, nutritional value, fatty acids profile by ^{1}H-NMR) and carbon-13 isotopic composition. The data were used to statistically discriminate the cultivars according to the composition, geographical area of origin and year of harvest, comparing the Romanian cultivars, selections, and hybrids, with the internationally available ones.

2. Results and Discussion

2.1. Geographical and Year of Harvest Discrimination

Chemical composition of the kernel, oil, and residue for the walnut samples is highlighted in Table 1. The samples are grouped according to the year of harvest and geographical origin of the walnut cultivar, with an emphasis on the Romanian cultivars and the selections and hybrids developed at the SCDP Râmnicu Vâlcea (Romania).

The walnut kernels contained mainly fats (69.0%), with a mean concentration of crude protein of 20.3% and a 10.7% content of carbohydrates, resulting in a 700 kcal/100 g kernel energy. The extracted oil, with a mean of 874 kcal energy, was composed mainly of linolenic acid (as sum of ω-6 fatty acids)—55.6% molar, followed by 22.7% monounsaturated fatty acids (mainly oleic), 11.1% linoleic acid (as sum of ω-3 fatty acids), and 10.6% saturated fatty acids, giving a mean unsaturation degree (IV–iodine value) of 139.

Table 1. Composition of the walnut kernel, oil, and residue.

Geographical Origin/Composition	All Samples (n = 64)	Germplasm Collection of Fruit Growing Research Station (SCDP) Vâlcea, Grown to RO-VL1							RO-VL2 Local Selections (n = 3)	RO-DB Red Kernel (n = 1)	GR (n = 2)		PRC (n = 2)	
		USA Cultivars (n = 18)	FR Cultivars (n = 7)	RO Cultivars (n = 20)	RO Selections (n = 5)	RO Hybrids (n = 6)	Samples Collected in Both 2015 and 2016 (n = 30)				GR-1	GR-2	PRC-1	PRC-2
							2015	2016						
kernel														
Total fat-k (%)	69.0 ± 7.6	66.6 ± 7.1	73.6 ± 6.1	71.0 ± 9.3	66.3 ± 2.3	66.5 ± 3.9	66.4 ± 4.0	73.4 ± 10.7	66.7 ± 4.0	79.6	72.0	60.6	79.4	61.1
Protein-k (%)	20.3 ± 6.4	19.8 ± 7.9	19.7 ± 5.9	20.1 ± 6.3	22.8 ± 4.9	26.0 ± 2.2	26.2 ± 4.6	13.3 ± 2.2	18.1 ± 4.5	16.7	12.9	17.4	14.6	15.6
Carbohydrates-k (%)	10.7 ± 8.0	13.5 ± 9.8	6.7 ± 1.8	8.9 ± 8.3	10.8 ± 5.9	7.5 ± 5.5	7.3 ± 5.7	10.2 ± 0.9	15.2 ± 0.9	3.7	15.1	22.1	6.0	23.2
Energy-k (kcal/100 g)	700 ± 22	703 ± 24	699 ± 19	699 ± 22	693 ± 9	702 ± 33	696 ± 21	709 ± 21	693 ± 18	713	730	698	704	711
oil														
SFA-o (%molar)	10.6 ± 0.9	10.2 ± 0.8	10.9 ± 0.7	10.7 ± 1.0	10.5 ± 0.7	10.8 ± 0.8	10.7 ± 0.7	10.2 ± 0.9	11.1 ± 0.7	13.4	11.0	12.3	8.9	9.3
Oleic-o (%molar)	22.7 ± 5.2	23.6 ± 5.5	21.3 ± 3.4	23.1 ± 4.2	24.7 ± 5.6	18.8 ± 3.5	19.9 ± 3.2	26.1 ± 5.2	16.1 ± 2.4	20.9	19.9	20.2	38.5	31.2
Linoleic-o (%molar)	55.6 ± 4.6	55.2 ± 5.0	56.3 ± 2.1	55.1 ± 4.0	54.8 ± 3.9	59.2 ± 5.3	58.0 ± 3.3	52.7 ± 4.4	60.5 ± 1.5	54.1	55.2	55.7	43.4	50.6
Linolenic-o (%molar)	11.1 ± 1.6	11.1 ± 1.6	11.5 ± 1.7	11.0 ± 1.4	10.1 ± 2.0	11.2 ± 2.1	11.4 ± 2.0	11.0 ± 1.0	12.4 ± 0.4	11.6	14.0	11.8	9.3	8.9
IV-o	139 ± 4	138 ± 4	140 ± 4	138 ± 4	137 ± 6	142 ± 2	141 ± 4	136 ± 4	144 ± 1.3	134	141	138	126	131
PUFA (%molar)	66.7 ± 4.8	66.2 ± 5.0	67.8 ± 3.3	66.2 ± 4.0	64.9 ± 5.2	70.4 ± 3.4	69.4 ± 3.3	63.7 ± 4.7	72.9 ± 1.7	65.7	69.2	67.5	52.6	59.5
UFA/SFA	8.49 ± 0.83	8.9 ± 0.8	8.2 ± 0.6	8.38 ± 0.78	8.58 ± 0.62	8.30 ± 0.66	8.43 ± 0.63	8.87 ± 0.96	8.07 ± 0.6	6.46	8.13	7.12	10.2	9.8
PUFA/SFA	6.32 ± 0.55	6.5 ± 0.5	6.3 ± 0.5	6.21 ± 0.62	6.20 ± 0.39	6.55 ± 0.57	6.55 ± 0.57	6.26 ± 0.51	6.61 ± 0.3	4.90	6.31	5.48	5.92	6.43
Energy-o (kcal/100 g)	874 ± 69	900 ± 56	848 ± 60	854 ± 71	892 ± 11	886 ± 11	886 ± 18	857 ± 104	887 ± 9	816	903	903	744	902
residue														
Protein-r (%)	41.3 ± 6.7	39.8 ± 6.8	41.7 ± 2.6	40.2 ± 7.6	43.7 ± 5.3	40.7 ± 6.0	42.5 ± 7.9	38.7 ± 6.0	52.3 ± 6.2	34.7	39.2	44.1	45.3	46.2
Energy-r (kcal/100 g)	303 ± 17	307 ± 23	293 ± 6	303 ± 18	300 ± 8	310 ± 16	312 ± 23	295 ± 11	304 ± 7	312	286	289	297	300

k—kernel, o—oil, r—residue.

The fatty acids composition allowed the estimation of the different nutritional fractions: PUFA—polyunsaturated fatty acids, UFA/SFA—unsaturated fatty acids/ saturated fatty acids, PUFA/SFA—polyunsaturated fatty acids/saturated fatty acids. The results as a general composition for the walnuts and walnuts oil and for specific cultivars were in agreement with other published data [1,17,20,22,23,27].

Considering the walnuts grown at the Fruit Growing Research Station (SCDP) Vâlcea, the Romanian selections and hybrids tend to present the extreme values for this set of samples. The Romanian hybrids registered high protein content (26.0%) and low fats (66.5%) for the kernel and oil, with a low composition in oleic acid (18.8% molar), and high linoleic acid (59.2% molar), that led to increased values of PUFA (70.4 % molar) and IV (142) compared to the other groups of walnuts.

The Romanian selections were also found to have a low-fat content (66.3%), with a higher oleic concentration (24.7% molar) and lower linoleic (54.8% molar) and linolenic (10.1% molar) concentrations, resulting in low PUFA. The international and Romanian cultivars had similar compositions, the values being intermediate between the Romanian hybrids and the Romanian selections, with the exception of the higher fatty content for the French cultivars (mean of 73.6%) and higher PUFA/SFA and UFA/SFA ratios for the USA cultivars.

For harvest year, discriminations were considered only among the cultivars from the Fruit Growing Research Station (SCDP) Vâlcea sampled both in 2015 and 2016, namely 15 cultivars, from which 10 were international and five Romanian ones. The classification of the samples was 100% correct, with the $\delta^{13}C$ of the residue (which is the carbon-13 fingerprint of the proteins and carbohydrates), protein content of the kernel and polyunsaturated fatty acids concentration, and the iodine value as the most important factors for discrimination. The walnuts from 2015 had a lower content of fats with a higher degree of unsaturation than the ones harvested in 2016, together with a higher content of proteins. Also, since in 2015 there were higher mean temperatures and precipitation in the maturation period (July–September) than in 2016 [28], a possible explanation could be drawn that increased temperatures and precipitation are conductive to the formation of proteins in the kernels and lower quantities of fat with higher concentrations of polyunsaturated fatty acids.

Discriminant analysis of walnuts grown in different locations is presented in Figure 1. For all samples, the first two most important factors summed 83.59% of the variance (F1 = 46.25% and F2 = 37.35%), with 100% correct classification of the samples.

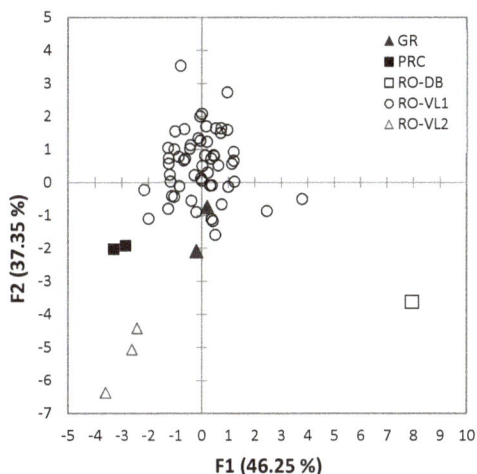

A

Figure 1. *Cont.*

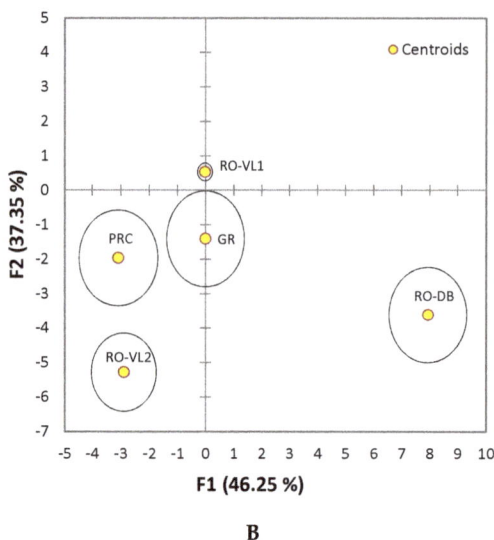

B

Figure 1. Discriminant analysis of walnuts according to the geographical origin of the walnut samples: (**A**) samples distribution by groups; (**B**) group centroids as mean discriminant scores for each group in the dependent variable for each of the discriminant functions.

The most important parameters for the geographical discrimination of walnuts were the $\delta^{13}C$ of the kernel, oil, and residue, the SFA, oleic concentrations, together with the indices IV, UFA/SFA, and PUFA/SFA. Since the number of samples grown in different locations than RO-VL1 was low, no general conclusions could be drawn. As a general observation, the two samples from China presented high oleic composition (38.5 and, respectively, 31.2% molar) compared with the other cultivars, resulting in a low IV of 129.

Other distinct compositions were observed for the walnut with red kernels (RO-DB), which had high fatty content (79.6%) with higher SFA (13.4% molar), and for the Romanian selections RO-VL2 ($n = 3$) which had low oleic and high linoleic/linolenic concentrations (mean of 60.5% and, respectively, 12.4% molar) resulting in a high IV (144).

2.2. Stable Carbon Isotopic Composition

The carbon stable isotopes are important for biological, cultivar and climatic differentiation [24,29,30]. Variation of $\delta^{13}C$ values between different plants and plant products may be attributed, besides the metabolic pathway, to geographical origin, environmental factors (temperature, humidity, rainfall, light exposure, water availability), cultivar, growth conditions and harvest time [26].

The variation in carbon-13 abundance in food is the result of isotopic fractionation that takes place in plants [25]. During the photosynthesis, the light (^{12}C) and heavier (^{13}C) isotopes of carbon will be discriminated, the processes being reflected in the isotopic compositions of the plant components. Carbon dioxide-fixing pathways in plants are the C3 (Calvin cycle), C4 (Hatch-Stack cycle), and CAM (Crassulacean acid metabolism), with the mention that all trees use the C3 metabolism. Since walnut is a C3 plant, the isotopic fingerprint of the kernel and oil should be in the range −34‰ to −22‰ [24]. Studies done on walnut and walnut oil samples produced in France and China, reported values of −28.67 ± 0.25‰ [25] and, respectively, −27.5‰ [24–26] for the $\delta^{13}C$.

In this study, the range of variation for the $\delta^{13}C$ of the kernel was of 7.6 delta units (between −29.9 and −22.3‰), while for the oil and residue was about six delta units, between −30.1 and −23.9‰ and, respectively, between −28.3 and −21.9‰. The oil was depleted in carbon-13 compared to the kernel with approx. 0.6‰, while the residue was enriched in carbon-13 compared to the kernel with approx. 1.9‰ (Table 2).

Table 2. Stable carbon composition (mean values) of the investigated walnut samples.

		δ^{13}C-Kernel (‰)	δ^{13}C-Oil (‰)	δ^{13}C-Residue (‰)
All samples ($n = 64$)		-26.8 ± 1.5	-27.4 ± 1.2	-24.9 ± 1.4
Samples collected in both years	2015 ($n = 15$)	-27.3 ± 1.4	-27.6 ± 0.8	-25.6 ± 1.3
	2016 ($n = 15$)	-26.9 ± 1.4	-27.1 ± 0.7	-24.5 ± 0.7
Samples growth to SCDP	USA cultivars ($n = 18$)	-27.4 ± 1.2	-27.5 ± 0.8	-25.3 ± 1.1
	RO selections ($n = 5$)	-27.1 ± 1.3	-28.3 ± 1.2	-25.2 ± 1.3
	RO cultivars ($n = 20$)	-27.0 ± 1.2	-27.4 ± 0.9	-25.2 ± 1.4
	RO hybrids ($n = 6$)	-26.9 ± 0.9	-28.0 ± 1.3	-24.5 ± 1.4
	FR cultivars ($n = 7$)	-26.8 ± 1.4	-27.8 ± 0.8	-25.0 ± 0.7
Samples from Greece	GR-2 2016	-26.3	-27.3	-24.1
	GR-1 2015	-26.3	-25.8	-22.7
Samples from China	PRC-2 (Zhong Lin no.1)	-25.6	-25.2	-22.1
	PRC-1 (Jin Bo Feng no.1)	-24.0	-26.0	-22.4
Samples from Romania (other than SCDP)	RO-DB red kernel selection ($n = 1$)	-25.1	-25.4	-23.6
	RO-VL ($n = 3$)	-23.3 ± 1.0	-24.8 ± 0.8	-22.7 ± 0.7
Samples classified by groups	C1 ($n = 23$)	-27.0 ± 1.7	-27.6 ± 1.3	-25.4 ± 1.6
	C2 ($n = 25$)	-26.6 ± 1.2	-27.4 ± 1.2	-24.7 ± 1.3
	C3 ($n = 6$)	-26.8 ± 1.4	-27.4 ± 1.1	-24.8 ± 1.1
	C4 ($n = 8$)	-26.3 ± 1.5	-26.8 ± 0.7	-24.1 ± 1.0
	C5 ($n = 2$)	-28.2	-26.5	-24.0

In photosynthesis, C3 plants transform the carbon dioxide (δ^{13}C = -8‰) in carbohydrates which will be further used to form other classes of compounds, including proteins and fatty acids, processes that will take place as a general rule with a 1‰ depletion in carbon-13 for proteins and 6‰ for lipids, since the ^{12}C isotope is favored for the reactions that take place [31]. The study done by Guo et al. [26] went in even more details and showed the differences between the δ^{13}C of individual fatty acids, which decreased slightly (0.7‰) in the elongation of $C_{18:0}$ from $C_{16:0}$, increased significantly (1.3‰) in the first desaturation (from $C_{18:0}$ to $C_{18:1}$) and kept stable in the further desaturation from $C_{18:1}$ to $C_{18:2}$.

The statistical analysis of the data (both isotopic and compositional) showed that the δ^{13}C of the kernel, oil, and residue were important in the geographical discrimination (with p-values < 0.0001), δ^{13}C of the residue (containing mostly the proteins and carbohydrates of the walnut) was important in differentiating the year of harvest (p-value = 0.004), while no carbon-13 fingerprint was important in discriminating the walnuts according to composition. These correlations indicate an influence of the location and year of harvest on the carbon isotopic composition of the walnuts. It can be mentioned that if the discrimination is done between cultivars from a restricted area, the isotopic composition is influenced by the biological parameters (metabolism) rather than the climatic factors, though no strong correlations were found between the isotopic fingerprint and composition of the walnuts.

The mean carbon-13 values of the different walnut groups are also presented in Table 2. The samples from 2016 were slightly more enriched in carbon-13 than the ones produced in 2015, with differences of about 0.4, 0.5 and 1.1‰ for the kernels, oil and residue. The trend is opposite than in other findings in which the isotopic fingerprint is enriched in the warmer years, though the two samples from Greece (Tripoli area) showed a more enriched isotopic composition for the oil and residue in the warmer year (2015).

Concerning the geographical origin of the cultivars, a trend of isotope enrichment was observed, from the walnuts produced in the Fruit Growing Research Station (SCDP) Vâlcea (kernel of -27‰, oil of -28‰ and residue of -25‰), to the ones produced in Greece, followed by the red kernel walnut from Romania–Dambovița county and walnuts from China, ending with the three walnut selections from Vâlcea county that registered the highest δ^{13}C values (-23.3‰ for kernel, -24.8‰ for oil, and -22.7‰ for residue). The data are also presented as a distribution in Figure 2.

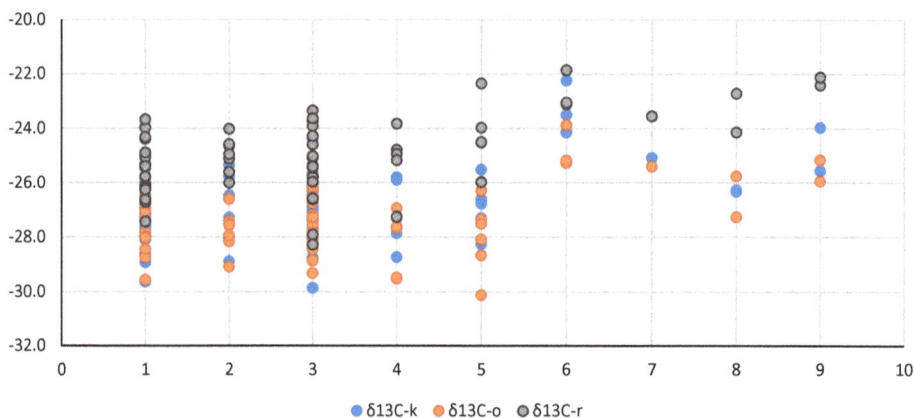

Figure 2. Carbon-13 composition of the kernel, oil and residue taking into account the geographical origin of the cultivars (1 = SCDP-USA cultivars, 2 = SCDP-FR cultivars, 3 = SCDP-RO cultivars, 4 = SCDP-RO selections, 5 = SCDP-RO hybrids, 6 = RO-VL, 7 = RO-DB, 8 = GR, 9 = PRC).

Using only the isotopic values, the walnuts can be differentiated according to the geographical origin, with the exception of the two samples from Greece which were classified in the RO–SCDP Vâlcea group, while the year of harvest discrimination gave only a 73.33% correct classification of the samples (11 out of 15). If the chemical composition is taken into account, groups C5 and C4 were the most enriched in carbon-13 (with the exception of C5 kernel), while groups C1, C2, and C3 were the most depleted.

2.3. Composition Discrimination

Nonspecific statistical analysis of the composition and isotopic data by agglomerative hierarchical clustering (AHC) classified the groups as presented in Table 3. The most important factors to differentiate these groups were the fat content, the energy of the oil, the oleic and linoleic concentrations, and the iodine value.

Groups C1 and C2 were the most similar, with the next level of similarity clustering groups C1 and C2 with C3, followed by group C5, while group C4 had the most distinct composition. Groups C1–C3 had the most unsaturated composition (high polyunsaturated acids concentrations resulting in high IV values), with group C1 being distinct due to the high protein content and group C3 due to the higher SFA content; group C5 (formed of Hartley and Vina cultivars harvested in 2016) had walnuts with low fat and protein compositions, but high in SFA and linolenic acid. Group C4, where Adams 10, Lara, one Chinese cultivar (Jin Bo Feng no.1), and other five Romanian cultivars all harvested in 2016 were found, had a walnut composition high in fats with increased oleic acid concentrations.

Group C1 is comprised mainly from samples of 2015 harvest; groups C3–C5 are formed of walnuts harvested in 2016, while group C2 is a mixture of 2015 and 2016 samples. For the cultivar Wilson Franquette, the walnuts sampled in both years of harvest were classified in the same group, indicating a metabolism that is less influenced by the environmental conditions. For the other cultivars, however, the walnuts obtained in different years were classified in different groups, more or less related to each other due to the difference in composition given by the production year.

Concerning the Romanian walnut cultivars, selections and hybrids, the majority of them were classified in the high unsaturation groups (C1–C3), with the exception of five of them produced in 2016 which had a composition high in fats and oleic acid (group C4).

Table 3. Classification of the walnuts according to composition.

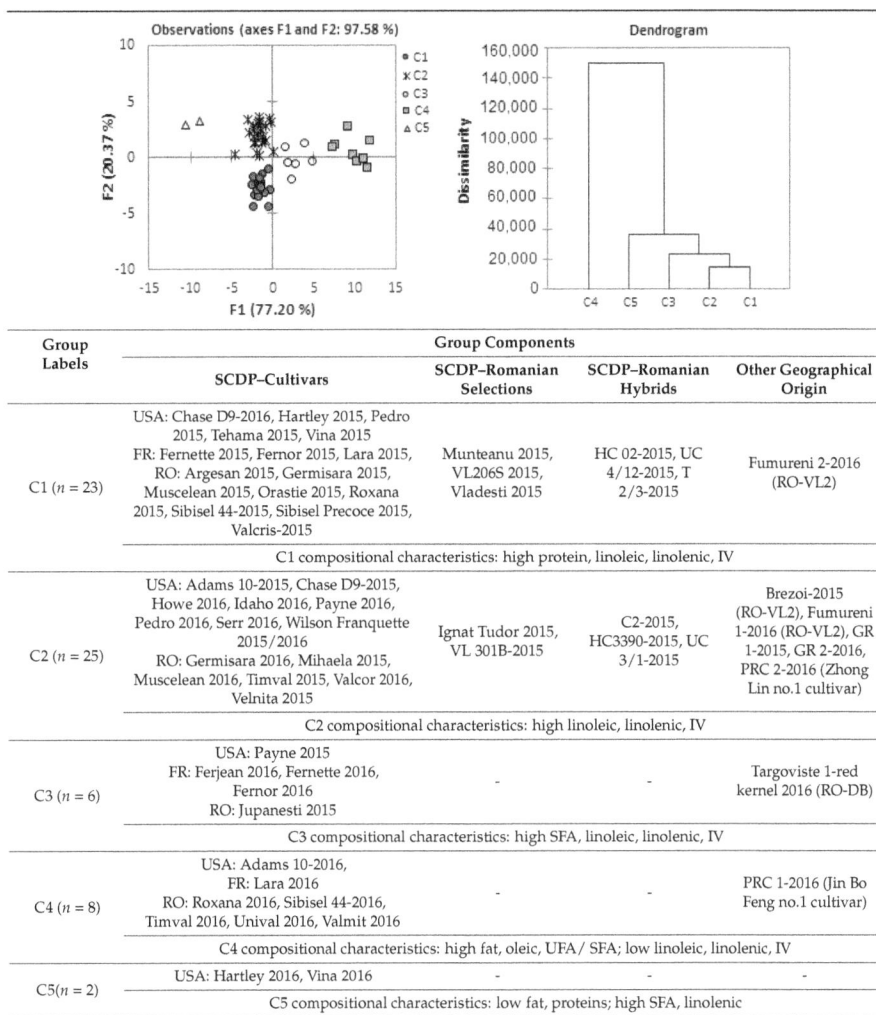

Group Labels	Group Components			
	SCDP–Cultivars	SCDP–Romanian Selections	SCDP–Romanian Hybrids	Other Geographical Origin
C1 (*n* = 23)	USA: Chase D9-2016, Hartley 2015, Pedro 2015, Tehama 2015, Vina 2015 FR: Fernette 2015, Fernor 2015, Lara 2015, RO: Argesan 2015, Germisara 2015, Muscelean 2015, Orastie 2015, Roxana 2015, Sibisel 44-2015, Sibisel Precoce 2015, Valcris-2015	Munteanu 2015, VL206S 2015, Vladesti 2015	HC 02-2015, UC 4/12-2015, T 2/3-2015	Fumureni 2-2016 (RO-VL2)
	C1 compositional characteristics: high protein, linoleic, linolenic, IV			
C2 (*n* = 25)	USA: Adams 10-2015, Chase D9-2015, Howe 2016, Idaho 2016, Payne 2016, Pedro 2016, Serr 2016, Wilson Franquette 2015/2016 RO: Germisara 2016, Mihaela 2015, Muscelean 2016, Timval 2015, Valcor 2016, Velnita 2015	Ignat Tudor 2015, VL 301B-2015	C2-2015, HC3390-2015, UC 3/1-2015	Brezoi-2015 (RO-VL2), Fumureni 1-2016 (RO-VL2), GR 1-2015, GR 2-2016, PRC 2-2016 (Zhong Lin no.1 cultivar)
	C2 compositional characteristics: high linoleic, linolenic, IV			
C3 (*n* = 6)	USA: Payne 2015 FR: Ferjean 2016, Fernette 2016, Fernor 2016 RO: Jupanesti 2015	-	-	Targoviste 1-red kernel 2016 (RO-DB)
	C3 compositional characteristics: high SFA, linoleic, linolenic, IV			
C4 (*n* = 8)	USA: Adams 10-2016, FR: Lara 2016 RO: Roxana 2016, Sibisel 44-2016, Timval 2016, Unival 2016, Valmit 2016	-	-	PRC 1-2016 (Jin Bo Feng no.1 cultivar)
	C4 compositional characteristics: high fat, oleic, UFA/ SFA; low linoleic, linolenic, IV			
C5(*n* = 2)	USA: Hartley 2016, Vina 2016	-	-	-
	C5 compositional characteristics: low fat, proteins; high SFA, linolenic			

3. Materials and Methods

3.1. Plant Material and Collection Site

The main biological material analyzed in this study consisted of walnut kernels of fruits harvested from the germplasm collection of the Fruit Growing Research Station (SCDP) belonging to the University of Craiova and located in Râmnicu Vâlcea, Romania. The aim of SCDP is to improve the walnut assortment in Southwest Romania through selection of local walnut population and controlled hybridization. The walnut cultivar collection field is located north of Râmnicu Vâlcea city, at 45°8′22.35″N and 24°22′36.85″E. The climate of the area is temperate, of Cfb Köppen-Geiger type [32], the average annual temperature is 10.2 °C and rainfall 715 mm. The walnut collection was planted in 1997 on an alluvial soil with medium content of nitrogen, phosphorus, and potassium, and with a pH of 6.8. No irrigation was provided to the trees. Fruits were harvested at full maturity, in 2015 and/or 2016, from a number of 11 walnut representative cultivars originated from USA (seven from

California, three from Oregon and one from Idaho), four cultivars from France, and 14 from Romania. Additionally, five advanced selections and six hybrids from the breeding program were sampled. Another set of samples was represented by walnuts grown in other locations than SCDP: Two cultivars from both Greece (Tripoli City area) and China (Taigu county and Taiyuan City from Shanxi Province), and the other four local selections from Romania (three from Vâlcea county and one with red kernel from Dâmbovița county). A total of 64 samples were investigated, from which 15 were taken both in 2015 and 2016 (see Table 4).

Table 4. Description of walnut samples collection.

Geographical Origin of the Cultivar	Description	Name	Genetic Origin	Harvest Year
Fruit Growing Location-Research Station (RO-VL1), Romania				
USA (Oregon)	Cultivar	Adams 10	Open pollinated seedling	2015 and 2016
	Cultivar	Chase D9	Open pollinated seedling	2015 and 2016
	Cultivar	Howe	Chance seedling	2016
USA (California)	Cultivar	Tehama	'Waterloo' × 'Payne'	2015
	Cultivar	Hartley	Open pollinated seedling	2015 and 2016
	Cultivar	Payne	Chance seedling	2015 and 2016
	Cultivar	Pedro	'Conway Mayette' × 'Payne'	2015 and 2016
	Cultivar	Vina	'Franquette' × 'Payne'	2015 and 2016
	Cultivar	Wilson Franquette	Selection of 'Franquette'	2015 and 2016
	Cultivar	Serr	'Payne' × PI 159568	2016
USA (Idaho)	Cultivar	Idaho	Selection from local populations	2016
France	Cultivar	Fernette	'Franquette' × 'Lara'	2015 and 2016
	Cultivar	Fernor	'Franquette' × 'Lara'	2015 and 2016
	Cultivar	Lara	Chance seedling of 'Payne'	2015 and 2016
	Cultivar	Ferjean	Grosvert' × 'Lara'	2016
Romania (Argeş)	Cultivar	Argesan	Selection from local populations	2015
	Cultivar	Jupâneşti	Selection from local populations	2015
	Cultivar	Mihaela	Selection from local populations	2015
	Selection	Ignat Tudor	Selection from local populations	2015
	Selection	Munteanu	Selection from local populations	2015
	Selection	Vladesti	Selection from local populations	2015
	Cultivar	Muscelean	Selection from local populations	2015 and 2016
	Cultivar	Roxana	Selection from local populations	2015 and 2016
Romania (Hunedoara)	Cultivar	Sibisel Precoce	Selection from local populations	2015
	Cultivar	Orastie	Selection from local populations	2015
	Cultivar	Germisara	Selection from local populations	2015 and 2016
	Cultivar	Sibişel 44	Selection from local populations	2015 and 2016
Romania (Vâlcea)	Cultivar	Valcris (syn. VL202 PO)	Selection from local populations	2015
	Selection	VL 206 S	Selection from local populations	2015
	Cultivar	Timval (syn. VL 54 B)	Selection from local populations	2015 and 2016
	Cultivar	Unival	Selection from local populations	2016
	Cultivar	Valcor	Selection from local populations	2016
	Cultivar	Valmit	Selection from local populations	2016
Romania (Craiova)	Hybrid	C2	Open pollinated seedling of 'Ideal'	2015
	Hybrid	HC 02	Open pollinated seedling of 'Ideal'	2015
	Hybrid	HC 3390	Open pollinated seedling of 'Ideal'	2015
	Hybrid	T2/3	Open pollinated seedling of 'Ideal'	2015
	Hybrid	UC 3/1	Open pollinated seedling of 'Ideal'	2015
	Hybrid	UC 4/12	Open pollinated seedling of 'Ideal'	2015
Romania (Bucureşti)	Selection	VL 301 B	Selection from local populations	2015
Romania (Iaşi)	Cultivar	Velnița	Selection from local populations	2015
Fruit Growing Location-Vâlcea county (RO-VL2), Romania				
Romania (Vâlcea)	Selection	Brezoi 1	Selection from local populations	2016
	Selection	Fumureni 1	Selection from local populations	2016
	Selection	Fumureni 2	Selection from local populations	2016
Fruit Growing Location-Dâmbovița county (RO-DB), Romania				
Romania (Dâmbovița)	Selection	Târgovişte 1-red kernel	Selection from local populations	2016

Table 4. *Cont.*

Geographical Origin of the Cultivar	Description	Name	Genetic Origin	Harvest Year
Fruit Growing Location-Tripoli (GR), Greece				
Greece (Tripoli)	Cultivar	Nut sample 1	'Franquette' × 'Hartley' x 'Chandler'	2015
Greece (Tripoli)	Cultivar	Nut sample 2	× 'Meylannaise'	2016
Fruit Growing Location-PRC				
China (Fruit Growing Institute in Taigu-Shanxi)	Cultivar	Jin Bo Feng no.1	Unknown	2016
China (Taiyuan market, Shanxi)	Cultivar	Zhong Lin no.1	Unknown	2016

3.2. Sample Preparation

After harvest, the in-shell walnuts were naturally dried. Nuts were cracked, and kernels were extracted from each cultivar sample. The mean sample consisted of 30 g of healthy kernels selected to be representative for the lot, grounded to a fine homogenous mix. From this, 1 g was dried by lyophilization and C/H/N composition and δ^{13}C-IRMS were performed. Another 5 g of mean grounded walnut kernels were used to extract oil for 16 h with 100 mL of petroleum ether using a Buchi B-811 Soxhlet extractor (Büchi Labortechnik AG, Flawil, Switzerland). The oil resulted after the removal of the solvent and filtration represented the fat content of the kernels and its composition was determined–fatty acids content by NMR, C/H/N and δ^{13}C-IRMS. The dried residue after oil extraction (consisting mainly of the proteins and carbohydrates of the walnuts) was also analyzed for the C/H/N composition and δ^{13}C-IRMS.

3.3. C/H/N Composition

The elemental composition (C/H/N) of the walnut, oil, and residue was determined in triplicate using a Flash2000 Thermo (Thermo Fisher Scientific, Leicestershire, UK) with a TCD detector (Thermo Fisher Scientific, UK) [33]. Analysis conditions: Chromatographic columns—(i) molecular sieve 5A for separating flue gases after combustion, gas type CO, and (ii) PoraPLOT Q (30 cm, 20 µm, 65 °C) for separating flue gases after combustion, gas type NO_2, CO_2, H_2O, SO_2, furnace temperature: 950 °C, TCD detector: 1000 mA, measuring time—12 min, sample mass 2.5 mg. Linear regression analysis, based on seven points created using certified reference material, traceable at ISO GUIDE 34/17025, was used to generate the calibration equation for each element. The condition of a correct measurement was to draw a calibration curve with an RSD < 0.5% and the correlation coefficients r^2 > 0.99. To verify the conformity of the results or if the calibration curve did not suffer changes, before every sample injection, a quality control (QC) gas mixture, rich in permanently gases and hydrocarbons, was used. The measured values were used to calculate the energy values for walnut, oil and residue, and the proximate composition, crude protein (N × 5.30) (AOAC 950.48) and carbohydrate content.

3.4. NMR Analysis

The walnut oil samples for the NMR analysis were prepared in duplicate by dissolving 70 mg of oil in 700 µL of CDCl3 (99.96% at.) with TMS (NMR grade) as internal standard. The reagents were purchased from Aldrich (St. Louis, MO, USA) and, respectively, Sigma-Aldrich (Darmstadt, Germany). All NMR experiments were recorded at 300 K using a Bruker AvanceIII400 spectrometer (Bruker France SAS, Wissembourg, France) operating at 9.4 T, equipped with a 5 mm BBO probe, observing ^{1}H at 400.2 MHz and ^{13}C at 100.6 MHz. ^{1}H-NMR spectra were acquired using spectral width of 8224 Hz; 65536 data points; pulse width of 10.5 µs; relaxation delay of 1.0 s; acquisition time of 4 s and 16 scans. To verify the spectra quality, the solvent peak was examined on the ^{1}H spectrum for each sample. Since the same quantity of deuterated chloroform was always added for the NMR analysis, the resolution of the peak at 7.3 ppm was used to attest the correct adjustment of the shims. Spectra were processed by applying an exponential line broadening of 0.3 Hz for sensitivity enhancement before Fourier transforms and were accurately phased and baseline adjusted. The determination of the fatty acids composition of the walnut oil was done according to Guillén and Ruiz [34].

3.4.1. δ^{13}C Measurement

The overall carbon-13 composition of the walnut kernel, oil, and residue was obtained using a Thermo DeltaVPlus Spectrometer (Thermo Fisher Scientific, Bremen, Germany) coupled with an 1112 Elemental Analyser equipped with an autosampler for solids. The isotopic composition is expressed as δ-values (‰), which refers the isotope ratio of the sample (S) to that of the international reference PDB:

$$\delta^{13}C = 1000 \cdot \left[\frac{\left(^{13}C/^{12}C\right)_S - \left(^{13}C/^{12}C\right)_{PDB}}{\left(^{13}C/^{12}C\right)_{PDB}} \right]$$

The δ^{13}C isotopic values were calibrated against an international reference material, sugar IAEA-CH-6 (IAEA—International Atomic Energy Agency, Vienna, Austria), and three other laboratory standards, namely, L-Alanine IA-R041 (Iso-Analytical Laboratory Standard), Beet Sugar IA-R005, and Cane Sugar IA-R006.

During measurements, a reference was injected at regular intervals as a working standard to check the accuracy of the results. Each sample was analyzed in triplicate and the results were validated if the difference was below 0.4‰.

3.5. Data Analysis

The data were evaluated by statistical methods. Descriptive statistical analysis (mean, standard deviation, etc.), agglomerative hierarchical clustering (AHC), and discriminant analysis (DA) were performed using commercial software packages as Microsoft Excel 2013 (Microsoft, Redmond, WA, USA) and XLSTAT Addinsoft 2014.5.03 version (Addinsoft Inc, Long Island, NY, USA).

4. Conclusions

The determined parameters (fat, protein, nutritional value, fatty acids profile by ^{1}H-NMR, and carbon-13 isotopic composition) for the walnuts and walnut oil were used to differentiate with certain degrees of success the cultivars grown in the same location or different locations and to show the composition differences for the same cultivar in different years of production. The study allowed the comparison of the Romanian cultivars, selections and hybrids with the internationally available ones. The data also showed that in order to enhance the discrimination of walnut and walnut oil samples a wider range of parameters is needed, for example to couple the isotopic fingerprint (of bulk, fractions, or individual compounds) with the composition data or vice versa.

Author Contributions: Conceptualization, R.P. and R.E.I.; methodology, R.P. and R.E.I.; software, R.P. and O.R.B.; validation, R.P., O.R.B. and D.C.; formal analysis, R.P., D.C., F.B. and E.I.G.; investigation, R.P.; resources, R.E.I.; data curation, Y.F.J.'A. and M.B.; writing—original draft preparation, R.P.; writing—review and editing, R.E.I.; visualization, M.B.; supervision, R.P. and R.E.I.; project administration, R.P.; funding acquisition, R.E.I.

Funding: This research was funded by the Romanian Ministry of Research and Innovation, grant number PN 18 12/2018, under the project "Research on the development of advanced analytical methods to investigate organic food of plant or animal origin".

Conflicts of Interest: The authors declare no conflict of interest.

References

1. Pereira, J.A.; Oliverira, I.; Sousa, A.; Ferreira, I.C.F.R.; Bento, A.; Estevihno, L. Bioactive properties and chemical composition of six walnut (*Juglans regia* L.) cultivars. *Food Chem. Toxicol.* **2008**, *46*, 2103–2111. [CrossRef]

2.   Cosmulescu, S.; Baciu, A.; Achim, G.; Botu, M.; Trandafir, I. Mineral Composition of fruits in different walnut (*Juglans regia* L.) cultivars. *Notulae Botanicae Horti Agrobotanici Cluj-Napoca* **2009**, *37*, 156–160.

3.   Li, L.; Tsao, R.; Yang, R.; Kramer, J.K.G.; Hernandez, M. Fatty acids profiles, tocopherol contents, and antioxidant activities of heartnut (Juglans ailanthiofolia Var. cordiformis) and Persian walnut (*Juglans regia* L.). *J. Agric. Food Chem.* **2007**, *55*, 1164–1168. [CrossRef]

4.   Tapia, M.I.; Sárnchez-Morgado, J.R.; Garcia-Para, J.; Ramirez, R.; Hernández, T.; Gonzáles Gómez, D. Comparative study of the nutritional and bioactive compounds content of four walnut (*Juglas regia* L.) cultivars. *J. Food Compost. Anal.* **2013**, *31*, 232–237. [CrossRef]

5.   Sabaté, J.; Fraser, G.E.; Burke, K.; Knutsen, S.F.; Bennett, H.; Lindsted, K.D. Effects of walnuts on serum lipid levels and blood pressure in normal men. *N. Engl. J. Med.* **1993**, *328*, 603–607. [CrossRef]

6.   Lavedrine, F.; Zmirou, D.; Ravel, A.; Balducci, F.; Alary, J. Blood cholesterol and walnut consumption: A cross-sectional survey in France. *Prev. Med.* **1999**, *28*, 333–339. [CrossRef] [PubMed]

7.   Hayes, D.; Angove, M.J.; Tucci, J.; Dennis, C. Walnuts (*Juglas regia* L.) chemical composition and research in human health. *Crit. Rev. Food Sci. Nutr.* **2016**, *56*, 1231–1241. [CrossRef]

8.   Molnar, T.J.; Zaurov, D.E.; Capik, J.M.; Eisenman, S.W.; Ford, T.; Nikolyi, L.V.; Funk, C.R. Persian Walnut (*Juglas regia* L.) in central Asia. *Annu. Rep. North. Nut Growers Assoc.* **2011**, *101*, 56–69.

9.   Mitra, S.K.; Rathore, D.S.; Bose, T.K. Walnut. In *Temperate Fruits*; Horticulture and Allied Publishers: Calcutta, India, 1991; pp. 377–414.

10.   Food and Agriculture Organisation of United Nations. Available online: http://www.fao.org/faostat (accessed on 9 April 2018).

11.   Marcone, M.F.; Wang, S.; Albabish, W.; Nie, S.; Somnarain, D.; Hill, A. Diverse food-based applications of nuclear magnetic resonance (NMR) technology. *Food Res. Int.* **2013**, *51*, 729–747. [CrossRef]

12.   Popescu, R.; Costinel, D.; Dinca, O.R.; Marinescu, A.; Stefanescu, I.; Ionete, R.E. Discrimination of vegetable oils using NMR spectroscopy and chemometrics. *Food Control* **2015**, *48*, 84–90. [CrossRef]

13.   Spangenberg, J.E.; Ogrinc, N. Authentication of vegetable oils by bulk and molecular carbon isotope analyses with emphasis on olive oil and pumpkin seed oil. *J. Agric. Food Chem.* **2001**, *49*, 1534–1540. [CrossRef]

14.   Paolini, M.; Bontempo, L.; Camin, F. Compund-specific δ^{13}C and δ^{2}H analysis of olive oil fatty acids. *Talanta* **2017**, *174*, 38–43. [CrossRef]

15.   Guyader, S.; Thomas, F.; Portaluri, V.; Jamin, E.; Akoka, S.; Silvestre, V.; Remaud, G. Authentication of edible fats and oils by non-targeted ^{13}C INEPT NMR spectroscopy. *Food Control* **2018**, *91*, 216–224. [CrossRef]

16.   Esteki, M.; Vander Heyden, Y.; Farajmand, B.; Kolahderazi, Y. Qualitative and quantitative analysis of peanut adulteration in almond powder samples using multi-elemental fingerprinting combined with multivariate data analysis methods. *Food Control* **2017**, *82*, 31–41. [CrossRef]

17.   Amaral, J.S.; Casal, S.; Pereira, J.A.; Seabra, R.M.; Oliveira, B.P.P. Determination of sterol and fatty acid compositions, oxidative stability, and nutritional value of six walnut (*Juglans regia* L.) cultivars grown in Portugal. *J. Agric. Food Chem.* **2003**, *51*, 7698–7702. [CrossRef] [PubMed]

18.   Batun, P.; Bakkalbasi, E.; Kazankaya, A.; Cavioglu, I. Fatty acid profiles and mineral contents of walnuts from different provinces of Van Lake. *GIDA* **2017**, *42*, 155–162. [CrossRef]

19.   Arranz, S.; Cert, R.; Perez-Jimenez, J.; Cert, A.; Saura-Calixto, F. Comparison between free radical scavenging capacity and oxidative stability of nut oils. *Food Chem.* **2008**, *110*, 985–990. [CrossRef]

20.   Bujdoso, G.; Konya, E.; Berki, M.; Nagy-Gasztonyi, M.; Bartha-Szughi, K.; Marton, B.; Izsepi, F.; Adanyi, N. Fatty acid composition, oxidative stability, and antioxidant properties of some Hungarian and other Persian walnut cultivars. *Turk. J. Agric. For.* **2016**, *40*, 160–168. [CrossRef]

21.   Al-Bachir, M. Effect of gamma irradiation on fungal load, chemical and sensory characteristics of walnuts (*Juglans regia* L.). *J. Stored Prod. Res.* **2004**, *40*, 355–362. [CrossRef]

22.   Crews, C.; Hough, P.; Godward, J.; Brereton, P.; Lee, M.; Guiet, S.; Winkelmann, W. Study of the main constituents of some authentic walnut oils. *J. Agric. Food Chem.* **2005**, *53*, 4853–4860. [CrossRef]

23.   Christopoulos, M.; Tsantili, E. Oil composition in stored walnut cultivars–quality and nutritional value. *Eur. J. Lipid Sci. Technol.* **2014**, *116*, 1–11. [CrossRef]

24.   Spangenberg, J.E.; Macko, S.A.; Hunziker, J. Characterization of olive oil by carbon isotope analysis of individual fatty acids: implications for authentication. *J. Agric. Food Chem.* **1998**, *46*, 4179–4184. [CrossRef]

25.   Morrison, D.J.; Dodson, B.; Slater, C.; Preston, T. ^{13}C natural abundance in the British diet: implications for ^{13}C breath tests. *Rapid Commun. Mass Spectrom.* **2000**, *14*, 1321–1324. [CrossRef]

26. Guo, L.X.; Xu, X.M.; Yuan, J.P.; Wu, C.F.; Wang, J.H. Characterization and authentication of significant Chinese edible oilseed oils by Stable Carbon Isotope Analysis. *J. Am. Oil Chem. Soc.* **2010**, *87*, 839–848. [CrossRef]

27. Martinez, M.L.; Labuckas, D.O.; Lamarque, A.L.; Maestri, D.M. Walnut (*Juglans regia* L.): Genetic resources, chemistry, by-products. *J. Sci. Food Agric.* **2010**, *90*, 1959–1967. [CrossRef] [PubMed]

28. National Meteorological Administration-Climate Monitoring. Available online: http://www.meteoromania.ro/anm2/clima/monitorizare-climatica/ (accessed on 9 April 2018).

29. Iacumin, P.; Bernini, L.; Boschetti, T. Climatic factors influencing the isotope composition of Italian olive oils and geographic characterization. *Rapid Commun. Mass Spectrom.* **2009**, *23*, 448–454. [CrossRef]

30. Costinel, D.; Tudorache, A.; Ionete, R.E.; Vremera, R. The impact of grape varieties to wine isotopic characterization. *Anal. Lett.* **2011**, *44*, 2856–2864. [CrossRef]

31. Hoefs, J. *Stable Isotope Geochemistry*; Springer: Berlin, Germany, 1980; Volume 9.

32. Rubel, F.; Kottek, M. Observed and projected climate shifts 1901-2100 depicted by world maps of the Köppen-Geiger climate classification. *Meteorol. Z.* **2010**, *19*, 135–141. [CrossRef]

33. Constantinescu, M.; Oancea, S.; Bucura, F.; Ciucure, C.; Ionete, R.E. Evaluation of the fuel potential of sewage sludge mixtures with beech sawdust and lignite. *J. Renew. Sustain. Ener.* **2018**, *10*, 053106. [CrossRef]

34. Guillén, M.D.; Ruiz, A. Rapid simultaneous determination by proton NMR of unsaturation and composition of acyl groups in vegetable oils. *Eur. J. Lipid Sci. Technol.* **2003**, *105*, 688–696. [CrossRef]

Sample Availability: Samples of the compounds are not available from the authors.

molecules

MDPI

Article

Chemical Composition, Antioxidant and Antihyperglycemic Activities of the Wild *Lactarius deliciosus* from China

Zhou Xu [1,†]**, Liang Fu** [2,†]**, Shiling Feng** [1]**, Ming Yuan** [1]**, Yan Huang** [1]**, Jinqiu Liao** [1]**, Lijun Zhou** [1]**, Hongyu Yang** [1] **and Chunbang Ding** [1,*]

[1] College of Life Sciences, Sichuan Agricultural University, Yaan 625014, China; xzhsicau@163.com (Z.X.); fengshilin@outlook.com (S.F.); yuanming@sicau.edu.cn (M.Y.); shirley11hy@163.com (Y.H.); liaojinqiu630@sicau.edu.cn (J.L.); zhouzhou124@126.com (L.Z.); yhy4868135@163.com (H.Y.)
[2] Dazhou Institute of Agricultural Sciences, Dazhou 635000, China; fuliangrain@126.com
[*] Correspondence: dcb@sicau.edu.cn; Tel.: +86-083-562-5014
[†] These authors contributed equally to this paper.

Academic Editors: Alessandra Gentili and Chiara Fanali
Received: 13 March 2019; Accepted: 1 April 2019; Published: 6 April 2019

check for updates

Abstract: The wild mushroom *Lactarius deliciosus* from China was studied for the first time to obtain information about its chemical composition, antioxidant, and antihyperglycemic activities. Nutritional value, dietary fiber, fatty acids, metal elements, free sugars, free amino acids, organic acids, flavor 5′-nucleotides, and volatile aroma compounds were determined. Potential antioxidant and antihyperglycemic activities were also tested by investigating 1,1-diphenyl-2-picrylhydrazyl (DPPH) and 2,2′-Azino-bis(3-ethylbenzothiazoline-6-sulfonic acid) (ABTS) radicals scavenging activities, ferric ion reducing activity, as well as α-amylase and α-glucosidase inhibitory activities using ethanol and aqueous extracts. The results showed that *L. deliciosus* was a good wild mushroom with high protein, carbohydrate, and dietary fiber contents, while low in fat and calorie, extensive unsaturated fatty acids contents, with negligible health risks about harmful metal elements. Twenty kinds of free amino acids were detected with a total content 3389.45 mg per 100 g dw. Flavor 5′-nucleotides including 5′-CMP, 5′-UMP, 5′-IMP, and 5′-AMP were 929.85, 45.21, 311.75, and 14.49 mg per 100 g dw, respectively. Mannitol (7825.00 mg per 100 g dw) was the main free sugar, and quininic acid (729.84 mg per 100 g dw) was the main organic acid. Twenty-five kinds of volatile aroma compounds were identified, acids (84.23%) were the most abundant compounds based on content, while aldehydes (15 of 25) were the most abundant compounds based on variety. In addition, both ethanol and aqueous extracts from *L. deliciosus* exhibited excellent antioxidant activity. While in antihyperglycemic activity tests, only ethanol extracts showed inhibitory effects on α-amylase and α-glucosidase.

Keywords: *Lactarius deliciosus*; chemical composition; antioxidant; antihyperglycemic

1. Introduction

Edible mushrooms consist of Basidiomycota and Ascomycota members [1]. Abundant protein, essential amino acids, mineral elements, dietary fiber, flavor 5′-nucleotides, and volatile aroma components endowed mushrooms with great nutritional value and unique flavor [2,3]. Thus, mushrooms have been consumed as popular food stuff and flavoring for centuries, and consumer demand has continued to increase in recent years. Nowadays, dozens of cultivated mushrooms such as *Agaricus bisporus*, *Lentinula edodes*, *Pleurotus ostreatus*, *Flammulina velutipes*, and *Auricularia auricula* occupy the main consumer market, but there are far more mushrooms that cannot be artificially cultivated called wild mushrooms that only come from field acquisition [4,5]. Although difficult to collect, wild mushrooms are still pursued by innumerable gluttons due to their unique flavors, and *Tuber melanosporum*, *Tricholoma matsutake* and *Collybia albuminosa*, etc., are regarded as gems [6,7]. Moreover, in the east, traditional medicine theory holds the concept that "drug homologous food", hence, mushrooms also serve as indispensable raw materials for pharmaceuticals [8].

Lactarius Persoon is an ectomycorrhizal group belonging to the family of Russulaceae and widespread in temperate, subtropical and tropical forests. To date, ~500 species of *Lactarius* Per. have been discovered around the world and *Lactarius deliciosus* is one of them [9]. Owing to its excellent taste, texture, and flavor, *L. deliciosus* is popular in China, with some chemical substances and bioactivities of *L. deliciosus* being reported. Research has shown that sesquiterpenoids contribute to the unique color of *L. deliciosus*, some of which possess potent biological activities such as antimicrobial and anticancer [10–12]. Ding et al. and Hou et al. reported that *L. deliciosus* polysaccharides exhibited significant anti-tumor activity in mice in vivo and immunomodulatory effects through proliferative growth of both B cell and macrophages cell in vitro [13,14]. Meanwhile, other *L. deliciosus* extracts also have various biological activities such as enzymes inhibition, antioxidant, antimicrobial, and anti-inflammatory [15–17].

As an essential class of food raw material, the nutritional value of mushrooms is comprehensively decided by proteins, carbohydrates, fats, minerals, etc., and this is crucial consumer focus point [18]. Other factors such as free sugars, free amino acids, organic acids, flavor 5′-nucleotides, and volatile aroma components could dramatically affect taste and flavor, and should not be ignored [19,20]. In China, although *L. deliciosus* is a popular wild edible mushroom, comprehensive studies of this fungi are relatively lacking. Based on the above, we collected wild *L. deliciosus* from Sichuan Province (South-west of China), and chemical compositions including proteins, carbohydrates, fats, dietary fiber, minerals, free sugars, free amino acids, organic acids, flavor 5′-nucleotides, and volatile aroma components were analyzed. In addition, we evaluated bioactivities by assaying antioxidant and anti-hyperglycemic activities of ethanol and aqueous extracts.

2. Results and Discussion

2.1. General Nutritional Value

Generally, mushrooms are considered as a valuable health food since they have perfect proportions of protein, fat, and carbohydrate. As summarized in Table 1, fresh *L. deliciosus* was preponderantly moist, and dry matter content was relatively low (8.00%). This result was in accordance with previous studies demonstrating that dry weight content of fresh mushrooms was generally 5–15% [21,22]. In the dried fruiting body, carbohydrate was the most abundant substance with 66.61 g per 100 g dw, followed by protein (17.19 g per 100 g dw), ash (8.62 g per 100 g dw), and fat (4.82 per 100 g dw). Moreover, 31.81 g per 100 g dw total dietary fiber was detected in *L. deliciosus*, indicating that consuming this mushroom is a great way for dietary fiber intake. On the whole, *L. deliciosus* is a good food that can meet the low-calorie requirements.

Table 1. Proximate composition, energetic value, dietary fiber, and fat composition of wild *L. deliciosus*.

Component	*L. deliciosus*
Moisture (g per 100 g)	92.00 ± 0.64
Dry matter (g per 100 g)	8.00 ± 0.64
Total carbohydrate (g per 100 g dw)	66.61 ± 1.02
Crude fat (g per 100 g dw)	4.82 ± 0.15
Crude Protein (g per 100 g dw)	17.19 ± 0.06
Ash (g per 100 g dw)	8.62 ± 0.25
Energy (kcal per 100 g dw)	378.60 ± 2.74
Total dietary fiber (g per 100 g dw)	31.81 ± 1.51
Insoluble dietary fiber (g per 100 g dw)	26.51 ± 1.54
Soluble dietary fiber (g per 100 g dw)	5.30 ± 0.36
C16:0 (% of total fatty acids)	5.17 ± 0.30
C18:0 (% of total fatty acids)	16.96 ± 0.19
C18:1 (% of total fatty acids)	48.37 ± 0.62
C18:2 (% of total fatty acids)	29.49 ± 0.55

Four fatty acids were identified in the crude fat and their constituents were as follows: palmitic acid (C16:0, 5.17%), stearic acid (C18:0, 16.96%), oleic acid (C18:1, 48.37%), and linoleic acid (C18:2, 29.49%) (Table 1 and Figure S1). Unsaturated fatty acids (C18:1 and C18:2) were dominant in *L. deliciosus* fat, which was in agreement with previous research on wild edible mushrooms [23,24]. Moreover, recent research has documented that unsaturated fatty acids may lower the risk of cardiovascular disease, type two diabetes, and cancer [25,26]. Nevertheless, considering the low level of fat in *L. deliciosus*, the health effects of various fatty acids are very limited.

Compared to green plants, the metal content in mushrooms is higher due to their effective mechanism of easily accumulating metals from the ecosystem [27]. Thus, wild edible mushrooms are regarded as an excellent choice for dietary mineral requirement. Concurrently, a hidden danger is arising with excess heavy metal ingestion. As summarized in Table 2, magnesium and calcium contents of *L. deliciosus* were 1244.29 and 247.07 mg per kg dw, this result was in agreement with a previous conclusion that calcium content in mushrooms is 100–500 mg/kg dw, and magnesium is 800–1800 mg per kg dw, based on data collected from over 1000 samples of 400 mushroom species [28]. Among trace elements, iron content (197.01 mg per kg dw) was notably the highest. Of note, compared with numerous studies, copper content (1.28 mg per kg dw) in this sample was relatively low, which might be related to the physiological property of this species due to *L. deliciosus* always possessing lower copper content in other comparative studies [29,30]. Usually, daily intake is 300 g of fresh mushroom, which contains ~30 g of dry matter [31]. Compared with recommended dietary allowance (RDA) and adequate intake (AI) for females and males (aged from 19 to 30) recommended by the Institute of Medicine, consumption of *L. deliciosus* does not provide significant contribution to calcium and copper supplementation, while good contribution of magnesium, zinc, manganese, iron, and chromium [32,33]. Notably, although a large daily intake value of chromium was observed (up to 344.57% for male and 482.40% for female of AI%), this poses no risk to the human body, with reference to the tolerable upper intake level recommended by the Institute of Medicine [33]. Moreover, toxic arsenic, cadmium, and plumbum were also detected in *L. deliciosus*. EU scientific committee standards stipulated provisional tolerable daily intake values for arsenic, cadmium, and plumbum for adults (of 60 kg body weight) were 0.13, 0.06, and 0.21 mg, respectively. Therefore, the intake of heavy metals (arsenic, cadmium, and plumbum) via consuming *L. deliciosus* is risk-free for the consumers.

Table 2. Contents and daily intake estimations of metal elements of wild *L. deliciosus*.

Element	Content (mg per kg dw)	Daily Intake (mg/day) [a]	RDA or AI (mg/d) [d]		RDA or AI % [d]	
			Male	Female	Male	Female
Magnesium	1244.29 ± 42.16	37.33	400 [b]	310 [b]	9.33	12.04
Calcium	247.07 ± 4.23	7.41	1000 [c]	1000 [c]	0.74	0.74
Zinc	52.34 ± 2.68	1.57	11 [b]	8 [b]	14.28	19.63
Manganese	23.12 ± 0.75	0.69	2.3 [c]	1.8 [c]	30.16	38.53
Iron	197.01 ± 13.14	5.91	8 [b]	18 [b]	73.88	32.84
Chromium	4.02 ± 0.69	0.12	0.035 [c]	0.025 [c]	344.57	482.40
Copper	1.28 ± 0.02	0.04	0.9 [b]	0.9 [b]	4.27	4.27
Arsenic	0.75 ± 0.04	0.02	-	-	-	-
Cadmium	1.91 ± 0.05	0.06	-	-	-	-
Plumbum	0.85 ± 0.04	0.03	-	-	-	-

[a] Daily element intake values calculated using 30 g *L. deliciosus* dry matter; [b] RDA recommended by the Institute of Medicine; [c] AI recommended by the Institute of Medicine; [d] RDA or AI for both males and females aged 19–30; - Toxic elements without RDA and AI.

2.2. Non-Volatile Compounds Relating to Special Flavour

Edible mushrooms are considered a valuable food for their abundance of nutrients and desirable complex delicious taste. The taste of mushrooms is primarily ascribed to abundant soluble non-volatile taste components, such as free amino acids, flavor 5′-nucleotides, free sugars, and organic acids [34,35]. As shown in Table 3 and Figure S2, twenty-two kinds of amino acids have been measured, of which, twenty kinds were found in the free amino acids of *L. deliciosus*, and the content of total free amino acids (TAA) was 3389.45 mg per 100 g dw. Among these amino acids, glutamic acid, glutamine, histidine, and alanine were found in relatively high concentrations. Moreover, the content of nine kinds of essential amino acids (EAA) in *L. deliciosus* was 1026.29 mg per 100 g dw, accounting for 30.28% of total free amino acids. This proportion was approximate to a previous report by Sun et al., in which they found the ratios of EAA/TAA in *Tricholomopsis lividipileata*, *Boletinus pinetorus*, and *Amanita hemibapha* were 26.82, 27.94, and 34.75%, respectively [36]. Free amino acids can be classified into four groups based on their taste characteristics. Aspartic acid and glutamic acid were monosodium glutamate-like (MSG-like) components responsible for umami taste. The level of MSG-like free amino acids in *L. deliciosus* was 415.71 mg per 100 g dw, which was at low-level (< 500 mg per 100 g dw) of MSG-like components according to the standard defined by Yang et al. [37]. Threonine, serine, glycine, and alanine were sweet taste amino acids, amounting to 734.82 mg per 100 g dw. Valine, methionine, isoleucine, leucine, phenylalanine, histidine, arginine, tryptophan, and tyrosine were classified as bitter amino acids. In *L. deliciosus*, the amount of bitter amino acids was 1033.29 mg per 100 g dw. Although bitter amino acids might bring bitterness, while sweet taste amino acids and soluble sugars could mask this unpleasant taste. Asparagine, glutamine, citrulline, proline, and lysine did not have effects on taste.

The chromatograms concerning flavor 5′-nucleotides, free sugars, and organic acids are presented in Figures S3–S5, and the compositions of these components are summarized in Table 4. The amount of total flavor 5′-nucleotides in *L. deliciosus* was 1301.30 mg per 100 g dw, and the individual contents of 5′-CMP, 5′-UMP, 5′-IMP, and 5′-AMP were 929.85, 45.21, 311.75, and 14.49 mg per 100 g dw, respectively, while 5′-GMP and 5′-XMP were not detected. Based on the amount of total flavor 5′-nucleotides, Yang et al. divided flavor 5′-nucleotides into three ranges, low (<100 mg per 100 g dw), medium (100–500 mg per 100 g dw), and high (>500 mg per 100 g dw) [37]. Accordingly, the amount of flavor 5′-nucleotides in *L. deliciosus* fell within the high category. Regarding free sugars, trehalose and mannitol have been detected, and their contents were 4990.09 and 7825.00 mg per 100 g dw, respectively. Compared with other wild mushrooms, the content of total free sugars in *L. deliciosus* was in the mid range, which would contribute a moderate sweet taste perception [38–40]. In the context of organic acids, quininic acid predominated in *L. deliciosus*, followed by L-malic acid and fumaric acid.

Oxalic acid was not found in *L. deliciosus* in this study, in which was however identified in another report of *L. deliciosus* from Portugal [41].

Table 3. Free amino acids composition of wild *L. deliciosus*.

Free Amino Acid	Content (mg per 100 g dw)
Aspartic acid [b]	39.43 ± 0.87
Glutamic acid [b]	376.29 ± 4.12
Asparagine	79.41 ± 0.84
Serine [c]	136.35 ± 0.66
Glutamine	794.06 ± 54.85
Histidine [a,d]	278.57 ± 23.62
Glycine [c]	91.32 ± 0.59
Threonine [a,c]	131.76 ± 3.45
Citrulline	16.26 ± 0.59
Arginine [d]	155.70 ± 11.34
Alanine [c]	375.38 ± 2.59
Tyrosine [d]	91.53 ± 1.38
Cystine	-
Valine [a,d]	122.17 ± 4.93
Methionine [a, d]	25.41 ± 0.62
Tryptophan [a, d]	68.95 ± 0.43
Phenylalanine [a, d]	133.61 ± 0.45
Isoleucine [a, d]	47.82 ± 0.95
Leucine [a, d]	109.53 ± 1.61
Lysine [a]	208.46 ± 16.92
Hydroxyproline	-
Proline	107.43 ± 12.44
Total free amino acids (TAA)	3389.45 ± 38.13

[a] Essential amino acids; [b] Monosodium glutamate-like (MSG-like) amino acids; [c] sweet amino acids; [d] bitter amino acids; - Not detected.

Table 4. Flavor 5′-nucleotides, free sugars and organic acids composition of wild *L. deliciosus*.

Component	Content (mg per 100 g dw)
5′-CMP	929.85 ± 42.33
5′-UMP	45.21 ± 6.72
5′-IMP	311.75 ± 13.43
5′-AMP	14.49 ± 3.37
Trehalose	4990.09 ± 307.95
Mannitol	7825.00 ± 466.72
Quininic acid	729.84 ± 71.80
L-Malic acid	415.63 ± 87.44
Fumaric acid	120.71 ± 11.45

Food flavor is a comprehensive concept that contains sweet, sour, bitter, spicy, astringent, umami, etc. Umami is especially important for edible mushrooms because they are usually used as natural freshness-enhancing materials. Equivalent umami concentration (EUC) value calculated using an equation from sensory evaluation, based on the report of Yamaguchi et al., has often been used to evaluate umami-like taste characteristics of mushrooms [42]. For *L. deliciosus*, the EUC value was 145.32 g per 100 g dw. Mau (2005) grouped mushrooms EUC values into four levels: the first level, >1000 g per 100 g dw, the second level, 100–1000 g per 100 g dw; the third level, 10–100 g per 100 g dw; and the fourth level, <10 g per 100 g dw [43]. Thus, the EUC value for *L. deliciosus* belonged to the second level.

2.3. Volatile Aroma Compounds Relating to Special Flavor

Although mushroom tastes depend on water-soluble non-volatile compounds, the function of volatile aroma compounds could not be ignored due to the fact that they directly influence consumer acceptability. Typical volatile aroma compounds in mushrooms originated mostly from chemical or enzymatic oxidation of unsaturated fatty acids and further interactions with proteins, peptides, and free amino acids [44]. In this research, volatile aroma compounds of *L. deliciosus* were estimated by headspace solid phase micro-extraction GS-MS combining library catalog. As shown in Table 5 and Figure S6, 25 compounds were identified, including 15 aldehydes, six acids, two alkanes, one alcohol, and one ester. Based on qualitative and quantitative analysis, acids were confirmed to be the most important aroma volatile compounds accounting for a 84.23% total chromatographic area, followed by aldehydes with 14.77%. Acids dominated in *L. deliciosus* aroma volatile compounds, which was in accordance with a previous study on *L. edodes* and *Pleurotus sajor-caju* conducted by Çağlarırmak [45]. However, other studies have shown that alcohols are important aromatic substances in mushrooms, which differed to our findings in the present research, this might be caused by the drying process, since Tian also found that drying leads to a sharp drop in alcohols while acids and aldehydes increase in *L. edodes* [44]. Moreover, volatile aroma compounds of mushrooms were also affected by growth conditions and genetic differences.

Table 5. Aroma volatile compounds of wild *L. deliciosus*.

Component	Composition (%)
Heptanal	2.03 ± 0.15
Benzaldehyde	2.23 ± 0.09
Hexanoic acid	0.67 ± 0.02
Octanal	1.19 ± 0.07
Benzyl alcohol	0.08 ± 0.00
2-Octenal	0.58 ± 0.11
Nonanal	1.58 ± 0.16
2-Nonenal	0.83 ± 0.05
Dodecane	0.29 ± 0.04
Decanal	0.79 ± 0.05
2-Decenal	1.62 ± 0.29
Undecanal	0.46 ± 0.02
2,4-Decadienal	0.29 ± 0.01
2-Undecenal	1.48 ± 0.18
2-Butyl-2-octenal	0.16 ± 0.02
n-Decanoic acid	0.76 ± 0.05
Decanoic acid, ethyl ester	0.24 ± 0.01
Tetradecane	0.38 ± 0.06
Dodecanal	0.47 ± 0.08
2-Dodecenal	0.65 ± 0.03
Tridecanal	0.41 ± 0.10
n-Hexadecanoic acid	12.82 ± 0.67
9,12-Octadecadienoic acid	3.11 ± 0.21
9-Octadecenoic acid	60.57 ± 2.89
Octadecanoic acid	6.30 ± 0.74

2.4. Antioxidant Activity

Free radicals induce cell-damage, which can cause DNA mutation, proteins damage, lipid peroxidation, and low-density lipoproteins modification. Free radicals can also cause several diseases including diabetes, cancer, neurodegenerative and cardiovascular diseases [46,47]. Usually, food or food extracts can act as antioxidant to counteract damage caused by free radicals. However, the antioxidant capacity of food is determined by complicated factors with various action mechanisms [48]. Thus, several evaluation methods are always used simultaneously to evaluate the antioxidant capacity of food. Due to this effect, we evaluated the antioxidant activity of *L. deliciosus* through DPPH and ABTS radical scavenging and ferric ion reducing (FRAP) assays using ethanol and aqueous extracts. The antioxidant capacity was expressed as trolox equivalent antioxidant capacity (TEAC) values. As shown in Table 6, the TEAC values of ethanol extracts with DPPH, ABT, S and FRAP assays were 18.38, 20.07, and 10.72 μmol Trolox/g dw, respectively. For aqueous extracts, these values were 45.63, 48.05, and 22.28 μmol Trolox/g dw, respectively. Aqueous extracts showed 2–3 fold higher antioxidant capacity compared to ethanol extracts, which might be related to the content of phenols in two extracts, since the total phenol content in aqueous extract was 3.01 fold higher than in ethanol extract.

Table 6. Antioxidant capacity and total phenols in ethanol and aqueous extracts obtained from wild *L. deliciosus*.

	TEAC $_{DPPH}$ (μmolTrolox/g dw) [a]	TEAC $_{ABTS}$ (μmolTrolox/g dw) [b]	TEAC $_{FRAP}$ (μmolTrolox/g dw) [c]	Total Phenols Content (mg GAE/g dw) [d]
Ethanol extract	18.38 ± 1.31	20.07 ± 1.75	10.72 ± 1.04	4.55 ± 0.24
Aqueous extract	45.63 ± 4.40	48.05 ± 3.37	22.28 ± 3.25	13.68 ± 0.26

[a] Trolox equivalent antioxidant capacity (TEAC), 1,1-diphenyl-2-picrylhydrazyl (DPPH); [b] 2,2'-Azino-bis(3-ethylbenzothiazoline-6-sulfonic acid) diammonium salt (ABTS); [c] Ferric ion reducing antioxidant power (FRAP); [d] Gallic acid equivalent weight (GAE).

2.5. Antihyperglycemic Activity

α-Amylase and α-glucosidase are key enzymes in the digestive system that catalyze carbohydrate hydrolysis to enhance blood glucose concentration. A proportion of the population, especially diabetics, suffer from hyperglycemia. A therapeutic approach to hyperglycaemia is to retard the absorption of glucose by inhibiting carbohydrate-hydrolyzing enzymes [49]. Thus, effective and nontoxic inhibitors of α-amylase and α-glucosidase are crucial for the treatment of hyperglycemia. As shown in Figure 1, ethanol extracts exhibited a dose-dependent increase in both α-amylase and α-glucosidase inhibitory assays, while compared with acarbose, the antienzyme activity of ethanol extract was weaker. At 5.0 mg/mL, ethanol extracts exhibited 29.53 and 52.36% on α-amylase and α-glucosidase inhibition, respectively. However, regarding aqueous extracts, no inhibitory effects on α-amylase and α-glucosidase were observed.

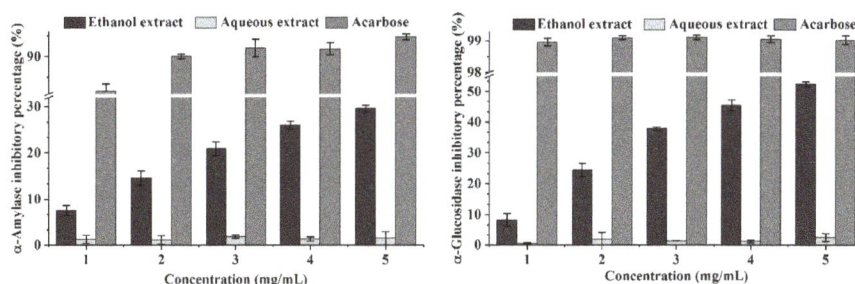

Figure 1. Antihyperglycemic activity of ethanol extract and aqueous extract obtained from wild *L. deliciosus*.

3. Materials and Methods

3.1. Mushroom Species

Wild growing *Lactarius deliciosus* fruiting bodies were collected from Dazhou (Southwest China) pine forests, in the autumn of 2016. All the samples were selected and authenticated at the Dazhou Institute of Agricultural Sciences based on their microscopic and macroscopic characteristics.

3.2. Standards and Reagents

1,1-diphenyl-2-picrylhydrazyl (DPPH), 2,2′-Azino-bis(3-ethylbenzothiazoline-6-sulfonic acid) diammonium salt (ABTS) were purchased from Sigma Chemical Co. (St. Louis, MO, USA). Ferric-tripyridyltriazine (Fe^{3+}-TPTZ) is purchased from Beyotime Biotechnology (Shanghai, China). Chromatographic grade analytical standards asparaginic acid, glutamic acid, asparagine, serine, glutamine, histidine, glycine, threonine, citrulline, arginine, alanine, tyrosine, cystine, valine, methionine, tryptophan, phenylalanine, isoleucine, leucine, lysine, hydroxyproline, proline, 5′-CMP, 5′-UMP, 5′-GMP, 5′-IMP, 5′-XMP, 5′-AMP, trehalose, mannitol, quininic acid, L-malic acid and fumaric acid purchased from Solarbio Science & Technology Co., Ltd. (Beijing, China). α-Amylase, α-glucosidase, 4-nitrophenyl α-d-glucopyranoside (PNPG), and soluble starch were purchased from Shanghai Yuanye Bio-Technology Co., Ltd. (Shanghai, China). All other reagents were analytical grade and obtained from Chengdu Kelong Chemical Factory (Chengdu, China).

3.3. Nutritional Value Assay

Moisture, ash, crude fat, crude proteins, and dietary fiber of the research sample were analyzed by the Association of Official Analytical Chemists methods [50]. Briefly, moisture content was measured by hot air heating at 105 °C until constant weight; ash content was measured by calcination at 600 ± 15 °C using a muffle furnace; crude fat was measured by Soxhlet extraction method with petroleum; crude protein content was estimated by the macro-Kjeldahl method using convert coefficient as 4.38; dietary fiber content was estimated by enzymatic hydrolysis method. Total carbohydrate content was measured by the phenol-sulfuric acid method after test sample completely hydrolyzed by hydrochloric acid. Total energy contribution was calculated according to the following equation [40]:

$$Energy\ (kcal) = 4 \times (g\ proteins + g\ carbohydrate) + 9 \times (g\ fat) \tag{1}$$

3.4. Fatty Acids Composition Assay

Fatty acids composition of crude fat was measured by GC-MS method. Briefly, 20 mg crude fat was methylated with 1 mL of sodium hydroxide (1 M) methanol solution. Then, 2 mL *n*-hexane was added to the mixture. Finally, the *n*-hexane phase was filtered through a 0.22-μm membrane filter before GC-MS analysis. The chromatographic analysis was performed on an Agilent 7890B-5977A GC-MS system equipped with an HP-5MS column (30 m × 0.25 mm × 0.25 μm). Helium was used as carrier gas at the flow rate of 1 mL/min. The injection volume was 1.0 μL with a split ratio of 1:50 at 250 °C. The column temperature was programmed as follow: initial temperature at 50 °C (held for 1 min), increased to 160 °C at 20 °C/min (held for 1 min), increased to 200 °C at 20 °C/min (held for 1 min), increased to 250 °C at 5 °C/min (held for 5 min). Mass spectrometry conditions: interface temperature 250 °C, ion source temperature 230 °C, MS quadrupoles temperature 150 °C, electron energy 70 eV, and m/z scanned area 35–550.

3.5. Metal Elements Assay

Metal elements of *L. deliciosus* were analyzed by ICP-MS method. Briefly, 0.2 g lyophilized sample powder was put into digestion tank, 5 mL HNO_3 was added and fully digested in a Multiwave Pro microwave digester (Anton Paar GmbH, Graz, Austria). Then, excess acid was expelled under 150 °C, and residual liquid was diluted with deionized water to 25 mL. Finally, metal elements analysis was performed on a PerkinElmer NexION350D system (PerkinElmer Co., Waltham, MA, USA). The working parameters were as following: Radio-frequency power 1500 W, nebulizer flow rate 0.8 L/min, coolant gas flow 15 L/min, scanning mode peak hopping, sampling depth 10, and isotopes of selected $^{24}Mg^+$, $^{43}Ca^+$, $^{53}Cr^+$, $^{55}Mn^+$, $^{57}Fe^+$, $^{63}Cu^+$, $^{66}Zn^+$, $^{75}As^+$, $^{111}Cd^+$, $^{208}Pb^+$.

3.6. Free Sugars and Free Amino Acids Assay

Free sugars and free amino acids were analyzed according to previous methods [23,36] with slight modification. Briefly, 0.2 g lyophilized sample powder was suspended in 20 mL aqueous ethanol (80%, *v/v*), and extracted 30 min at 80 °C. The clear supernatant was obtained through centrifuging at 10,000× *g* for 10 min. The above extraction process repeated again, and supernatant was collected together. Then, the supernatant was analyzed by an Agilent 1260 HPLC system equipped with a Hi-Plex Ca column (300 × 7.7 mm, 8.0 µm) and a Refractive Index Detector (RID) for free sugars assay. The analysis conditions of free sugars were as follows: injection volume: 5 µL; mobile phase: H_2O; flow rate of mobile phase: 0.5 mL/min; column temperature: 80 °C; detector temperature: 40 °C. Furthermore, free amino acids were analyzed by HPLC equipped with a Zorbax Eclipse AAA column (150 × 4.6 mm, 5.0 µm) and a Fluorescence Detector (FLD) (Agilent Technologies, Inc., Santa Clara, CA, USA) using above supernatant by online pre-column derivatization high performance liquid chromatography that has been detailed reported by Sun et al. [36].

3.7. Flavor 5′-Nucleotides Assay

Flavor 5′-nucleotides were analyzed according to the previous method [51] with slight modification. Briefly, 0.5 g lyophilized sample powder was mixed with 20 mL distilled water and extracted by boiling water bath for 1 min. Then, the clear supernatant was obtained through centrifuging at 10,000× *g* for 10 min. The above extraction process repeated again, and the supernatant was collected together. Finally, flavor 5′-nucleotides were analyzed by HPLC equipped with a Zorbax SB-C18 column (150 × 4.6 mm, 5.0 µm) and a Diode Array Detector (DAD) (Agilent Technologies, Inc., Santa Clara, CA, USA) using above supernatant. The analysis conditions were as follows: injection volume: 10 µL; mobile phase A: H_2O; mobile phase B: 0.01 M KH_2PO_4 (pH = 4.0; containing 1.45 M tetrabutylammonium hydrogen sulfate); mobile phase C: Acetonitrile; mobile phase gradient: 80-80-77-77% of mobile phase B and 3-3-6-6% of mobile phase C along with analysis time linear increased from 0-9.5-13-15 min; flow rate of mobile phase: 1.0 mL/min; column temperature: 25 °C; detector wavelength: 254 nm.

3.8. Organic Acids Assay

Organic acids were analyzed by SPE-HPLC method. Briefly, 0.15 g lyophilized sample powder was extracted twice by 20 mL methanol at 40 °C for 30 min. The supernatant was collected through centrifugation and dried under negative pressure condition. Then, the residue was re-dissolved by 1 mL distilled water and purified by SAX solid phase extraction column. Finally, the purified liquid was diluted to 100 mL and analyzed by HPLC equipped with a Zorbax SB-C18 column (150 × 4.6 mm, 5.0 µm) and a Diode Array Detector. The analysis conditions were as follows: injection volume: 10 µL; mobile phase A: 0.1% phosphoric acid solution; mobile phase B: methanol; mobile phase ratio: mobile phase A: mobile phase A = 97.5:2.5; flow rate of mobile phase: 1.0 mL/min; column temperature: 40 °C; detector wavelength: 210 nm.

3.9. Equivalent Umami Concentration

The equivalent umami concentration (EUC, g monosodium glutamate (MSG) per 100 g) was used to reflect the umami intensity of *L. deliciosus*, and is represented by the following addition equation according to Yamaguchi report [42].

$$Y = \sum a_i b_i + 1218 \left(\sum a_i b_i \right) \left(\sum a_j b_j \right) \tag{2}$$

where Y is the EUC of the mixture in terms of g MSG/100 g; a_i is the concentration (g per 100 g dw) of each umami amino acid (aspartic acid or glutamic acid); a_j is the concentration (g per 100 g dw) of each umami 5′-nucleotide (5′-IMP, 5′-GMP, 5′-XMP or 5′-AMP); b_i is the relative umami concentration for each umami amino acid to MSG (aspartic acid, 0.077; glutamic acid, 1); b_j is the relative umami concentration for umami 5′-nucelotide to 5′-IMP (5′-IMP, 1; 5′-GMP, 2.3; 5′-XMP, 0.61; 5′-AMP, 0.18); and 1218 is a synergistic constant based on the concentration g per 100 g used.

3.10. Volatile Aroma Components Assay

Volatile aroma components were analyzed using the headspace SPME-GC-MS method. Briefly, 2.0 g lyophilized sample powder was put into a sample bottle, a polydimethylsiloxane SPME fiber (100 µm, Supelco, Bellefonte PA) was used to adsorb statically for 40 min at 50 °C. Then, the volatile aroma components were released at 250 °C and analyzed on an Agilent 7890B-5977C GC-MS system equipped with DB-5MS column (30 m × 0.25 mm × 0.25 µm). Helium was used as carrier gas at the flow rate of 1 mL/min. The column temperature was programmed as follow: initial temperature at 40 °C (held for 1 min), increased to 70 °C at 10 °C/min (held for 2 min), increased to 105 °C at 3 °C/min (held for 1 min), increased to 180 °C at 5 °C/min (held for 1 min), increased to 220 °C at 10 °C/min (held for 5 min). Mass spectrometry conditions: interface temperature 280 °C, ion source temperature 230 °C, MS quadrupoles temperature 150 °C, electron energy 70 eV, and m/z scanned area 35–550.

3.11. Ethanol and Aqueous Extracts Preparation

The lyophilized sample was extracted with ethanol at 50 °C for 1 h three times, the supernatant was collected, concentrated and dried to obtain ethanol extracts. Then, the residue was extracted with distilled water at 80 °C for 1 h three times to obtain aqueous extract.

3.12. Antioxidant Activity Assay

3.12.1. DPPH Radical Scavenging Activity Assay

DPPH radical scavenging activity was measured according to the previous method described in [52] with slight modification. Briefly, 70 µL sample solution was mixed with 140 µL DPPH-ethanol solution. Then, the mixture was incubated in the dark for 30 min at room temperature, and the absorbance at 517 nm was measured using a spectrophotometric micro-plate reader. Trolox solution was used as a positive control and the antioxidant properties of samples were expressed as µmol trolox per g dry weight extract (µmol Trolox/g dw).

3.12.2. ABTS Radical Scavenging Activity Assay

ABTS radical scavenging activity was measured according to previous method described in [53] with slight modification. Briefly, 100 µL sample solution was mixed with 100 µL ABTS solution. Then, the mixture was incubated in the dark for 6 min at room temperature, and the absorbance at 734 nm was measured. Trolox solution was used as a positive control and the antioxidant properties of samples were expressed as µmol trolox per g dry weight extract (µmol Trolox/g dw).

3.12.3. Ferric Ion Reducing Activity Assay

Ferric ion reducing activity was measured according to previous method described in [54] with slight modification. Briefly, 100 µL sample solution was mixed with 100 µL Ferric-tripyridyltriazine (Fe^{3+}-TPTZ) solution. Then, the mixture was incubated in the dark for 10 min at room temperature, and the absorbance at 593 nm was measured. Trolox solution was used as a positive control and the antioxidant properties of samples were expressed as µmol trolox per g dry weight extract (µmol Trolox/g dw).

3.12.4. Total Polyphenols Content Assay

Total polyphenols contents of ethanol and aqueous extracts were determined by the Folin-Ciocalteu assay [55] with some modifications. Briefly, 1 mL sample solution was mixed with 2 mL Folin-Ciocalteu reagent. After incubated for two minutes, 2 mL Na_2CO_3 (10%, *w/v*) was added and the resulting mixture was incubated for 15 min at 50 °C. Finally, absorbance at 775 nm was measured and polyphenol contents was expressed as mg gallic acid equivalents per g dry weight extract (mg GAE/g dw).

3.13. Antihypertensive Activity Assay

3.13.1. α-Amylase Inhibition Activity Assay

α-Amylase inhibition activity was carried out according to previous method [56] with slight modification. Briefly, α-amylase (1 U/mL), soluble starch (1%) and a series concentration of sample were dissolved by 0.02 M sodium phosphate buffer (pH = 6.9, containing 0.0067 M sodium chloride). Sample solution (100 μL) and 100 μL α-amylase solution were mixed and incubated at 37 °C for 10 min. After adding 500 μL soluble starch solution, the mixture was incubated at 37 °C for 10 min. Then, the reaction was terminated by 100 μL hydrochloric acid solution (5 M). After diluting above mixture six times with distilled water, 50 μL liquid was taken, added into 96-well plate, and 150 μL of iodine solution was added as color developing reagent. The absorbance at 660 nm was measured using a spectrophotometric micro-plate reader. Finally, the α-amylase inhibition activity was calculated as the following formula:

$$\alpha - Amylase\ inhibition\ activity\ (\%) = \frac{Abs_1}{Abs_2} \times 100 \tag{3}$$

where Abs_1 was the absorbance of sample solution mixed with α-amylase solution and soluble starch solution; and Abs_2 was the absorbance of sample solution mixed with soluble starch solution (α-amylase solution was replaced by sodium phosphate buffer).

3.13.2. α-Glucosidase Inhibition Activity Assay

α-Glucosidase inhibition activity was carried out according to previous method described in [57] with slight modification. Briefly, α-glucosidase (0.5 U/mL), 4-nitrophenyl α-d-glucopyranoside (PNPG, 3 mM) a series concentration of sample were dissolved by 0.1 M sodium phosphate buffer (pH = 6.8). Sample solution (100 μL) and 100 μL α-glucosidase solution were mixed and incubated at 37 °C for 10 min. After adding 100 μL PNPG solution, the mixture was incubated at 37 °C for 10 min. Then, the reaction was terminated by 300 μL sodium carbonate solution (0.2 M). Finally, the absorbance at 405 nm was measured using a spectrophotometric micro-plate reader, and the α-glucosidase inhibition activity was calculated as the following formula:

$$\alpha - Glu\cos idase\ inhibition\ activity\ (\%) = \left[1 - \frac{(Abs_1 - Abs_2)}{Abs_0}\right] \times 100 \tag{4}$$

where Abs_0 was the absorbance of the α-glucosidase solution mixed with PNPG solution; Abs_1 was the absorbance of sample solution mixed with α-glucosidase solution and PNPG solution; and Abs_2 was the absorbance of sample solution mixed with PNPG solution (α-glucosidase solution was replaced by sodium phosphate buffer).

3.14. Statistical Analysis

All assays were carried out in triplicate, and the results are expressed as mean ± standard deviation (SD).

4. Conclusions

In conclusion, chemical analysis, GC-MS, ICP-MS, HPLC, SPE-HPLC, HS-SPME-GC-MS, radicals scavenging assays, ferric ion reducing activity assay, and enzymes inhibitory assays were conducted to evaluate nutritional value, non-volatile flavor compounds, volatile aroma compounds, potential antioxidant and anti-hyperglycemic activities of wild mushroom *L. deliciosus* from China for the first time. Experimental data indicated that *L. deliciosus* is a good wild edible mushroom with high nutritional value, low calorie level, and extensive flavor compounds. Moreover, *L. deliciosus* also have potential as natural antioxidant and anti-hyperglycemic agents in food and pharmaceutical industry in the future.

Supplementary Materials: The following are available online, Figure S1: Gas chromatography mass spectrum of fatty acids of *L. deliciosus* fat; Figure S2: High performance liquid chromatography of amino acids analyzed by online OPA-FMOC derivation; Figure S3: High performance liquid chromatography of flavor 5′-nucleotides; Figure S4: High performance liquid chromatography of organic acids; Figure S5: High performance liquid chromatography of free sugars; and Figure S6: Headspace solid phase micro-extraction gas chromatography mass spectrum of aroma volatile compounds from *L. deliciosus*.

Author Contributions: Z.X. and C.D. designed the research; Z.X., L.F., S.F. and M.Y. performed the experiments; Z.X., L.F., Y.H., J.L., L.Z. and H.Y. analyzed the data and wrote the draft manuscript; and C.D. provided financial support. All authors revised the manuscript.

Funding: This research received no external funding.

Conflicts of Interest: The authors declare no conflict of interest.

References

1. Okoro, I.O. Antioxidant activities and phenolic contents of three mushroom species, *Lentinus squarrosulus* Mont. *Volvariella esculenta* (Massee) Singer and *Pleurocybella porrigens* (Pers.) Singer. *Int. J. Nutr. Metab.* **2012**, *4*, 72–76.

2. Ouzouni, P.K.; Petridis, D.; Koller, W.D.; Riganakos, K.A. Nutritional value and metal content of wild edible mushrooms collected from West Macedonia and Epirus, Greece. *Food Chem.* **2009**, *115*, 1575–1580. [CrossRef]

3. Finimundy, T.C.; Dillon, A.J.P.; Henriques, J.A.P.; Ely, M.R. A review on general nutritional compounds and pharmacological properties of the Lentinula edodes mushroom. *Food Nutr. Sci.* **2014**, *5*, 1095–1105.

4. Reis, F.S.; Barros, L.; Martins, A.; Ferreira, I.C.F.R. Chemical composition and nutritional value of the most widely appreciated cultivated mushrooms: An inter-species comparative study. *Food Chem. Toxicol.* **2012**, *50*, 191–197. [CrossRef] [PubMed]

5. Miles, P.G.; Chang, S.T. *Mushrooms: Cultivation, Nutritional Value, Medicinal Effect, and Environmental Impact*, 2nd ed.; CRC Press: Boca Raton, FL, USA, 2004.

6. Culleré, L.; Ferreira, V.; Chevret, B.; Venturini, M.E.; Sánchez-Gimeno, A.C.; Blanco, D. Characterisation of aroma active compounds in black truffles (*Tuber melanosporum*) and summer truffles (*Tuber aestivum*) by gas chromatography–olfactometry. *Food Chem.* **2010**, *122*, 300–306. [CrossRef]

7. Cho, I.H.; Namgung, H.J.; Choi, H.K.; Kim, Y.S. Volatiles and key odorants in the pileus and stipe of pine-mushroom (*Tricholoma matsutake* Sing.). *Food Chem.* **2008**, *106*, 71–76. [CrossRef]

8. Mizuno, T. Bioactive biomolecules of mushrooms: Food function and medicinal effect of mushroom fungi. *Food Rev. Int.* **1995**, *11*, 5–21. [CrossRef]

9. Verbeken, A.; Nuytinck, J. Not every milkcap is a *Lactarius*. *Scr. Bot. Belg.* **2013**, *51*, 162–168.

10. Feussi Tala, M.; Qin, J.; Ndongo, J.T.; Laatsch, H. New azulene-type sesquiterpenoids from the fruiting bodies of *Lactarius deliciosus*. *Nat. Prod. Bioprospect.* **2017**, *7*, 269–273. [CrossRef]

11. Anke, H.; Bergendorff, O.; Sterner, O. Assays of the biological activities of guaiane sesquiterpenoids isolated from the fruit bodies of edible lactarius species. *Food Chem. Toxicol.* **1989**, *27*, 393–397. [CrossRef]

12. Bergendorff, O.; Sterner, O. The sesquiterpenes of *Lactarius deliciosus* and *Lactarius deterrimus*. *Phytochemistry* **1988**, *27*, 97–100. [CrossRef]

13. Ding, X.; Hou, Y.; Hou, W. Structure feature and antitumor activity of a novel polysaccharide isolated from *Lactarius deliciosus* Gray. *Carbohyd. Polym.* **2012**, *89*, 397–402. [CrossRef]

14. Hou, Y.; Liu, L.; Ding, X.; Zhao, D.; Hou, W. Structure elucidation, proliferation effect on macrophage and its mechanism of a new heteropolysaccharide from *Lactarius deliciosus* Gray. *Carbohyd. Polym.* **2016**, *152*, 648–657. [CrossRef]

15. Vetter, J. Trypsin inhibitor activity of basidiomycetous mushrooms. *Eur. Food Res. Technol.* **2000**, *211*, 346–348. [CrossRef]

16. Kosanić, M.; Ranković, B.; Rančić, A.; Stanojković, T. Evaluation of metal concentration and antioxidant, antimicrobial, and anticancer potentials of two edible mushrooms *Lactarius deliciosus* and *Macrolepiota procera*. *J. Food Drug Anal.* **2016**, *24*, 477–484. [CrossRef]

17. Moro, C.; Palacios, I.; Lozano, M.; D'Arrigo, M.; Guillamón, E.; Villares, A.; Martínez, J.A.; García-Lafuente, A. Anti-inflammatory activity of methanolic extracts from edible mushrooms in LPS activated RAW 264.7 macrophages. *Food Chem.* **2012**, *130*, 350–355. [CrossRef]

18. Rathore, H.; Prasad, S.; Sharma, S. Mushroom nutraceuticals for improved nutrition and better human health: A review. *Pharma Nutr.* **2017**, *5*, 35–46. [CrossRef]

19. Miyazawa, M.; Takahashi, T.; Horibe, I.; Ishikawa, R. Two new aromatic compounds and a new D-arabinitol ester from the mushroom *Hericium erinaceum*. *Tetrahedron* **2012**, *68*, 2007–2010. [CrossRef]

20. Myrdal Miller, A.; Mills, K.; Wong, T.; Drescher, G.; Lee, S.; Sirimuangmoon, C.; Schaefer, S.; Langstaff, S.; Minor, B.; Guinard, J.X. Flavor-enhancing properties of mushrooms in meat-based dishes in which sodium has been reduced and meat has been partially substituted with mushrooms. *J. Food Sci.* **2014**, *79*, S1795–S1804. [CrossRef]

21. Beluhan, S.; Ranogajec, A. Chemical composition and non-volatile components of Croatian wild edible mushrooms. *Food Chem.* **2011**, *124*, 1076–1082. [CrossRef]

22. Wang, X.M.; Zhang, J.; Wu, L.H.; Zhao, Y.L.; Li, T.; Li, J.Q.; Wang, Y.Z.; Liu, H.G. A mini-review of chemical composition and nutritional value of edible wild-grown mushroom from China. *Food Chem.* **2014**, *151*, 279–285. [CrossRef]

23. Barros, L.; Baptista, P.; Correia, D.M.; Casal, S.; Oliveira, B.; Ferreira, I.C.F.R. Fatty acid and sugar compositions, and nutritional value of five wild edible mushrooms from Northeast Portugal. *Food Chem.* **2007**, *105*, 140–145. [CrossRef]

24. Yilmaz, N.; Solmaz, M.; Türkekul, İ.; Elmastaş, M. Fatty acid composition in some wild edible mushrooms growing in the middle Black Sea region of Turkey. *Food Chem.* **2006**, *99*, 168–174. [CrossRef]

25. Orsavova, J.; Misurcova, L.; Ambrozova, J.; Vicha, R.; Mlcek, J. Fatty acids composition of vegetable oils and its contribution to dietary energy intake and dependence of cardiovascular mortality on dietary intake of fatty acids. *Int. J. Mol. Sci.* **2015**, *16*, 12871–12890. [CrossRef]

26. Calder, P.C. Functional roles of fatty acids and their effects on human health. *J. Parenter. Enter.* **2015**, *39*, 18S–32S. [CrossRef]

27. Ouzouni, P.K.; Veltsistas, P.G.; Paleologos, E.K.; Riganakos, K.A. Determination of metal content in wild edible mushroom species from regions of Greece. *J. Food Compos. Anal.* **2007**, *20*, 480–486. [CrossRef]

28. Kalač, P. Chemical composition and nutritional value of European species of wild growing mushrooms: A review. *Food Chem.* **2009**, *113*, 9–16. [CrossRef]

29. Çayır, A.; Coşkun, M.; Coşkun, M. The heavy metal content of wild edible mushroom samples collected in Canakkale province, Turkey. *Biol. Trace Elem. Res.* **2010**, *134*, 212–219. [CrossRef]

30. Mendil, D.; Uluözlü, Ö. D.; Hasdemir, E.; Çağlar, A. Determination of trace elements on some wild edible mushroom samples from Kastamonu, Turkey. *Food Chem.* **2004**, *88*, 281–285. [CrossRef]

31. Kalač, P.; Svoboda, L. A review of trace element concentrations in edible mushrooms. *Food Chem.* **2000**, *69*, 273–281. [CrossRef]

32. Food and Nutrition Board; Institute of Medicine. *Dietary Reference Intakes for Calcium, Phosphorus, Magnesium, Vitamin D, and Fluoride*; National Academy Press: Washington, DC, USA, 1997.

33. Trumbo, P.; Yates, A.A.; Schlicker, S.; Poos, M. Dietary reference intakes: Vitamin A, vitamin K, arsenic, boron, chromium, copper, iodine, iron, manganese, molybdenum, nickel, silicon, vanadium, and zinc. *J. Acad. Nutr. Diet.* **2001**, *101*, 294–301.

34. Pei, F.; Shi, Y.; Gao, X.; Wu, F.; Mariga, A.M.; Yang, W.; Zhao, L.; An, X.; Xin, Z.; Yang, F.; Hu, Q. Changes in non-volatile taste components of button mushroom (*Agaricus bisporus*) during different stages of freeze drying and freeze drying combined with microwave vacuum drying. *Food Chem.* **2014**, *165*, 547–554. [CrossRef] [PubMed]

35. Yin, C.; Fan, X.; Fan, Z.; Shi, D.; Yao, F.; Gao, H. Comparison of non-volatile and volatile flavor compounds in six *Pleurotus* mushrooms. *J. Sci. Food Agric.* **2018**. [CrossRef] [PubMed]

36. Sun, L.; Liu, Q.; Bao, C.; Fan, J. Comparison of free total amino acid compositions and their functional classifications in 13 wild edible mushrooms. *Molecules* **2017**, *22*, 350. [CrossRef] [PubMed]

37. Yang, J.H.; Lin, H.C.; Mau, J.L. Non-volatile taste components of several commercial mushrooms. *Food Chem.* **2001**, *72*, 465–471. [CrossRef]

38. Vieira, V.; Barros, L.; Martins, A.; Ferreira, I. Nutritional and biochemical profiling of *Leucopaxillus candidus* (Bres.) Singer wild mushroom. *Molecules* **2016**, *21*, 99. [CrossRef]

39. Mocan, A.; Fernandes, A.; Barros, L.; Crisan, G.; Smiljkovic, M.; Sokovic, M.; Ferreira, I. Chemical composition and bioactive properties of the wild mushroom *Polyporus squamosus* (Huds.) Fr: A study with samples from Romania. *Food Funct.* **2018**, *9*, 160–170. [CrossRef]

40. Toledo, C.V.; Barroetaveña, C.; Fernandes, Â.; Barros, L.; Ferreira, I.C.F.R. Chemical and antioxidant properties of wild edible mushrooms from native Nothofagus spp. forest, Argentina. *Molecules* **2016**, *21*, 1201. [CrossRef]

41. Barros, L.; Pereira, C.; Ferreira, I.C.F.R. Optimized analysis of organic acids in edible mushrooms from Portugal by ultra fast liquid chromatography and photodiode array detection. *Food Anal. Method.* **2013**, *6*, 309–316. [CrossRef]

42. Yamaguchi, S.; Yoshikawa, T.; Ikeda, S.; Ninomiya, T. Measurement of the relative taste intensity of some L-α-amino acids and 5′-nucleotides. *J. Food Sci.* **1971**, *36*, 846–849. [CrossRef]

43. Mau, J. The umami taste of edible and medicinal mushrooms. *Int. J. Med. Mushrooms* **2005**, *7*, 119–126. [CrossRef]

44. Tian, Y.; Zhao, Y.; Huang, J.; Zeng, H.; Zheng, B. Effects of different drying methods on the product quality and volatile compounds of whole shiitake mushrooms. *Food Chem.* **2016**, *197*, 714–722. [CrossRef]

45. Çağlarırmak, N. The nutrients of exotic mushrooms (*Lentinula edodes* and *Pleurotus* species) and an estimated approach to the volatile compounds. *Food Chem.* **2007**, *105*, 1188–1194. [CrossRef]

46. Ziegler, D.V.; Wiley, C.D.; Velarde, M.C. Mitochondrial effectors of cellular senescence: Beyond the free radical theory of aging. *Aging cell* **2015**, *14*, 1–7. [CrossRef] [PubMed]

47. Liochev, S.I. Reactive oxygen species and the free radical theory of aging. *Free Radic. Biol. Med.* **2013**, *60*, 1–4. [CrossRef] [PubMed]

48. Moo-Huchin, V.M.; Moo-Huchin, M.I.; Estrada-León, R.J.; Cuevas-Glory, L.; Estrada-Mota, I.A.; Ortiz-Vázquez, E.; Betancur-Ancona, D.; Sauri-Duch, E. Antioxidant compounds, antioxidant activity and phenolic content in peel from three tropical fruits from Yucatan, Mexico. *Food Chem.* **2015**, *166*, 17–22. [CrossRef]

49. Bhandari, M.R.; Jong-Anurakkun, N.; Hong, G.; Kawabata, J. α-Glucosidase and α-amylase inhibitory activities of Nepalese medicinal herb Pakhanbhed (*Bergenia ciliata*, Haw.). *Food Chem.* **2008**, *106*, 247–252. [CrossRef]

50. Hasan, M.T. *Official Methods of Analysis*, 15th ed.; Association of Official Analytical Chemists: Artington, VR, USA, 2015.

51. Sommer, I.; Schwartz, H.; Solar, S.; Sontag, G. Effect of gamma-irradiation on flavour 5′-nucleotides, tyrosine, and phenylalanine in mushrooms (*Agaricus bisporus*). *Food Chem.* **2010**, *123*, 171–174. [CrossRef]

52. Xiang, C.; Xu, Z.; Liu, J.; Li, T.; Yang, Z.; Ding, C. Quality, composition, and antioxidant activity of virgin olive oil from introduced varieties at Liangshan. *LWT Food Sci. Technol.* **2017**, *78*, 226–234. [CrossRef]

53. Li, X.; Lin, J.; Gao, Y.; Han, W.; Chen, D. Antioxidant activity and mechanism of Rhizoma Cimicifugae. *Chem. Cent. J.* **2012**, *6*, 140. [CrossRef]

54. Xia, Q.; Wang, L.; Xu, C.; Mei, J.; Li, Y. Effects of germination and high hydrostatic pressure processing on mineral elements, amino acids and antioxidants in vitro bioaccessibility, as well as starch digestibility in brown rice (*Oryza sativa* L.). *Food Chem.* **2017**, *214*, 533–542. [CrossRef] [PubMed]

55. Liu, K.; Xiao, X.; Wang, J.; Chen, C.Y.O.; Hu, H. Polyphenolic composition and antioxidant, antiproliferative, and antimicrobial activities of mushroom Inonotus sanghuang. *LWT Food Sci. Technol.* **2017**, *82*, 154–161. [CrossRef]

56. He, M.; Zeng, J.; Zhai, L.; Liu, Y.; Wu, H.; Zhang, R.; Li, Z.; Xia, E. Effect of in vitro simulated gastrointestinal digestion on polyphenol and polysaccharide content and their biological activities among 22 fruit juices. *Food Res. Int.* **2017**, *102*, 156–162. [CrossRef] [PubMed]

57. Hemalatha, P.; Bomzan, D.P.; Sathyendra Rao, B.V.; Sreerama, Y.N. Distribution of phenolic antioxidants in whole and milled fractions of quinoa and their inhibitory effects on α-amylase and α-glucosidase activities. *Food Chem.* **2016**, *199*, 330–338. [CrossRef]

Sample Availability: Lyophilized powder of *L. deliciosus* is available from the authors.

molecules

MDPI

Article

A UHPLC-UV Method Development and Validation for Determining Kavalactones and Flavokavains in *Piper methysticum* (Kava)

Yijin Tang * and Christine Fields *

Applied Food Sciences, Inc., 2500 Crosspark Road, Coralville, IA 52241, USA
* Correspondence: ytang@appliedfoods.com (Y.T.); cfields@appliedfoods.com (C.F.); Tel.: +1-309-716-5833 (Y.T.)

Academic Editors: Alessandra Gentili and Chiara Fanali
Received: 2 March 2019; Accepted: 27 March 2019; Published: 30 March 2019

check for updates

Abstract: An ultra-high-performance liquid chromatographic (UHPLC) separation was developed for six kava pyrones (methysticin, dihydromethysticin (DHM), kavain, dihydrokavain (DHK), desmethoxyyangonin (DMY), and yangonin), two unidentified components, and three Flavokavains (Flavokavain A, B, and C) in *Piper methysticum* (kava). The six major kavalactones and three flavokavains are completely separated ($R_s > 1.5$) within 15 min using a HSS T3 column and a mobile phase at 60 °C. All the peaks in the LC chromatogram of kava extract or standard solutions were structurally confirmed by LC-UV-MS/MS. The degradations of yangonin and flavokavains were observed among the method development. The degradation products were identified as cis-isomerization by MS/MS spectra. The isomerization was prevented or limited by sample preparation in a non-alcoholic solvent or with no water. The method uses the six kava pyrones and three flavokavains as external standards. The quantitative calibration curves are linear, covering a range of 0.5–75 µg/mL for the six kava pyrones and 0.05–7.5 µg/mL for the three flavokavains. The quantitation limits for methysticin, DHM, kavain, DHK, DMY, and yangonin are approximately 0.454, 0.480, 0.277, 0.686, 0.189, and 0.422 µg/mL. The limit of quantification (LOQs) of the three flavokavains are about 0.270, 0.062, and 0.303 µg/mL for flavokavain C (FKC), flavokavain A (FKA), and flavokavain B (FKB). The average recoveries at three different levels are 99.0–102.3% for kavalactones (KLs) and 98.1–102.9% for flavokavains (FKs). This study demonstrates that the method of analysis offers convenience and adequate sensitivity for determining methysticin, DHM, kavain, DHK, yangonin, DMY, FKA, FKB, and FKC in kava raw materials (root and CO_2 extract) and finished products (dry-filled capsule and tablet).

Keywords: *Piper methysticum* (kava); kavalactones; flavokavains; UHPLC-UV; mass spectra; isomerization; single-laboratory validation; quality control

---

1. Introduction

Kava (*Piper methysticum*) has been used for a traditional beverage in the Pacific islands, from ancient times, for its relaxant and anxiolytic effects [1,2].

The active constituents in kava root have been reported as a group of structurally related lipophilic lactone derivatives, kavalactones (KLs), with an arylethylene-α-pyrone skeleton. More than 18 kavalactones have been isolated from kava [3–5], including six major KLs (Figure 1A) as follows: Kavain, 5,5-dihydrokavain (DHK), methysticin, dihydromethysticin (DHM), yangonin, and desmethoxy-yangonin (DMY). Other types of compounds identified in kava include alkaloids, chalcones (flavokavains A, B, and C, Figure 1B), avanones (pinostrobin, 5,7-dimethoxy avanone), cinnamic acid derivatives (bornyl ester of 3,4-methylene dioxycinnamic acid, cinnamic acid bornyl ester), long-chain fatty acids and alcohols, and sterols [6–8].

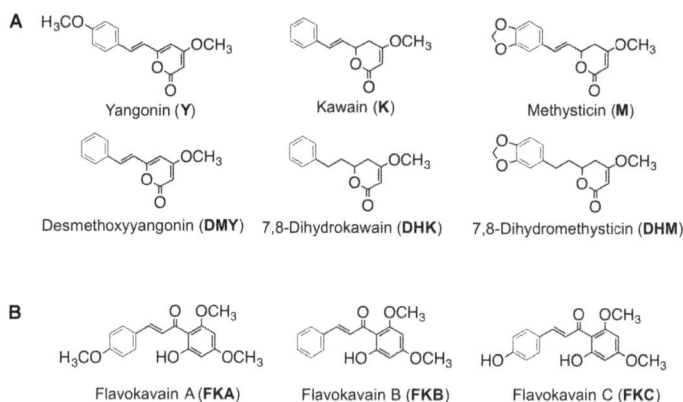

Figure 1. Chemical structures of (**A**) six major kavalactones and (**B**) three flavokavains.

Over a decade ago, potential safety issues with kava applications arose when several cases of liver damage were associated with kava consumption [9–11]. Given the concerns around potential liver toxicity with kava usage, the possible mechanisms for hepatotoxicity were investigated [12–14]. However, the cause of hepatotoxicity from kava was never clearly associated with one key factor. Kava continues to be consumed on a regular basis within local cultures, with no recent cases of hepatotoxicity reported. Recently, the World Health Organization revisited many of the reported kava-associated hepatotoxicity cases, evaluating the potential causes and providing guidance for safe use as well as recommendations for further studies to identify the potentially hepatotoxic compounds. The chalcone-based flavokavains A (FKA), B (FKB) and C (FKC) were studied with their potential risks and benefits. The studies suggested that the chalcones (FKA, FKB, and FKC), especially FKB, caused hepatotoxicity [15–17]. The possible toxicities of flavokavains (FKs) indicated that FKs should be limited to a very low level in the finished products being consumed.

As the main psychoactive components of kava, the contents and chemotypes of six major KLs, are the main criteria of kava beverage quality. Considering the possible hepatotoxicity of FKs, the reliable qualitative and/or quantitative determination of kava compounds is important for the best usage and purification techniques of kava. The RP-HPLC method was considered as an effective method with a good separation and a high accuracy, among many existing analytical methods for determining KLs [18–22]. Most recently, the HPLC method from Meissner and Häberlein [18] and the rapid HPLC method from Brown [22] were proposed for quantifying FKs, along with KLs. The two methods either had a long run time (50 min) or sacrificed the resolution among some key compounds. The UPLC-MS/MS method [23] was also reported for analysis of KLs, however, the linear range of UPLC-MS/MS was limited.

The main goal of this study was to develop and validate a more effective (U)HPLC method for quick and reliable quantification of the six major KLs and three FKs in kava raw roots and rhizomes, as well as finished products, based on the most prevalent instrumentation and standards available. During the method development of this study, it was observed that improper sample preparation of (standard) samples could lead the degradation of yangonin, as well the possible degradations of FKs, utilizing LC-UV-MS. The (U)HPLC methods were fully validated based on the AOAC Guideline [24] for Single-Laboratory Validation of Chemical Methods and ICH Q2 Guideline [25].

2. Results and Discussions

2.1. Identification of Isomerization of Yangonin and Flavokavains (A, B, and C), along with Sample Preparation

(U)HPLC-UV is still a more applicable method for quantitation of KLs and FKs in most of kava products. In this study, both standards and kava CO_2 extract samples were applied for the analysis and method development on LC-UV. The chromatographic separation was extended from the previous HPLC study [26] and achieved under a gradient separation at 60 °C. Optimum separation of KLs and FKs was achieved using an UHPLC column (Acquity HSS T3, 100 mm × 2.1 mm, 1.8 μm). Gradient elution was performed using water (100%, no addition of buffer or acids) as solvent A and isopropanol (100%, no addition of buffer or acids) as solvent B, with the gradient program listed in Table 1. The structures of KL or FK compounds were confirmed by the MS/MS spectra (Support Figure S1).

Table 1. Gradient UHPLC elution profile.

Time/min	A (H$_2$O, %)	B (IPA, %)	Flow Rate (mL/min)
0.00	95	5	0.50
0.35	90	10	0.50
0.39	78	22	
2.00	78	22	0.43
7.00	78	22	0.41
7.50	71	29	0.41
10.50	25	75	0.41
11.50	0	100	0.41
12.80	0	100	0.41
13.00	95	5	0.41
15.00	95	5	0.44
15.50		End	0.50

The two unknown compounds (U1 and U2) reported in a previous study [26] were also presented in the kava CO$_2$ extract (Figure 2). The (M + H)$^+$ was 261.1 Da for U1 and 263.1 Da for U2, from MS/MS spectra. The MS spectra suggest that the two compounds are very likely to be 5,6-dihydroyangonin (DHY) for U1 and 5,6,7,8-tetrahydroyangonin (THY) for U2, as the literature reported [2]. The structures of THY and DHY are associated with the mass fragmentations of the (M + H)$^+$ as 145.1 Da for U1 and 147.1 and 121.1 Da for U2. The two compounds are found as minor KLs in all-natural kava products. The quantitation of the major KLs could be interfered by the two minor components under a low-resolution LC.

The individual stock standard solutions were initially prepared in methanol. Then, the mixed standard solution was diluted with 50% H$_2$O/MeOH for LC analysis. Four unknown compounds were observed for the mixed standard solution over several weeks after the preparation. The unknown compounds were with the same molecular weight as both yangonin and flavokavains, but eluted significantly earlier (Figure 3). The fragmentation of the (M + H)$^+$ (parent ion) gives an identical product ion spectrum as that of yangonin and flavokavains. The isomerization from yangonin to cis-yangonin in kava standards or extract solutions was reported, especially if aqueous or alcoholic solutions were used [27,28]. MS/MS spectra suggested one of the unknown compounds as cis-yangonin, an isomer of yangonin. MS/MS spectra also indicated the other three compounds related to the isomerization of flavokavains A, B, and C. Those compounds are very likely to be the cis-isomer of flavokavains A, B, and C. For practical reasons, it was assumed that cis products could be formed with an alcoholic solvent or water [27,28]. Many LC methods were attempted for a good resolution (R$_s$ > 1.5) for all KLs, especially for the separation of M and DHM. It is very difficult to achieve a good chromatograph separation (R$_s$ > 1.5) for all KLs and FKs with the additions of cis-yangonin and cis-FKs. It is suggested that during validation of the proposed method, the stock standard solutions of KLs and FKs are prepared in a non-alcoholic solvent, like acetonitrile. The kava working standard solution and sample solution should be freshly prepared for LC analysis, to prevent or minimize the isomerization of yangonin and FKs.

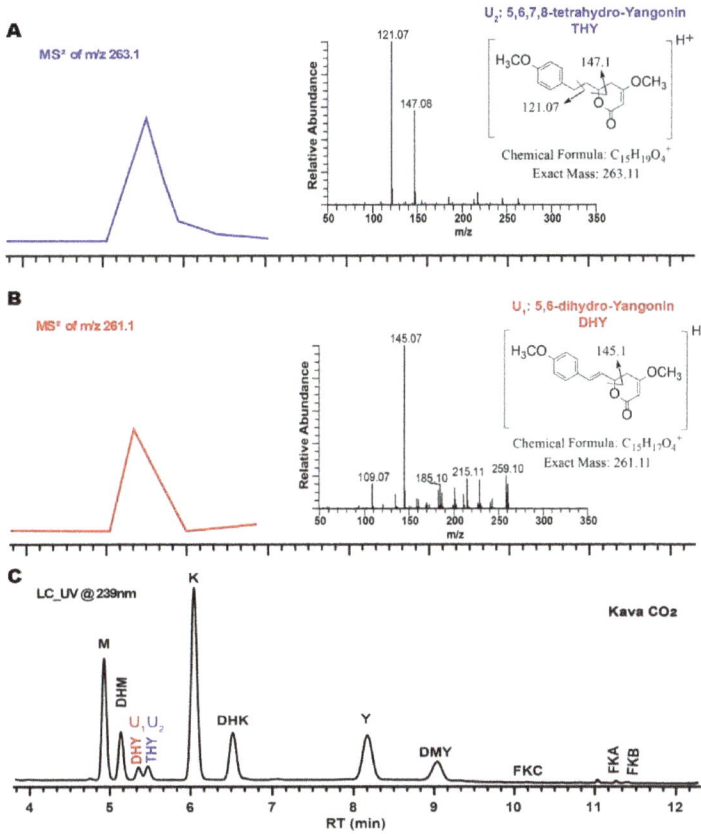

Figure 2. UHPLC-UV-MS data of kava CO_2 extract as follows: (**A**) MS/MS^2 trace of m/z 263.1 and MS^2 spectrum of peak U$_2$ (5,6,7,8-Tetrahydroyangonin, THY); (**B**) MS/MS^2 trace of m/z 261.1 and MS^2 spectrum of peak U$_1$ (5,6-Dihydroyangonin, DHY); (**C**) UV trace at the wavelength of 239 nm; M, methysticin; DHM, 7,8-Dihydromethysticin; K, kavain; DHK, 7,8-Dihydrokavain; Y, yangonin; DMY, desmethoxyyangonin; FKA (B or C) flavokavains A (B or C).

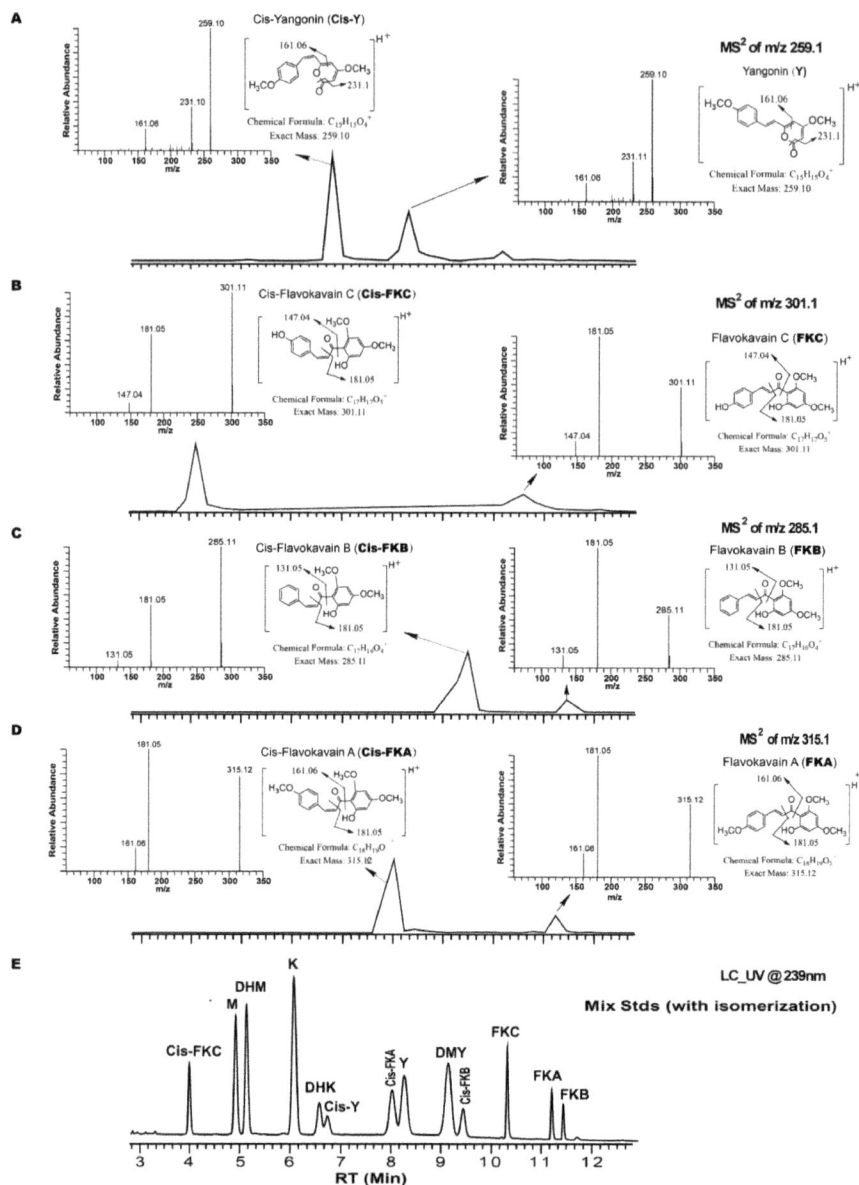

Figure 3. UHPLC-UV-MS data of degraded kava standard mixture, as follows: (**A**) MS/MS^2 trace of *m/z* 259.1, and MS^2 spectra of peaks of Y and cis-Y; (**B**) MS/MS^2 trace of *m/z* 301.1, and MS^2 spectra of peaks of FKC and cis-FKC; (**C**) MS/MS^2 trace of *m/z* 285.1, and MS^2 spectra of peaks of FKB and cis-FKB; (**D**) MS/MS^2 trace of *m/z* 315.1 and MS^2 spectra of peaks of FKA and cis-FKA; (**E**) UV trace at the wavelength of 239 nm; M, methysticin; DHM, 7,8-Dihydromethysticin; K, kavain; DHK, 7,8-Dihydrokavain; Y, Yangonin; Cis-Y, cis-yangonin; DMY, Desmethoxyyangonin; FKA (B or C), flavokavains A (B or C); cis-FKA (B, or C), flavokavains A (B or C).

2.2. Method Validation

2.2.1. Specificity/Resolution

Each individual reference standard was injected into the HPLC-UV to compare with the standard mixture (Support Figure S2). Identification of KLs in the test materials was determined by comparing peak retention times and UV spectra to the reference standards. Representative chromatograms of the standard mixture and kava products are displayed in Figure 4. Under the chromatographic conditions used in the present study, all six major KLs and three FKs were eluted separately following this following order: Methysticin, DHM, kavain, DHK, yangonin, DMY, FKC, FKA, and FKB. Values for the relative retention times (α), retention factors (k'), and chromatographic resolutions (R_s), calculated from the LC analysis of the kava standard mixtures (Figure 4), are summarized in Table 2. Retention factors (k') were calculated as $k' = (t_r - t_0)/t_0$, where t_r is the retention time of the analyte and t_0 is the retention time of unretained compounds (solvent front). The k' values were within the optimum range ($k' > 2$) for satisfactory chromatographic elution. An excellent chromatographic specificity was observed with the good resolution of the peaks ($R_s > 1.5$) and with no significant interfering peaks for all compounds in the mixed standard sample. The total chromatography run time was 15.5 min.

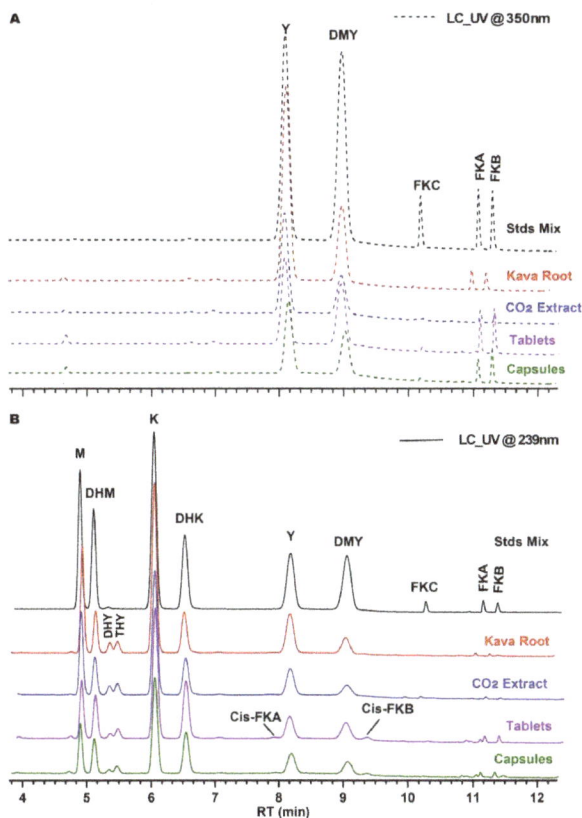

Figure 4. UHPLC chromatography of kava standard mixture and kava products. (**A**) UV trace at the wavelength of 350 nm; (**B**) UV trace at the wavelength of 239 nm; M, methysticin; DHM, 7,8-Dihydromethysticin; DHY, 5,6-Dihydroyangonin; THY, 5,6,7,8-Tetrahydroyangonin; K, kavain; DHK, 7,8-Dihydrokavain; DMY, Desmethoxyyangonin; Y, Yangonin; FKA (B or C), flavokavains A (B or C); cis-FKA (B), cis-flavokavains A (B).

Table 2. Parameters of kavalactones and flavokavains.

Compound	Retention Factor (k')	Relative RT (α)	Chromatographic Resolution (R_s)
Methysticin (1)	8.01	–	–
Dihydromethysticin (2)	8.41	$\alpha_{1/2} = 1.045$	$R_{1/2} = 1.78$
Kavain (3)	10.15	$\alpha_{2/3} = 1.184$	$R_{2/3} = 6.61$
Dihdyrokavain (4)	11.07	$\alpha_{3/4} = 1.082$	$R_{3/4} = 3.02$
Yangonin (5)	14.12	$\alpha_{4/5} = 1.253$	$R_{4/5} = 8.20$
Desmethoxyyangonin (6)	15.76	$\alpha_{5/6} = 1.109$	$R_{5/6} = 3.62$
Flavokavain C (7)	18.18	$\alpha_{6/7} = 1.144$	$R_{6/7} = 7.67$
Flavokavain A (8)	19.86	$\alpha_{7/8} = 1.088$	$R_{7/8} = 11.92$
Flavokavain B (9)	20.29	$\alpha_{8/9} = 1.020$	$R_{8/9} = 3.04$

2.2.2. Standard Linearity

To evaluate the linearity of the calibration curves, calibration standard mixed solutions, at the target concentrations, were prepared as described above. Each peak area of the chromatograms was recorded as the UV responses at 239 nm for methysticin, DHM, kavain, and DHK, and at 350nm for yangonin, DMY, and FKs. Calibration curves were plotted by the peak area vs. the concentration of the standard compounds (Support Figure S3). Regression analyses were processed by Lab Solution software. Calibration curves were linear over the concentration range used, with an $R^2 > 0.999$. The normalized intercept/slope of the regression line and the correlation coefficient were calculated for the whole data set. The method was evaluated by determining the coefficient of linearity and the intercept values, as summarized in Table 3.

Table 3. Calibration parameters for kavalactones and flavokavains from three different calibration curves.

	Calibration Range (µg/mL)	Slope (±SD) [a]	Y-Intercept (±SD) [a]	r^2 (±SD) [a]	LOD (µg/mL)	LOQ (µg/mL)
Methysitcin		106708.0 ± 700.0	-3443.1 ± 4846.8	0.99996 ± 0.00005	0.150	0.454
DHM		80132.1 ± 502.3	6666.6 ± 3847.5	0.99997 ± 0.00002	0.158	0.480
Kavain	$0.50\sim75.0$	176782.0 ± 1450.0	-5987.0 ± 4889.8	0.99996 ± 0.00004	0.091	0.277
DHK		80884.5 ± 381.5	11433.5 ± 5551.5	0.99995 ± 0.00002	0.226	0.686
Yangonin		167271.0 ± 1964.9	3729.7 ± 3167.9	0.99994 ± 0.00003	0.062	0.189
DMY		164787.0 ± 2352.4	-12618.3 ± 6952.1	0.99995 ± 0.00003	0.139	0.422
FKC		135720.0 ± 3576.4	-3194.0 ± 3663.5	0.99970 ± 0.00017	0.089	0.270
FKA	$0.05\sim7.50$	143817.0 ± 2873.9	768.8 ± 889	0.99980 ± 0.00018	0.020	0.062
FKB		150461.0 ± 3339.0	-1374.6 ± 4555.5	0.99988 ± 0.00017	0.100	0.303

Note: Calibration curves were performed three times on different days and established by measuring the concentration vs. the corresponding peak area. The given values are the mean of three replicates ± standard deviation [a].

2.2.3. Limit of Detection and Limit of Quantification

The limit of detection (LOD) and the limit of quantification (LOQ) of the kava standard assay were determined by the standard deviation of the y-intercepts over the slope of the regression lines from three replicated calibration curves on different days. The LODs and LOQs for six major KLs and three FKs were calculated and summarized in Table 3.

2.2.4. Recovery

A spike recovery study based on spiking six kavalactones and three flavokavains into a kava test material (kava root) at high, medium, and low levels, as well as non-spiked, were completed and shown in Table 4. The amounts of kavalactones and flavokavains were compared before and after spiking. As the AOAC guidelines suggest for single-laboratory validation, the recoveries were calculated in two ways, as follows: (1) Total recovery based on recovery of the native plus added analyte, and (2) marginal recovery based only on the added analyte (the native analyte is subtracted from both the numerator and denominator). The total recovery is used for the native analytes presented in amounts greater than about 10% of the amount added, otherwise, the marginal recovery is applied. The average recoveries for each analyte at each level were 99.0–102.3% for KLs and 98.1–102.9% for FKs, within the ranges of recovery limits from 95% to 102% at 10% concentration, 92% to 105% at 1% concentration, 90% to 108% at 0.1% concentration, and 85% to 110% for 0.01 at 0.01% concentration.

Table 4. Spike recovery results of HPLC-UV method for determination of kavalactones and flavokavains.

	Methysticin	DHM	Kavain	DHK	DMY	Yangonin	FKC	FKA	FKB
Spiked Level				3.0%				~0.3%	
Native (μg/mL) [a]	40.16 ± 0.62	22.14 ± 0.34	47.80 ± 0.74	29.75 ± 0.46	36.94 ± 0.57	14.52 ± 0.22	0.296 ± 0.005	1.27 ± 0.02	1.03 ± 0.02
Spiked (μg/mL) [b]	60.02 ± 0.80	58.73 ± 0.78	59.28 ± 0.79	59.23 ± 0.79	59.36 ± 0.79	57.73 ± 0.77	5.67 ± 0.08	7.99 ± 0.11	5.76 ± 0.08
After Spiked (μg/mL) [b]	100.2 ± 0.2	80.87 ± 0.44	107.1 ± 0.1	88.99 ± 0.33	96.30 ± 0.22	72.24 ± 0.54	5.97 ± 0.07	9.25 ± 0.09	6.79 ± 0.06
Detected (μg/mL) [a]	101.5 ± 0.3	81.43 ± 0.51	109.6 ± 0.6	89.63 ± 0.44	98.21 ± 0.41	73.42 ± 0.36	5.86 ± 0.16	9.48 ± 0.14	6.92 ± 0.11
Marginal recovery (%) [a]	102.1 ± 0.1	101.0 ± 0.1	104.2 ± 1.0	101.1 ± 0.4	103.2 ± 0.6	102.0 ± 0.4	**98.1 ± 1.6** [c]	**102.9 ± 0.6**	**102.2 ± 0.8**
Total Recovery (%) [a]	**101.3 ± 0.1**	**100.7 ± 0.1**	**102.3 ± 0.5**	**100.7 ± 0.3**	**102.0 ± 0.4**	**101.6 ± 0.3**	98.2 ± 1.5	102.5 ± 0.5	101.3 ± 0.1
Spiked Level				1.5%				~0.15%	
Native (μg/mL) [a]	41.04 ± 0.81	22.63 ± 0.45	48.86 ± 0.97	30.41 ± 0.60	37.76 ± 0.75	14.84 ± 0.29	0.302 ± 0.006	1.29 ± 0.03	1.05 ± 0.02
Spiked (μg/mL) [b]	30.21 ± 0.15	29.55 ± 0.15	29.83 ± 0.15	29.81 ± 0.15	29.87 ± 0.15	29.05 ± 0.14	2.86 ± 0.014	4.02 ± 0.02	2.90 ± 0.01
After Spiked (μg/mL) [b]	71.25 ± 0.66	52.18 ± 0.30	78.69 ± 0.82	60.22 ± 0.46	67.63 ± 0.60	43.89 ± 0.15	3.15 ± 0.01	5.31 ± 0.01	3.95 ± 0.01
Detected (μg/mL) [a]	70.50 ± 0.74	51.73 ± 0.42	78.02 ± 1.17	59.63 ± 0.51	67.21 ± 0.83	44.12 ± 0.36	3.10 ± 0.02	5.36 ± 0.02	3.97 ± 0.02
Marginal recovery (%) [a]	97.4 ± 2.5	99.0 ± 1.6	93.8 ± 2.5	98.6 ± 1.1	97.8 ± 2.9	100.8 ± 0.9	98.4 ± 0.5	101.0 ± 0.5	100.3 ± 0.9
Total Recovery (%) [a]	**99.0 ± 1.1**	**99.1 ± 0.9**	**99.2 ± 1.0**	**99.0 ± 0.5**	**99.4 ± 1.3**	**100.5 ± 0.6**	**98.1 ± 0.5**	**100.8 ± 0.4**	**100.3 ± 0.7**
Spiked Level				0.25%				~0.025%	
Native (μg/mL) [a]	40.83 ± 0.91	22.51 ± 0.50	48.60 ± 1.09	30.25 ± 0.68	37.56 ± 0.84	14.76 ± 0.33	0.300 ± 0.007	1.29 ± 0.03	1.05 ± 0.02
Spiked (μg/mL) [b]	5.01 ± 0.07	4.90 ± 0.07	4.95 ± 0.07	4.94 ± 0.07	4.95 ± 0.07	4.82 ± 0.07	0.473 ± 0.007	0.67 ± 0.01	0.48 ± 0.01
After Spiked (μg/mL) [b]	45.84 ± 0.85	27.41 ± 0.44	53.55 ± 1.02	35.20 ± 0.62	42.51 ± 0.78	19.58 ± 0.27	0.774 ± 0.004	1.95 ± 0.02	1.53 ± 0.02
Detected (μg/mL) [a]	45.51 ± 1.06	27.45 ± 0.51	53.47 ± 1.30	35.25 ± 0.78	42.21 ± 1.06	19.66 ± 0.32	0.779 ± 0.002	1.96 ± 0.03	1.53 ± 0.03
Marginal recovery (%) [a]	93.6 ± 4.8	100.7 ± 1.4	98.5 ± 5.7	101.1 ± 3.6	93.9 ± 5.6	101.7 ± 1.1	101.0 ± 0.7	101.7 ± 1.5	99.9 ± 3.2
Total Recovery (%) [a]	**99.3 ± 0.5**	**100.1 ± 0.2**	**99.9 ± 0.5**	**100.1 ± 0.5**	**99.3 ± 0.7**	**100.4 ± 0.3**	**100.6 ± 0.4**	**100.6 ± 0.5**	**100.0 ± 1.0**

Note: The given values are the mean of three replicated measurements ± standard deviation [a]. The values for all the spiked and after-spiked concentrations were calculated as the mean of the three replicated plus the standard deviation [b]. The bold recoveries were applied for the final evaluation [c].

2.2.5. Precision

The precision and accuracy of the method was assessed by determining the intraday precisions (n = 5) from repeating the analysis of the kava samples on the same day; and the interday precisions (n = 3 × 5, overall 15) from analyzing the same kava samples over the different days. Both intraday and interday precisions were calculated as RSD_r (%) = (standard deviation)/(mean) × 100. As shown in Table 5, all nine analytes had adequate precision in each of the four different solid matrices at different concentrations. The level of KLs were about 10–100 mg/g for CO_2 extract and root. The level of FKs in the kava CO_2 extract and the root were about 0.13–0.30 mg/g for FKC, 0.60–1.5 mg/g for FKA, and 0.50–0.90 mg/g for FKB. The intraday repeatability relative standard deviations (RSD_r) of the CO_2 extract and the root were 0.31% to 1.61% for KLs and 0.98% to 3.83% for FKs, which are <2% for KLs and <4% for FKs, as AOAC guidelines suggested. About 2–10 mg/g for KLs, 0.08–0.10 mg/g for FKC, 0.46–0.75 mg/g for FKA, and 0.60–1.00 mg/g for FKB were detected for kava products in tablets and capsules. The intraday RSD_r of kava products in tablets and capsules were 0.28% to 1.96% for KLs and 0.23% to 4.24% for FKs, which are also <3% for KLs and <6% for FKs, as AOAC guidelines suggested. Overall, the interday repeatability relative standard deviations (RSD_r) ranged from 0.50% to 2.56% for kavalactones and 2.44% to 5.52% for flavokavains. The Horwitz ratio (HorRat) values are used to evaluate method performance based on the ratios of actual precision to predicted precision. AOAC guidelines for single-laboratory validation accept a HorRat range from 0.5 to 2. In our method, the HorRat value for kavalactones ranged from 0.24 to 1.05 and for flavokavains ranged from 0.77 to 1.89. A HorRat value lower than 0.5 was considered acceptable, considering the analysis was performed under tightly controlled conditions.

Table 5. Precision summary of the HPLC-UV method for detecting kavalactones and flavokavains in kava products.

		Methysticin	DHM	Kavain	DHK	Yangonin	DMY	FKC	FKA	FKB
	LOQ (mg/g)	0.076	0.080	0.046	0.114	0.032	0.070	0.045	0.010	0.051
						Tablets				
D1	Mean ± SD (mg/g)	3.79 ± 0.03	3.94 ± 0.03	6.71 ± 0.07	7.17 ± 0.06	3.55 ± 0.03	2.76 ± 0.03	0.082 ± 0.001	0.477 ± 0.007	0.627 ± 0.008
	RSD_r (%) Intraday	0.86	0.80	1.06	0.82	0.98	1.08	1.28	1.39	1.32
D2	Mean ± SD (mg/g)	3.83 ± 0.01	3.96 ± 0.01	6.67 ± 0.03	7.05 ± 0.04	3.58 ± 0.04	2.78 ± 0.04	0.076 ± 0.001	0.452 ± 0.002	0.587 ± 0.002
	RSD_r (%) Intraday	0.34	0.34	0.45	0.50	1.22	1.30	1.64	0.49	0.28
D3	Mean ± SD (mg/g)	3.83 ± 0.02	3.96 ± 0.01	6.68 ± 0.02	7.06 ± 0.04	3.64 ± 0.03	2.75 ± 0.02	0.078 ± 0.001	0.459 ± 0.001	0.592 ± 0.002
	RSD_r (%) Intraday	0.39	0.28	0.32	0.60	0.81	0.70	0.77	0.23	0.49
	Mean ± SD (mg/g)	3.81 ± 0.03	3.95 ± 0.02	6.69 ± 0.05	7.09 ± 0.07	3.59 ± 0.05	2.77 ± 0.03	0.079 ± 0.003	0.463 ± 0.011	0.602 ± 0.019
	RSD_r (%) Interday	0.73	0.54	0.68	1.04	1.40	1.05	3.75	2.44	3.20
	HorRat	0.32	0.24	0.32	0.50	0.60	0.43	0.91	0.77	1.05
						Capsules				
	LOQ (mg/g)	0.151	0.160	0.092	0.229	0.063	0.141	0.090	0.021	0.101
D1	Mean ± SD (mg/g)	5.87 ± 0.07	5.80 ± 0.05	9.09 ± 0.09	9.48 ± 0.11	5.85 ± 0.09	3.74 ± 0.05	0.103 ± 0.002	0.702 ± 0.008	0.920 ± 0.010
	RSD_r (%) Intraday	1.14	0.82	0.97	1.20	1.50	1.33	2.18	1.16	1.08
D2	Mean ± SD (mg/g)	5.98 ± 0.07	5.93 ± 0.08	9.33 ± 0.12	9.73 ± 0.14	6.06 ± 0.10	3.83 ± 0.07	0.094 ± 0.004	0.773 ± 0.011	0.999 ± 0.019
	RSD_r (%) Intraday	1.10	1.31	1.29	1.45	1.66	1.74	4.24	1.39	1.93
D3	Mean ± SD (mg/g)	5.99 ± 0.12	5.98 ± 0.11	9.37 ± 0.12	9.85 ± 0.13	6.07 ± 0.10	3.87 ± 0.06	0.101 ± 0.002	0.796 ± 0.011	1.012 ± 0.012
	RSD_r (%) Intraday	1.96	1.82	1.32	1.32	1.59	1.58	2.04	1.43	1.21
	Mean ± SD (mg/g)	5.95 ± 0.10	5.90 ± 0.11	9.26 ± 0.17	9.69 ± 0.20	6.00 ± 0.14	3.81 ± 0.08	0.099 ± 0.004	0.757 ± 0.042	0.977 ± 0.044
	RSD_r (%) Interday	1.67	1.84	1.80	2.06	2.26	2.06	4.51	5.56	4.49
	HorRat	0.77	0.85	0.89	1.03	1.05	0.89	1.13	1.89	1.59
						CO_2 Extract				
	LOQ (mg/g)	0.454	0.480	0.277	0.686	0.189	0.422	0.270	0.062	0.303
D1	Mean ± SD (mg/g)	42.39 ± 0.41	26.75 ± 0.29	50.5 ± 0.7	36.2 ± 0.5	32.5 ± 0.3	14.0 ± 0.1	0.29 ± 0.01	1.34 ± 0.02	0.84 ± 0.02
	RSD_r (%) Intraday	0.97	1.08	1.28	1.39	1.06	0.99	2.13	1.84	2.11
D2	Mean ± SD (mg/g)	42.81 ± 0.21	27.0 ± 0.2	50.6 ± 0.3	36.2 ± 0.4	33.1 ± 0.2	14.2 ± 0.1	0.29 ± 0.01	1.43 ± 0.02	0.85 ± 0.01
	RSD_r (%) Intraday	0.50	0.54	0.55	0.97	0.56	0.38	2.09	1.34	1.22

Table 5. *Cont.*

		Methysticin	DHM	Kavain	DHK	Yangonin	DMY	FKC	FKA	FKB
D3	Mean ± SD (mg/g)	42.75 ± 0.53	27.1 ± 0.3	50.8 ± 0.5	36.9 ± 0.4	33.3 ± 0.2	14.1 ± 0.1	0.270 ± 0.004	1.47 ± 0.02	0.89 ± 0.03
	RSD$_r$ (%) Intraday	1.25	1.08	0.88	1.18	0.74	0.68	1.59	1.15	3.83
	Mean ± SD (mg/g)	42.65 ± 0.42	27.0 ± 0.3	50.6 ± 0.5	36.4 ± 0.5	33.0 ± 0.4	14.1 ± 0.1	0.28 ± 0.01	1.41 ± 0.06	0.86 ± 0.03
	RSD$_r$ (%) Interday	0.99	1.04	0.92	1.49	1.29	0.85	3.92	4.27	3.37
	HorRat	0.62	0.60	0.59	0.91	0.78	0.45	1.15	1.60	1.17
						Root				
	LOQ (mg/g)	0.151	0.160	0.092	0.229	0.063	0.141	0.090	0.021	0.101
D1	Mean ± SD (mg/g)	20.2 ± 0.1	11.05 ± 0.05	25.9 ± 0.4	14.7 ± 0.1	18.8 ± 0.1	7.31 ± 0.05	0.128 ± 0.001	0.64 ± 0.01	0.52 ± 0.01
	RSD$_r$ (%) Intraday	0.58	0.49	1.61	0.95	0.63	0.75	1.11	0.98	1.10
D2	Mean ± SD (mg/g)	20.3 ± 0.1	11.11 ± 0.04	25.3 ± 0.2	14.8 ± 0.2	18.6 ± 0.2	7.33 ± 0.07	0.139 ± 0.003	0.70 ± 0.01	0.57 ± 0.01
	RSD$_r$ (%) Intraday	0.31	0.38	0.87	1.14	1.01	0.97	2.14	1.51	1.39
D3	Mean ± SD (mg/g)	20.00 ± 0.12	10.89 ± 0.06	24.68 ± 0.18	14.65 ± 0.08	18.55 ± 0.12	7.23 ± 0.06	0.137 ± 0.001	0.63 ± 0.01	0.514 ± 0.003
	RSD$_r$ (%) Intraday	0.60	0.54	0.73	0.57	0.63	0.88	0.95	0.81	0.66
	Mean ± SD (mg/g)	20.19 ± 0.17	11.02 ± 0.11	25.30 ± 0.59	14.74 ± 0.15	18.62 ± 0.16	7.29 ± 0.07	0.135 ± 0.006	0.66 ± 0.03	0.53 ± 0.03
	RSD$_r$ (%) Interday	0.86	0.96	2.32	1.00	0.88	1.02	4.13	4.64	5.20
	HorRat	0.48	0.49	1.33	0.53	0.48	0.49	1.09	1.55	1.69

3. Discussion

As a more applicable method, the RP-HPLC method was reported for quantitation analysis of kava products [18–22,26]. Over many studies, Alexander H. Schmidt and Imre Molnar [26] applied computer-assisted optimization in the development to achieve a great resolution for the separations of all the major KLs. The study also included the two minor KLs (DHY and THY), but not FKs. As the above description, the kava root (~ 10% KLs), the kava CO_2 extract product (~ 20% KLs), and commercial tablet and capsule kava products were analyzed for the quantitation of each kava compound, based on the standard responses of M, DHM, K, and DHK (at 239 nm) and Y, DMY, and FKs (at 350 nm). The calculated concentrations are shown in Table 5. In this study, the percentage of each KL and FK over the total KLs and FKs, calculated as follows:

$$\frac{x}{M + DHM + K + DHK + Y + DMY + FKs} \times 100 \tag{1}$$

The kavalactone and flavokavain profile were normalized to the percentage of total kavalactone content and presented in Figure 5. Although there is a large difference in the total content of kava lactones, the relative intensities of the individual compounds differ only slightly. AFS kava root and the CO_2 extract contain a very similar profile of KLs at different levels, about 100 mg KLs per gram in the root and at about 200 mg KLs per gram in the CO_2 extract. The capsule and tablet products contain about 40 and 30 mg KLs per gram.

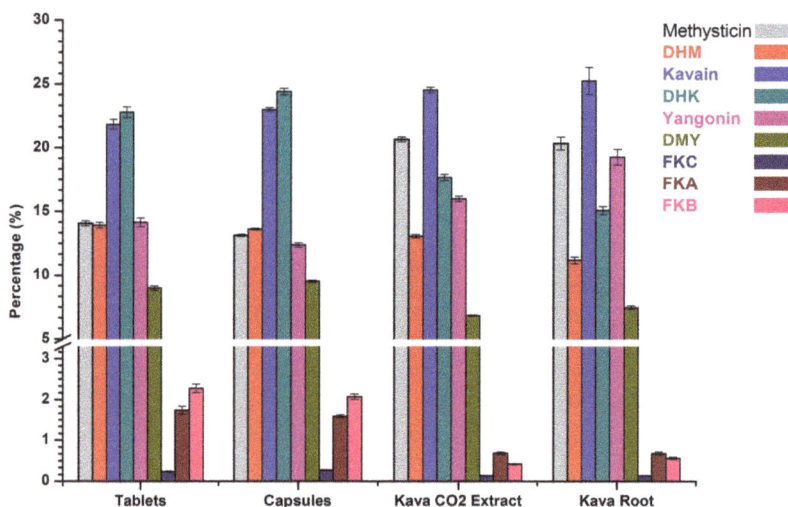

Figure 5. The profiles of kavalactones and flavokavains for kava products.

The quality of kava products was considered as a safety issue for consumers. The key criteria for the quality of kava product are the contents and chemotypes of six major KLs. Different quantity and ratio of KLs can impact their physiological action and safety [29]. The chemotypes of kava products were identified as noble or non-noble varieties following the simple system described by Lebot and Lévesque [30,31]. The six major KLs are used to define the chemotype (1 = DMY; 2 = DHK; 3 = yangonin; 4 = kavain; 5 = DHM; and 6 = methysticin). The different chemotypes of kava products are coded by listing, in decreasing order of proportion, the KLs. The noble cultivars have chemotypes rich in kavain, like 423561 or 423651. The chemotypes of 521634, 526341, or 254631 represent non-noble cultivars with very high proportions of DHM or DHK.

In this study, the chemotypes of kava products were determined by the percentage of each KL. The results show that the chemotypes of kava root (462351) were very similar to the CO_2 extract (463251), while the chemotypes of the capsule (245631) were very similar to the tablet (246531).

Due to the possible hepatotoxicities of FKs, another key criterion for the quality of kava products is to limit the amount of FKs, as FKs/KLs $\leq$ 0.29 and FKB% < 0.15% [16,17]. The ratios of FKs/KLs for kava products in this study were detected as 0.013 for the CO_2 extract, 0.014 for the root, 0.041 for the capsule, and 0.044 for the tablet. FKB was 0.086%, 0.055%, 0.06% and 0.096% for the kava CO_2 extract, the root, the tablets, and the capsules, respectively. In this study, it was noticed that cis-isomers of FKA and FKB were present at different levels in the powdered samples (Figure 4). The level of cis-isomers of FKA and FKB were found to be very low in both the kava CO_2 extract and the kava root, while a small amount of cis-isomers of FKA and FKB were observed in the capsule and tablet kava products.

4. Materials and Methods

4.1. Chemicals and Materials

The 2-propanol (HPLC grade), acetonitrile (ACN, HPLC grade), water (H2O, LC-MS grade), and methanol (MeOH, LC-MS grade) were purchased from Fisher Scientific (Hampton, NH, USA). The reference standard compounds of D, L-kavain (purity: 95% HPLC), yangonin (purity: 95% HPLC), flavokavain A (purity: 95% HPLC), flavokavain B (purity: 95% HPLC), and flavokavain C (purity: 95% HPLC) were purchased from Extrasynthese (Genay, France). The standard compounds of methysticin (purity: 99.63% HPLC), dihydromethysticin (purity 98.68% HPLC), desmethoxyyangonin (purity: 97.84% HPLC), and dihydrokavain (purity: 99.04% HPLC) products were from PhytoLab (Vestenbergsgreuth, Germany) and purchased from Cerilliant (Round Rock, TX, USA). Ultrapure (18 MΩ) water was produced using a Barnstead™ GenPure™ Pro Water Purification System from Thermo Fisher Scientific (Waltham, MA, USA).

4.2. Instrumentation

Method development and validation studies were performed on a Shimadzu Nexera-X2 UHPLC system (Shimadzu Scientific Instruments, Columbia, MD, USA), equipped with a LC-30AD pump, a SIL-30AC autosampler with a thermostated unit, a thermostated column compartment, and an SPD-M30A PDA detector. The UHPLC system was also interfaced with tandem Q-Exactive Orbitrap mass spectrometer (Thermo Fisher Scientific Inc., San Jose, CA, USA). High-resolution MS and MS^2 spectra were obtained on the Q-Exactive Orbitrap mass spectrometer equipped with a heated electrospray ionization, operated in both positive and negative ion mode. The optimized parameters were set as follows: Capillary voltage, 3.0 kV; sheath gas flow rate, 35 arbitrary unit; auxiliary gas flow rate, 5 arbitrary unit; sweep gas flow rate, 5 arbitrary unit; capillary temperature, 325 °C; and sheath gas heater temperature, 200 °C. MS scans were recorded in a mass range of m/z 100–1500 at a resolution of 70,000 with an AGC target of 3×10^6. After each MS scan, up to 5 of the most abundant multiply charged ions were selected for fragmentation. MS^2 scans were recorded in a mass range of m/z 50 to the parent ion at a resolution of 17,500, with an AGC target of 1×10^5 and a maximum fill time of 50 ms, using the stepped NCE of 25 and 35 for fragmentation in the HCD cell. Data were acquired from 50 to 1500 Da with dd-MS^2 or MS^2 in centroid mode. Raw data were acquired and processed using the Xcalibur software (Version 2.3.1, Thermo Electron Corporation, San Jose, CA, United States).

4.3. Chromatographic Condition

The chromatographic separation was extended from the previous HPLC study [24] and achieved under a gradient separation at 60 °C. Optimum separation of KLs and FKs was achieved using an UHPLC column (Acquity HSS T3, 100 mm × 2.1 mm, 1.8 µm). Gradient elution was performed using water (solvent A) and isopropanol (solvent B) with the gradient program listed in Table 1.

4.4. Standard Preparation

Individual standard stock solutions of 1000 µg/mL of each kava standard compound were accurately prepared by weighing about 10 mg of each compound and dissolving them into a 10 mL volumetric flask using acetonitrile. Volumetric flasks were sonicated for 10 min and wrapped with aluminum foil to protect them from light. Stock solutions were kept refrigerated. Working standard solutions were prepared fresh on a daily basis by pipetting aliquots of stock solutions and serial dilutions with 50% ACN were made at concentrations ranging from 0.5 µg/mL to 75 µg/mL for KLs and from 0.05 µg/mL to 7.5 µg/mL for FKs.

4.5. Test Materials and Sample Preparation

The kava CO_2 extract and root were obtained from Applied Food Sciences Inc. (AFS, Austin, TX, USA). Kava commercial products (dry-filled capsules and formulated tablets) were purchased from Foods market. All the samples were analyzed more than triplicate, unless stated otherwise.

4.5.1. Kava CO_2 Extract and Root Powders

A total of 100 mg of kava CO_2 extract or 300 mg of the root powder were first extracted with 15 mL ACN and sonicated for 30 min at 40 °C in a Fisher sonication bath. Following a 10-min centrifugation at 12,000× *g*, the supernatant was transferred to a 50 mL volumetric flask. The remaining residue was re-extracted twice with 15 mL ACN, following the same procedure. The volumetric flask was filled to the mark with ACN in the end. The samples were freshly diluted 1× with 18MW water and the extracts were filtered through a 3 mm syringe fitted with a 0.22 µm nylon filter (VWR) into an amber glass HPLC vial and readied for LC analysis.

4.5.2. Capsules and Phytocaps

The content of the 20 capsules or phytocaps were combined and mixed thoroughly. Two hundred milligrams of the capsules or phytocaps content were extracted with ACN and acetone following the same procedure as above.

4.6. Method Validation Parameters

This method was validated following the AOAC and ICH (Q2) guidelines [24,25] for conducing single-laboratory validation. For all the standards, 1000 µg/mL stock solutions were prepared by dissolving individual reference materials in ACN in volumetric flasks. The stock solutions of the reference materials were stored at −20 °C for long-term storage. The stock solution for each standard was mixed to the kava working standard solution at the concentration 100 µg/mL for KLs and 10 µg/mL for FKs, then diluted to the appropriate concentration to establish the retention time and combined at different concentration levels for external calibration.

4.6.1. Specificity/Resolution

The mixed reference standard was injected into the HPLC-UV to establish the selectivity of the method. The resolution for each reference standard was calculated. The value of $R_s > 1.5$ between closely eluting components was considered acceptable for FKs and major KLs.

4.6.2. Linearity

The linearity for the reference standard was determined by seven-point standard calibration curves. The standard curve for the six KLs ranged from 0.5 µg/mL to 75 µg/mL (0.5, 1.0, 5.0, 10.0, 25.0, 50.0, and 75.0 µg/mL). The standard curve for the three FKs ranged from 0.05 µg/mL to 7.5 µg/mL (0.05, 0.1, 0.5, 1.0, 2.5, 5.0, and 7.5 µg/mL). A simple linear regression was used to calculate R^2 value, the slope, and the y-intercept of each curve for each analyte. An $R^2 \geq 99.9\%$ value was considered acceptable. The calibration standards of seven KLs and three FKs were triplicated at the seven concentrations and analyzed over three days.

4.6.3. LOD and LOQ

The limit of detection (LOD) and limit of quantification (LOQ) of the kava standard assay were determined from the calibration curve method, as ICH Q2 (R1) recommendations [25], analyzing at least three replicates of the calibration standards. The LOD and LOQ of the proposed method were calculated using the following equations:

$$LOD = \frac{3.3 \times \text{Stdev y--intercept of Calibration Curve}}{\text{Slope of calibration curve } (A_{ave})} \tag{2}$$

$$LOQ = \frac{10 \times \text{Stdev y--intercept of Calibration Curve}}{\text{Slope of calibration curve } (A_{ave})} \tag{3}$$

4.6.4. Recovery

Spike recovery experiments were performed at three levels (high, 30 µg/mg; medium, 15 µg/mg; and low, 2.5 µg/mg) for KLs and three levels (high, 3 µg/mg; medium, 1.5 µg/mg; and low, 0.25 µg/mg) for FKs. Powdered kava root material was analyzed for KLs and FKs prior to the standards being spiked. The appropriate amount of reference standards was used to spike the powdered kava root material, followed by the extraction process. Considering the cost of the reference standards, for high-level spike recovery experiments a 5 mg sample was extracted with 5 mL extraction solvent (ACN). For medium-and low-level spike recovery experiments, a 10 mg sample was extracted with 10 mL extraction solvent (ACN). Three replicates were performed at each level and the mean recovery was calculated.

4.6.5. Precision

Four independent replicates of the same sample were prepared and analyzed on three separate days (n = 5 × 3). The within-day, between-day, overall precision for all nine target compounds were calculated for single-laboratory validation.

5. Conclusions

In conclusion, the UHPLC-UV method described herein for the determination of six major KLs and three FKs in kava raw materials and finished products was validated based on AOAC Guidelines for Single-Laboratory Validation of Chemical Methods for Dietary Supplements and Botanicals. This method maximized efficiency and chromatographic resolution under short analysis times. Both DHY and THY were found in all the kava products. The two minor KLs were well-separated from the major KLs under the excellent LC resolution. The isomerizations of yangonin and FKs were prevented or limited by the usage of non-alcoholic solvents, like acetonitrile, for sample preparation in this method. This suggested method of kava analysis is free of interference from minor KLs, like THY & DHY, and at a very low level of interferences from cis-isomers of yangonin and FKs. The results of the study demonstrate that this UHPLC-UV analytical method is a successful approach to determine methysticin, DHM, kavain, DHK, yangonin, DMY, FKA, FKB, and FKC in kava raw materials (kava root powder and kava CO_2 extract) and finished products (dry-filled capsules or formulated tablets) under a quick analysis time of 15 min and it therefore expands the scope to analyze a broad variety of market samples.

Supplementary Materials: The supplementary materials are available online.

Author Contributions: Y.T.—Study concept and design, acquisition of data, analysis and interpretation of data, and drafting of the manuscript. C.F.—Study concept and design, analysis and interpretation of data, and critical review of the manuscript.

Funding: Applied Food Sciences, Inc. (Austin, TX) funded this study and provided samples (kava root and KAVOA™ Kava CO_2 extract).

Conflicts of Interest: The authors declare no conflict of interest.

References

1. Singh, Y.N. Kava-an Overview. *J. Ethnopharmacol.* **1992**, *37*, 13–45. [CrossRef]
2. Pluskal, T.; Torrens-Spence, M.P.; Fallon, T.R.; de Abreu, A.; Shi, C.H.; Weng, J. The biosynthetic origin of psychoactive kavalactones in kava. *Febs. Open Bio.* **2018**, *8*, 87.
3. Zhang, N.; Wu, L.; Liu, X.; Shen, X. Plant-Derived Kavalactones and Their Bioactivities. *Med. Res.* **2018**, 2. [CrossRef]
4. Naumov, P.; Dragull, K.; Yoshioka, M.; Tang, C.S.; Ng, S.W. Structural characterization of genuine (−)-pipermethystine, (−)-epoxypipermethystine, (+)-dihydromethysticin and yangonin from the kava plant (Piper methysticum). *Nat. Product Commun.* **2008**, *3*, 1333–1336. [CrossRef]
5. Dragull, K.; Yoshida, W.Y.; Tang, C.S. Piperidine alkaloids from Piper methysticum. *Phytochemistry* **2003**, *63*, 193–198. [CrossRef]
6. Wu, D.; Nair, M.G.; DeWitt, D.L. Novel compounds from Piper methysticum Forst (*Kava Kava*) roots and their effect on cyclooxygenase enzyme. *J. Agric. Food Chem.* **2002**, *50*, 701–705. [CrossRef] [PubMed]
7. Dharmaratne, H.R.W.; Nanayakkara, N.P.D.; Khan, I.A. Kavalactones from Piper methysticum, and their C-13 NMR spectroscopic analyses. *Phytochemistry* **2002**, *59*, 429–433. [CrossRef]
8. World Health Organization. *Assessment of the Risk of Hepatotoxicity with Kava Products*; World Health Organization: Geneva, Switzerland, 2007.
9. Furbee, R.B.; Barlotta, K.S.; Allen, M.K.; Holstege, C.P. Hepatotoxicity associated with herbal products. *Clin. Lab. Med.* **2006**, *26*, 227. [CrossRef]
10. Ulbricht, C.; Basch, E.; Boon, H.; Ernst, E.; Hammerness, P.; Sollars, D.; Tsourounis, C.; Woods, J.; Bent, S. Safety review of kava (*Piper methysticum*) by the Natural Standard Research Collaboration. *Expert. Opin. Drug Saf.* **2005**, *4*, 779–794. [CrossRef]
11. Schmidt, M. Are kavalactones the hepatotoxic principle of kava extracts? The pitfalls of the glutathione theory. *J. Altern. Complement. Med.* **2003**, *9*, 183–187. [CrossRef]
12. Wu, D.; Yu, L.; Nair, M.G.; DeWitt, D.L.; Ramsewak, R.S. Cyclooxygenase enzyme inhibitory compounds with antioxidant activities from Piper methysticum (*kava kava*) roots. *Phytomedicine* **2002**, *9*, 41–47. [CrossRef] [PubMed]
13. Mathews, J.M.; Etheridge, A.S.; Black, S.R. Inhibition of human cytochrome P450 activities by kava extract and kavalactones. *Drug Metab. Dispos.* **2002**, *30*, 1153–1157. [CrossRef] [PubMed]
14. Zhou, P.; Gross, S.; Liu, J.H.; Yu, B.Y.; Feng, L.L.; Nolta, J.; Sharma, V.; Piwnica-Worms, D.; Qiu, S.X. Flavokawain B, the hepatotoxic constituent from kava root, induces GSH-sensitive oxidative stress through modulation of IKK/NF-kappaB and MAPK signaling pathways. *FASEB J.* **2010**, *24*, 4722–4732. [CrossRef] [PubMed]
15. Jhoo, J.W.; Freeman, J.P.; Heinze, T.M.; Moody, J.D.; Schnackenberg, L.K.; Beger, R.D.; Dragull, K.; Tang, C.S.; Ang, C.Y.W. In vitro cytotoxicity of nonpolar constituents from different parts of kava plant (*Piper methysticum*). *J. Agr. Food Chem.* **2006**, *54*, 3157–3162. [CrossRef]
16. Fratini Vergano Law Firm. Scientific and Legal assistance for the Development of a Quality and Safety Standard for Kava Production and Trade in the Pacific Region. Available online: http://www.acp-eu-tbt.org/pagenewsarc.cfm?id=279CC3E89126DCC6C78B90A8D1ED66FFD4E480FBF584FA6D1FCEE8E906FEFEC8DD (accessed on 30 November 2015).
17. Lebot, V.; Do, T.K.T.; Legendre, L. Detection of flavokavins (A, B, C) in cultivars of kava (*Piper methysticum*) using high performance thin layer chromatography (HPTLC). *Food Chem.* **2014**, *151*, 554–560. [CrossRef]
18. Meissner, O.; Haberlein, H. HPLC analysis of flavokavins and kavapyrones from Piper methysticum Forst. *J. Chromatogr. B Analyt. Technol. Biomed. Life Sci.* **2005**, *826*, 46–49. [CrossRef] [PubMed]
19. Murauer, A.; Ganzera, M. Quantitative Determination of Lactones in Piper methysticum (*Kava-Kava*) by Supercritical Fluid Chromatography. *Planta Med.* **2017**, *83*, 1053–1057. [CrossRef]

20. Wang, J.; Qu, W.Y.; Jun, S.J.; Bittenbender, H.C.; Li, Q.X. Rapid determination of six kavalactones in kava root and rhizome samples using Fourier transform infrared spectroscopy and multivariate analysis in comparison with gas chromatography. *Anal. Methods UK* **2010**, *2*, 492–498. [CrossRef]

21. Bilia, A.R.; Scalise, L.; Bergonzi, M.C.; Vincieri, F.F. Analysis of kavalactones from Piper methysticum (kava-kava). *J. Chromatogr. B Analyt. Technol. Biomed. Life Sci.* **2004**, *812*, 203–214. [CrossRef]

22. Liu, Y.; Lund, J.A.; Murch, S.J.; Brown, P.N. Single-Lab Validation for Determination of Kavalactones and Flavokavains in Piper methysticum (Kava). *Planta Med.* **2018**, *84*, 1213–1218. [CrossRef]

23. Wang, Y.; Eans, S.O.; Stacy, H.M.; Narayanapillai, S.C.; Sharma, A.; Fujioka, N.; Haddad, L.; McLaughlin, J.; Avery, B.A.; Xing, C. A stable isotope dilution tandem mass spectrometry method of major kavalactones and its applications. *PLoS ONE* **2018**, *13*, e0197940. [CrossRef]

24. Horwitz, W. *AOAC Guidelines for Single Laboratory Validation of Chemical Methods for Dietary Supplements and Botanicals*; AOAC International: Gaithersburg, MD, USA, 2002; p. 38.

25. Guideline, I.H.T. Validation of Analytical Procedures: Text and Methodology. In Proceedings of the International Conference on Harmonization, Geneva, Switzerland, November 2005.

26. Schmidt, A.H.; Molnar, I. Computer-assisted optimization in the development of a high-performance liquid chromatographic method for the analysis of kava pyrones in Piper methysticum preparations (vol 948, pg 51, 2002). *J. Chromatogr. A* **2006**, *1110*, 272. [CrossRef]

27. Smith, R.M.; Thakrar, H.; Arowolo, T.A.; Shafi, A.A. High-Performance Liquid-Chromatography of Kava Lactones from Piper-Methysticum. *J. Chromatogr.* **1984**, *283*, 303–308. [CrossRef]

28. Bobeldijk, I.; Boonzaaijer, G.; Spies-Faber, E.J.; Vaes, W.H.J. Determination of kava lactones in food supplements by liquid chromatography-atmospheric pressure chemical ionisation tandem mass spectrometry. *J. Chromatogr. A* **2005**, *1067*, 107–114. [CrossRef] [PubMed]

29. Lebot, V.; Merlin, M.; Lindstrom, L. *Kava: The Pacific Drug*; Yale University Press: New Haven, CT, USA, 1992; p. 255.

30. Lebot, V.; Le´vesque, J. The origin and distribution of kava (Piper methysticum Forst. f. Piperaceae): A phytochemical approach. *Allertonia* **1989**, *5*, 223–380.

31. Lebot, V.; Le´vesque, J. Genetic control of kavalactone chemo-types in Piper methysticum cultivars. *Phytochemistry* **1996**, *43*, 397–403. [CrossRef]

Sample Availability: Not Available.

Review

Analysis of Enantiomers in Products of Food Interest

Chiara Fanali [1,*], Giovanni D'Orazio [2], Alessandra Gentili [3] and Salvatore Fanali [4]

[1] Department of Medicine, University Campus Bio-Medico of Rome, Via Alvaro del Portillo 21, 00128 Rome, Italy

[2] Istituto per I Sistemi Biologici, Consiglio Nazionale delle Ricerche, Via Salaria km 29, 300-00015 Monterotondo, Italy; giovanni.dorazio@cnr.it

[3] Department of Chemistry, University of Rome "La Sapienza", Piazzale Aldo Moro 5, P.O. Box 34, Posta 62, 00185 Roma, Italy; alessandra.gentili@uniroma1.it

[4] Teaching Committee of Ph.D. School in Natural Science and Engineering, University of Verona, 37134 Verona, Italy; salvatore.fanali@gmail.com

* Correspondence: c.fanali@unicampus.it; Tel.: +39-06225419471

Received: 21 February 2019; Accepted: 20 March 2019; Published: 21 March 2019

Abstract: The separation of enantiomers has been started in the past and continues to be a topic of great interest in various fields of research, mainly because these compounds could be involved in biological processes such as, for example, those related to human health. Great attention has been devoted to studies for the analysis of enantiomers present in food products in order to assess authenticity and safety. The separation of these compounds can be carried out utilizing analytical techniques such as gas chromatography, high-performance liquid chromatography, supercritical fluid chromatography, and other methods. The separation is performed mainly employing chromatographic columns containing particles modified with chiral selectors (CS). Among the CS used, modified polysaccharides, glycopeptide antibiotics, and cyclodextrins are currently applied.

Keywords: chiral; chiral stationary phases; enantiomers; food; review

---

1. Introduction

In recent years, in the field of separation science, there has been great attention to analyzing products of food interest. All this has been done to meet the needs of both the industries operating in the field and the control laboratories belonging to the various national and international agencies. The use of modern and reliable analytical methods permits knowing the chemical composition of these products allowing, e.g., to determine their quality, to trace problems related to production, and storage processes, etc. In addition, the presence of dangerous pollutants such as pesticides present in foodstuff and/or additives can also be quantified.

A large number of compounds present in food products have been successfully examined, e.g., amino acids, proteins, carbohydrates, vitamins, lipids, mycotoxins, colorants, preservatives, herbicides, ionic compounds, fungicides, and enantiomers, etc.

Among all the classes of compounds analyzed so far, particular attention has been paid to the separation and determination of chiral compounds. Two enantiomers have the same chemical composition with, e.g., one asymmetric center, and due to the different spatial orientation of the substituent groups, they exhibit quite similar physical-chemical properties. However, due to the chirality, they can participate, with a different effect, to the various biochemical processes. In these processes, enantiomers could react differently with, e.g., enzymes, proteins, and peptides, etc., present in humans determining beneficial or dangerous effects. In addition, natural products must contain only one enantiomer as in the case of fruit juices where only the *L*-amino acid must be present. The existence of the antipode can indicate either an adulteration (addition of a racemate) or other problems related to poor storage. Therefore, it is very important to have reliable analytical methods able to determine qualitatively and quantitatively the enantiomeric forms present in samples of food interest.

Analytical techniques so far used for the analysis of enantiomers include: Gas chromatography (GC) [1], supercritical fluid chromatography (SFC) [2,3], thin layer chromatography (TLC) [4], high-performance liquid chromatography (HPLC) [5], capillary electrophoresis (CE), and capillary-/nano-liquid chromatography (CLC/nano-LC) [6,7], etc.

In this review, general principles of enantiomers separation utilizing chromatographic techniques and the most applied chiral stationary phases (CSPs) are reported and discussed. In addition, the most important applications of some analytical techniques to the analysis of chiral compounds in food matrices published in the last two years are presented.

2. General Separation Methods in Chiral Resolution

As previously mentioned, the very similar physical-chemical properties of two enantiomers are the main reasons for the difficulty in obtaining their resolution by using conventional methods such as, for example, reversed phase one in liquid chromatography. However, their separation can be easily achieved utilizing a chiral background.

Generally, two resolution methods can be used, namely indirect and direct ones. In the first procedure, the two compounds react with a chiral selector (CS) on forming stable diastereoisomers where strong bonds are formed. The two new compounds can be separated employing a non-chiral stationary phase (SP). The separation method offers some benefits, e.g., the introduction of additional groups useful for i) further interactions and ii) increase the sensitivity when UV and mass spectrometry (MS) detectors are applied. However, a high reagent purity is necessary and it is time-consuming [8].

The direct resolution method is the most applied in separation science. It is based on the use of a CSP where the two analytes interact continuously on forming diastereoisomeric complexes, and involving weak bonds (hydrogen, π-π, hydrophobic etc.). While some models have been proposed for chiral recognition, the "Three point" interaction could be considered in order to achieve compounds separation. However, enantiomers resolution can also be obtained when repulsion, in addition to interactions, are involved in the stereoselective mechanism [9].

Figure 1 shows a scheme of the "Three-point" interaction model.

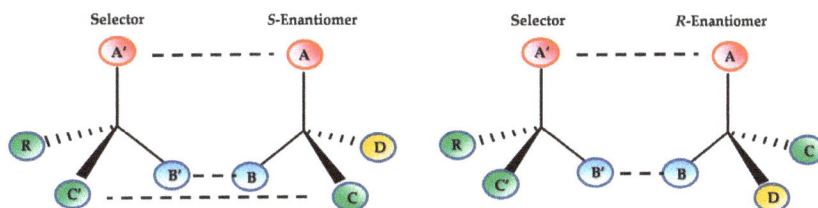

Figure 1. Scheme of "Three-point" interaction model.

Even if the number of CSs, applied to the resolution of enantiomers is quite large and varied, it is noteworthy to mention that it is not possible to find a universal one. They have been utilized employing different analytical techniques, e.g., GC, SFC, HPLC, CLC/nano-LC, and CE etc. [8,10–13].

Among all the chiral selectors used so far with the different analytical techniques, those that deserve special attention are peptides, chiral amino acids, cyclodextrins, polysaccharide derivatives, quinine-based, glycopeptides antibiotics, and chiral crown-ethers, etc.

3. The Most Used Chiral Stationary Phases for Enantiomers Separation

In this section, the most applied CSPs to the separation of enantiomers utilizing the analytical techniques above mentioned will be summarized. Some of them have been applied to the analysis of compounds of food interest. Over the years, a large number of CSPs has been prepared and studied for enantiomers separation. Several of them are commercially available under different trade name also containing the same CS. This is the case of Lux or Chiralpak by Phenomenex (USA) and Chiral Technologies (USA), respectively. They make use of modified polysaccharides either coated or immobilized on silica particles. In order to improve the enantioresolution capacity of this CS type, cellulose or amylose have been modified with phenylcarbamate groups containing substituents such as methyl, chloro, and bromo, etc., alone or in combination. The presence of these substituents with electron-withdrawing or electron-donating modifies the properties of the phenyl promoting different interactions with the enantiomers to be separated [14]. Concerning the interactions involved in the chiral recognition when polysaccharides are employed, hydrogen, π-π bonding have to be mentioned. The inert support is usually silica of different dimensions (sub-2 μm to 5μm) porous or recently core-shell [15,16]. Monolithic material has also been used [17,18]. These CSPs have been largely employed in HPLC [5,19], SFC [11,20], capillary electrochromatography (CEC) [21,22], and nano-LC [23].

Another CS, widely applied to enantiomers separation, deserving attention, includes cyclodextrins (CDs) or their derivatives. It can be either bonded to silica particles or monolithic material or rarely added to the mobile phase. In addition it can be coated on the capillary wall, as used in GC [24]. CDs or their derivatives have been widely used in capillary electrophoresis added to the background electrolyte [12]. CDs are oligosaccharides with a shape like a truncated cone with a hydrophobic cavity and a hydrophilic outside (presence of hydroxyl groups). The recognition mechanism is based on inclusion complexation where the two enantiomers fit into the cavity. Stereososelective interactions with hydroxyl or substituent groups (e.g., methyl, ethyl, and hydroxypropyl, etc.) take place. However, adsorption interactions could also be involved, especially if organic solvents are employed in the mobile phase. Schurig has showed a unified enantioselective approach for enantiomers separation. Here the same capillary column (50 μm i.d.) containing a permethylated-β-CD, thermally coated, has been employed in both open-GC, -SFC, -LC, and –CEC [25].

Among other CS used for enantiomers separation, chiral crown-ethers have a certain interest. These CSP type are commercially available, e.g., crownpakcr(+) and cr(−) (Chiral technologies, USA) or Larihc CF6- P (AZYP, USA). The second column contains cyclofructans groups forming a basket structure based on crown-ether. Therefore, enantiomers enter into the basket and on the contrary of CDs, the hydrophilic groups interact with the cavity. This CSP type has been used for the enantiomers separation of compounds containing amino groups in their structure (amino acids, amines, and derivatives). Recently Armstrong's group reported the screening of 119 primary amine-containing compounds achieving enantioresolution for 92% of analytes [26]. Finally, it is noteworthy to mention the use of ion-exchange type CSP [27–30] and those containing glycopeptides antibiotics (vancomycin, teicoplanin, etc.). All of them are commercially available columns offering good/high enantioselectivity towards a large number and type of compounds [31–35].

Glycopeptide antibiotics contain in their chemical structure a consistent number of asymmetric centers, amino, amide, aromatic, carboxylic groups offering different interaction types with the analyzed enantiomers. The recognition mechanism is based on the affinity interaction strongly influenced by the type of mobile phase, pH, ionic strength, organic solvent type, and concentration, etc. They have been used bonded to either porous or core-shell particles (1.7–5 μm diameter) [23,36].

4. Some Selected Applications to Enantiomers Separation in Food Chemistry

This section describes the recent data present in the literature reported in 2017–2018 and related to the enantiomers separation in food samples. Data on this topic, related to previous years, can be found in previous reviews [23,37].

4.1. Supercritical Fluid Chromatography

Although SFC is not a recent analytical technique, it has enormous potential for both analytical and preparative purposes. Among its main features, we can mention: Fast balance of the columns, high efficiency, and use of non-hazardous solvents. Making use of stationary phases, developed for HPLC applications, SFC has also been applied to the separation of enantiomers. Among these CSPs, those based on polysaccharides, cyclodextrins, vancomycin, and Pirkle-type are the most employed ones. The mobile phase contains carbon dioxide often modified with some organic solvents such as methanol, ethanol, isopropanol, and acetonitrile at relatively low concentrations. In addition to enhancing chiral resolution, other additives, e.g., formic acid (FA), acetic acid (HAc), trifluoroacetic acid (TFA), ammonia (NH_3), and diethylamine (DEA) have been used [37].

As can be observed in Table 1 six papers, dealing with enantiomers separation by SFC in samples of food interest, appeared in the considered last two years. In the developed analytical methods, chiral resolution was obtained utilizing CSP based on vancomycin and amylose or cellulose derivative.

Prothioconazole is a triazole fungicide widely used for its curative and protective action. Its enantiomers have been separated and analyzed by Jiang et al. [38] by using SFC. Analytes were extracted in a food sample (tomatoes) applying the Quick Easy Cheap Effective Rugged Safe (QuEChERS) sample preparation method and analyzed in a silica column coated with a modified cellulose derivative. Chiral separation was achieved in less than four minutes with good precision, accuracy, and recovery. In this study, method optimization was done by investigating some experimental parameters, e.g., the different type of polysaccharide columns, the composition of the mobile phase (carbon dioxide with additives such as methanol, ethanol, isopropanol-IPA). Best results were obtained using IPA as an additive at 30% (v/v) concentration. Another CS, namely amylose tris-(3,5-dimethylphenylcarbamate), coated on silica particles was used for the chiral separation of fenbuconazol and its metabolites by SFC. Analytes were also characterized by using MS/MS. Fenbuconazole is a compound with fungicidal activity, currently used for the treatment of fruits and vegetables. Two of its degradation lactone metabolites could contaminate groundwater and therefore their determination is of great interest. Very good enantiomers separation was obtained by SFC using a silica column coated with amylose tris-(3,5-dimethylphenylcarbamate) trough CO_2/ethanol applying a gradient elution mode. Sample preparation was done employing a QuEChERS method. The six stereoisomers were separated in less than four min. In order to increase the MS ionization, 0.1% (v/v) formic acid/methanol compensation solution was added. The optimized method was validated with various samples, e.g., tomatoes, cucumbers, apples, peaches, rice, and wheat. Application to a real sample (greenhouse cucumbers) revealed the presence of only the two enantiomers of fenbuconazol [39]. Figure 2 reports the chemical structure of the two enantiomers of fenbuconazol and its diastereoisomers.

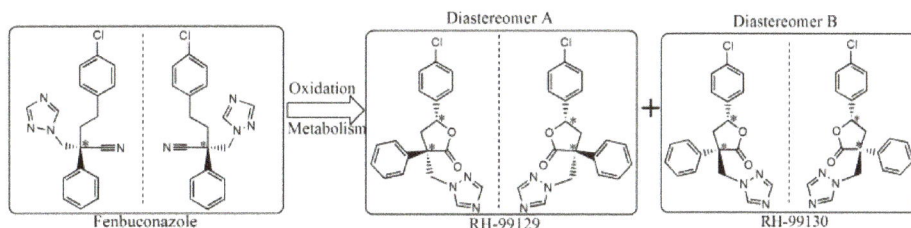

Figure 2. Chemical structure of fenbuconazol enantiomers and its metabolite diastereoisomers. Reproduced with permission of Elsevier from ref. [39].

Another polysaccharide based column (tris(3,5-dimethylphenylcarbamoyl) cellulose) was used for analysis of triticonazole enantiomers in cucumbers and tomatoes with the same method. The separation was achieved in three min (in HPLC, eighteen min) eluting with CO_2-ethanol (80:20, v/v) [40]. Four stereoisomers of propiconazole, a widely used triazole fungicide, were separated by SFC using different polysaccharide SPs. Among them, Chiralpak AD3 and Chiralpak IA3 allowed the separation of all stereoisomers in less than five min [41]. The two SPs contained amylose modified with tris(3,5-dimethylphenylcarbamate) coated and immobilized, respectively. Retention order was assessed by measuring the optical rotation of the four studied compounds. The optimized method was validated and applied to extract samples of wheat, grape, and soil matrices with good results concerning recovery.

4.2. High-Performance Liquid Chromatography

Selected applications to the analysis of chiral compounds utilizing HPLC are reported in Table 2 and described in this section.

Clenbuterol is a β_2 agonist that is used in the treatment of respiratory disorders in humans. However, this drug, in some countries, even if forbidden, is also administered to the animals. The two enantiomers have been recently separated utilizing both HPLC and SFC employing columns containing vancomycin or teicoplanin [42]. Baseline separation of the two enantiomers was obtained with both techniques, however, a shorter analysis time was observed utilizing SFC (3.5 and 6 min, respectively). The methods were applied to the analysis of meat samples and urines of humans after eating the meat. Cattle meat contained enriched R-(−)-clenbuterol. More recently, the two analytical techniques have also been used by Chen and Zhang [43] for the chiral resolution of some isobutylhydroxyamides. Different commercial columns, amylose-based have been employed. While analyzed compounds belonging to the same class, two different techniques and different columns had to be used. These compounds are quite important in the field of phytochemistry and traditional Chinese medicine; they are present in Sichuan pepper. In this study, authors separated and characterized several isobutylhydroxyamides enantiomers also studying protective effects (PC12 Cells), reporting that compounds with R configuration resulted protective on the contrary of the S form.

Vegetables such as spinach, tomatoes, and cucumber have been analyzed for the determination of R and S-titriconazol using HPLC. This compound belongs to the family of fungicides used in agriculture. In this study, the single enantiomers and the racemic mixture were used and the bioactivity was investigated. The analyzed vegetables contained a high concentration of titriconazol after foliar spraying. However, the concentration of the two enantiomers decreased along the time. The R-isomer was dissipated more than its antipode. Therefore, it can be concluded that using the single enantiomer (R-) would reduce potential health risk [44]. A different fungicide, pyrisoxazole was studied by Yang et al. [45]. Enantiomers were separated with a commercial cellulose tris(4-methylbenzoate) column by HPLC in less than 10 min. This fungicide contains two asymmetric centers and therefore two couples of enantiomers. A similar study was carried out by Wang et al. [46] to study the degradation of isofenphos-methyl, a pesticide, after treating cowpea, cucumber, and pepper.

Table 1. Selected applications of supercritical fluid chromatography (SFC) to Food analysis.

Samples	Matrix	Sample Preparation	Technique	Column and Chiral Stationary Phase	Mobile Phase	Detection	References
(R) and (S)-Prothioconazole	Tomatoes	QuEChERS	SFC	cellulose tris(3,5-dimethylphenylcarbamate)-silica coated, EnantioPak (China) (150 × 4.6 mm, 5 μm)	CO_2/2-propanol (80:20, v/v); 2.5 mL/min	UV, 254 nm	[38]
Fenbuconazole and metabolites	tomatoes, cucumbers, apples, peaches, rice and wheat,	QuEChERS	SFC	amylose tris-(3,5-dimethylphenylcarbamate)-coated chiral column	CO_2/ethanol; 1.8 mL/min	MS/MS	[39]
Triticonazole	cucumbers and tomatoes	QuEChERS	SFC	tris(3,5-dimethylphenylcarbamoyl) cellulose-coated silica gel EnantioPak OD column (China) (150 mm × 4.6 mm, 5 μm)	CO_2/ethanol (80:20, v/v)	-	[40]
Propiconazole	Wheat	Solid phase	SFC	Amylose tris(3,5-dimethylphenylcarbamate) and immobilized Amylose tris(3,5-dimethylphenylcarbamate), Chiralpak AD-3 and IA3, respectively (Daicel Chemical Industries, Japan) (150 mm × 4.6 mm i.d., 3 μm particle size)	CO_2/ethanol (93:7, v/v)	MS/MS	[41]
Clenbuterol	Meat	Liquid-liquid extractionand Solid phase	SFC or HPLC	Vancomycin, Astecchirobiotic V2 (150 × 4.6 mm or 2.1 mm, particle size 5 μm); Teicoplanin, Astecchirobiotic T (150 × 4.6 mm or 2.1 mm, particlesize 5 μm);	CO_2/ammonia or formic acid in SFC and MeOH (with water or ACN or 2-propanol) in HPLC	MS/MS	[42]
Isobutylhydroxyamides ZP'-amide A and ZP'-amide B	Pepper	Liquid-liquid extraction	SFC or HPLC	Amylose tris-(3,5-dimethylphenylcarbamate) coated, Chiralpak AD-H column (chiral column 1); Amylose tris(5-chloro-2-methylphenylcarbamate) coated, Chiralpak AY-H column (chiral column 2) (15 cm × 3 cm i.d., 5 μm; 15 cm × 3 cm i.d., 5 μm); Chiralpak AD-H column (chiral column 3); (25 cm × 0.46 cm i.d., 5 μm, 15 cm × 0.46 cm i.d.)	SFC, (40% MeOH, 3 mL/min); HPLC different MP (ethanol/diethylamine (100:0.1 v/v); 0.2 mL/min or n-hexane/isopropanol/ diethylamine(85:15:0.1 $v/v/v$); 1 mL/min or n-hexane/ethanol/ diethylamine (85:15:0.1 $v/v/v$); 1 mL/min)	MS	[43]

Table 2. Selected applications of HPLC to Food analysis.

Samples	Matrix	Sample Preparation	Technique	Column and Chiral Stationary Phase	Mobile Phase	Detection	References
(R,S)-triticonazole	Vegetables (tomatos, spinach, cucumber)	Liquid-liquid extraction and Solid phase	HPLC	cellulose tris (3-chloro-4-methylphenylcarbamate) (Lux Cellulose-2); 3-μm particles (250 mm or 4.6 mm i.d.)	-	UV	[44]
Isofenphos-methyl, (R,S)-O-methyl-O-(2-isopropoxcarbonyl)-phenyl-N-isopropylphosphora-midothioate	cowpea, cucumber, and pepper	QuEChERS	HPLC-MS/MS	Lux cellulose-3 chiral, column (250 mm × 4.6 mm i.d., 5 μm,	Isocratic, 0.80 mL/min	MS	[46]
Metconazole	Flour	QuEChERS	HPLC	Cellulose 3, 5-dimethylphenylcarbamate Enantiopak OD column	n-hexane-ethanol mixture (97:3, v/v) at the flow rate of 1.0 mL/min	UV, 220 nm	[47]
Thiols 3-sulfanylhexan-1-ol and its O-acetate, 3-sulfanylhexyl acetate	Wine	Solid phase	HPLC	Lux Amilose 1 and 2, cellulose 1	Gradient, 5 mM aqueous ammonium bicarbonate (A, pH 8.7) and acetonitrile (B)	MS/MS	[48]
8-O-4' type neolignans	Raspberry	-	HPLC	Chiralpak AD-H 250 × 4.6 mm, 5 μm	2-propanol-n-hexane (various ratios)	Polarimeter	[49]
Fungicide pyrisoxazole	Pakchoi, pepper, cabbage	QuEChERS	HPLC	cellulose tris(4-methylbenzoate)-Lux Cellulose-3, 150 mm × 2.0 mm, 3 μm	methanol and water (70:30 v/v), 0.35 mL/min	MS/MS	[45]
Chloramphenicol	Honey	Liquid-liquid extraction	HPLC	Chiralpak AGP, 3 × 5 mm, 5 μm	Gradient, water with 0.01% acetic acid (A) and methanol with 0.01% acetic acid (B)	MS/MS	[50]
α- and γ-Hexabromocyclododecanes	egg	Soxhlet extraction	HPLC	Permethylated-β-cyclodextrin–Chiral column Nucleosilβ-PM Macherey-Nagel(GmbH & Co., Düren, Germany), 20 cm × 4 mm × 5 μm	-	MS	[51]
Fluazifop-butyl and fluazifop	tomato, cucumber, pakchoi, rape	Liquid-liquid extraction	HPLC	Cellulose tris(3,5-dichlorophenylcarbamate) Chiralpak IC, 2504.6 mm I.D., 5μm particles	Not reported	MS/MS	[52]
Triacylglycerol	Chicken yolk and meat	Liquid-liquid extraction	HPLC	CHIRALCEL OD-3R, (Daicel Corporation, Tokyo, Japan) 4.6 mm i.d. ×150 mm,	Methanol, 0.5 mL min^{-1}	MS	[53]
Amino acids several	Vinegar, milk, kimchi, yogurt	Liquid-liquid extraction	HPLC	CROWNPAK CRIchiral column.	Acetonitrile/ethanol/water/TFA (80/15/5/0.5, $v/v/v/v$)	TOFMS	[54]
Amino acids	Chimchi (fermented vegetables)	Liquid-liquid extraction	HPLC	CROWNPAK CR-I(+) column, 3.0 mm i.d. 150 mm; particle size, 5 μm	Acetonitrile, ethanol, water, and TFA (80:15:5:0.5, $v/v/v/v$)	TOFMS	[55]

Table 2. *Cont.*

Samples	Matrix	Sample Preparation	Technique	Column and Chiral Stationary Phase	Mobile Phase	Detection	References
Amino acids	Vinegars	Liquid-liquid extraction	HPLC	CROWNPAK CR-I(+) and CR-I(−)(Daicel CPI, Osaka, Japan) (3.0 mm i.d. 150 mm, 5 μm)	acetonitrile, ethanol, water and TFA (80:15:5:0.5, $v/v/v/v$)	MS/MS	[56]
Triazole fungicide (paclobutrazol, myclobutanil, diniconazole, epoxiconazole)	Honey	Solid phase	HPLC	Chiralcel OD-RH column (150 mm × 4.6 mm, 5 μm, daicel, Japan)	ACN/2mM ammonium acetate, 50:42 (v/v)	MS/MS	[57]
Pesticides	cucumber, tomato, cabbage, grape, mulberry, apple and pear	Magnetic solid phase extraction	HPLC	Chiralpak IG column (250 mm × 4.6 mm, i.d. 5 μm, Daicel, Japan)	ACN/ water containing 5 mmol L^{-1} ammonium acetate and 0.1% (v/v) formic acid (65:35, v/v)	MS/MS	[58]
Fungicide prothioconazole and metabolites	Cucumber, pear	QuEChERS and Solid phase	UHPLC and HPLC	Cellulose-tris(4-methylbenzoate) Lux Cellulose-3, 2 or 3, 250 mm × 4.6 mm i.d., 5 μm and 150 mm × 2 mm i.d., 3 μm,	Acetonitrile:water	MS/MS	[59]
(R,S)-zoxamide	wine	-	UHPLC	Lux Amylose-2 chiral column (150 mm × 2 mm, 3 μm particle	acetonitrile and water (70:30, v/v), 0.5 mL/min	MS/MS	[60]
Zoxamide	Vegetable (tomato, cucumber), pepper, potato, grape, strawberry	-	UHPLC	Lux Amylose-2, 150 mm × 2 mm, 3 μm particle size	acetonitrile/water (70:30 v/v), 0.5 mL/min	MS/MS	[61]
Amino acids (derivatized with fluorescein isothiocyanate, FITC) (glutamic acid, aspartic acid, isoleucine, tryptophan, phenylalanine, tyrosine, histidine, proline)	Apple juice	-	Nano-LC (open tubular)	polymerization of 3-chloro-2-hydroxypropylmethacrylate (HPMA-Cl) and ethylene dimethacrylate (EDMA) with bonded β-cyclodextrin; 15 cm and i.d. 75 μm	acetonitrile:methanol:H2O at 0.1% v/v TFA (85:10:5, $v/v/v$); flow rate; 800 nL/min	UV, 214 nm	[62]

The two enantiomers were analyzed by HPLC and detected with MS/MS utilizing a commercial cellulose-based column (modified with 3-chloro-4-methylphenylcarbamate). Based on the presented results, the type of vegetable strongly influenced the degradation process, e.g., (*R*)-(−)-isofenphos-methyl was faster than (*S*)-(+)-isofenphos-methyl in cowpea and cucumber, while the opposite was observed in pepper. While showing interesting results, the authors did not further report about the reasons for the differences found.

Recently, an interesting HPLC method was studied utilizing cellulose or amylose 3, 5-dimethylphenylcarbamate for the separation of the four isomers of metconazole. An optimum resolution was achieved, after studying various mobile phases, employing the cellulose-based column and eluting with an organic solvent (hexane) with 3%, *v*/*v* of ethanol. The method was validated and applied to the analysis of the four compounds in flour. While the proposed method was carefully optimized studying the effect of various parameters on the compounds separation, the real sample was spiked and there was no finding of the transformation of the enantiomers/diastereoisomers [47].

The use of an HPLC method, employing polysaccharide-based columns (cellulose or amylose derivatives) allowed the chiral resolution of Thiols 3-sulfanylhexan-1-ol (**1**) and its O-acetate, 3-sulfanylhexyl acetate (**2**) after derivatization with 4-thiopyridine. The two compounds are quite important in assessing wine quality because of their potent aroma properties. In order to increase the MS signal and to have additional interaction groups for stereoselective interactions with the SP, the derivatization was helpful. Concerning enantiomers, the authors found that dry wines contained the two enantiomers of **1** and **2** at the same concentration, while in botrysed wines, elevated concentrations of *S*-form were found [48].

Chiralpak AD-H was used for semipreparative purposes for the enantiomers separation of some neolignans present in raspberry. A quantity of 2.5 mg of each enantiomer was obtained and the compounds investigated as potential inhibitors of β-amyloid aggregation. The study could be interesting in the field of nutraceutical research [49]. As reported above, all studies have been carried out using CSP polysaccharides bases, however, other CSs have also been used.

The enantioselective dissipation of an herbicide (fluazofop-butyl) was also studied in some vegetables (tomato, cucumber, pakchoi, and rape). The compound was degraded to fluazifop. Enantiomers separation was carried out in a commercial column packed with silica with immobilized cellulose tris(3,5-dichlorophenylcarbamate). As a result, the authors reported that the type of vegetable influenced the different behavior of the two chiral herbicides *S*-fluazifop-butyl dissipated faster than its antipode, while the contrary was observed in pakchoi, rape. The different behavior accounted for the presence of enzymes and to some chiral endogenous substances present in the plants [52].

α-, β-, and γ-hexabromocyclododecanes (HBCDs) are compounds used for different purposes, e.g., polystyrene foams, thermal insulation buildings, electric insulators, etc. Therefore they can be found in the environment and consequently in animals such as fishes, birds, chickens, etc. Enantiomers and diastereisomers, related to these compounds, have been separated by HPLC and the method applied to study the bioaccumulation in chicken tissues and eggs. Chiral compounds were separated using a permethylated-β-cyclodextrin silica column after recovering the different fractions subjected to sample preparation (soxhlet and gel permeation chromatography). It was observed that in adult chicken tissues (−)-α-HBCD and (+)-γ-HBCD were present at a higher concentration than the corresponding antipode [51].

A different chiral column containing α_1-acid glycoprotein (AGP) has been employed for the enantiomers and diastereoisomers separation of chloramphenicol (CAP) in honey by HPLC. The analysis of this antibiotic is very important because it has been found in some foodstuff causing health problems. Due to the presence of two asymmetric centers, four stereoisomers are exhibited (*RR*- and *SS*-CAP; *RS*- and *SR*-CAP) [50].

Usually, amino acid enantiomers present in the natural product are abundant in their L-form, however, in some fermented food, various D-amino acids could be present. Therefore, the enantiomers analysis is very important in order to assess food quality and properties. Amino acid chiral separation has been carried out utilizing a chiral crown-ether column in chimchi (Chinese cabbage) [55], in black vinegar and yogurt [54], in vinegar [56]. The chiral separation of a large number of amino acids has been obtained without derivatization. Finally, it has been found that the concentration of D-amino acids increased during the storage.

A few papers reported the enantiomers separation utilizing ultra high-performance liquid chromatography (UHPLC). This is a powerful methodology where compounds are separated in columns packed with particles of small diameter (<2–3 μm). High selectivity, high chromatographic efficiency, and short analysis time are the main advantages of this technique. However, it is noteworthy mentioning that increased back-pressure is obtained (high-pressure pumps are necessary) [59–61]. This methodology has been applied to the separation and analysis of a fungicide compound (prothioconazole) and its metabolite prothioconazole-desthio. The analytes have been base-line resolved in their enantiomers using a cellulose- tris(4-methylbenzoate) column by HPLC, while no baseline resolution was observed with UHPLC [59].

Among other recently developed analytical technique, one of them, namely nano-liquid chromatography (nano-LC open tubular) has been proposed for the enantiomers separation of some amino acids in apple juice. Amino acids have been derivatized with fluorescein isothiocyanate (FITC) and separated in a fused silica capillary containing, on the wall, a thin layer of polymeric material (monolithic) modified with a chiral selector (β-cyclodextrin). The optimized method was applied to the analysis of amino acid enantiomers in apple juice. While the technique can offer some advantages over conventional ones, there are some limitations especially considering the sample loading and the limited amount of chiral selector present in the capillary [62]. Table 2 summarizes the data related to the use of HPLC for enantiomers separation in food products. Some representative chiral separations of compounds in food matrices by HPLC-MS are reported in Figure 3.

4.3. Gas Chromatography

Table 3 reports the application to the analysis of enantiomers in food products achieved by gas chromatography.

Molecules **2019**, 24, 1119

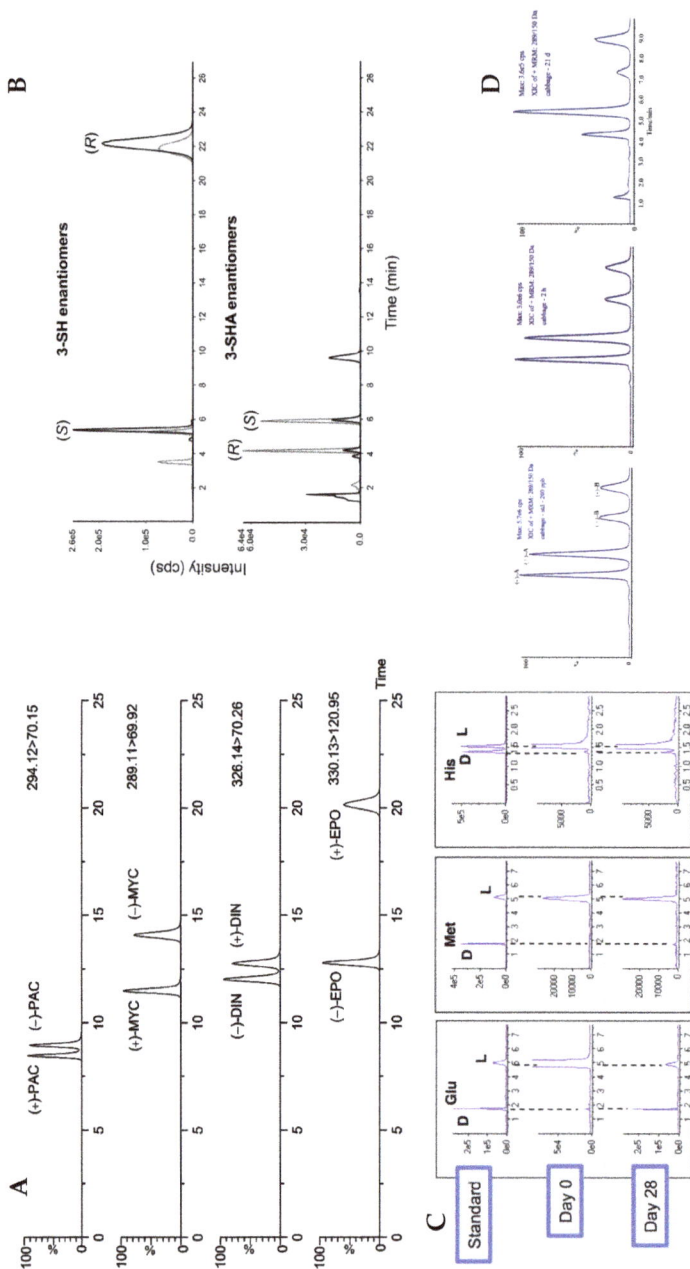

Figure 3. Typical enantioselective LC–MS/MS chromatograms: (**A**) chiral triazole fungicides in blank honey spiked using a Chiralcel OD-RH column modified with permission from [57]; (**B**) 3-sulfanylhexan-1-ol (3-SH 1) and 3-sulfanylhexyl acetate (3-SHA 2) isolated from a Sauvignon blanc wine on a CSP‑ Amylose-1 column; modified with permission from [48]. (**C**) Enantiomers resolution of some amino acids in kimchi stored for 25/28 days analyzed on a CSP‑ CROWNPAK CR-I by LC by LC‑TOF; modified with permission from [55]; and (**D**) Enantiomeric separation of fungicide pyrisoxazole by LC‑MS on a CSP‑ Lux Cellulose-3; modified with permission from Reference [45].

Table 3. Selected applications of GC to Food analysis.

Samples	Matrix	Sample Preparation	Technique	Column and Chiral Stationary Phase	Mobile Phase	Detection	References
TBECH enantiomers 1,2-dibromo-4-(1,2-dibromoethyl)-cyclohexane (α, β, γ, δ) HBCD 1,2,5,6,9,10-hexabromocyclododecane (α, β, γ)	Marine organisms, including 5 mollusk species, 6 crustacean species, and 19 fish species	pressurized fluid extraction	GC	CHIRALDEX B-TA capillary column (30 m × 0.25 mm i.d., 0.12 μm film thickness)	Carrier:40% methane in a helium carrier gas at a constant flow of 1.2 mL/min.	MS	[63]
α-pinene, β-pinene, borneol, camphene, carvone, linalool, limonene, α-terpineol, α-ionene, terpinen-4-ol	Juices (apple, pear, peach, carrot, lemon flesh, orange flesh, orange peel, tangerine flesh), and tangerine peel	Headspace Solid-Phase Microextraction (HS-SPME)	GC	30% 2,3-di-O-ethyl-6-O-*tert*-butyldimethylsilyl-β-cyclodextrin (diEt-CD) and 30% 2,6-dimethyl-3-O-pentyl-β-cyclodextrin (Pentyl-CD) coated	Hydrogen as carrier gas (1.25 mL/min)	MS	[24]
α-pinene, limonene, linalool, β-caryophyllene	Essential oil Thyme	Steam distillation	GC	non-bonded 2,3-di-O-methyl-6-O-butyl silyl derivative of β-cyclodextrin AstecChiraldex B-DM column (30 m length 0.25 mm internal, diameter 0.12 μm film thickness)	hydrogen as carrier gas (constant flow of 2.5 mL/min, 8 psistarting column head pressure	Electronic impact ionization and MS	[64]
Pyrethroid insecticide (α-cypermethrin)	tomato, cucumber, rape, cabbage, and pepper	Liquid-liquidextraction	GC	BGB-172 chiral column	N₂ 100–220 °C	ECD	[65]
Limonene, linalool, α-terpineol and 4-terpineol	Tea	Solid-Phase Microextraction	2D-GC	DB-WAX column (30 m × 0.25 mm i.d., 0.5 μm film thickness) (first) Cyclosil-B column (30 m × 0.25 mm i.d., 0.25 μm film thickness) (second)	helium at 1.2 mL/min	MS	[66]

Finally, gas chromatography (GC), widely used for enantiomers resolution, can offer high chromatographic efficiency. It is applied mainly to volatile compounds. The separation is performed in a capillary of thin i.d. where a chiral selector is coated or bonded on the wall. In the publications appeared during this time, polymeric material, β-cyclodextrin-based has been used as a chiral selector. Pollutants such as 1,2,5,6,9,10-hexabromocyclododecane (HBCD) (six diastereisomeric couple of enantiomers) were studied by GC-MS in order to verify the bioaccumulation in fishes [63]. In order to verify whether the processing would influence the enantiomeric ratio, chiral terpenoids present in juice industry by-products have been analyzed with GC-MS by Marsol-Vall et al. [24]. Among the results, authors observed that monoterpene alcohols (terpinen-4-ol and α-terpineol) exhibited a change on the enantiomeric ratio concluding that this was probably due to the heat applied during the drying process. The same technique has also been applied to the analysis of enantiomers present in essential oil from thyme. In this study, authors found the presence of (−)-linalool, (−)-borneol and (+)-limonene concluding that these compounds could be useful to assess origin and authenticity of the product [64]. A more advanced analytical technique, namely two dimensional-GC-MS(2D-GC-MS) has been applied to the analysis of chiral compounds present in tea samples [66]. The study aimed to investigate the chiral stability of lemon-flavored hard tea during storage. Samples were firstly analyzed by GC in a DB-Wax column and then target compounds heart-cutted and online submitted to a chiral column (cyclodextrin-based) for chiral analysis. Authors found that R/S ratio of limonene did not change during storage, while at high-temperature, S-limonene increased. Linalool enantiomers analysis, at the beginning of storage, revealed the presence of a higher amount of R-enantiomer. During storage, the R/S ratio decreased due to the conversion of S-isomer to R-one. Similar results have been obtained for α-terpineol.

5. Conclusions

The separation of chiral compounds has received great attention also in the last few years in different application fields including food chemistry. HPLC, SFC, and GC are the most utilized techniques because their features employing essentially the direct resolution method. Chiral columns, containing particles modified with selected CSs allowed the enantiomeric resolution of a large number of compounds. The majority of the enantioseparations have been obtained using polysaccharide-type CSs because of their very high chiral selectivity. The presented applications in the field of food analysis document the need of powerful analytical methods to assess the role of enantiomers for safety and product origin. Further studies are needed in order to (i) develop new CSPs (ii) study modern analytical techniques such as UHPLC and nano-LC offering high efficiency and reducing analysis time.

Funding: This research received no external funding.

Conflicts of Interest: The authors declare no conflict of interest.

References

1. Gil-Av, E. Present status of enantiomeric analysis by gas chromatography. *J. Mol. Evol.* **1975**, *6*, 131–144. [CrossRef] [PubMed]
2. Novell, A.; Méndez, A.; Minguillón, C. Effects of supercritical fluid chromatography conditions on enantioselectivity and performance of polyproline-derived chiral stationary phases. *J. Chromatogr. A* **2015**, *1403*, 138–143. [CrossRef] [PubMed]
3. Albals, D.; Vander Heyden, Y.; Schmid, M.G.; Chankvetadze, B.; Mangelings, D. Chiral separations of cathinone and amphetamine-derivatives: Comparative study between capillary electrochromatography, supercritical fluid chromatography and three liquid chromatographic modes. *J. Pharm. Biomed. Anal.* **2016**, *121*, 232–243. [CrossRef] [PubMed]
4. Armstrong, D.W.; Faulkner, J.R.J.; Han, S.M. Use of hydroxypropyl and hydroxyethyl-derivatized -cyclodextrins for the thin-layer chromatographic separation of enantiomers and diastereomers. *J. Chromatogr.* **1988**, *452*, 323–330. [CrossRef]

5. Khundadze, N.; Pantsulaia, S.; Fanali, C.; Farkas, T.; Chankvetadze, B. On our way to sub-second separations of enantiomers in high-performance liquid chromatography. *J. Chromatogr. A* **2018**, *1572*, 37–43. [CrossRef] [PubMed]

6. Si-Ahmed, K.; Aturki, Z.; Chankvetadze, B.; Fanali, S. Evaluation of novel amylose and cellulose-based chiral stationary phases for the stereoisomer separation of flavanones by means of nano-liquid chromatography. *Anal. Chim. Acta* **2012**, *738*, 85–94. [CrossRef]

7. Aturki, Z.; Rocco, A.; Rocchi, S.; Fanali, S. Current applications of miniaturized chromatographic and electrophoretic techniques in drug analysis. *J. Pharm. Biomed. Anal.* **2014**, *101*, 194–220. [CrossRef]

8. Fanali, S. Nano-liquid chromatography applied to enantiomers separation. *J. Chromatogr. A* **2017**, *1486*, 20–34. [CrossRef]

9. Davankov, V.A. The nature of chiral recognition:is it a three-point interaction? *Chirality* **1997**, *9*, 99–102. [CrossRef]

10. Cagliero, C.; Sgorbini, B.; Cordero, C.; Liberato, E.; Rubiolo, P.; Bicchi, C. Enantioselective Gas Chromatography with DerivatizedCyclodextrins in the Flavour and Fragrance Field. *Isr. J. Chem.* **2016**, *56*, 925–939. [CrossRef]

11. West, C. Enantioselective separations with supercritical fluids—Review. *Curr. Anal. Chem.* **2014**, *10*, 99–120. [CrossRef]

12. Chankvetadze, B. *Capillary Electrophoresis in Chiral Analysis*, 1st ed.; John Wiley & Sons: Chichester, UK, 1997; pp. 1–555.

13. Chankvetadze, B. Contemporary theory of enantioseparations in capillary electrophoresis. *J. Chromatogr. A* **2018**, *1567*, 2–25. [CrossRef] [PubMed]

14. Chankvetadze, B. Liquid chromatographic separation of enantiomers. In *Liquid Chromatography*, 2nd ed.; Fanali, S., Haddad, P.R., Poole, C.F., Riekkola, M.-L., Eds.; Elsevier Inc.: Amsterdam, The Netherlands, 2017; Volume 2, pp. 69–86.

15. Patel, D.C.; Waha, M.F.; Armstrong, D.W.; Breitbach, Z.S. Advances in high-throughput and high-efficiency chiral liquid chromatographic separations. *J. Chromatogr. A* **2016**, *1467*, 2–18. [CrossRef] [PubMed]

16. Lomsadze, K.; Jibuti, G.; Farkas, T.; Chankvetadze, B. Comparative high-performance liquid chromatography enantioseparations on polysaccharide based chiral stationary phases prepared by coating totally porous and core-shell silica particles. *J. Chromatogr. A* **2012**, *1234*, 50–55. [CrossRef] [PubMed]

17. Chankvetadze, B.; Yamamoto, C.; Okamoto, Y. Very Fast Enantioseparation in High-performance Liquid Chromatography Using Cellulose Tris(3,5-dimethylphenylcarbamate) Coated on Monolithic Silica Support. *Chem. Lett.* **2003**, *32*, 850–851. [CrossRef]

18. Chankvetadze, B.; Yamamoto, C.; Kamigaito, M.; Tanaka, N.; Nakanishi, K.; Okamoto, Y. High-performance liquid chromatographic enantioseparations on capillary columns containing monolithic silica modified with amylose tris(3,5- dimethylphenylcarbamate). *J. Chromatogr. A* **2006**, *1110*, 46–52. [CrossRef] [PubMed]

19. Khatiashvili, T.; Kakava, R.; Matarashvili, I.; Tabani, H.; Fanali, C.; Volonterio, A.; Farkas, T.; Chankvetadze, B. Separation of enantiomers of selected chiral sulfoxides with cellulose tris(4-chloro-3-methylphenylcarbamate)-based chiral columns in high-performance liquid chromatography with very high separation factor. *J. Chromatogr. A* **2018**, *1545*, 59–66. [CrossRef] [PubMed]

20. Pan, X.; Dong, F.; Xu, J.; Liu, X.; Chen, Z.; Zheng, Y. Stereoselective analysis of novel chiral fungicide pyrisoxazole in cucumber, tomato and soil under different application methods with supercritical fluid chromatography/tandem mass spectrometry. *J. Hazard. Mater.* **2016**, *311*, 115–124. [CrossRef] [PubMed]

21. D'Orazio, G.; Kakava, R.; Volonterio, A.; Fanali, S.; Chankvetadze, B. An attempt for fast separation of enantiomers in nano-liquid chromatography and capillary electrochromatography. *Electrophoresis* **2017**, *38*, 1932–1938. [CrossRef]

22. D'Orazio, G.; Fanali, C.; Karchkhadze, M.; Chankvetadze, B.; Fanali, S. Enantiomeric separation of some chiral analytes using amylose 3,5-dimethylphenylcarbamate covalently immobilized on silica by nano-liquid chromatography and capillary electrochromatography. *J. Chromatogr. A* **2017**, *1520*, 127–134. [CrossRef] [PubMed]

23. D'Orazio, G.; Fanali, C.; Asensio-Ramos, M.; Fanali, S. Chiral separations in food analysis. *TrAC Trends Anal. Chem.* **2017**, *96*, 151–171. [CrossRef]

24. Marsol-Vall, A.; Sgorbini, B.; Cagliero, C.; Bicchi, C.; Eras, J.; Balcells, M. Volatile composition and enantioselective analysis of chiral terpenoids of nine fruit and vegetable fibres resulting from juice industry by-products. *J. Chem.* **2017**, *2017*, 8675014. [CrossRef]

25. Schurig, V.; Mayer, S. Separation of enantiomers by open capillary electrochromatography on polysiloxane-bonded permethyl-beta-cyclodextrin. *J. Biochem. Biophys. Methods* **2001**, *48*, 117–141. [CrossRef]

26. Sun, P.; Armstrong, D.W. Effective enantiomeric separations of racemic primary amines by the isopropyl carbamate-cyclofructan6 chiral stationary phase. *J. Chromatogr. A* **2010**, *1217*, 4904–4918. [CrossRef] [PubMed]

27. Lajkó, G.; Grecsó, N.; Tóth, G.; Fülöp, F.; Lindner, W.; Ilisz, I.; Péter, A. Liquid and subcritical fluid chromatographic enantioseparation of Nα-Fmoc proteinogenic amino acids on Quinidine-based zwitterionic and anion-exchanger type chiral stationary phases. A comparative study. *Chirality* **2017**, *29*, 225–238. [CrossRef] [PubMed]

28. Ilisz, I.; Bajtai, A.; Lindner, W.; Péter, A. Liquid chromatographic enantiomer separations applying chiral ion-exchangers based on Cinchona alkaloids. *J. Pharm. Biomed. Anal.* **2018**, *159*, 127–152. [CrossRef]

29. Schmitt, K.; Woiwode, U.; Kohout, M.; Zhang, T.; Lindner, W.; Lämmerhofer, M. Comparison of small size fully porous particles and superficially porous particles of chiral anion-exchange type stationary phases in ultra-high performance liquid chromatography: Effect of particle and pore size on chromatographic efficiency and kinetic performance. *J. Chromatogr. A* **2018**, *1569*, 149–159. [CrossRef]

30. Patel, D.C.; Breitbach, Z.S.; Yu, J.; Nguyen, K.A.; Armstrong, D.W. Quinine bonded to superficially porous particles for high-efficiency and ultrafast liquid and supercritical fluid chromatography. *Anal. Chim. Acta* **2017**, *963*, 164–174. [CrossRef]

31. Wahab, M.F.; Wimalasinghe, R.M.; Wang, Y.; Barhate, C.L.; Patel, D.C.; Armstrong, D.W. Salient Sub-Second Separations. *Anal. Chem.* **2016**, *88*, 8821–8826. [CrossRef]

32. Barhate, C.L.; Regalado, E.L.; Contrella, N.D.; Lee, J.; Jo, J.; Makarov, A.A.; Armstrong, D.W.; Welch, C.J. Ultrafast Chiral Chromatography as the Second Dimension in Two-Dimensional Liquid Chromatography Experiments. *Anal. Chem.* **2017**, *89*, 3545–3553. [CrossRef]

33. Orosz, T.; Grecsó, N.; Lajkó, G.; Szakonyi, Z.; Fülöp, F.; Armstrong, D.W.; Ilisz, I.; Péter, A. Liquid chromatographic enantioseparation of carbocyclic β-amino acids possessing limonene skeleton on macrocyclic glycopeptide-based chiral stationary phases. *J. Pharm. Biomed. Anal.* **2017**, *145*, 119–126. [CrossRef] [PubMed]

34. Ismail, O.H.; Ciogli, A.; Villani, C.; De Martino, M.; Pierini, M.; Cavazzini, A.; Bell, D.S.; Gasparrini, F. Ultra-fast high-efficiency enantioseparations by means of a teicoplanin-based chiral stationary phase made on sub-2μm totally porous silica particles of narrow size distribution. *J. Chromatogr. A* **2016**, *1427*, 55–68. [CrossRef] [PubMed]

35. Ismail, O.H.; Antonelli, M.; Ciogli, A.; Villani, C.; Cavazzini, A.; Catani, M.; Felletti, S.; Bell, D.S.; Gasparrini, F. Future perspectives in high efficient and ultrafast chiral liquid chromatography through zwitterionic teicoplanin-based 2-μm superficially porous particles. *J. Chromatogr. A* **2017**, *1520*, 91–102. [CrossRef]

36. Rocchi, S.; Fanali, C.; Fanali, S. Use of a Novel Sub-2 μm Silica Hydride Vancomycin Stationary Phase in Nano-Liquid Chromatography. II. Separation of Derivatized Amino Acid Enantiomers. *Chirality* **2015**, *27*, 767–772. [CrossRef] [PubMed]

37. Harps, L.C.; Joseph, J.F.; Parr, M.K. SFC for chiral separations in bioanalysis. *J. Pharm. Biomed. Anal.* **2019**, *162*, 47–59. [CrossRef] [PubMed]

38. Jiang, Y.; Fan, J.; He, R.; Guo, D.; Wang, T.; Zhang, H.; Zhang, W. High-fast enantioselective determination of prothioconazole in different matrices by supercritical fluid chromatography and vibrational circular dichroism spectroscopic study. *Talanta* **2018**, *187*, 40–46. [CrossRef]

39. Tao, Y.; Zheng, Z.; Yu, Y.; Xu, J.; Liu, X.; Wu, X.; Dong, F.; Zheng, Y. Supercritical fluid chromatography–tandem mass spectrometry-assisted methodology for rapid enantiomeric analysis of fenbuconazole and its chiral metabolites in fruits, vegetables, cereals, and soil. *Food Chem.* **2018**, *241*, 32–39. [CrossRef] [PubMed]

40. Tan, Q.; Fan, J.; Gao, R.; He, R.; Wang, T.; Zhang, Y.; Zhang, W. Stereoselective quantification of triticonazole in vegetables by supercritical fluid chromatography. *Talanta* **2017**, *164*, 362–367. [CrossRef]

41. Cheng, Y.; Zheng, Y.; Dong, F.; Li, J.; Zhang, Y.; Sun, S.; Li, N.; Cui, X.; Wang, Y.; Pan, X.; et al. Stereoselective analysis & dissipation of propiconazole in wheat, grapes, & soil by supercritical fluid chromatography-tandem mass spectrometry. *J. Agric. Food Chem.* **2017**, *65*, 234–243. [CrossRef] [PubMed]

42. Parr, M.K.; Blokland, M.H.; Liebetrau, F.; Schmidt, A.H.; Meijer, T.; Stanic, M.; Kwiatkowska, D.; Waraksa, E.; Sterk, S.S. Distinction of clenbuterol intake from drug or contaminated food of animal origin in a controlled administration trial–the potential of enantiomeric separation for doping control analysis. *Food Addit. Contam. Part A* **2017**, *34*, 525–535. [CrossRef] [PubMed]

43. Chen, J.; Zhang, T.; Zhang, Q.; Liu, Y.; Li, L.; Si, J.; Zou, Z.; Hua, H. Isobutylhydroxyamides from Sichuan Pepper and Their Protective Activity on PC12 Cells Damaged by Corticosterone. *J. Agric. Food Chem.* **2018**, *66*, 3408–3416. [CrossRef]

44. Zhang, Q.; Zhang, Z.; Tang, B.; Gao, B.; Tian, M.; Sanganyado, E.; Shi, H.; Wang, M. Mechanistic Insights into Stereospecific Bioactivity and Dissipation of Chiral Fungicide Triticonazole in Agricultural Management. *J. Agric. Food Chem.* **2018**, *66*, 7286–7293. [CrossRef]

45. Yang, X.; Qi, P.; Wang, X.; Wang, Z.; Sun, Y.; Wang, L.; Xu, X.; Xu, H.; Wang, Q.; Wang, X.; et al. Stereoselective Analysis and Degradation of Pyrisoxazole in Cabbage, Pakchoi, and Pepper by Liquid Chromatography Tandem Mass Spectrometry. *J. Agric. Food Chem.* **2017**, *65*, 8295–8301. [CrossRef] [PubMed]

46. Wang, L.; Wang, X.; Di, S.; Qi, P.; Sun, Y.; Yang, X.; Zhao, C.; Wang, X. Enantioselective analysis and degradation of isofenphos-methyl in vegetables by liquid chromatography-tandem mass spectrometry. *Environ. Sci. Pollut. Res.* **2018**, *25*, 18772–18780. [CrossRef]

47. He, R.; Fan, J.; Tan, Q.; Lai, Y.; Chen, X.; Wang, T.; Jiang, Y.; Zhang, Y.; Zhang, W. Enantioselective determination of metconazole in multi matrices by high-performance liquid chromatography. *Talanta* **2018**, *178*, 980–986. [CrossRef] [PubMed]

48. Chen, L.; Capone, D.L.; Jeffery, D.W. Chiral analysis of 3-sulfanylhexan-1-ol and 3-sulfanylhexyl acetate in wine by high-performance liquid chromatography–tandem mass spectrometry. *Anal. Chim. Acta* **2018**, *998*, 83–92. [CrossRef] [PubMed]

49. Zhou, L.; Lou, L.L.; Wang, W.; Lin, B.; Chen, J.N.; Wang, X.B.; Huang, X.X.; Song, S.J. Enantiomeric 8-O-4′ type neolignans from red raspberry as potential inhibitors of β-amyloid aggregation. *J. Funct. Foods* **2017**, *37*, 322–329. [CrossRef]

50. Rimkus, G.G.; Hoffmann, D. Enantioselective analysis of chloramphenicol residues in honey samples by chiral LC-MS/MS and results of a honey survey. *Food Addit. Contam. Part A* **2017**, *34*, 950–961. [CrossRef] [PubMed]

51. Zheng, X.; Qiao, L.; Sun, R.; Luo, X.; Zheng, J.; Xie, Q.; Sun, Y.; Mai, B. Alteration of Diastereoisomeric and Enantiomeric Profiles of Hexabromocyclododecanes (HBCDs) in Adult Chicken Tissues, Eggs, and Hatchling Chickens. *Environ. Sci. Technol.* **2017**, *51*, 5492–5499. [CrossRef]

52. Qi, Y.; Liu, D.; Liu, C.; Liang, Y.; Zhan, J.; Zhou, Z.; Wang, P. Enantioselective behaviour of the herbicide fluazifop-butyl in vegetables and soil. *Food Chem.* **2017**, *221*, 1120–1127. [CrossRef]

53. Nagai, T.; Ishikawa, K.; Yoshinaga, K.; Yoshida, A.; Beppu, F.; Gotoh, N. Homochiral asymmetric triacylglycerol isomers in egg yolk. *J. Oleo Sci.* **2017**, *66*, 1293–1299. [CrossRef]

54. Konya, Y.; Taniguchi, M.; Fukusaki, E. Novel high-throughput and widely-targeted liquid chromatography–time of flight mass spectrometry method for D-amino acids in foods. *J. Biosci. Bioeng.* **2017**, *123*, 126–133. [CrossRef]

55. Taniguchi, M.; Konya, Y.; Nakano, Y.; Fukusaki, E. Investigation of storage time-dependent alterations of enantioselective amino acid profiles in kimchi using liquid chromatography-time of flight mass spectrometry. *J. Biosci. Bioeng.* **2017**, *124*, 414–418. [CrossRef]

56. Nakano, Y.; Konya, Y.; Taniguchi, M.; Fukusaki, E. Development of a liquid chromatography-tandem mass spectrometry method for quantitative analysis of trace D-amino acids. *J. Biosci. Bioeng.* **2017**, *123*, 134–138. [CrossRef]

57. Ye, X.; Ma, S.; Zhang, L.; Zhao, P.; Hou, X.; Zhao, L.; Liang, N. Trace enantioselective determination of triazole fungicides in honey by a sensitive and efficient method. *J. Food Compos. Anal.* **2018**, *74*, 62–70. [CrossRef]

58. Zhao, P.; Wang, Z.; Gao, X.; Guo, X.; Zhao, L. Simultaneous enantioselective determination of 22 chiral pesticides in fruits and vegetables using chiral liquid chromatography coupled with tandem mass spectrometry. *Food Chem.* **2019**, *277*, 298–306. [CrossRef]

59. Zhang, Z.; Zhang, Q.; Gao, B.; Gou, G.; Li, L.; Shi, H.; Wang, M. Simultaneous Enantioselective Determination of the Chiral Fungicide Prothioconazole and Its Major Chiral Metabolite Prothioconazole-Desthio in Food and Environmental Samples by Ultraperformance Liquid Chromatography-Tandem Mass Spectrometry. *J. Agric. Food Chem.* **2017**, *65*, 8241–8247. [CrossRef]

60. Pan, X.; Dong, F.; Liu, N.; Cheng, Y.; Xu, J.; Liu, X.; Wu, X.; Chen, Z.; Zheng, Y. The fate and enantioselective behavior of zoxamide during wine-making process. *Food Chem.* **2018**, *248*, 14–20. [CrossRef]

61. Pan, X.; Dong, F.; Chen, Z.; Xu, J.; Liu, X.; Wu, X.; Zheng, Y. The application of chiral ultra-high-performance liquid chromatography tandem mass spectrometry to the separation of the zoxamide enantiomers and the study of enantioselective degradation process in agricultural plants. *J. Chromatogr. A* **2017**, *1525*, 87–95. [CrossRef]

62. Aydoğan, C. Chiral separation and determination of amino acid enantiomers in fruit juice by open-tubular nano liquid chromatography. *Chirality* **2018**, *30*, 1144–1149. [CrossRef]

63. Ruan, Y.; Zhang, X.; Qiu, J.W.; Leung, K.M.Y.; Lam, J.C.W.; Lam, P.K.S. Stereoisomer-Specific Trophodynamics of the Chiral Brominated Flame Retardants HBCD and TBECH in a Marine Food Web, with Implications for Human Exposure. *Environ. Sci. Technol.* **2018**, *52*, 8183–8193. [CrossRef]

64. Cutillas, A.B.; Carrasco, A.; Martinez-Gutierrez, R.; Tomas, V.; Tudela, J. Thyme essential oils from Spain: Aromatic profile ascertained by GC–MS, and their antioxidant, anti-lipoxygenase and antimicrobial activities. *J. Food Drug Anal.* **2018**, *26*, 529–544. [CrossRef] [PubMed]

65. Yao, G.; Gao, J.; Zhang, C.; Jiang, W.; Wang, P.; Liu, X.; Liu, D.; Zhou, Z. Enantioselective degradation of the chiral alpha-cypermethrin and detection of its metabolites in five plants. *Environ. Sci. Pollut. Res.* **2019**, *26*, 1558–1564. [CrossRef] [PubMed]

66. He, F.; Qian, Y.L.; Qian, M.C. Flavor and chiral stability of lemon-flavored hard tea during storage. *Food Chem.* **2018**, *239*, 622–630. [CrossRef] [PubMed]

molecules

MDPI

Article

Development of an Impedimetric Aptasensor for Label Free Detection of Patulin in Apple Juice

Reem Khan [1,2,3], Sondes Ben Aissa [1,4]ORCID, Tauqir A. Sherazi [2], Gaelle Catanante [1]ORCID, Akhtar Hayat [3,*] and Jean Louis Marty [1,*]

[1] BAE: Biocapteurs-Analyses-Environnement, Universite de Perpignan Via Domitia, 52 Avenue Paul Alduy, 66860 Perpignan CEDEX, France; Kreemjadoon@gmail.com (R.K.); sondes.benaissa@fst.utm.tn (S.B.A.); gaelle.catanante@univ-perp.fr (G.C.)
[2] Department of chemistry, COMSATS University Islamabad, Abbottabad Campus, Abbottabad 22060, Pakistan; sherazi@cuiatd.edu.pk
[3] Interdisciplinary Research Centre in Biomedical Materials (IRCBM), COMSATS University Islamabad, Lahore Campus, Lahore 54000, Pakistan
[4] Université de Tunis El Manar, Faculté des Sciences de Tunis, Laboratoire de Chimie Analytique et Electrochimie (LR99ES15), Sensors and Biosensors Group, Campus Universitaire de Tunis El Manar, Tunis 2092, Tunisia
* Correspondence: akhtarhayat@cuilahore.edu.pk (A.H.); jlmarty@univ-perp.fr (J.L.M.); Tel.: +92-3317648291 (A.H.); +33-468662257 (J.L.M.)

Received: 6 February 2019; Accepted: 9 March 2019; Published: 13 March 2019

check for updates

Abstract: In the present work, an aptasensing platform was developed for the detection of a carcinogenic mycotoxin termed patulin (PAT) using a label-free approach. The detection was mainly based on a specific interaction of an aptamer immobilized on carbon-based electrode. A long linear spacer of carboxy-amine polyethylene glycol chain (PEG) was chemically grafted on screen-printed carbon electrodes (SPCEs) via diazonium salt in the aptasensor design. The NH_2-modified aptamer was then attached covalently to carboxylic acid groups of previously immobilized bifunctional PEG to build a diblock macromolecule. The immobilized diblocked molecules resulted in the formation of long tunnels on a carbon interface, while the aptamer was assumed as the gate of these tunnels. Upon target analyte binding, the gates were assumed to be closed due to conformational changes in the structure of the aptamer, increasing the resistance to the charge transfer. This increase in resistance was measured by electrochemical impedance spectroscopy, the main analytical technique for the quantitative detection of PAT. Encouragingly, a good linear range between 1 and 25 ng was obtained. The limit of detection and limit of quantification was 2.8 ng L^{-1} and 4.0 ng L^{-1}, respectively. Selectivity of the aptasensor was confirmed with mycotoxins commonly occurring in food. The developed apta-assay was also applied to a real sample, i.e., fresh apple juice spiked with PAT, and toxin recovery up to 99% was observed. The results obtained validated the suitability and selectivity of the developed apta-assay for the identification and quantification of PAT in real food samples.

Keywords: impedimetric aptasensor; screen-printed interface; bifunctional polymer arms; PAT detection; apple juice

---

1. Introduction

Effective detection of low molecular weight toxic molecules is vital in areas such as environmental monitoring and the food industry [1–3]. Mycotoxins are small agro-based food contaminants that pose serious health threats to humans and animals [4]. Patulin (PAT) is a low molecular weight mycotoxin produced by various fungal species. This small mycotoxin is commonly present in fruit- and vegetable-based products, especially apples [5,6]. The antibiotic properties of PAT were firstly reported in 1940. Some studies have also shown the adverse health effects of PAT on higher plants and animals [7]. Extensive studies have been carried out to highlight the health risks of PAT, and it has been concluded that PAT exhibits several chronic, acute, and cellular level toxic effects both on human and animals [8]. PAT has neurotoxic, immunosuppressive, and mutagenic effects on several animal species and cause severe damage to the intestinal epithelium, which leads to degeneration, inflammation, ulceration, and bleeding [8,9]. Due to such pronounced toxicities, different authorities have set regulatory limits for PAT in food. The European Union (EU) is the first organization that defined a maximum admissible limit, i.e., 50 μg L^{-1} [10].

To monitor the level of PAT in food, several analytical techniques have been established including HPLC, thin layer chromatography with mass spectrometry and UV detection [11], gas chromatography coupled with mass spectrometry [12], micellar electrokinetic chromatography (MEKC), and colorimetry [13–16]. Despite their high reliability and accuracy, these techniques suffer many disadvantages such as complex instrumentation, high running cost, and complicated handling methods that make on-site analysis unsuitable. Owing to such drawbacks, there is an immense need of reliable, fast, and easy-to-operate screening methods.

Biosensors, being simple, reliable, and fast screening tools, have emerged as an attractive alternative to classical analytical methods. Among different types of biosensors, electrochemical aptasensors offer significant advantages over the others. [17]. Change in the unique three-dimensional structure of aptamers upon target binding provides great flexibility in developing electrochemical aptasensors. In most cases, an aptamer is tagged with a redox label such as ferrocene or methylene blue [18]. These types of biosensor designs have been successfully employed to detect different targets, including mycotoxins. However, their main problem is that they require large conformational changes in the aptamer structure to tune the distance of the electro-active label from the transducer surface. Moreover, the process of labeling aptamers is expensive, and labeling sometimes decreases the activity of the modified biomolecules, which results in a more complex, laborious, and time-consuming assay design. [19]. As an alternative, interest in the development of simpler and cheaper label-free aptasensors has recently increased. These platforms are generally based on electrochemical impedance spectroscopy (EIS) as a powerful technique, allowing sensitive detection of the smallest variations in electron transfer processes. Still, in the case of small analytes, target binding events do generate an easily measurable impedimetric signal, because physical hindrance induced by these low molecular weight analytes is often insufficient to create detectable variation. However, ultrasensitive detection of low weight targeted analytes by a simple aptasensor design is quite difficult. Different methods have been used for signal amplification, such as strand displacement amplification, rolling circle amplification, and the use of nanomaterials [20,21]. Although these approaches offer advantages in signal intensification, they are costly, are complex, and have difficult operating conditions. These problems have resulted in the exploration of novel aptasensor designs to ensure the sensitive detection of low molecular weight targets. Integration of a spacer in aptasensor design could be a simple and effective method of signal amplification. In this work, we present this strategy for the detection of a small mycotoxin, i.e., PAT.

Furthermore, considering the mycotoxin analysis, multiple efforts have been made to design and develop biosensors for aflatoxins and ochratoxins, but very few reports are available for PAT detection [22,23]. In the present work, a novel impedimetric aptasensor was designed for the label-free detection of PAT. A long spacer, i.e., carboxy-amine PEG, was used to facilitate the label-free detection and enhance the sensitivity of the developed aptasensor. Disposable screen-printed carbon electrodes (SPCEs) were used as transducers. The SPCEs were based on a conventional three-electrode system with a working electrode made up of graphite (with a circular disk 4 mm in diameter), a counter electrode (curved line: 16 × 1.5 mm), and Ag/AgCl (straight line: 16 × 1.5 mm) as a pseudo reference electrode. All parameters were carefully optimized. The developed aptasensor was then assessed for a real sample, i.e., apple juice. To the best of our knowledge, no such impedimetric aptasensor has yet been reported for the detection of PAT.

2. Materials and Methods

2.1. Design and Working Principle of Impedimetric Aptasensor

A schematic representation of the working principle and the stepwise design strategy of the developed impedimetric aptasensor for PAT detection is shown in Scheme 1. As indicated, three major steps were carried out for the fabrication process. In the proposed aptasensor, a diblock macromolecule (carboxy-amine PEG + anti-PAT-aptamer) was chemically grafted onto the surface of SPCEs to form clusters of long spacer arms on the working electrode. Aminobenzoic acid was first electrochemically grafted onto the surface of SPCEs through linear sweep voltammetry (LSV). This technique is based on the electrochemical reduction of diazonium salt, leading to aryl-centered radicals by the spontaneous release of nitrogen. The voltammogram in Figure S1 shows the successful reduction of diazonium cations via a one-electron process. An insulating layer of carboxyphenyl was covalently formed on the surface.

Diazonium deposition was then followed by the chemical grafting of carboxy-amine PEG through the amide bond. The long spacer arms of heterobifunctional PEG acted like tunnels for the electrons of the redox marker to reach the surface of the electrode. The second part of the diblock, i.e., the anti-PAT aptamer, acted as gates for these tunnels. In the absence of a targeted analyte, the aptamer remained unfolded, and it is assumed that the gates were opened, thereby permitting the electron transfer from the redox probe to the surface of the electrode. Meanwhile, in the presence of the targeted analyte, 3D conformational changes in the aptamer structure resulted in the formation of an aptamer-PAT quadruplex complex that locked the gates of tunnels, resulting in blockage of the electron flow toward the electrode surface [24]. Furthermore, the conformational changes in the aptamer structure also resulted in the exposition of the negatively charged backbone of the aptamer. Consequently, steric hindrance after the formation of the quadruplex complex and the electrostatic repulsion between the anionic ferri/ferrocyanide redox system and the negatively charged aptamer collectively increased the impedance of the system. The higher the concentration of PAT is, the higher the impedance will be. This concept makes the quantitative detection of PAT feasible.

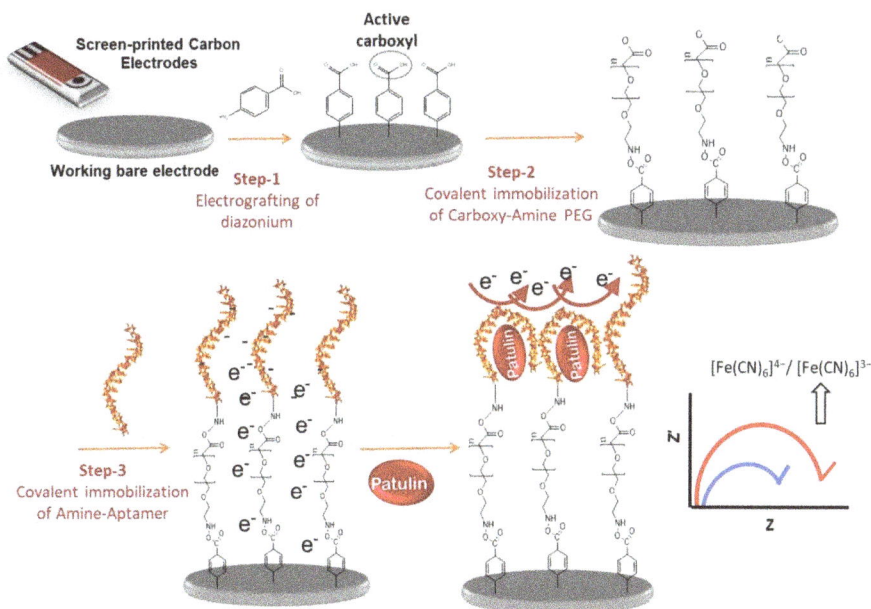

Scheme 1. Design and working principle of the proposed aptasensor.

2.2. Electrochemical Characterization of the Aptasensors

Cyclic voltammetry (CV) and EIS are widely used techniques to characterize the electrode/electrolyte interfacial properties at different modification steps of aptasensor fabrication. Thus, the different preparation stages were investigated by recording the impedance spectra and cyclic voltammograms of the modified electrode in the presence of the reversible [Fe (CN)$_6$]$^{4-/3-}$ redox system, shown in Figure 1.

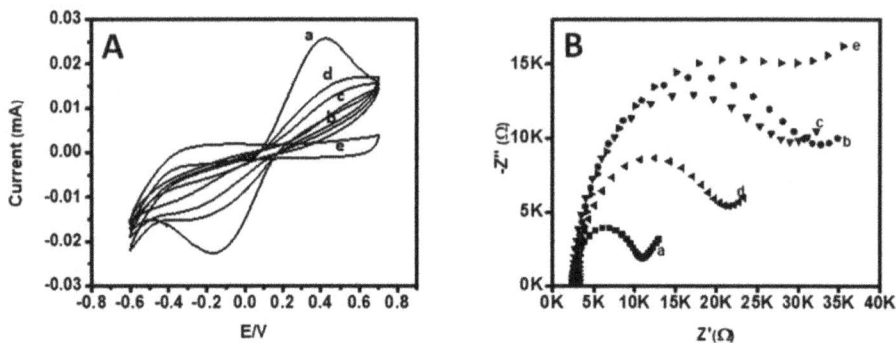

Figure 1. Characterization of modification steps during aptasensor fabrication. (**A**) Cyclic voltammogram obtained in 1 mM Ferri/Ferrocyanide after each modification step during aptasensor fabrication: (a) bare SPCE; (b) carboxyphenyl modified SPCE; (c) after immobilization of carboxy-amine PEG; (d) anti-patulin (anti-PAT) aptamer modified electrode; (e) after incubation of 50 ng L^{-1} patulin (PAT). (**B**) Nyquist plots of a 1 mM [Fe(CN)$_6$]$^{3-/4-}$ probe in PBS, pH 7.4.

2.2.1. Cyclic Voltammetry Characterization

CV in the ferri/ferrocyanide solution is a useful tool for examining the electrochemical behavior of functionalized electrodes, as electron transfer is more or less feasible depending on the hindrance of steric barriers. Therefore, this technique was used to investigate and characterize the electrode surface modifications. Figure 1A shows the cyclic voltammograms obtained after each modification of the electrode surface during aptasensor fabrication. Bare SPCEs exhibited a symmetric oxidation and reduction peaks (Figure 1A(a)), but the insulating layer of 4-aminobenzoic acid (ABA) blocked electron transfer from the probe to the electrode surface, so no oxidation or reduction took place, as we can see in Figure 1A(b). After that, bifunctional PEG was deposited through the amide bond. The long linear spacer arms of carboxy-amine PEG allow the electron to move from the probe solution to the electrode surface. Hence, an increase in redox peaks of cyclic voltammogram can be observed in Figure 1A(c). The immobilization of the aptamer over carboxy-amine PEG further decreased the resistance of the system and facilitated the charge transfer from the solution to the electrode surface, resulting in an increase in the redox shape of the voltammogram (Figure 1A(d)). After the completion of aptasensor fabrication, PAT was incubated over the aptasensor. Changes in the 3D conformation of the aptamer structure offered steric hindrance and resulted in the exposition of the negative charges of the aptamer that strongly repelled the redox probe. No redox reaction occurred at the electrode surface, as shown in Figure 1A(e).

2.2.2. Impedance Characterizations

EIS is a powerful and a useful tool for depicting the barrier properties of the modified electrode. The R$_{et}$ values obtained after fitting the curves to a Randles equivalent circuit are shown in Table S1 for various stages of aptasensor fabrication. Faradic impedance spectra presented in the form of Nyquist plots are shown in Figure 1B for the stepwise modification process. Bare SPCEs exhibited a small semicircle, as shown in Figure 1B(a), indicating a very low resistance in electron transfer from the redox probe to the electrode surface. After the electro-grafting of ABA via the diazonium ion, the resistance of the surface significantly increased (Figure 1B(b)). This increase in resistance was due to the presence of the COO⁻ surface, which acted as a barrier layer and hindered the electron transfer from the anionic ferri/ferrocyanide redox probe to the electrode due to strong electrostatic repulsion. The second modification was the covalent attachment of carboxy-amine PEG. For this purpose, the COOH group was first activated with EDC-NHS chemistry. After the activation of COOH carboxy-amine PEG was covalently attached via amide coupling between the terminal NH_2 group of carboxy-amine PEG and the COO⁻ group of the carboxyphenyl layer, there was a significant increase in R_{et} (Figure 1B(c)). Contrary to the expected results, a small decrease in impedance was observed, as compared to that obtained after ABA electrodeposition, although in both cases COO⁻ was exposed on the surface. This comparative decrease in R_{et} can be attributed to the formation of tunnels on the electrode surface after the deposition of heterobifunctional PEG, which facilitates the electron flow from the probe solution to the electrode. In the case of a carboxyphenyl layer, these tunnels were absent, so the substrate was completely and uniformly covered with negative charges. The electrostatic repulsion between COO⁻ did not allow any electrons to transfer from the probe solution to the electrode. Such behavior of modified electrodes demonstrates that the formation of tunnels on the electrode surface is due to the long linear spacer arms of carboxy-amine PEG [24]. After successful immobilization of bifunctional PEG, the exposed COOH group was again activated by EDC-NHS. Subsequently, amine-terminated anti-PAT aptamer was immobilized via amide bond formation. A decrease in R_{et} in Figure 1B(d), as compared to 1B(c), proved the effective immobilization of the aptamer, as the electrostatic repulsion between COO⁻ and the redox probe was reduced after aptamer immobilization. The incubation of PAT increased in R_{et} values, which can be seen in Figure 1B(e). The quadruplex formed upon target recognition, blocked the molecular gates and exposed the negatively charged backbone of the aptamer toward the anionic probe that in-turn increased the impedance of system.

2.3. Optimization of Experimental Parameters

The factors that can affect the electrochemical properties of the proposed aptasensor are pH, temperature, analyte incubation time, and the concentration of the aptamer immobilized on the electrode surface. Therefore, these parameters were optimized to obtain better analytical performance. Results are shown in Figure 2. The optimal response of the aptasensor was obtained at room temperature and pH 7.4. The concentration of carboxy-amine PEG was optimized to be 6 mg mL^{-1}.

Furthermore, the time required to immobilize the maximum concentration of aptamers on the active surface of the SPCEs was optimized. Variation in the Δ_{ratio} was recorded in the time range of 5–90 min. It was observed that maximum immobilization of the aptamer was achieved after 30 min (Figure 2A).

The concentration of the aptamer is another important parameter. We recorded the Δ_{ratio} for aptamer concentration ranging between 0.5 and 5.0 μM. Results in Figure 2B indicated that the maximum Δ_{ratio} was observed at a 2.0 μM concentration. Therefore, 2.0 μM was considered as the optimum concentration and was used in the following experiments.

The analyte incubation time is also an important parameter optimized during the aptasensor fabrication. Optimum incubation time for the analyte was found to be 45 min, as shown in Figure 2C.

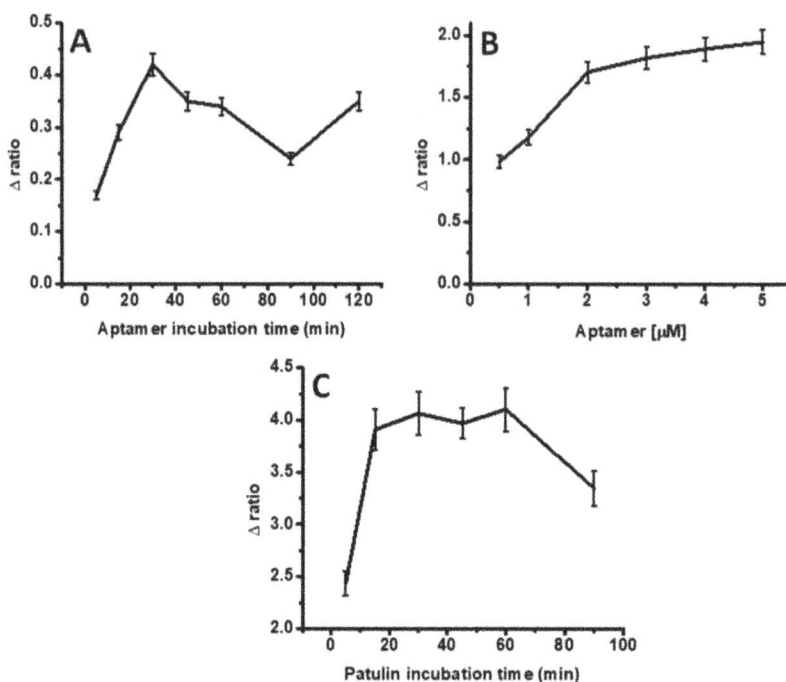

Figure 2. Optimization of experimental parameters. (**A**) Variation in Δ_{ratio} with increasing time after incubation with aptamer solution. (**B**) Variation in Δ_{ratio} with varying aptamer concentration. (**C**) Variation in Δ_{ratio} obtained at different intervals of time after the incubation of PAT.

2.4. Analytical Performance of the Aptasensors

To evaluate the analytical performance of the developed impedimetric aptasensor, different concentrations of PAT were incubated on the proposed aptasensing platform. A significant increase in the R_{et} was observed with increasing PAT concentration. A Nyquist plot for the electrochemical aptasensor in response to different concentrations of PAT is presented in Figure 3. The results indicated that the more PAT was bound to the electrode surface, the higher the R_{et} of the system was. This proportional increase in R_{et} indicated that the developed aptasensor can quantitatively detect PAT. In order to calculate a reliable estimation of increase or decrease in impedance, Δ_{ratio} was calculated for all aptasensors (different electrodes developed in triplicate).

Figure 3B shows that the Δ_{ratio} of the aptasensor gradually increases with an increasing concentration of PAT.

Limit of detection and limit of quantification were determined by calculating the signal-to-noise ratio in the blank. The formulae are

$$LOD = 3 * \text{standard deviation of blank / S}$$

$$LOQ = 10 * \text{standard deviation of blank / S}$$

where "S" is the slope of calibration curve.

The limit of detection obtained from the impedimetric aptasensor is 2.8 ng L^{-1}, while the limit of quantification is 4.0 ng L^{-1}.

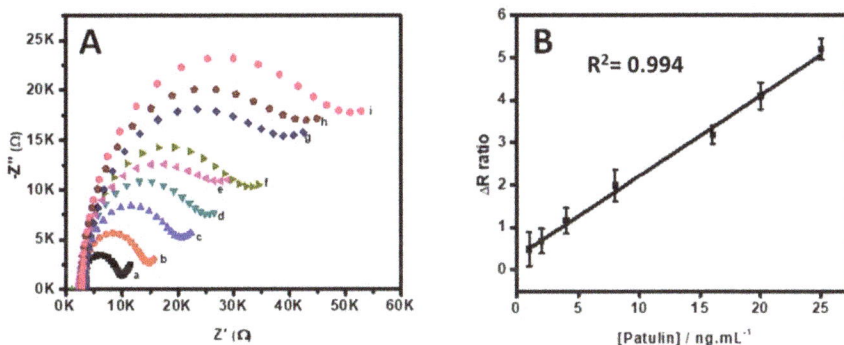

Figure 3. (**A**) Nyquist plot of the aptasensor after incubation with different PAT concentrations: (a) bare, (b) 1 ng mL^{-1}, (c) 2 ngmL^{-1}, (d) 4 ng mL^{-1}, (e) 8 ng mL^{-1}, (f) 16 ng mL^{-1}, (g) 20 ng mL^{-1}, (h) 25 ng mL^{-1}, and (i) 50 ng mL^{-1}. (**B**) Calibration plot of Δ_{ratio} with increasing PAT concentration.

2.5. Selectivity of the Aptasensors

Selectivity of the aptasensor against PAT was checked by incubating the aptasensor with different interfering analytes such as ochratoxin A, ochratoxin B, and aflatoxins. Obtained results are presented in Figure 4. No significant increase in R_{et} values can be seen in the case of mycotoxins other than PAT, which confirms the high selectivity and specificity of the aptasensor for PAT. In the same context, a control single-strand DNA was immobilized on the transducer surface in a fashion similar to that described for immobilization of the aptamer. No significant change in R_{et} was observed in the presence of PAT. These results also prove that there were no non-specific binding sites on the transducer surface.

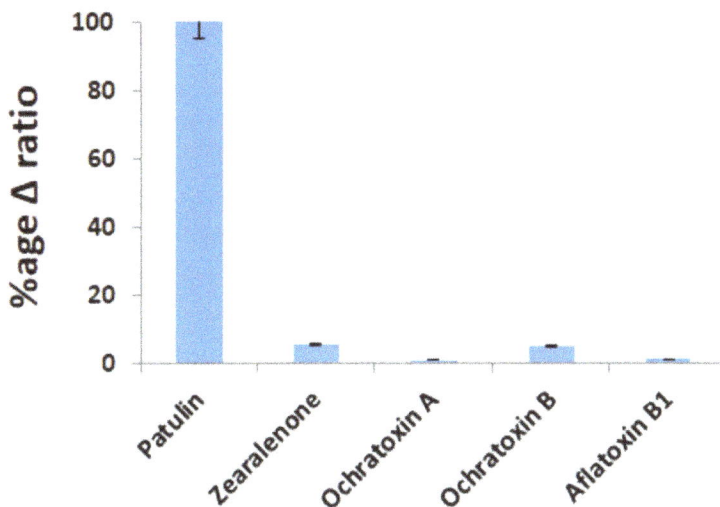

Figure 4. Selectivity test for the developed aptasensor.

2.6. Application to Spiked Apple Juice Sample

The developed impedimetric aptasensor was evaluated for real sample analysis. For a real sample application, PAT was detected in apple juice. Apple juice samples were spiked with three different concentrations of PAT, i.e., 4, 10, and 20 ng mL^{-1}. Controlled experiments were performed by detecting the same concentrations of PAT in phosphate binding buffer. The electrochemical response in apple juice exhibits the same phenomenon as observed in buffer. Results are presented in Table 1. This experiment confirms the feasibility of the aptasensor for field applications. The recovery values were calculated based on the integration of the Δ_{ratio} value (electrochemical impedimetric response) of the spiked value in the equation of the calibration curve.

$$Y = 0.3128 + 0.1901 * X$$

The precision and accuracy of the proposed aptasensor was confirmed by interday and intraday assay analysis. A relative standard deviation (RSD) of 2.8–4.0% ($n = 3$) was calculated for interday analysis, which indicated very good reproducibility of the assay. Similarly, for intraday analysis, RSD was 1.34%, which confirmed that results are reproducible.

Table 1. Recovery percentages obtained with designed electrochemical aptasensor for PAT monitoring in apple juice.

PAT Added (ng/mL)	PAT Found (ng/mL)	RSD %	RE %	R %
4	3.7	3.7	7.5	92.5
10	9.6	4.1	4	96
20	18.8	4.8	6	94

RSD % = relative standard deviation percentage; RE % = relative error percentage; R % = recovery percentage.

3. Experimental

3.1. Chemicals

Amine-terminated aptamer sequence for PAT was purchased from Microsynth (Balgach, Switzerland) The aptamer sequence was developed by Shijia Wu et al. by a graphene oxide-assisted SELEX (systematic evolution of ligands by exponential enrichment) process. Initially, eight aptamer sequences were shortlisted, and an aptamer named PAT-11 was then selected as the final recognition element based on its low K_d (dissociation constant) value, i.e., 21.83 nM, and the highest binding affinity and selectivity for PAT. The potential applicability of the selected aptamer sequence was further confirmed by developing an enzyme-chromogenic substrate system-based colorimetric aptasensor for the selective and sensitive detection of PAT in a real matrix [25]:

Sequence of PAT aptamer:

$5'$GGC CCG CCA ACC CGC ATC ATC TAC ACT GAT ATT TTA CCT T-NH$_2$-$3'$.

The sequence of control aptamers was as follows:

($5'$ GCA GTT GAT CCT TTG GAT ACC CTG G$3'$)-NH$_2$.

PAT, sodium phosphate dibasic Na$_2$HPO$_4$, potassium chloride KCl, potassium phosphate monobasic KH$_2$PO$_4$, calcium chloride CaCl$_2$, magnesium chloride MgCl$_2$, 4-aminobenzoic acid, carboxy-amine PEG (3400) ethyl-dimethyl-amino-propyl carbodiimide, *N*-hydroxy succinimide and ferricyanide, ferrocyanide, sodium nitrite, and 4-aminobenzoic acid were obtained from Sigma-Aldrich (Lyon, France). Sterilized water was used for aptamer solution preparation, while distilled water was used throughout the experimental procedure for reagent preparation. For real sample analysis, fresh apple juice was bought from a supermarket in Perpignan, France.

Home-made SPCEs were used for the electrochemical experiments.

3.2. Instrumentation

Pre-treatment of the aptamer (heating at 90 °C for 8 min followed by immediate cooling to 4 °C for 4 min then 15 min at RT) was done on thermocycler mastercycler (Eppendorf, Le pecq, France).

All the electrochemical measurements were performed on an Electrochemical Workstation "Biologics" equipped with EC-Lab software.

Lyophilized aptamer was diluted in phosphate binding buffer (PBB 50 mM). PBB was composed of 2 mM KH_2PO_4, 8mM Na_2HPO_4, 135 mM KCl, 60 mM NaCl, and 5 mM $MgCl_2$ in deionized water. The pH of the buffer was adjusted to 7.4.

3.3. Experimental Protocol

3.3.1. Covalent Immobilization of Aptamer onto the SPCEs

Before use, all electrodes were electrochemically washed with washing buffer (0.5 M H_2SO_4 diluted in 0.1 M KCl) with 8–10 CV scans between 1 and −1 V at 100 mV.s^{-1}

Diazonium salt was generated by mixing 2 mM ABA in 0.5 M HCl with 2 mM sodium nitrite. The solution was left to react for 5 min for maximum generation of diazonium salt. After 5 min, 150–200 µL of this salt solution was electrodeposited on SPCEs via LSV with a potential ranging between 0.6 and −0.8 V and a scan rate of 50 mV.s^{-1} [26]. Furthermore, the exposed COOH group of ABA was activated through EDC/NHS chemistry (by immersing the electrode for 60 min in solution of 100 mM EDC and 25 mM NHS solution in 100 mM MES buffer pH = 5.5). After activation of COOH, 6 mM carboxy-amine PEG solution was then incubated on a working electrode for 40 min to attach the spacer covalently with aminobenzoic acid via an amide bond. The terminal COOH group of carboxy amine PEG was again activated through carbodiimide chemistry. Afterward, 2 µM NH2-modified PAT-aptamer was cast (20 µL) on the activated surface of the working electrode for an optimized immobilization time of 30 min. The NH_2 group of the aptamer made a covalent bond with a terminal COOH group of PEG. Finally, the working area was treated with 1% BSA solution to block non-reactive sites, therefore avoiding the possibility of non-specific adsorption.

All the modification steps were characterized by CV and EIS. The CV was performed within a potential range from 0.7 to −0.6 V at a 50 mV.s^{-1} scan rate. EIS measurements were performed using 1 mM [Fe (CN)$_6$]$^{3-/4-}$ solution as a free redox probe containing 10 mM KCl.

The prepared electrode was stored at 4 °C and employed as a ready-to-use device during the present work. The ready-to-use electrodes did not show any variation in the electrochemical response over the entire period, suggesting the stability of the aptasensor.

3.3.2. Detection of Patulin

A stock solution of PAT was prepared in acetonitrile, and further dilutions were prepared in PBB. For the detection of PAT, selected concentrations of PAT were incubated for 1 h with a fabricated aptasensor, followed by rinsing with binding buffer, and electrochemical measurements were then carried out.

3.3.3. Apple Juice Sample Preparation

In a 5 mL falcon tube, apple juice was vortex-mixed with ethyl acetate solution (1:1) and centrifuged at 10m000 rpm for 5 min. The supernatant was dried in a nitrogen environment. Afterward, the obtained dry powder was dissolved in PBS buffer after filtering with a nylon syringe filter (0.22 mm). Different PAT concentrations (a 4, 10, and 20 ng mL^{-1} final concentration) were then added to 50 µL of the pre-treated apple juice samples in order to perform standard additions and recovery experiments. The pH of the apple juice sample was adjusted to 7.4.

3.4. Impedimetric Measurements

Impedance measurements were obtained at an applied potential of 100 mV (vs. Ag/AgCl reference electrode), within a frequency range from 10 KHz to 0.5 Hz, an AC amplitude of 10 mV, and a sampling rate of 50 points. All measurements were performed in PBB containing a redox probe (1:1 mixture of 2 mM K_3[Fe(CN)$_6$]/K_4[Fe(CN)$_6$]). Impedance data was registered after each modification step of electrode surface in the following order: (i) bare electrode; (ii) diazonium salt deposition; (iii) carboxy-amine PEG immobilization; (iv) aptamer immobilization steps; and (v) PAT detection. The impedance spectra were presented as Nyquist plots, ($-Z_{im}$ vs. Z_{re}) and fitted to the Randles equivalent circuit model, as shown in Figure S1, with EC Lab software. The parameter R_s, corresponds to the solution resistance. R_{et} is the resistance in electron transfer between the electrode surface and the redox marker, whereas C is the double-layer capacitance and W (Warburg parameter) is associated with the diffusion of the redox probe. For all fittings, performed on EC-Lab software, the chi-square goodness-of-fit test was carefully checked to verify calculations (with 80.73 at a 92% confidence level). In this work, we focused on the variation in electron transfer resistance (R_{et}). In order to minimize the electrode to electrode variations and to obtain reproducible and independent results, relative and normalized signals were required. Thus, the Δ_{ratio} parameter was calculated for all the electrodes. Δ_{ratio} is the ratio of R_{et} of the bare aptasensor to the R_{et} of the aptasensor with a particular amount of analyte. The R_{et} of the bare aptasensor was subtracted from the final values to ignore the electrode-to-electrode variation.

$$\Delta_{ratio} = R_{et\,(after\,PAT)} \text{ - } R_{et\,(bare)}\ /\ R_{et\,(before\,PAT)} - R_{et\,(bare)}$$

where $R_{et\,(after\,PAT)}$ is the value obtained incubating the aptasensor with the PAT, $R_{et\,(before\,PAT)}$ is the electron transfer resistance value of the bare electrode, and $R_{et\,(bare)}$ is the resistance in electron transfer of the blank electrode and buffer.

4. Conclusions

In summary, we successfully developed a field-portable impedimetric aptasensor for selective, sensitive, and label-free detection of the carcinogenic mycotoxin known as PAT. Electrochemical impedimetric aptasensing for label-free detection of PAT was based on diazonium chemistry and bifunctional polyethylene glycol as spacer. The carboxy-amine PEG proved to be an effective spacer that resulted in tunnel formation for the electron transfer from the redox probe to the electrode surface. The LOD and LOQ obtained were 2.8 ng L^{-1} and 4.0 ng L^{-1}, respectively. Real sample analysis and interference study confirms that the developed aptasensor has the potential for PAT detection in the real matrix.

Supplementary Materials: The following are available online at http://www.mdpi.com/1420-3049/24/6/1017/s1. Figure S1: Linear sweep voltammogram for electrodeposition of the in situ generated 4-carboxyphenyl diazonium salt in the diazotization mixture at SPCE, Figure S2: Randles Circuit for fitting of Nyquist Plot, Table S1: R_{et} Values obtained after each modification step of Aptasensor fabrication Electrode.

Author Contributions: Conceptualization, A.H. and J.L.M.; methodology, R.K. and S.B.A.; software, S.B.A.; validation and formal analysis, R.K., T.A.S. and S.B.A.; investigation, R.K.; resources, G.C.; data curation, R.K.; writing—original draft preparation, R.K. and S.B.A.; writing—review and editing, A.H.; visualization, T.A.S.; supervision, A.H.; project administration, A.H.; funding acquisition, A.H.

Funding: This research was funded by HEC (Higher Education Commission of Pakistan)-Technology Development Program (TDF), grant number. 028, 2017".

Acknowledgments: Akhtar Hayat would like to acknowledge the project; HEC (Higher Education Commission of Pakistan)-Technology Development Program (TDF) project; No. 028, 2017. S.B.A. acknowledges the University of Tunis El Manar (Tunisia) for the mobility grant "Bourse d'alternance", as well as the Tunisian Ministry of Higher Education and Scientific Research for LR99ES15 Lab funding.

References

1. Xia, X.; He, Q.; Dong, Y.; Deng, R.; Li, J. Aptamer-based Homogeneous Analysis for Food Control. *Curr. Anal. Chem.* **2018**, *14*, 1–9. [CrossRef]
2. Patra, J.K.; Mahato, D.K.; Kumar, P. Biosensor Technology—Advanced Scientific Tools, With Special Reference to Nanobiosensors and Plant- and Food-Based Biosensors. In *Nanomaterials in Plants, Algae and Microorganisms*; Elsevier Academic Press: Cambridge, MA, USA, 2019; pp. 287–303.

3. Vasilescu, A.; Marty, J.L. Electrochemical aptasensors for the assessment of food quality and safety. *Trends Anal. Chem.* **2016**, *79*, 60–70. [CrossRef]

4. Laganà, A. *Introduction to the Toxins Special Issue on LC-MS/MS Methods for Mycotoxin Analysis*; Multidisciplinary Digital Publishing Institute: Basel, Switzerland, 2017.

5. Moss, M. Fungi, quality and safety issues in fresh fruits and vegetables. *J. Appl. Microbiol.* **2008**, *104*, 1239–1243. [CrossRef] [PubMed]

6. Ioi, J.D.; Zhou, T.; Tsao, R.; Marcone, M.F. Mitigation of patulin in fresh and processed foods and beverages. *Toxins* **2017**, *9*, 157. [CrossRef]

7. Iwahashi, Y.; Hosoda, H.; Park, J.-H.; Lee, J.-H.; Suzuki, Y.; Kitagawa, E.; Murata, S.M.; Jwa, N.-S.; Gu, M.-B.; Iwahashi, H. Mechanisms of patulin toxicity under conditions that inhibit yeast growth. *J. Agric. Food Chem.* **2006**, *54*, 1936–1942. [CrossRef] [PubMed]

8. Glaser, N.; Stopper, H. Patulin: Mechanism of genotoxicity. *Food Chem. Toxicol.* **2012**, *50*, 1796–1801. [CrossRef] [PubMed]

9. Mahfoud, R.; Maresca, M.; Garmy, N.; Fantini, J. The mycotoxin patulin alters the barrier function of the intestinal epithelium: Mechanism of action of the toxin and protective effects of glutathione. *Toxicol. Appl. Pharmacol.* **2002**, *181*, 209–218. [PubMed]

10. Commission, E. Commission Regulation (EC) No 1881/2006 of 19 December 2006 setting maximum levels for certain contaminants in foodstuffs. *Off. J. Eur. Union* **2006**, *364*, 5–24.

11. Zheng, Y.; Wang, X.; He, S.; Gao, Z.; Di, Y.; Lu, K.; Li, K.; Wang, J. Aptamer-DNA Concatamer-Quantum Dots Based Electrochemical Biosensing Strategy for Green and Ultrasensitive Detection of Tumor Cells via Mercury-Free Anodic Stripping Voltammetry. *Biosens. Bioelectron.* **2018**, *126*, 261–268. [CrossRef] [PubMed]

12. Sadok, I.; Szmagara, A.; Staniszewska, M.M. The validated and sensitive HPLC-DAD method for determination of patulin in strawberries. *Food Chem.* **2018**, *245*, 364–370. [CrossRef]

13. Shephard, G.S.; Leggott, N.L. Chromatographic determination of the mycotoxin patulin in fruit and fruit juices. *J. Chromatogr. A* **2000**, *882*, 17–22. [CrossRef]

14. Takino, M.; Daishima, S.; Nakahara, T. Liquid chromatography/mass spectrometric determination of patulin in apple juice using atmospheric pressure photoionization. *Rapid Commun. Mass Spectrom.* **2003**, *17*, 1965–1972. [CrossRef] [PubMed]

15. Gimeno, A. Thin layer chromatographic determination of aflatoxins, ochratoxins, sterigmatocystin, zearalenone, citrinin, T-2 toxin, diacetoxyscirpenol, penicillic acid, patulin, penitrem A. *J. Assoc.* **1979**, *62*, 579–585.

16. Beltrán, E.; Ibáñez, M.; Sancho, J.V.; Hernández, F. Determination of patulin in apple and derived products by UHPLC–MS/MS. Study of matrix effects with atmospheric pressure ionisation sources. *Food Chem.* **2014**, *142*, 400–407. [CrossRef]

17. Grabowska, I.; Sharma, N.; Vasilescu, A.; Iancu, M.; Badea, G.; Boukherroub, R.; Ogale, S.; Szunerits, S. Electrochemical Aptamer-Based Biosensors for the Detection of Cardiac Biomarkers. *ACS Omega* **2018**, *3*, 12010–12018. [CrossRef]

18. Goud, K.Y.; Hayat, A.; Catanante, G.; Satyanarayana, M.; Gobi, K.V.; Marty, J.L. An electrochemical aptasensor based on functionalized graphene oxide assisted electrocatalytic signal amplification of methylene blue for aflatoxin B1 detection. *Electrochim. Acta* **2017**, *244*, 96–103. [CrossRef]

19. Rhouati, A.; Catanante, G.; Nunes, G.; Hayat, A.; Marty, J.-L. Label-free aptasensors for the detection of mycotoxins. *Sensors* **2016**, *16*, 2178. [CrossRef]

20. Wu, L.; Xiong, E.; Zhang, X.; Zhang, X.; Chen, J. Nanomaterials as signal amplification elements in DNA-based electrochemical sensing. *Nano Today* **2014**, *9*, 197–211. [CrossRef]

21. Huang, L.; Wu, J.; Zheng, L.; Qian, H.; Xue, F.; Wu, Y.; Pan, D.; Adeloju, S.B.; Chen, W. Rolling chain amplification based signal-enhanced electrochemical aptasensor for ultrasensitive detection of ochratoxin A. *Anal. Chem.* **2013**, *85*, 10842–10849. [CrossRef]

22. He, B.; Dong, X. Aptamer based voltammetric patulin assay based on the use of ZnO nanorods. *Microchim. Acta* **2018**, *185*, 462. [CrossRef] [PubMed]

23. Wu, Z.; Xu, E.; Jin, Z.; Irudayaraj, J. An ultrasensitive aptasensor based on fluorescent resonant energy transfer and exonuclease-assisted target recycling for patulin detection. *Food Chem.* **2018**, *249*, 136–142. [CrossRef] [PubMed]

24. Hayat, A.; Andreescu, S.; Marty, J.-L. Design of PEG-aptamer two piece macromolecules as convenient and integrated sensing platform: Application to the label free detection of small size molecules. *Biosens. Bioelectron.* **2013**, *45*, 168–173. [CrossRef] [PubMed]

25. Wu, S.; Duan, N.; Zhang, W.; Zhao, S.; Wang, Z. Screening and development of DNA aptamers as capture probes for colorimetric detection of patulin. *Anal. Biochem.* **2016**, *508*, 58–64. [CrossRef] [PubMed]

26. Mahouche-Chergui, S.; Gam-Derouich, S.; Mangeney, C.; Chehimi, M.M. Aryl diazonium salts: A new class of coupling agents for bonding polymers, biomacromolecules and nanoparticles to surfaces. *Chem. Soc. Rev.* **2011**, *40*, 4143–4166. [CrossRef] [PubMed]

molecules

MDPI

Article

Fatty Acid, Lipid Classes and Phospholipid Molecular Species Composition of the Marine Clam *Meretrix lyrata* (Sowerby 1851) from Cua Lo Beach, Nghe An Province, Vietnam

Quoc Toan Tran [1,2], Thi Thanh Tra Le [2,3], Minh Quan Pham [1,2], Tien Lam Do [1], Manh Hung Vu [4], Duy Chinh Nguyen [5], Long Giang Bach [5,6], Le Minh Bui [5,*] and Quoc Long Pham [1,2,*]

[1]   Institute of Natural Products Chemistry, Vietnam Academy of Science and Technology, Hanoi 122100, Vietnam; tranquoctoan2010@gmail.com (Q.T.T.); minhquanaries@gmail.com (M.Q.P.); dotienlam198@gmail.com (T.L.D.)
[2]   Graduate University of Science and Technology, Vietnam Academy of Science and Technology, Hanoi 122100, Vietnam; traltt@wru.vn
[3]   Department of Chemical Engineering, Faculty of Environment, Thuy loi University, Hanoi 122100, Vietnam
[4]   Institute of Marine Environment and Resources, Vietnam Academy of Science and Technology, Hanoi 122100, Vietnam; hungvm@imer.vast.vn
[5]   NTT Hi-Tech Institute, Nguyen Tat Thanh University, Ho Chi Minh City 700000, Vietnam; ndchinh@ntt.edu.vn (D.C.N.); blgiang@ntt.edu.vn (L.G.B.)
[6]   Center of Excellence for Biochemistry and Natural Products, Nguyen Tat Thanh University, Ho Chi Minh City 700000, Vietnam
*   Correspondence: blminh@ntt.edu.vn (L.M.B.); mar.biochem@fpt.vn (Q.L.P.); Tel.: +84-1900-2039 (L.M.B.)

Received: 15 January 2019; Accepted: 1 March 2019; Published: 4 March 2019

check for updates

Abstract: This study aims to analyze compositions of fatty acids and phospholipid molecular species in the hard clams *Meretrix lyrata* (Sowerby, 1851) harvested from Cua Lo beach, Nghe An province, Viet Nam. Total lipid of hard clams *Meretrix lyrata* occupied $1.7 \pm 0.2\%$ of wet weight and contained six classes: hydrocarbon and wax (HW), triacylglycerol (TAG), free fatty acids (FFA), sterol (ST), polar lipid (PoL), and monoalkyl diacylglycerol (MADAG). Among the constituents, the proportion of PoL accounted was highest, at 45.7%. In contrast, the figures for MADAG were lowest, at 1.3%. Twenty-six fatty acids were identified with the ratios of USAFA/SAFA was 2. The percentage of n-3 PUFA (ω-3) and n-6 PUFA (ω-6) was high, occupying 38.4% of total FA. Among PUFAs, arachidonic acid (AA, 20:4n-6), eicosapentaenoic acid (EPA, 20:5n-3), docosapentaenoic acid (DPA, 22:5n-3), and docosahexaenoic acid (DHA, 22:6n-3) accounted for 3.8%, 7.8%, 2.2% and 12.0% of total lipid of the clam respectively. Phospholipid molecular species were identified in polar lipids of the clams consisting six types: phosphatidylethalnolamine (PE, with 28 molecular species), phosphatidylcholine (PC, with 26 molecular species), phosphatidylserine (PS, with 18 molecular species), phosphatidylinositol (PI, with 10 molecular species), phosphatidylglycerol (PG, with only one molecular species), and ceramide aminoethylphosphonate (CAEP, with 15 molecular species). This is the first time that the molecular species of sphingophospholipid were determined, in *Meretrix lyrata* in particular, and for clams in general. Phospholipid formula species of PE and PS were revealed to comprise two kinds: Alkenyl acyl glycerophosphoethanolamine and Alkenyl acyl glycerophosphoserine occupy 80.3% and 81.0% of total PE and PS species, respectively. In contrast, the percentage of diacyl glycero phosphatidylcholine was twice as high as that of PakCho in total PC, at 69.3, in comparison with 30.7%. In addition, phospholipid formula species of PI and PG comprised only diacyl glycoro phospholipids. PE 36:1 (p18:0/18:1), PC 38:6 (16:0/22:6), PS 38:1 (p18:0/20:1), PI 40:5 (20:1/20:4), PG 32:0 (16:0/16:0) and CAEP 34:2 (16:2/d18:0) were the major molecular species.

Keywords: hard clams; *Meretrix lyrata*; lipid classes; fatty acids; phospholipids; molecular species of phospholipid; high resolution mass spectrometry

---

1. Introduction

Investigation on marine and freshwater two-part shell mollusks has been growing rapidly due to their rich nutritional value, variety of biofunctions, potent activities, and most importantly, ease of exploitation [1–4]. Poly-unsaturated fatty acids, mainly omega-3 fatty acids, have been a valuable nutrient group and the nutraceutical of interest. In almonds (*Prunus dulcis*), it is shown that the content of linoleic acid, a poly-unsaturated fatty acid, ranged from 10 to 31% depending on the origin of the sample [5]. The high content of PUFA is suggested to be associated with the cardio protective effect. The mollusk, especially bivalve mollusks, is widely consumed in Asian countries, providing high protein content, sugars, lipids, especially omega-3-fatty acids and essential amino acids, vitamin B12 and essential elements including iron, zinc and copper. Among these constituents, phospholipids and fatty acids play a major role in the functions of the immune system and the maintenance all hormonal systems of the organism. These important functionalities promote the development of new and more efficient extraction methods for mass production and commercialization of clams.

Clams, cockles, and arkshells, occupying the largest share in worldwide production of shellfish (about 6.16 million tons in 2016), are widely used as an important nutriment source [6]. Investigations of several clam species such as *Meretrix lusoria* (Asian hard clam), *Meretrix meretrix, Cyclina sinensis,* and *Chamelea gallina* (venus clam) have revealed that phospholipid (PL) and ω-3 long-chain polyunsaturated fatty acids (LC-PUFA) are abundantly found in clams [7–10]. Beneficial fatty acids in ω-3 LC-PUFAs include eicosapentaenoic acid (EPA, 20:5n-3), docosahexaenoic acid (DHA, 22:6n-3) and docosapentaenoic acid (DPA, 22:5n-3) [11–13]. Such dietary fatty acids have been studied extensively, showing a wide range of positive clinical effects including improvement in treatment of heart diseases, effective intervention of insulin sensitivity, inhibition of tumors, inflammation and metastasis. The FDA (Food and Drug Administration) has approved a qualified health claim for conventional foods and dietary supplements that contain EPA and DHA [14]. It states, "Supportive but not conclusive research shows that consumption of EPA and DHA omega-3 fatty acids may reduce the risk of coronary heart disease." The FDA also specifies that the labels of dietary supplements should not recommend a daily intake of EPA and DHA higher than 2 g per day for health benefits [14]. For patients who need to lower their triglyceride levels, the American Heart Association recommends 2–4 g/day of EPA plus DHA under the care of a physician [15]. Several prescription omega-3 preparations are also available to treat hypertriglyceridemia [16].

Clams could be potentially a good source of PL and PUFAs. Much evidence regarding lipid composition in various clam species has been presented. For example, in the striped Venus Clam (*Chamelea gallina*), it was shown that n-3 LC-PUFA constituted a large quantity in total fatty acids of, ranging from 33.7 to 41.9%. In addition, EPA (8.2–20.0% of total fatty acids) and DHA (12.5–20.3% of total fatty acids) were also found [8]. This is also similar to the Asian hard clam (*Meretrix lusoria*), where PUFA (46.8–49.2% of total fatty acids), DHA (13.3–16.5% of total fatty acids) and EPA (4.8–7.1% of total fatty acids) were found in high levels [9]. In the *Ruditapes philippinarum* clam, PLs were the principal lipids (57–75% of total lipids) [7]. In the lipid composition of the *Calyptogena phaseoliformis* clam, major lipids are the n-4 family non-methylene interrupted polyunsaturated fatty acids (NMI-PUFA) including 20:3n-4,7,15, 20:4n-1,4,7,15, and 21:3n-4,7,16, with significant levels of 20:2n-7,15 and 21:2n-7,16 as non-methylene interrupted n-7 dienes [17]. The major fatty acids of lipids in *Meretrix lamarckii* and *Ruditapes philippinarum* were 14:0, 16:0, 18:0, 16:1n-7, 18:1n-9, 18:1n-7, 20:4n-6, 20:5n-3, and 22:6n-3, while those of *Mesolinga soliditesta* were 16:0, 18:0, 16:1n-7, 18:1n-7, 20:1n-7, 20:1n-13, 20:2n-7, 15 (Δ5,13-20:2), and 22:2n-7,15 (Δ7,15-22:2) [18].

Due to the immensity of the sea aquaculture reservoir, the Vietnamese fishery industry was ranked very highly. In terms of fish landings from marine fishing areas, Vietnam occupied the 8th position in 2015 and 2016, reaching a volume of 2.607 and 2.678 million tons respectively [19]. Among Vietnam's highly valued aquatic organisms, the *Meretrix lyrata* (Sowerby 1851) clam, belonging to the *Veneridae* genus, also known as the hard clam, could be found on seashores and in estuarine areas. Apart from being an export product of high economic value, *Meretrix lyrata* also acts as a protein source for domestic consumption [20,21]. Considering the limited data on plasmalogens of bivalve species and the prevalence of the hard clam in Vietnamese diet, our study, for the first time, aims to report the fatty acids, lipid classes and phospholipid molecular species of the hard clams *Meretrix lyrata* S. The sample was collected in Cua Lo Beach, Nghe An Province, located in the North Central Coast region of Vietnam and populated with diversified aquaculture and ample clam species, in particular. We showed that the lipids from clams contained high percentages of PL. In addition, most of the predominant glycerophospholipid (GP) molecular species, such as phosphatidylethalnolamine (PE), phosphatidylcholine (PC), phosphatidylserine (PS), phosphatidylinositol (PI) and phosphatidylglycerol (PG) and sphingophospholipid (ceramide aminoethylphosphonate—CAEP), were identified.

2. Results and Discussion

2.1. Total Lipid

Total lipid (TL) constituted 1.7 ± 0.2% of wet weight of the clams. The composition and content of TL of *Meretrix lyrata* (Figure 1 and Table 1) was similar to that of other clams investigated previously [18,22–24]. Overall, hard clam species contained common lipids hydrocarbon and wax (HW), triacylglycerol (TAG), free fatty acids (FFA), sterol (ST), polar lipid (PoL) and monoalkyl diacylglycerol (MADAG). In contrast with lipid compositions of cnidarians and coral in which MADAG accounts for a significant proportion, MADAG component of the hard clam only represents a marginal content of 1.3 ± 0.2 of TL.

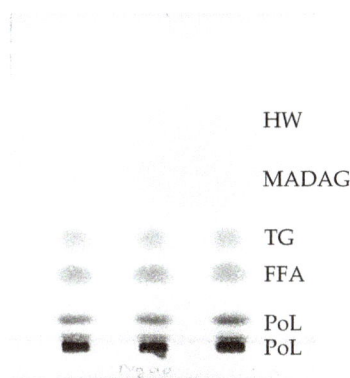

Figure 1. TLC determination of lipid classes of the clams.

Table 1. Composition of main lipid classes (% of total lipid) of the clams.

No.	Lipid Class	Content (%)
1	Hydrocarbon and wax (HW)	2.6 ± 0.1
2	Monoalkyl diacylglycerol (MADAG)	1.3 ± 0.2
3	Triacylglycerol (TG)	11.7 ± 0.3
4	Free fatty acids (FFA)	18.6 ± 0.8
5	Sterols (ST)	20.2 ± 0.5
6	Polar lipid (PoL)	45.7 ± 0.6

The findings for PoL levels in the total lipids of the hard clam are similar to those of other studies which have investigated lipid profiles and phospholipids. In addition, this result also highlights the seasonal variation of PoL content in clam. Specifically, the analyzed hard clams, which were collected in summer, exhibited a lower PoL content in comparison to those of other clam species collected in other seasons [25,26].

Regarding other lipid classes, TAG, FFA and ST were the major lipid classes in the non-polar lipid, representing 11.7%, 18.6% and 20.2% of TL respectively. This observation was in good agreement with previous publications of other clam species [22–24]. Notably, the content of FFA was found to be high in hard clam, suggesting possible involvement of lipase enzymes in the intestinal organs in the hydrolysis of TAG into FFA. The presence of sterols in hard clam is also consistent with the ubiquity of the lipid class in most marine organisms, which plays many key roles, such as presenting in membrane composition [26,27].

2.2. Fatty Acids in Total Lipids

Twenty-six fatty acids in the lipid sample of hard clam were found with the number of carbon atoms ranging from 14 to 22 (Table 2). Major FA were Acid 16:0, 16:1n-7, 18:0, 18:1n-9, 20:4n-6 (arachidonic acid AA), 20:5n-3 (EPA) and 22:6n-3 (DHA), respectively accounting for 16.0, 6.3, 5.0, 4.7, 3.8, 7.8, and 12.0% of the total FA content. Saturated fatty acids occupied 26.3% of total FA content.

Table 2. Fatty acid composition (% of total) of clams.

	R.t (min)	Area	Content (%)
Σ SAFAs		1,971,006	26.3
14:0 *	3.748	291,352.4	3.9
16:0 *	5.491	1,197,827	16.0
17:0 *	6.812	105,581.2	1.4
18:0 *	8.686	376,245.5	5.0
Σ MUFAs		1,191,078	15.9
16:1n-7 *	5.863	469,554.2	6.3
18:1n-9 *	9.228	355,129.6	4.7
18:1n-7 *	9.378	147,985.6	2.0
20:1n-11	15.324	99,722.7	1.3
20:1n-9 *	15.511	42,995.4	0.6
20:1n-7 *	15.875	75,690.3	1.0
Σ PUFAs		2,831,606	37.7
16:3n-3	7.292	56,506.2	0.8
18:2n-6 *	10.411	63,571.7	0.8
18:3n-3	12.383	77,148.8	1.0
18:4n-3 *	13.518	80,259.3	1.1
20:2n-6 *	17.775	108,771.3	1.4
20:3n-6	19.233	31,176.5	0.4
20:4n-6 *	20.617	283,411	3.8
20:4n-3 *	23.19	36,235.8	0.5
20:5n-3 *	24.929	588,733	7.8
21:3n-3	28.661	85,752.1	1.1
21:5n-6	29.156	144,854.4	1.9
21:5n-3	33.552	53,585.8	0.7
22:4n-6 *	36.426	115,438.2	1.5
22:5n-6	39.584	96,337.1	1.3
22:5n-3 *	44.249	161,865.8	2.2
22:6n-3 *	48.476	901,545.1	12.0
Octadecanal [a]	7.882	808,630.9	10.8
Other [b]		702,316.4	9.4

* Fatty acids were identified by GCMS and GC. [a] Fatty aldehyde was determined as its dimethylacetal derivatives by GC and GCMS. [b] 15:0, i-16:0, 16:1n-5, i-17:0, a-17:0, i-18:0, 17:3n-6, a-19:0, 19:0, 18:3n-6, 19:1n-9, 19:1n-7, 20:0, i-21:0, a-21:0, 20:3n-3, hexadecanal, hexadecenal, octedecenal. SAFAs: saturated fatty acids. MUFAs: monounsaturated fatty acids. PUFAs: poliunsaturated fatty acids.

Monounsaturated fatty acids (MUFA) accounted for 15.9% of total FA content and 29.6% of unsaturated fatty acids. Two major MUFA in the composition were C16 and C18 MUFA. Regarding polyunsaturated fatty acids (PUFA), content of PUFA represented 37.7% and 70.7% of total FA content and total unsaturated fatty acids respectively. Among PUFA, three C18 PUFA were found including18:2n-6, 18:3n-3 and 18:4n-3. However, C18 PUFA only accounted for 2.9% of the total FA content. In addition, the percentages of n-6 PUFA (ω-6) were 11.2% of total FA. The figure for n-3 PUFA (ω-3) was approximately 26.5%. Highly unsaturated fatty acids (HUFA) accounted for 32.1% in total and 59.9% in unsaturated fatty acids.

These results were similar to those of various reports [23,28,29]. In the composition, several odd-chain acids were found in small quantities and methyl-branched FA including i-16:0, i-17:0, a-17:0, i-18:0 and i-21:0 were detected as trace amounts. Among many long-chain DMA produced by plasmalogen PL under acid methanolysis of clam lipids, octadecanal dimethylacetal was the major DMA and amounted to 10.8% of FA composition. Three FAs with 21 carbon atoms were identified, including 21:3n-3, 21:5n-6 and 21:5n-3 and accounted for 1.1, 1.9, and 0.7% of FA, respectively. According to a previous study [26], 6,9,12,15,18-Heneicosapentaenoic acid (21:5n-3) (HPA), was shown to exhibit stronger inhibition for arachidonic acid synthesis from alpha-linoleic acid and dihomo-gamma-linolenic acid in hepatoma cells.

2.3. Fatty Acids in Fractions of Phospholipid

As shown in Table 3, GC analysis indicated that the fractions of phospholipid of the clams contained a high proportion of SAFAs. Saturated fatty acids exist dominantly in CAEP, at 75.8%. This figure was more than twice as much as of PS + PI. The figures for PE + PG and PC were lower, at 40.5, and 52.2% respectively. The abundance of SAFAs in the FA composition of CAEP was consistent with the molecular species content of 16:0, 17:0, and 18:0 acid, which were reported later in this study and totaled at 67.1% of CAEP from hard clam (Table S6).

Compared with the lipids recovered from the other clam species [17], the FA composition results of hard clam were also similar regarding DMA in PE and PC fraction. To be specific, both studies suggested that DMAs were non-existent in FA composition of PC in investigated clam species. In addition, present results also showed that CAEP fraction contains no DMA and PE is the only fraction which contains all three DMAs including of 16:0, 17:0 and 18:0 DMA, with corresponding contents of 0.4, 1.2, and 11.3. In fractions of PS + PI, only 17:0 DMA and 18:0 DMA exists at 1.5% and 14.6% respectively. Thus, two kinds of phospholipid formula species contained in PE and PS were plasmenyl and diacyl phospholipids and two phospholipid formula species of PC includes plasmanyl and diacyl glycero phospholipids.

Table 3. Fatty acid composition (% of each type phospholipid) of clams.

Fraction	PE + PG	PC	CAEP	PS + PI
Σ SAFAs	40.5	52.2	75.8	32.5
14:0	4.3	nd	nd	0.7
16:0	22.0	36.9	57.0	12.9
17:0	3.2	4.1	4.3	2.1
18:0	11.0	9.9	13.0	16.2
20:00	nd	1.3	1.5	0.4
Σ MUFAs	25.3	14.4	11.3	22.5
16:1n-7	10.9	3.6	nd	1.2
18:1n-9	8.4	7.9	8.5	9.5
18:1n-7	3.5	1.5	0.4	0.7
20:1n-11	1.1	0.5	0.3	7.4
20:1n-9	0.6	0.5	0.7	1.2
20:1n-7	0.8	0.4	1.4	2.5

Table 3. *Cont.*

Fraction	PE + PG	PC	CAEP	PS + PI
Σ PUFAs	14.9	29.9	9.1	22.7
16:3n-3	0.9	0.8	nd	nd
18:2n-6	0.9	1.0	2.1	1.5
18:3n-3	1.2	0.5	0.1	nd
18:4n-3	0.8	0.6	nd	nd
20:2n-6	nd	1.2	0.2	nd
20:3n-6	nd	0.7	0.3	nd
20:4n-6	1.7	2.4	1.7	5.8
20:4n-3	nd	0.3	nd	nd
20:5n-3	2.9	3.8	1.2	1.4
21:3n-6	nd	2.6	0.8	nd
21:3n-3	2.2	0.4	0.2	4.9
21:5n-6	nd	0.8	0.3	nd
21:5n-3	nd	0.1	1.1	nd
22:4n-6	nd	1.0	nd	0.7
22:5n-6	1.1	0.9	nd	1.5
22:5n-3	nd	1.8	nd	1.3
22:6n-3	3.2	11.0	1.0	5.5
Σ DMAs	12.9			16.0
16:0	0.4	nd	nd	nd
17:0	1.2	nd	nd	1.5
18:0	11.3	nd	nd	14.6
Other	6.3	3.5	3.8	6.3

nd: not detected or under 0.1% of fatty acid composition in each type phospholipid. SAFAs: saturated fatty acids. MUFAs: monounsaturated fatty acids. PUFAs: poliunsaturated fatty acids. DMAs: dimethylaxetal. PE: Phosphatidylethanolamine. PC: Phosphatidylcholine. PS: Phosphatidylserine. PI: Phosphatidylinositol. PG: Phosphatidylglycerol. CAEP Ceramide aminoethylphosphonate.

2.4. Molecular Species of Phospholipid

Molecular species of polar lipids from the hard clam were determined following the previously described HRMS fragmentations of PL standards [30]. TL of Clams contained 48.96% of polar lipids (PoL) (Table 1). PoL have got glycolipid and phospholipid. Especially in the hard clams, phospholipid dominated, with six types of phospholipids, including phosphatidylethanolamine (PE), phosphatidylcholine (PC), phosphatidylserine (PS), phosphatidylinositol (PI) and phosphatidylglycerol (PG) and Ceramide aminoethylphosphonate (CAEP). The molecular species and qualification of each phospholipid were performed by HPLC-HRMS [31].

Among phosphatidylethanolamine (PE) of clams, we determined 28 constituents. Alkenyl acyl glycerophosphoethanolamine (ethanolamine plasmalogen PlsEtn) was the major component in PE, with 80.3% of total PE species (Table S1), followed by PE 40:6 (m/z (M+) 776.5556), at 17.13%, Plasmalogen PE 36:1 (m/z (M+) 730.5794), at 15.4%, and PE 40:2 (m/z (M+) 784.6161), at 10.3%.

PlsEtn can be characterized according to their MS and MS/MS data. The signals of positive quasi-molecular ions $[M + H]^+$, cluster ions $[M + H + C_6H_{15}N]^+$ and negative quasi-molecular ions $[M - H]^-$ were observed in the HRMS spectra of all components of formula species of PE. For instance, we detected signals of negative quasi-molecular ions $[M - H]^-$ at m/z 728.5544 ($[C_{41}H_{79}NO_7P]^-$, calculated 728.5594, different 0.00556), positive quasi-molecular ions $[M + H]^+$ at m/z 730.5794 ($[C_{41}H_{81}NO_7P]^+$, calculated 730.5751, different 0.00488), and positive cluster ions $[M + C_6H_{15}N]^+$ at m/z 831.6822 ($[C_{47}H_{96}N_2O_7P]^+$, calculated 831.6955, different 0.01277) of PE 36:1 (Figure 2).

From the the MS^{2-} spectrum of the ions $[M - H]^-$ of PE 36:1, one signal corresponding to one carboxylate anion of 18:1 was detected at m/z 281.2466. For the plasmalogen, the fatty acid (FA) anion ($[RCOO]^-$) can only be liberated from the *sn*-2 position due to the ankenyl linkages at the *sn*-1 position [32]. Therefore, three possible molecular species including plasmenyl p18:0/18:1. Component PE 36:1 was identified as alkenyl acyl glycerophosphoethanolamine, p18:0/18:1 PlsEtn (Figure S1).

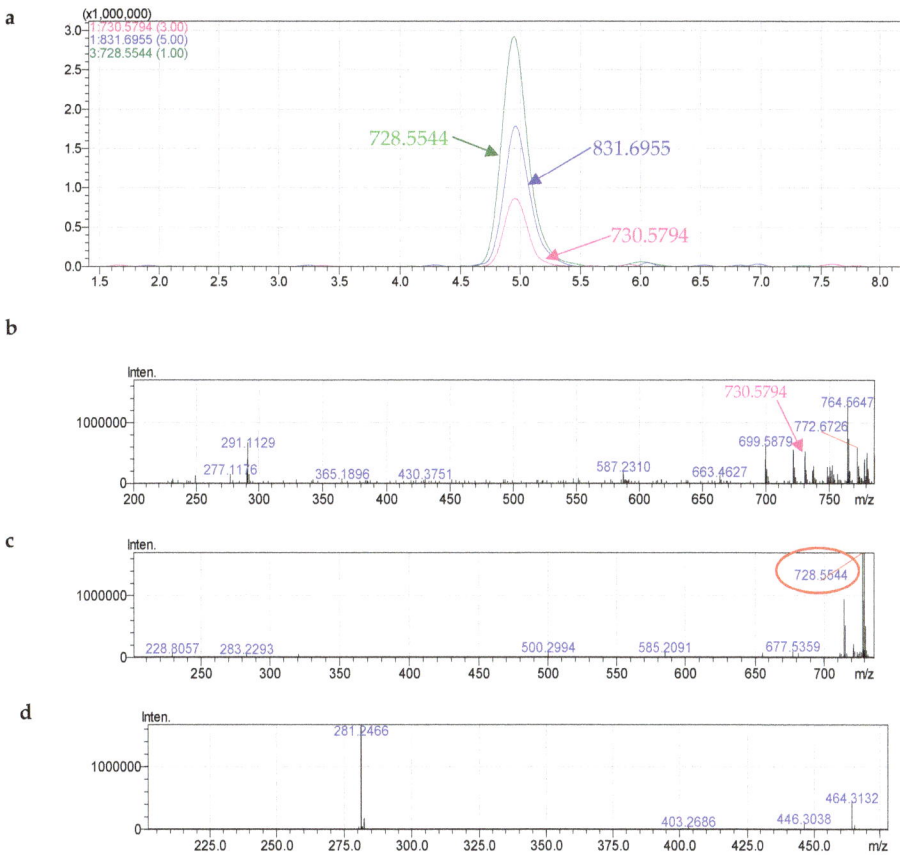

Figure 2. HPLC–HRMS (**a**) and fragmentation of PE 36:1(**b**—MS$^+$, **c**—MS$^-$, **d**—MS^{2-}).

Diacyl glycerophosphoethanolamine only accounted for 19.7% of total PE species. For an unknown PL containing two FAs, MS and MS/MS data also are needed to determine the two individual FAs esterified at the *sn*-1 and *sn*-2 positions of the glycerol backbone.

For example, for the unknown PE with measured *m/z* of 730.5411 (quasi-molecular ion ([M + H]$^+$)), PE 35:2 could tentatively be deduced according to the previously described formulae [30,31]. The MS^{2-} spectrum of the ions [M − H]$^-$ of component PE 35:2 contained the signal of two carboxylate anion of 18:1 at *m/z* 281.2466 and 17:1 at *m/z* 267.2315 (Figure 3 and Figure S2). The component PE 35:2 was determined as diacy glycerophosphoethanolamine, 18:1/17:1.

The choline glycerophospholipids (ChoGpl) of clams is summarized in Table S2. Among 26 components, 9 of which constituted more than 51% of total ChoGpl. The percentage of diacyl glycero phosphatidylcholine was twice as abundant as PakCho in total PC, at 69.3% in comparison with 30.7%.

The signals of positive quasi-molecular ions [M + H]$^+$, negative formate molecular ions [M + HCOO]$^-$ and cluster ions [M − CH$_3$]$^-$ were observed in the HRMS spectra of all components of formula species of PC. Formate molecular ions ([M + HCOO]$^-$) of each component lost methyl formate at the MS^{2-} stage (Figure 4). For instance, with the highest percentage at 16.0%, PC 38:6 formed negative acetylated molecular ions [M + HCOO]$^-$ at *m/z* 850.5571, positive quasi-molecular ions [M + H]$^+$ at *m/z* 806.5637 and cluster ions [M − CH$_3$]$^-$ at 790.5373 corresponding to composition [C$_{47}$H$_{81}$NO$_{10}$P]$^-$ (calculated850.5598, different 0.00326), [C$_{46}$H$_{81}$NO$_8$P]$^+$ (calculated806.5700, different 0.00573) and [C$_{45}$H$_{77}$NO$_8$P]$^-$ (calculated 790.5387, different 0.00193), respectively (Figure 4).

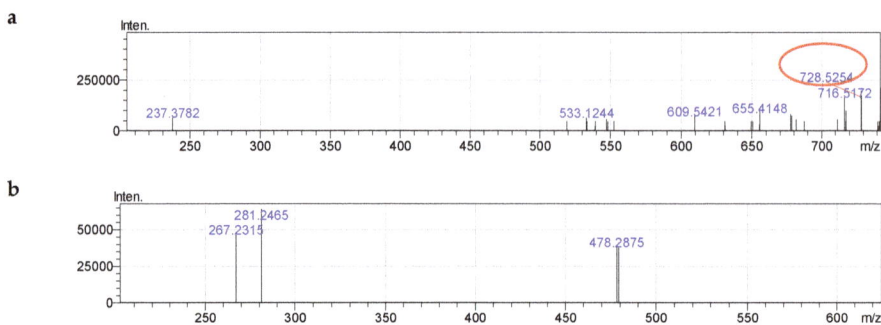

Figure 3. Fragmentation of PE (35:2): **a**—MS$^-$, **b**—MS^{2-}.

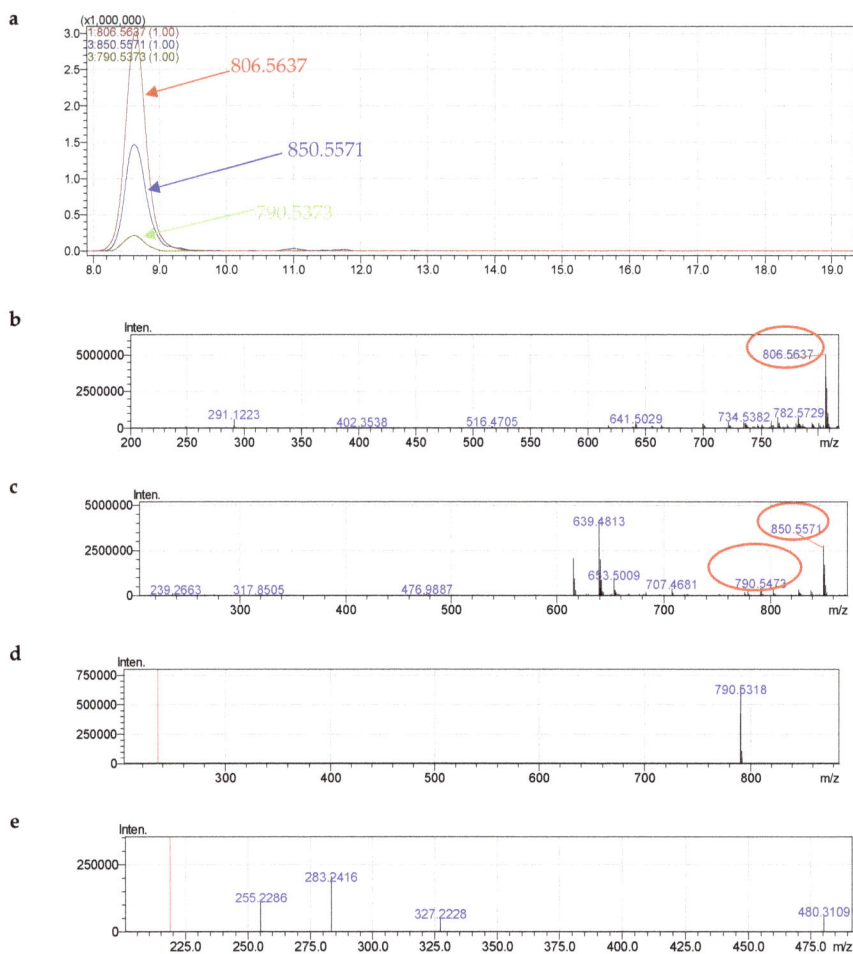

Figure 4. Fragmentation of PC (38:6): **a**—HPLC-HRMS; **b**—MS$^+$; **c**—MS$^-$ of fragmentation 850.5490 and 790.5301; **d**, **e**—MS^{2-} of fragmentation 850.5490 and 790.5301.

At the MS^{2-} stage, the ions at m/z 850.5571 eliminated a molecule of $C_2H_4O_2$ (methyl formate) and formed ions at m/z 790.5303, suggesting the formation of the quasi-molecular negative ions by the addition of formate ion to the lipid molecule. On MS^{2-} of component lost methyl formate, two carboxylate anion of 22:6 at m/z 327.2228 (calculated 327.2324, different 0.0105) and 16:0 at m/z 255.2286 (calculated 255.2324, different 0.00435) were observed (Figure 4 and Table S3). Normally, the *sn*-2 of PL is the preferred position for PUFAs [32]. Therefore, PC 38:6 was characterized as diacy glycerophosphocholine 16:0/22:6.

Among phosphotidylserine PS of clams, we determined 18 constituent components (Table S3). Alkenyl acyl glycerophosphoserine was the major component in PS, with 11 components occupying 81.0% of total PS species.

All components of formula species of PS had a signal of negative quasi-molecular ions $[M - H]^-$. On the other hand, positive quasi-molecular ions did not form. The MS^{2-} spectrum of $[M - H]^-$ of each component contained a signal of characteristic ion $[M–H–C_3H_5NO_2]^-$ corresponding to the loss of serine group (Figure 5). This was a specific fragmentation difference from the fragmentations of negative quasi-molecular ions of PE.

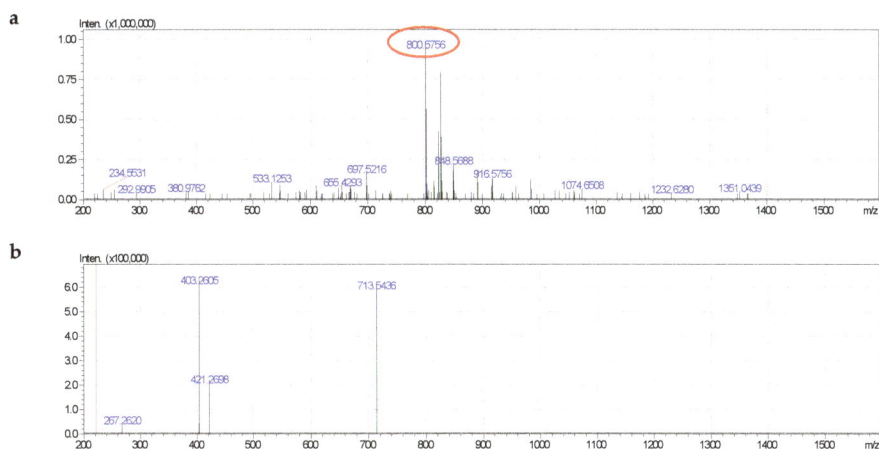

Figure 5. Fragmentation of PS (38:1): **a**—MS^-, **b**—MS^{2-}.

For example, PS 38:1 formed negative quasi-molecular ion $[M - H]^-$ at 800.5756 ($[C_{44}H_{83}NO_9P]^-$, calculated 800.5805, different 0.00549). On MS^{2-}, the absence of serine (713.6436) and serine and acyl groups (403.2605) was observed. However, signal of fatty acid was not detected (Figure S4). The component PS 38:1 was determined as alkenyl acyl glycerophosphoserine, p18:0/20:1.

Diacyl glycerophosphoserines only accounted for 19% of total PS species. On the MS^{2-} of signal negative quasi-molecular ion $[M-H]^-$ at 836.5402 ($[C46H79NO10P]-$, mass 836.5402, different 0.00451), the signal lost serine (749.4945) and lost simultaneously serine and acyl groups (419.2543) and one carboxylate anion of fatty acid 18:0 at m/z 283.2521 (Figure 6 and Figure S5). The component PS 40:5 was determined as diacy glycerophosphoserine, 18:0/22:5.

Among phosphotidylinositol PI of clams, we determined 10 components (Table S4). In addition, Alkenyl acyl glycerophosphoinositol was absent in PI. All ten components PI were diacyl glycerophosphoinositol with fatty acids 16:0. 17:0, 18:0, 19:0, 20:1, 20:4, 20:5.

The molecular species of clam PtdIns indicated the formation of both negative and positive quasi-molecular ions $[M - H]^-$. In addition, MS^{2-} fragmentation of the ions $[M - H]^-$ of PtdIns also reveals the presence of several characteristic ions such as PI 40:5 (Figure 7 and Figure S6). The fragmentation 911.5593 had formula $[C_{49}H_{84}O_{13}P]^-$, calculated 911.5650 (different 0.00620). The MS^{2-} spectra of component PI 40:5 also contained signals corresponding to the quasi-molecular ion, whose inositol, acyl groups and carboxylate anion of fatty acid have been simultaneously lost. In addition, fragmentation 297.0467 was quasi-molecular ion that lost diacyl groups. This fragmentation was important in determining the molecular species of PI.

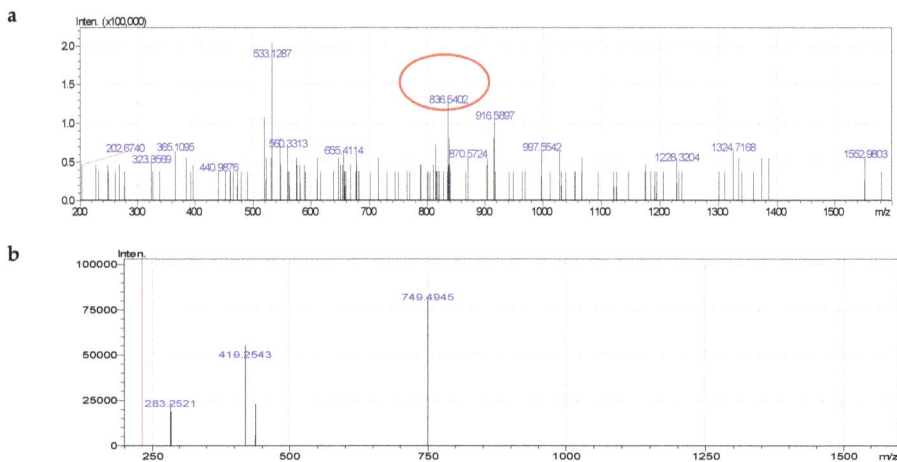

Figure 6. Fragmentation of PS (40:5): **a**—MS$^-$, **b**—MS^{2-}.

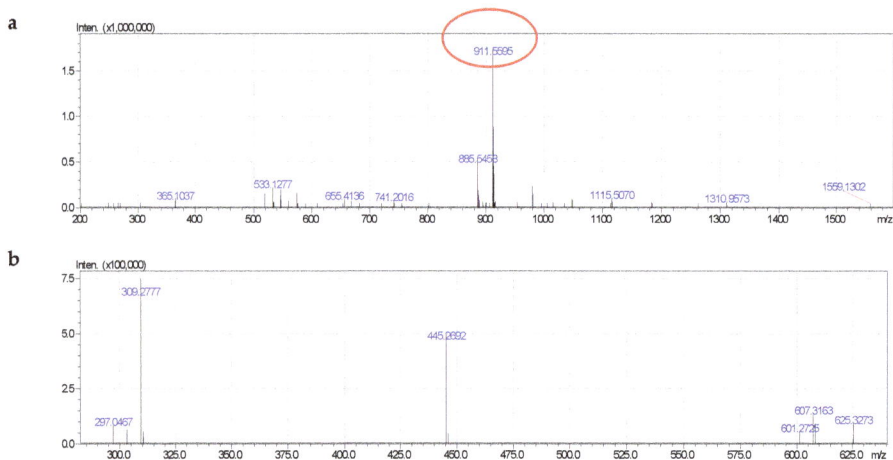

Figure 7. Fragmentation of PI (40:5): **a**—MS$^-$, **b**—MS^{2-}.

The component PI 40:5 (Figure S6) was determined as diacy glycerophosphoinositol 20:1/20:4.

We determined a single component, which was isomers of PG 32:0 constituting phosphatidylglycerol (PG) (Table S5).

PG 32:0 was described as follows (Figure 8 and Figure S7). The MS^{2-} spectra of component PG 32:0 contained signals corresponding to the negative quasi-molecular ion 721.4975 ([C$_{38}$H$_{74}$O$_{10}$P]$^-$, calculated 721.5020, different 0.00501). On MS^{2-}, we obtained fragmentations of carboxylate anion of fatty acid 16:0 (255.2304) with the simultaneous absence of glycerol and acyl groups (391.2255).

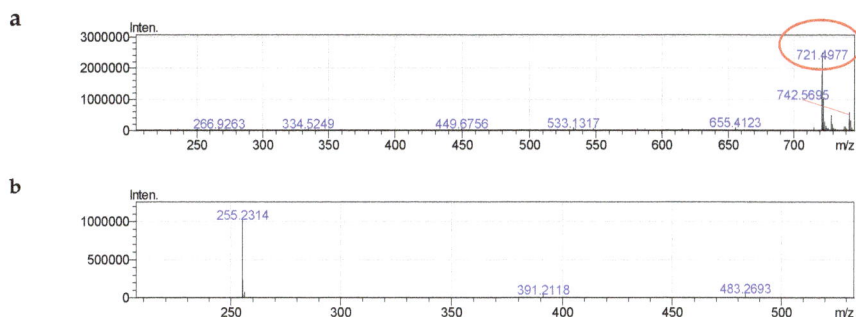

Figure 8. Fragmentation of PG (32:0): **a**—MS$^-$, **b**—MS^{2-}.

The component PG 32:0 (Figure S7) was determined as diacy glycerophosphoglycerol 16:0/16:0.

Fifteen molecular species of CAEP were determined (Table S6). In mass spectra of all molecular species, we observed signals of negative quasi-molecular ions [M − H]$^-$, positive quasi-molecular ions [M + H]$^+$ and positive quasi-molecular ions [M + Na]$^+$.

For example, we had signals of negative quasi-molecular ions [M − H]$^-$, at *m/z* 685.5565 ([C$_{39}$H$_{78}$N$_2$O$_5$P]$^-$, calculated 685.5648, different 0.00888), positive quasi-molecular ions [M + H]$^+$ at *m/z* 687.5830 ([C$_{39}$H$_{80}$N$_2$O$_5$P]$^+$, calculated 687.5805, different 0.00306), and positive adductions [M + Na]$^+$ at *m/z* 709.5627 ([C$_{39}$H$_{79}$N$_2$O$_5$NaP]$^+$, calculated 709.5619, different 0.00082) of CAEP 37:1 (Figure 9).

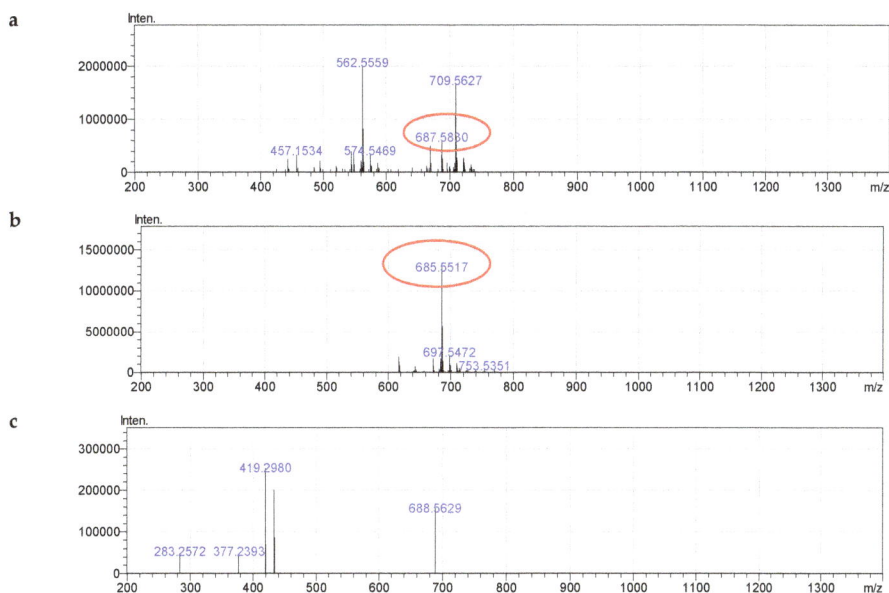

Figure 9. Fragmentation of CAEP (37:1): **a**—MS$^+$, **b**—MS$^-$, **c**—MS^{2-}.

The MS^{2-} spectrum of [M − H]$^-$ of CAEP which was sphingolipid signified a specific fragmentation similar to the fragmentations of negative quasi-molecular ions of glycerophospholipids (Figure S8). On MS^{2-}, we obtained carboxylate anion fragmentation of fatty acid 18:0 (283.2572). In addition, [M − H]$^-$ lost a neutral fragment amounted to 266.2567 and formed a single ion at 419.2980. Other fragmentations are also explained in Figure S8.

3. Materials and Methods

3.1. Material

The colonies of the clams were collected in May 2017 in Cua Lo beach, Cua Lo town, Nghe An provine, Vietnam.

The samples were transported immediately to the Institute of Natural Products Chemistry, Vietnam Academy of Science and Technology. Then, the shells and the meat were separated and stored at −5 °C.

3.2. Total Lipid Extraction

Soft tissue of clams was crushed and total lipid (TL) was extracted using modified Blight–Dyer extraction technique [31]. To be specific, lipids were extracted by homogenizing with the $CHCl_3/CH_3OH$ solution (1:2, v/v) (30 mL per 10 g of clams wet weight) in 6 h, at 30 °C. The obtained homogenate was filtered, and the residue was repeatedly extracted (6 h, 30 °C) in $CHCl_3$ (20 mL) a second time. After extraction, the homogenates were then mixed and separated into layers by adding 20 mL of H_2O. The lower layer was evaporated and the obtained TL obtained was dissolved in $CHCl_3$ and stored at −5 °C.

3.3. Analyses of Lipid Classes

To determine lipid class compositions, the extracted lipids were first dissolved in chloroform (10 mg/mL) and then spotted onto the one-dimensional thin-layer chromatography (TLC) using the pre-coated silica gel plates (6 cm × 6 cm) Sorbfil PTLC-AF-V (Sorbfil, Krasnodar, Russia). The first development of the plate was carried out with n-hexane/diethyl ether/acetic acid (85:15:1, $v/v/v$) for full length. Then, the plate was redeveloped with $CHCl_3/CH_3OH$ (2:1, v/v) for 5% length. Subsequently, the plates were air-dried, sprayed with 10% H_2SO_4/CH_3OH and heated at 240 °C for 10 min. An image scanner (Epson Perfection 2400 Photo, Epson, Suwa, Japan) operating in grayscale mode was employed to obtain chromatograms. For determination of lipid percentages, band intensity was evaluated using an image analysis program (Sorbfil TLC Video densitometer, Krasnodar, Russia).

3.4. Separation of Phospholipid Classes

The TLC plates used in this study were 10 × 10 cm glass-backed HPTLC Silica gel 60 plates (Merck, Darmstadt, Germany). Prior to sample loading, the plates were activated at 110 °C for at least 90 min and allowed to cool to room temperature in a vacuum desiccator. One dimensional TLC was employed to analyze phospholipids with the solvent system of chloroform–methanol–25% aqueous ammonia–benzene (65:30:6:10, $v/v/v/v$) [33]. Phosphomolybdate spray reagent for detection was prepared following a previous study [34].

Spots from TLC plates was eluted by the following procedure. First, for each phospholipid, corresponding band was scraped from the plate developed twice in 1st direction with the aforementioned solvent system. Second, elution was performed with chloroform–methanol (1:1, v/v).

3.5. Analyses of Fatty Acids

To obtain fatty acid methyl esters (FAME), the lipids were treated with 2% H_2SO_4 in CH_3OH in a screw-capped vial (2 h, 80 °C) under air and purified by TLC development in hexane–diethyl ether (95:5, v/v). FAME were analyzed with GC at 210 °C and identified with reference to authentic standards and a table of equivalent chain lengths [35]. Injector and detector temperatures were 240 °C.

To confirm structures of FA, corresponding FAME were analyzed with GC–MS and resulted spectra were matched with the NIST library and FA mass spectra archive [36]. Temperature of the GC–MS analysis initiated at 160 °C and then progressed at 2 °C/min to 240 °C which was kept constant for 20 min. Injector and detector temperatures were 250 °C.

3.6. Analysis of Molecular Species of Phospholipids

Phospholipids were analyzed with high performance liquid chromatography–high resolution mass spectrometry (HPLC–HRMS) to identify and quantify chemical structures of molecular species of phospholipids.

The high performance liquid chromatography (HPLC) separation of PL was performed at the constant content of $(C_2H_5)_3N$/acid formic (0.08:1, v/v) in the solvent system [37] that allowed carrying out efficient ionization in ESI conditions and obtaining a stable ion signal by the simultaneous registration of positive and negative ions. For polar lipids, HPLC separation was performed using the binary solvent gradient consisted of solvent mixture A: *n*-hexane/2-propanol/acid formic/$(C_2H_5)_3N$ (82:17:1:0.08, $v/v/v/v$) and mixture B: 2-propanol/H_2O/acid formic/$(C_2H_5)_3N$ (85:14:1:0.08, $v/v/v/v$). The gradient started at 5% of mixture B, and its percentage was increased to 80% over 25 min. This composition was maintained for 1 min before being returned to 5% of mixture B over 10 min and maintained at 5% for another 4 min (the total run time was 40 min). The flow rate was 0.2 mL/min. Polar lipids were detected by high resolution mass spectrometry (HRMS) and identified by a comparison with authentic standards using a Shimadzu LCMS Solution control and processing software (v.3.60.361, Shimadzu, Kyoto, Japan). The quantification of individual molecular species within each polar lipid class was carried out by calculating the peak areas for the individual extracted ion chromatograms [38].

3.7. Instrumental Equipment

The GC analysis was carried out on a Shimadzu GC-2010 chromatography (Kyoto, Japan) with a flame ionization detector on a SUPELCOWAX 10 (Supelco, Bellefonte, PA, USA) capillary column (30 m × 0.25 mm × 0.25 μm). Carrier gas was He at 30 cm/s. The GC–MS analysis was performed with a Shimadzu CMS-QP5050A instrument (Kyoto, Japan) (electron impact at 70 eV) with a MDN-5s (Supelco, Bellefonte, PA, USA) capillary column (30 m × 0.25 mm ID). Carrier gas was He at 30 cm/s.

The HPLC–HRMS analysis of polar lipids was performed with a Shimadzu Prominence liquid chromatograph equipped with two LC-20AD pump units, a high pressure gradient forming module, CTO-20A column oven, SIL-20A auto sampler, CBM-20A communications bus module, DGU-20A3 degasser, and a Shim-Pack diol column (50 mm × 4.6 mm ID, 5 μm particle size) (Shimadzu, Kyoto, Japan). Lipids were detected by a high resolution tandem ion trap–time of flight mass spectrometry with a Shimadzu LCMS-IT-TOF instrument (Kyoto, Japan) operating both at positive and negative ion mode during each analysis at electrospray ionization (ESI) conditions. Ion source temperature was 200 °C, the range of detection was m/z 200–1600, and potential in the ion source was −3.5 and 4.5 kV for negative and positive modes, respectively. The drying gas (N_2) pressure was 200 kPa. The nebulizer gas (N_2) flow was 1.5 L/min.

4. Conclusions

Lipid analysis of the present study revealed that the total lipid in hard clam (*Meretrix lyrata*) occupies 1.7 ± 0.2% of wet weight and that six lipid classes consisting of hydrocarbon and wax (HW), triacylglycerol (TAG), free fatty acids (FFA), sterol (ST), polar lipid (PoL) and monoalkyl diacylglycerol (MADAG) were detected. Among these classes, the proportion of PoL and MADAG accounted for the highest, at 45.7%, and the lowest, at 1.3% of total lipid, respectively. In addition, the ratios of PUFA/SAFA was 1.4. The total percentage of n-3 PUFA (ω-3) and n-6 PUFA (ω-6) was relatively high, at 38.4% of total FA.

To our knowledge, this is the first report to determine chemical structures and quantities of molecular species of phospholipid, in general, and phosphoethanolamines in particular, on the hard clams *Meretrix lyrata* from Cua Lo beach, Nghe An province, Viet Nam. Five types glycerophospholipid were identified: PE, PC, PS, PI and PG. One type sphingophospholipid was identified to be CAEP. Ninety-eight molecular species were identified in polar lipids of the clams. Alkenyl acyl forms of glycerophospholipids predominated in the molecular species determined. PE 36:1 (p18:0/18:1), PC 38:6 (16:0/22:6), PS 38:1 (p18:0/20:1), PI 40:5 (20:1/20:4), PG 32:0 (16:0/16:0) and CAEP 34:2 (16:2/d18:0) were the major molecular species.

Supplementary Materials: The supplementary materials are available online.

Author Contributions: Investigation, T.T.T.L., M.Q.P., T.L.D., M.H.V. and L.G.B.; Supervision, L.M.B. and Q.L.P.; Writing—original draft, Q.T.T.; Writing—review & editing, D.C.N.

Funding: This research was funded by Ministry of Science and Technology, Vietnam grant number KC.09.23/16-20.

Acknowledgments: We are grateful to Andrey B. Imbs and his group at Laboratory of Comparative Biochemistry A.V. Zhirmunsky Institute of Marine Biology, Far-Eastern Branch of the Russian Academy of Sciences, 17 Palchevskogo str., Vladivostok 690041, Russian Federation.

Conflicts of Interest: The authors declare no conflict of interest.

Abbreviations

AA	Arachidonic acid
CAEP	Ceramide aminoethylphosphonate
ChoGpl	Choline glycerophospholipids
DHA	Docosahexaenoic acid
DMA	Dimethylacetal
DPA	Docosapentaenoic acid
EPA	Eicosapentaenoic acid
FFA	Free fatty acids
GC	Gas chromatography
GCMS	Gas chromatography—mass spectrometry
HPA	Heneicosapentaenoic acid
HPLC-HRMS	High performance liquid chromatography–high resolution mass spectrometry
HRMS	High resolution mass spectrometry
HUFA	Highly unsaturated fatty acid
HW	Hydrocarbon and wax
LC-PUFA	Long-chain polyunsaturated fatty acid
MADAG	Monoalkyl diacylglycerol
MS	Mass spectrometry
MS/MS	Mass spectrometry / Mass spectrometry (or tandem mass spectrometry)
MUFA	Monounsaturated fatty acid
NMI-PUFA	Non-methylene interrupted polyunsaturated fatty acid
PakCho	1-*O*-alkyl-2-acyl-*sn*-glycero-3-phosphocholine
PC	Phosphatidylcholine
PE	Phosphatidylethalnolamine
PG	Phosphatidylglycerol
PI	Phosphatidylinositol
PL	Phospholipid
PlsEtn	Ethanolamine plasmalogen
PoL	Polar lipid
PS	Phosphatidylserine
PtdIns	Phosphatidylinositol
PUFA	Poly-unsaturated fatty acid
SAFA	Saturated fatty acid
ST	Sterol
TAG	Triacylglycerol
TL	Total lipid
TLC	Thin-layer chromatography
USAFA	Unsaturated fatty acid

References

1. Loureiro, D.R.P.; Soares, J.X.; Costa, J.C.; Magalhães, A.F.; Azevedo, C.M.G.; Pinto, M.M.M.; Afonso, C.M.M. Structures, Activities and Drug-Likeness of Anti-Infective Xanthone Derivatives Isolated from the Marine Environment: A Review. *Molecules* **2019**, *24*, 243. [CrossRef] [PubMed]

2. Huang, K.C.; Wu, W.T.; Yang, F.L.; Chiu, Y.H.; Peng, T.C.; Hsu, B.G.; Liao, K.W.; Lee, R.P. Effects of Freshwater Clam Extract Supplementation on Time to Exhaustion, Muscle Damage, Pro/Anti-Inflammatory Cytokines, and Liver Injury in Rats after Exhaustive Exercise. *Molecules* **2013**, *18*, 3825–3838. [CrossRef] [PubMed]

3. Greetham, D.; Zaky, A.; Makanjuola, O.; Du, C. A brief review on bioethanol production using marine biomass, marine microorganism and seawater. *Curr. Opin. Green Sustainable Chem.* **2018**, *14*, 53–59. [CrossRef]

4. Patil, M.P.; Kim, G.D. Marine microorganisms for synthesis of metallic nanoparticles and their biomedical applications. *Colloids Surf. B* **2018**, *172*, 487–495. [CrossRef] [PubMed]

5. Amorello, D.; Orecchio, S.; Pace, A.; Barreca, S. Discrimination of almonds (Prunus dulcis) geographical origin by minerals and fatty acids profiling. *Nat. Prod. Res.* **2016**, *30*, 2107–2110. [CrossRef] [PubMed]

6. FAO Yearbook of Fishery and Aquaculture Statistics 2018. Available online: http://www.fao.org/fishery/statistics/en (accessed on 20 February 2019).

7. Fernández, M.J.; Labarta, U.; Albentosa, M.; Pérez-Camacho, A. Lipid composition of Ruditapes philipponarum spat: Effect of ration and diet quality. *Comp. Biochem. Physiol.* **2006**, *144*, 229–237.

8. Orban, E.; Di, L.G.; Nevigato, T.; Casini, I.; Caproni, R.; Santaroni, G.; Giulini, G. Nutritional and commercial quality of the striped venus clam, Chamelea gallina, from the Adriatic sea. *Food Chem.* **2007**, *101*, 1063–1070. [CrossRef]

9. Karnjanapratum, S.; Benjakul, S.; Kishimura, H.; Tsai, Y. Chemical compositions and nutritional value of Asian hard clam (Meretrix lusoria) from the coast of Andaman Sea. *Food Chem.* **2013**, *141*, 4138–4145. [CrossRef] [PubMed]

10. Boselli, E.; Pacetti, D.; Lucci, P.; Frega, N. Characterization of phospholipid molecular species in the edible parts of bony fish and shellfish. *J. Agric. Food Chem.* **2012**, *60*, 3234–3245. [CrossRef] [PubMed]

11. Janssen, C.I.; Kiliaan, A.J. Long chain polyunsaturated fatty acids (LC-PUFA) from genesis to senecscence: The influence of LC-PUFA on neural development, aging and neurodegeneration. *Prog. Lipid Res.* **2014**, *5*, 1–7. [CrossRef] [PubMed]

12. Lorente-Cebrian, S.; Costa, A.; Navas-Carretero, S.; Zabala, M.; Martinez, J.; Moreno-Aliaga, M. Role of omega-3 fatty acids in obesity, metabolic syndrome and cardiovascular diseases: A review of the evidence. *J. Physiol. Biochem.* **2013**, *69*, 633–651. [CrossRef] [PubMed]

13. Cottin, S.; Sanders, T.A.; Hall, W.L. The differential effects of EPA and DHA on cardiovascular risk factors. *Proc. Nutr. Soc.* **2011**, *70*, 215–231. [CrossRef] [PubMed]

14. Edelstein, S. *Food Science: An Ecological Approach*; Jones & Bartlett Publishers: Burlington, MA, USA, 2014; pp. 70–71.

15. Miller, M.; Stone, N.J.; Ballantyne, C.; Bittner, V.; Criqui, M.H.; Ginsberg, H.N. Triglycerides and cardiovascular disease: A scientific statement from the American Heart Association. *Circulation* **2011**, *123*, 2292–2333. [CrossRef] [PubMed]

16. Weintraub, H.S. Overview of prescription omega-3 fatty acid products for hypertriglyceridemia. *Postgrad Med.* **2014**, *126*, 7–18. [CrossRef] [PubMed]

17. Saito, H. Identification of novel n-4 series polyunsaturated fatty acids in a deep-sea clam, Calyptogena phaseoliformis. *J. Chromatogr. A* **2007**, *1163*, 247–259. [CrossRef] [PubMed]

18. Saito, H.; Murata, M.; Hashimoto, J. Lipid characteristics of a seep clam, Mesolinga soliditesta: Comparison with those of two coastal clams, Meretrix lamarckii and Ruditapes philippinarum. *Deep Sea Res. Part I* **2014**, *94*, 150–158. [CrossRef]

19. FAO Yearbook. Fishery and Aquaculture Statistics 2016. Available online: http://www.fao.org/fishery/static/Yearbook/YB2016_USBcard/navigation/index_intro_e.htm (accessed on 20 February 2019).

20. Hamli, H.; Idris, M.H.; Rajaee, A.H.; Kamal, A.H.M. Reproductive Cycle of Hard Clam, Meretrix lyrata Sowerby, 1851 (Bivalia: Veneridae) from Sarawak, Malaysia. *Trop Life Sci. Res.* **2015**, *26*, 59–72. [PubMed]

21. Thao, N.T.; Nha le, Q. Effects of glucose and probiotic supplementation in nursing juvenile clam, Meretrix lyrata. *Commun. Agric. Appl. Biol. Sci.* **2013**, *78*, 304–307. [PubMed]

22. Klingensmith, J.S.; Stillway, L.W. Lipid composition of selected tissues of the hardsell clam, Mercenaria mercenaria. *Comp. Biochem. Physiol.* **1981**, *71B*, 111–112.

23. Caers, M.; Coutteau, P.; Sorgeloos, P. Impact of starvation and of feeding algal and artificial diets on the lipid content and composition of juvenile oysters (Crassostrea gigas) and clams (Tapes philippinarum). *J. Mar. Biol.* **2000**, *136*, 891–899. [CrossRef]

24. Aouini, F.; Ghribi, F.; Boussoufa, D.; Bejaoui, S.; Navarro, J.; Cafsi, M.E. Changes in lipid classes, fatty acid composition and lipid peroxidation in the gills of the clam Donax Trunculus after permethrin exposure. *Research Gate* **2016**, *186*.

25. Beninger, P.G. Seasonal variation of the mayor lipid classes in relation to the reproductive activity of two species of clams raised in a common habitat: Tapes decssatus L. (Jeffreys, 1863) and Philippinraum (Adams and Reeve), 1850. *J. Exp. Mar. Bio. Ecol.* **1984**, *79*, 79–90. [CrossRef]

26. Prato, E.; Danieli, A.; Michele, M.; Francesca, B. Lipid and Fatty Acid Compositions of Mytilus galloprovincialis Cultured in the Mar Grande of Taranto (Southern Italy): Feeding Strategies and Trophic Relationships. *Zool. Stud.* **2010**, *49*, 211–219.

27. Bernsdorff, C.; Winter, R. Differential properties of the sterols cholesterol, ergosterol, β-sitosterol, trans-7-dehydrocholesterol, stigmasterol and lanosterol on DPPC bilayer order. *J. Phys. Chem. A* **2003**, *107*, 10658–10664. [CrossRef]

28. Chen, T.C.C. A comparison of the behavior of cholesterol, 7-dehydrocholesterol and ergosterol in phospholipid membranes. *Biochim. Biophys. Acta* **2012**, *1818*, 1673–1681. [CrossRef] [PubMed]

29. Dennis, F.; Riccardo, G.; Agnese, P.; Deborah, P.; Barrara, Z.; Luciiano, B.; Enrico, M. Comparison of eleven extraction methods for quantitative evalution of total lipids and fatty acids in the clam Anadara Inaequivalvis (Bruguiere). *J. Shellfish Res.* **2013**, *32*, 285–290.

30. Andray, B.I.; Ly, P.D.; Viacheslav, G.R.; Vasily, I.S. Fatty acid, Lipid class, and Phospholipid Molecular Species Composition of the Soft Coral Xenia sp. (Nha Trang Bay, the South China Sea, Vietnam). *Springer* **2015**, *50*, 575–589.

31. Liu, Z.Y.; Zhou, D.-Y.; Wu, Z.-X.; Yin, F.-W.; Zhao, Q.; Xie, H.-K.; Zhang, J.-R.; Qin, L.; Shahidi, F. Extraction and detailed characterization of phospholipid-enriched oils from six species of edible clams. *Food Chem.* **2017**, *239*, 1175–1181. [CrossRef] [PubMed]

32. Peterson, B.L.; Cummings, B.S. A review of chromatographic methods for the assessment of phospholipids in biological samples. *Biomed. Chromatogr.* **2006**, *20*, 227–243. [CrossRef] [PubMed]

33. Bligh, E.G.; Dyer, W.J. A rapid method of total lipid extraction and purification. *Can. J. Biochem. Phys.* **1959**, *37*, 911–917. [CrossRef]

34. Vaskovsky, V.E.; Terekhova, T.A. HPTLC of phospholipid mixtures containing phosphatidylglycerol. *J. High. Resolut. Chromatogr. Chromatogr. Commun.* **1979**, 671–672. [CrossRef]

35. Vaskovsky, V.E.; Kostetsky, E.Y.; Vasendin, I.M. A universal reagent for phospholipid analysis. *J. Chromatogr.* **1975**, *114*, 129–141. [CrossRef]

36. Christie, W. Equivalent chain lengths of methyl ester derivatives of fatty acids on gas chromatography a reappraisal. *J. Chromatogr.* **1988**, *447*, 305–314. [CrossRef]

37. Mass Spectrometry of Fatty Acid Derivatives. Available online: http://lipidlibrary.aocs.org/ms/masspec.html (accessed on 10 August 2014).

38. Harrabi, S.; Herchi, W.; Kallel, H.; Mayer, P.; Boukhchina, S. Liquid chromatographic-mass spectrometric analysis of glycerophospholipids in corn oil. *Food Chem.* **2009**, *114*, 712–716. [CrossRef]

Sample Availability: Not available.

molecules

MDPI

Article

Morphology and Molecular Identification of Twelve Commercial Varieties of Kiwifruit

Qiaoli Xie [1], Hongbo Zhang [1], Fei Yan [2], Chunxia Yan [1], Shuguang Wei [1], Jianghua Lai [1], Yunpeng Wang [1,*] and Bao Zhang [1,*]

[1] School of Forensic Medicine, Xi'an Jiaotong University, 76 Yanta West Road, Xi'an 710061, China; sunshineqiaoer@xjtu.edu.cn (Q.X.); zhanghb@mail.xjtu.edu.cn (H.Z.); yanchunxia@mail.xjtu.edu.cn (C.Y.); weisg@mail.xjtu.edu.cn (S.W.); laijh1011@mail.xjtu.edu.cn (J.L.)

[2] School of Energy and Power Engineering, Chongqing University, 174 Shapingba Main Street, Chongqing 400030, China; yanfei0506@cqu.edu.cn

* Correspondence: wyp033@xjtu.edu.cn (Y.W.); zhangbao_814@mail.xjtu.edu.cn (B.Z.); Tel.: +86-029-82655113 (Y.W.); +86-029-82655113 (B.Z.)

Received: 22 December 2018; Accepted: 28 February 2019; Published: 3 March 2019

check for updates

Abstract: The quality and safety of food are important guarantees for the health and legal rights of consumers. As an important special fruitcrop, there are frequently shoddy practices in the kiwifruit (*Actinidia chinensis*) market, which harms the interests of consumers. However, there is lack of rapid and accurate identification methods for commercial kiwifruit varieties. Here, twelve common commercial varieties of kiwifruit were morphologically discriminated. DNA barcodes of chloroplast regions *psbA-trnH*, *rbcL*, *matK*, *rpoB*, *rpoC1*, *ycf1b*, *trnL* and *rpl32_trnL(UAG)*, the nuclear region *At103* and intergenic region *ITS2* were amplified. Divergences and phylogenetic trees were used to analyze the phylogenetic relationship of these twelve commercial kiwifruit varieties. The results showed that *matK*, *ITS2* and *rpl32_trnL(UAG)* can be utilized as molecular markers to identify CuiYu, JinYan, HuangJinGuo, ChuanHuangJin, HuaYou, YaTe, XuXiang and HongYang. This provides experimental and practical basis to scientifically resolve kiwifruit-related judicial disputes and legal trials.

Keywords: food safety; kiwifruit (*Actinidia chinensis*); molecular identification; phylogeny; DNA barcode

1. Introduction

The quality and safety of food are important guarantees for the health and legal rights of consumers. Kiwifruit (*Actinidia chinensis*), also called 'the king of fruits', is an important economical crop because of its exceedingly high content of ascorbic acid (vitamin C), dietary fiber, nutritional minerals compositions and other health beneficial metabolites [1]. China, the origin of kiwifruit, possesses the largest planted area of kiwifruit in the world. According to statistics, in 2016, the national kiwifruit cultivation area reached 0.365 million acres, and the output was 2.15 million tons (China Industry Report Network) [2] Since the commercial establishment of kiwifruit, its classification has been controversial. These kiwifruits, named with geographical indications, cannot represent the species of kiwifruit. Besides, many commercial varieties vary greatly in market demand and price due to differences in taste and nutritional value. Therefore, there are frequent problems in the kiwifruit industry, such as false labelling, and lack of origin confirmation and identification. In order to protect consumer rights, we are looking for ways to quickly and accurately identify the commercial varieties of kiwifruit in the market and scientifically resolve judicial disputes and legal trials. However, morphological-based identification methods have great difficulties for non-professionals, and the methods of omics or chemistry are complex, time consuming and susceptible to environmental factors [3,4].

The DNA barcoding technique is a quick and effective molecular marker technology for classification and identification of organisms by using a standard gene region [5]. Since Hebert et al. recommended mitochondrial *cox1* gene as a DNA barcode for animal species identification in 2003 [5], it has been widely and effectively applied in the classification, identification, and phylogenetic analysis of thousands of species [6,7]. DNA barcode technology is an ideal identification method because of its accurate identification and simple operation. To date, the Consortium for the Barcode of Life (CBOL) formally proposed chloroplast markers *rbcL* and *matK* as the core barcodes for plant species identification [8]. Noncoding intergenic spacer *psbA-trnH* barcode was used to identify species of medicinal pteridophytes and members of Dendrobium of Orchid [9,10]. *ITS2* was selected as a standard barcode for identifying medical plants [11]. Two chloroplast genome markers, coding *rpoB* and *rpoC1* were utilized to discriminate 92 species in 32 diverse genera of land plants [12]. *ycf1b* was reported as the most variable plastid genome region and can serve as a core barcode of land plants [13]. *trnL* was chosen as the barcoding gene for reference library constructing and high-throughput sequencing for wetland plants [14]. *At103* (Mgprotoporphyrin IX monomethyl ester cyclase) developed by Li et al. [15] as universally amplifiable marker for phylogenetic reconstructions and that together with *matK* to be used to distinguish toxic hybrids form parental species [16]. Fu et al. [17] found that *rpl32_trnL (UAG)* had a greater degree of variation and could be used as the core barcode sequence of cherry plants. These genes have the potential to be as powerful as mitochondrial CO1 gene in identifying species [10]. However, as far as we have been concerned, there have been no reports on DNA barcode for common kiwifruit commercial varieties in the market. Previous studies have provided us with reference to the feasibility of using DNA barcodes to identify kiwifruit commercial varieties. Lee et al. made use of *ITS2* to identify the varieties and provenances of Taiwan's domestic and imported made teas [18]. Enan and Ahmed using chloroplast DNA barcode *psbK-psbI* spacers for identification of Emirati date palm (*Phoenix dactylifera* L.) varieties (cultivar-level) [19]. Jaakola et al. successfully identified the blueberry varieties "Northcountry" and "Northblue" using DNA barcode technology combined with high-resolution dissolution profiles [20]. Through DNA barcodes, He et al. authenticated cultivars of *Angelica anomala* Ave'-Lall [21].

Single nucleotide polymorphism (SNP) is a single nucleotide variation with a specific and determined genetic location in at least 1% of the population [22]. SNP is one of the richest and most stable genetic polymorphisms in the genome, which is suitable for solving the differences between closely related species [23]. SNP typing has been successfully used to conveniently and accurately identify plant origin [24], medicinal plants [25], and bacteria [26].

In this study, we attempted to use the morphological method combined with molecular biological methods for rapid cultivar identification of twelve kiwifruit commercial varieties in China. SNP typing method was evaluated by using ten candidate DNA barcoding markers (chloroplast genome *psbA-trnH*, *rbcL*, *matK*, *rpoB*, *rpoC1*, *ycf1b*, *trnL* and *rpl32_trnL(UAG)*, the nuclear region *At103* and intergenic region *ITS2*), which were used for molecular identification. This study will not only lay a foundation for phylogenetic analysis but also provides experimental and practical basis for the rapid identification of kiwifruit.

2. Results

2.1. Morphological Identification

In order to enable consumers to visually identify the commercial varieties of kiwifruit in the first place, by referring to the classification criteria of "Flora of China", the morphology of 12 commercial kiwifruit were analyzed. We mainly made statistics on fruit type, fruit shape, fruit size, peel color, peel spots, hair presence, hair length, hair softness and hardness, hair shedding, pulp color, pulp taste, seed color, seed number and shape, beak prominence, beak diameter, fruit picking time and so on. The results are shown in Figure 1 and Supplementary Table S1. The 72 samples of the 12 commercial varieties of kiwifruit were harvested from 150–160 days after pollination. Their fruits are all bacca.

The shape of the fruit is mostly cylindrical and spherical. HuangJinGuo and ChuanHuangJin are long oval. JinYan is cylindrical. HongYang is short cylindrical and CuiYu is oblate cone. Statistical analysis of fruit size showed that the fruit size of FengXianLou is the largest, followed by HuangJinGuo, CuiXiang, and HongYang is the smallest. For the peel, only CuiYu has no spots on the peel surface, other commercial varieties all have spots. The color of mostly commercial varieties' peel is brown. The fruit peel of HuangJinGuo, ChuanHuangJin and JinYan is yellower. The peel of HongYang is greenish. Hair analysis found that the skin of HuangJinGuo, ChuanHuangJin, CuiYu and HongYang have no hair, other commercial varieties all have hair. QinMei and HaiWoDe have very dense hair on the surface. HuaYou has few hair on the surface. The surface of FengXianLou has the longest hair, soft and easy to fall off. HuaYou's surface hair is the shortest and easy to fall off.

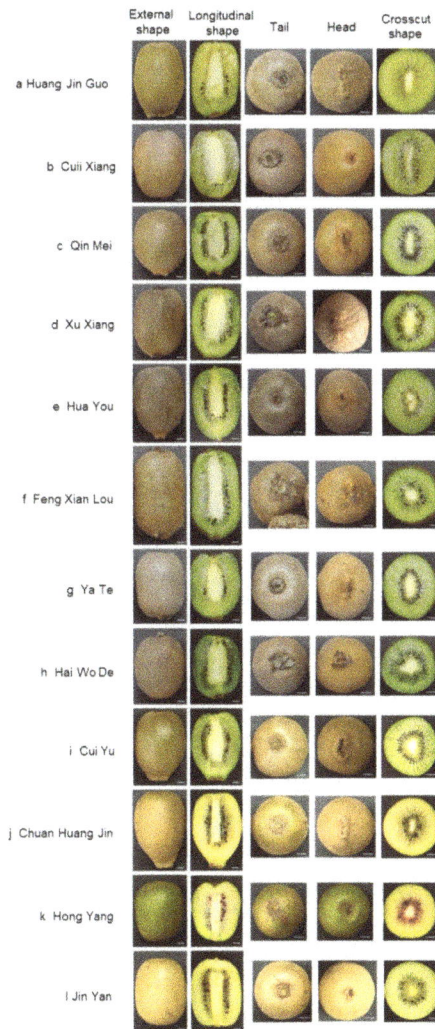

Figure 1. External and internal forms of twelve commercial kiwifruit. (**a**) HuangJinGuo, (**b**) CuiXiang, (**c**) QinMei, (**d**) XuXiang, (**e**) HuaYou, (**f**) FengXianLou, (**g**) YaTe, (**h**) HaiWoDe, (**i**) CuiYu, (**j**) ChuanHuangJin, (**k**) HongYang, (**l**) JinYan. Each variety contains external shape, longitudinal shape, head, tail, crosscut shape. Scale bar represents 6 mm.

All statistical results are in Supplementary Table S1. For the color of the flesh, the HongYang is the easiest to distinguish, which has a radial red color in the pulp center. The flesh color of HuangJinGuo, ChuanHuangJin and JinYan are yellow. The flesh color of HuaYou is yellow-green and other commercial varieties are all green. The fruit of HuangJinGuo, ChuanHuangJin, JinYan, CuiXiang, HuaYou, CuiYu are sweet, no sour taste, and other commercial varieties are sweet and sour. Most of the seeds are dark brown and flat oval. HaiWoDe has the largest number of seeds, and HuangJinGuo has the fewest seeds. Statistical results are in Supplementary Table S1. Kiwifruit has a beak at the top. Except CuiXiang and HongYang, the beaks of other commercial varieties are prominent. The longest beak exists on HuangJinGuo, followed by QinMei, and HongYang has the smallest beak.

The above analysis shows that HongYang is the most discernible variety by the peel color and flesh color. JinYan is easy to identify because of its yellow peel and flesh. The shape of CuiYu is similar to that of HuangJinGuo, while there are differences between them. The surface of CuiYu has no spots and its flesh is green. The surface of HuangJinGuo is spotted and its flesh is yellow. Therefore, CuiYu is easier to be distinguished. The hair of Huayou is short and small and its flesh is yellow-green, which made it easier to be discriminated. FengXianLou is a long cylindrical shape with many hairs easy to fall off makes it easy to be identified. It is not easy to distinguish HuangJinGuo and ChuanHuangJin. CuiXiang, QinMei, XuXiang, YaTe and HaiWoDe all have green pulp, but the shape of CuiXiang is slightly flat, and its beak is the smallest and not prominent. CuiXiang is sweet, while QinMei, XuXiang, YaTe and HaiWoDe are sour and sweet. YaTe's skin is white and brown compared with QinMei, XuXiang and HaiWoDe. QinMei is greener than XuXiang and HaiWoDe. For ordinary consumers, XuXiang and HaiWoDe are relatively difficult to be differentiated.

2.2. Analysis of Variable Sites in Different Commercial Varieties

The sequencing results showed that the product length of *rbcL*, *matK*, *psbA-trnH*, *ITS2*, *rpoB*, *rpoC1*, *trnL* and *rpl32_trnL(UAG)* are 743 bp, 889 bp, 502 bp, 491 bp, 512 bp, 529 bp, 193bp and 1010 bp, respectively. Sequences of *rbcL*, *psbA-trnH*, *rpoB*, *rpoC1* and *trnL* have no difference in these twelve kiwi commercial varieties (Supplementary Figure S1). *ycf1b* and *At103* get no results of amplification. The amplification efficiency of 72 *ITS2* sequences is 100%. Sequencing results showed that the sequences of 6 repetitive samples from each commercial variety are consistent and stable. There are two haplotypes in *ITS2* sequence. Bases at 115 bp, 132 bp and 310 bp of HuangJinGuo, ChuanHuangJin and HongYang are "C", that of the other nine commercial varieties are "T". Bases at 206 bp and 215 bp of HuangJinGuo, ChuanHuangJin and HongYang are "G", that of the other nine commercial varieties are "A" (Figure 2a, Table 1 and Supplementary Table S2).

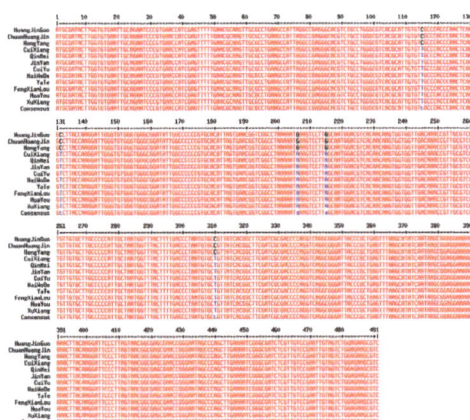

(a)

Figure 2. *Cont.*

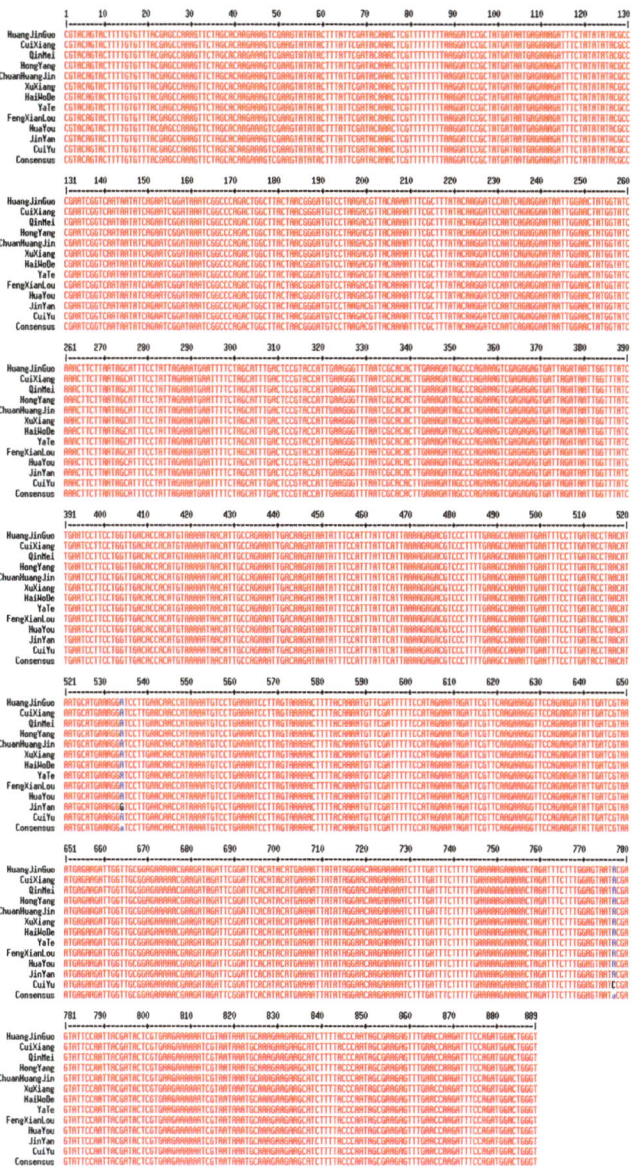

(b)

Figure 2. *Cont.*

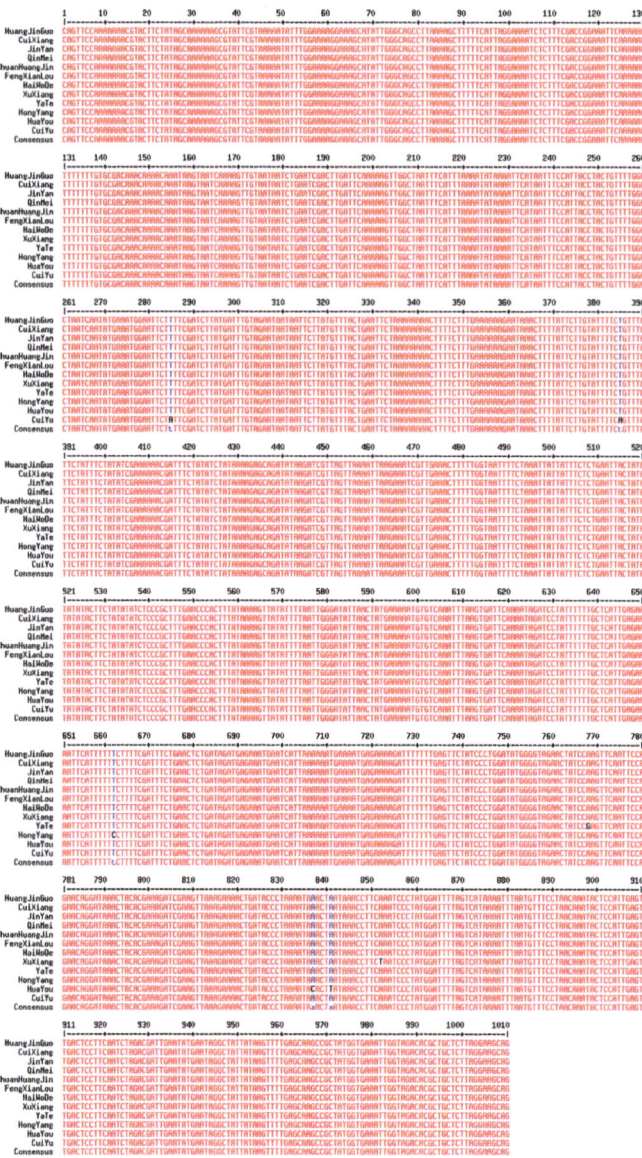

(c)

Figure 2. DNA barcode sequence alignment of 12 commercial kiwifruits. (**a**) *ITS2* sequence alignment 491 bp; (**b**) *matK* sequence alignment 889 bp; (**c**) *rpl32_trnL(UAG)* sequence alignment 1010 bp.

Table 1. SNP site information for two DNA barcodes.

Variety	SNP Site Information (bp)												
	ITS2					matK					rpl32			
	115	132	206	215	310	534	777	285	385	662	768	837	841	852
HuangJinGuo	C	C	G	G	C	A	A	T	T	T	A	A	A	A
ChuanHuangJin	C	C	G	G	C	A	A	T	T	T	A	A	A	A
HongYang	C	C	G	G	C	A	A	T	T	C	A	A	A	A
CuiiXiang	T	T	A	A	T	A	A	T	T	T	A	A	A	A
QinMei	T	T	A	A	T	A	A	T	T	T	A	A	A	A
JinYan	T	T	A	A	T	G	A	T	T	T	A	A	A	A
CuiYu	T	T	A	A	T	A	C	A	A	T	A	A	A	A
HaiWoDe	T	T	A	A	T	A	A	T	T	T	A	A	A	A
YaTe	T	T	A	A	T	A	A	T	T	T	G	A	A	A
FengXianLou	T	T	A	A	T	A	A	T	T	T	A	A	A	A
HuaYou	T	T	A	A	T	A	A	T	T	T	A	C	T	A
XuXiang	T	T	A	A	T	A	A	T	T	T	A	A	A	T

For *matK*, the amplification efficiency of 72 *matK* sequences is 100%. Sequencing results showed that the sequences of 6 repetitive samples of each commercial variety are consistent and stable. There are three haplotypes in *matK* sequences. Bases at 534 bp of JinYan are "G", which of the other eleven commercial varieties are "A", and bases at 777 bp of CuiYu are "C", the other eleven commercial varieties are "A" (Figure 2b, Table 1 and Supplementary Table S3). The amplification efficiency of 72 *rpl32_trnL(UAG)* sequences is 100%. There are six haplotypes in *rpl32_trnL(UAG)* sequence. Bases at 285 bp and 385 bp of CuiYu are "A", that of the other 11 commercial varieties are "T". Bases at 662 bp of HongYang are "C", that of the other 11 commercial varieties are "T". Bases at 768 bp of YaTe are "G", that of the other 11 commercial varieties are "A". Bases at 837 bp and 841 bp of HuaYou are "C" and "T", respectively, that of the other 10 commercial varieties are "A". Bases at 852 bp of XuXiang are "T", that of the other 11 commercial varieties are "A". (Figure 2c, Table 1 and Supplementary Table S4).

2.3. K2P Genetic Distance Analysis

Genetic distances of all samples were calculated by using the MEGA 7.0 [27] software (Pennsylvania State University, USA). The results of *ITS2*, *matK*, *rpl32_trnL(UAG)* and *ITS2+matK+rpl32_trnL(UAG)* are shown in Supplementary Table S6. For *ITS2*, there is no difference among HuangJinGuo, ChuanHuangJin and HongYang. The genetic distance between them and nine other commercial varieties is 0.01. There is no difference among the nine commercial varieties (Supplementary Table S5a). For *matK*, the genetic distance between JinYan and CuiYu is 0.002, and between JinYan and other 10 commercial varieties is 0.001. The genetic distance between CuiYu and other 10 cultivars is 0.001, and there is no difference among the other 10 cultivars (Supplementary Table S5b). For *rpl32_trnL(UAG)*, the genetic distance between HuaYou and HongYang, YaTe, XuXiang is 0.003. The genetic distance between HuaYou and JinYan, CuiXiang, FengXianLou, ChuanHuangJin, HaiWoDe, QinMei and HuangJinGuo is 0.002. The genetic distance between HongYang and CuiYu is 0.003. The genetic distance between HongYang and YaTe, XuXiang is 0.002. The genetic distance between HongYang and JinYan, CuiXiang, FengXianLou, ChuanHuangJin, HaiWoDe, QinMei and HuangJinGuo is 0.001. The genetic distance between CuiYu and YaTe, XuXiang is 0.003. The genetic distance between CuiYu and JinYan, CuiXiang, FengXianLou, ChuanHuangJin, HaiWoDe, QinMei and HuangJinGuo is 0.001. The genetic distance between JinYan and YaTe, XuXiang is 0.001. There is no difference between JinYan and CuiXiang, FengXianLou, ChuanHuangJin, HaiWoDe, QinMei and HuangJinGuo. The genetic distance between CuiXiang and YaTe, XuXiang is 0.001. There is no difference between CuiXiang and FengXianLou, ChuanHuangJin, HaiWoDe, QinMei and HuangJinGuo. The genetic distance between YaTe and XuXiang is 0.002. There is no difference between YaTe and FengXianLou, ChuanHuangJin, HaiWoDe, QinMei and HuangJinGuo. The genetic distance between FengXianLou and XuXiang is 0.001. There is no difference between FengXianLou and ChuanHuangJin, HaiWoDe, QinMei and HuangJinGuo. The genetic distance between ChuanHuangJin and XuXiang is 0.001. There is no difference between ChuanHuangJin and HaiWoDe, QinMei and HuangJinGuo. The genetic distance between HaiWoDe and XuXiang is 0.001. There is no difference between HaiWoDe and QinMei and HuangJinGuo. The genetic distance between QinMei and XuXiang is 0.001. There is no difference between QinMei and HuangJinGuo. The genetic distance between HuangJinGuo and XuXiang is 0.001 (Supplementary Table S5c). For *ITS2+matK+rpl32_trnL(UAG)*, the genetic distance between YaTe and JinYan is 0.001. The genetic distance between XuXiang and YaTe, JinYan is 0.001. There is no difference between QinMei and YaTe, JinYan, XuXiang. There is no difference between HaiWoDe and YaTe, JinYan, XuXiang, QinMei. There is no difference between CuiXiang and YaTe, JinYan, XuXiang, QinMei, HaiWoDe. The genetic distance between HuaYou and YaTe, JinYan, XuXiang, QinMei, HaiWoDe, CuiXiang is 0.001. There is no difference between FengXianLou and YaTe, JinYan, XuXiang, QinMei, HaiWoDe and CuiXiang. The genetic distance between FengXianLou and HuaYou is 0.001. The genetic distance between CuiYu and YaTe, JinYan, XuXiang, HuaYou is 0.002. The genetic distance between CuiYu and QinMei, HaiWoDe, CuiXiang, FengXianLou is 0.001. The genetic distance between HuangJinGuo and YaTe, JinYan, XuXiang, HuaYou, CuiYu is 0.003. The genetic distance between HuangJinGuo and QinMei, HaiWoDe and CuiXiang, FengXianLou is 0.002. The genetic distance between HongYang and YaTe, JinYan, XuXiang, QinMei, HaiWoDe, CuiXiang, HuaYou, FengXianLou is 0.003. The genetic distance between HongYang and CuiYu is 0.004. There is no difference between HongYang and CuiYu. The genetic distance between ChuanHuangJin and YaTe, JinYan, XuXiang, HuaYou, CuiYu is 0.003. The genetic distance between ChuanHuangJin and QinMei, HaiWoDe and CuiXiang, FengXianLou is 0.002. There is no difference between ChuanHuangJin and HuangJinGuo, HongYang (Supplementary Table S5d).

2.4. Phylogenetic Analysis

ITS2, *matK* and *rpl32_trnL(UAG)* sequences and *ITS2+matK+rpl32_trnL(UAG)* combination sequences were tested 1000 times by bootstrap method to build NJ phylogenetic tree (Figure 3 and Supplementary Figure S2). Phylogenetic analysis of *ITS2* gene sequences (Figure 3a and Supplementary Figure S2a) showed that HuangJinGuo, ChuanHuangJin and HongYang were clustered into one group, and the other nine commercial varieties were grouped into another category. For the *matK* gene sequences, JinYan and CuiYu were separately distinguished, and the other ten commercial varieties were clustered into one category (Figure 3b and Supplementary Figure S2b). Figure 3c and Supplementary Figure S2c indicated that, using *rpl32_trnL(UAG)*, HuaYou, YaTe, XuXiang, HongYang, CuiYu could be distinguished from the other seven commercial varieties, respectively.

For *ITS2+matK+rpl32_trnL(UAG)*, CuiYu, JinYan, YaTe, XuXiang, HuaYou, HongYang, HuangJinGuo and ChuanHuangJin can be distinguished from the other four commercial varieties, respectively(Figure 3d).

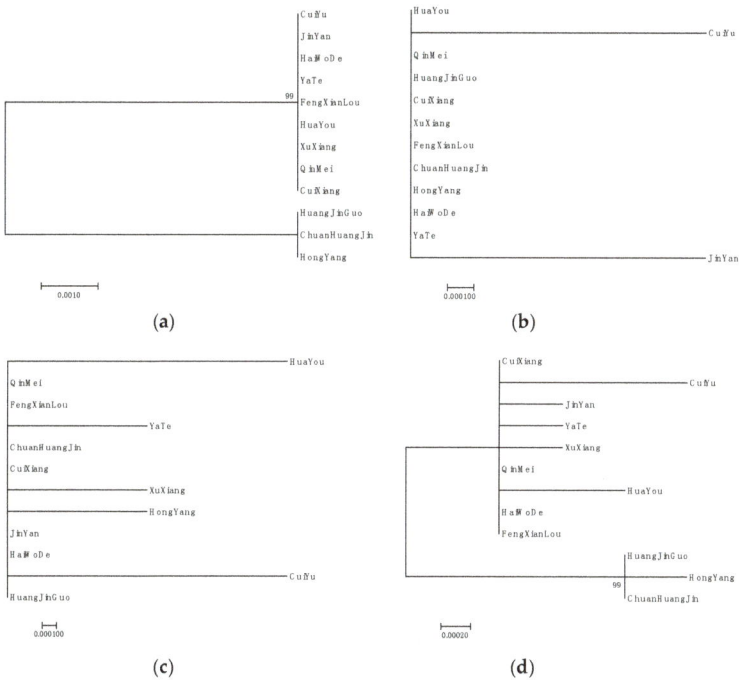

Figure 3. Phylogenetic analysis of 12 kiwifruit commercial varieties. (**a**) Analysis of *ITS2* sequence fragments; (**b**) Analysis of *matK* sequence fragments; (**c**) Analysis of *rpl32_trnL(UAG)* sequence fragments; (**d**) Analysis of *ITS2+matK+rpl32_trnL(UAG)* sequence fragments.

3. Discussion

The accurate classification of kiwifruit is one of the problems in the current kiwifruit plant resource utilization and market regulation. It is very difficult to classify kiwifruit only by morphology, because many shapes of kiwifruit are very close that requires the participation of professionals. If the suitable DNA barcode can be screened for kiwifruit, it is very meaningful to classification and identification, resource protection, variety selection and market regulation of kiwifruit.

In this study, 72 samples of 12 commercial varieties were collected in the market. We firstly conducted a morphological analysis of twelve kiwifruit commercial varieties that are easily confused in the market. Suggestions for morphological identification are proposed. By referring to the classification criteria of "Flora of China (http://foc.iplant.cn/)", the morphology of 12 commercial kiwifruit was analyzed in Supplementary Table S1. We mainly made statistics on fruit type, fruit shape, fruit size, peel color, peel spots, hair presence, hair length, hair softness and hardness, hair shedding, pulp color, pulp taste, seed color, seed number and shape, beak prominence, beak diameter, fruit picking time (Supplementary Table S1). The analysis results showed that HongYang is most discernible by the peel color and flesh color. Therefore, the expensive HongYang, whose price is 4 to 5 times than that of QinMei, HuaYou, YaTe and HaiWoDe in the market, can be identified morphologically. JinYan whose price is 3 to 4 times than that of QinMei, HuaYou, YaTe and HaiWoDe in the market, is easy to identify because of its yellow peel and flesh. The shape of CuiYu is similar to that of HuangJinGuo, while there are differences between them. The surface of HuangJinGuo is spotted and its flesh is yellow. Therefore, CuiYu is easier to be distinguished. The hair of Huayou is short and small and its flesh is yellow-green, which made it easier to be discriminated. FengXianLou is a long cylindrical shape with many hairs easy to fall off makes it easy to be identified. It is not easy to distinguish HuangJinGuo and ChuanHuangJin. CuiXiang, QinMei, XuXiang, YaTe and HaiWoDe all have green pulp, but the shape of CuiXiang is slightly flat, and its beak is the smallest and not prominent. CuiXiang is sweet, while QinMei, XuXiang, YaTe and HaiWoDe are sour and sweet. YaTe's skin is white and brown compared with QinMei, XuXiang and HaiWoDe. QinMei is greener than XuXiang and HaiWoDe. For ordinary consumers, XuXiang and HaiWoDe are relatively difficult to be differentiated. As the identification of morphology is very demanding on the professional, there are limitations in the identification of morphology.

Recently, DNA barcodes are often used as a standard for identifying plant species [28,29]. The DNA region, as an effective barcode, should contain sufficient variability for identification, contain conserved regions for the development of universal primers and be short enough to be sequenced in one reaction. In animals, a portion of mitochondrial cytochrome C oxidation enzyme I gene sequence is often used as a general barcode and is also used in forensic identification [16,28]. Currently, there is no common area available in plants. In the nuclear and plastid genomes, there are multiple loci were selected as DNA barcodes in plants [12,30–32].

Ten DNA barcoding *rbcL*, *matK*, *psbA-trnH*, *rpoB*, *rpoC1*, *ITS2*, *ycf1b*, *trnL*, *rpl32_trnL(UAG)* and *At103* were explored to quickly and accurately identify twelve kiwifruit commercial varieties in this study. The primer sequences used for amplification of these ten regions, the length of the products, and the amplification procedures are listed in Table 2. The *psbA-trnH* sequence of the intergenic region is one of the most mutated non-coding regions in the plant chloroplast genome and is often used as a barcode for material identification [32–34]. However, the results of this experiment found that the *psbA-trnH* sequences of the 12 commercial kiwifruits are identical, which means that the *psbA-trnH* is not suitable for identifying the 12 commercial kiwifruits. Intergenic region *ITS2* (internal transcribed spacer 2) and the cp gene *rbcL* may have potential as universal plant barcodes [34]. Three regions *matK*, *rpoB* and *rpoC1* were outlined as viable markers for land plant barcoding [33]. CBOL formally proposed chloroplast markers *rbcL* and *matK* as the core barcodes for plant species identification [8]. The results, in this study, showed that the *rbcL*, *rpoB* and *rpoC1* sequences of the 12 commercial kiwifruits are unanimous, which indicates the three barcodes sequences are improper for the identification of the 12 commercial kiwifruits. *ycf1b* was reported as the most variable plastid genome region and can serve as a core barcode of land plants [13]. *At103* together with *matK* to be used to distinguish toxic hybrids form parental species [16]. Our results showed that these two DNA barcodes could not be amplified in the 12 commercial varieties of kiwifruit, indicating they are not suitable to be DNA barcodes of kiwifruit. *trnL* was chosed as the barcoding gene for reference library constructing for wetland plants [14], whose amplification results were indistinguishable among the 12 commercial varieties of kiwifruit. Therefore, *trnL* could not be used as the barcode to discriminate these 12 commercial varieties of kiwifruit.

There are two SNP sites in the *matK* sequences of 12 commercial kiwifruits. For *matK*, bases at 534 bp of Jin Yan are "G", that of the other eleven commercial varieties are "A", and bases at 777 bp of Cui Yu are "C", that of the other eleven commercial varieties are "A" (Figure 2b, Table 1 and Supplementary Table S3). It can be applied to distinguish Jin Yan, CuiYu and other 10 commercial varieties. Genetic distance analysis discovered that the genetic distance between JinYan and CuiYu is 0.002, and between JinYan and 10 other commercial varieties is 0.001. The genetic distance between CuiYu and other 10 cultivars is 0.001, and there is no difference among the other 10 cultivars (Supplementary Table S5b). The phylogenetic analysis also found that Jin Yan, CuiYu and other ten commercial varieties is clustered into three branches. JinYan and CuiYu is separately distinguished from other ten commercial varieties, respectively (Figure 3b).

Table 2. Primers and amplification procedures used for PCR in this study.

Gene Amplification Region	Primer Name	Sequence (5′ →3′)	Length of Amplified Fragment	Amplification Procedure
*rbc*L	*rbc*L-F *rbc*L-R	ATGTCACCACAAACAGA TCGCATGTACCTGCAGTA	743 bp	94 °C 5 min; [94 °C 30S, 54.5 °C 30S, 72 °C 45S]* 35 cycles; 72 °C 10 min; 4 °C 10 min
*mat*K	*mat*K-F *mat*K-R	CGTACAGTACTTTTGTGTTTAC ACCCAGTCCATCTGGAAATC	889 bp	94 °C 5 min; [94 °C 30S, 56 °C 30S, 72 °C 55S]* 35 cycles; 72 °C 10 min; 4 °C 10 min
*psbA-trn*H	*psbA-trn*H-F *psbA-trn*H-R	TATGCATGAACGTAATGCT GCATGGTGGATTCACAAT	502 bp	94 °C 5 min; [94 °C 30S, 55 °C 30S, 72 °C 30S]* 35 cycles; 72 °C 10 min; 4 °C 10 min
ITS2	ITS2-F ITS2-R	ATGCGATACTTGGTGTG GACGCTTCTCCAGACTA	491 bp	94 °C 5 min; [94 °C 30S, 55 °C 30S, 72 °C 30S]* 35 cycles; 72 °C 10 min; 4 °C 10 min
*rpo*B	*rpo*B-F *rpo*B-R	AAGTGCATTGTTGGAACTGG CCCAGCATCACAATTCC	512 bp	94 °C 5 min; [94 °C 30S, 55 °C 30S, 72 °C 35S]* 35 cycles; 72 °C 10 min; 4 °C 10 min
rpoC1	*rpoC1*-F *rpoC1*-R	CAAAGAGGGAAGAT TAAGCATATCTTGAGT	529 bp	94 °C 5 min; [94 °C 30S, 54.5 °C 30S, 72 °C 40S]* 35 cycles; 72 °C 10 min; 4 °C 10 min
rpl32_trnL(UAG)	*rpl32* *trnL(UAG)*	CAGTTCCAAAAAAACGTACTTC CTGCTTCCTAAGAGCAGCGT	1010 bp	94 °C 5 min; [94 °C 30S, 57.8 °C 30S, 72 °C 45S]* 35 cycles; 72 °C 10 min; 4 °C 10 min
*trn*L	*trn*L-F *trn*L-R	CGAAATCGGTAGACGCTACG CCATTGAGTCTCTGCACCTATC	193 bp	94 °C 5 min; [94 °C 30S, 56.5 °C 30S, 72 °C 30S]* 35 cycles; 72 °C 10 min; 4 °C 10 min
ycf1b	*ycf1b*-F *ycf1b*-R	TCTCGACGAAAATCAGATTGTTGTGAAT ATACATGTCAAAGTGATGGAAAA	No result	94 °C 5 min; [94 °C 30S, 56.5 °C 30S, 72 °C 45S]* 35 cycles; 72 °C 10 min; 4 °C 10 min 56.5 °C
At103	*At103*-F *At103*-R	CTTCAAGCCMAAGTTCATCTTCTA ATCATTGAGGTACATNGTMACATA	No result	94 °C 5 min; [94 °C 30S, 54 °C 30S, 72 °C 45S]* 35 cycles; 72 °C 10 min; 4 °C 10 min 56.5 °C

In addition, *ITS2* sequence analysis of 12 commercial kiwifruit results showed that compared to *rbcL*, *matK*, *psbA-trnH*, *rpoB*, *rpoC1* and *trnL*, the *ITS2* sequence difference is better. It was found that there are five SNPs in the 491 bp of *ITS2* sequence, bases at 115 bp, 132 bp and 310 bp of HuangJinGuo, ChuanHuangJin and HongYang are "C", that of the other nine commercial varieties are "T". Bases at 206 bp and 215 bp of HuangJinGuo, ChuanHuangJin and HongYang are "G", that of the other nine commercial varieties are "A" (Figure 2a, Table 1 and Supplementary Table S2). It can be used to distinguish HuangJinGuo, ChuanHuangJin, HongYang and other 9 commercial varieties. Genetic distance analysis discovered that for *ITS2*, there is no difference among HuangJinGuo, ChuanHuangJin and HongYang. The genetic distance between them and other 9 commercial varieties was 0.01. There is no difference among the nine commercial varieties (Supplementary Table S5a). *ITS2* showed more variable sites than the plastid regions *matK* and *rbcL*. This result confirmed that *ITS2* showed greater discriminatory power than plastid regions [35]. Phylogenetic analysis also showed that HuangJinGuo, ChuanHuangJin and HongYang were clustered into one group with the *ITS2* sequence, and the other nine commercial varieties were grouped into another category (Figure 3a).

rpl32_trnL(UAG) had a greater degree of variation and could be used as the core barcode sequence of cherry plants [17]. *rpl32_trnL(UAG)* sequence analysis of 12 commercial kiwifruit results showed that compared to *rbcL*, *matK*, *psbA-trnH*, *rpoB*, *rpoC1*, *trnL* and *ITS2*, the *rpl32_trnL(UAG)* sequence difference is the maximum. It was found that there are six SNPs in the 1010bp of *rpl32_trnL(UAG)* sequence. Bases at 285 bp and 385 bp of CuiYu are "A", that of the other 11 commercial varieties are "T". Bases at 662 bp of HongYang are "C", that of the other 11 commercial varieties are "T". Bases at 768 bp of YaTe are "G", that of the other 11 commercial varieties are "A". Bases at 837 bp and 841 bp of HuaYou are "C" and "T", respectively, that of the other 10 commercial varieties are "A". Bases at 852 bp of XuXiang are "T", that of the other 11 commercial varieties are "A". (Figure 2c, Table 1 and Supplementary Table S4). It can be used to distinguish CuiYu, JinYan, YaTe, XuXiang, HuaYou, HongYang, HuangJinGuo and ChuanHuangJin from the other four commercial varieties, respectively. Genetic distance analysis discovered that, for *rpl32_trnL(UAG)*, the genetic distance between HuaYou and JinYan, CuiXiang, FengXianLou, ChuanHuangJin, HaiWoDe, QinMei and HuangJinGuo is 0.002. The genetic distance between HongYang and CuiYu is 0.003. The genetic distance between HongYang and YaTe, XuXiang is 0.002. The genetic distance between HongYang and JinYan, CuiXiang, FengXianLou, ChuanHuangJin, HaiWoDe, QinMei and HuangJinGuo is 0.001. The genetic distance between CuiYu and YaTe, XuXiang is 0.003. The genetic distance between CuiYu and JinYan, CuiXiang, FengXianLou, ChuanHuangJin, HaiWoDe, QinMei and HuangJinGuo is 0.001. The genetic distance between JinYan and YaTe, XuXiang is 0.001. There is no difference between JinYan and CuiXiang, FengXianLou, ChuanHuangJin, HaiWoDe, QinMei and HuangJinGuo. The genetic distance between CuiXiang and YaTe, XuXiang is 0.001. There is no difference between CuiXiang and FengXianLou, ChuanHuangJin, HaiWoDe, QinMei and HuangJinGuo. The genetic distance between YaTe and XuXiang is 0.002. There is no difference between YaTe and FengXianLou, ChuanHuangJin, HaiWoDe, QinMei and HuangJinGuo. The genetic distance between FengXianLou and XuXiang is 0.001. There is no difference between FengXianLou and ChuanHuangJin, HaiWoDe, QinMei and HuangJinGuo. The genetic distance between ChuanHuangJin and XuXiang is 0.001. There is no difference between ChuanHuangJin and HaiWoDe, QinMei and HuangJinGuo. The genetic distance between HaiWoDe and XuXiang is 0.001. There is no difference between HaiWoDe and QinMei and HuangJinGuo. The genetic distance between QinMei and XuXiang is 0.001. There was no difference between QinMei and HuangJinGuo. The genetic distance between HuangJinGuo and XuXiang is 0.001 (Supplementary Table S5c). *rpl32_trnL(UAG)* showed more variable sites than the other plastid regions *matK* and *rbcL*. This result confirmed that *rpl32_trnL(UAG)* showed better discriminatory power than other plastid regions. Phylogenetic analysis further indicated that HuaYou, YaTe, XuXiang, HongYang, CuiYu can be distinguished from the other seven commercial varieties, respectively (Figure 3c).

Multi-site combination methods based on plastid (chloroplast) sequences are considered as effective strategies in the identification of plant species [31,33,34]. Therefore, we comprehensively analyzed the sequences of *ITS2*, *matK* combined with *rpl32_trnL(UAG)*, and then found that, for *ITS2+matK+rpl32_trnL(UAG)*, there is no difference between QinMei and YaTe, JinYan, XuXiang. There is no difference between HaiWoDe and YaTe, JinYan, XuXiang, QinMei. There is no difference between CuiXiang and YaTe, JinYan, XuXiang, QinMei, HaiWoDe. The genetic distance between YaTe and JinYan is 0.001. The genetic distance between XuXiang and YaTe, JinYan is 0.001. The genetic distance between HuaYou and YaTe, JinYan, XuXiang, QinMei, HaiWoDe, CuiXiang are 0.001. There is no difference between FengXianLou and YaTe, JinYan, XuXiang, QinMei, HaiWoDe and CuiXiang. The genetic distance between FengXianLou and HuaYou is 0.001. The genetic distance between CuiYu and YaTe, JinYan, XuXiang, HuaYou is 0.002. The genetic distance between CuiYu and QinMei, HaiWoDe, CuiXiang, FengXianLou is 0.001. The genetic distance between HuangJinGuo and YaTe, JinYan, XuXiang, HuaYou, CuiYu is 0.003. The genetic distance between HuangJinGuo and QinMei, HaiWoDe and CuiXiang, FengXianLou is 0.002. The genetic distance between HongYang and YaTe, JinYan, XuXiang, QinMei, HaiWoDe, CuiXiang, HuaYou, FengXianLou is 0.003. The genetic distance between HongYang and CuiYu is 0.004. There is no difference between HongYang and CuiYu. The genetic distance between ChuanHuangJin and YaTe, JinYan, XuXiang, HuaYou, CuiYu is 0.003. The genetic distance between ChuanHuangJin and QinMei, HaiWoDe and CuiXiang, FengXianLou is 0.002. There is no difference between ChuanHuangJin and HuangJinGuo, HongYang (Supplementary Table S5d). These results suggest that CuiYu, JinYan, YaTe, XuXiang, HuaYou, HongYang, HuangJinGuo and ChuanHuangJin can also be differentiated by the genetic distance of *ITS2+matK+rpl32_trnL(UAG)* combination. We found that the genetic distance between samples of HaiWoDe, FengXianLou, QinMei and CuiXiang in this study is 0.000, which means that most of them have a unique *ITS2*, *matK* and *rpl32_trnL(UAG)* sequences. This feature is thus useful for identifying these twelve commercial varieties and related species, which is similar to previous study [36].

The analysis of SNP loci is an important tool for analyzing small differences between populations [37–39]. In this study, by analyzing the sequence variable sites of 8 DNA barcodes including *ITS2*, *psbA-trnH*, *rbcL*, *matK*, *rpoB*, *rpoC1* and *rpl32_trnL(UAG)* in 12 commercial kiwifruit, stable SNP loci of DNA barcode *ITS2* are found between HuangJinGuo, ChuanHuangJin, HongYang and the other nine commercial varieties. Likewise, DNA barcode *matK* has stable SNP loci between JinYan, CuiYu and the other 10 commercial varieties. *rpl32_trnL(UAG)* has stable SNP loci between CuiYu, JinYan, YaTe, XuXiang, HuaYou, HongYang, HuangJinGuo and ChuanHuangJin and the other four commercial varieties. The presence of these SNP loci makes the identification of CuiYu, JinYan, YaTe, XuXiang, HuaYou, HongYang, HuangJinGuo and ChuanHuangJin kiwifruits possible. These results indicate that intergenic region *ITS2* could distinguish HuangJinGuo and HongYang from the nine other commercial varieties, and *matK* could be utilized to distinguish JinYan and CuiYu from 10 the other commercial varieties. *rpl32_trnL(UAG)* could be utilized to distinguish HuaYou, YaTe, XuXiang, HongYang, CuiYu from the other seven commercial varieties. SNPs of *ITS2*, *matK* and *rpl32_trnL(UAG)* combining with the current plant DNA barcode systems could be used to accurately identify kiwifruit, which comprises samples of closely-related or subspecies from different cultivation areas. These results are good agreements with previous reports [38,39].

In conclusion, morphologically, JinYan, CuiYu, HuangJinGuo and HongYang can be discriminated. At the molecular level, SNPs based on DNA barcoding intergenic region *ITS2* and cp gene *matK* and *rpl32_trnL(UAG)* combined together can distinguish CuiYu, JinYan, YaTe, XuXiang, HuaYou, HongYang, HuangJinGuo and ChuanHuangJin from the other four commercial varieties, respectively, that is consistent with the results of morphology and better than the results of morphology. These results will provide an experimental basis for the classification and identification of commercially available kiwifruits and will offer a theoretical support for scientific trials and forensic botany of such disputes.

4. Materials and Methods

4.1. Samples Collection and DNA Extraction

Seventy two samples from twelve kiwifruit commercial varieties were collected in this study. Their names are HuangJinGuo, CuiXiang, QinMei, XuXiang, HuaYou, FengXianLou, YaTe, HaiWoDe, CuiYu, ChuanHuangJin, HongYang and JinYan. Six replicates were randomly selected for each commercial variety in the local market. According to our survey, the price of HongYang is 4 to 5 times than that of QinMei, HuaYou, YaTe and HaiWoDe in the market. Similarly, ChuanHuangJin, HuangJinGuo and JinYan, whose price are 3 to 4 times than that of QinMei, HuaYou, YaTe and HaiWoDe. HuangJinGuo, CuiXiang, QinMei, XuXiang, HuaYou, FengXianLou, YaTe and HaiWoDe were collected from Shaanxi province of China. CuiYu, ChuanHuangJin, HongYang and JinYan were collected from Si Chuan province of China (Detailed information in Supplementary Table S6). Among the 12 commercial varieties, HongYang is diploid, HaiWoDe is hexaploid, HuangJinGuo, CuiXiang, XuXiang, HuaYou, CuiYu, ChuanHuangJin and JinYan are tetraploid. QinMei, FengXianLou and YaTe are unknown, which is needed further research. DNA of each kiwifruit was extracted by Plant Genomic DNA Extraction Kit (Omega, Norcross, GA, USA) according to the manufacturer's instructions.

4.2. PCR Amplification, Cloning, and Sequencing

PCR primers of DNA barcodes (*rbcL*, *matK*, *psbA-trnH*, *ITS2*, *rpoB*, *rpoC1*, *ycf1b*, *trnL*, *rpl32_trnL(UAG)* and *At103*) were designed based on DNA barcodes [9–11,16], CBOL Plant Working Group (2009), China Bole Group (2011) as well as reported chloroplast and genomic sequences of kiwifruit [40–43]. Primer sequences are listed in Table 1. They were synthesized in BGI (Beijing Genomics Institute, Beijing, China). Polymerase chain reaction made use of the same reaction system. Amplification procedures of ten different primer pairs are shown in Table 1. The PCR products were separated by 1.2~1.5% agarose gel electrophoresis. The remaining PCR products were recovered and purified. Then, the purified products were ligated into pMD18-T vector (Takara, DaLian, China) and transformed into DH-5α *E. coli*. Positive clones were screened and sequenced in Beijing Genomics Institute.

4.3. Sequence Alignments and Analysis

Multiple sequence alignment was performed by using the results of *psbA-trnH*, *rbcL*, *rpoB*, *rpoC1* and *trnL* amplification by http://multalin.toulouse.inra.fr/multalin/. (Supplementary Figure S1). Multi-sequence alignment of *ITS2*, *matK*, *rpl32_trnL(UAG)* and *ITS2+matK+rpl32_trnL(UAG)* were carried out by ClustalX 1.8 [44] software (Conway Institute UCD, Dublin, Ireland), respectively. Then, the phylogenetic trees of these twelve kiwifruit commercial varieties were reconstructed. Neighbor-joining (NJ) method of MEGA 7.0 is applied to construct the phylogenetic tree. One thousand bootstrap resamplings were performed to assess the reliability. The steps are as follows: Phylogeny/construct/Test Neighbor-joining tree/Nucleotide Sequences/Statistical method: Neighbor-joining, No. of bootstrap repetitions: 1000, Model: Kimura 2-parameter model, Missing Data Treatment: Pairwise deletion/compute. The K2P distances between different genetic resources were calculated by MEGA7.0. Variation of different commercial varieties was analyzed according to the genetic distance of each barcode fragment.

Supplementary Materials: The following are available online at http://www.mdpi.com/1420-3049/24/5/888/s1, Figure S1: *psbA-trnH*, *rbcL*, *rpoB*, *rpoC1* and *trnL* sequences alignment of 12 commercial kiwifruits. Figure S2: Phylogenetic analysis of 12 kiwifruit commercial varieties. Table S1: The information of the experimental materials, Table S2: Phenotypic characteristics of twelve commercial varieties of kiwifruit, Table S3: *ITS2* sequence of 72 kiwifruit samples, Table S4: *matK* sequence of 72 kiwifruit samples. Table S5: *rpl32_trnL (UAG)* sequence of 72 kiwifruit samples. Table S6: Genetic distance of *ITS2*, *matK*, *rpl32_trnL(UAG)* and *ITS2+matK+rpl32_trnL(UAG)* sequences of 12 kiwifruit commercial varieties.

Author Contributions: Conceptualization, Q.X., Y.W. and B.Z.; methodology, Q.X. and H.Z.; validation, H.Z., C.Y., S.W. and J.L.; data curation, Q.X. and F.Y.; writing—original draft preparation, Q.X., Y.W., H.Z. and F.Y.; writing—review and editing, C.Y., J.L. and S.W.; visualization, Q.X. and F.Y.; supervision, C.Y., J.L. and S.W.; project administration, Y.W. and B.Z.; funding acquisition, Q.X.

Funding: This research was supported by National Natural Science Foundation of China (No. 31801870), China Postdoctoral Science Foundation (No. 2016M590951), Shaanxi Natural Science Foundation (No. 2016JQ8018), the Fundamental Research Funds for the Central Universities (No. xjj2016091) and Shaanxi Postdoctoral Science Foundation (No. 2016BSHEDZZ120).

Acknowledgments: In this section you can acknowledge any support given which is not covered by the author contribution or funding sections. This may include administrative and technical support, or donations in kind (e.g., materials used for experiments).

Conflicts of Interest: The authors declare no conflict of interest.

References

1. Hunter, D.C.; Greenwood, J.; Zhang, J.L.; Skinner, M.A. Antioxidant and 'Natural Protective' Properties of Kiwifruit. *Curr. Top. Med. Chem.* **2011**, *11*, 1811–1820. [CrossRef] [PubMed]
2. Analysis of global kiwifruit yield and production area distribution. Available online: http://www.chyxx.com/industry/201602/389018.html (accessed on 22 December 2018).
3. Garcia, C.V.; Quek, S.Y.; Stevenson, R.J.; Winz, R.A. Characterisation of bound volatile compounds of a low flavour kiwifruit species: Actinidia eriantha. *Food Chem.* **2012**, *134*, 655–661. [CrossRef] [PubMed]
4. Nishiyama, I.; Yamashita, Y.; Yamanaka, M.; Shimohashi, A.; Fukuda, T.; Oota, T. Varietal difference in vitamin C content in the fruit of kiwifruit and other Actinidia species. *J. Agric. Food Chem.* **2004**, *52*, 5472–5475. [CrossRef] [PubMed]
5. Hebert, P.D.N.; Cywinska, A.; Ball, S.L.; DeWaard, J.R. Biological identifications through DNA barcodes. *Proc. R. Soc. B-Biol. Sci.* **2003**, *270*, 313–321. [CrossRef] [PubMed]
6. Jiang, Y.; Ding, C.B.; Zhang, L.; Yang, R.W.; Zhou, Y.H.; Tang, L. Identification of the genus Epimedium with DNA barcodes. *J. Med. Plants Res.* **2011**, *5*, 6413–6417. [CrossRef]
7. Liu, J.X.; Shi, L.C.; Han, J.P.; Li, G.; Lu, H.; Hou, J.Y.; Zhou, X.T.; Meng, F.Y.; Downie, S.R. Identification of species in the angiosperm family Apiaceae using DNA barcodes. *Mol. Ecol. Resour.* **2014**, *14*, 1231–1238. [CrossRef] [PubMed]

8. Hollingsworth, P.M.; Forrest, L.L.; Spouge, J.L.; Hajibabaei, M.; Ratnasingham, S.; van der Bank, M.; Chase, M.W.; Cowan, R.S.; Erickson, D.L.; Fazekas, A.J.; et al. A DNA barcode for land plants. *Proc. Natl. Acad. Sci. USA* **2009**, *106*, 12794–12797.

9. Yao, H.; Song, J.Y.; Ma, X.Y.; Liu, C.; Li, Y.; Xu, H.X.; Han, J.P.; Duan, L.S.; Chen, S.L. Identification of Dendrobium Species by a Candidate DNA Barcode Sequence: The Chloroplast psbA-trnH Intergenic Region. *Planta Med.* **2009**, *75*, 667–669. [CrossRef] [PubMed]

10. Ma, X.Y.; Xie, C.X.; Liu, C.; Song, J.Y.; Yao, H.; Luo, K.; Zhu, Y.J.; Gao, T.; Pang, X.H.; Qian, J.; et al. Species Identification of Medicinal Pteridophytes by a DNA Barcode Marker, the Chloroplast psbA-trnH Intergenic Region. *Biol. Pharm. Bull.* **2010**, *33*, 1919–1924. [CrossRef] [PubMed]

11. Chen, S.; Yao, H.; Han, J.; Liu, C.; Song, J.; Shi, L.; Zhu, Y.; Ma, X.; Gao, T.; Pang, X.; et al. Validation of the ITS2 region as a novel DNA barcode for identifying medicinal plant species. *PLoS ONE* **2010**, *5*, e8613. [CrossRef] [PubMed]

12. Fazekas, A.J.; Burgess, K.S.; Kesanakurti, P.R.; Graham, S.W.; Newmaster, S.G.; Husband, B.C.; Percy, D.M.; Hajibabaei, M.; Barrett, S.C. Multiple multilocus DNA barcodes from the plastid genome discriminate plant species equally well. *PLoS ONE* **2008**, *3*, e2802. [CrossRef] [PubMed]

13. Dong, W.; Xu, C.; Li, C.; Sun, J.; Zuo, Y.; Shi, S.; Cheng, T.; Guo, J.; Zhou, S. ycf1, the most promising plastid DNA barcode of land plants. *Sci. Rep.* **2015**, *5*, 8348. [CrossRef] [PubMed]

14. Yang, Y.; Zhan, A.; Cao, L.; Meng, F.; Xu, W. Selection of a marker gene to construct a reference library for wetland plants, and the application of metabarcoding to analyze the diet of wintering herbivorous waterbirds. *PeerJ* **2016**, *4*, e2345. [CrossRef] [PubMed]

15. Li, M.G.; Wunder, J.; Bissoli, G.; Scarponi, E.; Gazzani, S.; Barbaro, E.; Saedler, H.; Varotto, C. Development of COS genes as universally amplifiable markers for phylogenetic reconstructions of closely related plant species. *Cladistics* **2008**, *24*, 727–745. [CrossRef]

16. Bruni, I.; De Mattia, F.; Galimberti, A.; Galasso, G.; Banfi, E.; Casiraghi, M.; Labra, M. Identification of poisonous plants by DNA barcoding approach. *Int. J. Leg. Med.* **2010**, *124*, 595–603. [CrossRef] [PubMed]

17. Fu, T.; Wang, Z.L.; Lin, L.J.; Lin, L.; Li, W.; Yuan, D.M. Molecular Phylogenetic Analysis of Wild Cerasus Plants in South China. *J. Nucl. Agric. Sci.* **2018**, *32*, 2126–2134.

18. Lee, S.C.; Wang, C.H.; Yen, C.E.; Chang, C. DNA barcode and identification of the varieties and provenances of Taiwan's domestic and imported made teas using ribosomal internal transcribed spacer 2 sequences. *J. Food Drug Anal.* **2017**, *25*, 260–274. [CrossRef] [PubMed]

19. Enan, M.R.; Ahmed, A. Cultivar-level phylogeny using chloroplast DNA barcode psbK-psbI spacers for identification of Emirati date palm (Phoenix dactylifera L.) varieties. *Genet. Mol. Res.* **2016**, *15*. [CrossRef] [PubMed]

20. Jaakola, L.; Suokas, M.; Haggman, H. Novel approaches based on DNA barcoding and high-resolution melting of amplicons for authenticity analyses of berry species. *Food Chem.* **2010**, *123*, 494–500. [CrossRef]

21. He, Y.; Hou, P.; Fan, G.; Song, Z.; Arain, S.; Shu, H.; Tang, C.; Yue, Q.H.; Zhang, Y. Authentication of Angelica anomala Ave-Lall cultivars through DNA barcodes. *Mitochondrial DNA* **2012**, *23*, 100–105. [CrossRef] [PubMed]

22. de Paula Careta, F.; Paneto, G.G. Recent patents on high-throughput single nucleotide polymorphism (SNP) genotyping methods. *Recent Pat. DNA Gene Seq.* **2012**, *6*, 122–126. [CrossRef] [PubMed]

23. Yamamoto, T.; Nagasaki, H.; Yonemaru, J.; Ebana, K.; Nakajima, M.; Shibaya, T.; Yano, M. Fine definition of the pedigree haplotypes of closely related rice cultivars by means of genome-wide discovery of single-nucleotide polymorphisms. *BMC Genom.* **2010**, *11*, 267. [CrossRef] [PubMed]

24. Yuan, Y.; Jiang, C.; Liu, L.; Yu, S.; Cui, Z.; Chen, M.; Lin, S.; Wang, S.; Huang, L. Convenient, sensitive and high-throughput method for screening botanic origin. *Sci. Rep.* **2014**, *4*, 5395. [CrossRef] [PubMed]

25. Wang, H.T.; Kim, M.K.; Kim, Y.J.; Lee, H.N.; Jin, H.Z.; Chen, J.; Yang, D.C. Molecular authentication of the Oriental medicines Pericarpium Citri Reticulatae and Citri Unshius Pericarpium using SNP markers. *Gene* **2012**, *494*, 92–95. [CrossRef] [PubMed]

26. Maiden, M.C.J.; van Rensburg, M.J.J.; Bray, J.E.; Earle, S.G.; Ford, S.A.; Jolley, K.A.; McCarthy, N.D. MLST revisited: The gene-by-gene approach to bacterial genomics. *Nat. Rev. Microbiol.* **2013**, *11*, 728–736. [CrossRef] [PubMed]

27. Kumar, S.; Stecher, G.; Tamura, K. MEGA7: Molecular Evolutionary Genetics Analysis Version 7.0 for Bigger Datasets. *Mol. Biol. Evol.* **2016**, *33*, 1870–1874. [CrossRef] [PubMed]

28. Kress, W.J.; Wurdack, K.J.; Zimmer, E.A.; Weigt, L.A.; Janzen, D.H. Use of DNA barcodes to identify flowering plants. *Proc. Natl. Acad. Sci. USA* **2005**, *102*, 8369–8374. [CrossRef] [PubMed]

29. Ferri, G.; Alu, M.; Corradini, B.; Licata, M.; Beduschi, G. Species Identification Through DNA "Barcodes". *Genet. Test. Mol. Biomark.* **2009**, *13*, 421–426. [CrossRef] [PubMed]

30. Chase, M.W.; Salamin, N.; Wilkinson, M.; Dunwell, J.M.; Kesanakurthi, R.P.; Haider, N.; Savolainen, V. Land plants and DNA barcodes: Short-term and long-term goals. *Philos. Trans. R. Soc. B-Biol. Sci.* **2005**, *360*, 1889–1895. [CrossRef] [PubMed]

31. Pennisi, E. Taxonomy—Wanted: A barcode for plants. *Science* **2007**, *318*, 190–191. [CrossRef] [PubMed]

32. Newmaster, S.G.; Fazekas, A.J.; Steeves, R.A.D.; Janovec, J. Testing candidate plant barcode regions in the Myristicaceae. *Mol. Ecol. Resour.* **2008**, *8*, 480–490. [CrossRef] [PubMed]

33. Chase, M.W.; Cowan, R.S.; Hollingsworth, P.M.; van den Berg, C.; Madrinan, S.; Petersen, G.; Seberg, O.; Jorgsensen, T.; Cameron, K.M.; Carine, M. A proposal for a standardised protocol to barcode all land plants. *Taxon* **2007**, *56*, 295–299. [CrossRef]

34. Kress, W.J.; Erickson, D.L. A two-locus global DNA barcode for land plants: The coding rbcL gene complements the non-coding trnH-psbA spacer region. *PLoS ONE* **2007**, *2*, e508. [CrossRef] [PubMed]

35. Hollingsworth, P.M.; Graham, S.W.; Little, D.P. Choosing and using a plant DNA barcode. *PLoS ONE* **2011**, *6*, e19254. [CrossRef] [PubMed]

36. Feng, S.G.; Jiao, K.L.; Zhu, Y.J.; Wang, H.F.; Jiang, M.Y.; Wang, H.Z. Molecular identification of species of Physalis (Solanaceae) using a candidate DNA barcode: The chloroplast psbA-trnH intergenic region. *Genome* **2018**, *61*, 15–20. [CrossRef] [PubMed]

37. Chen, X.C.; Liao, B.S.; Song, J.Y.; Pang, X.H.; Han, J.P.; Chen, S.L. A fast SNP identification and analysis of intraspecific variation in the medicinal Panax species based on DNA barcoding. *Gene* **2013**, *530*, 39–43. [CrossRef] [PubMed]

38. Huang, Q.; Duan, Z.; Yang, J.; Ma, X.; Zhan, R.; Xu, H.; Chen, W. SNP typing for germplasm identification of Amomum villosum Lour. Based on DNA barcoding markers. *PLoS ONE* **2014**, *9*, e114940. [CrossRef] [PubMed]

39. Murphy, B.P.; Tranel, P.J. Identification and Validation of Amaranthus Species-Specific SNPs within the ITS Region: Applications in Quantitative Species Identification. *Crop. Sci.* **2018**, *58*, 304–311. [CrossRef]

40. Huang, S.X.; Ding, J.; Deng, D.J.; Tang, W.; Sun, H.H.; Liu, D.Y.; Zhang, L.; Niu, X.L.; Zhang, X.; Meng, M. Draft genome of the kiwifruit Actinidia chinensis. *Nat. Commun.* **2013**, *4*, 2640. [CrossRef] [PubMed]

41. Pilkington, S.M.; Crowhurst, R.; Hilario, E.; Nardozza, S.; Fraser, L.; Peng, Y.Y.; Gunaseelan, K.; Simpson, R.; Tahir, J.; Deroles, S.C. A manually annotated Actinidia chinensis var. chinensis (kiwifruit) genome highlights the challenges associated with draft genomes and gene prediction in plants. *BMC Genom.* **2018**, *19*, 257. [CrossRef] [PubMed]

42. Lin, M.; Qi, X.; Chen, J.; Sun, L.; Zhong, Y.; Fang, J.; Hu, C. The complete chloroplast genome sequence of Actinidia arguta using the PacBio RS II platform. *PLoS ONE* **2018**, *13*, e0197393. [CrossRef] [PubMed]

43. Yao, X.; Tang, P.; Li, Z.; Li, D.; Liu, Y.; Huang, H. The First Complete Chloroplast Genome Sequences in Actinidiaceae: Genome Structure and Comparative Analysis. *PLoS ONE* **2015**, *10*, e0129347. [CrossRef] [PubMed]

44. Thompson, J.D.; Gibson, T.J.; Plewniak, F.; Jeanmougin, F.; Higgins, D.G.J.N. The CLUSTAL_X windows interface: Flexible strategies for multiple sequence alignment aided by quality analysis tools. *Nucleic Acids Res.* **1997**, *25*, 4876–4882. [CrossRef] [PubMed]

molecules

MDPI

Article

Alternative Ultrasound-Assisted Method for the Extraction of the Bioactive Compounds Present in Myrtle (*Myrtus communis* L.)

Ana V. González de Peredo [1], Mercedes Vázquez-Espinosa [1], Estrella Espada-Bellido [1], Marta Ferreiro-González [1], Antonio Amores-Arrocha [2], Miguel Palma [1], Gerardo F. Barbero [1,*] and Ana Jiménez-Cantizano [2]

[1] Department of Analytical Chemistry, Faculty of Sciences, University of Cadiz, Agrifood Campus of International Excellence (ceiA3), IVAGRO, 11510 Puerto Real, Cadiz, Spain; ana.velascogope@uca.es (A.V.G.d.P.); mercedes.vazquez@uca.es (M.V.-E.); estrella.espada@uca.es (E.E.-B.); marta.ferreiro@uca.es (M.F.-G.); miguel.palma@uca.es (M.P.)

[2] Department of Chemical Engineering and Food Technology, Faculty of Sciences, University of Cadiz, Agrifood Campus of International Excellence (ceiA3), IVAGRO, 11510 Puerto Real, Cadiz, Spain; antonio.amores@uca.es (A.A.-A.); ana.jimenezcantizano@uca.es (A.J.-C.)

* Correspondence: gerardo.fernandez@uca.es; Tel.: +34-956-01-6355

Academic Editors: Alessandra Gentili and Chiara Fanali
Received: 5 February 2019; Accepted: 26 February 2019; Published: 2 March 2019

check for updates

Abstract: The bioactive compounds in myrtle berries, such as phenolic compounds and anthocyanins, have shown a potentially positive effect on human health. Efficient extraction methods are to be used to obtain maximum amounts of such beneficial compounds from myrtle. For that reason, this study evaluates the effectiveness of a rapid ultrasound-assisted method (UAE) to extract anthocyanins and phenolic compounds from myrtle berries. The influence of solvent composition, as well as pH, temperature, ultrasound amplitude, cycle and solvent-sample ratio on the total phenolic compounds and anthocyanins content in the extracts obtained were evaluated. The response variables were optimized by means of a Box-Behnken design. It was found that the double interaction of the methanol composition and the cycle, the interaction between methanol composition and temperature, and the interaction between the cycle and solvent-sample ratio were the most influential variables on the extraction of total phenolic compounds (92.8% methanol in water, 0.2 s of cycle, 60 °C and 10:0.5 mL:g). The methanol composition and the interaction between methanol composition and pH were the most influential variables on the extraction of anthocyanins (74.1% methanol in water at pH 7). The methods that have been developed presented high repeatability and intermediate precision (RSD < 5%) and the bioactive compounds show a high recovery with short extraction times. Both methods were used to analyze the composition of the bioactive compounds in myrtle berries collected from different locations in the province of Cadiz (Spain). The results obtained by UAE were compared to those achieved in a previous study where microwave-assisted extraction (MAE) methods were employed. Similar extraction yields were obtained for phenolic compounds and anthocyanins by MAE and UAE under optimal conditions. However, UAE presents the advantage of using milder conditions for the extraction of anthocyanins from myrtle, which makes of this a more suitable method for the extraction of these degradable compounds.

Keywords: anthocyanins; bioactive compounds; Box–Behnken design; ultrasound-assisted extraction; myrtle; *Myrtus communis* L.; phenolic compounds

Molecules **2019**, *24*, 882

1. Introduction

People's diet is currently improving with a growing demand for healthy food, such as vegetables and fruit. Some berries have been particularly demanded mainly due to their phenolic composition [1,2]. Phenolic compounds are quite prone to oxidation due to their high content in double bonds and hydroxyl groups [3]. This characteristic provides them with a substantial capacity to prevent the oxidation of free radicals, i.e., chemically unstable species that may damage lipid cells, proteins and DNA [4]. Therefore, numerous studies support a positive association between the consumption of berries, which are rich in phenolic compounds, and the prevention against some diseases, such as cardiovascular or neurodegenerative diseases [5]. For that reason, the phenolic compounds extracted from berries are being used by the food, cosmetic and pharmaceutical industries to replace synthetic antioxidants [6,7].

Myrtle (*Myrtus communis* L.), is a widely spread plant throughout the Mediterranean area and the Middle East [8]. This evergreen shrub produces dark blue edible berries of different shapes in most of the characterized ecotypes [9]. These berries are rich in antioxidant compounds, in a considerably greater degree than most other fruit types [10]. Specifically, myrtle berries have a high content of phenolic compounds and anthocyanins [11]. The major phenolic compounds which can be identified in myrtle berries are quercetin 3-*O*-galactoside, quercetin 3-*O*-rhamnoside, myricetin 3-*O*-rhamnoside, quercetin 3-*O*-glucoside, ellagic acid and myricetin [12,13]. The major anthocyanins which can be identified in myrtle berries are delphinidin 3,5-*O*-diglucoside, delphinidin 3-*O*-glucoside, cyanidin 3-*O*-galactoside, cyanidin 3-*O*-glucoside, cyanidin 3-*O*-arabinoside, petunidin 3-*O*-glucoside, delphinidin 3-*O*-arabinoside, peonidin 3-*O*-glucoside, malvidin 3-*O*-glucoside, petunidin 3-*O*-arabinoside and malvidin 3-*O*-arabinoside [14,15]. These compounds, as above mentioned, have exhibited some potentially positive properties for human health, such as antidiabetic, anti-inflammatory, anti-cancer and antioxidant properties [16]. These medicinal properties have meant new fields where myrtle can be used. Consequently, it is currently used in the perfume, cosmetics, healthcare and food industries [17]. In spite of myrtle's broad potential and the fact that it grows in many extensive areas, its intensive exploitation takes place mainly in Sardinia. In this area, "Mirto" liquor is produced by macerating myrtle leaves and berries [18].

In order to determine the quality control of these beneficial compounds, the development of fast and efficient methods of extraction and analysis is required. Several studies have been carried out on the antioxidant activity and phenolic compounds in the extracts obtained from myrtle berries [19]. Most of these studies use maceration as the extraction technique, while few studies have been found that employ new extraction methods to obtain the extract from myrtle berries [20]. Compared to the commonly used extraction methods, some novel extraction techniques, such as ultrasound-assisted extraction (UAE), microwave-assisted extraction (MAE) or pressurized-liquid extraction (PLE) greatly reduce time, costs and volume of solvent, and also improve the quality of the extracts [21]. In a previous study recently published by the authors, MAE was used for the determination of total phenolic compounds and total anthocyanins in myrtle berries [22]. Based on the results obtained, it was concluded that MAE is an eco-friendlier and easier to use a technique for the extraction of both, phenolic compounds and anthocyanins, from myrtle berries. In this study, UAE is presented as an alternative extraction method, since UAE offers extraction yields comparable to those obtained by means of MAE, but exceeds MAE in terms of the number of solvent used, easiness, and economic cost. The use of ultrasounds is supported by the phenomenon of cavitation. This phenomenon makes of UAE a widely used method, since it breaks cell walls and releases the target compounds out of their natural matrices [23]. For the reason, UAE has been used for the extraction of the antioxidant compounds found in fruits matrices, such as papayas [24], mulberries [25], oranges [26] or sugarcane [27]. With regard to myrtle, some articles have been found in literature where UAE is used to extract biological compounds from myrtle berries [28,29], but it has not yet been carried out a thorough optimization and development of this technique for the specific extraction of phenolic compounds and anthocyanins. This would be very convenience, since the correct applicability would improve the analysis of the raw material, which would enhance the quality, for example, of liquor, as the main product obtained from myrtle. Moreover, the correct applicability of the method would allow to track down the evolution of the fruit's chemical composition during its maturity process. Reliable maturity indices would be useful to establish optimum harvesting periods, etc.

The aim of the present study is, therefore, to determine optimum conditions for the efficient extraction methods to obtain the greatest possibly yields of substances with antioxidant activities, anthocyanins and phenolic compounds from myrtle berries. Furthermore, the UAE results were compared to those obtained by MAE.

2. Results and Discussion

2.1. Development of the UAE Method

The Box-Behnken design was applied to the optimization of the variables that mainly influence UAE with regards to total phenolic and total anthocyanins yields. The main variables that affect UAE efficiency are solvent composition, temperature, amplitude of the ultrasound, cycle, pH, and solvent-sample ratio [30]. Table 1 shows such influential variables and their corresponding values studied in this work.

Table 1. Influential variables with their corresponding values studied in this work.

Variables	Studied Ranges	
	Phenolic Compounds	Anthocyanins
Temperature (°C)	10, 35, 60	10, 35, 60
Amplitude (%)	30, 50, 70	30, 50, 70
Cycle (s)	0.2, 0.45, 0.7	0.2, 0.45, 0.7
pH	2, 4.5, 7	2, 4.5, 7
Solvent-sample ratio (mL/0.5 g)	10, 15, 20	10, 15, 20
Solvent composition (% methanol in water)	50, 75, 100	25, 50, 75

Analysis of variance (ANOVA) was applied to the set of results in order to evaluate the effect of the different factors on their response and the possible interactions between them. Tables 2 and 3 show the results obtained from this analysis.

Table 2. ANOVA for the quadratic model adjusted to the extraction of total phenolic compounds.

	Source	Coefficient	Sum of Squares	Degrees of Freedom	Mean Square	F-Value	p-Value
	Model		2940.04	27	108.89	2.31	0.0179
Methanol	X_1	−1.15	31.97	1	31.97	0.6787	0.4175
Temperature	X_2	−1.01	24.50	1	24.50	0.5202	0.4772
Amplitude	X_3	0.0166	0.0066	1	0.0066	0.0001	0.9906
Cycle	X_4	−4.17	418.01	1	418.01	8.88	0.0062
pH	X_5	−0.8371	16.82	1	16.82	0.3571	0.5553
Ratio	X_6	0.5498	7.26	1	7.26	0.1541	0.6979
Methanol × Temperature	$X_1 X_2$	6.75	364.53	1	364.53	7.74	0.0099
Methanol × Amplitude	$X_1 X_3$	1.01	8.14	1	8.14	0.1729	0.6809
Methanol × Cycle	$X_1 X_4$	−2.00	63.93	1	63.93	1.36	0.2545
Methanol × pH	$X_1 X_5$	4.38	153.56	1	153.56	3.26	0.0826
Methanol × Ratio	$X_1 X_6$	−0.7961	5.07	1	5.07	0.1077	0.7454
Temperature × Amplitude	$X_2 X_3$	−0.7882	4.97	1	4.97	0.1055	0.7479
Temperature × Cycle	$X_2 X_4$	0.6373	3.25	1	3.25	0.0690	0.7949
Temperature × pH	$X_2 X_5$	2.87	132.23	1	132.23	2.81	0.1058

Table 2. *Cont.*

Source		Coefficient	Sum of Squares	Degrees of Freedom	Mean Square	F-Value	p-Value
Temperature × Ratio	X_2X_6	−2.36	44.72	1	44.72	0.9495	0.3388
Amplitude × Cycle	X_3X_4	−4.26	144.96	1	144.96	3.08	0.0911
Amplitude × pH	X_3X_5	−1.14	10.47	1	10.47	0.2222	0.6413
Amplitude × Ratio	X_3X_6	1.87	56.01	1	56.01	1.19	0.2855
Cycle × pH	X_4X_5	−2.64	55.80	1	55.80	1.18	0.2863
Cycle × Ratio	X_4X_6	6.71	359.66	1	359.66	7.64	0.0104
pH × Ratio	X_5X_6	1.68	22.66	1	22.66	0.4812	0.4940
Methanol × Methanol	X_1^2	−8.06	668.48	1	668.48	14.19	0.0009
Temperature × Temperature	X_2^2	−0.2020	0.4197	1	0.4197	0.0089	0.9255
Amplitude × Amplitude	X_3^2	1.79	32.81	1	32.81	0.6967	0.4115
Cycle × Cycle	X_4^2	1.87	36.12	1	36.12	0.7669	0.3892
pH × pH	X_5^2	2.96	90.01	1	90.01	1.91	0.1786
Ratio × Ratio	X_6^2	−2.20	49.57	1	49.57	1.05	0.3144
Residual		45.85	1224.48	26	47.10		
Lack of Fit			1075.70	21	51.22	1.72	0.2858
Pure Error			148.78	5	29.76		
Total			4164.52	53			

Table 3. ANOVA for the quadratic model adjusted to the extraction of total anthocyanins.

Source		Coefficient	Sum of Squares	Degrees of Squares	Mean Square	F-Value	p-Value
Model			2514.45	27	93.13	1.47	0.1631
Methanol	X_1	4.80	553.02	1	553.02	8.75	0.0065
Temperature	X_2	−2.05	100.49	1	100.49	1.59	0.2185
Amplitude	X_3	−1.88	85.15	1	85.15	1.35	0.2563
Cycle	X_4	−2.47	146.43	1	146.43	2.32	0.1400
pH	X_5	1.13	30.92	1	30.92	0.4892	0.4905
Ratio	X_6	−0.1120	0.3008	1	0.3008	0.0048	0.9455
Methanol × Temperature	X_1X_2	−5.59	250.28	1	250.28	3.96	0.0572
Methanol × Amplitude	X_1X_3	−4.68	174.94	1	174.94	2.77	0.1082
Methanol × Cycle	X_1X_4	−2.18	76.24	1	76.24	1.21	0.2821
Methanol × pH	X_1X_5	6.51	338.73	1	338.73	5.36	0.0288
Methanol × Ratio	X_1X_6	0.0864	0.0598	1	0.0598	0.0009	0.9757
Temperature × Amplitude	X_2X_3	−0.1454	0.1691	1	0.1691	0.0027	0.9591
Temperature × Cycle	X_2X_4	−1.16	10.77	1	10.77	0.1704	0.6831
Temperature × pH	X_2X_5	−3.14	157.76	1	157.76	2.50	0.1262
Temperature × Ratio	X_2X_6	0.8116	5.27	1	5.27	0.0834	0.7751
Amplitude × Cycle	X_3X_4	3.73	111.16	1	111.16	1.76	0.1963
Amplitude × pH	X_3X_5	−1.63	21.28	1	21.28	0.3368	0.5667
Amplitude × Ratio	X_3X_6	−1.04	17.31	1	17.31	0.2739	0.6052
Cycle × pH	X_4X_5	−0.7232	4.18	1	4.18	0.0662	0.7990
Cycle × Ratio	X_4X_6	0.2305	0.4249	1	0.4249	0.0067	0.9353
pH × Ratio	X_5X_6	−0.9561	7.31	1	7.31	0.1157	0.7365
Methanol × Methanol	X_1^2	−1.95	39.12	1	39.12	0.6191	0.4385
Temperature × Temperature	X_2^2	2.23	50.98	1	50.98	0.8068	0.3773
Amplitude × Amplitude	X_3^2	2.89	86.14	1	86.14	1.36	0.2536

Table 3. *Cont.*

	Source	Coefficient	Sum of Squares	Degrees of Squares	Mean Square	F-Value	p-Value
Cycle × Cycle	X_4^2	−1.17	14.04	1	14.04	0.2221	0.6414
pH × pH	X_5^2	2.03	42.24	1	42.24	0.6683	0.4211
Ratio × Ratio	X_6^2	−1.64	27.72	1	27.72	0.4386	0.5136
Residual		23.34	1643.02	26	63.19		
Lack of Fit			1469.28	21	69.97	2.01	0.2248
Pure Error			173.75	5	34.75		
Total			4157.47	53			

This information was supplemented with Pareto Charts (Figure 1). Pareto charts show each effect and combination of effects by a bar in decreasing order of significance. From a graphical point of view, this allows to visualize the influencing variables and their degree of influence.

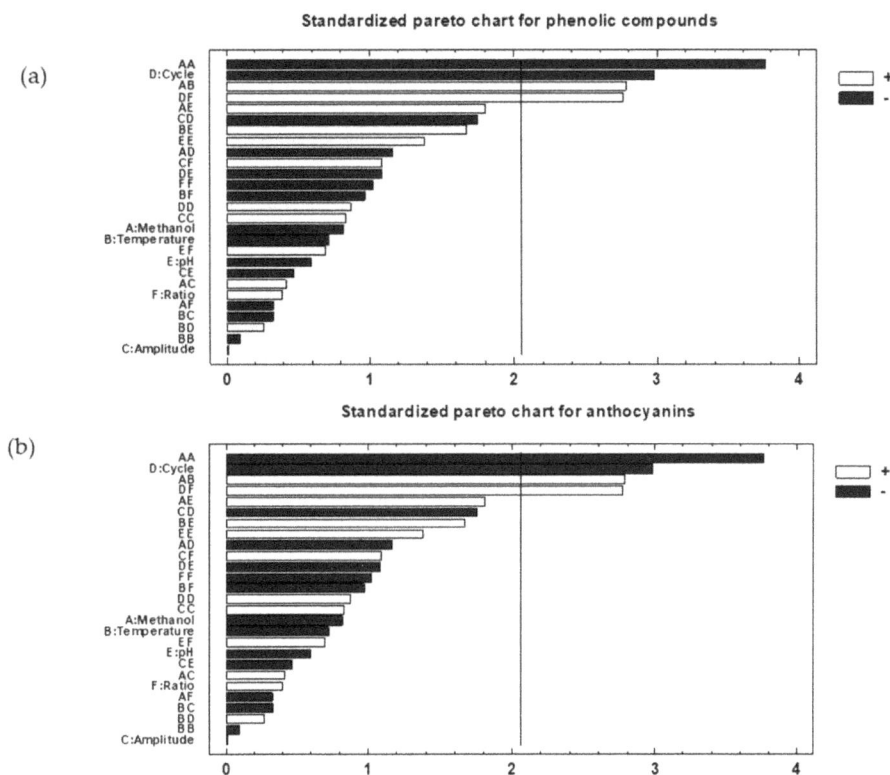

Figure 1. Pareto charts of the standardized effects: (**a**) Total phenolic compounds and (**b**) total anthocyanins.

In the case of total phenolic compounds (Figure 1a), cycle (X_4) was the only linear term which had a significant effect, with a *p*-value lower than 0.01. Its effect on the response variable was negative ($b_4 = -4.17$). Numerous studies show that the cycle is an influential variable since the use of ultrasound cycles (pulse) improves the extraction of certain compounds of interest, such as the phenolic compounds in natural matrices [31–33]. The negative effect means that a decrease in the cycle increases the extraction of phenolic compounds. This may be due to the negative chemical and physical effects of cavitation [34]. The negative effect is often due to the reactions of free radicals formed during the sonication with molecules in the medium [35], which accelerates the degradation process of phenolic compounds. In addition, the solvent composition had a significant quadratic influence (X_1^2) on the response variable (*p*-value < 0.01). The solvent composition is an important variable since it is necessary to extract the phenolic compounds with solvents of similar polarity [36]. Specifically, X_1^2 showed a negative effect ($b_{11} = -8.06$). With regards to interactions between factors, minor interactions between methanol and temperature (X_1X_2) (*p*-value < 0.01) and between cycle and solvent-sample ratio (X_4X_6) (*p*-value < 0.05) were observed. Both interactions showed positive coefficients ($b_{12} = 6.75$ and $b_{46} = 6.70$).

In the case of anthocyanins (Figure 1b), solvent composition (X_1) was the only linear term that had a significant effect, with a *p*-value lower than 0.01. Its effect on the response variable was positive ($b_1 = 4.80$), which indicates that an increase in the methanol percentage in the solvent favored the anthocyanins content in the extract. Many pieces of research have been found in the literature which shows that hydroalcoholic mixtures are more efficient than pure solvents for the extraction of moderately polar molecules, such as phenolic compounds [36]. Phenolic compounds have a moderate polarity, so they are not extracted adequately when pure water mixtures are used (high polarity). The use of methanol increases the solubility of phenolic compounds and the use of water in a lower percentage helps the desorption of the solute from the sample [37]. With regard to the interactions between factors, a minor interaction between methanol and pH (X_1X_5) (*p*-value < 0.05) was observed with a positive coefficient ($b_{15} = 6.51$). These results agree with the bibliographic data [38] where the concentration of organic solvent used and the pH are influential variables on the extraction and stability of anthocyanins from vegetable matrices. Which regard to quadratic effects, non-significant interactions were obtained (*p*-value > 0.05).

The polynomial Equations (1) and (2) for anthocyanins and total phenolic compounds were obtained from the coefficients of the effects and interactions (Tables 2 and 3). Therefore, two second-order mathematical models were obtained to predict the Y_{TA} and Y_{TP} response values as a function of the independent variables. Lack of fit test showed *p*-values greater than 0.05 for phenolic compounds and for anthocyanins which means that both models fit well.

$$\begin{aligned}
Y_{TA}\ (\mathrm{mg \cdot g^{-1}}) =\ & 23.3422 + 4.80026 \cdot X_1 - 2.04623 \cdot X_2 - 1.88357 \cdot X_3 - 2.47009 \cdot X_4 + 1.13497 \cdot X_5 - \\
& 0.111961 \cdot X_6 - 1.95032 \cdot X_1^2 - 5.59331 \cdot X_1X_2 - 4.67624 \cdot X_1X_3 - 2.18294 \cdot X_1X_4 + 6.50698 \cdot X_1X_5 + \\
& 0.0864269 \cdot X_1X_6 + 2.22638 \cdot X_2^2 - 0.145373 \cdot X_2X_3 - 1.16027 \cdot X_2X_4 - 3.14005 \cdot X_2X_5 + 0.811562 \cdot X_2X_6 \\
& + 2.89386 \cdot X_3^2 + 3.72761 \cdot X_3X_4 - 1.63114 \cdot X_3X_5 - 1.04011 \cdot X_3X_6 - 1.16818 \cdot X_4^2 - 0.72317 \cdot X_4X_5 + \\
& 0.230451 \cdot X_4X_6 + 2.02637 \cdot X_5^2 - 0.956103 \cdot X_5X_6 - 1.64156 \cdot X_6^2
\end{aligned} \tag{1}$$

$$\begin{aligned}
Y_{TP}\ (\mathrm{mg \cdot g^{-1}}) =\ & 45.8533 - 1.15407 \cdot X_1 - 1.0103 \cdot X_2 + 0.0166375 \cdot X_3 - 4.17338 \cdot X_4 - 0.837125 \cdot X_5 + \\
& 0.549829 \cdot X_6 - 8.06168 \cdot X_1^2 + 6.75029 \cdot X_1X_2 + 1.00895 \cdot X_1X_3 - 1.99896 \cdot X_1X_4 + 4.38119 \cdot X_1X_5 \\
& -0.796125 \cdot X_1X_6 - 0.201989 \cdot X_2^2 - 0.78815 \cdot X_2X_3 + 0.637275 \cdot X_2X_4 + 2.87481 \cdot X_2X_5 - 2.3643 \cdot X_2X_6 \\
& + 1.78604 \cdot X_3^2 - 4.25675 \cdot X_3X_4 - 1.14384 \cdot X_3X_5 + 1.87097 \cdot X_3X_6 + 1.87383 \cdot X_4^2 - 2.6411 \cdot X_4X_5 + \\
& 6.70505 \cdot X_4X_6 + 2.95825 \cdot X_5^2 + 1.68306 \cdot X_5X_6 - 2.19536 \cdot X_6^2
\end{aligned} \tag{2}$$

Both mathematical models can be reduced by omitting the insignificant terms (*p*-value > 0.05). The Equations (3) and (4) of the two reduced models were expressed as follows:

$$Y_{TA}\ (\mathrm{mg \cdot g^{-1}}) = 23.3422 + 4.80026 \cdot X_1 - 6.60698 \cdot X_1X_5, \tag{3}$$

$$Y_{TP} \text{ (mg·g}^{-1}) = 45.8533 - 4.17338 \cdot X_4 - 8.06168 \cdot X_{12} + 6.75029 \cdot X_1 X_2 + 6.70505 \cdot X_4 X_6 \qquad (4)$$

The trends outlined above were recorded in three-dimensional (3D) surface plots using the fitted model in order to improve our understanding of both, the main and the interaction effects, of the most influential parameters. The combined effects of cycle-methanol, methanol–temperature and cycle-ratio on the total phenolic compounds recovery are represented in Figure 2a–c. The combined effect of solvent composition and pH on the total anthocyanins recovery is represented in Figure 2b.

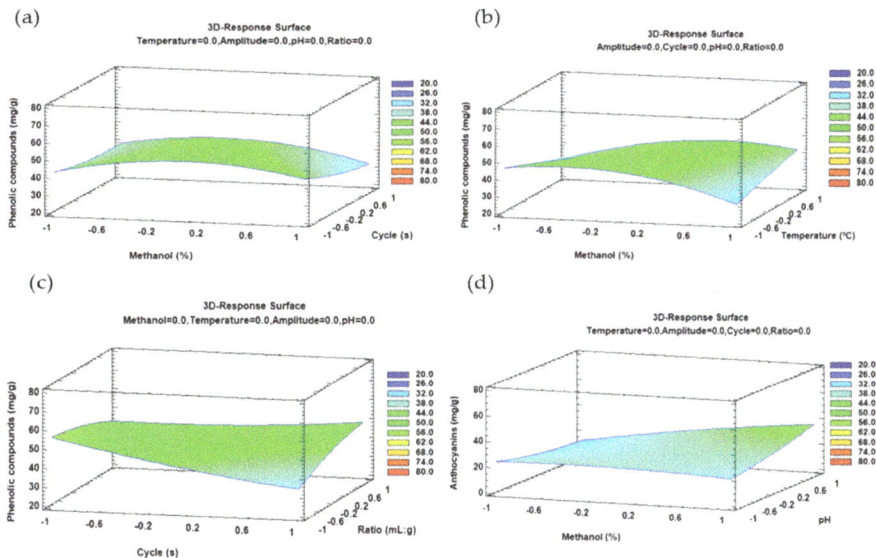

Figure 2. 3D-surface plots of the Box–Behnken design to represent the influence of: (**a**) Solvent composition and cycle on the total phenolic compounds; (**b**) solvent composition-temperature on the total phenolic compounds; (**c**) cycle-ratio on the total phenolic compounds; (**d**) solvent composition-pH on the total anthocyanins.

2.2. Optimal Conditions

According to the experimental design, the ideal UAE conditions to extract the phenolic compounds were as follows: 92.8% methanol in water as a solvent, 60 °C extraction temperature, 65.48% ultrasound amplitude, 0.2 s cycles, pH 6.8, and 10:0.5 mL:g solvent-sample ratio. With regard to the temperature, no higher temperatures were verified, since they might imply a greater degradation of the compounds of interest and a high loss of methanol that would affect the solvent-sample ratio [39]. With respect to pH, an almost neutral value was determined as optimal, since different research shows that acidified solvents may enhance the formation of free radicals in aqueous solutions because of their higher concentration of H$^+$ or thermal treatment [32], which would hinder the recovery of the phenolic compounds [35].

With regards to anthocyanins, optimum UAE conditions were as follows: 74.1% methanol in water solvent, 10 °C extraction temperature, 30% ultrasound amplitude, 0.3 s cycles, pH 7, and 18:0.5 mL:g as the solvent-sample ratio. With respect to temperature, the lowest end of the range studied (10 °C) was determined as the optimum value. Although anthocyanins are also phenolic compounds, they are more thermally sensitive than other phenolic compounds. High temperatures can diminish the recovery of anthocyanins due mainly to oxidation, cleavage of covalent bonds or an increase in oxidation reactions as a result of the thermal treatment [40]. With respect to the solvent pH, neutral pH was found to be optimum for the extraction of anthocyanins. Although pH between 1 and 3 usually generates stable conformation for anthocyanins, there are many articles in the literature where the highest extraction yields take place with a higher pH (3–7) [34]. This behavior might be the effect of different factors on anthocyanins stability (light, temperature, extraction time, etc.), which may turn them into other compounds [41]. Specifically, some authors affirm that ultrasound can promote the degradation of the anthocyanins because of the radical hydroxyl (OH$^\bullet$) and *hydrogen peroxide* (H_2O_2) produced inside the cavitation bubbles when subjected to conditions, such as high ultrasonic power, high amplitude, low temperature and long treatment time [42]. No higher pH was checked for anthocyanins since this may cause unstable structures as a result of basic hydrolysis [43].

In conclusion, for both, phenolic compounds and anthocyanins, maximum extractions were obtained when the solvent had a high percentage of methanol and neutral pH. Specifically, for the extraction of phenolic compounds a higher range of percentages was required.

2.3. Extraction Time

Once the effects of the variables on the extraction methods and the optimal values were known, the kinetics of the extractions was studied. Several extractions were carried out under optimal ultrasound conditions while extraction time varied between 2, 5, 10, 15, 20, and 25 min. The average results obtained ($n = 3$) for phenolic compounds and for anthocyanins are represented in Figure 3.

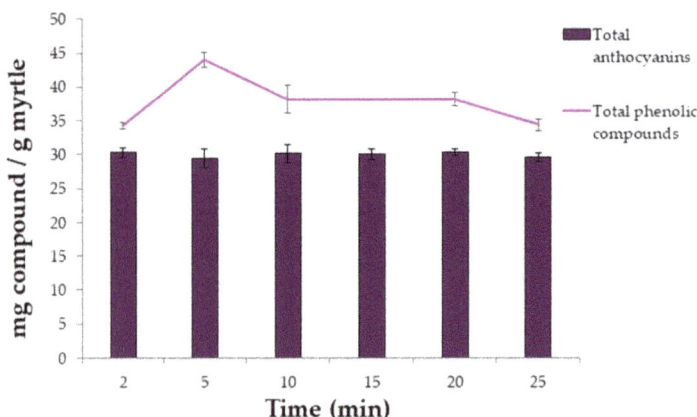

Figure 3. Recovery of anthocyanins (mg·g^{-1}) and total phenolic compounds (mg·g^{-1}) using different extraction times ($n = 3$).

It can be seen that large recoveries are achieved for both types of bioactive compounds and that long extraction times are not required. The Phenolic compounds present their maximum extraction at 5 min. However, longer extraction times led to lower recoveries, probably due to degradation of the phenolic compounds [25]. With respect to the anthocyanins, 2 min was determined as their optimum extraction time, since it exhibited the same yields as with longer times, while saving both, time and costs.

2.4. Repeatability and Intermediate Precision of UAE Methods

The precision of the extraction methods was evaluated in terms of repeatability (intra-day) and intermediate precision (inter-day). Repeatability was evaluated by performing 10 extractions under the same conditions on the same day. Intermediate precision was evaluated by performing 10 additional extractions on each one of the following two days. Altogether, 30 extractions were carried out under optimal extraction conditions to evaluate the precision of the extraction method for phenolic compounds and for anthocyanins. This method is employed in numerous studies [25,44]. The results were expressed by the coefficient of variation (CV) of the means. The repeatability results obtained were: 2.95% for phenolic compounds and 2.23% for anthocyanins. The intermediate precision results were: 4.66% for phenolic compounds and 4.15% for anthocyanins. As it can be seen, all the results are within acceptable limits ($\pm$10%) according to AOAC [45] and supported the accuracy—with diversions lower than 5.0%—of the extraction methods for total anthocyanins and total phenolic compounds.

2.5. Application of the Developed Methods to Ecotypes from Two Locations

In order to determine the applicability of the developed methods, once they had been optimized, they were applied to a new set of samples. Specifically, 14 ecotypes of myrtle were evaluated. 8 ecotypes from Puerto Real region (My-1, My-2, My-3, My-4, My-5, My-6, My -7, and My-8) and 6 from San José del Valle region (My-9, My-10, My-11, My-12, My-13, and My-14). The phenolic compounds were extracted from the 14 ecotypes in duplicate by applying the UAE method according to the optimum conditions previously determined. This should ensure the greatest possible yields. The quantification of the total phenolic compounds content in the extracts was carried out by Folin-Ciocalteau reagent.

The anthocyanins compounds were also extracted from the 14 samples according to the optimal conditions determined for the developed UAE method. The anthocyanins content in the extracts was quantified by UHPLC-UV-vis. The total anthocyanins content is the result of adding up each separate anthocyanin content. The average extraction and quantification results are shown in Table 4.

2.6. Analysis by Conglomerate

As a consequence of the chemical results obtained, it can be observed that there are differences between the average values obtained from the different myrtle ecotypes. To objectively study if these visual differences are related to the origin of the ecotypes, a comparative chemometric study was carried out using all the average values. Specifically, the data matrix $D_{13 \times 14}$ ($D_{variables \times ecotypes}$) (Table 4) was evaluated using an exploratory tool, i.e., Hierarchical Cluster Analysis (HCA). Ward's method and square Euclidean distance were employed and the variables for the differentiation were: The total phenolic compounds ($mg \cdot g^{-1}$) from each experiment; each individual anthocyanin content, 11 anthocyanins ($mg \cdot g^{-1}$), and total anthocyanins ($mg \cdot g^{-1}$). Myrtle is a shrub that grows better in warm and humid areas, and requires rich, humid soils. These characteristics match those of Puerto Real, which is near the sea with humid and sandy soils. San José del Valle has drier climate conditions and its clay soil is not so fertile [46,47]. These differences may lead to differences between the maturation processes of the autochthonous ecotypes in Puerto Real and San José del Valle, and consequently, to variations in their bioactive composition. The results of the analysis are graphically represented as a dendrogram in Figure 4. An obvious differentiation of the samples into two groups can be observed: Cluster A, includes only the ecotypes from Puerto Real, and Cluster B, only includes the ecotypes from San José del Valle. Therefore, based on their tendency to fall into a particular group in accordance with their origin, it can be said that phenolic compounds and anthocyanins contents in each ecotype is related to the berries' geographical area of origin. Specifically, the ecotypes in Cluster A, Puerto Real, present a total phenolic compounds and anthocyanins content greater than the ecotypes in Cluster B, from San José del Valle. The differences can be attributed to the different climatic and soil conditions above mentioned.

Table 4. Extract concentrations (mg·g⁻¹) of total anthocyanins and total phenolic compounds ($n = 3$) obtained from different myrtle ecotypes by means of ultrasound-assisted method (UAE).

Compounds [1]	Myrtle ecotypes from Puerto Real									Myrtle ecotypes from San José del Valle				
	MY-1	MY-2	MY-3	MY-4	MY-5	MY-6	MY-7	MY-8	MY-9	MY-10	MY-11	MY-12	MY-13	MY-14
D-3,5-diGl	0.166 ± 0.006	0.415 ± 0.016	0.404 ± 0.015	0.504 ± 0.020	0.377 ± 0.014	0.375 ± 0.016	0.183 ± 0.008	0.103 ± 0.0034	0.127 ± 0.004	0.065 ± 0.002	0.079 ± 0.0023	0.176 ± 0.0067	0.113 ± 0.003	0.134 ± 0.004
Del-3-Glu	10.552 ± 0.161	16.309 ± 0.534	10.305 ± 0.358	11.114 ± 0.384	12.557 ± 0.481	12.258 ± 0.432	8.315 ± 0.326	5.929 ± 0.165	3.667 ± 0.152	1.566 ± 0.008	2.145 ± 0.084	3.823 ± 0.124	2.987 ± 0.121	3.346 ± 0.135
Cy-3-Ga	0.150 ± 0.006	0.382 ± 0.008	0.216 ± 0.009	0.359 ± 0.007	0.319 ± 0.012	0.371 ± 0.0135	0.176 ± 0.007	0.392 ± 0.012	0.050 ± 0.001	0.052 ± 0.002	0.068 ± 0.0018	0.1757 ± 0.005	0.077 ± 0.002	0.036 ± 0.001
Cy-3-Gl	1.794 ± 0.084	3.004 ± 0.15	1.126 ± 0.023	1.241 ± 0.036	1.613 ± 0.03	1.313 ± 0.0578	1.946 ± 0.034	0.791 ± 0.043	0.852 ± 0.024	0.293 ± 0.01	0.301 ± 0.016	0.512 ± 0.025	0.452 ± 0.013	0.786 ± 0.032
Cy-3-Ar	0.092 ± 0.003	0.126 ± 0.002	0.076 ± 0.002	0.130 ± 0.002	0.112 ± 0.004	0.170 ± 0.006	0.061 ± 0.14	0.081 ± 0.0025	0.170 ± 0.003	0.040 ± 0.001	0.410 ± 0.021	0.657 ± 0.027	0.549 ± 0.23	0.188 ± 0.006
Pet-3-Gl	6.217 ± 0.110	11.007 ± 0.38	7.173 ± 0.314	8.691 ± 0.301	8.329 ± 0.297	10.242 ± 0.392	5.538 ± 0.0423	5.082 ± 0.176	6.810 ± 0.235	1.580 ± 0.018	1.676 ± 0.065	2.123 ± 0.085	1.823 ± 0.067	6.643 ± 0.114
Del-3-Ara	0.840 ± 0.005	2.378 ± 0.102	1.635 ± 0.062	1.775 ± 0.067	1.917 ± 0.064	2.009 ± 0.081	1.335 ± 0.021	1.562 ± 0.035	1.833 ± 0.076	0.367 ± 0.006	0.324 ± 0.009	0.679 ± 0.021	0.454 ± 0.014	1.234 ± 0.033
Peo-3-Gl	1.030 ± 0.017	1.179 ± 0.043	0.659 ± 0.019	0.689 ± 0.021	0.610 ± 0.023	0.816 ± 0.029	0.755 ± 0.34	0.481 ± 0.019	0.473 ± 0.015	0.305 ± 0.017	0.357 ± 0.012	0.921 ± 0.032	0.564 ± 0.013	0.446 ± 0.012
Mal-3-Gl	9.834 ± 0.314	16.691 ± 0.641	16.050 ± 0.65	18.301 ± 0.614	12.748 ± 0.768	23.817 ± 1.032	8.244 ± 0.273	9.072 ± 0.348	6.987 ± 0.246	5.195 ± 0.068	5.453 ± 0.185	7.679 ± 0.23	6.456 ± 0.241	5.979 ± 0.263
Pet-3-Ar	0.269 ± 0.009	0.551 ± 0.034	0.448 ± 0.014	0.587 ± 0.013	0.482 ± 0.028	0.751 ± 0.0013	0.332 ± 0.012	0.583 ± 0.015	0.438 ± 0.017	0.197 ± 0.002	0.206 ± 0.008	0.657 ± 0.023	0.454 ± 0.012	0.446 ± 0.012
Mal-3-Ar	0.226 ± 0.004	0.302 ± 0.009	0.402 ± 0.012	0.436 ± 0.121	0.296 ± 0.015	0.598 ± 0.019	0.240 ± 0.009	0.395 ± 0.014	0.185 ± 0.005	0.221 ± 0.001	0.244 ± 0.012	0.846 ± 0.032	0.679 ± 0.031	0.165 ± 0.006
Total anthocyanins	31.170 ± 1.023	52.346 ± 1.124	38.493 ± 3.516	43.827 ± 2.557	43.274 ± 1.904	60.252 ± 0.002	35.207 ± 1.175	32.903 ± 1.162	21.591 ± 0.92	10.381 ± 0.037	11.263 ± 0.431	18.249 ± 2.30	14.608 ± 1.964	19.403 ± 1.97
Total Phenolic compounds	70.747 ± 1.433	86.439 ± 3.125	64.7586 ± 0.914	60.034 ± 0.44	72.155 ± 3.174	76.124 ± 1.170	76.070 ± 3.417	69.815 ± 2.947	49.824 ± 0.362	52.66 ± 0.038	45.789 ± 1.824	51.278 ± 2.051	49.103 ± 0.192	50.897 ± 1.564

[1] Del-3,5-diGl, delphinidin 3,5-O-diglucoside; Del-3-Glu, delphinidin 3-O-glucoside; Cy-3-Ga, cyanidin 3-O-galactoside; Cy-3-Gl, cyanidin 3-O-glucoside; Cy-3-Ar, cyanidin 3-O-arabinoside; Pet-3-Gl, petunidin 3-O-glucoside; Del-3-Ara, delphinidin 3-O-arabinoside; Peo-3-Gl, peonidin 3-O-glucoside; Mal-3-Gl, malvidin 3-O-glucoside; Pet-3-Ar, petunidin 3-O-arabinoside; Mal-3-Ar, malvidin 3-O-arabinoside.

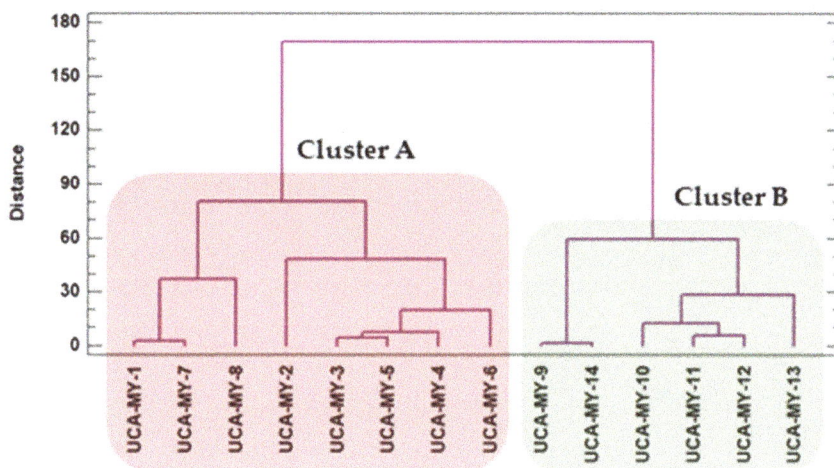

Figure 4. Dendrogram representing the bioactive compounds in 14 myrtle ecotypes according to the matrix of the results from Hierarchical Cluster Analysis (HCA) (Table 3).

2.7. Comparison Study: UAE vs. MAE

As above mentioned, the greatest phenolic compounds and anthocyanins yields using UAE were obtained under the following optimal conditions: methanol:water solvent ratio 92.8% *v/v* at 60 °C for phenolic compounds and methanol:water solvent ratio 74.1% *v/v* at 10 °C for anthocyanins.

In comparison with traditionally used methods for the extraction of the phenolic compounds in myrtle berries, UAE achieves a greater recovery of the compounds of interest, while using less solvent and in a shorter time, with the consequent cost reduction [11,17]. This increased effectiveness with greater extraction yields of both, total phenolic compounds and total anthocyanins, could be based on the phenomenon of cavitation, which breaks cell walls and releases the compounds of interest from the myrtle berries' matrices.

With the purpose of rounding up this study, the results obtained using UAE were compared to those achieved by MAE in previous work. For that purpose, the same number of samples were run under optimum conditions and later on analyzed [22]. The total phenolic compounds and total anthocyanins content extracted from myrtle berries at different times using UAE and MAE are shown in Figure 5.

With respect to the phenolic compounds (Figure 5a), a similar trend is observed in both extraction methods. The phenolic compounds yield increases until the maximum extraction value is reached at 5 min for UAE and at 15 min for MAE. From then on, the quantity of the extract begins to decrease. The optimum time for UAE, 5 min, indicates that UAE degrades phenolic compounds faster than MAE and the recovery is also lower. When compared to UAE, as recently reported by Ghafoor et al. [48,49], MAE obtains extracts with a substantially greater content of phenolic compounds than the one obtained by UAE.

With respect to the anthocyanins (Figure 5b), their content levels are very similar in MAE and UAE extracts. In addition, the extraction time required to achieve good yields (2 min) is low for both methods, which would considerably reduce costs when operating at an industrial scale. At the optimal time of two minutes, the anthocyanins extraction is slightly higher when MAE is used. However, at longer times, MAE extracts have lower anthocyanins content [50]. MAE optimal operating temperature (50 °C) makes anthocyanins, thermally labile, begin to degrade before.

Additionally, both techniques were applied to different myrtle ecotypes (Figure 6). As above noted, MAE stands out as a more efficient method for the extraction of the phenolic compounds in myrtle berries. With respect to anthocyanins, although some particular ecotypes produced greater yields by MAE, most of them produced greater yields when UAE was employed. Anthocyanins are extremely susceptible to degradation and the combination of high pressure and temperature that is employed for MAE would enhance such degradation and affect negatively their recovery.

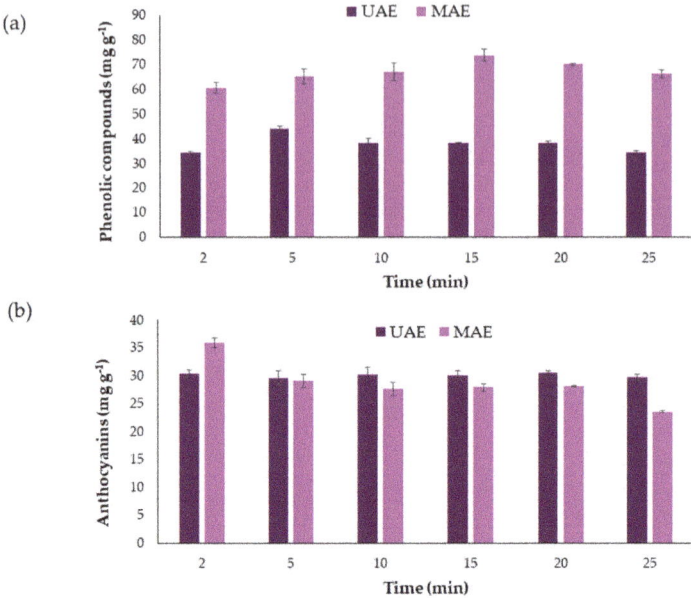

Figure 5. Comparison between the implementation of UAE and microwave-assisted extraction (MAE) to myrtle berries: (**a**) To extract total phenolic compounds; (**b**) to extract anthocyanins.

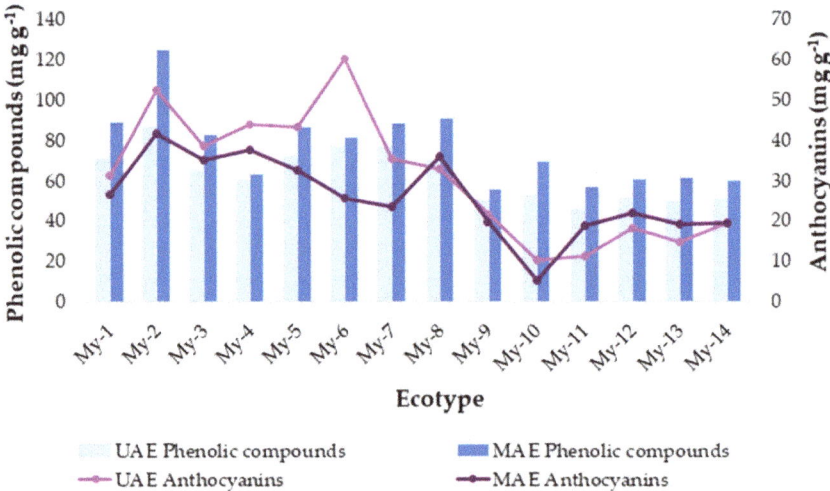

Figure 6. Concentration of total phenolic compounds and total anthocyanins ($n = 3$) in UAE and MAE extracts from several myrtle berries ecotypes.

Therefore, the UAE should be seriously considered as the preferred method for the extraction of the anthocyanins in myrtle berries.

3. Materials and Methods

3.1. Plant Materials

The biological materials used for this study were different myrtle berries (14 ecotypes) collected by the authors from two different areas in Cadiz province during their optimum ripeness stages in 2016. The first collection area was Puerto Real (eight ecotypes). This area is characterized by its humid climate due to its proximity to the sea. The second collection area was San José del Valle (6 ecotypes), also within Cadiz province, but located inland at 50 km from the coast. This location has a drier climate and its soils have a lower water content. The guidelines described by M., Mulas and M.R., Cani [47] were applied to characterize the morphology of both, leaves and berries, to confirm that the samples had been collected from different ecotypes. The samples were subjected to a pretreatment to improve the contact surface with the solvent [51]. First, the seeds were separated from the pulp. Secondly, the pulp was lyophilized in a Virtis Benchtop K freeze dryer (SP Cientific, New York, United States) and crushed by means of a regular spice grinder. Finally, the samples were stored in a freezer at −20 °C prior to analysis.

3.2. Chemicals and Solvents

The solvents used for the extraction were a mix of methanol and water at different concentration levels and with different pH. The methanol (Fischer Scientifics, Loughborough, United Kingdom) was HPLC grade. Ultra-pure water was obtained from a Milli-Q water purification system (EMD Millipore Corporation, Bedford, MA, United States). The pH adjustment of the solvents was done by means of hydrochloric acid and sodium hydroxide, both analytical grade and purchased from Panreac Química S.A.U. (Castellar del Valles, Barcelona, Spain). For the Folin-Ciocalteau spectrophotometric method, anhydrous sodium carbonate (Panreac Química S.A.U., Castellar del Valles) and Folin–Ciocalteu (Merck KGaA, EMD Millipore Corporation, Darmstadt, Germany) were employed. For the HPLC analyses, methanol (Fischer Scientific, Loughborough, United Kingdom) and formic acid (Scharlau, Barcelona, Spain) were used. These solvents were degassed and filtered through a 0.22 µm membrane (Nylon Membrane Filter, FILTER-LAB, Barcelona, Spain) before being used. The standard for the phenolic compounds was gallic acid and the standard for anthocyanins was cyanidin chloride. Both standards were purchased from Sigma-Aldrich Chemical Co. (St. Louis, MO, USA).

3.3. Ultrasound-Assisted Extraction Procedure

To extract the total phenolic compounds and the total anthocyanins from the myrtle berries, UAE was used. A UP200S probe (Hielscher Ultrasound Technology, Berlin, Germany) was employed, coupled to a processor that allows adjusting the amplitude and the cycle. For the adjustment of the temperature, a thermostatic bath (Frigiterm-10, Selecta, Barcelona, Spain) was employed. The temperature, the cycle, and the amplitude were selected for each extraction according to the experiment. About 0.5 g of the lyophilized and homogenized sample was weighed in a Falcon tube and the corresponding volume of solvent was added depending on the experiment. The Falcon tube was placed in a double vessel through which the water from the thermostatic bath circulated. The initial extraction time set was 10 min, followed by a sample cooling time. After that time, the extracts were centrifuged (7500 rpm, 5 min) and the supernatants were placed in a 25 mL volumetric flask. The precipitates from the extraction were redissolved in 5 mL of the same extraction solvent and centrifuged again under the same conditions. The new supernatants were placed in the volumetric flask and it was completed with the same solvent. The final extracts were stored at −20 °C for their correct conservation until further analysis. The UAE conditions set for the extractions were: Solvent composition (50–100% methanol in water for phenolic compounds and 25–75% for anthocyanins), temperature (10–60 °C), amplitude (30–70%), cycle (0.2–0.7 s), pH (2–7) and solvent-sample ratio (10:0.5–20:0.5 mL:g).

3.4. Determining the Content of Total Phenolic Compounds by Folin-Ciocalteau Essay

The total phenolic compounds content in myrtle berry was determined by adapted/modified Folin-Ciocalteau (FC) method [52]. This method has been previously used by many researchers to determine the total phenolic compounds content [53,54]. It is based on a redox reaction in a basic medium that gives rise to a complex of blue coloration with a wide absorption up to 765 nm. The extracts were filtered using a 0.45 μm nylon syringe filter (Membrane Solutions, Dallas, United States). The protocol of the method is the following: 250 μL of the previously filtered extract was transferred to a 25 mL volumetric flask. After this, 12.5 mL of water, 1.25 mL of Folin-Ciocalteau reagent and 5 mL of a 20% aqueous sodium carbonate solution were added. Finally, the flask was made up with water, and after 30 min the absorbance was measured at the maximum. All the extracts were analyzed in duplicate. The range of absorbance obtained for the studied samples was 0.4-1.4. The equipment used to measure the absorbance was a Helios Gamma (γ) Unicam UV-vis Spectrophotometer (Thermo Fisher Scientific, Waltham, MA, United States). The calibrated curve for the quantification was constructed based on a reference standard gallic acid pattern under the same conditions as the extracts [55]. The computer application used to process the data was Microsoft Office Excel 2013. The following regression equation y = 0.0010x + 0.0059 and the following correlation coefficient R^2 = 0.9999 were obtained. The linear range of work was 100–2600 mg L^{-1}. The results are expressed in milligrams of gallic acid equivalent per gram of lyophilized weight.

3.5. Identification of Anthocyanins by UHPLC-QToF-MS

Ultra-high performance liquid chromatography (UHPLC) coupled to quadrupole-time-of-flight mass spectrometry (QToF-MS) (Xevo G2 QToF, Waters Corp., Milford, MA, United States) was used to identify the anthocyanins in the UAE extracts. The column employed was a reverse-phase C18 analytical column with 1.7 μm particle size, 2.1 mm × 100 mm (ACQUITY UPLC CSH C18, Waters). The mobile phase was 2% formic acid–water solution (phase A) and methanol solution (phase B). The studied bioactive compounds were determined by employing the UHPLC-QToF-MS method described in a previously research [22]. The individual anthocyanins were identified based on their retention time and molecular weight. The following eleven anthocyanins were identified in the samples: delphinidin 3,5-*O*-diglucoside (*m/z* = 627.1561), delphinidin 3-*O*-glucoside (*m/z* = 465.1033), cyanidin 3-*O*-galactoside (*m/z* = 449.1084), cyanidin 3-*O*-glucoside (*m/z* = 449.1084), cyanidin 3-*O*-arabinoside (*m/z* = 419.0978), petunidin 3-*O*-glucoside (*m/z* = 479.1189), delphinidin 3-*O*-arabinoside (*m/z* = 435.0927), peonidin 3-*O*-glucoside (*m/z* = 463.1240), malvidin 3-*O*-glucoside (*m/z* = 493.1346), petunidin 3-*O*-arabinoside (*m/z* = 449.1084) and malvidin 3-*O*-arabinoside (*m/z* = 463.1240). Before their identification, all the UAE extracts were filtered through a 0.20 μm nylon syringe filter (Membrane Solutions, Dallas, TX, United States). The anthocyanins structures are shown in Figure 7.

3.6. Determination of Anthocyanins by UHPLC-UV-Vis System

For the separation and quantification of the anthocyanins present in UAE extracts from myrtle berries, an Elite UHPLC LaChrom Ultra System (Hitachi, Tokyo, Japan) was used. The UHPLC system consists of an L-2420U UV-Vis detector, an L-2200U autosampler, an L-2300 column oven set at 50 °C and two L-2160 U pumps. The column used was a "Fused Core" C18 with 2.6 μm particle size, 2.1 mm × 100 mm (Phenomenex Kinetex, Torrance, CA, United States). The mobile phase consisted of a 5% formic acid–water solution (phase A) and a methanol solution (phase B). The studied bioactive compounds were determined by employing the UHPLC-UV-Vis method described in previous research [22]. Before their analysis, all the UAE extracts were filtered through a 0.20 μm nylon syringe filter (Membrane Solutions, Dallas, TX, United States) and diluted in Milli-Q water. The individual anthocyanins present in myrtle extracts were quantified in cyanidin equivalents by means of a regression curve of anthocyanidin standard, cyanidin chloride

($y = 252640.4136x - 28462.4337$; $R^2 = 0.9999$). The standards with a known concentration were prepared between 0.06 and 35 mg·L^{-1}. The limit of detection (LOD) (0.196 mg·L^{-1}) and the limit of quantification (LOQ) (0.653 mg·L^{-1}) were calculated as three and ten times respective to the standard deviation of the blank divided by the slope of the calibration curve. Assuming that the 11 anthocyanins have similar absorbance, and taking into account the molecular weight of each anthocyanin, a calibration curve was plotted for each anthocyanin present in myrtle, which allowed to quantify the compounds of interest. All the analyses were carried out in duplicate. Figure 8 shows the HPLC chromatogram that represents the eleven anthocyanins detected in the analyses.

Anthocyanins present in myrtle berries	R1	R2	R3	R4
Delphinidin 3,5-O-diglucoside	OH	OH	$C_6H_{11}O_5$	$C_6H_{11}O_5$
Delphinidin 3-O-glucoside	OH	OH	$C_6H_{11}O_5$	H
Cyanidin 3-O-galactoside	OH	H	$C_6H_{11}O_5$	H
Cyanidin 3-O-glucoside	OH	H	$C_6H_{11}O_5$	H
Cyanidin 3-O-arabinoside	OH	H	$C_5H_9O_4$	H
Petunidin 3-O-glucoside	OH	H	$C_{12}H_{21}O_9$	H
Delphinidin 3-O-arabinoside	OH	OH	$C_5H_9O_4$	H
Peonidin 3-O-glucoside	OCH$_3$	H	$C_6H_{11}O_5$	H
Malvidin 3-O-glucoside	OCH$_3$	OCH$_3$	$C_6H_{11}O_5$	H
Petunidin 3-O-arabinoside	OCH$_3$	OH	$C_5H_9O_4$	H
Malvidin 3-O-arabinoside	OCH$_3$	OCH$_3$	$C_5H_9O_4$	H

Figure 7. Radicals of the different anthocyanins present in myrtle berries.

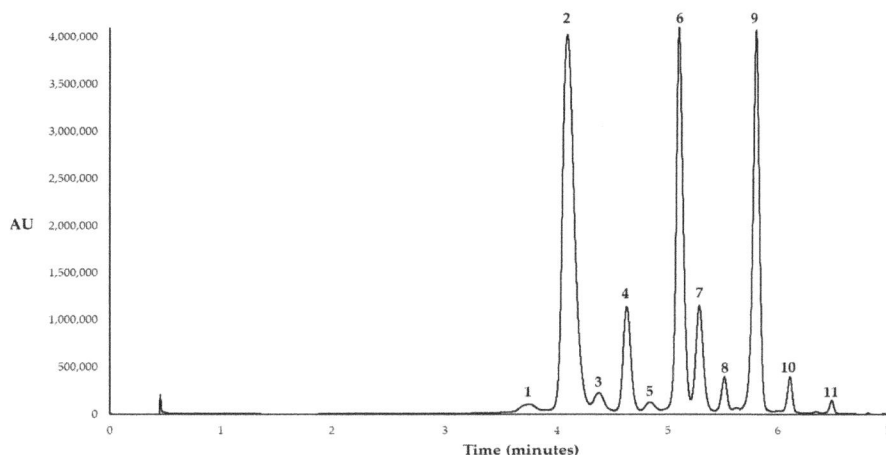

Figure 8. Chromatogram of the 11 anthocyanins identified in the UAE extracts from myrtle berries. Peak assignment: (1) Delphinidin 3,5-O-diglucoside; (2) delphinidin 3-O-glucoside; (3) cyanidin 3-O-galactoside; (4) cyanidin 3-O-glucoside; (5) cyanidin 3-O-arabinoside; (6) petunidin 3-O-glucoside; (7) delphinidin 3-O-arabinoside; (8) peonidin 3-O-glucoside; (9) malvidin 3-O-glucoside; (10) petunidin 3-O-arabinoside; (11) malvidin 3-O-arabinoside.

3.7. Application of Box-Behnken Design (BBD) to the Optimization of the Extraction Methods

In order to optimize the extraction variables, a response surface experiment (RSM) known as Box-Behnken (BBD) was carried out [56]. Box-Behnken design (BBD) is an independent rotatable quadratic design with no embedded factorial or fractional factorial points. The variable combinations are at the midpoint of the edges and at the center of the space [57]. It is useful because it allows one to avoid carrying out experiments under extreme conditions and, therefore, the possibility of deceiving results [58]. When this statistical experiment design is employed in conjunction with a response surface methodology (RSM) the effects of six independent factors on each response can be studied. The independent factors studied were: Solvent composition (% methanol in water) (X_1), solvent pH (X_2), extraction temperature (X_3), ultrasound amplitude (X_4), cycle (X_5), and a solvent-sample ratio (X_6). For each variable, there are three levels, coded as −1 (low), 0 (central point or middle), and +1 (high). Specifically, the studied ranges were as follows: Solvent composition: 50, 75, 100% for phenolic compounds and 25, 50, 75% for anthocyanins; temperature: 10, 35, 60 °C; amplitude: 30, 50, 70%; cycle: 0.2, 0.45, 0.7 s; pH: 2, 4.5, 7 and solvent-sample ratio: 10:0.5, 15:0.5, 20:0.5 mL:g. The ranges for the study were selected taking into account previous experiences by the research team. The response variables studied were: The experimental results for total phenolic compounds (Y_{TP}, mg·g^{-1}) and the experimental results for total anthocyanins (Y_{TA}, mg·g^{-1}). The design consisted of 54 treatments with six repetitions at the center point. All the trials were performed in random order. The whole experimental design matrix used can be seen in Table 5. My-9 from San Jose del Valle was the myrtle sample used for the optimization procedure.

Table 5. Experimental and predicted values for total phenolic compounds and total anthocyanins contents based on Box–Behnken design.

Run	Factors						Responses			
	X_1	X_2	X_3	X_4	X_5	X_6	Y_{TP} (mg·g^{-1})		Y_{TA} (mg·g^{-1})	
							Experimental	Predicted	Experimental	Predicted
1	0	0	−1	0	−1	−1	55.4367	51.0831	20.7595	23.8541
2	0	0	1	0	−1	−1	48.6499	49.6621	29.8821	25.4294
3	0	0	−1	0	1	−1	51.4343	48.3304	29.0606	31.2985
4	0	0	1	0	1	−1	41.6667	42.334	27.9193	26.3493
5	0	0	−1	0	−1	1	46.1649	45.0747	27.6777	27.6226
6	0	0	1	0	−1	1	47.6107	51.1375	25.6501	25.0375
7	0	0	−1	0	1	1	50.4893	49.0542	28.4150	31.2426
8	0	0	1	0	1	1	45.7652	50.5417	23.6023	22.1329
9	0	−1	0	−1	−1	0	57.1168	57.3752	28.5256	24.7846
10	0	1	0	−1	−1	0	44.7735	48.3304	26.8602	29.2928
11	0	−1	0	1	−1	0	41.9739	53.0361	27.755	23.6113
12	0	1	0	1	−1	0	41.3703	46.5404	26.6112	23.4784
13	0	−1	0	−1	1	0	57.8351	55.2335	27.0548	34.781
14	0	1	0	−1	1	0	71.3186	57.688	27.1787	26.729
15	0	−1	0	1	1	0	41.3184	40.33	28.5542	30.715
16	0	1	0	1	1	0	48.1604	45.3335	18.8744	18.0219
17	−1	0	−1	−1	0	0	37.8811	41.5155	14.3354	19.5394
18	1	0	−1	−1	0	0	39.2436	41.1874	19.2532	42.8583
19	−1	0	1	−1	0	0	48.7908	48.0444	18.0553	17.6695
20	1	0	1	−1	0	0	43.5972	51.7521	21.1111	22.2834
21	−1	0	−1	1	0	0	54.6703	45.6802	17.6946	11.5099
22	1	0	−1	1	0	0	35.7746	37.3562	20.6987	26.097
23	−1	0	1	1	0	0	37.9611	35.1821	10.1680	24.5504
24	1	0	1	1	0	0	33.6931	30.8939	20.6241	20.4326
25	0	−1	−1	0	0	−1	39.6751	44.4043	25.5771	30.4887
26	0	1	−1	0	0	−1	42.5789	48.6886	26.0578	25.0639
27	0	−1	1	0	0	−1	43.7849	42.272	27.3046	29.0925
28	0	1	1	0	0	−1	42.0191	43.4037	24.5787	23.0862
29	0	−1	−1	0	0	1	47.4523	46.4907	27.6042	30.7219
30	0	1	−1	0	0	1	39.3819	41.3177	28.7060	28.5433
31	0	−1	1	0	0	1	58.3748	51.8422	25.7965	25.1653
32	0	1	1	0	0	1	48.6688	43.5167	28.9420	22.4052
33	−1	−1	0	0	−1	0	58.0268	57.5556	22.9897	19.5292
34	1	−1	0	0	−1	0	41.2425	32.9846	19.5373	27.3024

Table 5. *Cont.*

Run	\multicolumn{6}{c}{Factors}	\multicolumn{4}{c}{Responses}
	X_1	X_2	X_3	X_4	X_5	X_6	Y_{TP} (mg·g^{-1}) Experimental	Predicted	Y_{TA} (mg·g^{-1}) Experimental	Predicted
35	−1	1	0	0	−1	0	38.5507	36.2848	24.6490	32.9035
36	1	1	0	0	−1	0	46.8626	38.7149	20.2517	18.3034
37	−1	−1	0	0	1	0	35.7902	41.3694	17.7105	15.0653
38	1	−1	0	0	1	0	34.6257	34.3231	17.7143	48.8664
39	−1	1	0	0	1	0	20.7714	31.5978	19.0511	15.8794
40	1	1	0	0	1	0	48.5130	51.5526	19.2534	27.3073
41	−1	0	0	−1	0	−1	49.3362	46.1577	18.9945	14.4979
42	1	0	0	−1	0	−1	49.0123	49.4397	21.8380	28.2914
43	−1	0	0	1	0	−1	30.0802	28.3987	16.72389	13.4627
44	1	0	0	1	0	−1	24.1848	23.6849	20.7428	18.5244
45	−1	0	0	−1	0	1	35.7748	35.4395	16.4343	13.6402
46	1	0	0	−1	0	1	33.0203	35.537	19.5057	27.7794
47	−1	0	0	1	0	1	44.0929	44.5007	14.9678	13.5268
48	1	0	0	1	0	1	34.2591	36.6024	19.4501	18.9343
49	0	0	0	0	0	0	37.2240	45.8533	27.1667	23.3422
50	0	0	0	0	0	0	52.1910	45.8533	27.2028	23.3422
51	0	0	0	0	0	0	47.5538	45.8533	17.7051	23.3422
52	0	0	0	0	0	0	50.5994	45.8533	14.0658	23.3422
53	0	0	0	0	0	0	43.5576	45.8533	26.5943	23.3422
54	0	0	0	0	0	0	43.9939	45.8533	27.3184	23.3422

The results for total phenolic compounds and total anthocyanins contents were entered into a polynomial equation. The response of the total phenolic compounds and the anthocyanins obtained in each of the experiments was entered into a second-order polynomial equation in order to correlate the relationship between the independent variables and the response (Equation (5)):

$$Y = \beta_0 + \sum_{i=1}^{k} \beta_i X_i + \beta_{ii} X_i^2 + \sum_{i}\sum_{i=1}^{k} \beta_{ij} X_i X_j + r \tag{5}$$

where Y is the predicted response (Y_{TP} and Y_{TA}); β_0 is the model constant; X_i and X_j are the independent variables; β_i are the linear coefficients; β_{ij} are the coefficients corresponding to the interactions; β_{ii} are the quadratic coefficients and r is the pure error sum of squares.

Design Expert software 11 (Trial Version, Stat-Ease Inc., Minneapolis, MN, USA) was the software employed for experimental design, the data analysis, and the model building. The statistical significance of the model, lack of fit, and regression terms were evaluated based on the analysis of variance (ANOVA).

The results of applying the extraction method to different myrtle ecotypes were studied using a multivariate analysis, hierarchical clustering analysis (HCA). Ward's method and the Euclidean square distance, were employed. Statgraphic Centurion XVII (Statgraphics Technologies, Inc., The Plains, VA, United States) was the software used.

4. Conclusions

This work has successfully developed quick and effective methods to extract bioactive compounds, such as anthocyanins and total phenolics from myrtle (Myrtus communis L.) pulp. A thorough search in the relevant literature showed that this is the first study in which UAE has been optimized for the extraction of phenolic compounds and anthocyanins from myrtle berries. The following optimal UAE conditions have been determined to extract the phenolic compounds: 92.8% methanol in water, 6.8 pH, 60 °C temperature, 65.48% ultrasound amplitude, 0.2 s cycle, and 10:0.5 as the optimum solvent-sample ratio. With regards to anthocyanins, optimal UAE conditions were: 74.1% methanol in water, 7 pH, 10 °C temperature, 30% ultrasound amplitude, 0.3 s cycle, and 18:0.5 as the optimum solvent-sample ratio. The optimum extraction times were only 5 and 2 min for phenolic compounds and anthocyanins, respectively. Both extraction methods presented satisfactory intra-day repeatability and inter-day repeatability (CV < 5%). The methods were applied to 14 different myrtle ecotypes.

The hierarchical cluster analysis (HCA), showed a correlation between the bioactive composition (total phenolic compounds and total and individual anthocyanins contents in the extracts) and the ecotypes' geographical area of origin. In conclusion, the results have indicated that UAE is a feasible alternative to conventional methods for the extraction of valuable components from myrtle berries. These results would mean a substantial improvement at the industrial level, since they would allow the manufacturers to quickly determine the quality of the raw materials and save costs. Furthermore, UAE results were compared to those achieved by MAE. The proposed UAE method proved to be an effective procedure to extract the bioactive compounds in myrtle berries, and a particularly efficient alternative for the extraction of anthocyanins.

Author Contributions: Conceptualization, E.E.-B. and G.F.B.; methodology, A.V.G.d.P. and M.V.-E.; software, M.F.-G.; formal analysis, A.V.G.d.P., M.V.-E and A.A.-A.; investigation, A.V.G.d.P. and M.V.-E.; resources, A.J.-C. and M.P.; data curation, E.E.-B., M.F.-G. and G.F.B.; writing—original draft preparation, A.V.G.d.P. and M.V.-E.; writing—review and editing, G.F.B. and E.E.-B.; supervision, E.E.-B. and G.F.B.; project administration, G.F.B. and E.E.-B.

Funding: This research was funded by the University of Cadiz and V. la Andaluza (Project OT2017/032).

Conflicts of Interest: The authors declare no conflict of interest.

References

1. Tian, Y.; Puganen, A.; Alakomi, H.L.; Uusitupa, A.; Saarela, M.; Yang, B. Antioxidative and antibacterial activities of aqueous ethanol extracts of berries, leaves, and branches of berry plants. *Food Res. Int.* **2018**, *106*, 291–303. [CrossRef] [PubMed]

2. Li, D.; Ye, Q.; Jiang, L.; Luo, Z. Effects of nano-TiO₂ packaging on postharvest quality and antioxidant activity of strawberry (Fragaria × ananassa Duch.) stored at low temperature. *J. Sci. Food Agric.* **2017**, *97*, 1116–1123. [CrossRef] [PubMed]

3. Łata, B.; Trampczynska, A.; Paczesna, J. Cultivar variation in apple peel and whole fruit phenolic composition. *Sci. Hortic.* **2009**, *121*, 176–181. [CrossRef]

4. Ahmad, N.; Zuo, Y.; Lu, X.; Anwar, F.; Hammed, S. Anwar. Characterization of free and conjugated phenolic compounds in fruits of selected wild plants. *Food Chem.* **2016**, *190*, 80–89. [CrossRef] [PubMed]

5. Bamba, B.S.B.; Shi, J.; Tranchant, C.C.; Xue, S.J.; Forney, C.F.; Lim, L.-T. Influence of Extraction Conditions on Ultrasound-Assisted Recovery of Bioactive Phenolics from Blueberry Pomace and Their Antioxidand Activity. *Molecules* **2018**, *23*, 1685. [CrossRef] [PubMed]

6. Wannes, A.; Mhamdi, B.; Sriti, J.; Ben Jemia, M.; Ouchikh, O.; Hamdaoui, G.; Kchouk, M.E.; Marzouk, B. Antioxidant activities of the essential oils and methanol extracts from myrtle (*Myrtus communis* var. italica L.) leaf, stem and flower. *Food Chem. Toxicol.* **2010**, *48*, 1362–1370. [CrossRef] [PubMed]

7. Limwachiranon, J.; Jiang, L.; Huang, H.; Sun, J.; Luo, Z. Improvement of phenolic compounds extraction from high-starch lotus (*Nelumbo nucifera* G.) seed kernels using glycerol: New insights to amylose/amylopectin—Phenolic relationships. *Food Chem.* **2019**, *274*, 933–941. [CrossRef] [PubMed]

8. Pereira, P.C.; Cebola, M.-J.; Bernardo-Gil, M.G. Evolution of the Yields and Composition of Essential Oil from Portuguese Myrtle (*Myrtus comunis* L.) through the Vegetative Cycle. *Molecules* **2009**, *14*, 3094–3105. [CrossRef] [PubMed]

9. Yildirim, F.; San, B.; Yildirim, A.N.; Polat, M.; Ercişli, S. Mineral Composition of Leaves and Fruit in Some Myrtle (*Myrtus communis* L.). *Genotypes. Erwerbs Obstbau* **2015**, *57*, 149–152. [CrossRef]

10. Kuskoski, E.M.; Asuero, A.G.; Troncoso, A.M.; Mancini-Filho, J.; Fett, R. Aplicación de diversos métodos químicos para determinar actividad antioxidante en pulpa de frutos. *Ciênc. Tecnol. Aliment.* **2005**, *25*, 726–732. [CrossRef]

11. Maldini, M.; Chessa, M.; Petretto, G.L.; Montoro, P.; Rourke, J.P.; Foddai, M.; Nicoletti, M.; Pintore, G. Profiling and Simultaneous Quantitative Determination of Anthocyanins in Wild *Myrtus communis* L. Berries from Different Geographical Areas in Sardinia and their Comparative Evaluation. *Phytochem. Anal.* **2016**, *27*, 249–256. [CrossRef] [PubMed]

12. Babou, L.; Hadidi, L.; Grosso, C.; Zaidi, F.; ValentÃ£o, P.; Andrade, P.B. Study of phenolic composition and antioxidant activity of myrtle leaves and fruits as a function of maturation. *Eur. Food Res. Technol.* **2016**, *242*, 1447–1457. [CrossRef]

13. Tuberos, C.I.G.; Rosa, A.; Bifulco, E.; Melis, M.P.; Atseri, A.; Pirisi, F.M.; Dessi, M.A. Chemical composition and antioxidant activities of *Myrtus communis* L. berries extracts. *Food Chem.* **2010**, *123*, 1242–1251. [CrossRef]

14. Scorrano, S.; Lazzoi, M.R.; Mergola, L.; Di Bello, M.P.D.; Del Sole, R.; Vasapollo, G. Anthocyanins Profile by Q-TOF LC/MS in *Myrtus communis* Berries from Salento Area. *Food Anal. Methods* **2017**, *10*, 2404–2411. [CrossRef]

15. Tuberoso, C.I.G.; Melis, M.; Angioni, A.; Pala, M.; Cabras, P. Myrtle hydroalcoholic extracts obtained from different selections of *Myrtus communis* L. *Food Chem.* **2007**, *101*, 806–811. [CrossRef]

16. Dahmoune, F.; Nayak, B.; Moussi, K.; Remini, H.; Madani, K. Optimization of microwave-assisted extraction of polyphenols from *Myrtus communis* L. leaves. *Food Chem.* **2015**, *166*, 585–595. [CrossRef] [PubMed]

17. Messaoud, C.; Boussaid, M. *Myrtus communis* berry color morphs: A comparative analysis of essential oils, fatty acids, phenolic compounds, and antioxidant activities. *Chem. Biodiv.* **2011**, *8*, 300–310. [CrossRef] [PubMed]

18. Serreli, G.; Jerković, I.; Gil, K.A.; Marijanović, Z.; Pacini, V.; Tuberoso, C.I.G. Phenolic Compounds, Volatiles and Antioxidant Capacity of White Myrtle Berry Liqueurs. *Plant Foods Hum. Nutr.* **2017**, *72*, 205–210. [CrossRef] [PubMed]

19. Aidi Wannes, W.; Marzouk, B. Differences between myrtle fruit parts (*Myrtus communis* var. *italica*) in phenolics and antioxidant contents. *J. Food Biochem.* **2013**, *37*, 585–594. [CrossRef]

20. Pereira, P.; Cebola, M.J.; Oliveira, M.C.; Bernardo-Gil, M.G. Supercritical fluid extraction vs conventional extraction of myrtle leaves and berries: Comparison of antioxidant activity and identification of bioactive compounds. *J. Supercrit. Fluids.* **2016**, *113*, 1–9. [CrossRef]

21. Picó, Y. Ultrasound-assisted extraction for food and environmental samples. *TrAC Trends Anal. Chem.* **2013**, *43*, 84–99. [CrossRef]

22. González de Peredo, A.V.; Vázquez-Espinosa, M.; Espada-Bellido, E.; Jiménez-Cantizano, A.; Ferreiro-González, M.; Amores-Arrocha, A.; Barbero, G.F. Development of new analytical microwave-assisted extraction methods for bioactive compounds from myrtle (*myrtus communis* L.). *Molecules* **2018**, *23*, 2992. [CrossRef] [PubMed]

23. Mason, T.J.; Paniwnyk, L.; Lorimer, K.P. The uses of ultrasound in food technology. *Ultrason. Sonochem.* **1996**, *3*, S253–S260. [CrossRef]

24. Briones-Labarca, V.; Plaza-Morales, M.; Giovagnoli-Vicuña, C.; Jamett, F. High hydrostatic pressure and ultrasound extractions of antioxidant compounds, sulforaphane and fatty acids from Chilean papaya (*Vasconcellea pubescens*) seeds: Effects of extraction conditions and methods. *LWT Food Sci. Technol.* **2015**, *60*, 525–534. [CrossRef]

25. Espada-Bellido, E.; Ferreiro-González, M.; Carrera, C.; Palma, M.; Barroso, C.G.; Barbero, G.F. Optimization of the ultrasound-assisted extraction of anthocyanins and total phenolic compounds in mulberry (*Morus nigra*) pulp. *Food Chem.* **2017**, *219*, 23–32. [CrossRef] [PubMed]

26. Khan, M.K.; Abert-Vian, M.; Fabiano-Tixier, A.S.; Dangles, O.; Chemat, F. Ultrasound-assisted extraction of polyphenols (*flavanone glycosides*) from orange (*Citrus sinensis* L.) peel. *Food Chem.* **2010**, *119*, 851–858. [CrossRef]

27. Feng, S.; Luo, Z.; Tao, B.; Chen, C. Ultrasonic-assisted extraction and purification of phenolic compounds from sugarcane (*Saccharum officinarum* L.) rind. *LWT Food Sci. Technol.* **2015**, *60*, 970–976. [CrossRef]

28. Sarais, G.; D'Urso, G.; Lai, C.; Pirisi, F.M.; Pizza, C.; Montoro, P. Targeted and untargeted mass spectrometric approaches in discrimination between *Myrtus communis* cultivars from Sardinia region. *J. Mass Spectrom.* **2016**, *51*, 704–715. [CrossRef] [PubMed]

29. Pereira, P.; Cebola, M.J.; Oliveira, M.C.; Bernardo Gil, M.G. Antioxidant capacity and identification of bioactive compounds of *Myrtus communis* L. extract obtained by ultrasound-assisted extraction. *J. Food Sci. Technol.* **2017**, *54*, 4362–4369. [CrossRef] [PubMed]

30. Bridgers, E.N.; Chinn, M.S.; Truong, V.D. Extraction of anthocyanins from industrial purple-fleshed sweetpotatoes and enzymatic hydrolysis of residues for fermentable sugars. *Ind. Crops Prod.* **2010**, *32*, 613–620. [CrossRef]

31. Hashemi, S.M.B.; Michiels, J.; Joris, A.Y.; Hosseini, M. Kolkhoung (*Pistacia khinjuk*) kernel oil quality is affected by different parameters in pulsed ultrasound-assisted solvent extraction. *Ind. Crops Prod.* **2015**, *70*, 28–33. [CrossRef]

32. Kazemi, M.; Karim, R.; Mirhosseini, H.; Abdul Hamid, A. Optimization of pulsed ultrasound-assisted technique for extraction of phenolics from pomegranate peel of Malas variety: Punicalagin and hydroxybenzoic acids. *Food Chem.* **2016**, *206*, 156–166. [CrossRef] [PubMed]

33. You, Q.; Yin, X.; Ji, C. Pulsed counter-current ultrasound-assisted extraction and characterization of polysaccharides from *Boletus edulis*. *Carbohydr. Polym.* **2014**, *101*, 379–385. [CrossRef] [PubMed]

34. Machado, A.P.D.F.; Pereira, A.L.D.; Barbero, G.F.; Martínez, J. Recovery of anthocyanins from residues of *Rubus fruticosus*, *Vaccinium myrtillus* and *Eugenia brasiliensis* by ultrasound assisted extraction, pressurized liquid extraction and their combination. *Food Chem.* **2017**, *231*, 1–10. [CrossRef] [PubMed]

35. Kidak, R.; Ince, N.H. Ultrasonic destruction of phenol and substituted phenols: A review of current research. *Ultrason. Sonochem.* **2006**, *13*, 195–199. [CrossRef] [PubMed]

36. Machado, A.P.D.F.; Pasquel-Reátegui, J.L.; Barbero, G.F.; Martínez, J. Pressurized liquid extraction of bioactive compounds from blackberry (*Rubus fruticosus* L.) residues: A comparison with conventional methods. *Food Res. Int.* **2015**, *77*, 675–683. [CrossRef]

37. Mustafa, A.; Turner, C. Pressurized liquid extraction as a green approach in food and herbal plants extraction: A review. *Anal. Chim. Acta* **2011**, *703*, 8–18. [CrossRef] [PubMed]

38. Cavalcanti, R.N.; Santos, D.T.; Meireles, M.A.A. Non-thermal stabilization mechanisms of anthocyanins in model and food systems-An overview. *Food Res. Int.* **2011**, *44*, 499–509. [CrossRef]

39. Carrera, C.; Ruiz-Rodriguez, A.; Palma, M.; Barroso, C.G. Ultrasound assisted extraction of phenolic compounds from grapes. *Anal. Chim. Acta* **2012**, *732*, 100–104. [CrossRef] [PubMed]

40. Pereira, D.T.V.; Tarone, A.G.; Cazarin, C.B.B.; Barbero, G.F.; Martínez, J. Pressurized liquid extraction of bioactive compounds from grape marc. *J. Food Eng.* **2019**, *240*, 105–113. [CrossRef]

41. Tiwari, B.K.; O'Donnell, C.P.; Cullen, P.J. Effect of non thermal processing technologies on the anthocyanin content of fruit juices. *Trends Food Sci. Technol.* **2009**, *20*, 137–145. [CrossRef]

42. Tiwari, B.K.; Patras, A.; Brunton, N.; Cullen, P.J.; O'Donnell, C.P. Effect of ultrasound processing on anthocyanins and color of red grape juice. *Ultrason. Sonochem.* **2009**, *17*, 598–604. [CrossRef] [PubMed]

43. Fleschhut, J.; Kratzer, F.; Rechkemmer, G.; Kulling, S.E. Stability and biotransformation of various dietary anthocyanins in vitro. *Eur. J. Nutr.* **2006**, *45*, 7–18. [CrossRef] [PubMed]

44. Stipcovich, T.; Barbero, G.F.; Ferreiro-González, M.; Palma, M.; Barroso, C.G. Fast analysis of capsaicinoids in Naga Jolokia extracts (*Capsicum chinense*) by high-performance liquid chromatography using fused core columns. *Food Chem.* **2018**, *239*, 217–224. [CrossRef] [PubMed]

45. AOAC International (Ed.) AOAC Peer Verified Methods Program. In *Manual on Policies and Procedures*; AOAC International: Arlingt, MD, USA, 1998.

46. González-Varo, J.P.; Albaladejo, R.G.; Aparicio, A. Mating patterns and spatial distribution of conspecific neighbours in the Mediterranean shrub *Myrtus communis* (Myrtaceae). *Plant Ecol.* **2009**, *203*, 207–215. [CrossRef]

47. Mulas, M. Germplasm Evaluation of Spontaneous Myrtle (*Myrtus communis* L.) for Cultivar Selection and Crop Development. *J. Herbs Spices Med. Plants* **1999**, *6*, 31–49. [CrossRef]

48. Ghafoor, K.; Choi, Y.H.; Jeon, J.Y.; Jo, I.H. Optimization of ultrasound- assisted extraction of phenolic compounds, antioxidants, and anthocyanins from grape (*Vitis vinifera*) seeds. *J. Agric. Food Chem.* **2009**, *57*, 4988–4994. [CrossRef] [PubMed]

49. Casazza, A.; Aliakbarian, B.; Mantegna, S.; Cravotto, G.; Perego, P. Extraction of phenolics from Vitis vinifera wastes using non-conventional techniques. *J. Food Eng.* **2010**, *100*, 50–55. [CrossRef]

50. Wijngaard, H.; Hossain, M.B.; Rai, D.K.; Brunton, N. Techniques to extract bioactive compounds from food by-products of plant origin. *Food Res. Int.* **2012**, *46*, 505–513. [CrossRef]

51. Tušek, A.J.; Benković, M.; Cvitanović, A.B.; Valinger, D.; Jurina, T.; Kljusurić, J.G. Kinetics and thermodynamics of the solid-liquid extraction process of total polyphenols, antioxidants and extraction yield from Asteraceae plants. *Ind. Crops Prod.* **2016**, *91*, 205–214. [CrossRef]

52. Singleton, V.L.; Orthofer, R.; Lamuela-Raventós, R.M. Analysis of total phenols and other oxidation substrates and antioxidants by means of folin-ciocalteu reagent. *Methods Enzymol.* **1999**, *299*, 152–178. [CrossRef]

53. Vallverdú-Queralt, A.; Medina-Remón, A.; Andres-Lacueva, C.; Lamuela-Raventos, R.M. Changes in phenolic profile and antioxidant activity during production of diced tomatoes. *Food Chem.* **2011**, *126*, 1700–1707. [CrossRef] [PubMed]

54. Li, Y.; Skouroumounis, G.K.; Elsey, G.M.; Taylor, D.K. Microwave-assistance provides very rapid and efficient extraction of grape seed polyphenols. *Food Chem.* **2011**, *129*, 570–576. [CrossRef] [PubMed]

55. Batista, Â.G.; Ferrari, A.S.; Da Cunha, D.C.; da Silva, J.K.; Cazarin, C.B.B.; Correa, L.C.M.; Prado, M.A.; de Carvalho-Silva, L.B.; Esteves, E.A.; Junior, M.R.M. Polyphenols, antioxidants, and antimutagenic effects of *Copaifera langsdorffii* fruit. *Food Chem.* **2016**, *197*, 1153–1159. [CrossRef] [PubMed]

56. Zhou, Y.; Zheng, J.; Gan, R.-Y.; Zhou, T.; Xu, D.-P.; Li, H.-B. Optimization of Ultrasound-Assisted Extraction of Antioxidants from the Mung Bean Coat. *Molecules* **2017**, *22*, 638. [CrossRef] [PubMed]

57. Maran, J.; Manikandan, S.; Thirugnanasambandham, K.; Vigna Nivetha, C.; Dinesh, R. Box-Behnken design based statistical modeling for ultrasound-assisted extraction of corn silk polysaccharide. *Carbohydr. Polym.* **2013**, *92*, 604–611. [CrossRef] [PubMed]

58. Ferreira, S.L.C.; Bruns, R.E.; Ferreira, H.S.; Matos, G.D.; David, J.M.; Brandao, G.C.; da Silva, E.G.P.; Portugal, L.A.; dos Reis, P.S.; Souza, A.S.; et al. Box-Behnken design: An alternative for the optimization of analytical methods. *Anal. Chim. Acta* **2007**, *597*, 179–186. [CrossRef] [PubMed]

Sample Availability: Samples with the compounds delphinidin 3,5-*O*-diglucoside, delphinidin 3-*O*-glucoside, cyanidin 3-*O*-galactoside, cyanidin 3-*O*-glucoside, cyanidin 3-*O*-arabinoside, petunidin 3-*O*-glucoside, delphinidin 3-*O*-arabinoside, peonidin 3-*O*-glucoside, malvidin 3-*O*-glucoside, petunidin 3-*O*-arabinoside and malvidin 3-*O*-arabinoside are available from the authors.

molecules

MDPI

Article

Flavor Profile Evolution of Bottle Aged Rosé and White Wines Sealed with Different Closures

Meng-Qi Ling [1,2], Han Xie [1,2], Yu-Bo Hua [3], Jian Cai [4], Si-Yu Li [1,2], Yi-Bin Lan [1,2],
Ruo-Nan Li [1,2], Chang-Qing Duan [1,2] and Ying Shi [1,2,*]

[1] Center for Viticulture & Enology, College of Food Science and Nutritional Engineering,
    China Agricultural University, Beijing 100083, China; knighhtt@cau.edu.cn (M.-Q.L.);
    m18800135515@163.com (H.X.); siyuli@cau.edu.cn (S.-Y.L.); clanyibin@gmail.com (Y.-B.L.);
    liruonan0011@163.com (R.-N.L.); chqduan@cau.edu.cn (C.-Q.D.)
[2] Key Laboratory of Viticulture and Enology, Ministry of Agriculture and Rural Affairs, Beijing 100083, China
[3] Shandong Taila Winery Co., Ltd., Weihai 264500, China; huayubo-111@163.com
[4] College of Biological Resources and Food Engineering, Qujing Normal University, Qujing 655011, China;
    caijian928@outlook.com
*   Correspondence: shiy@cau.edu.cn; Tel.: +86-1062-7373-04

check for
updates

Received: 12 February 2019; Accepted: 22 February 2019; Published: 27 February 2019

Abstract: Bottle aging is the final stage before wines are drunk, and is considered as a maturation time when many chemical changes occur. To get a better understanding of the evolution of wines' flavor profile, the flavor compounds (phenolic and volatile compounds), dissolved oxygen (DO), and flavor characters (OAVs and chromatic parameters) of rosé and dry white wines bottled with different closures were determined after 18 months' bottle aging. The results showed the main phenolic change trends of rosé wines were decreasing while the trends of white wines were increasing, which could be the reason for their unique DO changing behaviors. Volatile compounds could be clustered into fluctuating, increasing, and decreasing groups using k-means algorithm. Most volatile compounds, especially some long-chain aliphatic acid esters (octanoates and decanoates), exhibited a lower decrease rate in rosé wines sealed with natural corks and white wines with screw caps. After 18 months of bottle aging, wines treated with natural corks and their alternatives could be distinguished into two groups based on flavor compounds via PLS-DA. As for flavor characters, the total intensity of aroma declined obviously compared with their initial counterparts. Rosé wines exhibit visual difference in color, whereas such a phenomenon was not observed in white wines.

Keywords: rosé wines; white wines; bottle aging; flavor profile; closures

1. Introduction

Wine flavor is composed of a wide variety of compounds with different organoleptic properties, which will slightly evolve during bottle aging due to the limited quantities of oxygen penetrating through the closures [1–4]. The increase of dissolved oxygen (DO) in wines means the replenishment of oxygen is higher than the consumption of oxygen by antioxidants (such as phenolic compounds), while the decrease of DO represents a relatively higher consumption of oxygen [5]. Different types of closures exhibit different abilities in preventing oxygen penetration due to their structural differences [6,7]. Natural corks are the traditional choice of closure in the wine industry, but other types of closures are also used by wine producers. Agglomerated corks and technical corks are made of offcuts of oak wood, thus can be named oak-based corks as opposed to natural corks. In comparison with oak-based corks, polymer synthetic plugs and screw caps are more economical and less dependent on the raw material limitation.

Bottle aging is an important period when wine's flavor characters must be preserved as much as possible. The main function of closures is to ensure a good seal and to prevent any organoleptic deterioration of wine during storage. However, during bottle aging, various reactions may occur, such as oxidation, hydrolysis, and reactions caused by charge transfer and formation of covalent bonds, which will influence wine flavor evolution [6]. Wine aroma quality like the fruity and floral perception usually decreases due to the diminishment of critical aroma compounds including long-chain aliphatic acid ethyl esters, terpenes, and norisoprenoids [8,9]. Wine astringency also decreases because of a decline in the mean polymerization degree of tannins [10], while the hue and color stability usually increase due to the formation of stable orange-yellow pigments such as pyranoanthocyanins [11]. So far, most research observing wine flavor changes during bottle aging has focused on dry red wines [1,12], or is only concerned about the flavor quality, determined either by volatile compounds or non-volatile compounds [3,4,13,14]; comprehensive investigations monitoring the evolution of rosé and dry white wines' flavor profiles and oxidation patterns during bottle aging are quite limited.

The objectives of our research were: (1) understanding the evolution of flavor compounds of rosé and white wines during an 18-month bottle aging; (2) observing the oxidation pattern differences of rosé and white wines during an 18-month bottle aging and finding out the possible reasons that caused the differences; (3) comparing the effects of natural cork and its alternatives on the flavor profiles of rosé and white wines after an 18-month bottle aging.

2. Results and Discussion

2.1. Evolution of Phenolic Compounds and Dissolved Oxygen during Bottle Aging

Seventeen phenolic compounds were quantified in rosé wines (Figure 1a) and only four non-anthocyanin compounds were quantified, which might be due to the short maceration time during winemaking. Phenolic compounds in all rosé wine samples were clustered into two groups using k-means algorithm after normalization process (dividing the concentration by the maximum value of each compounds among all samples). A boxplot based on the normalized data was carried out to exhibit the change trend of phenolic compounds of each cluster, the fluctuating group (Cluster 1) and the decreasing group (Cluster 2) (Figure 1b). The majority of phenolic compounds belonged to Cluster 2, which was similar to previous reports on red wines [11,15]. Considering the limited non-anthocyanin phenolics detected in rosé wines, the decrease of major anthocyanins should not be strongly correlated with the formation of polymeric pigments [10]. Therefore, these compounds should participate in reactions such as oxidation and degradation [16]. Only caffeic acid and 4-hydroxycinnamic acid were detected in white wines, which increased from 1.07 ± 0.17 mg/L to 1.98 ± 0.26 mg/L and 0.97 ± 0.04 mg/L to 1.6 ± 0.11 mg/L, respectively (see Table S1), due to hydrolysis of their tartaric esters [10].

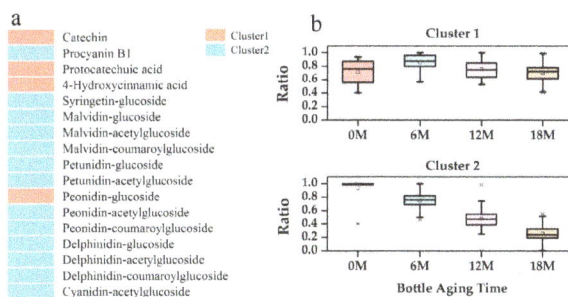

Figure 1. Statistical clustering for change trends of phenolic compounds in all rosé wines during an 18-month bottle aging. (**a**) General phenolic profile of rosé wines; (**b**) phenolic compounds' change trends.

The level of dissolved oxygen (DO) in wines mainly depends on oxygen concentration in the headspace at bottling and the ingress rate of oxygen into the bottle through closures [9]. During bottling, wines are exposed to air and thus have a chance to absorb oxygen. The oxidation substrates, generally phenolic compounds in wines, are positively correlated to wines' oxygen absorption capacity [6]. Rosé and white wines have unique DO evolution behaviors (Figure 2). In all rosé wines, the concentrations of DO decreased constantly during the first 10 months of bottle aging (Figure 2a), possibly because their major antioxidants, the phenolic compounds, were decreasing due to oxidation (Figure 1). However, in all white wines, the change trends of DO showed a slight decrease in the first two months, and then a drastic fluctuation (Figure 2b). The accumulation of DO meant the replenishment of oxygen was higher than the consumption. The oxygen dissolved in white wines probably came from the headspace oxygen, which was much higher in white wines (5.12 ± 0.76 mg/L) than in rosé wines (1.57 ± 0.73 mg/L). Moreover, caffeic acid and 4-hydroxycinnamic acid in white wines are good antioxidants. However, they might react slowly or not react at all until their concentrations reach a certain degree due to the hydrolysis of their tartaric esters, just like the formation of pinotins, a group of pyranoanthocyanins slowly formed by reactions between hydroxycinnamic acids and anthocyanins during wine aging [17]. After a four-month bottle aging, the accumulation of caffeic acid and 4-hydroxycinnamic acid was sufficient to trigger oxidation reactions, leading to a drastic decrease in DO (Figure 2b). Furthermore, white wines sealed with a screw cap had the lowest concentration of DO during bottle aging, which means a screw cap might provide better preservation for white wines. The DO concentrations were below the detection limit after 10 months of bottle aging for all rosé and white wines, indicating that the oxygen replenished from outside of the closures was far less than the oxygen consumption potential of wines. According to the results, we believe that although closures act as an oxygen barrier [18], DO's change trends during preservation depend more on the wine type and also the initial oxygen dissolved during the bottling process.

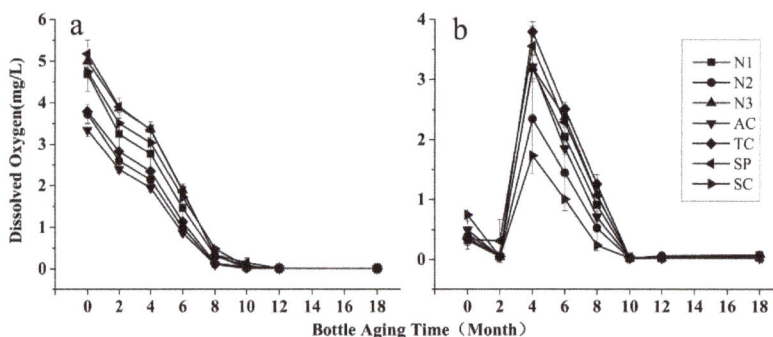

Figure 2. Change trends of DO in rosé (**a**) and white wines (**b**) during an 18-month bottle aging. (Symbols and abbreviations: N1: natural cork-1; N2: natural cork-2; N3: natural cork-3; AC: agglomerated cork; TC: '1+1' technical cork; SP: polymer synthetic plug; SC: screw cap.)

2.2. Evolution of Volatile Compounds during Bottle Aging

There were 72 volatile compounds quantified in this study, including 15 alcohols, 32 esters, five aliphatic acids, seven terpenes, three norisoprenoids and 10 other volatile compounds (Figure 3a). Detailed information is shown in Table S2. These two types of wines shared similar aromatic profiles, except there were more terpenes only quantified in whites (terpinolene, α-terpineol, and β-farnesene) to represent floral notes (Figure 3a). (E)-2-hexen-1-ol, 1-heptanol, benzyl alcohol, heptyl acetate, ethyl 2-hydroxy-4-methylpentanoate, benzaldehyde, and guaiacol were only quantified in rosé wines. Some volatile compounds, such as ethyl decanoate, n-decanoic acid, and dodecanoic acid, were reported to be associated with the oxidation of wines [8]. Wine oxidation was responsible for some unpleasant sensory descriptors ('green apple,' 'cooked potato,' and 'curry'), and could also interact with remaining pleasant aroma compounds, leading to the suppression of certain positive attributes [19]. The evolution of volatile compounds in wines during bottle aging was analyzed by k-means algorithm. It was clear that all compounds could be classified into three groups, namely, Cluster 1, Cluster 2, and Cluster 3, shown in the boxplot based on the normalized data from k-means algorithm (Figure 3b,c). Cluster 1 compounds stayed stable and were mostly alcohols, some ethyl esters, aliphatic acids, and a few terpenes (linalool and terpinen-4-ol). Compounds in Cluster 2 and 3 exhibited an increasing and decreasing trend, respectively, which indicated that they were the main changing compounds. Overall, more volatile compounds had a decreasing trend than an increasing trend during bottle aging. Almost half of the esters detected in our study were found in Cluster 3, which corroborates previous studies' finding that esters decreased during ageing due to hydrolysis [2,20]. By calculating the absolute values of the differences between final concentrations (after an 18-month bottle aging) and initial concentrations of those changing compounds, we used clustering analysis to differentiate wines sealed with natural corks and their alternatives (Figure 4).

For rosé wines, wines treated with similar closures (natural corks or alternatives of natural corks) did not show much consistency based on Cluster 2 compounds (Figure 4a), while Cluster 3 compounds could precisely distinguish wines with natural corks from their alternatives (Figure 4b). In Figure 4b, the difference between final concentration and initial concentration of Cluster 3 compounds showed less of a heat response in rosé wines sealed with natural corks than those sealed with alternative closures, which meant natural corks could better prevent aroma loss, especially for acetates (heptyl acetate, hexyl acetate, 3-hexen-1-ol, phenethyl acetate). β-Damascenone and (6E)-nerolidol decreased more in rosé wines sealed with a '1+1' technical cork and agglomerated cork, while the compounds in wines with a polymer synthetic plug seemed to have the largest decrease rate during bottle aging.

As for white wines, agglomerated cork and '1+1' technical cork were mixed with natural corks in a clustering analysis for compounds in Cluster 2 and 3, mainly because they were all oak-based closures (Figure 4c,d). In Cluster 2, α-ionone and TDN had a relatively lower increase rate in wines with natural corks compared to their alternatives, which might be due to the 'scalping phenomenon' of corks. Some studies have showed that cork and synthetic closures may scalp several aroma compounds from wine, such as TDN and methoxypyrazines [18] (Figure 4c). In Figure 4d, white wines with a screw cap had the most compounds that exhibited less heat response among all samples, which meant the screw cap was better ay preserving aroma quality in white wines. This might be due to there being the lowest level of DO concentration in white wines sealed with a screw cap (Figure 2b). It was those long-chain aliphatic acid esters that decreased less in both rosé and white wines sealed with a screw cap, such as ethyl decanoate, ethyl 9-decenoate, ethyl dodecanoate, ethyl hexadecanoate, isoamyl decanoate, isoamyl octanoate, and isobutyl octanoate, which could differentiate a screw cap from other closures. The result corresponded to a previous study that found that long-chain aliphatic acid esters would decrease with a higher oxidation level [8], while a screw cap seemed to be the best choice in terms of minimizing wine oxidation [1,9]. On the contrary, agglomerated cork and polymer synthetic plug could lead to more loss in the aroma compounds (Figure 4d).

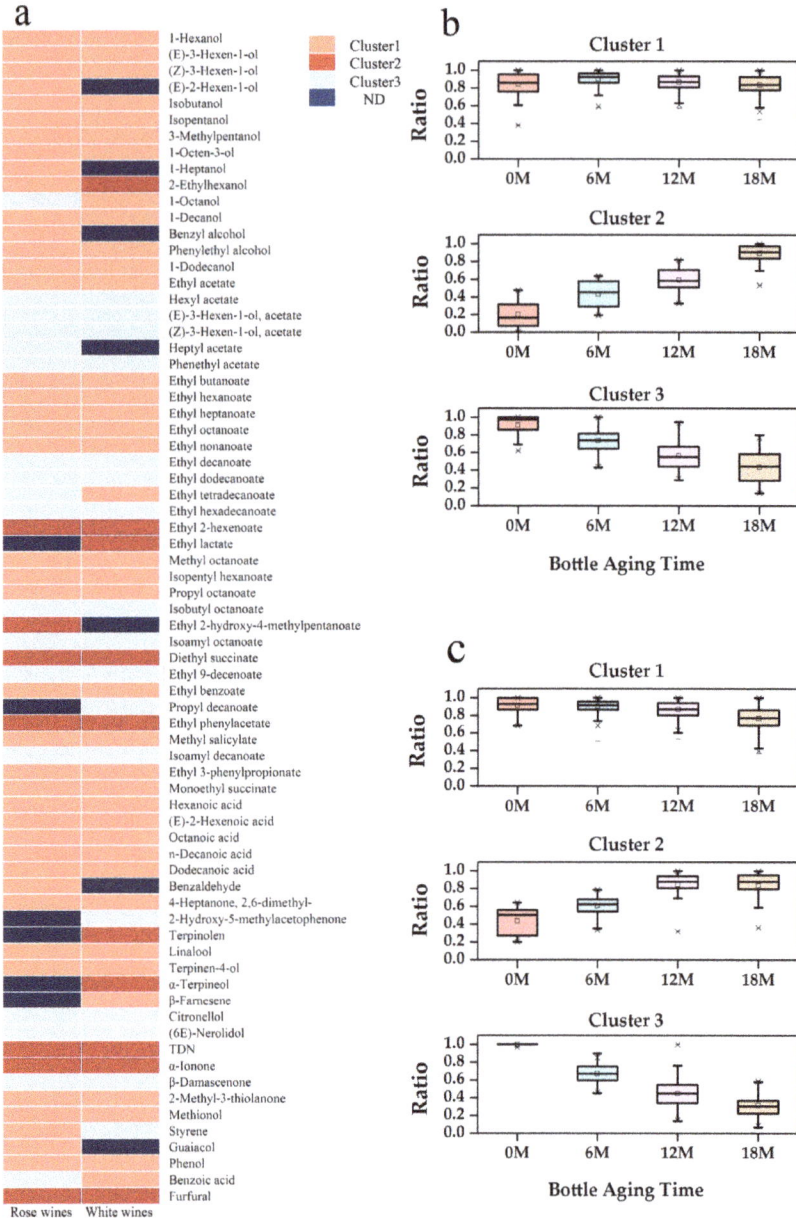

Figure 3. Change trends of volatile compounds during an 18-month bottle aging. (**a**) General volatile profile of rosé and white wines; (**b**) volatile compounds' change trends in rosé wines; (**c**) volatile compounds' change trends in white wines.

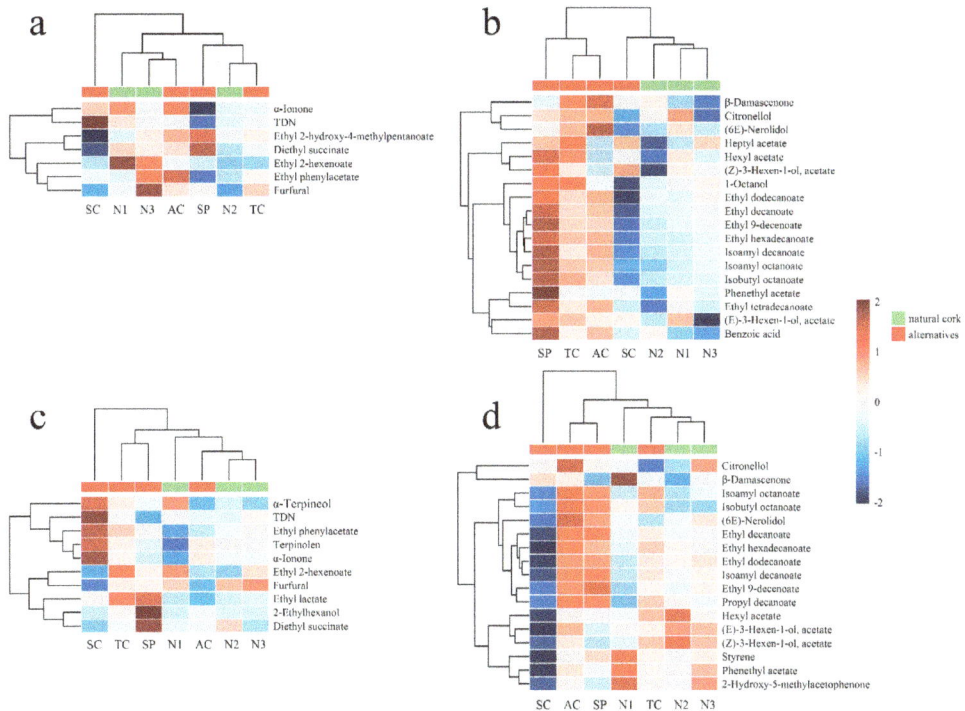

Figure 4. Clustering analysis of wines sealed with natural corks and their alternatives based on absolute values of the differences between final concentrations and initial concentrations. (**a**) Difference value of increasing volatile compounds in rosé wines; (**b**) difference value of decreasing volatile compounds in rosé wines; (**c**) difference value of increasing volatile compounds in white wines; (**d**) difference value of decreasing volatile compounds in white wines.

2.3. Flavor Profile Analysis of Wines after an 18-Month Bottle Aging

During bottle aging, wines are expected to maintain their original organoleptic characters as much as possible. In this section, partial least squares-discriminate analysis (PLS-DA) of different flavor compounds (phenolic and volatile compounds) was conducted in wines after an 18-month bottle aging. Wines sealed with three natural corks (N1, N2, N3) were set up as a group, while wines sealed with other closures (AC, TC, SP, SC) were set up as a single group of natural cork alternatives to see the effect of natural corks and their alternatives on wines' flavor profile (Figure 5). The use and interpretation of the VIP (variable importance in projection) values obtained from the PLS-DA model make it possible to determine potential flavor markers in the classes selected. Besides calculating the VIP values in PLS-DA, a one-way ANOVA test was also applied to check the statistical differences between different flavor compounds and guarantee a more statistically reliable selection of critical flavor markers (Table 1). Moreover, by calculating OAVs and chromatic parameters in all initial wines (prior to bottle aging) and final wines (after an 18-month bottle aging), the aromatic and chromatic characters of these wines were also compared.

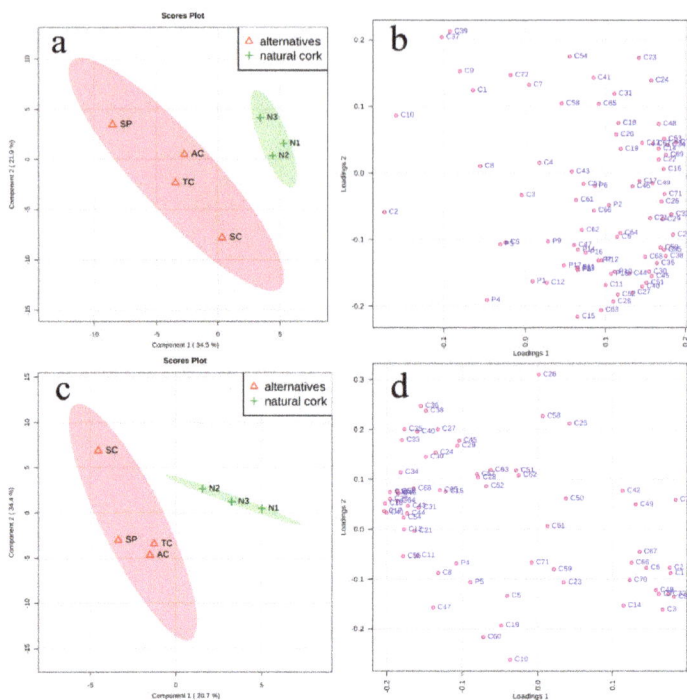

Figure 5. PLS-DA of flavor compounds in 18-month bottle-aged wines treated with natural corks and their alternatives. (**a**) PLS-DA model for rosé wine differentiation; (**b**) scattering plot of PLS-DA model for rosé wines; (**c**) PLS-DA model for white wine differentiation; (**d**) scattering plot of PLS-DA model for white wines. Note: Flavor compounds' numbers in (**b**,**d**) are provided in Tables S1,S2.

Table 1. Differentiated flavor compounds in rosé and white wines (VIP > 1, p < 0.05).

No. [a]	CAS	Compounds [b]	'Cabernet Sauvignon' Rosé Wines			'Chardonnay' Dry White Wines		
			VIP1	VIP2	p values	VIP1	VIP2	p values
C10	104-76-7	2-Ethylhexanol	1.145	1.055	**0.024**	1.045	1.229	**7.07×10^{-7}**
C12	112-30-1	1-Decanol	0.286	0.617	0.001	1.443	1.022	**0.035**
C24	106-30-9	Ethyl heptanoate	1.876	1.720	**0.022**	0.348	1.025	0.992
C29	124-06-1	Ethyl tetradecanoate	1.404	1.245	**0.022**	0.155	0.944	0.029
C33	111-11-5	Methyl octanoate	1.266	1.187	**0.021**	0.711	0.954	0.954
C38	2035-99-6	Isoamyl octanoate *	1.136	1.111	**0.002**	0.117	1.44	0.678
C48	142-62-1	Hexanoic acid **	1.655	1.464	**0.022**	0.717	0.917	0.975
C55	1450-72-2	2-Hydroxy-5-methylacetophenone	Trace	Trace	Trace	1.673	1.218	**0.013**
C60	18794-84-8	β-Farnesene	Trace	Trace	Trace	1.053	0.963	**0.000**
C64	8013-90-9	α-Ionone *	0.341	0.851	0.003	1.464	1.039	**0.037**
C65	23726-93-4	β-Damascenone **	1.301	1.274	**0.014**	0.428	0.883	0.749
C68	100-42-5	Styrene	0.691	0.934	0.001	1.121	0.847	**0.010**
C69	90-05-1	Guaiacol *	1.777	1.575	**0.04**	Trace	Trace	Trace

The data in bold were the most differentiated compounds (VIP>1, p < 0.05); [a.] The No. was in accordance with Table S2; [b.] * OAV > 0.1, ** OAV > 1.0.

2.3.1. Flavor Compounds

In rosé wines, principal component 1 (PC1) and principal component 2 (PC2) accounted for 56.4% of the total variance (Figure 5a). According to variable importance in projection values (VIP > 1) of the first two PCs from the PLS-DA model and P values (p < 0.05) from a one-way ANOVA test, differentiated flavor compounds in rosé wines sealed with natural corks and their alternatives were selected (Table 1). Closures seemed to have more effect on volatile compounds than phenolic compounds because those differentiated flavor compounds were all volatile compounds. Only rosé wines with natural corks were in the quadrant where PC1 and PC2 were both positive. PC1, which accounted for 34.5% of the total variance, could distinguish wines sealed with natural corks from those with a polymer synthetic plug (SP) (Figure 5a). As shown in Figure 5b, the positive direction of PC1 was mostly driven by esters, such as ethyl octanoate (C25), methyl octanoate (C33), isopentyl hexanoate (C34), and isoamyl octanoate (C38). Except for 2-ethylhexanol (C10), the other differentiated compounds all lay in the positive part in PC1. Ethyl heptanoate (C24), hexanoic acid (C48), β-damascenone (C65), and guaiacol (C69) had a higher concentration in wines with natural corks (see Table S2). All phenolic compounds, especially 4-hydroxycinnamic acid (P4) and catechin (P1), drove PC2 in the negative direction, where wines with a screw cap lay.

As for white wines, PC1 and PC2 accounted for 55.1% of the total variance (Figure 5c). Since phenolic compounds are quite limited in white wines, aroma properties play a more important role from an organoleptic perspective. According to the loading plot, the quadrant where PC2 was positive and PC1 was negative contained the most types of volatile compounds (alcohols, esters, aliphatic acids, terpenes, and norisoprenoids) (Figure 5d). Wines sealed with a screw cap lay in this quadrant. The most differentiated compounds (VIP > 1, $P < 0.05$) in white wines were 2-ethylhexanol (C10), 1-decanol (C12), 2-hydroxy-5-methylacetophenone (C55), β-farnesene (C60), α-ionone (C64), and styrene (C68). Those volatile compounds characterized the negative direction of PC1 (Table 1, Figure 5d) to distinguish wines sealed with a screw cap from those with natural corks. Wines sealed with agglomerated cork (AC), '1+1' technical cork (TC), and polymer synthetic plug (SP) were together in the third quadrant, close to the 'zero' of plot score.

Comparing the flavor compounds in rosé and white wines, wines with natural corks could be distinguished from their alternatives by the results of PLS-DA. The only differentiated flavor compound rosé and white wines had in common was 2-ethylhexanol, which was related to the citrus odor. The quadrant where rosé wines with natural corks lay contained more compounds, indicating that natural corks could perform better in preventing flavor compounds from loss during bottle aging in rosé wines. Whereas, in white wines, it was wines with a screw cap that stood out with more volatile compounds, which might be due to having the best sealing environment. This result was also in accordance with a previous study in Semillon wines that bottling with a screw cap tends to lead to a higher fruity, citrus sensory score and less of an oxidized aroma [9].

2.3.2. Flavor Characters

To learn more about the aroma profile in rosé and white wines, Duncan's multiple range tests based on OAVs were applied to identify which types of closures had a different effect on wines after an 18-month bottle aging (Table 2). Overall, the rosé and white wines were both fruity, floral, and sweet in odor, which makes sense given that most of the volatile compounds quantified in this study were esters and the thresholds of terpenes and norisoprenoids were usually low. White wines exhibited more 'berry' and 'sweet' odors, mainly due to the high level of ethyl esters, such as ethyl acetate, ethyl octanoate, ethyl decanoate, and ethyl dodecanoate (Table S2). During an 18-month bottle aging, the total intensity of aroma in both types of wines declined prominently, especially 'tropical fruity,' 'floral,' 'berry,' and 'sweet' aroma descriptions. The results indicated that the period of bottle aging was likely to contribute to the loss of aroma properties. The decrease was more obvious in white wines, as the studies showed that white wines were very unstable and lost their desirable fresh and fruity characters over time [20]. According to the results of Duncan's multiple range tests, the difference between different closures mainly concerned the 'berry' and 'sweet' odors. Ethyl acetate, ethyl 2-hydroxy-4-methylpentanoate, ethyl decanoate, and β-damascenone were representative compounds of those odorant series. Wine samples sealed with a screw cap contained a higher concentration of long-chain aliphatic acid esters, which might be the reason why it performed better in preventing 'berry' and 'sweet' odors from declining. There was no significant difference between rosé wines sealed with three natural corks (N1, N2, N3) and a screw cap (SC), except for the 'berry' aroma description.

On the contrary, OAVs in rosé wines sealed with a polymer synthetic plug (SP) were significantly lower than those sealed with other closures after an 18-month bottle aging. White wines with agglomerated cork (AC) and polymer synthetic plug (SP) had the lowest OAVs among other types of closures, indicating that those two types of closures were not appropriate for the preservation of white wines during bottle aging.

Table 2. OAVs in all initial wines (0M) and final wines (after an 18-month bottle aging).

Aroma Description	0M	N1	N2	N3	TC	AC	SP	SC
				'Cabernet Sauvignon' rosé wines				
Tropical Fruity	56.01 ± 5.02	33.11 ± 1.07ab	31.9 ± 1.13ab	33.68 ± 2.62a	29.96 ± 0.65b	30.51 ± 0.11ab	30.26 ± 0.67b	31.77 ± 1.49ab
Floral	35.17 ± 3.08	12.98 ± 0.06ab	12.18 ± 0.5cd	13.48 ± 0.2a	11.49 ± 0.08de	11.28 ± 0.32e	12.44 ± 0.2bc	12.34 ± 0.37bc
Berry	14.86 ± 1.33	10.95 ± 0.65b	11.15 ± 0.8b	10.68 ± 1.08bc	9.74 ± 0.37bc	9.34 ± 0.21c	7.51 ± 0.2d	13.57 ± 0.62a
Herbaceous/Vegetal	1.27 ± 0.07	1.27 ± 0.05ab	1.31 ± 0.06a	1.27 ± 0.08ab	1.24 ± 0.03ab	1.24 ± 0.01ab	1.18 ± 0.02b	1.3 ± 0.06a
Chemical	8.57 ± 0.8	8.94 ± 0.43ab	8.47 ± 0.31ab	8.69 ± 0.94ab	7.83 ± 0.2b	8.09 ± 0.19b	6.65 ± 0.18c	9.32 ± 0.5a
Fatty	9.66 ± 0.27	9.01 ± 0.32a	9.04 ± 0.58a	9.01 ± 0.36a	8.8 ± 0.14a	8.58 ± 0.16a	7.24 ± 0.32b	8.99 ± 0.19a
Sweet	26.05 ± 2.21	21.02 ± 1.18ab	21.14 ± 1.16ab	20.97 ± 2.41ab	19.07 ± 0.65bc	19.15 ± 0.45bc	17.02 ± 0.43c	23.01 ± 1.12a
				'Chardonnay' dry white wines				
Tropical Fruity	44.8 ± 1.19	28.13 ± 0.21a	26.77 ± 1.12a	28.44 ± 1.13a	26.62 ± 2.71a	27.02 ± 0.5a	27.86 ± 2.97a	28.61 ± 1.03a
Floral	19.81 ± 0.35	8.93 ± 0.21b	9.66 ± 0.04a	9.41 ± 0.11a	9.47 ± 0.27a	9.43 ± 0a	9.6 ± 0.24a	9.57 ± 0.17a
Berry	21.12 ± 0.49	8.73 ± 0.39ab	8.13 ± 0.12bc	8.11 ± 0.61bc	7.56 ± 0.65bc	6.66 ± 0.42c	6.72 ± 1.16c	10.16 ± 0.54a
Herbaceous/Vegetal	1.3 ± 0.02	1.01 ± 0.17a	0.89 ± 0.03a	0.88 ± 0.04a	0.85 ± 0.04a	0.83 ± 0.02a	0.88 ± 0.1a	0.93 ± 0.05a
Chemical	10.68 ± 0.4	6.43 ± 0.02ab	5.99 ± 0.3ab	6.43 ± 0.52ab	5.82 ± 0.73b	5.68 ± 0.3b	5.33 ± 0.95b	7.14 ± 0.32a
Fatty	12.39 ± 0.36	11.74 ± 0.62ab	11.45 ± 0.48ab	10.77 ± 0.16b	11.46 ± 0.61ab	11.14 ± 0.34b	12.9 ± 0.74b	12.9 ± 0.94a
Sweet	34.94 ± 0.81	20.58 ± 0.15ab	18.68 ± 0.94ab	20.01 ± 1.09ab	18.34 ± 2.24ab	17.87 ± 0.59b	18.58 ± 2.87ab	22.09 ± 1.11a

Odor activity values (OAVs) were shown through average ± standard error. The odor thresholds were taken from [21–27]. The details are shown in Table S2. Different letters in the same row indicate significant differences at $p < 0.05$ by Duncan's multiple-range test.

As for chromatic characters, due to the decrease of anthocyanins in rosé wines during bottle aging, a^* values were much lower, while b^* values were higher in final wines than in initial wines (Figure 6a), indicating that the hue tended to undergo a yellowing phenomenon after 18 months of bottle aging. Although the ΔE values of rosé wines in our research ranged from 1.62 to 2.37 and the chromatic difference in wines was perceivable by human eyes when the ΔE value was above 2.8 [28], judging from the direct comparison in Figure 6a, visual differences existed in all rosé wines after an 18-month bottle aging compared to the initial wines. This may be due to the extreme brightness caused by low color substances in rosé wines, since it was reported that an ΔE value >1.0 in model solutions was visually perceivable by the human eye [28]. The ΔE value of white wines ranged from 0.08 to 0.33, corresponding to no visual differences in all white wines after an 18-month bottle aging (Figure 6b).

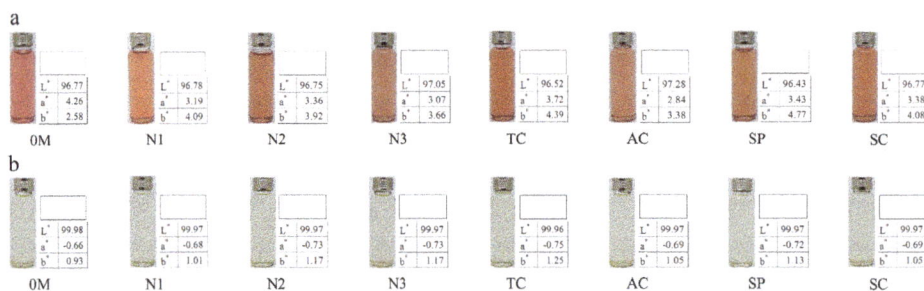

Figure 6. Sample photos and chromatic parameters of rosé (**a**) and white wines (**b**) prior to and after an 18-month bottle aging.

3. Materials and Methods

3.1. Wine Samples

'Cabernet Sauvignon' rosé wines and 'Chardonnay' dry white wines were made strictly according to the local published winemaking standards in October 2014, at Shandong Taila Winery Co., Ltd.

(Rushan, Shandong, China). Detailed information about the two original wine samples is given in Table S3. Bottling was performed at the packaging line of this winery to guarantee a 750 mL volume of each bottle. Seven types of closures were used for treatment, including three different natural corks [named natural cork-1 (N1), natural cork-2 (N2), and natural cork-3 (N3)] and four natural cork alternatives ['1+1' technical cork (TC), agglomerated cork (AC), polymer synthetic plug (SP) and screw cap (SC)]. The physical indexes of these closures are provided in Table S4. For each closure treatment, 30 bottles of wine were collected as samples. After bottling, wine samples were stored in a cellar, with an average temperature of 16 ± 1 °C and a relative humidity of $65 \pm 5\%$. The length of the experiment was 18 months.

3.2. Oxygen Measurements

The dissolved oxygen (DO) levels were recorded by noninvasive oxygen sensors (5 mm sensor spots PSt3, NomaSense O_2 P6000, Yantai Vinventions Co., Ltd., Yantai, China). For each closure treatment, three bottles of wine were selected randomly to equip one sensor positioned at mid height and one in the neck of the bottle. Wine samples were analyzed every two months for oxygen data acquisition.

3.3. Flavor Compounds Detection

Phenolic compounds were detected via high-performance liquid chromatography/triple-quadrupole tandem mass spectrometry (HPLC-QqQ-MS/MS) method using an Agilent series 1200 instrument fitted with a Poroshell 120 EC-C18 column (150×2.1 mm, 2.7 µm, Agilent Technologies, Santa Clara, CA, USA). The method was established and validated in our earlier work, and is reliable for detecting and quantifying 45 non-anthocyanin and 95 anthocyanin compounds [29,30]. Volatile compounds were detected using headspace solid phase micro-extraction (HS-SPME) with a 2 cm DVB/CAR/PDMS 50/30 µm SPME fiber (Supelco, Bellefonte, PA, USA) and an Agilent 7890 gas chromatography kit equipped with an Agilent 5975 mass spectrometer (GC–MS). The details were given in our previous study [31]. Retention indices (RI) [adjected by C7–C24 n-alkane series (Supelco)] were compared with those in the NIST11 database via an Automatic Mass Spectral Deconvolution & Identification System (AMDIS) for identification. The ChemStation software (Agilent Technologies, Inc.) was used to calculated the peak areas and quantification was conducted on the basis of the calibration curve of volatile compound standards purchased from Sigma-Aldrich (St. Louis, MO, USA).

Three bottles of wine samples from each closure treatment were collected randomly every six months for flavor compound detection. Each sample was analyzed in duplicate.

3.4. Flavor Analysis

By calculating the ratio of the volatile compound concentration to its odor perception threshold, the odor activity value (OAV) was obtained. This analysis was used to assess the potential contribution of individual volatile compounds to wine aroma.

Chromatic parameters were measured in triplicate using a spectrophotometer Shimadzu UV-Vis 2600 (Shimadzu, Kyoto, Japan) [30] and analyzed via CIELab system to detect the lightness (L^*), reddish attribute (a^*), and yellowish attribute (b^*) of each wine [32]. The visual change in wine, described as ΔE, was calculated using the following equation [28]:

$$\Delta E^* = [(\Delta L^*)^2 + (\Delta a^*)^2 + (\Delta b^*)^2]^{1/2} \tag{1}$$

3.5. Statistical Analysis

K-means algorithm and clustering analysis were conducted respectively by 'kmeans' function and 'pheatmap' function in R environment (3.4.0) (http://www.r-project.org/). Partial least squares-discriminate analysis (PLS-DA) was conducted by MetaboAnalyst 4.0 (http://www.

metaboanalyst.ca/). One-way ANOVA tests and Duncan's multiple range tests were carried out using SPSS software version 20.0 for Windows (SPSS Inc., Chicago IL, USA).

4. Conclusions

In conclusion, by assessing the dissolved oxygen (DO), phenolic and volatile compounds, OAV and chromatic parameters, a comprehensive understanding of the evolution of oxidation patterns and flavor profiles in rosé and dry white wines during an 18-month bottle aging was established. Those two types of wines had different change trends of DO because of their unique phenolic compositions and the initial concentration of headspace oxygen, which makes it important for winemakers to control the bottling process. Many volatile compounds exhibited a decreasing trend during bottle aging and the total intensity of aroma declined obviously compared with their counterparts prior to bottling in all wines. Rosé wines sealed with natural corks and screw caps exhibited no significant difference after an 18-month bottle aging, except for in the 'berry' aroma description, while a polymer synthetic plug could lead to a greater loss of aroma quality in rosé wines during bottle aging. Agglomerated corks and polymer synthetic plugs were not appropriate for the preservation of white wines, while screw caps behaved better than any other closures in maintaining aroma compounds in white wines. After an 18-month bottle aging, rosé wines exhibit a difference in color, whereas this phenomenon was not observed in white wines. Our study has extended the research into the evolution of flavor profiles in rosé and white wines; further attention should be given to the wines' flavor chemistry and quality control during bottle aging.

Supplementary Materials: Supplementary Materials are available online. Table S1: Phenolic compounds identified in this work; Table S2: Volatile compounds identified in this work and their aroma parameters; Table S3: Detailed information of two original wine samples; Table S4: Physical index of various bottle closures used in this work.

Author Contributions: C.-Q.D. and Y.S. designed the study. M.-Q.L., H.X., Y.-B.H., J.C., S.-Y.L., Y.-B.L. and R.-N.L. conducted the experiments and collected the data. M.-Q.L. analyzed the data and drafted the manuscript.

Funding: This research was funded by the China Agriculture Research System (Grant No. CARS-29).

Conflicts of Interest: The authors declare no conflict of interest.

References

1. Kwiatkowski, M.J.; Skouroumounis, G.K.; Lattey, K.A.; Waters, E.J. The impact of closures, including screw cap with three different headspace volumes, on the composition, colour and sensory properties of a Cabernet Sauvignon wine during two years' storage. *Aust. J. Grape Wine R* **2007**, *13*, 81–94. [CrossRef]
2. Guaita, M.; Petrozziello, M.; Motta, S.; Bonello, F.; Cravero, M.C.; Marulli, C.; Bosso, A. Effect of the closure type on the evolution of the physical-chemical and sensory characteristics of a Montepulciano d'Abruzzo *Rosé* wine. *J. Food Sci.* **2013**, *78*, C160–C169. [CrossRef] [PubMed]
3. Liu, D.; Xing, R.-R.; Li, Z.; Yang, D.-M.; Pan, Q.-H. Evolution of volatile compounds, aroma attributes, and sensory perception in bottle-aged red wines and their correlation. *Eur. Food Res. Technol.* **2016**, *242*, 1937–1948. [CrossRef]
4. Gao, Y.; Tian, Y.; Liu, D.; Li, Z.; Zhang, X.X.; Li, J.M.; Huang, J.H.; Wang, J.; Pan, Q.H. Evolution of phenolic compounds and sensory in bottled red wines and their co-development. *Food Chem.* **2015**, *172*, 565–574. [CrossRef] [PubMed]
5. Ugliano, M. Oxygen contribution to wine aroma evolution during bottle aging. *J. Agric. Food Chem.* **2013**, *61*, 6125–6136. [CrossRef] [PubMed]
6. Karbowiak, T.; Gougeon, R.D.; Alinc, J.-B.; Brachais, L.; Debeaufort, F.; Voilley, A.; Chassagne, D. Wine oxidation and the role of cork. *Crit. Rev. Food Sci. Nutr.* **2009**, *50*, 20–52. [CrossRef]
7. Lopes, P.; Silva, M.A.; Pons, A.; Tominaga, T.; Lavigne, V.; Saucier, C.; Darriet, P.; Teissedre, P.L.; Dubourdieu, D. Impact of oxygen dissolved at bottling and transmitted through closures on the composition and sensory properties of a Sauvignon Blanc wine during bottle storage. *J. Agric. Food Chem.* **2009**, *57*, 10261–10270. [CrossRef] [PubMed]

8. Lee, D.H.; Kang, B.S.; Park, H.J. Effect of oxygen on volatile and sensory characteristics of Cabernet Sauvignon during secondary shelf life. *J. Agric. Food Chem.* **2011**, *59*, 11657–11666. [CrossRef] [PubMed]

9. Godden, P.; Francis, L.; Field, J.; Gishen, M.; Coulter, A.; Valente, P.; HØJ, P.; Robinson, E. Wine bottle closures: Physical characteristics and effect on composition and sensory properties of a Semillon wine 1. Performance up to 20 months post-bottling. *Aust. J. Grape Wine R* **2001**, *7*, 64–105. [CrossRef]

10. Wirth, J.; Morel-Salmi, C.; Souquet, J.; Dieval, J.; Aagaard, O.; Vidal, S.; Fulcrand, H.; Cheynier, V. The impact of oxygen exposure before and after bottling on the polyphenolic composition of red wines. *Food Chem.* **2010**, *123*, 107–116. [CrossRef]

11. Vazallo-Valleumbrocio, G.; Medel-Marabolí, M.; Peña-Neira, Á.; López-Solís, R.; Obreque-Slier, E. Commercial enological tannins: Characterization and their relative impact on the phenolic and sensory composition of Carménère wine during bottle aging. *LWT-Food Sci. Technol.* **2017**, *83*, 172–183. [CrossRef]

12. Han, G.; Ugliano, M.; Currie, B.; Vidal, S.; Dieval, J.B.; Waterhouse, A.L. Influence of closure, phenolic levels and microoxygenation on Cabernet Sauvignon wine composition after 5 years' bottle storage. *J. Sci. Food Agric.* **2015**, *95*, 36–43. [CrossRef] [PubMed]

13. He, J.; Zhou, Q.; Peck, J.; Soles, R.; Qian, M.C. The effect of wine closures on volatile sulfur and other compounds during post-bottle ageing. *Flavour Frag. J.* **2013**, *28*, 118–128. [CrossRef]

14. Skouroumounis, G.K.; Kwiatkowski, M.J.; Francis, I.L.; Oakey, H.; Capone, D.L.; Duncan, B.; Waters, E.J. The impact of closure type and storage conditions on the composition, colour and flavour properties of a Riesling and a wooded Chardonnay wine during five years' storage. *Aust. J. Grape Wine R* **2005**, *11*, 369–377. [CrossRef]

15. Puértolas, E.; Saldaña, G.; Condón, S.; Álvarez, I.; Raso, J. Evolution of polyphenolic compounds in red wine from Cabernet Sauvignon grapes processed by pulsed electric fields during aging in bottle. *Food Chem.* **2010**, *119*, 1063–1070. [CrossRef]

16. Waterhouse, A.L.; Laurie, V.F. Oxidation of Wine Phenolics: A Critical Evaluation and Hypotheses. *Am. J. Enol. Viticult.* **2006**, *57*, 306–313.

17. Ivanova-Petropulos, V.; Hermosín-Gutiérrez, I.; Boros, B.; Stefova, M.; Stafilov, T.; Vojnoski, B.; Dörnyei, Á.; Kilár, F. Phenolic compounds and antioxidant activity of Macedonian red wines. *J. Food Compos. Anal.* **2015**, *41*, 1–14. [CrossRef]

18. Silva, M.A.; Julien, M.; Jourdes, M.; Teissedre, P.-L. Impact of closures on wine post-bottling development: A review. *Eur. Food Res. Technol.* **2011**, *233*, 905–914. [CrossRef]

19. Coetzee, C.; Van Wyngaard, E.; Suklje, K.; Silva Ferreira, A.C.; du Toit, W.J. Chemical and sensory study on the evolution of aromatic and nonaromatic compounds during the progressive oxidative storage of a Sauvignon blanc wine. *J. Agric. Food Chem.* **2016**, *64*, 7979–7993. [CrossRef] [PubMed]

20. Makhotkina, O.; Pineau, B.; Kilmartin, P.A. Effect of storage temperature on the chemical composition and sensory profile of Sauvignon Blanc wines. *Aust. J. Grape Wine R* **2012**, *18*, 91–99. [CrossRef]

21. Peinado, R.A.; Mauricio, J.C.; Moreno, J. Aromatic series in sherry wines with gluconic acid subjected to different biological aging conditions by *Saccharomyces cerevisiae* var. capensis. *Food Chem.* **2006**, *94*, 232–239. [CrossRef]

22. Franco, M.; Peinado, R.A.; Medina, M.; Moreno, J. Off-vine grape drying effect on volatile compounds and aromatic series in must from Pedro Ximénez grape variety. *J. Agric. Food Chem.* **2004**, *52*, 3905–3910. [CrossRef] [PubMed]

23. Ferreira, V.; López, R.; Cacho, J.F. Quantitative determination of the odorants of young red wines from different grape varieties. *J. Sci. Food Agric.* **2000**, *80*, 1659–1667. [CrossRef]

24. Bao, J.; Zhenwen, Z. Volatile compounds of young wines from Cabernet Sauvignon, Cabernet Gernischet and Chardonnay varieties grown in the Loess Plateau Region of China. *Molecules* **2010**, *15*, 9184–9196. [CrossRef]

25. Sacks, G.L.; Gates, M.J.; Ferry, F.X.; Lavin, E.H.; Kurtz, A.J.; Acree, T.E. Sensory threshold of 1,1,6-trimethyl-1,2-dihydronaphthalene (TDN) and concentrations in young Riesling and non-Riesling Wines. *J. Agric. Food Chem.* **2012**, *60*, 2998–3004. [CrossRef] [PubMed]

26. Li, H.; Tao, Y.-S.; Wang, H.; Zhang, L. Impact odorants of Chardonnay dry white wine from Changli County (China). *Eur. Food Res. Technol.* **2007**, *227*, 287–292. [CrossRef]

27. Zea, L.; Moyano, L.; Moreno, J.; Cortes, B.; Medina, M. Discrimination of the aroma fraction of Sherry wines obtained by oxidative and biological ageing. *Food Chem.* **2001**, *75*, 79–84. [CrossRef]

28. Habekost, M. Which color differencing equation should be used. *Int. Circ. Graph. Educ. Res.* **2013**, *6*, 20–33.

29. Li, S.-Y.; He, F.; Zhu, B.-Q.; Xing, R.-R.; Reeves, M.J.; Duan, C.-Q. A systematic analysis strategy for accurate detection of anthocyanin pigments in red wines. *Rapid Commun. Mass Spectrom.* **2016**, *30*, 1619–1626. [CrossRef] [PubMed]

30. Li, S.-Y.; He, F.; Zhu, B.-Q.; Wang, J.; Duan, C.-Q. Comparison of phenolic and chromatic characteristics of dry red wines made from native Chinese grape species and *Vitis vinifera*. *Int. J. Food Prop.* **2017**, *20*, 2134–2146. [CrossRef]

31. Lan, Y.B.; Qian, X.; Yang, Z.J.; Xiang, X.F.; Yang, W.X.; Liu, T.; Zhu, B.Q.; Pan, Q.H.; Duan, C.Q. Striking changes in volatile profiles at sub-zero temperatures during over-ripening of 'Beibinghong' grapes in Northeastern China. *Food Chem.* **2016**, *212*, 172–182. [CrossRef] [PubMed]

32. Ayala, F.; Echávarri, J.F.; Negueruela, A.I. A New Simplified Method for Measuring the Color of Wines. III. All Wines and Brandies. *Am. J. Enol. Vitic.* **1999**, *50*, 359–363.

Sample Availability: Samples of the compounds are not available from the authors.

molecules

MDPI

Article

Analysis of 27 β-Blockers and Metabolites in Milk Powder by High Performance Liquid Chromatography Coupled to Quadrupole Orbitrap High-Resolution Mass Spectrometry

Jian-Qiao Cheng [1,2,†], Tong Liu [1,†], Xue-Mei Nie [1], Feng-Ming Chen [1], Chuan-Sheng Wang [2] and Feng Zhang [1,*]

[1]   Institute of Food Safety, Chinese Academy of Inspection & Quarantine, Beijing 100176, China; 13352479167@163.com (J.-Q.C.); liutongyes@163.com (T.L.); niexuemei_00@163.com (X.-M.N.); chenfengmingok@163.com (F.-M.C.)
[2]   College of Applied Chemistry, Shenyang University of Chemical Technology, Shenyang 110142, China; wchsh18@163.com
*   Correspondence: fengzhang@126.com; Tel.: +86-13651290763
†   These authors contributed equally to this work.

Academic Editors: Alessandra Gentili and Chiara Fanali
Received: 31 January 2019; Accepted: 19 February 2019; Published: 25 February 2019

check for
updates

Abstract: This paper presents an application of high performance liquid chromatography coupled with quadrupole orbitrap high-resolution mass spectrometry (HPLC-Q-Orbitrap HRMS) for the analysis of 27 β-blockers and metabolites in milk powder. Homogenized milk power samples were extracted by acetonitrile and purified by using Oasis PRiME HLB solid-phase extraction cartridges. The Ascentis® C8 chromatographic column was used to separate the analytes. The quantification was achieved by using matrix-matched standard calibration curves with carazolol-d_7 and propranolol-d_7 as the internal standards. The results show an exceptional linear relationship with the concentrations of analytes over wide concentration ranges (0.5–500 µg kg^{-1}) as all the fitting coefficients of determination r^2 are > 0.995. All the limits of detection (LODs) and quantitation (LOQs) values were within the respective range of 0.2–1.5 µg kg^{-1} and 0.5–5.0 µg kg^{-1}. Overall average recoveries were able to reach 66.1–100.4% with the intra- and inter-day variability under 10%. This method has been successfully applied to the screening of β-blockers and metabolites in commercial milk powders. At the same time, the corresponding characteristic fragmentation behavior of the 27 compounds was explored. The characteristic product ions were determined and applied to the actual samples screening.

Keywords: β-blockers; metabolites; milk powder; Q-Orbitrap

1. Introduction

β-blockers (BBS) are structurally analogous to the catecholamines, which can act as non-specific β-adrenergic receptor blocking agents. They play an extremely important role in the treatment of cardiovascular diseases such as coronary heart disease, hypertension, arrhythmia and cardiac insufficiency. However, improper use of β-blockers can cause an increase of myocardial oxygen consumption, vascular resistance, oxygen free radicals and myocardial cell apoptosis, etc. [1]. β-blockers are usually used in animals to reduce morbidity and mortality during transportation (to the slaughterhouse or livestock farm), mating, childbirth and in other stressful situations. Such stress usually results in a poor quality of meat, or even in the premature death of the animal [2–4]. The illegal use of β-blockers gives rise to drug residues in edible animal tissue, which can be metabolized in the body. Several metabolites of β-blockers are pharmacologically active and also harmful to the body. For example, 4-hydroxyphenyl carvedilol (the metabolite of carvedilol) exhibits an approximately thirteen-fold higher-adrenoreceptor blocking potency compared to carvedilol itself [5]. In order to protect public health, many countries and organizations began to establish regulations. For example, carazolol has maximum residue limits (MRLs) in animal-based foods. The European Union and the International Codex Alimentarius Commission have asked for MRLs of carazolol in edible animal tissues, with MRLs of 25 μg kg^{-1} for porcine kidney and 15 μg kg^{-1} for bovine kidney. The European Union has also asked for a maximum residue limit of 1.0 μg kg^{-1} of carazolol in the milk powder [1,6]. However, no restrictions have been placed upon their metabolites (4-hydroxyphenyl carvedilol, etc.). Due to the fact that eating food containing high levels of carazolol and other β-blockers can be harmful to consumer health (especially to infants and children), control of β-blockers is required [3]. Therefore, it is necessary to establish a high-throughput analytical method for β-blockers and their metabolites.

Many approaches for the detection of β-blockers have been reported, such as enzyme-linked immunosorbent assay (ELISA) [7], gas chromatography coupled with mass spectrometry (GC-MS) [8–11], liquid chromatography with fluorescence detection (LC-UV) [12], and liquid chromatography mass spectrometry (LC-MS) [13–17]. Although these methods play important roles in the detection of β-blockers, they also have some drawbacks. For example, the quantification of the ELISA method is not accurate and the operation is troublesome, and GC-MS involves derivative steps before chromatographic separation, which are time-consuming and which increase the possibility of contamination [18]. And the LC-UV method is limited by poor sensitivity, so it cannot meet the requirements for ultra-trace analyses. LC-MS was the most widely used method for the qualitative and quantitative detection of β-blockers multiple residues. Liquid chromatography is generally used, coupled with a low resolution mass spectrometry (LRMS) analyzer such as triple-quadrupole (QqQ). Orbitrap is the newest HRMS analyzer. Most identification and determination studies of β-blockers were undertaken using the LTQ Orbitrap (linear ion trap quadrupole Orbitrap high resolution mass spectrometry), achieving LODs below 2 μg kg^{-1} and 5 ng mL^{-1} [19,20]. The Q-Orbitrap (Q-Exactive™, hybrid quadrupole-orbitrap mass spectrometer, Thermo Fisher Scientific, Bremen, Germany) combines high-performance quadrupole precursor selection with high resolution and accurate mass (HR/AM) Orbitrap detection, which has great potential to avoid both false positive and negative results in residue analyses. Compared with the LTQ Orbitrap, the Q-Orbitrap has higher sensitivity, and its use has become widespread in the confirmation and quantification of drugs residues in food [21,22]. In addition, it can realize real-time positive and negative switches; therefore, the time spent on the preparation process and method optimization is significantly reduced. However, analysis of β-blockers and their metabolites using the Q-Orbitrap has not been reported.

In this study, a high-throughput, high performance liquid chromatography coupled to quadrupole Orbitrap high-resolution mass spectrometry (HPLC-Q-Orbitrap HRMS) has been developed for the screening of 27 analytes, including 21 β-blockers and 6 metabolites in milk powder samples. In addition to this, the corresponding characteristic fragmentation behavior and the product ions of the 27 compounds are described in detail. They will provide a basis for the target-free screening of these drugs and the identification markers of the newly-emerging β-blockers residues.

2. Results and Discussion

2.1. LC Parameters Optimization

Chromatographic conditions were studied in order to obtain the best separation and retention for the compounds. Four HPLC columns (Waters ACQUITY UPLC® BEH C18 (1.7 μm, 50 × 2.1 mm, Milford, MA, USA), Thermo Accucore aQ (2.6 μm, 150 × 2.1 mm, Bellefonte, PA, USA), ALDRICH Ascentis® C8 (3 μm, 10 cm × 4.6 mm, Bellefonte, PA, USA), Waters ACQUITY UPLC™ BEH Phenyl (1.7 μm, 50 × 2.1 mm, Milford, MA, USA) were evaluated in (0.1% FA) H$_2$O-MeCN in their appropriate gradient elution at 0.5 mL min^{-1}. Similar separation performances for total analytes were observed by the first three columns; however, the Ascentis® C8 column provided better shape and retention for hydroxyatenolol, as shown in Figure 1. Therefore, the Ascentis® C8 column was chosen.

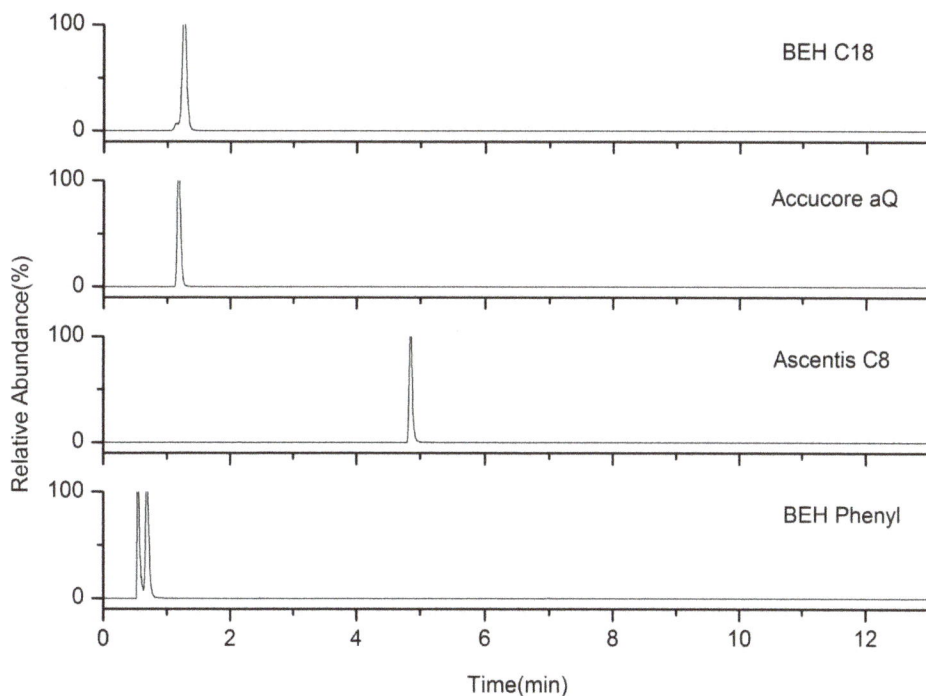

Figure 1. The effect of the columns on the chromatographic separation of hydroxyatenolol.

Several mobile phases were tested using MeCN or MeOH as an organic solvent and water as a polar solvent with FA addition (from 0% to 0.5%). As shown in Figure 2A, the mixture of MeOH and water showed higher responses and better separation for isomers of practolol and atenolol. The proton donor tendency of MeOH contributed to the formation of positive adducts. The better volatility and lower surface tension of MeOH can also improve desolvation of the droplets. Therefore, MeOH was chosen as the organic solvent. On the other hand, the addition of FA could improve the phenomenon of peak tailing and the response, which is probably because the excess silanols in the stationary phase combine with the acid rather than with their targets. The retention time of isomers of atenolol and practolol were greatly affected by pH, so a good chromatographic separation could not obtained until 0.1% FA was added (mobile phase pH = 2.58) (Figure 2B). Therefore, 0.1% FA-H_2O was chosen as the water phase.

In addition, injection volumes of 2 μL to 5 μL were evaluated using the aforementioned conditions. Taking the isomers of atenolol and practolol as examples, the experimental results showed that the separation factor decreased from 1.41 to 0.77 with increasing injection volume. So, 2 μL was chosen as the experimental injection volume.

Furthermore, several elution gradient profiles were also optimized to obtain better chromatographic separation and less analysis time (within 15 min). Other parameters (flow rate, column temperature) were also characterized to achieve better target separation and peak shapes. Under these conditions (see Section 3.2), the retention times (RT) of these 27 analytes were constant, and ranged ranging from 3.85 min to 9.01 min.

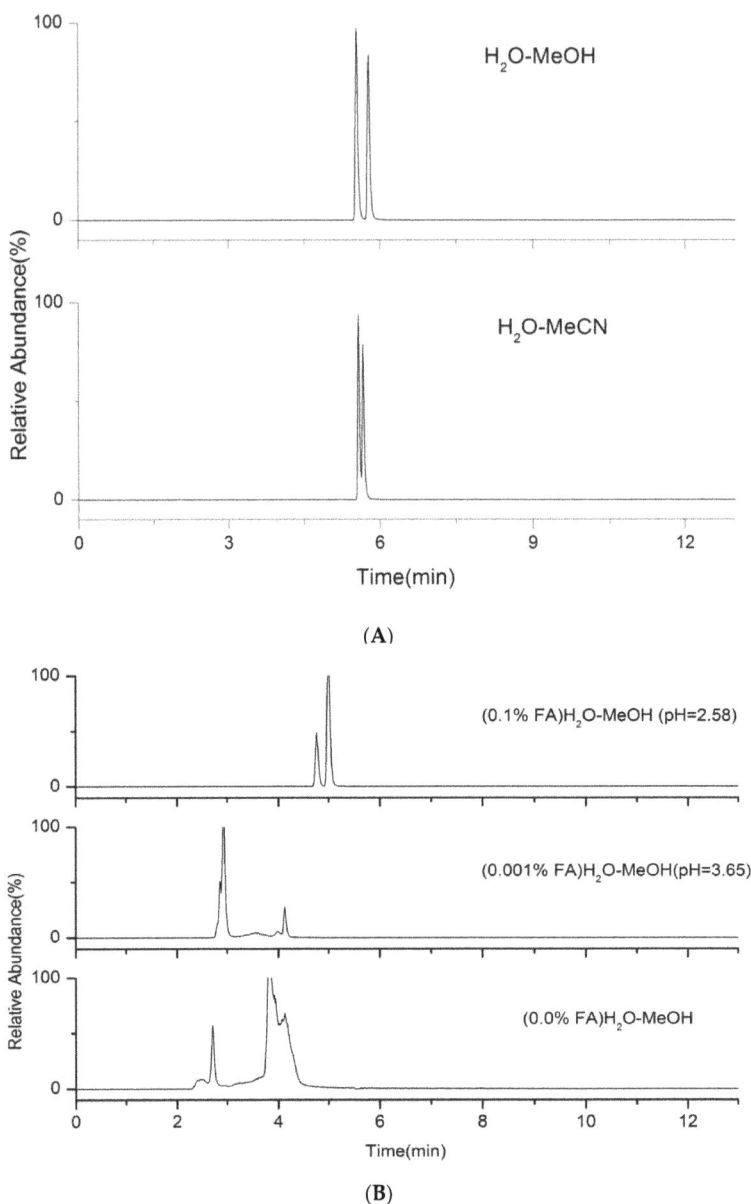

Figure 2. The effect of different mobile phases on the chromatographic separation of the isomers of atenolol and practolol. (**A**) Different organic solvents (atenolol:practolol = 2:1); (**B**) Different concentrations of FA in water phase (atenolol:practolol = 1:1).

2.2. Optimization of the Mass Spectrometric Parameters

The optimization of MS parameters was performed by infusing a standard solution of 100 μg L^{-1} of each β-blockers in methanol-water (50:50, v/v) as the mobile phase under Full scan mode (Full MS). The precursor ions were selected in both positive and negative modes. Consistent with previous report [2], β-blockers and metabolites tend to form [M + H]$^+$ adduct ions in positive mode. The Full

MS/ddMS^2 scan mode, which can achieve the non-target list qualitative and quantitative detection in a single run, was used for screening all samples. All the MS parameters were optimized to provide the best responses of the analytes. The optimized parameters values are summarized in Section 3.2.

2.3. The Proposed Fragmentation Pathways for 27 β-Blockers

In this study, secondary mass spectrometry data of all the substances were extracted, and the fragment ions occurring many times were selected to analyze the fragmentation pathway. β-blockers are mainly classified into three kinds of structures. Labetalol and sotalol have the structure of phenylethanolamine, and others have the structure of aryloxypropanolamine except for timolol and hydroxytimolol, which have special chemical structures. The principal structures of phenylethanolamine and aryloxypropanolamine compounds are shown in Figure 3.

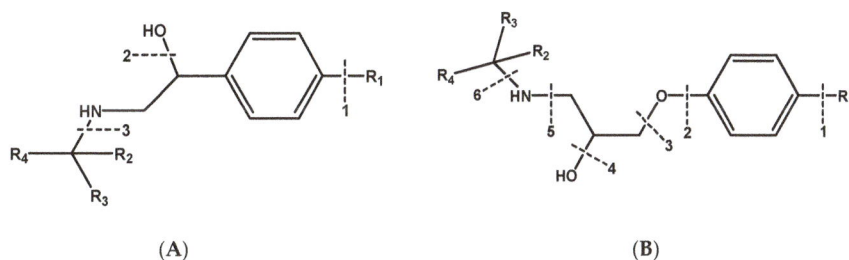

(A) **(B)**

Figure 3. Principal structure of (**A**) phenylethanolamine compounds; (**B**) aryloxypropanolamine compounds.

2.3.1. Phenylethanolamines Structure

For phenylethanolamine structure, there is a characteristic loss of one molecule of water at position 2 at first, and then a characteristic cleavage at positions 1 and 3, as shown in Figure 3A. This phenomenon was consistent with the fragmentation of β-agonists compounds with the phenylethanolamine structure that were studied in our laboratory [23].

2.3.2. Aryloxypropanolamines Structure

For the aryloxypropanolamine structure (Figure 3B), the proposed fragmentation pathways can be divided into three types, depending on the different substitution groups.

Type I (R$_6$ was H, R$_7$ and R$_8$ were methyl)

For type I, the bonds between carbon and oxygen were preferentially broken at position 2. At first, the phenyl structure (C$_6$H$_4$-R$_5$) was lost to form the fragment ion [C$_6$H$_{14}$NO$_2$ + H]$^+$ at m/z 133.06412. Then, the potential loss of [C$_3$NH$_9$]$^+$ or [OH]$^+$ gave m/z 74.06063 (formula C$_3$H$_6$O$_2$) or 116.10702 (formula C$_6$H$_{14}$NO) moieties. And the loss of one molecule of water and a series of fragmentations occurred. The suggested fragmentation pathway of type I is shown in Figure 4.

It is worth mentioning that the Υ-H of the amine structure with 1-propene (C$_6$H$_{12}$N at m/z 98.09663) rearranged to an unsaturated group, accompanied with the cleavage of the β-bond in the amine structure to produce the McLafferty Rearrangement. Imine structure of the fragment C$_3$H$_7$N (at m/z 57.07101) was produced by this reaction. The mechanism of the McLafferty Rearrangement reaction is shown in Figure 5.

Type II (R$_6$, R$_7$ and R$_8$ were methyl group)

For type II, the R$_6$ position of the aryloxypropanolamine structure was substituted with a methyl group. The methyl group can easily get lost to produce a type I structure, and then have a similar fragmentation pathway to the type I mentioned above.

Type III (R₆ or R₇ was an isophthalic ether structure, others were H)

For type III, the R_6 or R_7 position of the aryloxypropanolamine structure was substituted with an isophthalic ether structure, which has strong electronegativity. The bond between carbon and nitrogen at position 8 was easily broken, and it was difficult to form the fragment at m/z 116.10702 (formula $C_6H_{14}NO$).

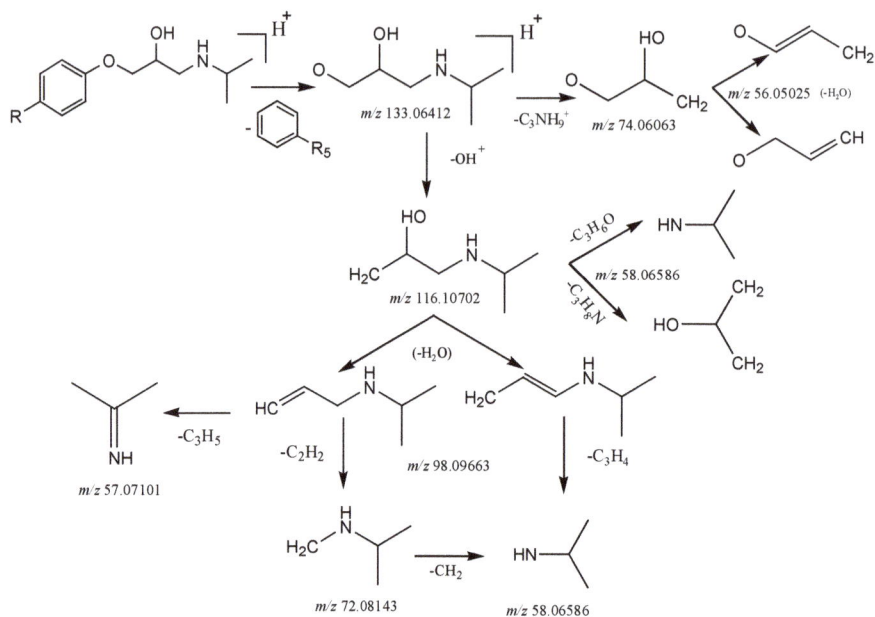

Figure 4. The suggested fragmentation pathways of aryloxypropanolamine compounds.

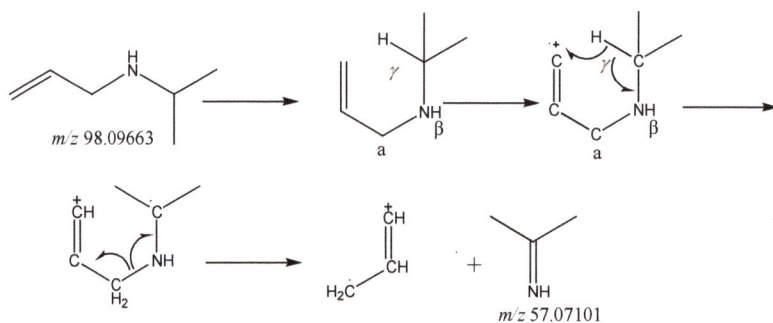

Figure 5. The McLafferty Rearrangement of m/z 98.09663.

2.3.3. Special Structures

For timolol and hydroxytimolol (structures see Figure 6.), they can be considered as β-blockers for newly-emerging aryloxypropanolamine compounds, which differ in structure from the others. These two compounds have characteristic fragments at m/z 74.06063, m/z 57.07101 and m/z 56.05025.

The structure types of the 21 β-blockers and 6 metabolites are listed in Table 1. The possible structures of the corresponding characteristic fragments are described and summarized in Table 2. By exploring the exact mass of these identification markers, it is possible to find newly-emerging β-blockers residues in a complex food matrix.

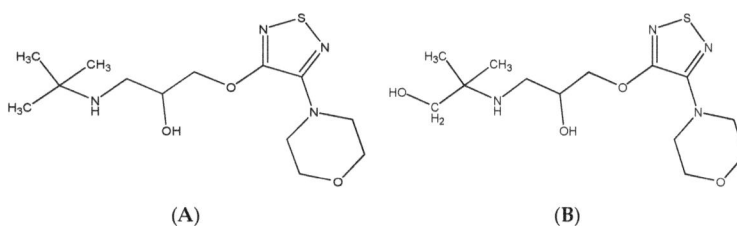

(A)   (B)

Figure 6. Chemical structures of timolol and hydroxytimolol, (**A**) Timolol, Theoretical m/z 317.16419 (**B**) Hydroxytimolol, Theoretical m/z 333.15910.

Table 1. Formula, ionization mode, theoretical mass, measured mass, mass accuracy and MS^2 data for 27 β-blockers.

Analytes	Formula	Theoretical Mass [m/z]	Measured Mass [m/z]	Accuracy [ppm]	MS^2	Structure Type
Carazolol	$C_{18}H_{22}N_2O_2$	299.17540	299.17484	1.87	222.09090 116.10712 98.09675 72.08148 56.05032	Type I
Oxprenolol	$C_{15}H_{23}NO_3$	266.17507	266.17496	0.41	133.06451 116.10696 98.09662 72.08138 56.05027	Type I
Propranolol	$C_{16}H_{21}NO_2$	260.16451	260.16373	3.00	183.07979 116.10689 98.09659 72.08135 58.06586	Type I
Alprenolol	$C_{15}H_{23}NO_2$	250.18016	250.17953	2.52	173.09550 116.10687 98.09660 72.08134 56.05024	Type I
Bisoprolol	$C_{18}H_{31}NO_4$	326.23258	326.23169	3.00	133.06441 116.10689 98.09659 74.06059 56.05026	Type I
Betaxolol	$C_{18}H_{29}NO_3$	308.22202	308.22174	0.91	133.06445 116.10691 98.09661 72.08135 56.05004	Type I
Sotalol	$C_{12}H_{20}N_2O_3S$	273.12674	273.12680	0.46	255.11484 213.06822 176.12991 133.07552 98.05713	phenylethanolamine
Pindolol	$C_{14}H_{20}N_2O_2$	249.15975	249.15961	0.56	172.07532 116.10711 98.09679 72.08147 58.06597	Type I
Nadolol	$C_{17}H_{27}NO_4$	310.20128	310.20084	1.42	354.13795 236.12750 201.09059 74.06068 56.05030	Type II
Timolol	$C_{13}H_{24}N_4O_3S$	317.16419	317.16367	1.64	261.10089 244.07440 188.04840 74.06068 57.07074	Special structure
Acebutolol	$C_{18}H_{28}N_2O_4$	337.21218	337.21310	2.73	218.11726 116.10712 98.09680 72.08150 56.05036	Type I
Celiprolol	$C_{20}H_{33}N_3O_4$	380.25438	380.25299	3.67	324.19070 307.16397 251.10155 74.06061 56.05026	Type II

Table 1. *Cont.*

Analytes	Formula	Theoretical Mass [*m/z*]	Measured Mass [*m/z*]	Accuracy [α] [ppm]	MS[2]	Structure Type
Labetalol	$C_{19}H_{24}N_2O_3$	329.18597	329.18613	0.49	311.17395 294.14755 207.11201 179.08063 162.05423	phenylethanolamine
Cloranolol	$C_{13}H_{19}Cl_2NO_2$	292.08656	292.08658	0.07	236.02318 218.01273 174.97054 74.06063 56.05020	Type II
Penbutolol	$C_{18}H_{29}NO_2$	292.22711	292.22672	1.33	236.16374 201.12683 133.06451 74.06063 57.07070	Type II
Practolol	$C_{14}H_{22}N_2O_3$	267.17032	267.16965	2.51	190.08589 116.10711 98.09682 72.08146 56.05036	Type I
Carvedilol	$C_{24}H_{26}N_2O_4$	407.19653	407.19565	2.16	283.14340 224.12755 100.07599 74.06063 56.05036	Type III
Bupranolol	$C_{14}H_{22}ClNO_2$	272.14118	272.14020	3.60	216.07790 198.06741 181.04089 74.06061 56.05027	Type II
Atenolol	$C_{14}H_{22}N_2O_3$	267.17032	267.16983	1.83	133.06412 116.10690 98.09663 74.06060 56.05026	Type I
Esmolol	$C_{16}H_{25}NO_4$	296.18563	296.18558	0.17	133.06467 116.10737 98.09705 72.08168 56.05050	Type I
Metoprolol	$C_{15}H_{25}NO_3$	268.19072	268.19028	1.64	133.06435 116.10693 98.09660 74.06057 56.05026	Type I
Diacetolol	$C_{16}H_{24}N_2O_4$	308.18088	308.18088	0.00	291.16943 116.10702 98.09670 72.08143 56.05031	Type I
α-hydroxymetoprolol	$C_{15}H_{25}NO_4$	284.18563	284.18472	3.20	133.06435 116.10691 98.09663 74.06059 56.05026	Type I
α-hydroxyatenolol	$C_{14}H_{22}N2O_4$	283.16523	283.16507	0.57	133.08632 116.10760 89.06059 74.06103 57.07010	Type I
(S)-Hydroxytimolol	$C_{13}H_{24}N_4O_4S$	333.15910	333.15823	2.61	261.10059 188.04814 146.11705 74.06059 56.05025	Special structure
7-Hydroxyproprenolol	$C_{16}H_{21}NO_3$	276.15942	276.15930	0.43	199.07463 116.1067 98.09663 74.06057 58.06586	Type I
4-Hydroxyphenylcarvedilol	$C_{24}H_{26}N_2O_5$	423.19145	423.19141	0.09	283.14267 240.12180 100.07578 74.06049 56.05022	Type III
1	56.05025	C_3H_4O	or			

$$^{\alpha}ppm = \frac{|m_{measured} - m_{theoretical}|}{m_{theoretical}} \times 10^6$$

Table 2. The possible structure of corresponding characteristic fragments.

No.	*m/z*	The Molecular Formula	The Possible Structure
2	57.07101	C_3H_7N	
3	58.06586	C_3H_6O or C_3H8N	or
4	72.08143	$C_4H_{10}N$	
5	74.06063	$C_3H_6O_2$	
6	98.09663	$C_6H_{12}N$	or
7	116.10702	$C_6H_{14}NO$	
8	133.06412	$C_6H_{15}NO_2{}^+$	

2.4. The Optimization of the Sample Preparation Procedure

A rapid enzymolysis method was chosen to ensure the processing flux and dissociate the possible bound residual drug. Neutral enzymatic environment (closing to pH 5.2) can significantly reduce matrix co-extraction. Na_2EDTA was added to the buffer to reduce the chelation between metal ions and strongly polar targets. The alkalized aqueous phase and salting out after the enzymolysis facilitate the target extraction into the organic solvent. pH at 9.0, 9.5, 10.0, 10.5, 11.0 and 12.0 were evaluated, and the best extraction efficiency was obtained at pH 10.0, which was consistent with the literature [1].

For food samples, MeCN was commonly used as the extraction solvent due to its protein precipitation ability. Since the acetic-buffer could be used to increase the recoveries of pH-dependent compounds, pure MeCN and different contents of acids (0.1% HOAc, 1% HOAc, 0.1% FA, 1% FA, *v/v*) in MeCN were compared for extraction efficiency in this study. For extraction solvents containing HOAc or FA, the recoveries of some analytes (such as sotalol, hydroxymetoprolol, labetalol, epractolol and hydroxytimoloven) were lower than 60%, as can be seen in Figure 7A. Pure MeCN provided better extraction efficiency with all analyte recoveries being higher than 65%; therefore, pure MeCN was found to be the most suitable extract solvent. Then, the solvent volume was investigated for optimization of the recoveries of the targets. It can be observed in Figure 7B that the recoveries of analytes increased with the solvent volume. When the solvent volume reached 15 mL, recoveries began to be stable. In order to ensure the stability of the recoveries, 20 mL extract solvent was chosen to extract all analytes.

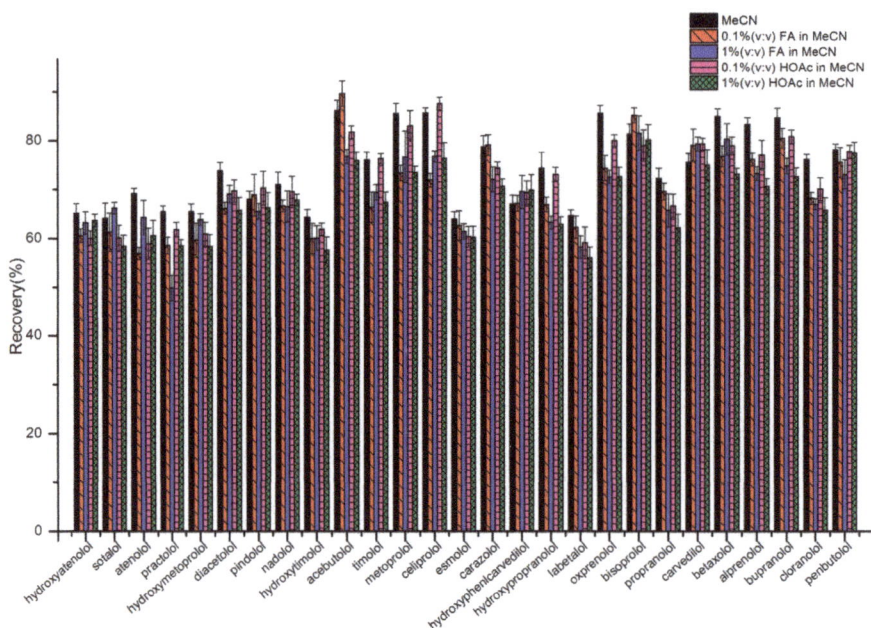

(**A**)

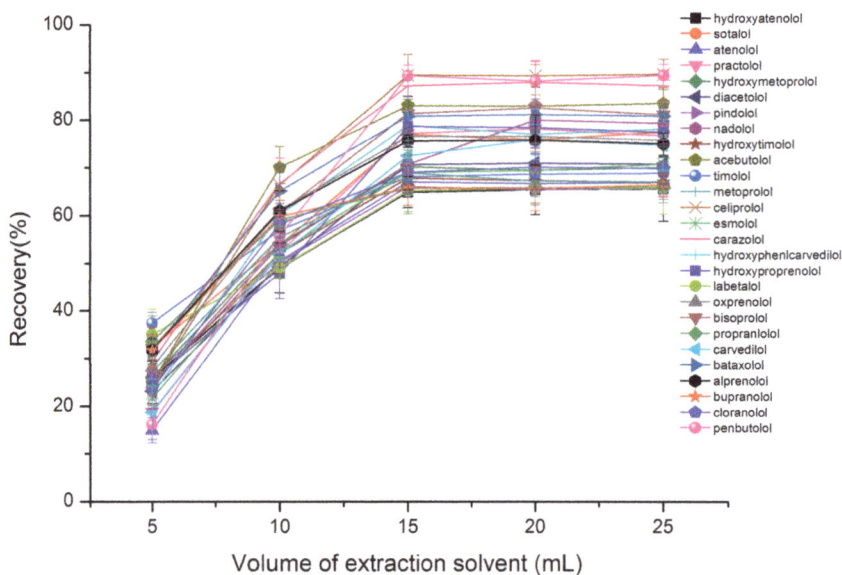

(**B**)

Figure 7. The effect of extracting solvents on the recovery of 27 analytes in milk powder. (**A**) The type of extract solvent, (**B**) the volume of extraction solvent (n = 3).

Taking into account the characteristics of the β-blockers and matrix interferences in milk powder samples, low temperature high-speed centrifugation, low temperature high-speed centrifugation + SPE (PRiME HLB column), and low temperature high-speed centrifugation + QuEChERS (quick, easy, cheap, effective, rugged and safe) methods were tested as purification steps. For the low temperature, high-speed centrifugation + QuEChERS method, the water removal step (using anhydrous magnesium sulfate) might take some water-soluble targets (such as sotalol) away, resulting in a low recovery rate. Compared with low temperature, high-speed centrifugation, PRiME HLB column used in SPE step could adsorb nonpolar interferences (some fats and phospholipids) in milk powder samples, which results in a smaller matrix effect and better target recoveries, as shown in Table 3. Therefore, the solid phase extraction PRiME HLB cartridge was selected for purification step.

2.5. Validation of the Proposed Method

2.5.1. Linearity and Sensitivity

The linearity of the proposed method was evaluated using matrix-matched spiked samples over the range of 0.5–500 µg kg^{-1}. Calibration curves resulted from the ratios of the peak area of the target compounds to the peak area of the isotope-labeled internal standards. The results showed a good linearity relationship with correlation coefficients (r^2) higher than 0.995 (Table 4). Limits of detection (LODs) and quantification (LOQs) are fundamental parameters used to evaluate the sensitivity of instructions and methods. The LODs were determined by the injection of a series of diluted standard solutions corresponding to a signal-to-noise (S/N) ratio of 3. The LOQs were determined by the injection of a series of spiked samples corresponding to a signal-to-noise (S/N) ratio of 10. Under the optimum condition, the LODs and LOQs were in the range of 0.2–1.5 µg kg^{-1} and 0.5–5.0 µg kg^{-1}, respectively, which allows the quantification of analytes presented at low content, indicating that good sensitivity was obtained.

Table 3. Validation parameters of the developed method.

NO.	Analyte	Matrix Effect C (%)		QC Concentration (μg kg^{-1})	Average Recovery (%)	Intra-Day Precision (%) (n = 5)	Inter-Day Precision (%) (n = 5)
		PRiME HLB	Centrifugation				
1	Atenolol	115.7	118.5	2	72.5	2	3.1
				4	76.3	3.1	2.2
				8	74.8	4.1	5
2	Sotalol	85.6	86.8	2	83.6	3.2	3.2
				4	87	5.6	1.9
				8	81.1	2.7	7.9
3	Pindolol	101.7	108.2	1	89.2	1.6	5.2
				2	100.4	4.8	2.7
				4	83.7	3.6	6.4
4	Nadolol	102.2	112.8	0.5	83.6	7.9	5.5
				1	93.2	2.1	3.1
				2	78.5	4.4	5.3
5	Metoprolol	120.9	140	1	80.4	3.9	5
				2	90	6.2	3.5
				4	84.2	2.1	7.7
6	Timolol	116.9	133.7	1	76.5	2.1	7.4
				2	83.8	5.2	3.9
				4	78.8	3.8	3.2
7	Acebutolol	129.2	155.3	0.5	95.6	2.2	3.2
				1	89.3	5.7	5
				2	92.1	2.4	7.5
8	Oxprenolol	109.9	123.3	1	69.6	1.9	4.4
				2	89.7	6.7	7.5
				4	84.8	3.3	3.9
9	Celiprolol	165.1	181.5	1.5	98.5	7.3	5.6
				3	87.4	3.5	5.4
				6	93.3	2.1	7.7
10	Bisoprolol	134.6	156.3	0.5	93.8	7.1	7.2
				1	90.4	4.4	3.6
				2	84.6	2.5	5.6
11	Labetalol	91.2	102.1	0.5	91.7	2.1	2.8
				1	86.6	7.5	6.4
				2	83.4	3.3	6.5

Table 3. *Cont.*

NO.	Analyte	Matrix Effect C (%)		QC Concentration (µg kg^{-1})	Average Recovery (%)	Intra-Day Precision (%) (n = 5)	Inter-Day Precision (%) (n = 5)
		PRiME HLB	Centrifugation				
12	Alprenolol	102.1	117	0.5	74.8	5.5	4.2
				1	81.3	7.2	8.9
				2	81.1	4.1	5.4
13	Propranolol	98.5	120.7	0.5	80.2	8.7	2.3
				1	83.5	3.5	1.8
				2	80.4	6.4	5.6
14	Betaxolol	117.6	146.5	2	79.8	7.1	4.5
				4	91.5	2.7	3.7
				8	85.2	5.6	2.5
15	Cloranolol	109.2	126.1	2	72.1	8.1	3.5
				4	75.5	3.5	5.4
				8	76.6	3.1	7.2
16	Penbutolol	109.4	134.3	1	85.5	3.4	6.5
				2	97.6	2.6	4.4
				4	76.4	1.7	2.5
17	Practolol	115.7	120.6	0.5	71.9	4.4	2.6
				1	73.6	3.4	7.5
				2	75.2	5	5.8
18	Carazolol	80.1	96.1	0.5	99.3	3.1	2.8
				1	85.3	4	5.7
				2	85.9	7.9	4.9
19	Carvedilol	82.7	105.6	2	79.5	5.7	2.5
				4	78	2.5	7.5
				8	84.6	3.4	5.6
20	Esmolol	101.8	106.9	3	72.5	5.1	3.6
				6	83.2	8.1	2.3
				12	73.4	4.3	7.5
21	Bupranolol	112.6	122.9	0.5	73.4	3.2	5.9
				1	79.4	5.2	3.5
				2	80	1.7	5.4
22	Diacetolol	134	152.2	1	81.7	3.8	3.5
				2	87.7	2.3	5.9
				4	82.9	6.9	2.5

Table 3. *Cont.*

NO.	Analyte	Matrix Effect C (%)		QC Concentration (μg kg⁻¹)	Average Recovery (%)	Intra-Day Precision (%) (n = 5)	Inter-Day Precision (%) (n = 5)
		PRiME HLB	Centrifugation				
23	α-Hydroxymetoprolol	107.5	113.8	1	85.8	2.4	7.1
				2	89.8	4.7	4.6
				4	84.5	5.6	5.6
24	α-Hydroxyatenolol	80.1	72.8	5	67.7	4.3	3.2
				10	66.1	3.4	5.4
				20	68.6	7.7	6.9
25	(s)-Hydroxytimolol	93.4	103.6	1	78.5	2.2	5.8
				2	91.3	5.5	5.7
				4	85.7	6.3	4.6
26	7-Hydroxypropranolol	84.7	99.9	1	69.8	5.6	3.5
				2	73.6	2.2	5.3
				4	78.8	3.2	7.2
27	4-Hydroxyphenlcarvedilol	85.6	103.1	2	73.5	3.7	1.8
				4	66	7	4.5
				8	67.4	6.3	6.9

Table 4. Regression data, Precision, LODs, and LOQs for the investigated compounds.

Analytes	Linear Equation	Linear Range (μg kg⁻¹)	Correlation Coefficient (r^2)	LOD (μg kg⁻¹)	LOQ (μg kg⁻¹)
Atenolol	$Y = -0.0196094 + 0.0399798X$	2–200	0.9994	0.6	2
Sotalol	$Y = -0.0302291 + 0.0364729X$	2–200	0.9995	0.6	2
Pindolol	$Y = 0.374297 + 0.109209X$	1–200	0.9967	0.3	1
Nadolol	$Y = 0.00696148 + 0.0330124X$	0.5–50	0.9987	0.2	0.5
Metoprolol	$Y = 0.250919 + 0.0925317X$	0.5–50	0.9975	0.3	1
Timolol	$Y = -0.0667935 + 0.0862132X$	1–100	0.9997	0.3	1
Acebutolol	$Y = 0.0484461 + 0.047836lX$	0.5–50	0.9977	0.2	0.5
Oxprenolol	$Y = -0.157711 + 0.0272438X$	1–100	0.9990	0.3	1
Celiprolol	$Y = -0.00691133 + 0.068192X$	2–200	0.9966	0.5	1.5
Bisoprolol	$Y = 0.130461 + 0.111734X$	0.5–50	0.9969	0.2	0.5
Labetalol	$Y = -0.0396352 + 0.0389041X$	0.5–50	0.9997	0.2	0.5
Alprenolol	$Y = -0.164222 + 0.588251X$	0.5–50	0.9997	0.2	0.5
Propranolol	$Y = 0.0870785 + 0.148381X$	0.5–50	0.9996	0.2	0.5
Betaxolol	$Y = 0.132601 + 0.130952X$	2–200	0.9987	0.6	2
Cloranolol	$Y = 0.214191 + 0.103506X$	2–200	0.9991	0.6	2

Table 4. *Cont.*

Analytes	Linear Equation	Linear Range (μg kg^{-1})	Correlation Coefficient (r^2)	LOD (μg kg^{-1})	LOQ (μg kg^{-1})
Penbutolol	Y = −0.0422939 + 0.063028X	1–100	0.9989	0.3	1
Practolol	Y = −0.0153134 + 0.114317X	0.5–50	0.9990	0.2	0.5
Carazolol	Y = −0.015998 + 0.0807354X	0.5–50	0.9998	0.2	0.5
Carvedilol	Y = −0.00662109 + 0.0671112X	2–200	0.9998	0.6	2
Esmolol	Y = −0.142646 + 0.159519X	5–500	0.9994	1	3
Bupranolol	Y = −0.126666 + 0.325598X	0.5–50	0.9995	0.3	0.5
Diacetolol	Y = 0.204209 + 0.0797893X	1–100	0.9973	0.3	1
α-Hydroxymetoprolol	Y = 0.0134667 + 0.105121X	1–100	0.9992	0.3	1
α-Hydroxyatenolol	Y = −0.0738747 + 0.0396782X	5–500	0.9993	1.5	5
(S)-Hydroxytimolol	Y = 0.168441 + 0.119248X	1–100	0.9989	0.3	1
7-Hydroxypropranolol	Y = 0.0104 + 0.167864X	1–100	0.9999	0.3	1
4-Hydroxyphenilcarvedilol	Y = 0.0993695 + 0.159881X	2–200	0.9991	0.6	2

Y: The ratio of the peak area of the target to the area of the isotope peak, X: Corresponding concentration (μg kg^{-1}).

2.5.2. Matrix effect

Suppression or enhancement of the target signal usually occurs in the HESI source, especially for complicated food matrices. With a matrix effect (ME) value was between 80% and 120%, signal suppression or enhancement effect can be considered tolerable. As shown in Table 3, many of the analytes did not significantly express the matrix effect, except metoprolol, acebutolol, celiprolol, bisoprolol, diacetolol and α-hydroxyatenolol, which showed a significant matrix effect (ME < 80%, or ME > 120%). In order to accurately quantify the compounds, the assay was quantified with matrix-matched internal standard calibration.

2.5.3. Trueness and Precision

Recovery experiments were performed to evaluate the trueness of the method due to the lack of certified reference materials (CRM). As shown in Table 3, recoveries at three spiking levels (LOQ, 2 × LOQ, 4 × LOQ) ranged from 66.1% to 100.4%. The precision was calculated in terms of intra-day repeatability and inter-day reproducibility, which were expressed as relative standard deviations (RSDs). The results of intra-day and inter-day analyses performed at three spiking levels are presented in Table 3. Repeatability and reproducibility were in the range of 1.6–8.7% and 1.8–8.9%, respectively. Consequently, these results indicated that the developed method in this study is quite reliable, accurate and reproducible for determining β-blockers and their metabolites in milk powder samples.

2.6. Real Samples Analysis

In order to estimate the reliability and practicability of the developed method, 30 samples of infant formula milk powder purchased at local markets were analyzed in this study. The samples were regarded as representative, since they were ranged from phase 1 to stage 4 produced by reputable manufactures. High accuracy parent ions and product ions were used for qualitative analysis simultaneously. Full MS data of this mode were used for quantitative analysis. None of the 27 targeted analytes were detected by the developed Q-Orbitrap high resolution mass spectrometry method. However, an unknown compound which has the same molecule mass (m/z 309.18005) but different retention time (7.15 min vs. 5.75 min) with diacetolol has been screened in one sample. As shown in Figure 8, under the MS^2 conditions, the detected unknown compound had high-accuracy product ions 98.09692 (m/z) and 72.08157 (m/z), which was similar to diacetolol, and also the characteristic fragment ion m/z 57.07080. So, it is reasonable to speculate that this unknown substance may be an isomer of diacetolol. Considering the possible structure of product ions, the detected diacetolol's isomer could have two possible chemical structures ($C_{16}H_{25}O_4N_2$), as shown in Figure 9. For the first one, the acetyl group on the phenyl ring is in the presence of an acetyl amino group, which is more likely. The reason for this is that intermediate isomers are produced during the synthesis process, resulting in the isomerism of the target compounds. The second possibility may be diastereoisomers. Imino NH, phenolic hydroxyl oxygen and carbonyl oxygen through hydrogen bonding make the nitrogen atom form a relatively stable chiral center, as shown in Figure 9B. The specific structure needs to be further confirmed by NMR or other techniques.

(A)

(B)

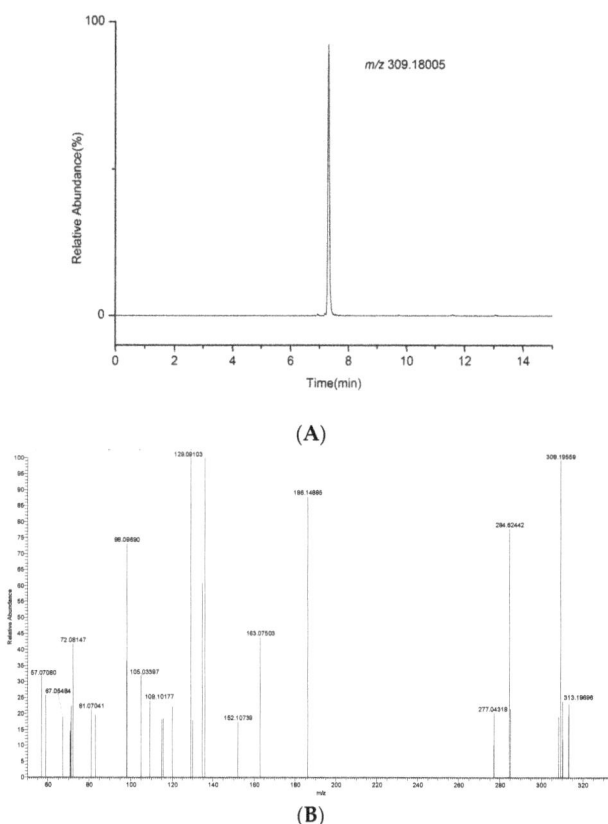

Figure 8. Extracted ion chromatogram and product ion spectrum of the suspected sample. (**A**) The extracted ion chromatogram of the suspected sample. (**B**) The product ion spectrum of the suspected sample.

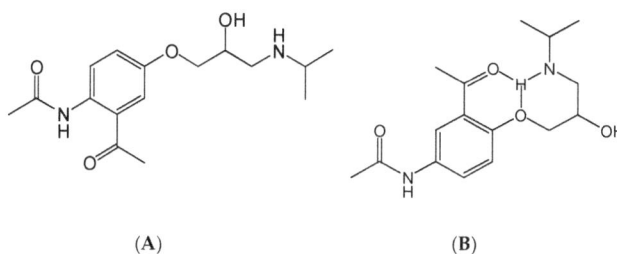

(A)   (B)

Figure 9. The possible chemical structure of diacetolol isomer detected in sample. (**A**) The first possible chemical structure. (**B**) The second possible chemical structure.

3. Materials and Methods

3.1. Chemicals and Reagents

Formic acid (FA), acetic acid (HOAc) were purchased from Sigma-Aldrich (Steinheim, Germany). Sodium chloride, ammonium acetate, ethylenediaminetetraacetic acid disodium salt (Na_2EDTA) were obtained from Beijing Chemical Company (Beijing, China). HPLC-grade methanol (MeOH) and acetonitrile (MeCN) were supplied by Fisher Scientific (Lough borough, UK). Ultra-pure water (H_2O) was obtained using a Milli-Q Ultrapure system (Millipore, Brussels, Belgium). β-glucuronidase/arylsulfatase was supplied by Merck (Darmstadt, Germany). Solid phase extraction cartridges Oasis PRiME HLB (500 mg, 6 cm^3) were obtained from Waters (Milford, MA, USA).

Standards of carazolol, timolol maleate, nadolol, sotalol hydrochloride, pindolol, atenolol, metoprolol, carazolol-d_7 (internal standard, ISTD), propranolol-d_7 (ISTD) (purity > 96%) were purchased from Dr. Ehrenstorfer (Augsburg, Germany). Acebutolol, carvedilol, penbutolol sulfate, propranolol, betaxolol hydrochloride, alprenolol, oxprenolol, celiprolol, bisoprolol fumarate, labetalol hydrochloride (purity > 96%) were purchased from U.S. Pharma-copoeia (Rockville, MD, USA). Diacetolol, cloranolol, esmolol, bupranolol, practolol, 7-hydroxypropranolol, hydroxytimolol, 4-hydroxyphenylcarvedilol, α-hydroxyatenolol (purity > 96%) were obtained from Toronto Research Chemicals (North York, Canada). α-hydroxymetoprolol (100 µg mL^{-1}, methanol) were obtained from AccuStandard (Chiron, Norway).

All standard stock solutions were prepared in MeOH at 100 µg mL^{-1}. The mixed working standard solutions were prepared daily via proportional dilution of the stock solutions. All of the standard solutions were stored at −20 °C in a dark amber bottle.

Matrix-matched standard working solutions were prepared in blank sample extracts, which were obtained from a commercial product purchased from a local market and affirmed in advance not to contain any of the tested analytes. All of the standard solutions were stored at −20 °C in a dark amber bottle.

Extraction solution: 37.5 g Na_2EDTA was dissolved in an ammonium acetate buffer produced by dissolving 15.4 g ammonium acetate in 1 L deionized water and then using acetic acid to adjust the pH to 5.2.

3.2. Instrument and Analytical Conditions

The UHPLC/HESI Q-Orbitrap system consisted of a Thermo UltiMate 3000 $UHPLC^+$ system coupled with a Q Exactive mass spectrometer (Thermo Fisher Scientific, Bremen, Germany). The system was controlled by Exactive Tune 1.1 and Xcalibur 2.2 software (Thermo Fisher Scientific, San Jose, CA, USA).

Chromatographic separation was achieved on an Ascentis® C8 chromatographic column (100 × 4.6 mm, 3 µm) (SUPELCO® Analytical, Bellefonte, PA, USA). The autosampler tray temperature, column oven temperature, flow rate and injection volume were set at 10 °C, 30 °C, 0.5 mL min^{-1} and 2 µL, respectively. The mobile phase consisted of water containing 0.1% FA (A) and MeOH (B). The gradient used for eluting analytes with mobile phase is as follows: 0–0.5 min, 5% B; 0.5–9 min, 5–95% B; 9–12.5 min, 95% B; 12.5–14 min, 95–5% B; 14–15 min, 5% B.

The Q-Orbitrap HRMS was equipped with a heated electrospray ionization (HESI, Waltham, MA, USA) source and the analysis was operated in the Full MS/dd-MS^2 (data-dependent MS^2) scanning mode, which includes a Full scan followed by MS/MS scan of precursors in the inclusion list. All analytes were measured in positive mode and precursor ion selected was $[M + H]^+$ in all cases. To keep a balance between the selectivity and the sensitivity with Full MS, a mass resolution of 70,000 FWHM was selected, and this turned out to be optimal for the majority of the analytes. For the dd-MS^2 scan, 35,000 FWHM was used for time-saving and to ensure sufficient scan points of the Full MS. The stepped normalized collision energy (NCE) was set to 15%, 25% and 35%. The spray voltage, capillary temperature, aux gas heater temperature were set to 3.0 kV, 350 °C and 350 °C, respectively. The sheath gas, auxiliary gas, sweep gas and S-lens RF level were set to 40, 10, 0 (arbitrary units) and 50 V, respectively. Main MS acquisitions parameters are listed in Table 1. All the extracted mass traces were based on a 2 ppm mass window (accuracy).

3.3. Sample Preparation

Two grams of each sample were precisely weighed in polypropylene centrifuge tube (50 mL). Then, 100 µL mixed ISTD solution (1µg mL^{-1}), 40 µL β-glucuronidase/arylsulfatase and 5 mL EDTA extract solution were added to the sample after being fully dissolved in 5 mL H_2O. The mixture was placed in a water bath shaker at 50 °C for 60 min after vortex-mixing for 1 min. After cooling to room temperature, the pH of the extract was adjusted to 10.0 with 3 mol L^{-1} NaOH solution. MeCN (20 mL) and NaCl (2.5 g) were added to the mixture and then shaken for 30 min. After that, the extract was centrifuged at 10,000 rpm at 4 °C for 10 min. The supernatant was decanted to another polypropylene centrifuge tube. The above procedure was repeated and combined with the supernatant. Next, 2 mL of the supernatant directly passed through the Oasis PRiME HLB (500 mg, 6 cm³) cartridge. After sample loading, the cartridge was washed with 2 mL H_2O/MeOH (95:5, *v/v*) and 2 mL MeOH/MeCN (1:9, *v/v*), and all of the effluent was collected. The mixture was evaporated with a gentle N_2 stream at 40 °C, and redissolved in 1 mL of H_2O/MeOH (1:1). The sample extract was vortexed for 0.5 min and filtered through a 0.22 µm nylon membrane, and was ready for Q-Orbitrap HRMS analysis.

3.4. Method Validation

Linearity, precision and recovery were carried out to validate the method. An internal standard method which using carazolol-d_7 and propranolol-d_7 as ISTD was utilized for quantification. A matrix-matched calibration cure was constructed by linear regression of the ratios of chromatographic peak areas of the standards and the ISTD. The linearity was discussed by the coefficient of determination (r^2).

Blank milk sample powders spiked at three concentration levels (LOQ, 2 × LOQ and 4 × LOQ) which were tested for the recovery experiments. Each level was analyzed in five replicates. Intra-day precision was performed by spiking blank milk at three concentration levels (LOQ, 2 × LOQ and 4 × LOQ) with five replicates in one day. To evaluate inter-day precision, the same concentration levels were performed during over consecutive days.

The matrix effect (ME) was calculated by comparing the response of analytes prepared in solvent and in extracted blank matrix at the same concentration, respectively. The value of matrix effect can be calculated as (Equation (1)):

$$ME\ (\%) = B/A \times 100 \tag{1}$$

A refers to the peak areas obtained from neat solution standards, while B refers to the corresponding peak areas of standards spiked after extraction from matrix [24,25].

4. Conclusions

In this study, a rapid HPLC-Q-Orbitrap HRMS method for simultaneous analyses of 27 compounds (21 β-blockers and 6 metabolites) in milk powder has been developed. Simultaneous qualitative and quantitative analysis of analytes were achieved using Full MS/dd-MS² acquisition mode of the Q-Orbitrap mass analyzer and the preparation procedure comprised a simple acetonitrile step, followed by a cleanup using cartridges. The method has been well validated, and is particularly effective and valuable for the routine screening of β-blockers and metabolites in infant formula milk powder. At the same time, the corresponding characteristic fragmentation behavior of the 27 compounds were explored, the characteristic product ions were determined and applied to the actual sample screening.

Author Contributions: J.-Q.C. and T.L. conceived and designed the experiments, performed the experiments, analyzed the data, wrote the paper; X.-M.N. and F.-M.C. prepared the sample, analyzed the data; C.-S.W. analyzed the data; F.Z. designed the experiment, wrote and revised the paper.

Funding: The authors gratefully acknowledge financial support from the project of National Certification and Accreditation Administration Committee (CNCA) (2016YFF0203903).

References

1. Xu, H.; Zhang, H.W.; Wang, F.M. Determination of 9 β-blockers residues in dairy products by liquid chromatography-tandem mass spectrometry. *J. Food Safety Qual.* **2014**, *12*, 3884–3890.
2. Zhang, J.; Shao, B.; Yin, J. Simultaneous detection of residues of β-adrenergic receptor blockers and sedatives in animal tissues by high-performance liquid chromatography/tandem mass spectrometry. *J. Chromatogr. B Anal. Technol. Biomed. Life Sci.* **2009**, *877*, 1915–1922. [CrossRef] [PubMed]
3. Mitrowska, K.; Posyniak, A.; Zmudzki, J. Rapid method for the determination of tranquilizers and a β-blocker in porcine and bovine kidney by liquid chromatography with tandem mass spectrometry. *Anal. Chim. Acta* **2009**, *637*, 185–192. [CrossRef] [PubMed]
4. Hao, J.; Jiang, J.; Shao, R.T.; Ding, X.Y.; Shi, N.; Lu, Y. Determination of Eleven β-Blocker residues in Animal Derived Foods by Ultra High Performance Liquid Chromatography-Tandem Mass Spectrometry with Molecularly Imprinted Solid Phase Extraction. *J. Instrum. Anal.* **2016**, *10*, 1278–1282.
5. Gehr, T.W.B.; Tenero, D.M.; Boyle, D.A.; Qian, Y.; Sica, D.A.; Shusterman, N.H. The pharmacokinetics of carvedilol and its metabolites after single and multiple dose oral administration in patients with hypertension and renal insufficiency. *Eur. J. Clin. Pharmaco.* **1999**, *55*, 269–277. [CrossRef]
6. The Council of the European Communities. Council Regulation 2377/90/EEC. *Off. J. Eur. Commun.* **1990**, *224*, 1.
7. Cooper, J.; Delahaut, P.; Fodey, T.L. Development of a rapid screening test for veterinary sedatives and the β-blocker carazolol in porcine kidney by ELISA. *Analyst* **2004**, *129*, 169–174. [CrossRef] [PubMed]
8. Delbeke, F.T.; Debackere, M.; Desme, T.N. Comparative study of extraction methods for the GC and GC-MS screening of urine for β-blocker abuse. *J. Pharmaceu. Biomed. Anal.* **1988**, *6*, 827–835. [CrossRef]
9. Ternes, T.A.; Hirsch, R.; Mueller, J. Methods for the determination of neutral drugs as well as β-blockers and β_2-sympathomimetics in aqueous matrices using GC/MS and LC/MS/MS. *Anal. Bioanal. Chem.* **1998**, *362*, 329–340.
10. Magiera, S.; Uhlschmied, C.; Rainer, M. GC–MS method for the simultaneous determination of β-blockers, flavonoids, isoflavones and their metabolites in human urine. *J. Pharmaceu. Biomed. Anal.* **2011**, *56*, 93–102. [CrossRef] [PubMed]
11. Amendola, L.; Molaioni, F.; Botrè, F. Detection of β-blockers in human urine by GC-MS-MS-EI: Perspectives for the antidoping control. *J. Pharmaceu. Biomed. Anal.* **2000**, *23*, 211–221. [CrossRef]
12. Delamoye, M.; Duverneuil, C.; Paraire, F. Simultaneous determination of thirteen β-blockers and one metabolite by gradient high-performance liquid chromatography with photodiode-array UV detection. *Forensic Sci. Int.* **2004**, *141*, 23–31. [CrossRef] [PubMed]
13. Zuppa, A.F.; Shi, H.; Adamson, P.C. Adamson, Liquid chromatography-electrospray mass spectrometry (LC-MS) method for determination of esmolol concentration in human plasma. *J. Chromatogr. B* **2003**, *796*, 293–301. [CrossRef]
14. Delahaut, P.; Levaux, C.; Eloy, P. Validation of a method for detecting and quantifying tranquillisers and a β-blocker in pig tissues by liquid chromatography–tandem mass spectrometry. *Anal. Chim. Acta* **2003**, *483*, 335–340. [CrossRef]
15. Lee, H.B.; Sarafin, K.; Peart, T.E. Determination of β-blockers and β_2-agonists in sewage by solid-phase extraction and liquid chromatography–tandem mass spectrometry. *J. Chromatogr. A* **2007**, *1148*, 158–167. [CrossRef] [PubMed]
16. Umezawa, H.; Lee, X.P.; Arima, Y. Simultaneous determination of β-blockers in human plasma using liquid chromatography–tandem mass spectrometry. *Biomed. Chromatogr.* **2008**, *22*, 702–711. [CrossRef] [PubMed]
17. Farré, M.; Gros, M.; Hernández, B. Analysis of biologically active compounds in water by ultra-performance liquid chromatography quadrupole time-of-flight mass spectrometry. *Rapid Commun. Mass Spectrom.* **2008**, *22*, 41–51. [CrossRef] [PubMed]
18. Cheng, Y.; Nie, X.M.; Wu, H.Q. A high-throughput screening method of bisphenols, bisphenols digycidyl ethers and their derivatives in dairy products by ultra-high performance liquid chromatography-tandem mass spectrometry. *Anal. Chim. Acta* **2017**, *950*, 98–107. [CrossRef] [PubMed]
19. Choi, J.H.; Lamshöft, M.; Zühlke, S. Determination of sedatives and adrenergic blockers in blood meal using accelerated solvent extraction and Orbitrap mass spectrometry. *J. Chromatogr. A* **2012**, *1260*, 111–119. [CrossRef] [PubMed]

Molecules **2019**, *24*, 820

20. Li, X.; Shen, B.; Jiang, Z.; Huang, Y.; Zhuo, X. Rapid screening of drugs of abuse in human urine by high-performance liquid chromatography coupled with high resolution and high mass accuracy hybrid linear ion trap-Orbitrap mass spectrometry. *J. Chromatogr. A* **2013**, *1302*, 95–104. [CrossRef] [PubMed]

21. Guo, C.; Shi, F.; Gong, L. Ultra-trace analysis of 12 β-agonists in pork, beef, mutton and chicken by ultrahigh-performance liquid-chromatography-quadrupole-orbitrap tandem mass spectrometry. *J. Pharmaceu. Biomed. Anal.* **2015**, *107*, 526–534. [CrossRef] [PubMed]

22. Hong, Y.H.; Xu, X.L.; Li, W.Q. A high-accuracy screening method of 44 cephalosporins in meat using liquid chromatography quadrupole-orbitrap hybrid mass spectrometry. *Anal. Methods* **2017**, *10*. [CrossRef]

23. Wang, X.J.; Zhang, F.; Li, W.Q. Simultaneous determination of 12 β-agonists in feeds by ultra-high-performance liquid chromatography-quadrupole-time-of-flight mass spectrometry. *J. Chromatogr. A* **2013**, *1278*, 82–88.

24. Matuszewski, B.K.; Constanzer, M.L.; Chavez-Eng, C.M. Strategies for the assessment of matrix effect in quantitative bioanalytical methods based on HPLC–MS/MS. *Anal. Chem.* **2003**, *75*, 3019–3030. [CrossRef] [PubMed]

25. Cappiello, A.; Famiglini, G.; Palma, P.; Pierini, E.; Termopoli, V.; Trufelli, Y. Overcoming matrix effects in liquid chromatography–mass spectrometry. *Anal. Chem.* **2008**, *80*, 9343–9348. [CrossRef] [PubMed]

Sample Availability: Samples of the compounds are not available from the authors.

molecules

MDPI

Article

Identification of the *Pol* Gene as a Species-Specific Diagnostic Marker for Qualitative and Quantitative PCR Detection of *Tricholoma matsutake*

Luying Shan [1,†], Dazhou Wang [1,†], Yinjiao Li [1], Shi Zheng [1], Wentao Xu [2] and Ying Shang [1,2,*]

[1] Yunnan Institute of Food Safety, Kunming University of Science and Technology, Yunnan 650500, China; shanluying1994@163.com (L.S.); wdz101545171@163.com (D.W.); 18487156903@163.com (Y.L.); zhengshi199201@163.com (S.Z.)

[2] Beijing Laboratory of Food Quality and Safety, College of Food Science and Nutritional Engineering, China Agricultural University, Beijing 100083, China; xuwentao@cau.edu.cn

*   Correspondence: shangying1986@163.com; Tel./Fax: +86-871-6592-0216

† These authors contributed equally to this work.

Received: 24 December 2018; Accepted: 24 January 2019; Published: 28 January 2019

check for updates

Abstract: *Tricholoma matsutake* is a rare, precious, and wild edible fungus that could not be cultivated artificially until now. This situation has given way to the introduction of fake *T. matsutake* commodities to the mushroom market. Among the methods used to detect food adulteration, amplification of species-specific diagnostic marker is particularly important and accurate. In this study, the *Pol* gene is reported as a species-specific diagnostic marker to identify three *T. matsutake* varieties and 10 other types of edible mushrooms through qualitative and quantitative PCR. The PCR results did not reveal variations in the amplified region, and the detection limits of qualitative and quantitative PCR were found to be 8 ng and 32 pg, respectively. Southern blot showed that the Pol gene exists as a single copy in the *T. matsutake* genome. The method that produced the purest DNA of *T. matsutake* in this study was also determined, and the high-concentration salt precipitation method was confirmed to be the most suitable among the methods tested. The assay proposed in this work is applicable not only to the detection of raw materials but also to the examination of processed products containing *T. matsutake*.

Keywords: *Tricholoma matsutake*; *Pol* gene; qualitative and quantitative PCR; DNA extraction

1. Introduction

Tricholoma matsutake is an ectomycorrhizal agaricomycete predominantly associated with pines and oaks. It is a commercially valuable edible mushroom [1] with great significance, not only because of its delicate flavor but also because of its diverse biological properties [2], which include multiple immunostimulatory, hematopoietic, antineoplastic, antimutation, and antioxidation activities [3].

The growing number of *T. matsutake* consumers has steadily expanded market demands for this mushroom. On account of the continued deterioration of ecological systems and the environment and excessive picking, the natural productivity of this gourmet mushroom has gradually declined. To date, no artificial cultivation method has yet been developed for *T. matsutake* due to the lack of information concerning precise soil requirements and cues for sporophore formation [4]. Therefore, wild *T. matsutake* and its products are in a short supply in the market, and, as such, its economic value has risen sharply. To address demands for the mushroom and reap higher profits, merchants frequently pass off counterfeit or adulterated *T. matsutake* products as genuine items, for example, sliced dried *Agaricus blazei* [5]. Establishing an effective and convenient method for identifying authentic *T. matsutake* is an urgent necessity.

Traditional methods of identifying *T. matsutake* are generally divided into two categories, namely, morphological [6] and physicochemical methods [7], and each category presents some limitations [8]. Morphological methods are effective in identifying fresh and integrated sporophores. Unfortunately, identification becomes much more difficult when applied to processed products on account of the destruction of the morphological characteristics of the mushroom. Physicochemical methods, such as gas chromatography, high-performance liquid chromatography, and mass spectrometry, require selection of a unique compound or fingerprint to represent *T. matsutake*, and the detection result is easily influenced by variety, habitat, and processing method [9,10]. Thus, two or more physicochemical means are commonly combined to obtain a valid result. If an isomeride exists in the analyte, detection becomes even more complicated. When the unique compound is added artificially, the physicochemical method may fail.

Rapid developments in modern biotechnology have enabled the wide use of molecular biological methods for authentication due to their accuracy, convenience, and speed. PCR-based methods, in particular, have been broadly adopted by many laboratories and fields [11,12]. Compared with conventional PCR, quantitative PCR technology has realized the leap of PCR from qualitative to quantitative, and it has higher specificity, effective resolution and higher degree of automation, and widely used in many fields such as gene expression research, transgenic research, drug efficacy assessment, pathogen detection and food composition analysis, especially. Therefore, real-time quantitative PCR detection is considered to be an easy-to-use, accurate, specific, sensitive, and quantitative method [13–15]. Furthermore, the species-specific diagnostic marker (also denoted as endogenous reference gene in some literatures) is a significant parameter during PCR amplification, which can evaluate the quality of the extracted DNA and provide the means to quantify the amount of the tested DNA substance in the processed food samples [15].

Identification of species-specific diagnostic marker, which requires species specificity, a consistent and low copy number, and low heterogeneity in the same species [13,14], has principally focused on crops. Numerous species-specific diagnostic marker of have been developed and reported [13], *vicK* for *Staphylococcus aureus* [15], Ribosomal Protein *L21* [16] for genetically modified (GM) wheat; *CruA* [17], *PEP* [18], *HMG-I/Y* [19], *FatA* [20], and *BnAccg8* [21] for canola; *lectin* [17,22], *β-actin* [17], and *hsp* [23] for soybean; *Cotton-ppi-PPF* [24], *ACP1* [25], *Sad1* [26], and *SAH7* [27] for cotton; *hmga* [28–30], *10kDa zein* [30–33], *Ivr1* [23,30,34,35], *zSSIIb* [22,36,37], and *Adh1* [30] for maize; and *gos9* [38], *RBE4* [39], and *SPS* [40] for rice. Appropriate reference genes under abiotic stress of annual ryegrass also was selected [41]. However, species-specific diagnostic markers for *T. matsutake* have yet to be established or reported.

Extracting high-quality DNA from *T. matsutake* fruiting bodies is crucial for downstream molecular experiments. At present, the cetyltrimethyl ammonium bromide (CTAB) method, which is especially suitable for plant DNA extraction, is commonly used to extract mushroom DNA [42]. However, high-quality DNA from *T. matsutake* fruiting bodies is difficult to acquire using this method because the mushroom contains thick cytoderms or capsules [43] and is rich in polyphenols and viscous polysaccharides. Thus, exploring a suitable DNA isolation method for *T. matsutake* fruiting bodies is an important endeavor.

In this work, the *Pol* gene, which encodes RNase H and integrase, was selected as a candidate species-specific diagnostic DNA marker for *T. matsutake* through sequence alignment and BLAST. The gene was verified to be a valid species-specific diagnostic marker given its features of appearance as a single copy in genomic DNA, good species specificity, and real-time quantitative detection limit as low as 32 pg. A suitable method for extracting the DNA of *T. matsutake*, i.e., high-concentration salt precipitation method, was also determined. The benefits of the present study are multifold: it provides a convenient and accurate approach for detecting *T. matsutake* and its related processed products, it aids efforts to improve food safety and detect food adulteration, and it develops an effective technique for detecting adulterations in the wild-mushroom market.

2. Results and Discussion

2.1. Comparison of DNA Extraction Methods

The DNA of *T. matsutake* fruiting bodies from Yunnan was obtained using the CTAB, SDS-CTAB, high-concentration salt precipitation, and kit methods. The results of 1% agarose gel for genomic DNA (Figure 1A) indicated that the four DNA extraction methods could successfully extract genomic DNA with good integrity. All of the OD260/280 and DNA concentrations obtained are listed in Table 1, there are significant differences in the DNA concentrations, and the OD260/280 ratio of the High-concentration salt precipitation method compared with other three extraction methods reveals that the purity of genomic DNA extracted by this method is the highest.

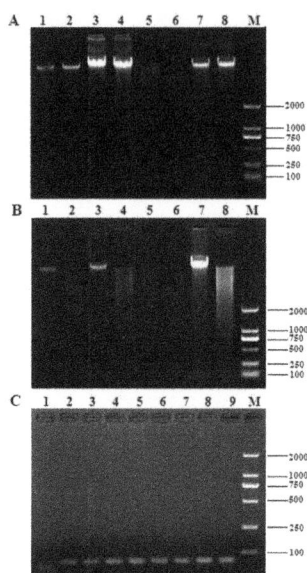

Figure 1. The effect of different DNA extraction methods. (**A**) The electrophoresis profile of *T. matsutake* genomic DNA; (**B**) the digestion result of *T. matsutake* genomic DNA. 1–2: cetyltrimethyl ammonium bromide (CTAB); 3–4: SDS-CTAB; 5–6: high-salt enzymolysis; 7–8: fungal genomic DNA isolation kit; M: DNA marker DL 2000; (**C**) the electrophoresis profile of PCR products of Tricholoma matsutake genomic DNA amplified by 18S universal primers. 1: negative control; 2–3: CTAB; 4–5: SDS-CTAB; 6–7: high-salt enzymolysis; 8-9: fungal genomic DNA isolation kit; M: DNA marker DL 2000.

Table 1. Comparison of purity and concentration of DNA extracted by four methods ($\bar{a} \pm$ SD, $n = 3$).

Method	OD260/280	Concentration (ng/µL)
CTAB	2.01 ± 0.02 [a]	983.33 ± 2.89 [a]
SDS-CTAB	1.97 ± 0.01 [a]	1138.33 ± 2.89 [b]
High-concentration salt precipitation	1.88 ± 0.02 [b]	676.67 ± 0.01 [c]
Kit method	1.66 ± 0.11 [c]	163.33 ± 2.89 [d]

[a-d] the upper letter of the OD260/280 and DNA concentrations indicates a significant difference (Duncan test, $p < 0.05$).

The restriction enzyme digestion result in Figure 1B showed that the genomic DNA extracted by the four methods could be digested by *EcoR* I, and the products were in a dispersive state. However, the digestibility of genomic DNA extracted by the CTAB, SDS-CTAB, and kit methods was lower than that of the high-concentration salt precipitation method, and there were a large number of large fragments, which showed brighter digestion products in the upstream part of the lanes, while the high-concentration salt precipitation method was more uniform. It revealed the high-concentration salt precipitation method was superior in quality and has less enzyme inhibitors.

The PCR products of the 18S rRNA region amplified with the fungal universal 18S primers were staining to identify whether the extracted *T. matsutake* genomic DNA can be used for subsequent PCR amplification. As shown in Figure 1C, the 18S rRNA fragments were efficiently amplified using the DNA extracted by the four methods.

In summary, by integrating the above four indicators shown in Table 1 and Figure 1, all genomic DNA extracted by the four methods could meet with the requirements of the PCR experiment, however, the high-concentration salt precipitation method was confirmed to be the most suitable DNA extraction way among the methods tested.

2.2. Species-Specific Diagnostic Marker of T. Matsutake

Among the genes selected, the *Pol* gene (*Tricholoma matsutake pol* gene for polyprotein encoding RNase H and integrase, Genbank No. AB016926) showed no homology with other genes from different varieties of the mushroom.

It has been reported that retrotransposons have been incorporated into the genome of their hosts and inherited to the host progenies since the earliest establishment of their parasitism. In a previous study, this *Pol* gene was a retroelement from *T. matsutake*, which related to RNase H and integrase of retrotransposons [44]. The reverse transcriptase domain was found in *T. matsutake* worldwide, this finding suggested that retroelements associate with ectomycorrhizal basidiomycetes and might be useful as genetic markers for identification, phylogenetic analysis, and mutagenesis of this fungal group [45]. After the BLAST analysis of *Pol* gene, it just confirmed the statement above. Hence, we chose the *Pol* gene as the candidate species-specific diagnostic marker of *T. matsutake*.

2.3. Species Specificity of Qualitative PCR aSSAYS

The genomic DNA isolated from 10 non-*T. matsutake* species (*R. virescens, A. deliciosus, B. speciosus, T. albuminosus, A. blazei, L. edodes, P. eryngii, F. velutipes,* and *P. ostreatus*) and three *T. matsutake* strains (Yunnan, Sichuan, and Jilin) was subjected to 18s rDNA amplification (Figure 2A). The 18s rDNA primer pair was used to confirm all the extracted DNA can be effectively amplified, and the quality of the extracted DNA clearly met the conditions of PCR (Figure 2A).

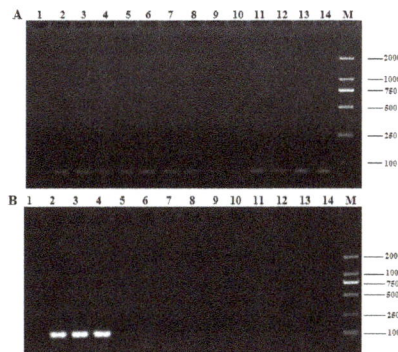

Figure 2. Specificity of the *Pol* gene detection in qualitative PCR. (**A**) The electrophoresis profiles of the DNA products amplified by the 18S rDNA; (**B**) the species specific identification of the *Pol* gene from *T. matsutake* by conventional PCR. 1: negative control, 2: Yunnan Shangri-La *T. matsutake*, 3: Sichuan Ganzi *T. matsutake*, 4: Jilin Yanbian *T. matsutake*, 5: *Russula virescens*, 6: *Agaricus deliciosus*, 7: *Boletus speciosus Forst*, 8: *Termitornyces albuminosus*, 9: *Agaricus blazei*, 10: *Agrocybe aegerita*, 11: *Lentinula edodes*, 12: *Pleurotus eryngii*, 13: *Flammulina velutipes*, 14: *Pleurotus ostreatus*, M: DNA marker DL 2000.

The primer pair *Pol-F/R* was applied to the qualitative PCR of the *Pol* gene; PCRs were also conducted using the genomic DNA of the 13 mushroom samples indicated above. Electrophoretic analysis of all qualitative PCR products (Figure 2B) revealed no objective product, except in the PCR products of the three *T. matsutake* samples. The results confirm that the qualitative PCR applied in this work are highly specific for *T. matsutake*.

2.4. Homology Analysis of the Pol Gene among Different T. Matsutake Varieties

An excellent species-specific diagnostic marker should have low heterogeneity and a consistent copy number in the same species. We carried out qualitative PCR using identical amounts of DNA from the three *T. matsutake* strains to determine whether the *Pol* gene undergoes any sort of variation. Amplification products with identical sizes and relative intensities were obtained for all varieties after qualitative PCR (Figure 2B, lines 2–4), and the slight differences, which were attributed to the quality of the isolated DNA [13], were considered negligible. As shown in Figure 3, the homologous similarity sequence identity between the PCR products and the reference *Pol* gene was 94.51%. These results indicate that the *Pol* gene did not show the sequence variation among the *T. matsutake* varieties studied.

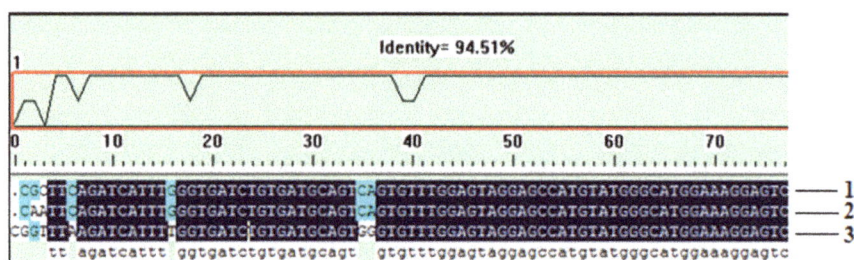

Figure 3. The sequencing result of *Pol* gene in different samples. 1, Yunnan Shangri-La *T. matsutake*; 2, Sichuan Ganzi *T. matsutake*; 3, Jilin Yanbian *T. matsutake*.

2.5. Confirmation of the Pol Gene Copy Number by Southern Blot

Besides species specificity, low sequence variation, and a consistent copy number, an excellent species-specific diagnostic marker is expected to possess a low copy number. Therefore, using Southern blot, we analyzed the copy number of the *Pol* gene in two *T. matsutake* varieties gathered from Yunnan and Jilin Provinces. Whether *Hind* III or *EcoR* I was used to digest the genomic DNA of *T. matsutake*, only one hybridization band was found in the nylon membrane (Figure 4), which means the *Pol* gene is only present as a single copy in the *T. matsutake* genome.

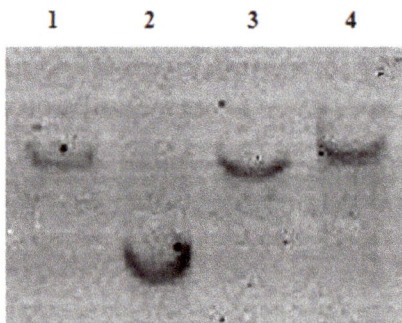

Figure 4. Southern blot result. 1, Jilin Yanbian *T. matsutake* digested by EcoR I; 2, Jilin Yanbian *T. matsutake* digested by Not I; 3, Yunnan Shangri-La *T. matsutake* digested by EcoR I; 4, Yunnan Shangri-La *T. matsutake* digested by Not I.

2.6. Sensitivity of the Qualitative and Taqman-Based Real-Time Quantitative PCR Assays

Genomic DNA from *T. matsutake* was diluted five times from 200 ng/μL to 12.8 pg/μL over a gradient, and the results (Figure 5A) showed a detection limit of 8 ng for qualitative PCR. With the same way, in Figure 5B, the sensitivity of Taqman quantitative PCR was found to be 32 pg. A standard curve of the *Pol* gene was then generated by using the proposed quantitative PCR system, and a linear relationship (R^2 = 0.993) with a slope of −3.081 was determined between the DNA quantities and Ct values (Figure 5C).

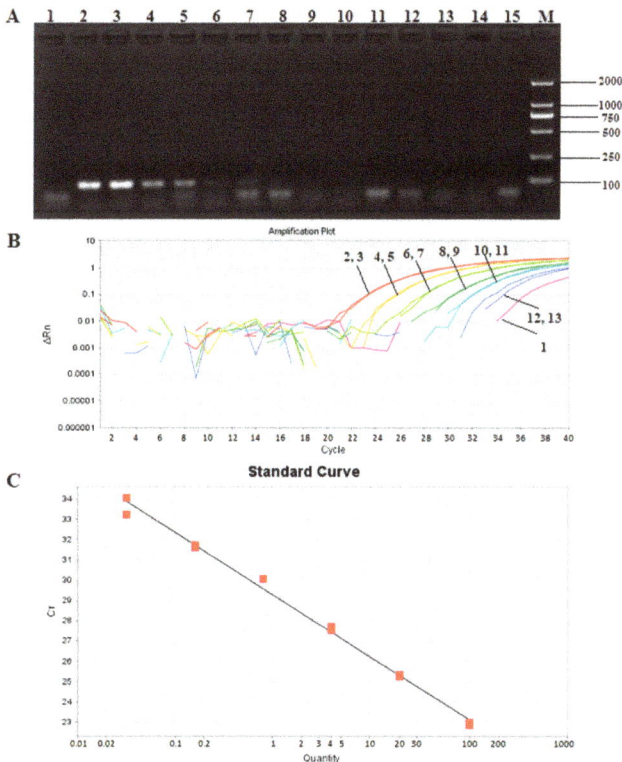

Figure 5. Sensitivity of the *Pol* gene detection in qualitative and Taqman quantitative PCR. (**A**) Sensitivity detection of *Pol* gene by conventional PCR. 1: negative control; 2–3: 200 ng; 4–5: 40 ng; 6–7: 8 ng; 8–9: 1.6 ng; 10–11: 0.32 ng; 12–13: 0.064 ng; 14–15: 0.0128 ng; M: DNA marker DL 2000; (**B**) the amplification curves of *Pol* gene by quantitative PCR. 1: negative control; 2–3: 100 ng; 4–5: 20 ng; 6–7: 4 ng; 8–9: 0.8 ng; 10–11: 160 pg; 12–13: 32 pg; (**C**) the standard curve of the sensitivity detection of Pol gene by qPCR.

2.7. Application of the Pol gene to Detect Processed T. Matsutake *Products*

We used the established Taqman-based PCR system to detect the source of *T. matsutake* in processed products, including mushroom biscuit, oil, and sauce. The target products could be amplified by employing Taqman-based quantitative PCR with the primer *Pol-F/R* and the probe *Pol-P*. The PCR results (Figure 6A) were consistent with the ingredients listed on the packaging of each products. The results of real-time quantitative PCR (Figure 6B) were in keeping with those of qualitative PCR. So, the results of both assays were prove that *T. matsutake* biscuits tested contained *T. matsutake* and the *Pol* gene was a practical and precise species-specific diagnostic marker for *T. matsutake* in highly processed foods, and affirmed its absence, as necessary.

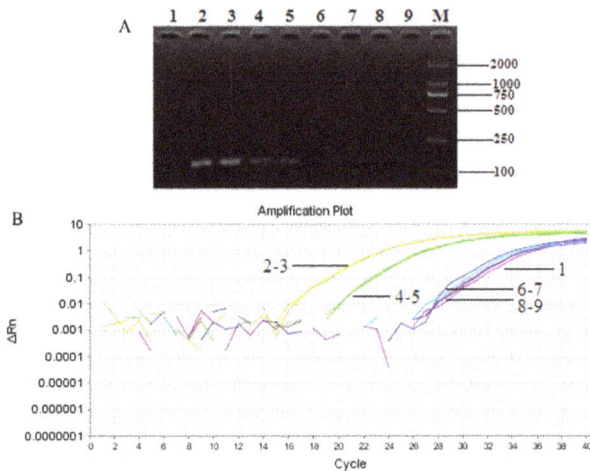

Figure 6. The detection of *T. matsutake* component in different kinds of food samples by qualitative (**A**) and real-time quantitative (**B**) PCR. 1: negative control; 2–3: positive control (Yunnan Shangri-La *T. matsutake*); 4–5: *T. matsutake* biscuits; 6–7: *T. matsutake* sauce; 8–9: Oil field *T. matsutake*; M: DNA marker DL 2000.

3. Materials and Methods

3.1. Materials

The following mushroom species were purchased from local farmers' markets: *Russula virescens, Agaricus deliciosus, Boletus speciosus, Termitornyces albuminosus, Agaricus blazei, Agrocybe cylindracea, Lentinula edodes, Pleurotus eryngii, Flammulina velutipes,* and *Pleurotus ostreatus.* Three *T. matsutake* varieties strains collected from the provinces of Yunnan, Sichuan, and Jilin, China, were kindly supplied by all China Federation of Supply and Marketing Cooperative's Kunming Institute of Edible Fungi. All of the samples were collected in the quantity of 200 g.

Three processed products labeled containing *T. matsutake* component, including mushroom biscuit, oil, and sauce, were gathered from Internet and local markets in Kunming, Yunnan province. They were used to verify the application of this selected species-specific diagnostic marker.

3.2. Genomic DNA Extraction

Four methods, including an improved CTAB method [46], the SDS-CTAB method [47], high-concentration salt precipitation method [48], and the kit method (Rapid Fungi Genomic DNA Isolation Kit, B518229, Sangon, Shanghai, China), were compared to determine the most suitable protocol for extracting the genomic DNA of *T. matsutake*. The *T. matsutake* fruiting bodies from Yunnan was used as the sample, each method was performed in triplicate. First, added the equal silica into 5 g mushroom sample, and then grounded them with liquid nitrogen. The target genomic DNA was isolated from the 0.2 g sample powder. All DNA extraction steps were performed in accordance with the references and manufacturer protocol. At last, the DNA was eluted in 50 μL EB buffer.

The subsequent DNA extraction was followed the selected optimal method. All of the OD260/280 and DNA concentrations were measured with a NanoDrop2000 spectrophotometer (Thermo Scientific, Waltham, MA, USA), the quality of genomic DNA was further analyzed on 1% agarose gel with ethidium bromide (0.1 μg /mL), and run with 1 × TAE buffer. Finally, the DNA solutions were stored at −20 °C.

3.3. Enzyme Digestion of T. Matsutake Genomic DNA

The DNA digestion was performed in a 20 µL reaction system containing 2 µL of 10× H Buffer, 1 µL *EcoR* I (15 U/µL) (TaKaRa Biotechnology Co. Ltd., Dalian, China), 3 µg genomic DNA, and added ultrapure water to 20 µL, and put the tube in a water bath (37 °C) for 2 h, after that added 2 µL 10× loading Buffer (Tiangen, Beijing, China) to stop the reaction. The digestion products was further analyzed on 1% agarose gel with ethidium bromide (0.1 µg /mL), and run with 1× TAE buffer.

3.4. Southern Blot

Complete enzyme cleavage of DNA from *T. matsutake* fruiting bodies obtained from Yunnan and Jilin was performed using *Hind* III and *EcoR* I, respectively, according to the manufacturer's instructions (TaKaRa) to ascertain the copy number of the *Pol* gene in the same species. The cleaved DNA was separated by 0.8% garose gel in 1×TAE at a constant voltage of 20 V overnight. Thereafter, the DNA fragments were transferred to a nylon membrane (Amersham Biosciences Shanghai Ltd., Darmstadt, Germany) from the 0.8% agarose gel. A 775 bp DNA fragment of the *Pol* gene was amplified with *Southern blot*-F/R and used as the hybridized probe, which was labeled with DIG-dUTP (DIG Hybridization Detection Kit, Mylab Co., Beijing, China). Pre-hybridization was carried out at 42 °C for 2 h, after which the pre-hybridization solution was poured off. The probe was denatured into single-strand DNA for 10 min at 100 °C and then cooled for 5 min. Exactly 4 µL of the denatured probe and 8 mL of the hybridization solution were added to a hybridization bag, and hybridization was performed at 42 °C overnight. Finally, the nylon membrane was washed twice with 3× SSC/0.5% SDS, and autoradiography was performed for 2–3 days.

3.5. Species-Specific Diagnostic Marker Selection of T. Matsutake

Genes belonging to *T. matsutake* were sought in Genbank, and several genes were preliminarily selected as detection targets according to their detailed gene information. Afterword, BLAST analysis was performed to determine the homology of the targets. The gene that with the lowest homology compared with all the DNA sequences in Genbank was selected as the candidate.

3.6. Primers and Probe

All primers and TaqMan fluorescent dye-labeled probes were designed using ABI Prism Primer Express version 3.0 software (Applied Biosystems, Foster City, CA, USA) and synthesized by Sangon Co. Ltd. (Shanghai, China).

The universal primer 18S rDNA-F/R was used to evaluate DNA quality. The *Pol* gene was assessed through qualitative PCR with the primer *pol*-F/R and through quantitative PCR with the primer *pol*-F/R, as well as the probe *pol*-P. Southern blot-F/R was used for Southern blot assay. The detailed sequences of these primers and probe are listed in Table 2.

Table 2. Primers used in qualitative and quantitative PCR.

Primer Name	Primer Sequence (5′→3′)	Length	Product Size (bp)	Reference
18S-F	CCTGAGAAACGGCTACCAT	19	80	
18S-R	ATCTTCACTACCTCCCCATTCTG	23		
pol-F	GACTCCCATACTGAAGCCAAT	21	107 (from 69 to 175 bp)	This study
pol-R	ACTCCTTTCCATGCCCATAC	20		
pol-probe	(FAM)-TGGCTCCTACTCCAAACACTGACAC-(TAMRA)			
Southern blot-F	CGTGATGGATGGAATACCTGT	21	775 (from 563 to 1157 bp)	
Southern blot-R	GTGTACCCCCCCTTAGACTGA	21		

3.7. PCR Conditions

Qualitative PCR was performed using an ABI SimpliAmp thermal cycler (Applied Biosystems) in a 25 µL reaction system containing 2.5 µL of 10×buffer, 0.2 mM dNTP, 0.4 µM of each primer, 2.5 units of Taq DNA Polymerase (TaKaRa Biotechnology Co. Ltd.), 100 ng DNA template, and 16.3 µL of ultrapure water. The 18S rDNA amplification was carried out to test DNA quality via the following program: 5 min at 95 °C; 40 cycles of 30 s at 95 °C, 30 s at 58 °C, and 30 s at 72 °C; and 10 min at 72 °C. During *Pol* gene amplification, the procedures applied for 18S rDNA amplification were followed, but annealing was optimized at 56 °C, 58 °C, 60 °C, 62 °C, and 64 °C separately. After comparison, the annealing temperature 60 °C was chosen.

The PCR amplified products were analyzed on 2% agarose gel with ethidium bromide (0.1 µg/mL), and run with 1× TAE buffer. The PCR products, that conducted the homology analysis, were sequenced using *pol*-F/R separately by Sangon (Shanghai, China) company.

Real-time quantitative PCR of the *Pol* gene was also conducted in an ABI StepOne Plus Real-Time System (Applied Biosystems, Thermo Fisher Scientific, Waltham, MA, USA) using the TaqMan probe methods. TaqMan probe-based qPCR was performed using a 20 µL reaction system containing 1×TaqMan Gene Expression Master Mix (TaKaRa, Dalian, China), 200 nM primers, 200 nM probe, and 50 ng of DNA. Each sample was quantified twice for each biological replicate.

3.8. Sensitivity of the Qualitative and Taqman-Based Real-Time Quantitative PCR Assays

Genomic DNA from *T. matsutake* was diluted seven times from 200 ng/µL to 12.8 pg/µL (five-fold serial dilutions) over a gradient using the non- *T. matsutake* DNA, and a series of PCRs with 1 µL DNA sample were conducted to determine the detection limit of qualitative PCR, and the products were analyzed using 2% agarose.

To evaluate the sensitivity of Taqman-based quantitative PCR, genomic DNA was serially diluted six times to final concentrations ranging from 100 ng/µL to 32 pg/µL (five-fold serial dilutions) using the non-*T. matsutake* DNA, and with 1 µL DNA sample, the detection limit of Taqman quantitative PCR was determined.

4. Conclusions

In conclusion, this work demonstrated that High-quality DNA from *T. matsutake* fruiting bodies was obtained following the high-concentration salt precipitation method, and the *Pol* gene was selected and validated as an ideal species-specific diagnostic marker for the PCR-based detection of *T. matsutake* sources after assay of its species specificity, copy number, high homology in different varieties, and sensitivity. The detection limit of Taqman-based quantitative PCR analysis was 32 pg, which means this method could be used to detect processed *T. matsutake* products containing low amounts of the target genomic DNA.

Author Contributions: Methodology, L.S.; formal analysis, D.W.; validation, Y.L. and S.Z.; writing—original draft preparation, L.S.; writing—review and editing, W.X. and Y.S.; supervision, Y.S.; funding acquisition Y.S.

Funding: This work was supported by the National Natural Science Foundation of China (31801635) and the Yunnan Provincial Department of Education Fund for Scientific Research Project (2018JS020).

Conflicts of Interest: The authors declare no conflict of interest.

References

1. Xu, J.P.; Sha, T.; Li, Y.C.; Zhao, Z.W.; Yang, Z.L. Recombination and genetic differentiation among natural populations of the ectomycorrhizal mushroom Tricholoma matsutake from southwestern China. *Mol. Ecol.* **2008**, *17*, 1238–1247. [CrossRef] [PubMed]
2. Geng, X.; Tian, G.; Zhao, Y.; Zhao, L.; Wang, H.; Ng, B.T. A Fungal α-Galactosidase from Tricholoma matsutake with Broad Substrate Specificity and Good Hydrolytic Activity on Raffinose Family Oligosaccharides. *Molecules* **2015**, *20*, 13550–13562. [CrossRef] [PubMed]
3. Yin, X.; You, Q.; Su, X. A comparison study on extraction of polysaccharides from Tricholoma matsutake by response surface methodology. *Carbohydr. Polym.* **2014**, *102*, 419–422. [CrossRef] [PubMed]

4. Vaario, L.-M.; Pennanen, T.; Lu, J.; Palmén, J.; Stenman, J.; Leveinen, J.; Kilpeläinen, P.; Kitunen, V. Tricholoma matsutake can absorb and accumulate trace elements directly from rock fragments in the shiro. *Mycorrhiza* **2015**, *25*, 325–334. [CrossRef] [PubMed]

5. Sun, Y.; Shi, Y. FT-TRS Analysis of Tricholoma matsutake Sing and Agaricus Blazei Murrill. *Phys. Test. Chem. Anal. (Part B Chem. Anal.)* **2013**, *49*, 1076–1078.

6. Zhu, Y.; Zhang, H.; Zhu, Y.; Pu, X. An Appraisal of Past and Present Methods for Identifying Poisonous Mushrooms. *J. Gansu Sci.* **2008**, *20*, 44–48.

7. Liu, Y.; Liu, N.; Zhang, Y. Research progress of food authentication technology. *Sci. Technol. Food Ind.* **2016**, *37*, 374–383.

8. Chen, Y.; Dong, W.; Wu, Y.; Yuan, F.; Huang, W.; Ge, Y. Research and application of food authentication system. *Sci. Technol. Food Ind.* **2008**, *29*, 216–218.

9. Qiang, L.; Li, Z.; Li, W.; Li, X.; Huang, W.; Hua, Y.; Zheng, L. Chemical compositions and volatile compounds of Tricholoma matsutake from different geographical areas at different stages of maturity. *Food Sci. Biotechnol.* **2016**, *25*, 71–77.

10. Ding, X.; Hou, Y.L. Identification of genetic characterization and volatile compounds of Tricholoma matsutake from different geographical origins. *Biochem. Syst. Ecol.* **2012**, *44*, 233–239. [CrossRef]

11. Yang, L.; Yang, Y.; Jin, W.; Zhang, X.; Li, X.; Wu, Y.; Li, J.; Li, L. Development and Interlaboratories Validation of Event-Specific Quantitative Real-Time PCR Method for Genetically Modified Rice G6H1 Event. *J. Agric. Food Chem.* **2018**, *66*, 8179. [CrossRef]

12. Liu, Y.; Jiang, D.; Lu, X.; Wang, W.; Xu, Y.; He, Q. Phage Medicated Immuno-PCR for Ultrasensitive Detection of Cry1Ac protein based on Nanobody. *J. Agric. Food Chem.* **2016**, *64*, 7882–7889. [CrossRef]

13. Xu, W.; Bai, W.; Guo, F.; Luo, Y.; Yuan, Y.; Huang, K. A papaya-specific gene, papain, used as an endogenous reference gene in qualitative and real-time quantitative PCR detection of transgenic papayas. *Eur. Food Res. Technol.* **2008**, *228*, 301–309. [CrossRef]

14. Shang, Y.; Zhu, P.; Huang, K.; Liu, W.; Tian, W.; Luo, Y.; Xu, W. A peach (Prunus persica)-specific gene, Lhcb2, used as an endogenous reference gene for qualitative and real-time quantitative PCR to detect fruit products. *LWT-Food Sci. Technol.* **2014**, *55*, 218–223. [CrossRef]

15. Liu, Z.-M.; Shi, X.-M.; Pan, F. Species-specific diagnostic marker for rapid identification of Staphylococcus aureus. *Diagn. Microbiol. Infect. Dis.* **2007**, *59*, 379–382. [CrossRef]

16. Liu, Y.K.; Li, H.P.; Huang, T.; Cheng, W.; Gao, C.S.; Zuo, D.Y.; Zhao, Z.X.; Liao, Y.C. Wheat-Specific Gene, Ribosomal Protein L21, Used as the Endogenous Reference Gene for Qualitative and Real-Time Quantitative Polymerase Chain Reaction Detection of Transgenes. *J. Agric. Food Chem.* **2014**, *62*, 10405–10413. [CrossRef]

17. James, D.; Schmidt, A.-M.; Wall, E.; Green, M.; Masri, S. Reliable Detection and Identification of Genetically Modified Maize, Soybean, and Canola by Multiplex PCR Analysis. *J. Agric. Food Chem.* **2003**, *51*, 5829–5834. [CrossRef]

18. Zeitler, R.; Pietsch, K.; Waiblinger, H.-U. Validation of real-time PCR methods for the quantification of transgenic contaminations in rape seed. *Eur. Food Res. Technol.* **2002**, *214*, 346–351. [CrossRef]

19. Weng, H.; Yang, L.; Liu, Z.; Ding, J.; Pan, A.; Zhang, D. Novel Reference Gene, High-mobility-group protein I/Y, Used in Qualitative and Real-Time Quantitative Polymerase Chain Reaction Detection of Transgenic Rapeseed Cultivars. *J. AOAC Int.* **2005**, *88*, 577–584.

20. Demeke, T.; Ratnayaka, I. Multiplex qualitative PCR assay for identification of genetically modified canola events and real-time event-specific PCR assay for quantification of the GT73 canola event. *Food Control* **2008**, *19*, 893–897. [CrossRef]

21. Hernández, M.; Río, A.; Esteve, T.; Prat, S.; Pla, M. A Rapeseed-Specific Gene, Acetyl-CoA Carboxylase, Can Be Used as a Reference for Qualitative and Real-Time Quantitative PCR Detection of Transgenes from Mixed Food Samples. *J. Agric. Food Chem.* **2001**, *49*, 3622–3627. [CrossRef]

22. Yoshimura, T.; Kuribara, H.; Matsuoka, T.; Kodama, T.; Iida, M.; Watanabe, T.; Akiyama, H.; Maitani, T.; Furui, S.; Hino, A. Applicability of the Quantification of Genetically Modified Organisms to Foods Processed from Maize and Soy. *J. Agric. Food Chem.* **2005**, *53*, 2052–2059. [CrossRef]

23. Van Duijn, G.; Biert, R.V.; Bleeker-Marcelis, H.; Peppelman, H.; Hessing, M. Detection methods for genetically modified crops. *Food Control* **1999**, *10*, 375–378. [CrossRef]

24. Chaouachi, M.; Giancola, S.; Romaniuk, M.; Laval, V.; Bertheau, Y.; Brunel, D. A Strategy for Designing Multi-Taxa Specific Reference Gene Systems. Example of Application—ppi Phosphofructokinase (ppi-PPF) Used for the Detection and Quantification of Three Taxa: Maize (*Zea mays*), Cotton (*Gossypium hirsutum*) and Rice (*Oryza sativa*). *J. Agric. Food Chem.* **2007**, *55*, 8003–8010. [CrossRef]

25. Lee, S.-H.; Kim, J.-K.; Yi, B.-Y. Detection Methods for Biotech Cotton MON 15985 and MON 88913 by PCR. *J. Agric. Food Chem.* **2007**, *55*, 3351–3357. [CrossRef]

26. Yang, L.; Chen, J.; Huang, C.; Liu, Y.; Jia, S.; Pan, L.; Zhang, D. Validation of a cotton-specific gene, Sad1, used as an endogenous reference gene in qualitative and real-time quantitative PCR detection of transgenic cottons. *Plant Cell Rep.* **2005**, *24*, 237–245. [CrossRef]

27. Baeumler, S.; Wulff, D.; Tagliani, L.; Song, P. A Real-Time Quantitative PCR Detection Method Specific to Widestrike Transgenic Cotton (Event 281-24-236/3006-210-23). *J. Agric. Food Chem.* **2006**, *54*, 6527–6534. [CrossRef]

28. Zimmermann, A.; Hemmer, W.; Liniger, M.; Lüthy, J.; Pauli, U. A Sensitive Detection Method For Genetically Modified MaisGardTMCorn Using a Nested PCR-system. *LWT-Food Sci. Technol.* **1998**, *31*, 664–667. [CrossRef]

29. Pardigol, A.; Guillet, S.; Pöpping, B. A simple procedure for quantification of genetically modified organisms using hybrid amplicon standards. *Eur. Food Res. Technol.* **2003**, *216*, 412–420. [CrossRef]

30. Hernández, M.; Duplan, M.-N.; Berthier, G.; Vaïtilingom, M.; Hauser, W.; Freyer, R.; Pla, M.; Bertheau, Y. Development and Comparison of Four Real-Time Polymerase Chain Reaction Systems for Specific Detection and Quantification of *Zea mays* L. *J. Agric. Food Chem.* **2004**, *52*, 4632–4637. [CrossRef]

31. Vaïtilingom, M.; Pijnenburg, H.; Gendre, F.; Brignon, P. Real-Time Quantitative PCR Detection of Genetically Modified Maximizer Maize and Roundup Ready Soybean in Some Representative Foods. *J. Agric. Food Chem.* **1999**, *47*, 5261–5266. [CrossRef] [PubMed]

32. Höhne, M.; Santisi, C.; Meyer, R. Real-time multiplex PCR: An accurate method for the detection and quantification of 35S-CaMV promoter in genetically modified maize-containing food. *Eur. Food Res. Technol.* **2002**, *215*, 59–64. [CrossRef]

33. Matsuoka, T.; Kuribara, H.; Takubo, K.; Akiyama, H.; Miura, H.; Goda, Y.; Kusakabe, Y.; Isshiki, K.; Toyoda, M.; Hino, A. Detection of Recombinant DNA Segments Introduced to Genetically Modified Maize (*Zea mays*). *J. Agric. Food Chem.* **2002**, *50*, 2100–2109. [CrossRef] [PubMed]

34. Brodmann, P.D.; Ilg, E.C.; Berthoud, H.; Herrmann, A. Real-Time Quantitative Polymerase Chain Reaction Methods for Four Genetically Modified Maize Varieties and Maize DNA Content in Food. *J. AOAC Int.* **2002**, *85*, 646–653. [PubMed]

35. Rønning, S.B.; Vaïtilingom, M.; Berdal, K.G.; Holst-Jensen, A. Event specific real-time quantitative PCR for genetically modified Bt11 maize (*Zea mays*). *Eur. Food Res. Technol.* **2003**, *216*, 347–354. [CrossRef]

36. Kuribara, H.; Shindo, Y.; Matsuoka, T.; Takubo, K.; Futo, S.; Aoki, N.; Hirao, T.; Akiyama, H.; Goda, Y.; Toyada, M.; et al. Novel Reference Molecules for Quantitation of Genetically Modified Maize and Soybean. *J. AOAC Int.* **2002**, *85*, 1077–1089.

37. Lee, S.-H.; Min, D.-M.; Kim, J.-K. Qualitative and Quantitative Polymerase Chain Reaction Analysis for Genetically Modified Maize MON863. *J. Agric. Food Chem.* **2006**, *54*, 1124–1129. [CrossRef]

38. Hernández, M.; Esteve, T.; Pla, M. Real-Time Polymerase Chain Reaction Based Assays for Quantitative Detection of Barley, Rice, Sunflower, and Wheat. *J. Agric. Food Chem.* **2005**, *53*, 7003–7009. [CrossRef]

39. Jeong, S.-C.; Pack, I.S.; Cho, E.-Y.; Youk, E.S.; Park, S.; Yoon, W.K.; Kim, C.-G.; Choi, Y.D.; Kim, J.-K.; Kim, H.M. Molecular analysis and quantitative detection of a transgenic rice line expressing a bifunctional fusion TPSP. *Food Control* **2007**, *18*, 1434–1442. [CrossRef]

40. Ding, J.; Jia, J.; Yang, L.; Wen, H.; Zhang, C.; Liu, W.; Zhang, D. Validation of a Rice Specific Gene, Sucrose Phosphate Synthase, Used as the Endogenous Reference Gene for Qualitative and Real-Time Quantitative PCR Detection of Transgenes. *J. Agric. Food Chem.* **2004**, *52*, 3372–3377. [CrossRef]

41. Liu, Q.; Qi, X.; Yan, H.; Huang, L.; Nie, G.; Zhang, X. Reference Gene Selection for Quantitative Real-Time Reverse-Transcriptase PCR in Annual Ryegrass (*Lolium multiflorum*) Subjected to Various Abiotic Stresses. *Molecules* **2018**, *23*, 172. [CrossRef] [PubMed]

42. Jiang, S.; Shao, B.; Chen, W.; Li, S.; Wang, Z.; Zhun, X.; Liao, J. Study on the Comparison of three Genomic DNA Extraction Methods for 15 Familiar Edulis and Medicincal Fungi. *Food Sci.* **2004**, *25*, 36–40.

Molecules **2019**, *24*, 455

43. González-Mendoza, D.; Argumedo-Delira, R.; Morales-Trejo, A.; Pulido-Herrera, A.; Cervantes-Díaz, L.; Grimaldo-Juarez, O.; Alarcón, A. A rapid method for isolation of total DNA from pathogenic filamentous plant fungi. *Genet. Mol. Res.* **2010**, *9*, 162–166. [CrossRef] [PubMed]

44. Murata, H.; Yamada, A.; Babasaki, K. Identification of Repetitive Sequences Containing Motifs of Retrotransposons in the Ectomycorrhizal Basidiomycete Tricholoma matsutake. *Mycologia* **1999**, *91*, 766–775. [CrossRef]

45. Murata, H.; Yamada, A. marY1, a Member of the gypsy Group of Long Terminal Repeat Retroelements from the Ectomycorrhizal Basidiomycete Tricholoma matsutake. *Appl. Environ. Microbiol.* **2000**, *66*, 3642–3645. [CrossRef]

46. Nan, Z.; Xu, W.; Bai, W.; Zhai, Z.; Luo, Y.; Yan, X.; Jing, H.; Huang, K. Event-specific qualitative and quantitative PCR detection of LY038 maize in mixed samples. *Food Control* **2011**, *22*, 1287–1295.

47. Shang, Y.; Yan, Y.; Xu, W.; Tian, W.; Huang, K. Research on Gene Mobility and Gene Flow Between Genetically Modified Mon 15985 Cotton and Pleurotus Ostreatus. *J. Food Saf.* **2016**, *36*, 423–432. [CrossRef]

48. Yan, Y.; Xu, W.; Su, C.; Luo, Y.; Wang, Y.; Gu, X.; Dai, Y.; Tian, H. Study on the methods of extraction of Pleurotus ostreatus genomic DNA. *Sci. Technol. Food Ind.* **2011**, 190–193.

Sample Availability: Samples of the compounds are not available from the authors.

molecules

MDPI

Article

Small Molecular Weight Aldose (D-Glucose) and Basic Amino Acids (L-Lysine, L-Arginine) Increase the Occurrence of PAHs in Grilled Pork Sausages

Wen Nie [1], Ke-zhou Cai [1,2,*](ORCID), Yu-zhu Li [1], Shuo Zhang [1], Yu Wang [1], Jie Guo [1], Cong-gui Chen [1,2] and Bao-cai Xu [1,3]

[1] School of Food Science and Engineering, Hefei University of Technology, Hefei 230009, Anhui Province, China; 17356511419@163.com (W.N.); 1932526946@163.com (Y.L.); 18856023156@163.com (S.Z.); wyll_ah92@163.com (Y.W.); hfutkaylee@163.com (J.G.); chencg1629@hfut.edu.cn (C.C.); baocaixu@163.com (B.X.)

[2] Key Laboratory on Deep Processing of Agricultural Products for Anhui Province, Hefei 230009, Anhui Province, China

[3] Engineering Research Centre of Bio-Process, Ministry of Education, Hefei 230009, Anhui Province, China

* Correspondence: kzcai@hfut.edu.cn; Tel.: +15305516697

Academic Editor: Alessandra Gentili
Received: 22 November 2018; Accepted: 16 December 2018; Published: 19 December 2018

check for updates

Abstract: (1) Background: Amino acids and carbohydrates are widely used as additives in the food industry. These compounds have been proven to be an influencing factor in the production of chemical carcinogenic compounds polycyclic aromatic hydrocarbons (PAHs). However, the effect of the properties of the amino acids and carbohydrates on the production of PAHs is still little known. (2) Methods: We added different (i) R groups (the R group represents an aldehyde group in a glucose molecule or a ketone group in a fructose molecule); (ii) molecular weight carbohydrates; (iii) polarities, and (iv) acid-base amino acids to pork sausages. The effects of the molecular properties of carbohydrates and amino acids on the formation of PAHs in grilled pork sausages were investigated. (3) Results: The results showed that a grilled sausage with aldehyde-based D-glucose was capable of producing more PAHs than a sausage with keto-based D-fructose. A higher PAH content was determined in the grilled pork sausage when the smaller molecular weight, D-glucose, was added compared with the sausage where the larger molecular weight, 4-(α-D-glucosido)-D-glucose and cellulose were added. The addition of basic amino acids (L-lysine, L-arginine) was capable of producing more PAHs compared with the addition of acidic amino acids (L-glutamic acid, L-aspartate). When amino acid containing a benzene ring was added, a smaller volume of PAHs was produced compared with the addition of other amino acids. (4) Conclusions: Our study suggests that systematic consideration of molecule properties is necessary when using food additives (amino acids and carbohydrates) for food processing.

Keywords: amino acids; carbohydrates; acidity; polarity; molecular weight

1. Introduction

Polycyclic aromatic hydrocarbons (PAHs), formed through incomplete combustion of wood or gasoline, are regarded as potentially genotoxic and carcinogenic to humans [1,2]. The occurrence of PAHs in food, such as edible oils, meat, and dairy products, has been regarded as a consequence of high-temperature processing [3]. Previous epidemiological studies have speculated that food consumption might contribute 88–98% of PAHs exposure. This is especially the case in populations that are non-smoking and not subject to occupational exposure [4]. Among the sources of PAHs exposure, meat products account for the largest proportion. The formation and control of PAHs during the processing of meat products is topical in current research.

The definite mechanism for the formation of PAHs is not well understood. Some researchers proposed that they might be formed through free radical reactions, intramolecular addition or the polymerization of small molecules [5–7]. In meat products, PAHs are formed during processing at a high temperature, such as by smoking, drying, roasting, and grilling [8–10]. Grilled sausages are a common food and can be contaminated with high concentrations of PAHs [11]. Amino acids, a kind of food additive, are often used to improve the sensory properties (color, taste, and texture) of meat products to meet the needs of consumers [12,13]. Some scholars have studied the pyrolysis of aliphatic α-amino acids, glutamine, glutamic acid, and aspartic acid and the formation of PAHs. The results show that single chemical components of glutamine, glutamic acid, and aspartic acid can pyrolyze to form PAHs at high temperatures. The author speculates that all three acids appear to break down to smaller "building blocks" and then pyrosynthesize the PAHs [14]. Other scholars have explored whether glucose can enhance the pyrolysis of proline to form PAHs and found that glucose can provide a low-temperature pathway for the decomposition of proline, thus, enhancing the pyrolysis of proline to form PAHs [15]. The above studies indicate that some amino acids can be pyrolyzed to form PAHs in a single simulated system, and glucose can promote the pyrolysis of proline to form PAHs. However, meat products are a complex matrix. The formation of PAHs in meat products by glucose and amino acids is significantly different from the formation of PAHs by the pyrolysis of glucose and amino acids in a single chemical system. There is still no information on the effect of amino acid types on the PAHs formation in cooked meat.

In addition to nutritional properties, carbohydrates can also be used as a sweetener, gel, thickener, or stabilizer in the meat industry. The physical and chemical properties, the degree of polymerization, solubility, viscosity, film formation, and gelation greatly influence the retention and release of flavor [16,17]. In addition, carbohydrates retain aromatic components and reduce flavor loss during meat processing [17]. Other studies considered using polysaccharide to promote water holding capacity and texture [18]. Although many carbohydrates have been successfully used to increase the flavor, color, and texture of meat, there is still limited research on their safety including the effect of PAHs production in cooked meat.

Therefore, it is of interest to specifically study the effects of amino acids and carbohydrate types on PAHs formation in cooked meat. In this paper, we tested the effects of carbohydrates with a different R-base (an aldehyde group in a glucose molecule and a ketone group in a fructose molecule) and different molecular weights, and the property of amino acids (with different polarities and a different acid-base) on the formation of twelve kinds of PAHs in grilled sausages. All of these PAHs are noted as potential food contaminants by the United States Environmental Protection Agency [19].

2. Results and Discussion

2.1. Validation

The detection limits, recoveries, and precision for twelve kinds of PAHs data are presented in Table 1. Linearity was determined using external standard plot method. The HPLC of all 12 PAHs in the master sample is presented in Figure 1. The high R^2 values indicated good linearity over the concentration range. As can be seen, the limit of detection (LOD) and limit of quantification (LOQ) of all PAHs ranged from 0.03 to 0.18 µg kg^{-1} and 0.10 to 0.91 µg kg^{-1} and met the Commission of the European Communities (2011b). Relative standard deviations (RSD) and recovery in our study ranged from 1.23 to 9.71% and 79.22 to 105.78%, respectively. According to the regulation [20], recovery sets should be in the range of 50–120%; recovery in the experiments varied between 79.22% and 105.78%, which is in good accordance with the regulation. The precision was adequate, with RSD < 10%. It has been well established that the presence of some impurities in meat samples, such as aliphatic hydrocarbons, fatty acids, phenols, and polycyclic organic compounds, may greatly reduce the extraction efficiency of PAHs [21].

Table 1. Parameters of validation of 12 polycyclic aromatic hydrocarbons (PAHs) in sausages.

PAHs	Linear Range (ng mL^{-1})	Correlation Coefficients (R^2)	LOD (μg kg^{-1})	LOQ (μg kg^{-1})	Recovery [a] (%)	RSD [b] (%)
NA	0.10–10.00	0.9999	0.05	0.17	105.78	1.23
AC	0.10–10.00	0.9996	0.18	0.60	93.46	1.46
FLU	0.20–10.00	0.9999	0.06	0.20	84.54	5.54
FLT	0.20–10.00	0.9993	0.16	0.53	83.83	7.89
BaA	0.25–5.00	0.9998	0.09	0.30	90.52	4.37
CHR	0.30–7.50	0.9999	0.10	0.91	92.56	3.42
BbFA	1.00–20.00	0.9995	0.06	0.20	96.74	1.90
BkFA	1.00–20.00	0.9999	0.08	0.27	95.33	2.05
BaP	0.20–10.00	0.9999	0.11	0.37	91.31	4.55
DBahA	0.10–10.00	0.9991	0.03	0.10	84.53	9.71
BPE	0.20–10.00	0.9998	0.04	0.20	88.34	4.56
IPY	0.30–10.00	0.9997	0.06	0.30	79.22	6.18

Explanations: LOD–limit of detection; LOQ–limit of quantification; RSD–relative standard deviation of repeatability; [a] mean recoveries of four different spiking levels in triplicate in the same day; [b] mean relative standard deviations (RSD) of two different spiking levels in triplicates on two different day; Naphthalene (NA), Acenaphthene (Ac), Fluorene (FLU), Fluoranthene (FLT), Benzo[a]anthracene (BaA), Chrysene (CHR), Benzo[b]fluoranthene (BbFA), Benzo[k]fluoranthene (BkFA), Benzo[a]pyrene (BaP), Dibenzo[a,h]anthracene (DBahA), Benzo[g,h,i]perylene (BPE), Indeno[1,2,3-c,d]pyrene (IPY).

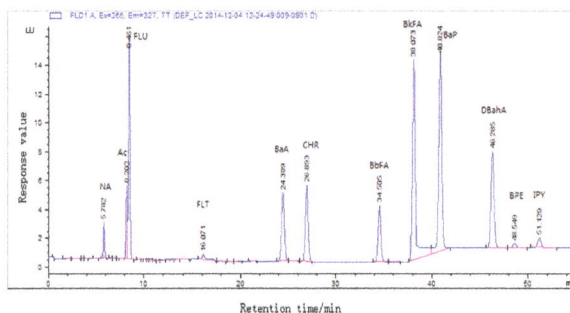

Figure 1. Standard chromatogram of 12 polycyclic aromatic hydrocarbons.

2.2. Effect of Carbohydrate Characteristics on PAHs Content

Carbohydrate, as an important nutrient of meat and a common additive in meat products, was reported to have a notable influence on PAHs production during meat processing [22,23]. In this study, the effects of aldose (D-glucose) and ketose (D-fructose) on PAHs production during meat processing were investigated by adding D-glucose and D-fructose to grilled pork sausages. The D-glucose, 4-(α-D-Glucosido)-D-glucose, and cellulose were used to explore the effect of carbohydrate molecular weight on the formation of PAHs during meat processing. The results showed there was no significant difference in the content of PAHs between the grilled pork sausage with D-fructose and the blank control sausage ($p > 0. 05$). However, compared with these two groups, the occurrence of PAHs in the grilled pork sausages with D-glucose was significantly increased ($p < 0.05$) (details are shown in Table 2). The reason for this phenomenon might be due to D-glucose containing free aldehyde groups, which can be decomposed to smaller molecular aldehyde compounds at high temperatures, and these aldehydes can be further subjected to complex cracking, polymerization or condensation reaction to ultimately produce PAHs. However, the ketone group in D-fructose needs to be converted into an aldehyde group at a higher temperature to further generate PAHs by a cleavage reaction [24].

Table 2. Concentrations ($n = 3$) of 12 PAHs produced as a consequence of adding different carbohydrates to grilled sausages.

Species	NA	Ac	FLU	FLT	BaA	CHR	BbFA	BkFA	DBahA	BPE	BaP	IPY	$\sum PAH_{12}$
Control group	17.93 ± 1.35 c	12.99 ± 1.02 c	2.79 ± 0.14 c	ND	2.21 ± 0.16 c	3.62 ± 0.10b c	ND	7.52 ± 0.36 b	0.71 ± 0.05 c	1.25 ± 0.08 c	0.88 ± 0.12 d	ND	49.90 ± 3.46 c
D-Fructose	18.36 ± 1.41 c	11.15 ± 1.14 c	3.13 ± 0.26 c	1.06 ± 0.15 c	3.04 ± 0.27 bc	2.47 ± 0.16 d	ND	6.01 ± 0.33 c	1.04 ± 0.11 b	1.37 ± 0.09 b	1.26 ± 0.06 c	ND	48.89 ± 4.05 c
D-Glucose	34.25 ± 1.22 a	23.45 ± 1.68 a	7.66 ± 0.23 a	4.24 ± 0.18 a	7.41 ± 0.36 a	6.75 ± 0.36 a	0.47 ± 0.03 a	13.76 ± 1.28 a	3.16 ± 0.30 a	3.07 ± 0.26 a	5.59 ± 0.22 a	ND	109.81 ± 6.49 a
4-(α-D-Glucosido)-D-glucose	25.17 ± 2.23 b	17.42 ± 1.35 b	5.06 ± 0.44 b	2.31 ± 0.18 b	4.01 ± 0.23 b	4.48 ± 0.24 b	0.29 ± 0.01 b	8.47 ± 0.31 b	1.35 ± 0.10 b	1.54 ± 0.10 b	3.46 ± 0.25 b	ND	73.56 ± 5.67 b
Cellulose	17.16 ± 1.27 c	11.56 ± 1.06 c	3.03 ± 0.21 c	ND	2.44 ± 0.17 c	3.53 ± 0.28 c	ND	6.46 ± 0.31 c	0.82 ± 0.03 bc	1.39 ± 0.11 b	1.12 ± 0.18 cd	ND	47.51 ± 3.38 c

Explanations: the results are expressed as the mean and standard deviation; ND—Means no PAHs detected; a, b, c, d, different letters indicate significant differences in PAHs content between any two rows in Table 2 ($p < 0.05$); Naphthalene (NA), Acenaphthene (Ac), Fluorene (FLU), Fluoranthene (FLT), Benzo[a]anthracene (BaA), Chrysene (CHR), Benzo[b]fluoranthene (BbFA), Benzo[k]fluoranthene (BkFA), Benzo[a]pyrene (BaP), Dibenzo[a,h]anthracene (DBahA), Benzo[g,h,i]perylene (BPE), Indeno[1,2,3-c,d]pyrene (IPY).

When comparing the influence of the carbohydrate's molecular weight on the content of PAHs in grilled pork sausages, it was found that the concentration of PAHs in sausages supplemented with D-glucose or 4-(α-D-Glucosido)-D-glucose was significantly higher than that of the blank control group. However, the content of PAHs in the grilled sausage to which cellulose had been added was not significantly different from that of the control group. In addition, the PAHs content in the grilled sausage with D-glucose was significantly higher than that of in the grilled pork sausage with 4-(α-D-Glucosido)-D-glucose ($p < 0.05$) (details are shown in Table 2). A higher PAHs content was determined in the grilled sausage with smaller molecular weight, D-glucose, when compared with the larger molecular weight, 4-(α-D-Glucosido)-D-glucose and cellulose. However, there was no significant difference in the content of PAHs in grilled pork sausages with the addition of small molecular weight D-fructose and large molecular weight cellulose ($p > 0.05$). This result may be due to the effect of carbohydrates on PAHs production being affected by both the R group and molecular weight, and the effect of R group on PAHs formation was greater than the effect of molecular weight. Some scholars explored the formation of PAHs from the pyrolysis of carbohydrates, amino acids, and fatty acids, and it was found that high-temperature pyrolysis of D-glucose produced more PAHs than starch, which result is consistent with our research [25]. Other scholars studied the mechanism of dehydration, carbonization, decarbonylation, decarboxylation, dehydrogenation, and cross-linking of cellulose at high temperatures when studying the mechanism of pyrolysis of cellulose to form PAHs. It was proposed that PAHs could be formed through transformation and rearrangement [26]. The study also found that the yield of PAHs produced by pyrolysis of D-glucose and sucrose was higher than that of cellulose. This result is consistent with our findings. However, the current mechanism of the molecular weight of carbohydrates on the formation of PAHs during high-temperature processing is still unclear.

2.3. Effect of Amino Acid Properties on PAHs Content

To explore the effects of amino acid polarity and acidity/alkalinity on the occurrence of PAHs in meat processing under high temperature, several different polar amino acids, non-polar amino acids, acidic amino acids, and basic amino acids were chosen to be added to grilled pork sausages. When investigating the effects of three non-polar amino acids on the content of PAHs in grilled pork sausages, it was found that the content of PAHs in sausages supplemented with L-proline was significantly increased compared with the blank control group ($p < 0.05$). However, the content of PAHs in grilled pork sausages added L-tryptophan or L-phenylalanine was not significantly different from that of the blank control group ($p > 0.05$) (details are shown in Table 3). The effects of three polar amino acids on the content of PAHs in grilled pork sausages were investigated. The result showed that the addition of L-threonine or L-serine significantly increased the PAHs content in grilled pork sausages when compared with the blank control group ($p < 0.05$). However, there was no significant difference between the sausage supplemented with L-tyrosine and the blank control group ($p > 0.05$) (details are shown in Table 3). The authors did not find obvious rules when comparing the effects of polar and non-polar amino acids on the formation of PAHs in grilled pork sausages. This indicates that the formation of PAHs in grilled pork sausages is not affected by the polarity of the amino acids.

Table 3. Concentrations ($n = 3$) of 12 PAHs produced as a consequence of adding different amino acids to grilled sausages.

	Species	NA	Ac	FLU	FLT	BaA	CHR	BbFA	BkFA	DBahA	BPE	BaP	IPY	∑PAH12
	Control group	17.93 ± 1.35 de	12.99 ± 1.02 cd	2.79 ± 0.14 cd	ND	2.21 ± 0.16 c	3.62 ± 0.10 c	ND	7.52 ± 0.36 c	0.71 ± 0.05 c	1.25 ± 0.08 cd	0.88 ± 0.12 b	ND	49.90 ± 3.46 d
Non-polar amino acid	L-proline	30.42 ± 2.26 c	27.10 ± 1.54 b	6.17 ± 0.33 b	3.16 ± 0.18 a	5.06 ± 0.27 ab	7.83 ± 0.24 b	1.01 ± 0.04 a	17.43 ± 1.33 a	1.64 ± 0.11 b	2.73 ± 0.13 b	2.23 ± 0.20 a	ND	104.78 ± 8.72 b
	L-tryptophan	18.35 ± 1.42 de	11.43 ± 1.24 d	3.01 ± 0.26 c	ND	1.88 ± 0.13 d	3.59 ± 0.21 c	ND	7.17 ± 0.45 c	0.75 ± 0.07 c	1.16 ± 0.11 d	1.01 ± 0.16 b	ND	48.35 ± 3.18 d
	L-phenylalanine	16.77 ± 1.37 e	13.21 ± 1.10 cd	2.85 ± 0.24 cd	ND	2.35 ± 0.17 c	3.81 ± 0.26 c	ND	8.02 ± 0.43 c	0.67 ± 0.05 c	1.36 ± 0.14 cd	0.85 ± 0.11 b	ND	49.89 ± 4.05 d
Polar amino acid	L-tyrosine	18.19 ± 1.52 de	12.15 ± 1.13 d	2.66 ± 0.22 d	ND	1.99 ± 0.14 cd	3.77 ± 0.28 c	ND	7.86 ± 0.46 c	0.69 ± 0.04 c	1.22 ± 0.11 d	0.94 ± 0.06 b	ND	49.47 ± 4.44 d
	L-threonine	39.45 ± 3.26 b	28.58 ± 2.19 ab	6.82 ± 0.31 b	ND	4.86 ± 0.20 b	7.96 ± 0.30 b	0.86 ± 0.03 a	16.54 ± 1.42a	1.56 ± 0.12 b	2.75 ± 0.21 b	1.96 ± 0.08 a	ND	111.34 ± 9.43 b
	L-serine	37.49 ± 3.33 b	34.01 ± 2.64 a	8.12 ± 0.36 a	3.72 ± 0.20 a	5.78 ± 0.26 ab	9.47 ± 0.82 ab	1.02 ± 0.01 a	19.68 ± 1.54 a	1.85 ± 0.11 b	3.27 ± 0.20 a	2.35 ± 0.17 a	ND	126.76 ± 10.17 a
Basic amino acid	L-lysine	50.48 ± 2.57 a	32.08 ± 1.88 a	8.74 ± 023 a	ND	8.76 ± 0.18 a	12.15 ± 0.32 a	ND	12.78 ± 1.14 b	3.21 ± 0.08 a	3.13 ± 0.10 a	1.52 ± 0.05 ab	ND	132.85 ± 6.13 a
	L-arginine	52.67 ± 3.36 a	30.91 ± 2.49 a	9.64 ± 0.33 a	ND	7.26 ± 0.21 a	11.61 ± 1.40 a	ND	17.89 ± 1.42 a	3.69 ± 0.33 a	3.98 ± 0.36 a	2.17 ± 0.14 a	ND	139.82 ± 8.45 a
Acidic amino acid	L-glutamic acid	20.96 ± 3.61 d	17.67 ± 2.20 c	2.37 ± 0.27 d	ND	5.45 ± 0.22 ab	7.27 ± 0.31 b	ND	10.17 ± 1.26 b	1.62 ± 0.11 b	1.86 ± 0.14 c	2.09 ± 0.10 a	ND	69.46 ± 9.27 c
	L-aspartate acid	22.76 ± 2.18 d	15.37 ± 1.06 c	3.54 ± 0.16 c	ND	3.03 ± 0.17 bc	4.59 ± 0.20 c	ND	9.54 ± 0.96 bc	2.92 ± 0.16 a	1.58 ± 0.10 c	1.14 ± 0.04 b	ND	64.47 ± 3.47 c

Explanations: the results are expressed as the mean and the standard deviation; ND—Means no PAHs detected; a, b, c, d, e, different letters indicate significant differences in PAHs content between any two rows in Table 3 ($p < 0.05$); Naphthalene (NA), Acenaphthene (Ac), Fluorene (FLU), Fluoranthene (FLT), Benzo[a]anthracene (BaA), Chrysene (CHR), Benzo[b]fluoranthene (BbFA), Benzo[k]fluoranthene (BkFA), Benzo[a]pyrene (BaP), Dibenzo[a,h]anthracene (DBahA), Benzo[g,h,i]perylene (BPE), Indeno[1,2,3-c,d]pyrene (IPY).

When investigating the effects of two basic amino acids on the content of PAHs in grilled pork sausage, it was found that the content of PAHs in the sausage with L-lysine and L-arginine was significantly higher than that in the blank control group. However, there was no significant difference in the content of PAHs in the grilled pork sausage with L-arginine or L-lysine ($p > 0.05$). When investigating the effects of two acidic amino acids on the content of PAHs in grilled pork sausage, it was found that the content of PAHs in the sausage to which L-glutamic acid and L-aspartic acid were added was significantly higher than that in the blank control group ($p < 0.05$). However, there was no significant difference in the PAHs content between the sausages supplemented with L-glutamic acid or L-aspartic acid ($p > 0.05$). In addition, when comparing the effects of acidic amino acids (L-lysine and L-arginine) and basic amino acids (L-glutamic acid and L-aspartate) on the formation of PAHs in grilled pork sausages, the authors found that the content of PAHs in sausages with basic amino acids was significantly higher than that in grilled pork sausages with acidic amino acids ($p < 0.05$) (details are shown in Table 3). The reason may be due to the basic amino acids promoting the Maillard reaction during the grilling of the sausages. Previous studies have shown that the Maillard reaction was increased with the increase of Ph [23], and the Maillard reaction has a great relationship with the formation of PAHs [23,27]. The pyrolysis of proteins produces free amino acids and these amino acids can react with reducing sugars (such as glucose) to form the Amalido compound (1-amino-1-deoxy-2-ketosaccharide) [14]. These compounds then undergo pyrolysis to produce PAHs. But the detailed relationship between Maillard reaction and PAHs production is still unclear.

2.4. Effect of Benzene Ring in Amino Acid on PAHs Content

The authors found a very interesting phenomenon when investigating the effects of three non-polar amino acids and three polar amino acids on the content of PAHs in grilled pork sausages. When investigating the effects of three non-polar amino acids on the content of PAHs in grilled pork sausages, it was found that the content of PAHs in sausages supplemented with L-tryptophan and L-phenylalanine was significantly lower than proline (as shown in Figure 2A). When investigating the effects of three polar amino acids on the content of PAHs in grilled pork sausages, the results showed that the content of PAHs in sausages supplemented with L-tyrosine was significantly lower than the sausage with L-serine or L-threonine ($p < 0.05$), but there was no significant difference compared with the control group ($p > 0.05$) (as shown in Figure 2B). Aspartic acid and proline degrade at high temperatures to produce PAHs, such as phenanthrene and aromatics. Tryptophan could not be degraded to produce PAHs when treated at high temperature [15]. This is consistent with the results of our present experiment. It is well known that L-tryptophan, L-phenylalanine, and L-tyrosine are all aromatic amino acids, and each contains a benzene ring. Therefore, it is reasonable to speculate that the benzene ring in the amino acid molecule is a significant reason that these aromatic amino acids hardly influence the production of PAHs. This may be due to the benzene ring in the amino acid molecule being structurally stable and not easily cleaved into small molecular substances, for example, an aldehyde, an enyne, etc. at a high temperature.

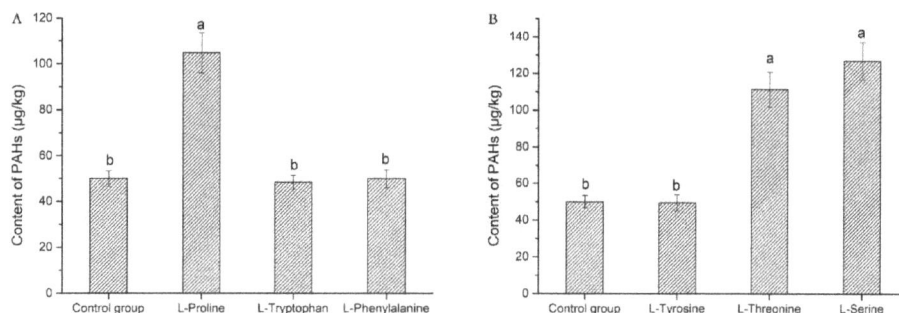

Figure 2. Effect of benzene ring in amino acid on polycyclic aromatic hydrocarbons (PAHs) content. (**A**) The effects of three non-polar amino acids on the content of PAHs in grilled pork sausages; (**B**) The effects of three polar amino acids on the content of PAHs in grilled pork sausages; a, b, different letters indicate significant differences in PAHs content between any two bar graphs in Figure 2A,B ($p < 0.05$).

3. Materials and Methods

3.1. Standards and Reagents

D-glucose (CAS number: 50-99-7), D-fructose (CAS number: 53188-23-1), 4-(α-D-Glucosido)-D-glucose (CAS number: 6363-53-7), cellulose (CAS number: 9004-34-6), Non-polar amino acids: L-proline (CAS number: 147-85-3), L-tryptophan (CAS number: 73-22-3) and L-phenylalanine (CAS number: 63-91-2); Polar amino acids: L-tyrosine (CAS number: 60-18-4), L-threonine (CAS number: 72-19-5) and L-serine (CAS number: 56-45-1); Basic amino acids: L-Lysine (CAS number: 56-87-1) and L-arginine (CAS number: 74-79-3); Acidic amino acids: L-glutamic acid (CAS number: 56-86-0) and L-aspartic acid (CAS No.: 56-84-8) (See Table 4 for specific information on the compounds studied above). The above reagents were all more than 99% pure and were purchased from Aladdin Reagent (Shanghai Aladdin Biochemical Technology Co., Ltd., Shanghai, China). Cyclohexane, dichloromethane (analytical pure) and acetonitrile (chromatographic pure) were purchased from Sinopharm Group. Twelve kinds of PAHs that were used as standards (Naphthalene (NA), Acenaphthene (Ac), Fluorene (FLU), Fluoranthene (FLT), Benzo[a]anthracene (BaA), Chrysene (CHR), Benzo[b]fluoranthene (BbFA), Benzo[k]fluoranthene (BkFA), Benzo[a]pyrene (BaP), Dibenzo[a,h]anthracene (DBahA), Benzo[g,h,i]perylene (BPE), and Indeno[1,2,3-c,d]pyrene (IPY)) were purchased at a purity of more than 99% from Toronto Research Chemicals, Brisbane, North York, ON, Canada.

3.2. Sample Preparation

Fresh hind legs and fat from pigs were purchased from a Carrefour supermarket in Hefei, China. After the visible connective tissues and fat had been removed, the meat was cut into small pieces and ground twice using an electric meat mincer (sieve plate aperture 4 mm; TJ12-H, Henglian, Guangdong, China). The fat was also ground twice using an electric meat mincer with a sieve plate aperture of 6 mm. Four thousand grams of ground meat, 1000 g of fat, 100 g of sodium chloride were thoroughly mixed, then divided into 14 parts, each about 360 g. One point eight grams of an amino acid or carbohydrate was added to each portion and then mixed well. After curing for 6 h at 6–10 °C, the two groups were stuffed into edible casings (Shengguan Casing Company, Guangxi, China) to form sausages. The diameter, length and weight of the sausages were 1.9 cm, 20 cm, and 100 ± 10 g, respectively. All experiments were carried out in triplicate.

Table 4. Molecular weight, pKa and logP of each study compound.

	Molecular Weight (g/mol)	pKa	logP
D-Glucose	180.16	12.43 (t = 18 °C)	−3.24
D-Fructose	180.16	12.06 (t = 18 °C)	−2.23
4-(α-D-Glucosido)-D-glucose	342.30	-	−5.03
Cellulose	>50000	-	-
L-Proline	115.13	10.64 (t = 25 °C)	−2.54
L-Tryptophan	204.22	7.38 (t = 25 °C)	−1.06
L-Phenylalanine	165.19	1.24 (t = 25 °C)	−1.38
L-Tyrosine	181.19	2.20 (t = 25 °C)	−2.26
L-Threonine	119.12	5.60 (t = 25 °C)	−2.94
L-Serine	105.09	2.21 (t = 25 °C)	−3.07
L-Lysine	146.19	3.12 (t = 0 °C)	−3.05
L-Arginine	174.20	2.24 (t = 0 °C)	−4.20
L-Glutamic acid	147.13	2.23 (t = 0 °C)	−3.69
L-Aspartic acid	133.10	2.01 (t = 0 °C)	−3.89

Explanations: pKa—The acid dissociation constant refers to a specific equilibrium constant to represent the ability of an acid to dissociate hydrogen ions; logP—The logP value refers to the logarithm of the partition coefficient of a substance in n-octane (oil) and water.

3.3. Grilling Tools and Cooking Procedures

In this study, the sausages were grilled using an electric oven (Electrolux, EOT5004K, Guang dong, China). The sausages were placed on a metal grid 12 cm below the heating source and baked for 20 min at 240 °C and flipped every 5 min.

3.4. Extraction and Clean-Up

The extraction and clean-up procedures were performed as described by Farhadian et al. [28]. After completion of the heat treatment, the sausage was removed and cooled to room temperature. Five grams of minced sausage was precisely weighed and placed into a 50 mL centrifuge tube containing 25 mL of cyclohexane. After shaking for 5 min with a quick mixer (SK-1, Jintan is Jairel Electric Co., Ltd., China), the centrifuge tube was subjected to ultrasonic treatment for 20 min (SB-5200D; Xinzhi Biotechnology Co., Ltd., Ningbo, China). After sonication, the solution was applied to a Florisil solid phase extraction (SPE) column previously treated with dichloromethane (3 mL) and cyclohexane (5 mL). The PAHs adsorbed by the SPE column were eluted with 9 mL of cyclohexane dichloromethane (3:1 *v/v*). The eluate was evaporated by rotary evaporation (40 °C, 30 rmp) until only 1–2 mL of concentrate remained. The concentrate was transferred into a 10 mL centrifuge tube and dried in a nitrogen atmosphere. A total of 2 mL of acetonitrile was then added to dissolve the extract, which was then filtered through a 0.22 μm membrane for analysis by HPLC.

3.5. HPLC Analysis of PAHs

The PAH analyses were performed based on the method described by Farhadian et al. [28]. PAH analysis was carried out using an HPLC apparatus (Agilent, Santa Clara, CA, USA) equipped with a 600 controller pump, fluorescence detector (G1312, Agilent, USA) and a 20 μL loop injector. A PAH column (250 mm × 4.6 mm, 5 μm particle size) (Agilent, Santa Clara, CA, USA) was used. The mobile phase was constituted of acetonitrile and water. The elution conditions applied were: 0–3 min, 60% of acetonitrile isocratic; 3–15 min, 60–100% of acetonitrile; 15–46 min, 100% of acetonitrile isocratic; 46–53, 100–60% of acetonitrile, gradient. The flow rate was 1.0 mL min⁻¹. A fluorescence detector operated at excitation/emission wavelength 265/327 nm for NAP, 285/320 nm for Ac, 256/300 for FLU, 275/450 nm for FLT, 274/382 nm for BaA, 260/360 for CHR, 283/430 for BbFA and BkFA, 285/410 for BaP, 284/395 for DBahA, 290/410 for BPE, and 301/480 for IPY. Separation was performed under isocratic conditions. Each PAH sample solution was passed through a 0.22 μm filter before injection into the HPLC system. The quantification of PAHs was performed using an external calibration curve method. The quantification of 12 PAHs was carried out through the external standard method.

3.6. Method Validation

The calibration curve, linearity (tested through square correlation coefficients R^2) and limits of detection and quantification (LOD and LOQ) were obtained using the signal-tonoise ratio of $S/N = 3$ and $S/N = 10$, respectively [29]. Accuracy was evaluated using spiked sausage samples at four levels of concentration (ranging from 0.25 to 20.0 µg kg^{-1}), and recoveries were calculated. The recoveries were calculated from the differences in total amounts of each PAHs between the spiked and unspiked samples. For precision, the sample was spiked with two levels (ranging from 0.25 to 8.0 µg kg^{-1}) on two different days, by a single analyst using the same equipment, and the relative standard deviation (RSD) was determined.

3.7. Statistical Analysis

The sample data are provided as the mean ± standard deviation. The data were subjected to analysis of variance (ANOVA) and Duncan's multiple range test for statistical significance ($p < 0.05$) using the SPSS 17.0 software system (SPSS Inc., Chicago, IL, USA).

4. Conclusions

This study tested the effects of molecule properties of carbohydrates and amino acids on PAHs formation in grilled pork sausage. The results showed that addition of aldehyde-based D-glucose could significantly increase the PAHs production when compared with that of the keto-based D-fructose group. The grilled sausage with smaller molecular weight, D-glucose, was determined to have more PAHs content as compared with the sausage with larger molecular weight, 4-(α-D-Glucosido)-D-glucose and cellulose. The addition of basic amino acids (L-lysine, L-arginine) was capable of producing more PAHs in grilled sausages than adding acidic amino acids (L-glutamic, L-aspartate). The addition of the amino acid containing a benzene ring produced fewer PAHs than other amino acids in the grilled sausages. Our study suggests that systematic consideration of the molecule properties is necessary when using food additives (amino acids and carbohydrates) for food processing.

Author Contributions: K.C., C.C., B.X. and W.N. conceived and designed the experiments. W.N., Y.L., Y.W., J.G. and S.Z. performed the experiments. W.N. wrote the manuscript. K.C. revised the manuscript.

Acknowledgments: Financial support of this research provided by the National Science Foundation of China [31501585], Major Science and Technology Program of Anhui, China [18030701159] and project of Yang sheng's student community. We also thank Xingyong Chen of Anhui Agricultural University for grammar correction of the manuscript.

Conflicts of Interest: The authors declare no competing financial interest.

References

1. Iwegbue, C.M.; Onyonyewoma, U.A.; Bassey, F.I.; Nwajei, G.E. Concentrations and health risk of polycyclic aromatic hydrocarbons in some brands of biscuits in the nigerian market. *Hum. Ecol. Risk Assess* **2015**, *21*, 338–357. [CrossRef]
2. Simon, R.; Gomez, J.A.; Von, H.C.; Wenzl, T.E. Results of a european inter-laboratory comparison study on the determination of eu priority polycyclic aromatic hydrocarbons (PAHs) in edible vegetable oils. *Anal. Bioanal. Chem.* **2008**, *391*, 1397–1408. [CrossRef]
3. Authority, E.F. Findings of the efsa data collection on polycyclic aromatic hydrocarbons in food. *Efsa J.* **2007**, *5*, 3–55. [CrossRef]
4. Alomirah, H.; Al-Zenki, S.; Al-Hooti, S.; Zaghloul, S.; Sawaya, W. Concentrations and dietary exposure to polycyclic aromatic hydrocarbons (PAHs) from grilled and smoked foods. *Food Control* **2011**, *22*, 2028–2035. [CrossRef]

5. Raj, A.; Prada, I.D.; Amer, A.A. A reaction mechanism for gasoline surrogate fuels for large polycyclic aromatic hydrocarbons. *Combust. Flame.* **2012**, *159*, 500–515. [CrossRef]

6. Richter, H.; Howard, J.B. Formation of polycyclic aromatic hydrocarbons and their growth to sootda review of chemical reaction pathways. *Prog. Energy Combust. Sci.* **2000**, *26*, 565–608. [CrossRef]

7. Mojica, M.; Francisco, M.A. The Diels-Alder cycloaddition reaction of substituted hemifullerenes with 1,3-butadiene: Effect of electron-donating and electron-withdrawing substituents. *Molecules* **2016**, *21*, 200. [CrossRef]

8. Farhadian, A.; Jinap, S.; Abas, F.S. Determination of polycyclic aromatic hydrocarbons in grilled meat. *Food Control* **2010**, *21*, 606–610. [CrossRef]

9. Saito, E.; Tanaka, N.; Miyazaki, A. Concentration and particle size distribution of polycyclic aromatic hydrocarbons formed by thermal cooking. *Food Chem.* **2014**, *153*, 285–291. [CrossRef]

10. Santos, C.; Gomes, A. Polycyclic aromatic hydrocarbons incidence in Portuguese traditional smoked meat products. *Food Chem. Toxicol.* **2011**, *49*, 2343–2347. [CrossRef]

11. Singh, L.; Agarwalauthor, T. Polycyclic aromatic hydrocarbons formation and occurrence in processed food. *Food Chem.* **2016**, *199*, 768–781. [CrossRef] [PubMed]

12. Zhou, C.; Li, J.; Tan, S. Effect of L-lysine on the physicochemical properties of pork sausage. *Food Sci. Biotechnol.* **2014**, *23*, 775–780. [CrossRef]

13. Zhou, C.; Li, J.; Tan, S. Effects of L-arginine on physicochemical and sensory characteristics of pork sausage. *Adv. J. Food Sci. Technol.* **2014**, *6*, 660–667. [CrossRef]

14. Phillip, F.; Britt, A.C.; Buchanan, C.V.; Owens, J.D.J. Glucose enhance the formation of nitrogen containing polycyclic aromatic compounds and polycyclic aromatic hydrocarbons in the pyrolysis of proline. *Fuel* **2004**, *83*, 1417–1432. [CrossRef]

15. Ramesh, K.S.; Geoffrey, C. Formation of low molecular weight heterocycles and polycyclic aromatic compounds(PACs) in the pyrolysis of a-amino acids. *J. Anal. Appl. Pyrolysis* **2003**, *66*, 97–121. [CrossRef]

16. Naknean, P.; Meenune, M. Factors affecting retention and release of flavour compounds in food carbohydrates. *Int. Food Res. J.* **2010**, *17*, 23–34. [CrossRef]

17. Goubet, I.; Le, Q.J.; Voilley, A.J. Retention of aroma compounds by carbohydrates: Influence of their physicochemical characteristics and of their physical state. A review. *J. Agric. Food. Chem.* **1998**, *46*, 1981–1990. [CrossRef]

18. Xiong, Y.L.; Noel, D.C.; Moody, W.G. Textural and sensory properties of low-fat beef sausages with added water and polysaccharides as affected by pH and salt. *J. Food Sci.* **2010**, *64*, 5. [CrossRef]

19. Appendix A to 40 CFR Part 423, 2010. United States Environmental Protection Agency (US-EPA). Available online: http://www.epa.gov/waterscience/methods/pollutants.htm (accessed on 1 July 2018).

20. COMMISSION REGULATION (EU) No 835 2011 of 19 August 2011 Amending Regulation (EC) No 18812006 as Regards Maximum Levels for Polycyclic Aromatic Hydrocarbons in Foodstuffs. Available online: http://www.docin.com/p-275090873.html (accessed on 19 August 2018).

21. Chen, B.H.; Lin, Y.S. Formation of polycyclic aromatic hydrocarbons during processing of duck meat. *J. Agric. Food Chem.* **1997**, *45*, 1394–1403. [CrossRef]

22. Bansal, V.; Kim, K.H. Review of PAH contamination in food products and their health hazards. *Environ. Inter.* **2015**, *84*, 26–38. [CrossRef]

23. Britt, P.F.; Buchanan, A.C.; Owens, C.V.; Todd, S.J. Formation of nitrogen containing polycyclic aromatic hydrocarbons from the co-pyrolysis of carbohydrates and amino acids. *Div. Fuel Chem.* **2002**, *47*, 400–403. [CrossRef]

24. Kandhro, A.; Sherazi, S.T.; Mahesar, S.A.; Bhanger, M.I.; Younis, T.M. GC-MS quantification of fatty acid profile including trans fa in the locally manufactured margarines of pakistan. *Food Chem.* **2008**, *109*, 207–211. [CrossRef]

25. Masuda, Y.; Mori, K.; Kuratsune, M. Polycyclic aromatic hydrocarbons formed by pyrolysis of carbohydrates, amino acids, and fatty acids. *Gan* **1967**, *58*, 69–74. [CrossRef] [PubMed]

26. Thomas, E.; McGrath, W.; Geoffrey, C. Low temperature mechanism for the formation of polycyclic aromatic hydrocarbons from the pyrolysis of cellulose. *J. Anal. Appl. Pyrolysis* **2003**, *66*, 51–70. [CrossRef]

27. Wanwisa, W.; Kanithaporn, V. Effects of oil types and pH on carcinogenic polycyclic aromatic hydrocarbons (PAHs) in grilled chicken. *Food Control* **2017**, *79*, 119–125. [CrossRef]

28. Farhadian, A.; Jinap, S.; Hanifah, H.N.; Zaidul, I.S. Effects of meat preheating and wrapping on the levels of polycyclic aromatic hydrocarbons in charcoal-grilled meat. *Food Chem.* **2010**, *124*, 141–146. [CrossRef]
29. Juan, C.; Zinedine, A.; Moltó, J.C.; Idrissi, L.; Mañes, J. Aflatoxins levels in dried fruits and nuts from rabat-salé area, morocco. *Food Control* **2008**, *19*, 849–853. [CrossRef]

Sample Availability: Samples are not available from the authors.

molecules

MDPI

Article

Spectrum-Effect Relationships between High-Performance Liquid Chromatography (HPLC) Fingerprints and the Antioxidant and Anti-Inflammatory Activities of Collagen Peptides

Junwen Wang [1,†], Dan Luo [2,†], Ming Liang [3], Ting Zhang [3], Xiquan Yin [3], Ying Zhang [3], Xiangliang Yang [1,2] and Wei Liu [1,2,*]

[1] College of Life Science and Technology, Huazhong University of Science and Technology, Wuhan 430074, China; wangjw@hust.edu.cn (J.W.); yangxl@hust.edu.cn (X.Y.)
[2] National Engineering Research Center for Nanomedicine, Wuhan 430075, China; Laurel565118@163.com
[3] Infinitus (China) Co., Ltd., Guangzhou 510665, China; Fiona.Liang@infinitus-int.com (M.L.); Tinca.Zhang@infinitus-int.com (T.Z.); Xiquan.Yin@infinitus-int.com (X.Y.); sophia.zhang@infinitus-int.com (Y.Z.)
[*] Correspondence: wliu@hust.edu.cn; Tel.: +86-136-5729-5678
[†] These authors contributed equally to this work.

Academic Editors: Alessandra Gentili and Chiara Fanali
Received: 13 November 2018; Accepted: 6 December 2018; Published: 10 December 2018

Abstract: A total of 13 batches of collagen peptide samples were extracted, isolated, and purified from chicken sternal cartilage under various process parameters. The fingerprint profiles of 13 batches of collagen peptides were established by high-performance liquid chromatography (HPLC). In addition, the amino acid profiles and molecular weight distributions of collagen peptides were investigated. The in vitro antioxidant activities of the peptide samples were measured using the 2,2′-Azinobis (3-ethylbenzothiazoline-6-sulphonic acid) diammonium salt (ABTS) assay, the 2,2-diphenyl-1-picrylhydrazyl (DPPH) assay, the ferric-reducing antioxidant power (FRAP) assay and an assay of the oxidative damage induced by hydrogen peroxide (H_2O_2) in the degenerative cartilage cells from the knee joint of rat C518 (C518 cell line). The anti-inflammatory activities of the peptide samples were assessed by measuring the inflammatory responses induced by lipopolysaccharides (LPS) in C518 cells. Subsequently, the spectrum-effect relationships between HPLC fingerprints and the antioxidant and anti-inflammatory activities of collagen peptides were investigated using grey relational analysis (GRA). Fifteen common peaks were obtained from the HPLC fingerprints of collagen peptides. Each collagen peptide sample had a characteristic set of amino acid types and contents. All of the hydrolysates of the collagen peptides were primarily composed of fractions II (500–1000 Da) and III (1000–3000 Da). Collagen peptides exhibited good scavenging activity on ABTS radical, DPPH radical, and ferric-reducing antioxidant power. Collagen peptides were also effective against H_2O_2-induced cellular oxidative damage in C518 cells. The antioxidant activity of collagen peptides was due to the low molecular weight and the presence of antioxidant and hydrophobic amino acid residues within its sequence. Collagen peptides significantly inhibited the secretion of inflammatory cytokines IL-1β, TNF-α, and PGE2 in C518 cells. The anti-inflammatory activity of collagen peptides may include increased synthesis of the key components of extracellular matrix (ECM) and inhibited apoptosis of chondrocytes. The GRA results showed that peaks 2, 3, and 8 were the main components contributing to the antioxidant activity of the collagen peptides, whereas peaks 11 and 14 were the main components contributing to the anti-inflammatory activity of the collagen peptides. The components of peaks 8 and 14 were identified as GPRGPPGPVGP and VAIQAVLSLYASGR by UPLC-MS/MS. Those identified collagen peptides offer a potential therapeutic strategy for the treatment of osteoarthritis (OA) due to their antioxidative stress and due to them disturbing the catabolism and anabolism processes in arthrodial cartilage.

Molecules **2018**, *23*, 3257

Keywords: collagen peptide; HPLC fingerprint; antioxidant; anti-inflammatory; spectrum-effect relationship

---

1. Introduction

Osteoarthritis (OA), the most common degenerative joint disease worldwide found in elderly individuals, is a leading cause of physical disability. Type II collagen (CII) is a principal component of the extracellular matrix (ECM) and constitutes 90–95% of the total protein content in the articular cartilage [1]. Repeated oral administrations of CII induce oral tolerance and inhibit the development of OA. The immunologic response to orally administered antigens occurs in the gut-associated lymphoid tissue (GALT), which is composed of epithelium, lamina propria, Peyer's patch (PP), and mesenteric lymph nodes (MLN). Orally taken native CII antigens interact with PP in the GALT, resulting in turning off the T-cell attack to CII in the cartilage. This desensitization process in PP, also known as oral tolerance, avoids the recognition of endogenous CII in the cartilage as an antigen by the immune system [2]. The effects of oral administration of CII obtained from chicken, bovine, and sheep sources have been evaluated for the treatment of arthritis [3]. Chicken sternal cartilage containing a high amount of collagen is recognized as a potential source of CII [4]. Naturally derived CII from chick sternal cartilage is thought to be a potential oral alternative for OA treatment.

Although the etiology and underlying mechanisms of OA are complicated, much evidence has suggested that the progression of OA in patients is significantly associated with oxidative stress and the inflammatory factor network [5,6].

Recently, oxidative stress has been one of the research hotspots in the field of OA. There is increasing evidence that oxidative stress due to chronic production of endogenous reactive oxygen species (ROS) plays an important role in the physiology and pathophysiology of OA. Oxidative stress regulates intracellular signaling processes and induces chondrocyte senescence and apoptosis, which is characterized by degradation of ECM and a decrease in the synthesis of proteoglycan and CII [7]. Moreover, increased levels of ROS can damage DNA, including mitochondrial DNA, thereby affecting cell viability and contributing to the disruption of ECM homeostasis. Increased levels of ROS may also contribute to the senescent secretory phenotype and the reduced sensitivity of chondrocytes to insulin-like growth factor I (IGF-I) [8].

Antioxidant properties should be evaluated by a variety of methods because most natural antioxidants are multifunctional. Several assays have been frequently used to estimate antioxidant capacities in natural antioxidants, including the 2,2'-Azinobis (3-ethylbenzothiazoline-6-sulphonic acid) diammonium salt (ABTS) assay, the 2,2-diphenyl-1-picrylhydrazyl (DPPH) assay, and the ferric-reducing antioxidant power (FRAP) assay [9]. H_2O_2 is a common ROS produced in cellular metabolism. Intracellular steady-state concentrations of H_2O_2 above 1 μM cause oxidative stress that induces growth arrest and cell death. H_2O_2 induces chondrocyte apoptosis through the regulation of phosphatidylinositol 3-kinase (PI3K)/protein kinase B (Akt) and c-Jun N-terminal kinase (JNK) signaling pathways [5]. In experimental models used to investigate oxidative stress responses of cells, or cytoprotection by antioxidant agents, cultured cells are often exposed to H_2O_2 added as a bolus into the culture medium [10]. Therefore, the ABTS assay, DPPH assay, FRAP assay and an assay of cellular antioxidant activity (the protective effect of collagen peptides against H_2O_2-induced cellular oxidative damage in C518 cells) were performed to evaluate the antioxidant activity of collagen peptides.

Inflammation is another promoting factor in the OA process, including chondrocyte and synovium inflammation. The group of inflammatory cytokines is the most important group of compounds participating in the pathogenesis of OA. Among the many representatives of this group, the greatest importance is attributed to IL-1β, TNF-α, and PGE2, which have been shown to modulate ECM turnover, to accelerate the degradation of cartilage and to induce chondrocyte apoptosis in the development of OA [6,11].

305

IL-1β demonstrates potent bioactivities in inhibiting ECM synthesis and promoting cartilage breakdown [12]. The matrix metalloproteinases (MMPs) play important roles in cartilage degradation in OA, and IL-1β significantly upregulates the expression of MMPs, aggravating degradation processes in OA [13]. Moreover, it has been suggested that IL-1β induces the expression of the TNF-α gene in chondrocytes and upregulates the surface expression of the TNF receptor (TNFR) [14]. The effect of TNF-α in most cases coincides with the action of IL-1β, and in the case of many phenomena occurring in the course of OA there is a marked synergism between the two cytokines. This effect is the result of activation of the same group of intracellular signaling pathways, which in turn triggers similar effects that increase the inflammation and catabolism in joint tissues [6]. PGE2 is considered to be the major contributor to inflammatory pain in arthritic conditions. IL-1β has been shown to stimulate and produce high levels of PGE2 that may induce pain and degeneration with OA [12]. Pharmacologic inhibitors to these mediators may potentially be used as biological treatments in the future. Accordingly, the protective effect of collagen peptides against lipopolysaccharide (LPS)-induced proinflammatory mediators IL-1β, TNF-α, and PGE2 in C518 cells was investigated.

In virtue of the extreme complexity of collagen peptides, the components contributing to their effects are still unclear. Traditional methods for analysis of the active components of peptides are complex and time-consuming, and the spectrum-effect relationships of collagen peptides have not been reported. The aim of this study was to develop an efficient method for screening the active components of CII by establishing spectrum-effect relationships.

2. Results and Discussion

2.1. High-Performance Liquid Chromatography (HPLC) Fingerprints of the Collagen Peptides

The HPLC fingerprints of the collagen peptides and the reference chromatogram generated by the software are shown in Figure 1. The HPLC fingerprints of collagen peptide samples S1–S13 are shown in Figures S1–S13. Fifteen common peaks were obtained from all chromatograms, which were marked 1–15 in the reference chromatogram. Peak 8 (t = 9.087 min) was selected as the reference peak because it was a medium peak in the middle of the chromatogram. The relative retention time (RRT), relative peak area (RPA), and coefficient of variance (C.V.%) of the peak area of 15 common characteristic peaks are shown in Table 1. The majority of the C.V.% values were greater than 18.54%, which showed that the content of each common constituent in the samples varied significantly. The content of the unknown component, which was represented by peaks 5 and 7, showed a particularly large degree of variation.

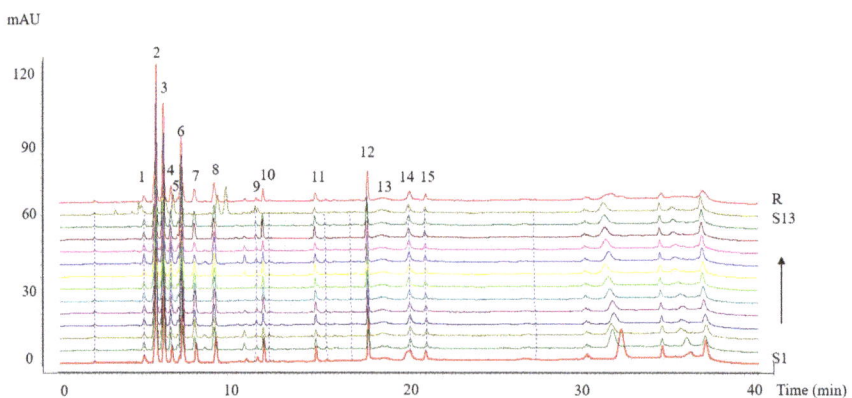

Figure 1. The high-performance liquid chromatography (HPLC) fingerprints of 13 batches of collagen peptide samples (S1–S13) and reference standard fingerprint (R).

Table 1. The relative retention time (RRT) and relative peak area (RPA) of the common peaks of the collagen peptides. C.V.%: coefficient of variance.

Peak No.	RRT	S1	S2	S3	S4	S5	S6	S7	S8	S9	S10	S11	S12	S13	C.V.%
1	0.538	0.333	0.376	0.241	0.329	0.307	0.364	0.372	0.355	0.400	0.371	0.348	0.337	0.499	18.54
2	0.620	4.766	4.808	5.099	4.516	4.653	4.700	4.821	4.745	5.031	4.780	4.769	4.543	1.761	22.17
3	0.669	3.585	3.401	4.693	3.406	3.429	3.467	3.502	3.478	3.449	3.420	3.535	3.395	2.997	13.60
4	0.720	0.533	0.617	0.453	0.570	0.559	0.594	0.608	0.590	0.750	0.625	0.527	0.575	0.880	20.92
5	0.768	0.182	0.201	0.098	0.183	0.156	0.178	0.193	0.191	0.226	0.209	0.180	0.176	0.000	37.45
6	0.789	2.972	2.862	2.683	2.828	2.825	2.857	2.893	2.898	3.043	2.964	2.950	2.834	0.081	31.60
7	0.875	0.751	0.547	0.791	0.664	0.845	0.675	0.696	0.598	0.315	0.414	0.703	0.655	0.000	41.04
8(s)	1.000	1.000	1.000	1.000	1.000	1.000	1.000	1.000	1.000	1.000	1.000	1.000	1.000	1.000	9.95
9	1.266	0.198	0.179	0.149	0.177	0.177	0.105	0.141	0.207	0.192	0.185	0.182	0.163	0.256	22.45
10	1.308	0.655	0.440	0.343	0.408	0.473	0.403	0.402	0.426	0.305	0.360	0.652	0.394	0.170	34.07
11	1.629	0.451	0.372	0.229	0.438	0.304	0.364	0.348	0.350	0.246	0.295	0.445	0.350	0.183	26.69
12	1.957	1.459	0.923	0.618	0.928	0.866	0.862	0.932	0.893	0.651	0.805	1.396	0.934	0.261	36.93
13	2.044	0.045	0.253	0.359	0.304	0.218	0.304	0.357	0.323	0.233	0.371	0.383	0.358	0.386	27.75
14	2.216	1.086	0.693	0.665	0.709	0.539	0.602	0.719	0.703	0.511	0.741	0.751	0.694	0.608	23.95
15	2.316	0.415	0.269	0.258	0.270	0.221	0.233	0.280	0.275	0.219	0.282	0.277	0.264	0.288	22.00

The similarity values ranged from 0.612 to 0.998 (Table 2). Compared to the reference chromatogram, the similarities between the reference fingerprint and the entire chromatographic profiles of 13 batches of collagen peptide samples were evaluated. The correlation coefficients of batches 1–13 were 0.777, 0.936, 0.972, 0.963, 0.977, 0.976, 0.969, 0.977, 0.972, 0.967, 0.961, 0.953, and 0.711, respectively. The similarity values of all batches exceeded 0.930, with the exception of S1 and S13, which indicated that there were relatively small differences among them.

Table 2. Similarities among collagen peptides from 13 origins.

	S1	S2	S3	S4	S5	S6	S7	S8	S9	S10	S11	S12	S13	R
S1	1.000													
S2	0.704	1.000												
S3	0.747	0.950	1.000											
S4	0.733	0.964	0.984	1.000										
S5	0.695	0.926	0.959	0.959	1.000									
S6	0.696	0.889	0.932	0.921	0.969	1.000								
S7	0.691	0.877	0.920	0.909	0.963	0.998	1.000							
S8	0.696	0.888	0.928	0.918	0.967	0.972	0.967	1.000						
S9	0.687	0.884	0.924	0.912	0.964	0.987	0.985	0.972	1.000					
S10	0.680	0.870	0.910	0.899	0.957	0.963	0.961	0.994	0.974	1.000				
S11	0.696	0.885	0.916	0.910	0.918	0.927	0.916	0.950	0.919	0.940	1.000			
S12	0.868	0.911	0.903	0.909	0.917	0.907	0.939	0.913	0.932	0.993	1.000			
S13	0.626	0.612	0.693	0.637	0.631	0.642	0.635	0.639	0.637	0.633	0.702	0.714	1.000	
R	0.777	0.936	0.972	0.963	0.977	0.976	0.969	0.977	0.972	0.967	0.961	0.953	0.711	1.000

2.2. HPLC Fingerprints of Amino Acids

The HPLC fingerprints of the amino acids and the reference chromatogram generated by the software are shown in Figure 2. Twenty-one common peaks were obtained from all chromatograms, which were marked 1–21 in the reference chromatogram. Peak 7 (Arg, t = 13.881 min) was selected as the reference peak because it was a medium peak in the middle of the chromatogram. The RRT, RPA, and C.V.% of the peak area of 21 common characteristic peaks are shown in Table 3.

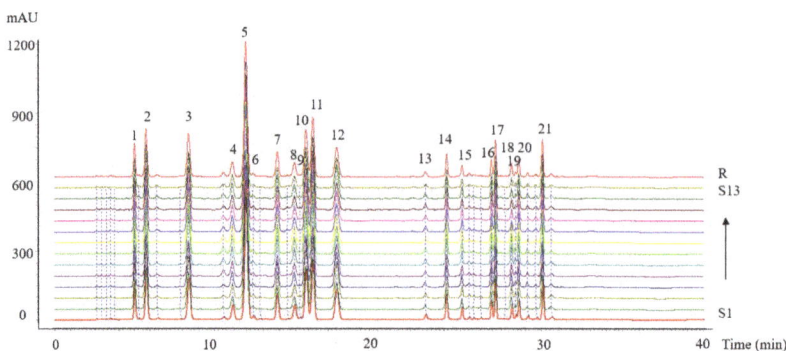

Figure 2. The HPLC fingerprints of the amino acids of 13 batches of collagen peptide samples (S1–S13) and amino acid reference standard fingerprint (R).

Table 3. The RRT and RPA of the common peaks of the amino acids.

Peak No.	RRT	S1	S2	S3	S4	S5	S6	S7	S8	S9	S10	S11	S12	S13	C.V.%
1	0.361	0.944	0.903	0.885	0.953	0.974	0.843	0.865	0.854	0.804	0.896	0.857	0.915	0.931	8.30
2	0.413	1.701	1.646	1.712	1.732	1.800	1.645	1.648	1.622	1.555	1.627	1.612	1.717	1.775	8.63
3	0.603	1.804	1.801	1.905	1.893	1.918	2.020	2.015	1.966	2.013	2.008	2.089	2.101	1.939	12.48
4	0.801	0.639	0.618	0.607	0.632	0.796	0.709	0.721	0.709	0.663	0.680	0.680	0.689	0.704	12.10
5	0.861	6.032	5.889	6.115	6.099	6.320	6.367	6.329	6.125	6.366	6.308	6.569	6.575	6.287	10.86
6	0.894	0.134	0.133	0.121	0.135	0.157	0.134	0.139	0.137	0.131	0.133	0.145	0.135	0.143	9.29
7(s)	1.000	1.000	1.000	1.000	1.000	1.000	1.000	1.000	1.000	1.000	1.000	1.000	1.000	1.000	10.39
8	1.075	0.631	0.611	0.599	0.626	0.705	0.650	0.659	0.641	0.627	0.632	0.666	0.656	0.661	9.96
9	1.098	0.065	0.061	0.068	0.067	0.063	0.058	0.052	0.050	0.052	0.057	0.064	0.070	0.081	10.37
10	1.125	2.148	2.076	2.126	2.137	2.204	2.096	2.114	2.047	2.082	2.083	2.120	2.136	2.143	9.40
11	1.156	2.398	2.372	2.440	2.441	2.539	2.529	2.524	2.461	2.566	2.551	2.683	2.661	2.549	10.60
12	1.263	1.638	1.608	1.948	1.833	2.146	1.391	1.322	1.339	1.443	1.556	1.545	1.926	1.974	10.40
13	1.649	0.212	0.197	0.173	0.191	0.205	0.182	0.187	0.188	0.181	0.187	0.216	0.198	0.152	12.71
14	1.741	0.732	0.665	0.627	0.670	0.736	0.636	0.641	0.609	0.590	0.601	0.662	0.639	0.624	9.65
15	1.809	0.328	0.328	0.334	0.361	0.379	0.335	0.348	0.343	0.354	0.335	0.353	0.355	0.328	10.43
16	1.938	0.479	0.481	0.398	0.455	0.498	0.448	0.453	0.429	0.425	0.021	0.484	0.462	0.438	30.24
17	1.957	0.943	0.920	0.833	0.922	0.982	0.914	0.936	0.905	0.897	0.426	0.969	0.944	0.943	18.25
18	2.027	0.366	0.345	0.349	0.357	0.394	0.448	0.445	0.442	0.252	0.008	0.289	0.255	0.318	40.55
19	2.045	0.111	0.110	0.125	0.128	0.135	0.147	0.126	0.123	0.079	0.284	0.101	0.104	0.144	39.35
20	2.061	0.476	0.464	0.443	0.474	0.528	0.480	0.485	0.472	0.470	0.473	0.515	0.513	0.447	10.51
21	2.166	1.026	0.985	0.949	0.985	1.090	0.942	0.966	0.951	0.955	0.951	1.003	0.971	1.059	7.48

Subsequently, eighteen common characteristic peaks in the chromatogram were identified by comparing the retention times and online UV spectra with those of the standards as follows: Peak 1 (Asp), peak 2 (Glu), peak 4 (Ser), peak 5 (Gly), peak 6 (His), peak 7 (Arg), peak 8 (Thr), peak 9 (HyP), peak 10 (Ala), peak 11 (Pro), peak 13 (Tyr), peak 14 (Val), peak 15 (Met), peak 16 (Ile), peak 17 (Leu), peak 19 (HyL), peak 20 (Phe), and peak 21 (Lys). Each collagen peptide sample had a characteristic set of amino acid types and contents. As a whole, the collagen peptide samples were rich in Gly, Ala, and Pro, which accounted for 23.98%, 16.20%, and 9.64% of the amino acid residues, respectively, whereas HyP (0.24%), HyL (0.51%), His (0.52%), and Tyr (0.73%) were relatively rare among the 18 quantified amino acids (Table 3).

It was found that the collagen peptide samples had a similar amino acid composition to the reference. The characteristic that collagen peptide samples had a high content of Gly and Pro was in accordance with previous findings, although the contents of Gly and Pro obtained in the current research were slightly lower than the values reported by Cao [4]. From these results it can be concluded that Gly, which constitutes approximately one-fourth of all residues in collagen, was present as a quarter residue in the sequence, and that high amounts of Pro could be accommodated while maintaining planar peptide bonds. These assumptions led to the construction of the correct model of sternal cartilage collagen as a (Gly-X-Y) n pattern. Indeed, it was the most commonly found triplet in collagen chains space, indicating that the collagen purified from chicken sternal cartilage was typical CII. Bioactive peptides usually contain 3–20 amino acid residues, and their activities are related to their amino acid compositions [15]. Generally, the antioxidant activities of the amino acid side chain, such as the imidazole group in His (peak 6), the phenolic hydroxyl group in Tyr (peak 13), and thioether in Met (peak 15), determine the antioxidant activities of peptides [16]. Additionally, antioxidative peptides containing aromatic amino acids at the C-terminus have high radical scavenging activities. Aromatic amino acids such as His (peak 6) and Tyr (peak 13) increase the antioxidant activities of peptides and protein hydrolysates because they easily donate protons to electron-deficient radicals and maintain their stabilities via resonance structures. The hydrophobic amino acids at the N-terminus of a peptide sequence, such as Gly (peak 5), Val (peak 14), and Leu (peak 17), are believed to contribute to antioxidative activity, as it is thought that hydrophobic amino acids can increase the abundance of peptides at the water–lipid interface, which may facilitate greater interaction between the peptide and fatty acids [17,18]. Collagen peptide samples S3 and S13 exhibited a lower abundance of Gly, His, Leu, Met, Tyr, and Val in comparison to the other samples. These findings were in accordance with the results of the ABTS and DPPH assays, which indicated that low antioxidant amino acid content in the collagen peptide samples might be related to low antioxidant activity.

The similarity values ranged from 0.995 to 1.000 (Table 4). Compared to the reference chromatogram, the similarities between the reference fingerprint and the entire chromatographic profiles of 13 batches of collagen peptide samples were evaluated: The correlation coefficients of batches 1–13 were 0.999, 1.000, 0.999, 0.999, 0.998, 0.999, 0.999, 0.999, 0.999, 1.000, 0.999, 0.999, and 0.999, respectively. The similarity values of all batches exceeded 0.998, which indicated that there were relatively small differences among them.

Table 4. Similarities among amino acids from 13 origins.

	S1	S2	S3	S4	S5	S6	S7	S8	S9	S10	S11	S12	S13	R
S1	1.000													
S2	1.000	1.000												
S3	0.999	0.999	1.000											
S4	0.999	1.000	1.000	1.000										
S5	0.998	0.998	0.999	0.999	1.000									
S6	0.998	0.998	0.997	0.997	0.995	1.000								
S7	0.998	0.998	0.996	0.997	0.995	1.000	1.000							
S8	0.998	0.998	0.996	0.997	0.995	1.000	1.000	1.000						
S9	0.998	0.998	0.997	0.997	0.995	0.999	0.999	0.999	1.000					
S10	0.999	0.999	0.998	0.999	0.997	0.999	0.999	0.999	1.000	1.000				
S11	0.998	0.998	0.997	0.998	0.996	0.999	0.999	0.999	1.000	1.000	1.000			
S12	0.998	0.999	0.999	0.999	0.998	0.998	0.997	0.998	0.999	0.999	0.999	1.000		
S13	0.999	0.999	1.000	1.000	0.999	0.997	0.996	0.997	0.997	0.998	0.998	0.999	1.000	
R	0.999	1.000	0.999	0.999	0.998	0.999	0.999	0.999	0.999	1.000	0.999	0.999	0.999	1.000

2.3. Molecular Weight Distribution of Cartilage Hydrolysates

The collagen peptides were fractionated according to molecular weight into four categories: I (below 500 Da), II (500–1000 Da), III (1000–3000 Da), and IV (heavier than 3000 Da). The relative content, obtained from their peak areas relative to the total peak area, of each peptide fraction is presented in Table 5. It was found that all of the hydrolysates were primarily composed of fractions II (500–1000 Da) and III (1000–3000 Da), which accounted for approximately 64.80% of the sample after enzymatic treatment.

Molecular weight reflects the hydrolysis of proteins and is highly correlated with the bioavailability and bioactivity of peptides [19,20]. Peptides of low molecular weight are more potent than those of high molecular weight because they cross the intestinal barrier more easily and exert biological effects. Moreover, short-chain peptides are more effective than long-chain peptides with regard to their effects on physiological processes, as they are less susceptible to gastrointestinal hydrolysis.

Table 5. Molecular weight distribution of collagen peptides.

Samples	<500 (Da)	500–1000 (Da)	1000–3000 (Da)	>3000 (Da)
S1	12.48	45.32	24.60	17.60
S2	8.89	39.18	30.77	21.16
S3	9.68	30.68	26.23	33.41
S4	12.89	32.34	25.76	29.01
S5	13.09	32.99	20.99	32.93
S6	12.37	34.21	22.21	31.21
S7	7.56	34.10	32.03	26.31
S8	7.03	33.55	30.48	28.94
S9	8.90	43.10	31.82	16.18
S10	9.51	44.95	24.41	21.13
S11	12.51	43.36	28.25	15.88
S12	10.24	42.43	31.54	15.79
S13	6.07	27.28	29.76	36.89

The peptide fraction with lower molecular weight was probably associated with higher antioxidant activity. Samples S1, S10, and S11 had greater percentages of fractions I (below 500 Da) and II (500–1000 Da) in comparison to the other samples. In contrast, S13 contained relatively low percentages of fractions I (below 500 Da) and II (500–1000 Da) (Table 5). These findings were consistent with observations from the in vitro antioxidant assay mentioned below and support the idea that the functional properties of antioxidative peptides are highly influenced by their molecular weight.

2.4. Antioxidant Activity

2.4.1. Comparative Antioxidant Activity of 13 Batches of Collagen Peptides

The ABTS radical scavenging activity of the collagen peptides varied from 34.91% to 48.14%, which represented a variation of approximately 1.4-fold. The results showed that the ABTS radical inhibitory capacity of the samples (at a concentration of 2 mg/mL) decreased in the following order: S1 > S11 > S7 > S8 > S5 > S12 > S2 > S4 > S10 > S6 > S9 > S3 > S13. S1 exhibited the most ABTS radical scavenging activity, followed by S11 and S7 (Table 6).

The DPPH radical scavenging activity of the collagen peptides varied from 18.46% to 55.06%, which represented a variation of approximately three-fold. The results showed that the DPPH radical inhibitory capacity of the samples (at a concentration of 25 mg/mL) decreased in the following order: S12 > S11 > S10 > S5 > S7 > S9 > S6 > S8 > S4 > S2 > S1 > S3 > S13. S12 exhibited the most DPPH radical scavenging activity, followed by S11 and S10 (Table 6).

The ferric radical scavenging activity of collagen peptides varied from 31.56% to 49.10%, which represented a variation of approximately 1.5-fold. The results showed that the ferric radical scavenging activity of the samples (at a concentration of 50 mg/mL) decreased in the following order: S10 > S9 > S3 > S11 > S6 > S1 > S2 > S7 > S12 > S8 > S4 > S5 > S13. S10 exhibited the most ferric radical scavenging activity, followed by S9 and S3 (Table 6).

The results of these experiments suggested that all of the collagen peptide samples possessed good in vitro antioxidant efficacy. Furthermore, previous results suggested that the chicken sternal cartilage hydrolysates had great oxygen radical absorption capacity (ORAC), with a half maximal inhibitory concentration (IC50) value of 0.48 ± 0.02 μM TE/mg peptides. The exposure of abundant antioxidant residues was responsible for their ORAC antioxidant activity [21]. It is interesting to note that S13 exhibited the lowest antioxidant capacity in all three antioxidant assays, although the antioxidant activity of the collagen peptides varied according to the assay used.

The enzyme/substrate (E/S) ratio, pH, and temperature of digestion had a more significant effect on the hydrolysis of collagen peptides compared to digestion time, as no significant difference of the DPPH radical scavenging activity of S12, S11, and S10 was observed with the extension of the hydrolysis time.

It was clear that enzyme dosage was an important parameter for collagen peptide hydrolysis, since there were relatively significant differences in the antioxidant activity of S10 and S13, which was particularly evident in the FRAP assay.

S11 exhibited the greatest antioxidant capacity in the ABTS assay, DPPH assay, and FRAP assay (the top 4). Thus, the optimal enzymatic hydrolysis conditions for collagen peptide, papain treatment at an E/S ratio of 0.50% at 50 °C and at pH 6.5 for 5 h, could be used as a basis for production scale-up. The results of the optimal enzymatic hydrolysis conditions were basically consistent with a previous study [22].

Table 6. The antioxidant activity of collagen peptides in the 2,2′-Azinobis (3-ethylbenzothiazoline-6-sulphonic acid) diammonium salt (ABTS) assay, the 2,2-diphenyl-1-picrylhydrazyl (DPPH) assay, and the ferric-reducing antioxidant power (FRAP) assay.

Samples	ABTS	DPPH	FRAP
S1	48.14 ± 0.15	44.38 ± 1.28	37.02 ± 0.57
S2	43.21 ± 0.19	45.94 ± 2.50	35.19 ± 0.72
S3	36.78 ± 0.25	40.16 ± 1.17	42.65 ± 1.00
S4	42.65 ± 1.23	46.05 ± 2.18	33.06 ± 0.51
S5	43.30 ± 1.38	53.50 ± 4.36	31.56 ± 0.17
S6	41.93 ± 0.43	49.83 ± 2.10	37.65 ± 0.22
S7	46.07 ± 0.39	51.50 ± 1.12	34.77 ± 0.34
S8	44.72 ± 0.17	49.28 ± 2.93	33.83 ± 0.25
S9	38.79 ± 0.35	50.39 ± 2.06	45.58 ± 0.54
S10	42.50 ± 0.37	53.62 ± 1.21	49.10 ± 0.56
S11	47.84 ± 0.21	54.23 ± 2.39	38.08 ± 0.47
S12	43.28 ± 0.18	55.06 ± 2.49	34.00 ± 0.35
S13	34.91 ± 0.22	18.46 ± 1.64	18.15 ± 0.90

2.4.2. Protective Effect of Collagen Peptides against H_2O_2-Induced Oxidative Damage

The results of the Cell Counting Kit-8 (CCK-8) assay showed that the collagen peptides were not cytotoxic ($P < 0.01$). Exposure to exogenous H_2O_2 decreased the viability of C518 cells, which indicated that exposure to H_2O_2 was cytotoxic for C518 cells. In contrast, a significant increase in C518 cell viability was observed in all samples, especially for S1, S2, and S9, when C518 cells were pretreated with collagen peptides (Figure 3). Taken together, these findings clearly show that collagen peptides conferred significant protection against H_2O_2-induced cellular oxidative damage and can thus be used as a natural antioxidant.

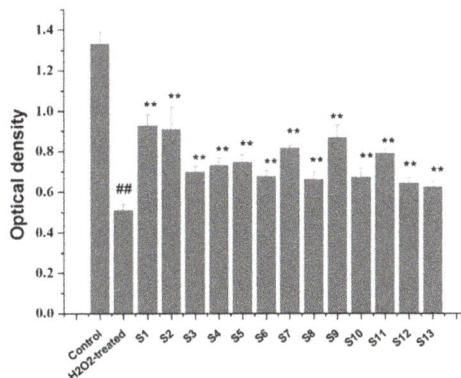

Figure 3. Collagen peptides prevented H_2O_2-induced oxidative damage. Data are expressed as means $\pm$ SD ($n = 3$); ** $p < 0.01$ versus H_2O_2-treated group; ## $p < 0.01$ versus the control group.

2.5. Inhibitory Effects of Collagen Peptides on LPS-Induced Release of Proinflammatory Mediators IL-1β, TNF-α, and PGE2

The effects of the collagen peptide samples on IL-1β secretion varied significantly. S2, S10, and S11 exhibited extremely strong inhibitory effects on IL-1β expression, whereas those of S1 and S12 were slightly weaker. S8, S9, and S13 showed strong inhibitory effects on IL-1β expression, whereas S3, S4, S5, S6, and S7 did not show anti-inflammatory activity (Table 7).

The levels of released TNF-α were increased in the LPS-stimulated culture media. Pretreatment with collagen peptides at a concentration of 2.0 mg/mL for 24 h substantially blocked TNF-α generation in C518 cells. S1, S2, and S3 showed the strongest inhibitory effects on TNF-α generation (Table 7).

All collagen peptide samples significantly attenuated LPS-induced PGE2 production, which was not consistent with the data from the experiments assessing IL-1β and TNF-α secretion. S1, S2, and S8 showed the strongest inhibitory effects on LPS-induced PGE2 production (Table 7).

Taken together, our findings show that collagen peptides significantly inhibited the secretion of IL-1β, TNF-α, and PGE2. Notably, the inhibitory effect of collagen peptides on PGE2 was particularly apparent. Furthermore, previous studies also confirmed that CII treatment dose-dependently inhibited the overproduction of inflammatory cytokines levels (IL-1β and TNF-α) in synoviocytes in the collagen-induced rat arthritis (CIA) model. In conclusion, CII extracted from chick sternal cartilage possessed antiarthritis activity, which may have been a result of its regulation of the humoral and cellular immune systems [23]. Thus, downregulation of proinflammatory mediators by collagen peptides may be a promising therapy for OA. Nevertheless, the molecular mechanisms underlying the anti-inflammatory effects of collagen peptides are not completely understood, so further investigation is required.

Table 7. Inhibitory effects of collagen peptides on lipopolysaccharide (LPS)-induced IL-1β, TNF-α, and PGE2 release in C518 cells.

Group	IL-1β (ng/L)	TNF-α (ng/L)	PGE2 (ng/L)
Control	69.66 ± 4.32	169.65 ± 6.85	90.35 ± 5.15
LPS-treated	86.62 ± 2.44 [#]	211.33 ± 11.43 [#]	115.10 ± 5.20 [#]
S1	61.69 ± 1.37 **	152.62 ± 7.55 *	40.79 ± 9.56 **
S2	44.42 ± 0.81 **	128.97 ± 3.09 *	47.10 ± 2.81 **
S3	78.73 ± 1.28	132.91 ± 10.76 *	59.56 ± 0.02 **
S4	81.94 ± 0.28	172.71 ± 14.85 *	52.57 ± 5.22 **
S5	79.22 ± 1.84	168.35 ± 8.69 *	74.69 ± 2.90 **
S6	85.00 ± 2.99	158.69 ± 5.57 *	49.68 ± 2.81 **
S7	75.99 ± 2.51	199.97 ± 0.78 *	55.84 ± 4.92 **
S8	69.37 ± 4.56 *	193.53 ± 8.59 *	40.79 ± 9.97 **
S9	72.15 ± 5.55 *	185.42 ± 7.68 *	52.57 ± 7.47 **
S10	42.46 ± 4.70 **	176.09 ± 1.32 *	55.84 ± 4.92 **
S11	54.56 ± 2.10 **	193.05 ± 5.25 *	49.68 ± 6.29 **
S12	59.48 ± 3.28 **	184.95 ± 6.76 *	52.57 ± 9.80 **
S13	74.5 ± 3.15 *	182.56 ± 7.17 *	68.82 ± 2.89 **

Data are expressed as means ± SD (n = 3); ** $p < 0.01$ and * $p < 0.05$ versus the LPS-treated group; [#] $p < 0.05$ versus the control group.

2.6. GRA Results

The ranking results of the GRA were as follows (the top 10): ABTS scavenging activity, 14 > 8 > 15 > 3 > 6 > 2 > 1 > 13 > 5 > 4; DPPH scavenging activity, 2 > 8 > 6 > 13 > 3 > 12 > 10 > 14 > 5 > 11; ferric reducing antioxidant power, 2 > 3 > 8 > 1 > 4 > 14 > 6 > 15 > 13 > 11; protection against H_2O_2-induced oxidative damage, 10 > 11 > 15 > 14 > 9 > 2 > 12 > 1 > 4 > 5; inhibited production of LPS-induced IL-1β, 11 > 12 > 10 > 14 > 15 > 9 > 3 > 8 > 5 > 1; inhibited production of LPS-induced TNF-α, 3 > 14 > 8 > 11 > 15 > 2 > 7 > 12 > 10 > 9; inhibited production of LPS-induced PGE2, 6 > 14 > 8 > 2 > 3 > 15 > 5 > 11 > 1 > 12 (Table 8). Overall, peaks 2, 3, and 8 were the main contributors to the antioxidant activity of collagen peptides, whereas peaks 11 and 14 were the main contributors to the anti-inflammatory activity of collagen peptides.

Table 8. The correlation coefficient between the common characteristic peaks and the efficacy of collagen peptides.

Peak No.	r						
	ABTS	DPPH	FRAP	Cell Viability	IL-1β	TNF-α	PGE2
1	0.7732	0.7338	0.7722	0.7315	0.6441	0.6492	0.7622
2	0.7788	0.8110	0.8102	0.7334	0.6435	0.6733	0.8079
3	0.7795	0.7744	0.8081	0.7065	0.6531	0.7025	0.7983
4	0.7497	0.7451	0.7584	0.7294	0.6303	0.6537	0.7368
5	0.7503	0.7467	0.6933	0.7168	0.6450	0.6313	0.7748
6	0.7795	0.7853	0.7456	0.7166	0.6349	0.6501	0.8282
7	0.6673	0.7068	0.6722	0.6819	0.6279	0.6617	0.7029
8	0.8040	0.7982	0.7783	0.7085	0.6507	0.6852	0.8159
9	0.7285	0.7181	0.6870	0.7345	0.6569	0.6555	0.7486
10	0.7008	0.7639	0.6960	0.7534	0.6802	0.6574	0.7271
11	0.7225	0.7461	0.6985	0.7517	0.6943	0.6827	0.7707
12	0.7161	0.7728	0.6950	0.7331	0.6811	0.6581	0.7486
13	0.7544	0.7842	0.7083	0.6555	0.6238	0.6039	0.7446
14	0.8080	0.7530	0.7561	0.7428	0.6729	0.6853	0.8254
15	0.7947	0.7428	0.7397	0.7492	0.6587	0.6789	0.7932

Generating peptides via proteolysis or peptidolysis results in complex combinations of peptides with different masses and physicochemical properties. Classical strategies commonly used to achieve isolation and subsequent characterization employ chromatographic techniques coupled to mass spectrometry, in particular liquid chromatography (LC) coupled to tandem MS (MS/MS) [24,25]. The accuracy and speed of peptide identification are some of the key features that set MS/MS apart from the other methodologies used to analyze protein mixtures [26]. Moreover, the probability-based search engine Mascot has found widespread use as a tool in correlating tandem mass spectra with peptides in a sequence database [27,28].

The amino acid sequences of the collagen peptides were tentatively identified using ultra-performance liquid chromatography-tandem mass spectrometry (UPLC-MS/MS), and all the resulting fragmentation spectra (MS/MS) were searched against the *Gallus gallus* database (NCBI) with the Mascot database searching software (version 2.1, Matrix Science, London, UK). In addition, we also compared the ultraviolet spectra and retention times of the peaks with those of standard peptides obtained by the solid phase peptide synthesis (SPPS) method, and they matched each other. The primary structures of the peptide fractions were determined as GPRGPPGPVGP (peak 8, 987.12 Da) and VAIQAVLSLYASGR (peak 14, 1447.67 Da). Peaks 8 and 14 probably each represented more than one peptide, but only the structure of the main component of each peak was identified in this study. This is the first report of the isolation and identification of these peptides from chicken sternal cartilage. However, further study is needed to identify the structures of other collagen peptide components, particularly the effective components.

3. Materials and Methods

3.1. Materials

HPLC-grade acetonitrile was purchased from Tedia (Fairfield, OH, USA). Amino acid standards, ABTS, DPPH, 2,4,6-tripyridyl-s-triazine (TPTZ), LPS, CCK-8, rat IL-1β, PGE2, and the TNF-α ELISA kit were purchased from Sigma-Aldrich (St. Louis, MO, USA). Dulbecco's Modified Eagle's Medium (DMEM), fetal bovine serum (FBS), phosphate-buffered saline (PBS), and 1% penicillin/streptomycin were bought from Gibco (Carlsbad, CA, USA). The synthesized peptides (purity ≥ 95%) were obtained from Bankpeptide Biological Technology Co., Ltd. (Hefei, China). Other reagents were of analytical grade.

3.2. Preparation of Collagen Peptides

Collagen peptide was prepared according to the methods reported by Xie with some modifications [22]. Fresh chicken sternal cartilage was removed from the periosteum and the calcified portion, cut into slices, minced into paste, and stored at −20 °C. Then, the paste was treated with 5 M guanidine hydrochloride in Tris-HCl (0.05 M, pH 7.5) at a sample/solution ratio of 1:10 (w/v), with gentle stirring for 24 h at 4 °C to remove proteoglycans [29]. The precipitate was washed using Tris-HCl (0.05 M, pH 7.5) and acetic acid (0.5 M). In the enzymatic hydrolysis process of collagen peptides, the four parameters including the enzyme (papain) to substrate ratio (E/S: 0.25%, 0.50%, and 0.75% w/w), pH (6.0, 6.5, and 7.0), temperature (50, 55, and 60 °C) and time (4, 5, and 6 h), were investigated in single-factor experiments (Table 9). At the end of the enzymatic hydrolysis, the solution was heated in a boiling water bath for 10 min to inactivate the enzymes. The extracts were collected by centrifugation (Avanti J-26XP, Beckman Coulter, Brea, CA, USA) at 10,000× g for 30 min at 4 °C. The supernatant was collected and salted out overnight at 4 °C by adding 2.5 M NaCl in Tris-HCl (0.05 M, pH 7.5) at a ratio of 1:20 (w/v). The resulting precipitate was collected by centrifugation (Avanti J-26XP, Beckman Coulter, USA) at 10,000× g for 20 min at 4 °C. Subsequently, the precipitate was dissolved with acetic acid (0.1 M) and allowed to reach dialysis equilibrium with NaCl (0.2 M). The dialysate was freeze-dried and referred to as "collagen peptides".

Table 9. The processing conditions of collagen peptides. E/S: enzyme to substrate ratio.

Samples	E/S (w/w)	pH	Temperature (°C)	Time (h)
S1	0.75%	6.0	50	6
S2	0.50%	6.0	50	6
S3	0.25%	6.0	50	6
S4	0.50%	6.0	55	6
S5	0.50%	6.5	55	6
S6	0.50%	7.0	55	6
S7	0.50%	7.0	50	4
S8	0.50%	7.0	55	4
S9	0.50%	7.0	60	4
S10	0.50%	6.5	50	6
S11	0.50%	6.5	50	5
S12	0.50%	6.5	50	4
S13	0.25%	6.5	50	6

3.3. HPLC Fingerprints

3.3.1. HPLC Conditions

Chromatographic analyses of collagen peptides were performed using a Shimadzu LC-20A HPLC system (Kyoto, Japan), including a quaternary solvent delivery system, online degasser, autosampler manager, column compartment and diode-array detector (DAD), as well as LCsolution software. The HPLC operating conditions were as follows: The column was an Inertsil ODS-SP C18 column (4.6 × 250 mm, 5 μm). A binary gradient elution system, comprising buffer (A) 0.05% TFA/acetonitrile and buffer (B) 0.1% TFA/water, was applied as follows: Initial, 98% B; 30 min, 40% B; 30.01 min, 98% B; 40 min, 98% B. The detection wavelength was set at 214 nm, the flow rate was 0.8 mL/min, the column temperature was maintained at 40 °C, the sample temperature was ambient, and the injection volume was 10 μL.

3.3.2. Preparation of the Sample Solution

Collagen peptides were dissolved in deionized water at a concentration of 10 mg/mL and then filtered through a 0.22 µm micropore film to yield the sample solution.

3.3.3. Analysis of HPLC Fingerprints

Validation of Methodology

The chromatographic fingerprinting methodology was validated to assess its precision, repeatability, and stability. Precision was evaluated by the analysis of six injections of the same testing sample consecutively. Repeatability was examined by determination of six different samples prepared from the same collagen peptide sample. Stability was examined by analysis of the sample solution at 0, 2, 4, 6, 8, 12, and 24 h. The methodology validation showed that the relative standard deviation (RSD) for precision was in the range of 1.28–1.75%, whereas that of reproducibility was less than 1.18% and that of storage stability was 1.46–1.83%. All results indicated that the HPLC fingerprint analysis method was valid and satisfactory.

Similarity Analysis

The HPLC fingerprints were matched automatically using the Similarity Evaluation System for Chromatographic Fingerprint of Traditional Chinese Medicine software developed by the Chinese Pharmacopoeia Committee (Version 2012) (Beijing, China). The reference fingerprint was generated using the median method [30], after which the similarity values between the reference fingerprint and the entire chromatographic profiles of 13 batches of collagen peptides were calculated.

3.4. Amino Acid Composition

Amino acid composition of collagen peptides was determined according to the method of Sun with a slight modification [31]. A high-performance liquid chromatography equipped with a PICO.TAG column (Waters, Milford, MA, USA) was used. The amino acid composition of collagen peptides was determined after hydrolysis at 150 °C for 1 h with 6 M hydrochloric acid prior to derivatization with phenyl isothiocyanate. A binary gradient elution system, comprising (A) 0.1 M ammonium acetate (the pH was adjusted to 6.5 with acetic acid)/acetonitrile (93:7 ratio) and (B) 80% acetonitrile/water, was applied as follows: Initial, 0% B; 15 min, 15% B; 18 min, 24% B; 25 min, 40% B; 30 min, 40% B; 30.01 min, 100% B; 40 min, 100% B. The amino acid standards included L-alanine (Ala), L-arginine (Arg), L-aspartic acid (Asp), L-glutamic acid (Glu), L-glycine (Gly), L-histidine (His), hydroxylysine (HyL), hydroxyproline (HyP), L-isoleucine (Ile), L-leucine (Leu), L-lysine (Lys), L-methionine (Met), L-phenylalanine (Phe), L-proline (Pro), L-serine (Ser), L-threonine (Thr), L-tyrosine (Tyr), and L-valine (Val).

3.5. Molecular Weight Distribution

Molecular weight distribution of collagen peptides was determined by gel permeation chromatography on a Superdex Peptide HR 10/300 GL (10 × 300 mm, Amersham Biosciences Co., Piscataway, NJ, USA) with UV detection at 214 and 280 nm. The mobile phase was 5 mM phosphate buffer containing 10 mM NaCl (pH 7.4), at a flow rate of 0.5 mL/min, which corresponded to an operating pressure of 1.8 MPa. A molecular weight calibration curve was obtained from the following standards: Glycine trimer (189 Da), oxidized glutathione (612 Da), vitamin B12 (1355 Da), aprotinin (6500 Da), cytochrome C (12500 Da), and bovine serum albumin (66430 Da) (Sigma Co., St. Louis, MO, USA). UNICORN 5.0 software (Amersham Biosciences Co., Piscataway, NJ, USA) was used to analyze the chromatographic data.

3.6. Antioxidant Activity Determination

3.6.1. ABTS Assay

The ABTS assay was conducted following the method of Arnao with some modifications [32]. The stock solutions included 7.0 mM ABTS$^+$ solution and 2.45 mM $K_2S_8O_2$ solution. The working solution was prepared by mixing the two stock solutions in equal quantities and allowing them to react for 12 h at room temperature in the dark. Before analysis, this stock solution was diluted with PBS to obtain an absorbance of 0.70 ± 0.02 at 734 nm. Next, 100 μL samples were allowed to react with 2 mL ABTS$^{.+}$ solution in the dark. After 6 min, the absorbance was recorded at 734 nm. Results were determined using the following Equation (1):

$$\text{ABTS radical scavenging capacity (\%)} = [(A_{blank} - A_{sample})/A_{blank}] \times 100\%, \tag{1}$$

where A_{sample} and A_{blank} are the absorbance of the test sample and blank sample, respectively.

3.6.2. DPPH Assay

The DPPH assay was performed according to the method of Brand–Williams with some modifications [33]. First, 1 mL of 0.2 mM DPPH solution was added to an equal volume of the sample solution. The mixture was vortexed for a few seconds and incubated for 30 min at room temperature, after which the absorbance of the mixture was measured at 517 nm. Results were calculated according to the following Equation (2):

$$\text{DPPH radical scavenging activity (\%)} = [(A_{blank} - A_{sample})/A_{blank}] \times 100\%, \tag{2}$$

where A_{sample} and A_{blank} are the absorbance of the test sample and blank sample, respectively.

3.6.3. FRAP Assay

The FRAP assay was performed according to the methods reported by Benzie with some modifications [34]. The FRAP solution was prepared by mixing 10 volumes of acetate buffer (300 mM, pH 3.6) with 1 volume of TPTZ (10 mM dissolved in 40 mM HCl) and 1 volume of $FeCl_3 \cdot 6H_2O$ (20 mM in water). Next, 200 μL of the sample were allowed to react with 2 mL FRAP solution in the dark for 30 min at 37 °C, and the absorbance of the sample was recorded at 593 nm. Aqueous standard solutions of $FeSO_4 \cdot 7H_2O$ (100–1000 μM) were used to generate a calibration curve. The final results were expressed as millimole $FeSO_4 \cdot 7H_2O$ equivalents per gram of collagen peptides (mM $FeSO_4 \cdot 7H_2O$/g, collagen peptides).

3.6.4. In Vitro H_2O_2-Induced Oxidative Damage Model

Cell Lines and Culture

C518 cells were purchased from Saiqi Biological Engineering Co., Ltd. (Shanghai, China). C518 cells were cultured and maintained in DMEM supplemented with 10% FBS, 100 U/mL penicillin, and 100 μg/mL streptomycin at 37 °C under a humidified atmosphere containing 5% CO_2.

Protective Effects of Collagen Peptides against H_2O_2-Induced Oxidative Stress

C518 cells were seeded in 96-well plates (2×10^4 cells/mL, 100 μL) and incubated for 24 h. They were pretreated with various collagen peptide samples (1.0 mg/mL) and cultured for 30 min. The control was pretreated with DMEM under the same conditions. Thereafter C518 cells were exposed to H_2O_2 (1 mM) in the presence or absence of collagen peptides for 24 h. Finally, cell viability was determined using the CCK-8 assay.

3.7. Anti-Inflammatory Activity Determination

C518 cells were seeded in 96-well plates (2×10^4 cells/mL, 100 μL) and incubated for 24 h. They were pretreated with various collagen peptide samples (2.0 mg/mL) and cultured for 1 h. The control was pretreated with DMEM under the same conditions. Thereafter C518 cells were exposed to LPS (10 μg/mL) in the presence or absence of collagen peptides for 24 h. Finally, the abundance of IL-1β, PGE2, and TNF-α in cell-free supernatants was determined using ELISA kits according to the manufacturer's instructions.

3.8. Spectrum-Effect Relationships

The spectrum-effect relationships between HPLC fingerprints and antioxidant and anti-inflammatory effects were established by GRA. Grey system theory is an interdisciplinary scientific area that was first introduced in the early 1980s [35], and it has been applied in decision making in an extremely wide range of multiple-attribute group decision-making problems [36].

3.8.1. Data Preprocessing

In GRA, initial data preprocessing is performed to transform the original data sequences, with different measurement units, into comparable sequences. In this study, the mean value normalization preprocessing method was applied in the data series treatment to obtain a dimensionless matrix, because this method can easily accommodate a wide range of units among the factors used [37].

3.8.2. Definition of Reference Sequences and Comparison Sequences

The antioxidant and anti-inflammatory activities of the collagen peptides were utilized as reference sequences, and the characteristic peak areas obtained by HPLC fingerprints were utilized as comparison sequences. The original reference sequences and comparison sequences are represented by $\{X_o(k)\}$ and $\{X_i(k)\}$, $i = 1, 2, \dots, m; k = 1, 2, \dots, n$ respectively, where m is the number of experiments and n is the total number of observations of data.

3.8.3. Grey Relational Grade Calculation

The grey relational coefficient was calculated from the deviation sequence using the following relation:

$$\gamma\{X_o(k), X_i(k)\} = \frac{\Delta\min + \xi\Delta\max}{\Delta_{oi}\min(k) + \xi\Delta\max} 0 < \gamma\{X_o(k), X_i(k)\} \ll 1, \tag{3}$$

where $\Delta_{oi}(k)$ is the deviation sequence of the reference sequence $\{X_o(k)\}$ and comparability sequence $\{X_i(k)\}$, and ξ is the resolution coefficient, usually $\xi \in (0,1)$. The resolution coefficient is typically chosen to be 0.5. A grey relational grade is the weighted average of the grey relational coefficient and is defined as follows:

$$\gamma_{oi} = \frac{1}{n}\gamma_{oi}\{X_o(k), X_i(k)\}, \tag{4}$$

3.8.4. Grey Relational Coefficient Calculation and Ranking

The grey relational grade between the reference sequence and comparison sequences was calculated. The grey relational grade was proportional to the similarity of the developing trends (i.e., the greater the grade, the more similarity) [38].

3.9. Purification and Identification of Collagen Peptides

3.9.1. Purification of Collagen Peptides by RP-HPLC

The collagen peptide was dissolved in distilled water and loaded onto a semipreparative C18 RP-HPLC column (10.0 mm × 250 mm, 5 μm, Agilent Technologies, Santa Clara, CA, USA). The HPLC 1200 system (Agilent Technologies, USA) was equipped with a quaternary pump solvent delivery system and a diode-array detector (DAD). The sample injection volume and concentration were 250 µL and 10 mg/mL, respectively. The column was eluted by a linear gradient of acetonitrile (2–60%) containing 0.1% TFA at a flow rate of 1.0 mL/min. The UV absorbance of the eluent was monitored at 214 nm. This step was repeated several times, and the different elution fractions were pooled, concentrated, and lyophilized for sequence identification.

3.9.2. Identification of Collagen Peptides by UPLC-MS/MS Analysis

A Waters Acquity UPLC system (Waters, Milford, MA, USA) coupled with a Thermo Q Exactive mass spectrometer (Thermo Fisher Scientific, Bremen, Germany) was used for peptide separation and identification. The UPLC-MS/MS operating conditions were as follows: The column was an Eksigent C18 trap column (75 µm × 250 mm, 3 µm). A binary gradient elution system, comprising (A) 0.1% TFA/acetonitrile and (B) 0.1% TFA/water, was applied as follows: Initial, 90% B; 60 min, 40% B. The detection wavelengths were 214 and 280 nm, the flow rate was 3.0 µL/min, the freeze-dried sample was dissolved in 0.1% aqueous formic acid, and the injection volume was 1.0 µL. The mass spectrometer was fitted with an electrospray ionization (ESI) source used in the positive ion mode.

3.10. Statistical Analysis

All assays were carried out using at least three independent sets of experiments, and all results were expressed as average values with their corresponding standard deviations. One-way analysis of variance by Tukey's test was performed using SPSS 22.0 (IBM Co., Armonk, NY, USA).

4. Conclusions

In this work, HPLC fingerprints and a series of in vitro antioxidant and anti-inflammatory assays were combined to investigate the spectrum-effect relationship of collagen peptides. The results showed that peaks 2, 3, and 8 were the main components contributing to the antioxidant activity of the collagen peptides, whereas peaks 11 and 14 were the main components contributing to the anti-inflammatory activity of the collagen peptides. Subsequently, the amino acid sequences of peaks 8 and 14 were identified as GPRGPPGPVGP and VAIQAVLSLYASGR by UPLC-MS/MS. These identified peptides may have potential as drugs or functional foods for the treatment or prevention of OA. This report establishes a new platform for identifying the functional components of collagen peptides by the spectrum-effect relationship, which may lead to the development of new directions in the utilization of bioactive peptides in the future. However, further studies are required to investigate the cellular and molecular mechanism by which peptide pretreatment exerts its effects in the animal OA models.

Supplementary Materials: The following are available online, Figures S1–S13: The HPLC fingerprints of collagen peptide samples S1–S13.

Author Contributions: J.W. and D.L. designed the research; J.W., D.L., M.L., T.Z., X.Y. (Xiquan Yin) and Y.Z. performed the study and analyzed the data; J.W., X.Y. (Xiangliang Yang) and W.L. drafted and revised the manuscript; all authors approved the final version.

Funding: This research was financially supported by the National High Technology Research and Development Program of China (SS2012AA022704) and the Natural Science Foundation of China (NSFC, 31470968).

Acknowledgments: The authors thank the analytical and testing center of the Huazhong University of Science and Technology for their help in the testing of HPLC-MS/MS analysis.

Conflicts of Interest: The authors declare no conflict of interest.

References

1. Jeevithan, E.; Zhang, J.Y.; Wang, N.P.; He, L.; Bao, B.; Wu, W.H. Physico-chemical, antioxidant and intestinal absorption properties of whale shark type-II collagen based on its solubility with acid and pepsin. *Process. Biochem.* **2015**, *50*, 463–472. [CrossRef]

2. Bakilan, F.; Armagan, O.; Ozgen, M.; Tascioglu, F.; Bolluk, O.; Alatas, O. Effects of native type II collagen treatment on knee osteoarthritis: A Randomized Controlled Trial. *Eurasian J. Med.* **2016**, *48*, 95. [CrossRef] [PubMed]
3. Jeevithan, E.; Wu, W.H.; Wang, N.P.; He, L.; Bao, B. Isolation, purification and characterization of pepsin soluble collagen isolated from silvertip shark (*Carcharhinus albimarginatus*) skeletal and head bone. *Process. Biochem.* **2014**, *49*, 1767–1777. [CrossRef]
4. Cao, H.; Xu, S.Y. Purification and characterization of type II collagen from chick sternal cartilage. *Food Chem.* **2008**, *108*, 439–445. [CrossRef] [PubMed]
5. Lepetsos, P.; Papavassiliou, A.G. ROS/oxidative stress signaling in osteoarthritis. *BBA-Mol. Basis Dis.* **2016**, *1862*, 576–591. [CrossRef]
6. Wojdasiewicz, P.; Poniatowski, Ł.A.; Szukiewicz, D. The role of inflammatory and anti-Inflammatory cytokines in the pathogenesis of osteoarthritis. *Mediat. Inflamm.* **2014**, *2014*, 1–19. [CrossRef] [PubMed]
7. Zhuang, C.; Wang, Y.J.; Zhang, Y.K.; Xu, N.W. Oxidative stress in osteoarthritis and antioxidant effect of polysaccharide from *angelica sinensis*. *Int. J. Biol. Macromol.* **2018**, *115*, 281–286. [CrossRef] [PubMed]
8. Heijink, A.; Vanhees, M.; van den Ende, K.; van den Bekerom, M.P.; van Riet, R.P.; van Dijk, C.N.; Eygendaal, D. Biomechanical considerations in the pathogenesis of osteoarthritis of the elbow. *Knee Surg. Sport Tr. A* **2012**, *20*, 423–435. [CrossRef] [PubMed]
9. Thaipong, K.; Boonprakob, U.; Crosby, K.; Cisneros-Zevallos, L.; Hawkins Byrne, D. Comparison of ABTS, DPPH, FRAP, and ORAC assays for estimating antioxidant activity from guava fruit extracts. *J. Food Compos. Anal.* **2006**, *19*, 669–675. [CrossRef]
10. Gulden, M.; Jess, A.; Kammann, J.; Maser, E.; Seibert, H. Cytotoxic potency of H_2O_2 in cell cultures: Impact of cell concentration and exposure time. *Free Radic. Biol. Med.* **2010**, *49*, 1298–1305. [CrossRef]
11. Shen, C.L.; Smith, B.J.; Lo, D.F.; Chyu, M.C.; Dunn, D.M.; Chen, C.H.; Kwun, I.S. Dietary polyphenols and mechanisms of osteoarthritis. *J. Nutr. Biochem.* **2012**, *23*, 1367–1377. [CrossRef] [PubMed]
12. Lee, A.S.; Ellman, M.B.; Yan, D.Y.; Kroin, J.S.; Cole, B.J.; van Wijnen, A.J.; Im, H.-J. A current review of molecular mechanisms regarding osteoarthritis and pain. *Gene* **2013**, *527*, 440–447. [CrossRef]
13. Gu, Y.T.; Chen, J.; Meng, Z.L.; Ge, W.Y.; Bian, Y.Y.; Cheng, S.W.; Xing, C.K.; Yao, J.L.; Fu, J.; Peng, L. Research progress on osteoarthritis treatment mechanisms. *Biomed. Pharmacother.* **2017**, *93*, 1246–1252. [CrossRef] [PubMed]
14. Qin, J.; Shang, L.; Ping, A.S.; Li, J.; Li, X.J.; Yu, H.; Magdalou, J.; Chen, L.B.; Wang, H. TNF/TNFR signal transduction pathway-mediated anti-apoptosis and anti-inflammatory effects of sodium ferulate on IL-1β-induced rat osteoarthritis chondrocytes in vitro. *Arthritis Res. Ther.* **2012**, *14*, R242. [CrossRef] [PubMed]
15. Roufik, S.; Gauthier, S.F.; Turgeon, S.L. In vitro digestibility of bioactive peptides derived from bovine β-lactoglobulin. *Int. Dairy J.* **2006**, *16*, 294–302. [CrossRef]
16. Matsui, R.; Honda, R.; Kanome, M.; Hagiwara, A.; Matsuda, Y.; Togitani, T.; Ikemoto, N.; Terashima, M. Designing antioxidant peptides based on the antioxidant properties of the amino acid side-chains. *Food Chem.* **2018**, *245*, 750–755. [CrossRef]
17. Chi, C.F.; Hu, F.Y.; Wang, B.; Li, Z.R.; Luo, H.Y. Influence of amino acid compositions and peptide profiles on antioxidant capacities of two protein hydrolysates from skipjack tuna (*Katsuwonus pelamis*) dark muscle. *Mar. Drugs* **2015**, *13*, 2580–2601. [CrossRef]
18. Sarmadi, B.H.; Ismail, A. Antioxidative peptides from food proteins: A review. *Peptides* **2010**, *31*, 1949–1956. [CrossRef]
19. Jin, J.; Ma, H.L.; Wang, B.; Aelg, Y.; Wang, K.; He, R.; Zhou, C. Effects and mechanism of dual-frequency power ultrasound on the molecular weight distribution of corn gluten meal hydrolysates. *Ultrason. Sonochem.* **2016**, *30*, 44–51. [CrossRef]
20. Yang, B.; Yang, H.S.; Li, J.; Li, Z.X.; Jiang, Y.M. Amino acid composition, molecular weight distribution and antioxidant activity of protein hydrolysates of soy sauce lees. *Food Chem.* **2010**, *124*, 551–555. [CrossRef]
21. Lin, X.L.; Yang, L.; Wang, M.; Zhang, T.; Liang, M.; Yuan, E.D.; Ren, J.Y. Preparation, purification and identification of cadmium-induced osteoporosis-protective peptides from chicken sternal cartilage. *J. Funct. Foods* **2018**, *51*, 130–141. [CrossRef]
22. Xie, Y.; Pang, X.; Zhao, R.B. Isolation and identification of type-II collagen peptide from chicken sternal cartilage. *Food Sci.* **2011**, *32*, 26–29.

23. Cao, H.; Xu, S.Y.; Ge, H.S.; Xu, F. Molecular characterisation of type II collagen from chick sternal cartilage and its anti-rheumatoid arthritis activity. *Food Agr. Immunol.* **2014**, *25*, 119–136. [CrossRef]
24. Maux, S.L.; Nongonierma, A.B.; Fitzgerald, R.J. Improved short peptide identification using HILIC-MS/MS: Retention time prediction model based on the impact of amino acid position in the peptide sequence. *Food Chem.* **2015**, *173*, 847–854. [CrossRef] [PubMed]
25. Zhu, C.Z.; Zhang, W.G.; Zhou, G.H.; Xu, X.L.; Kang, Z.L.; Yin, Y. Isolation and identification of antioxidant peptides from jinhua ham. *J. Agric. Food Chem.* **2013**, *61*, 1265–1271. [CrossRef] [PubMed]
26. Xu, C.J.; Ma, B. Software for computational peptide identification from MS-MS data. *Drug Discov. Today* **2006**, *11*, 595–600. [CrossRef]
27. Lam, H.; Deutsch, E.W.; Eddes, J.S.; Eng, J.K.; King, N.; Stein, S.E.; Aebersold, R. Development and validation of a spectral library searching method for peptide identification from MS/MS. *Proteomics* **2007**, *7*, 655–667. [CrossRef]
28. Weatherly, D.B.; Atwood, J.A.; Minning, T.A.; Cavola, C.; Tarleton, R.L.; Orlando, R. A heuristic method for assigning a false-discovery rate for protein identifications from Mascot database search results. *Mol. Cell. Proteom.* **2005**, *4*, 762–772. [CrossRef]
29. Wang, H. Zaocys type II collagen regulated the balance of Treg/Th17 cells in mice with collagen-induced arthritis. *J. South Med. Univ.* **2014**. [CrossRef]
30. Meng, J.M.; Jiang, H.M.; Lu, J.Q.; Zhou, K.; Xiao, Y.S. Preliminary Study on HPLC Fingerprint of *Thlaspi arvense* L. *Med. Plant* **2017**, *8*, 15–19. [CrossRef]
31. Sun, W.Z.; Zhao, H.F.; Zhao, Q.Z.; Zhao, M.M.; Yang, B.; Wu, N.; Qian, Y.L. Structural characteristics of peptides extracted from Cantonese sausage during drying and their antioxidant activities. *Innov. Food Sci. Emerg. Technol.* **2009**, *10*, 558–563. [CrossRef]
32. Arnao, M.B.; Cano, A.; Acosta, M. The hydrophilic and lipophilic contribution to total antioxidant activity. *Food Chem.* **2001**, *73*, 239–244. [CrossRef]
33. Brand-Williams, W.; Cuvelier, M.E.; Berset, C. Use of a free radical method to evaluate antioxidant activity. *LWT-Food Sci. Technol.* **1995**, *28*, 25–30. [CrossRef]
34. Benzie, I.F.; Strain, J.J. The ferric reducing ability of plasma (FRAP) as a measure of "antioxidant power": The FRAP assay. *Anal. Biochem.* **1996**, *239*, 70–76. [CrossRef] [PubMed]
35. Deng, J.L. Control problems of grey systems. *Syst. Control Lett.* **1982**, *1*, 288–294. [CrossRef]
36. Wu, H.H. A comparative study of using grey relational analysis in multiple attribute decision making problems. *Qual. Eng.* **2002**, *15*, 209–217. [CrossRef]
37. Kadier, A.; Abdeshahian, P.; Simayi, Y.; Ismail, M.; Hamid, A.A.; Kalil, M.S. Grey relational analysis for comparative assessment of different cathode materials in microbial electrolysis cells. *Energy* **2015**, *90*, 1556–1562. [CrossRef]
38. Kuo, Y.Y.; Yang, T.; Huang, G.W. The use of grey relational analysis in solving multiple attribute decision-making problems. *Comput. Ind. Eng.* **2008**, *55*, 80–93. [CrossRef]

Sample Availability: Samples of the compounds are available from the authors.

molecules

MDPI

Article

Phytochemical and Biological Characteristics of Mexican Chia Seed Oil

Yingbin Shen [1,†][ORCID], Liyou Zheng [2,†], Jun Jin [2][ORCID], Xiaojing Li [2], Junning Fu [1], Mingzhong Wang [3][ORCID], Yifu Guan [4,*] and Xun Song [5,*]

1   Department of Food Science and Engineering, Jinan University, Guangzhou 510632, China; shenybin412@gmail.com (Y.S.); juningf0313@163.com (J.F.)
2   State Key Laboratory of Food Science and Technology, Synergetic Innovation Center of Food Safety and Nutrition, School of Food Science and Technology, Jiangnan University, 1800 Lihu Road, Wuxi 214122, China; liyou890513@sina.com (L.Z.); zgzjjin@126.com (J.J.); Lixiaojing19900810@163.com (X.L.)
3   Shenzhen Kivita Innovative Drug Discovery Institute, Shenzhen 518110, China; mzwang2000@163.com
4   Research Institute for Marine Drugs, Guangxi University of Chinese Medicine, Nanning 530200, China
5   School of Pharmaceutical Science, Shenzhen University, Shenzhen 518060, China
*   Correspondence: 000917@gxtcmu.edu.cn (Y.G.); 13480243@life.hkbu.edu.hk (X.S.); Tel.: +86-755-86671931 (X.S.)
†   These authors contributed equally to this work and share first authorship.

Academic Editors: Alessandra Gentili and Chiara Fanali
Received: 11 November 2018; Accepted: 3 December 2018; Published: 6 December 2018

check for updates

Abstract: The purpose of this research was to investigate the chemical profile, nutritional quality, antioxidant and hypolipidemic effects of Mexican chia seed oil (CSO) in vitro. Chemical characterization of CSO indicated the content of α-linolenic acid (63.64% of total fatty acids) to be the highest, followed by linoleic acid (19.84%), and saturated fatty acid (less than 11%). Trilinolenin content (53.44% of total triacylglycerols (TAGs)) was found to be the highest among seven TAGs in CSO. The antioxidant capacity of CSO, evaluated with ABTS$^{\bullet+}$ and DPPH$^{\bullet}$ methods, showed mild antioxidant capacity when compared with Tocopherol and Catechin. In addition, CSO was found to lower triglyceride (TG) and low-density lipoprotein-cholesterol (LDL-C) levels by 25.8% and 72.9%respectively in a HepG2 lipid accumulation model. As CSO exhibits these chemical and biological characteristics, it is a potential resource of essential fatty acids for human use.

Keywords: Chia seed oil; polyunsaturated fatty acid; antioxidant; lipid-lowering effect

1. Introduction

Vegetable oils such as palm, canola, soybean, rapeseed, corn, olive, and sunflower oils are commonly used in cooking and industrial food manufacturing. The chemical constituents and content of cooking oils, particularly the type of fatty acids, are considered important criteria for quality of oils and their health benefits. Nutrition studies indicate that the type and content of dietary fat intake is closely related to various diseases, such as cardiovascular disease, diabetes, and depression [1]. Thus, consumption of healthier polyunsaturated fatty acids (PUFAs), which are considered 'good fat', is a vitally important prevention strategy for fat related health complications. However, as PUFAs cannot be synthesized in the human body, it is essential to supplement them through diet to meet daily needs [2]. Due to the huge market of PUFA rich oils, a number of industrial food manufacturing plants tend to seek alternative, plant-based sources of PUFAs.

Molecules **2018**, *23*, 3219; doi:10.3390/molecules23123219   321   www.mdpi.com/journal/molecules

Chia, botanically known as *Salvia hispanica* L., native to southern Mexico and northern Guatemala, is a 2 m tall herbaceous plant that belongs to the Lamiaceae family [3]. Currently, chia is grown on a commercial scale in the South America, Australia, Europe, and Southeast Asia, due to the high edible value in whole seeds, flour, and oil in food industry. In general, the major constituents in chia seeds include 25–40% oil, 17–24% protein, and 18–30% dietary fiber [4]. It also contains significant quantities of minerals and phytonutrients, including tocopherols, carotenoids and phytosterols, and phenolic compounds [5]. Chia seed oil is probably one of the healthiest oils on the market, containing predominantly essential fatty acids such as α-linolenic acid and linoleic acid. Numerous findings have confirmed that chia seed oil is a healthy oil for lowering the risk for cardiovascular disease, hepatoprotective effect, inflammation, and prevention of obesity-related disorders [6]. At present, the in vitro cancer cytotoxic properties of CSO have been reported by Ramzi et al., who showed that the CSO significantly inhibited the proliferation of human lymphoblastic leukemic cell lines, HeLa, and MCF-7 cells [7]. Up to now, there has been no evidence of adverse effects of chia seeds, and toxicological data on CSO from animal and controlled human trials on the safety and efficacy are still limited. However, experience gained from previous and current use of chia seeds for food purposes in developing countries can be regarded as supportive evidence to allow a positive conclusion on the safety of CSO [8].

Research findings by Ayerza et al. indicated that the content of the bioactive nutrients of chia seed can be affected by the geographical location and climate condition [9]. To date, research has been carried out on basic chemical and physical characteristics, including quantification of fatty acids, tocopherol, and polyphenols. However, to the best of our knowledge, sn-2 fatty acid composition and minor components of Mexican chia seed oil have not been extensively studied. The purpose of this research was to elaborate on most chemical compositions (fatty acids, triacylglycerol, tocopherols, sterols, polyphenols, metal elements, and PAHs) and functional values (antioxidant activity and lipid-lowering effects) of oil extracted from chia seeds grown in Mexico, to further evaluate its nutritional value and thus contribute towards identification of a potential food with medicinal and industrial applications.

2. Results and Discussion

2.1. Physical and Chemical Profiles

The physical and chemical properties analysis of CSO are presented in Table 1. The oil content of chia seed ranged from 31.39 to 32.39 g/100 g, much higher than that in soybean (17.6 to 25.4 g/100 g) reported by Dornbos et al. [10]. The result was consistent with the published data (31.2 g/100 g, on average, of Brazil chia seed) [11] and 28.5–32.70 g/100 g in chia seeds from Colombia, Argentina, Peru and Bolivia [12]. The results obtained in this study suggested that chia seed was a good source of crude oil based on the high yield of oil content.

Table 1. Physical and chemical characters of chia seed oil.

Oil Parameters	Content	
Oil content (g/100 g)	31.89 ± 0.50	
Oil stability (induction period)/h	0.68 ± 0.03	
Oil color/(units)		
	R	Y
	1.65 ± 0.07	13.00 ± 0.00

Table 1. *Cont.*

Oil Parameters	Content
[a] Fatty Acid Composition	
Palmitic acid C16:0	7.07 ± 0.01
Palmitoleic acid C16:1 (n-9)	0.06 ± 0.00
Trianoic acid C17:0	0.16 ± 0.01
Stearic acid C18:0	2.81 ± 0.04
Oleic acid C18:1 (n-9)	5.50 ± 0.01
Vaccenic acid C18:1 (n-7)	0.80 ± 0.01
Linoleic acid C18:2 (n-6)	19.84 ± 0.01
α-Linolenic C18:3 (n-3)	63.64 ± 0.06
Arachidic acid C20:0	0.12 ± 0.01
SFA	10.16 ± 0.06
PUFA	89.84 ± 0.07
PUFA/SFA	8.85 ± 0.06
n-3/n-6 FA ratio	3.21 ± 0.00
sn-2 Fatty Acid Composition	
C16:0	1.10 ± 0.05
C18:0	0.88 ± 0.09
C18:1	6.38 ± 0.12
C18:2	25.07 ± 0.06
C18:3	63.76 ± 0.66

[a] Values reported as means ± SD of three replicate analyses (n = 3). SFA = total saturated fatty acids, PUFA = total polyunsaturated fatty acids, n-6 = total omega-6, n-3 = total omega-3 fatty acids.

2.2. Oxidative Stability

The induction period, widely used to determine stability of edible oils, is an indicator of oxidative processes. The oxidative stability of oils was measured as the induction period in response to forced oxidation. The results in Table 1 indicate that the induction period of CSO was found to be 0.68 h, which is similar to the reported value for Camelina oil (0.63 h) [13], but much lower than that of soybean oil [14]. The possible reason for lower induction period is the high content of unsaturated fatty acids in CSO (89.84% of total lipids). However, as there is limited data available on the stability of CSO, there is rising need for investigation on the stability of CSO during processing and storage.

At present there aren't any color standards for CSO, and thus the Lovibond colorimeter was used firstly for color measurement. The color of CSO, represented by R and Y Lovibond scale, displayed more yellow (13 units) than red (1.65 units), which was consistent with that reported by Maira et al. [15]. The color ratio of CSO was also similar to that of linseed oil with yellow (70 units) and red (8.6 units) [16]. Nevertheless, CSO has different color than oils such as palm (yellow, 3.2 units; red, 27.4 units) and soy oil (yellow, 4.6 units; red, 10.6 units), which show more red than yellow [17]. Total pigment and carotenoid content are responsible for the natural color of vegetable oil, which could be an indicator of metamorphic oils. Significant changes of the R and Y values were found on the CSO during the metamorphic process, which could be used as a mark in distinguishing the degree of metamorphism.

2.3. Fat and Triacylglycerol Composition

As shown in Table 2, the most abundant fatty acid was α-linolenic acid (63.64% of total lipids), followed by linoleic acid (19.84% of total lipids), palmitic acid (7.07% of total lipids), and oleic acid (5.5% of total lipids). Among the fatty acids in CSO, the order of abundance of the identified components: α-linolenic acid > linoleic acid > palmitic acid > oleic acid > stearic acid > vaccenic acid > trianoic acid > arachidic acid > palmitoleic acid (Table 2 and Figure S1). This study showed that the major type of fatty acid in CSO was polyunsaturated fatty acid (PUFA, 89.84%). Interestingly, some research has reported that animal feed with chia seed of high PUFA content can increase the level of PUFA in meat fats, as well as aroma and flavor [18].

Table 2. Fat, TAGs and minor components of chia seed oil.

Oil Parameters	Percentages
Fat Compositions/%	
TAG	82.60 ± 0.15
1,3-DAG	0.82 ± 0.03
1,2(2,3)-DAG	0.74 ± 0.02
Total DAG	1.56 ± 0.02
FFA	15.18 ± 0.11
TAG Compositions/%	
aLnaLnaLn	53.44 ± 0.47
aLnaLnL	23.76 ± 0.22
aLnLL	8.22 ± 0.24
aLnaLnP	6.25 ± 0.05
aLnLO	1.80 ± 0.24
aLnOP	4.43 ± 0.28
aLnOO	2.10 ± 0.35
Di-UTAG	10.69 ± 0.33
Tri-UTAG	89.31 ± 0.34
Minor components	[a] Content (mg/kg)
Tocopherols	
α-tocopherol	5.10 ± 0.42
γ-tocopherol	70.38 ± 7.99
δ-tocopherol	1.48 ± 0.06
Total tocopherols	76.96 ± 8.47
Squalene	226.43 ± 38.19
Phytosterols	
Campesterol	387.77 ± 59.05
Stigmasterol	177.47 ± 31.57
β-Sitosterol	2433.56 ± 71.69
Total phytosterols	2998.80 ± 162.30
Mineral contents	
Boron	0.193 ± 0.012
Magnesium	3.566 ± 0.185
Aluminum	4.104 ± 0.644
Calcium	1.226 ± 0.082
Manganese	0.098 ± 0.010
Zinc	0.153 ± 0.017
Arsenic	0.014 ± 0.004
Strontium	0.071 ± 0.014

Tri-UTAG, triunsaturated triacylglycerols; Di-UTAG, diunsaturated triacylglycerols; P, palmitic acid; S, stearic; O, oleic; L, linoleic acid; Ln, linolenic acid; aLn, α- linolenic acid. [a] All the measurements are in terms of mean ± standard deviation.

CSO has very high contents of α-linolenic (63.64%) and linoleic acids (19.84%), together accounting for more than 83%. In addition, the level of palmitic acid was 7.07%. We found that the ratio n-3/n-6 in CSO was 3.21, in agreement with the range of 3.18–4.18 reported by Ixtaina et al. [19], which is much higher than that of most vegetable oils, such as canola oil (0.45), olive oil (0.13), soybean oil (0.15), and walnut oil (0.20) [20]. Furthermore, the sn-2 fatty acid composition of CSO is indicated in Table 1 and Figure S2. Obviously, α-linolenic, linoleic, and oleic acids in CSO were the major sn-2 fatty acids, which were responsible for more than 95%. This result is in line with the recognized principle that unsaturated fatty acids mainly occupy the sn-2 position.

Based on a daily nutrient criterion for linoleic acid and α-linolenic acid proposed by National Institutes of Health [21], CSO could be applied as a good dietary supplement. In addition, CSO was characterized by a high polyunsaturated fatty acid/saturated fatty acid (PUFA/SFA) ratio, which is highly favorable for the reduction of serum cholesterol and atherosclerosis, and the prevention of cardiovascular disease [22]. Many studies have shown that CSO plays a role in lowering serum triglyceride, and also raises the level of HDL-C in rats [23]. As shown in Table 1, the PUFA/SFA ratio of CSO was 8.85. Thus, the incorporation of CSO into the diet could bring great beneficial effects to the cardiovascular system due to the high content of PUFAs.

The compositions of acylglycerol and free fatty acids are indicated in Table 2. The main composition of CSO was triacylglycerol, with its content up to 82.60%. The content of total diacylglycerol and FFA were 1.56% (1,3-DAG = 0.82% and 1,2 or 2,3-DAG = 0.74%) and 15.18%, respectively. As shown in Table 2, seven different TAGs were found in CSO, including aLnaLnaLn, aLnaLnL, aLnLL, aLnaLnP, aLnLO, aLnOP and aLnOO (Figure S3). The main triacylglycerols in CSO were aLnaLnaLn, aLnaLnL, aLnLL and aLnaLnP, and their levels were 53.44%, 23.76%, 8.22% and 6.25%, respectively. We found that α-linolenic acid was present in all the measured TAGs, which was also discovered by Ixtaina and coworkers [19].

2.4. Tocopherol and Phytosterol Levels

Tocopherols are natural antioxidants which can stabilize oils. The α-, γ- and δ-tocopherols, and total tocopherol content, in CSO are shown in Table 2 and Figure S4. CSO contained 76.96 mg/kg of tocopherols, mainly γ-tocopherol (>91%) and α-tocopherol (6.6%). δ-Tocopherol was present in low concentration (1.48 mg/kg). However, β-tocopherol was not detected in CSO. The content of total tocopherols (76.96 ± 8.47 mg/kg) in CSO was much lower than that of research (238–427 mg/kg) reported by Matthäus [24]. This difference may be explained by different varieties, processing, and storage conditions.

Table 2 and Figure 1A also show the content and the composition of sterols in CSO. Three phytosterols including campesterol, stigmasterol, and β-sitosterol were identified, with β-sitosterol as the most abundant sterol, accounting for more than 81% of the total amount of sterols. β-Sitosterol was predominant (2433.56 mg/kg), followed by campesterol (387.77 mg/kg) and stigmasterol (177.47 mg/kg). The total phytosterols of the CSO amounted to 3 g/kg, lower than the previous data ranged from 7 to 17 g/kg [25], which is probably linked to the loss of sterols during oil refining. Compared with other common edible oils, CSO contained a higher total tocopherol content than olive, soybean, peanut, corn, sunflower, and canola oils (260–1000 mg/kg) [26], indicating that CSO probably has relatively better oxidative stability.

It is well known that squalene is an important contributor to reduction of cholesterol levels. In this study, squalene was found in CSO at 226.43 mg/kg, much higher than the concentrations in walnuts, almonds, peanuts, hazelnuts, and macadamia nuts revealed by Maguire et al. [27]. The range of squalene content in CSO was 50–500 mg/kg, dependent on the extraction method used [27]. The high content of squalene in CSO is probably an important contribution to the beneficial effects of cancer prevention and health in human diet.

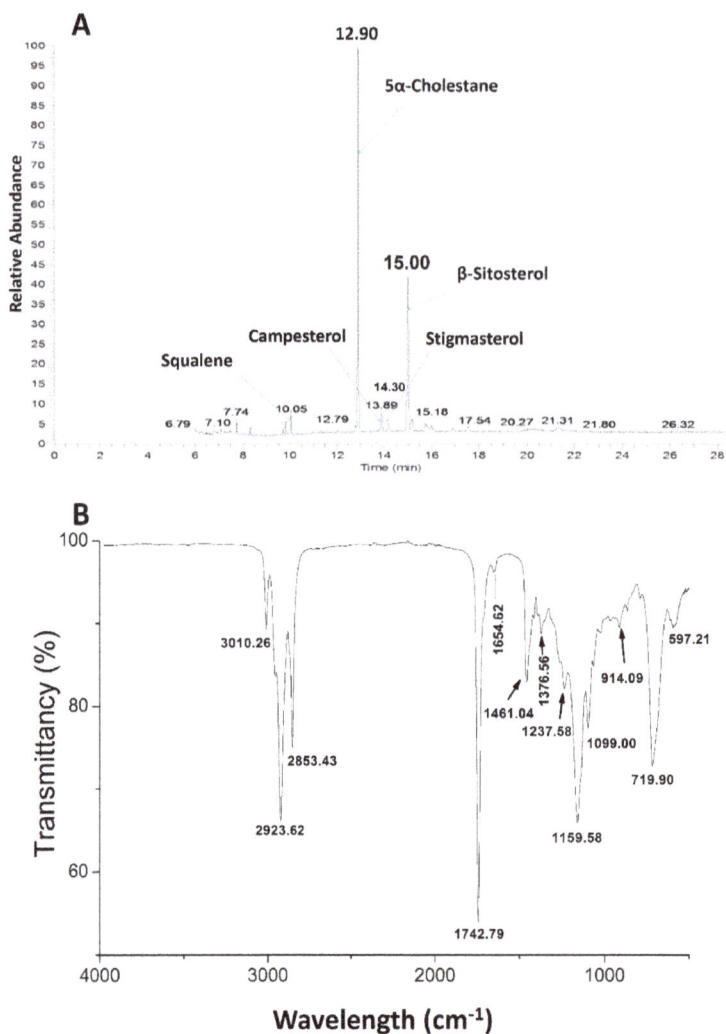

Figure 1. (**A**) Sterols in Mexican chia seed oil were analyzed using a gas chromatograph-mass spectrometer system by standards. (tR = 10.05 min), squalene; (tR = 12.90 min), 5-α-cholestan; (tR = 13.89 min), campesterol; (tR = 14.30 min), stigmasterol; (tR = 15.00 min), β-sitosterol. (**B**) FTIR spectra for chia seed oil extracted.

2.5. Determination of Trace Element Levels

Trace elements like manganese (Mn) increase the rate of oxidation of oil by the formation of free radicals of fatty acids and hydroperoxides [28], so they are undesirable in oils. The content of trace elements in CSO are shown in Table 2. Aluminum (Al), magnesium (Mg), calcium (Ca), boron (B), zinc (Zn), manganese (Mn), strontium (Sr) and arsenic (As) were detected in CSO. As indicated in Table 2, mineral composition of chia seed oil is low in elements, like all other oils (Ca = 1.226 mg/kg; Mg = 3.566 mg/kg; Zn = 0.153 mg/kg; Al = 4.104 mg/kg; Mn = 0.098 mg/kg; B = 0.193 mg/kg; Sr = 0.071 mg/kg). The most abundant trace element was aluminum (4.104 mg/kg). CSO is a rich source of minerals such as Ca, Mg and Zn. There were no toxic elements (Ni, Pb, Cd, Tl and Hg) detected except arsenic. However, mean levels of arsenic (0.014 mg/kg) do not raise concern. It is possible that concentrations of some elements were influenced by the growth, manufacturing process, and equipment. As arsenic was found in CSO, further investigation would be needed to determine its valence state, since different valence states vary in toxicity. All metals measured in CSO were lower than the maximum level accepted for virgin vegetable oils [29], and also lower than those in crude and degummed sunflower oils [30]. Thus, it could be used in food, medical and cosmetics industry.

2.6. Determination of Polycyclic Aromatic Hydrocarbons (PAHs)

PAHs constitute the critical hazards which are widespread in vegetable oil such as sunflower, olive, peanut, soybean, corn, canola, and palm oils. PAHs might occur from raw materials or the refining process [31]. Table 3 indicates the levels of the PAHs detected in CSO. Ten PAHs were revealed in CSO, including pyrene, fluoranthene, BaA, BbFlu, DBahA, Chr, BaP, BkFlu, IP and BghiP. Pyrene had the highest level (180.24 µg/100g) of the PAHs found in this oil. The concentrations of fluoranthene, BaA, and BbFlu were 84.72, 66.92, and 31.49 µg/100g, respectively. The total concentration of the other 6 PAHs was 45.18 µg/100g. The average sum of PAHs in CSO (41.06 µg/kg) was found to be higher than that in sunflower oil (17.36 µg/kg), and lower than that in soybean oil (65.33 µg/kg) [32]. One reason for PAHs introduced in vegetable oils is because of the heating-drying and extraction processes on oilseed or raw material [33]. Maximum concentration for total PAHs in edible oil in Regulation (European Union) is 10 µg/kg [34]. Thus, using uncontaminated raw materials, improving the refining process, and using activated carbon to remove PAHs during the oil refining process, are necessary requirements to decrease PAHs before market use [35].

Table 3. Polycyclic aromatic hydrocarbons (PAH) in chia seed oil.

PAHs	Concentration (µg/100 g)
Fluoranthene	84.72 ± 9.85
Pyrene	180.24 ± 18.84
Benzo (*a*) anthracene (BaA)	66.92 ± 10.05
Chrysene (Chr)	5.69 ± 0.98
Benzo (*b*) fluoranthene (BbFlu)	31.49 ± 8.84
Benzo (*k*) fluoranthene (BkFlu)	2.21 ± 0.95
Benzo (*a*) pyrene(BaP)	3.68 ± 1.02
Dibenzo (*a.h*) anthracene (DBahA)	30.04 ± 6.62
Indeno(1,2,3-*cd*) pyrene and Benzo (*g,h,i*) perylene (IP and BghiP)	3.56 ± 0.88

2.7. FT-IR Spectrum Analysis

The FT-IR spectrum of CSO in Figure 1B shows an absorption band at about 3000 cm^{-1} corresponding with C–H stretching, and we observed the =C–H, asymmetric and symmetric methyl groups at 3010.26, 2923.62 and 2853.43 cm^{-1}, respectively [36]. The frequencies at 1461.04 and 1742.79 cm^{-1} were assigned to bending vibration of lipid CH$_2$ groups and the ester carbonyl stretching (C=O) fatty acids, respectively. According to the study by Guillen and Cabo, the distinct weak peak at 1654 cm^{-1} refers to C=C stretching absorptions of disubstituted *cis*-olefins, which could be used to evaluate the level of total lipids and identify total unsaturation in the oil [37].

Absorption bands associated with (C–C(=O)–O) and (OC–C) coupled bonds stretching are usually very strong at 1159.58 and 1099 cm^{-1}. The peak at 1099 cm^{-1} is related to C–O–C stretch of triglyceride ester linkage. The bands at 719.9 cm^{-1} are due to a combined vibration of *cis*-disubstituted olefins having seven or more carbon atoms, and CH$_2$ out-of-plane rocking [36].

The FT-IR spectrum clearly indicates that CSO contains C=C, representing the unsaturated fatty acids, and ester functional groups. The high level of α-linolenic acid (~64% in CSO), with three double bonds, led to a high degree of unsaturation in the spectrum. It was interesting to note that characteristic absorption of OH at 3470 cm^{-1} is not found on the FT-IR spectrum, which suggests that the moisture content could be neglected in CSO.

FTIR is one of the most practical, non-destructive and relatively cost-efficient techniques to evaluate relative levels of unsaturated fatty acid in vegetable oils by their characteristic peaks. The integration area of absorption peaks at 3010 cm^{-1} and 1743 cm^{-1} were representative of the content of PUFAs and total lipid respectively. Furthermore, the percentage of unsaturated fatty acid could be calculated by the area ratio of 3010/1743 cm^{-1}. The percentage of unsaturated fatty acid found in this study indicates that CSO contains a high concentration of UFAs, which provided the unsaturation observed using individual feature of the peak at 3010 and 1654 cm^{-1} [36].

2.8. Antioxidant and Cytotoxicity

The antioxidant capability of CSO was evaluated by the inactivation of DPPH and ABTS assays. The radical inhibition decreased after treatment with CSO, as shown in Table 4. The IC_{50} value (IC_{50} value is the concentration of the sample required to inhibit 50% of radicals) of DPPH of the chia seed oil was 33.94 mg/mL, while the IC_{25} value (IC_{25} value is the concentration of the sample required to inhibit 25% of radicals) of ABTS was 28.51 mg/mL. Overall, catechin and tocopherol demonstrated superior scavenging activity than CSO. Both the ABTS and DPPH methods as investigated in our study gave similar results as the study by Xuan et al. [38] and by Scapin et al. for ethanol extract of Brazil chia seeds (3.84 mg/mL) [38], and by Scapin et al. for a Brazil chia seed ethanol extraction (3.84 mg/mL) obtained by solvent extraction [39]. Compared with sunflower, safflower, canola, and soybean oil, CSO presented the lowest DPPH and ABTS radical scavenging capability [38].

Table 4. Antioxidant activity and cytotoxicity of chia seed oil.

Samples	Antioxidant Activity		Cytotoxicity [c] IC_{25} (µg/mL)	
	DPPH [a] IC_{50} (mg/mL)	ABTS [b] IC_{25} (mg/mL)	LNcap	HepG2
Chia seed oil	33.94	28.51	580.12	889.68
Catechin	0.005	-	-	-
Tocopherol	-	0.004	-	-

[a] IC_{50} value is the concentration of the sample required to inhibit 50% of radical of DPPH, [b] IC_{25} value is the concentration of the sample required to inhibit 25% of radical of ABTS, [c] IC_{25} value is the concentration of the sample required to inhibit 25% of cell growth. IC_{25} value was much more reliable in ABTS and cytotoxicity assay because it was in the range of tested concentration.

Further, the cytotoxic property of chia seed oil was assessed in LNcap and HepG 2 cells. Our data indicated the IC_{25} values (IC_{25} value is the concentration of the sample required to inhibit 25% of cell growth) of CSO were up to 580 and 889.68 µg/mL in LNcap and HepG 2 cell lines for 48 h, separately, which means cell toxic effects can be excluded by this extract method on this cell line (Table 4).

2.9. The Effect of Chia Seed Oil on the Hepg2 Lipid Accumulation Model

Hepatic steatosis can be induced in HepG2 cells by exposing cells to pathophysiological levels of oleic-acid (OA) to mimic the influx of excess lipo-toxicity of free fatty acids in hepatocytes [40]. To date there have been no reports published indicating anti-hepatic lipogenic effect in OA-induced HepG2 Cells. Thus, this study was designed to evaluate the anti-hepatic lipogenic effect of chia seed oil in the selected in vitro model. In the present study, upon establishing this successful hepatic steatosis model, intracellular TG, TC, HDL-C and LDL-C levels were significantly increased (Figure 2) after 0.2 mM OA stimulation. CSO (500 µg/mL) exhibited no cytotoxicity on HepG2 cells (Table 4). As showed in Figure 2, significant decreases of TG and LDL-C levels by 25.8% and 72.9% were seen in the CSO treatment group, separately, compared with OA-group, similar to the tocopherol group. Moreover, the group treated with CSO (500 µg/mL) demonstrated a higher reduction of LDL-C than the tocopherol treated group. We have demonstrated that chia seed oil does not reduce the TC secreted in HepG2 cells, but increases the levels of cellular HDL-C. These results could be supplementary in vitro evidence to the in vivo studies by Ayerza et al., which indicated that chia seed diets for 4 weeks decreased TG levels and increased HDL-C cholesterol content in rat serum [41]. In terms of lipid lowering efficacy, this study indicates that chia seed oil is more effective at reducing triglycerides than total cholesterol, similar to tocopherol. Thus, chia seed oil has great potential as nutraceutical in the food industry with its TG and LDL-C reducing effects, and increasing effect on the secretion of HDL-C.

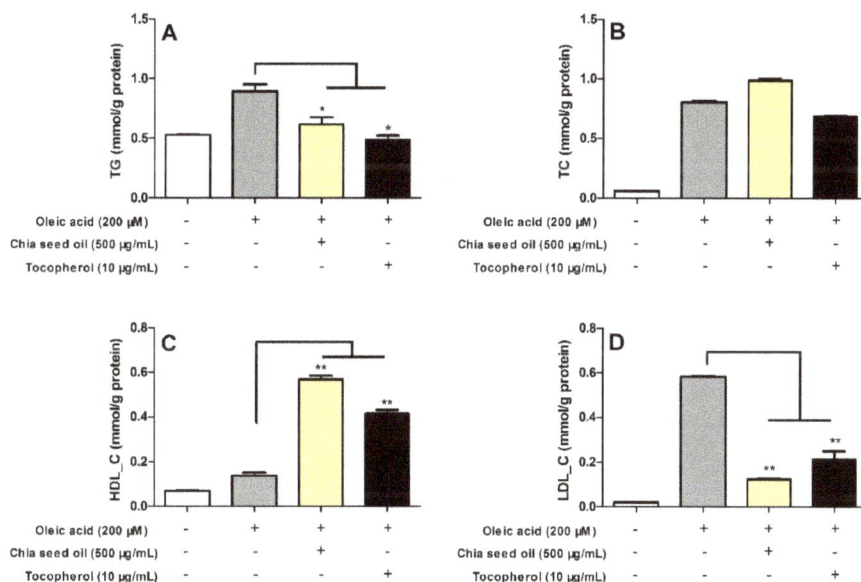

Figure 2. Effects of chia seed oil on accumulation of the intracellular (**A**) TG, (**B**) TC, (**C**) HDL-C, and (**D**) LDL-C in oleic acid (OA)-induced HepG2 cells. OA-induced accumulation of TG and LDL-C was significantly reduced by chia seed oil co-treatment in HepG2 cells. Vertical bars represent the mean ± SE of three independent experiments. * $p < 0.05$, ** $p < 0.01$ versus OA group.

3. Materials and Methods

3.1. Chemicals

Standards of tocopherols, phytosterols and squalene were purchased from Sigma Aldrich (St. Louis, MO, USA). Triacylglycerols were obtained from J&K Scientific (Beijing, China). Target standard compounds of polycyclic aromatic hydrocarbons (PAHs, dissolved in acetonitrile, 200 µg/mL), including benzo(*a*)anthracene (BaA), chrysene (Chr), benzo(*b*)fluoranthene BbFlu, benzo(*k*)fluoranthene (BkFlu), benzo(*a*)pyrene (BaP), dibenzo(*a.h*)anthracene (DBahA), benzo(*g.h.i*)perylene (BghiP), and indeno(1,2,3-*cd*)pyrene (IP), were purchased from Sigma-Aldrich (St. Louis, MO, USA). Chemical solvents were of HPLC grade. Commercial detection kits of triglyceride (TG), total cholesterol (TC), low-density lipoprotein-cholesterol (LDL-C) and high-density lipoprotein-cholesterol (HDL-C) were obtained from Nanjing Jiancheng Bioengineering Institute (Nanjing, China).

3.2. Extraction of Chia Seed Oil

Chia seeds (*S. hispanica* L.) were purchased from Guangzhou Yongheng Biotechnology Company Co., Ltd (Guangzhou, China). The chia seeds (100 g) were ground, and extracted with 0.5 L of *n*-hexane at 60 °C for 4 h in three repetitions. The obtained extracts were then combined and concentrated using a rotavapor apparatus. The resulting oil was stored at −20 °C for further analysis.

3.3. Oil Content of Chia Seeds

Determination of chia seed oil (CSO) content was adopted by the China National GB Standard for Oilseeds (GB/T 14488.1-2008) [42]. The oil content was reported as g/100 g seed.

3.4. Chemical Properties

3.4.1. Oxidative Stability

The oxidative stability of CSO was evaluated by the widely used Rancimat analysis [43]. Briefly, 3 g oil was added to the reaction vessel in triplicate, followed by heating to 120 °C. The samples were then dried under airflow at the rate of 20 L/h. A receiver with 60 mL of distilled water was used for collecting the effluent air, containing volatile organic acids from CSO. The conductivity of the water as the process of oxidation proceeded, and the induction period in hours (h) was automatically measured by the machine.

3.4.2. Color Determination

Color measurement was performed via official AOCS method (Cc 13b-45) [44]. In brief, CSO was loaded to a 25.4 mm cell, and the R/Y value was recorded following the instructions of Lovibond PFX 995 (Tintometer, Amesbury, UK).

3.4.3. Fatty Acid Composition

Fatty acid composition of CSO was detected as methyl esters by the following method [45]. Briefly, 100 mg CSO was dissolved in 2 mL potassium hydroxide solution (KOH, 10% w/v in MeOH) and placed in water at 85 °C for 45 min. As the solution cooled, 5 mL of H_2O and 5 mL of hexane were added and mixed thoroughly. Then mixed solution was subjected to liquid–liquid partitioning 3 times, and hexane fractions were combined. The hexane fractions with unsaponifiable substances were then washed with 10% alcohol until neutral pH was obtained, and finally dried by rotary evaporator. The hexane extract containing unsaponifiable substances was kept at −20 °C until further analysis for determination of sterols and squalene.

Saponified fatty acids were extracted from the aqueous layer (previously acidified with 2 M HCl) three times using hexane. The hexane layer containing saponified fatty acids was dried, and dissolved in 2 mL Boron trifluoride-methanol solution (14% methanol). The mixture reaction was then incubated at 60 °C for 45 min to obtain methylation of fatty acids. Hexane was used to extract fatty acid methyl esters (FAMEs) from the cooled mixture. Analysis of FAMEs was further carried out by gas chromatography (7820A, Agilent, Santa Clara, CA, USA), coupled with a FID and a Trace TR-FAME capillary column (i.d. 0.25 µm, 60 m × 0.25 mm, Thermo Fisher, Grand Island, NY, USA). The analysis condition was programmed as follows: the injector and FID temperatures were kept at 250 °C, and initial oven temperature was held at 80 °C for 3 min. It was then increased to 215 °C (15 °C/min), and finally up to 215 °C (20 min hold time). Nitrogen was used as carrier gas at 1 mL/min, with a split ratio of 1:20. Retention times (RT) of fatty acids were compared with those of authentic standards, and contents of individual components were expressed in relative percentages.

3.4.4. Sn-2 Fatty Acid Composition

Analysis of sn-2 fatty acids of CSO was conducted by the assay used in reference [46]. Briefly, 10 mg CSO was added to a reaction solution containing 1 mL of 1 M Tris buffer (pH 8.0), 0.25 mL of bile salts (0.05%), 0.1 mL of calcium chloride (2.2%), and 10 mg of pancreatic lipase. The reaction mixture was kept incubated at 40 °C, with shaking for 3 min. The reaction was stopped by adding 1 mL of 6 M HCl. Separation of above reaction product was performed by preparative TLC on silica gel, with hexane/diethyl ether/acetic acid (1/1/0.02, $v/v/v$) as the developing solvent system. The band responding to the 2-monoacylglycerol (2-MAG) on the TLC plate was scraped into a glass tube and extracted 3 times with diethyl ether. The 2-MAG was dried and methylated to esters following the same saponification process as above. The resultant product was finally analyzed by the same analytical method.

3.4.5. Triacylglycerol Composition

Analysis of triacylglycerol components of CSO was conducted by the official assay (AOCS Ce 5c-93), with HPLC-ELSD (Agilent 1200, Agilent Technologies, Santa Clara, CA, USA) equipped with a C18 column (150 mm × 4.6 mm, 5 µm, Varian, Palo Alto, CA, USA) [47]. CSO was prepared in hexane at 2 mg/mL and analyzed using the following conditions. Briefly, the mobile phase consisted of acetonitrile and isopropanol with a flow rate of 0.8 mL/min. The initial mobile phase was held at 60% acetonitrile for 40 min, changed linearly to 55% (40–80 min), returned to 60% (80–85 min) and maintained at 60% for 5 min (85–90 min). High sensitivity was obtained by using ELSD temperature set at 55 °C with a gain value of 8, and gas flow rate of 1.5 mL/min. Peaks were identified by comparing their retention times with standard reference compounds. Triacylglycerols were confirmed by comparison of RT with their standards. Relative content was reported as a percentage.

3.4.6. FT-IR Spectrum Analysis

The infrared spectrum of CSO was measured by Fourier transform infrared spectroscopy (FTIR) with a Vector 70 model FT-IR instrument (Bruker, Billerica, MA, USA) in the infrared region of $4000-400$ cm^{-1}. This enabled investigation of structural information with a resolution of 4 cm^{-1}. The assignment of functional groups corresponding to IR absorption bands was performed by comparison with distinctive bands of functional groups of the chemicals in edible oils.

3.5. Minor Components

3.5.1. Tocopherol and Phytosterol Levels

Analysis of the composition of tocopherols was performed by HPLC using Waters 2475 multi λ fluorescence detector (λ_{Ex} 293 nm; λ_{Em} 325 nm) (Waters, Milford, MA, USA) based on reported methods [48]. Briefly, the column used in this analysis was a DIOL column (100 mm × 3 mm, 7 µm, Varian, Palo Alto, CA, USA). CSO was dissolved in hexane at 0.25 mg/mL and injected directly into the HPLC column. The optimized conditions for the run were as follows: in a linear gradient elution with the flow rate of 0.5 mL/min, and solvents hexane/tetrahydrofuran (A, 98/2, v/v) and isopropanol (B) with the following gradient timetable: 0–40 min, 100% A; 40–45 min, 100–95% A; 50–51 min, 95–100% A; 51–60 min, 100% A. The identification and quantification of α-, β-, γ- and δ-tocopherols in CSO was determined by comparison with their retention times and the area of the standards of tocopherols, and the unit was expressed as mg/kg of CSO.

The analysis of sterols and squalene was carried out by GC-MS equipped with a flame ionization detector (FID) and using a DB-5MS capillary column (60 m × 0.25 mm, 0.25 µm, Agilent, Santa Clara, CA, USA). Gas chromatograph conditions were as follows: helium was used as carrier gas at the flow rate of 1 mL/min, then the split ratio was 1:100. Oven temperature program: initial temperature 200 °C for 1 min, then 200 °C to 300 °C at 1 °C/min, and maintained at 300 °C for 18 min; detector temperature set at 290 °C, injector temperature 290 °C, injection volume, 1.0 µL. Sterol and squalene were determined and quantified by the internal standard (5α-cholestane), and the contents were expressed as mg/kg.

3.5.2. Determination of Polycyclic Aromatic Hydrocarbons (PAH)

PAH levels were analyzed by LC/MS/MS. Briefly, CSO (0.5 g) was added a 10 ml Pyrex tube, and mixed with 5 µL of internal standard (2 mg/mL) and 3 mL of *n*-hexane by shaking. SupelMIP solid-phase extraction (SPE, Anpel, Shanghai, China) was applied to extract PAHs from the oil. The mixture contents were fully loaded to the SPE column, to allow adhesion of PAHs to the membrane. The column was then washed with hexane (5 mL), and PAHs were eluted with methylene dichloride (10 mL) at a low speed. Eluent was then dried and dissolved in acetonitrile (200 µL) for further experiments.

LC/MS/MS with APCI Source (Agilent, Santa Clara, CA, USA) was applied to detect PAHs in CSO. Separation was achieved using a C18 column (100 × 2.1 mm, 1.8 µm; Agilent, Santa Clara, CA, USA) with a mobile phase comprising of solvent A (0.1% formic acid in acetonitrile) and solvent B (0.1% formic acid in water). The column oven temperature was set at 25 °C, and flow rate was 0.5 mL/min. The optimized conditions for separation were as follow: linear gradient elution was started from 45% A and then increased from 45 to 100% A (1–15 min), held 100% A from 15 to 21 min, next decreased 100% to 45% A from 21–22 min, lastly maintaining 45% A for 3 min. The PAH concentrations in CSO were quantified with internal standard.

3.5.3. Determination of Trace Element Levels

The content of trace elements of CSO was determined by inductively coupled plasma-mass spectrometry (ICP-MS), based on previous methods with minor modifications [49]. Briefly, the oven temperature was programmed from 60 °C (3 min hold) to 175 °C at 5 °C/min, maintained at 175 °C for 15 min, then raised to 220 °C at 2 °C/min, and finally held at 220 °C for 10 min.

3.5.4. Antioxidant Activity In Vitro

The in vitro antioxidant effects of CSO were evaluated by DPPH and ABTS assays [50]. Briefly, 50 µL volumes of different concentrations of CSO were added to 150 µL of DPPH solution (200 µM) with shaking. After incubation of 40 min, optical density was recorded at 517 nm (OD_{517}). Tocopherol was used as a positive control. In the same manner, ABTS radical scavenging capability of CSO was measured at a wavelength of 734 nm. Catechin was used as the positive control in ABTS assay. The formula for calculating % scavenging effects (SE) by DPPH and ABTS was as follows:

$$SE(\%) = \frac{ODcontrol - ODsample}{ODcontrol} \times 100\%$$

3.6. Cell Cytotoxicity

The cytotoxic effects of the chia seed oil against LNcap cells and HepG2 were tested using a 3-(4,5-dimethyl-2-yl)-2, 5 diphenyltetrazolium bromide (MTT) assay [51]. Cells were obtained from Stem Cell Bank, Chinese Academy of Sciences, Shanghai, China. Cells were cultured in Dulbecco's modified Eagle's medium (DMEM; Thermo, Grand Island, NY, USA) containing 10% fetal bovine serum (FBS; Gibco, USA) in an incubator (37 °C, 5% CO2). The dried CSO was dissolved in DMSO to 20 mg/mL and sterilized by filtration through a 0.22 µm membrane filter (Millipore, Burlington, MA, USA) as a stock solution. To each well of a 96-microwell plate, 190 µL of LNcap cells or HepG2 (5 × 10⁴ cell/mL) was added with or without 10 µL of different concentrations of CSO. After 48 h incubation, 20 µL of MTT (5 mg/mL) was added and then incubated for an additional 4 h. The supernatant was then removed completely, and 150 µL of DMSO was added to each well. Lastly, the OD value at 570 nm was measured using a microplate reader (Bio Tek, San Diego, CA, USA). Cell viability was measured and calculated by the formula: cell viability (%) = [OD $_{570}$ (sample)/OD $_{570}$ (control)] × 100%.

3.7. CSO Inhibits OA-Induced Intracellular Lipid Level in HepG2 Cells

3.7.1. Cell Treatment

HepG2 was cultured in DMEM with high glucose (Gibco, USA) supplement with 10% FBS (Gibco, USA) at 37 °C with 5% CO_2 in a cell incubator. The oleic acid (OA) was dissolved in 10% bovine serum albumin (BSA, fatty acid free, Sigma, St. Louis, MO, USA), and diluted in a culture medium with 1%

BSA as the final concentration of 200 µM. CSO was prepared in DMEM as stock solution. The cells were seeded in 6-well plates and incubated for 24 h, before being cultured as follows: DMEM plus 1% BSA as a control group; DMEM with OA-BSA at 200 µM as a model group; CSO at final concentration of 500 µg/mL was applied with OA-BSA simultaneously as a co-treatment.

3.7.2. Measurement of Cellular Lipid Levels in HepG2 Cells

Contents of TG, TC, LDL-C and HDL-C in HepG2 cell model were measured by commercial kits (Nanjing Jiancheng Bioengineering Institute, Nanjing, China). Briefly, cells were collected and washed twice with pre-chilled PBS to remove culture medium after treatment for 24 h. Then, 200 µL of 2% Triton X-100 was added for lysing cells by sonication. The supernatant was collected for analysis of TG, TC, LDL-C and HDL-C. Results were normalized to protein concentration using the Bradford protein assay kit (Bio-Rad, Hercules, CA, USA).

3.8. Statistical Analysis

Statistical analysis was carried out using GraphPad Prism 7.0 (GraphPad, San Diego, CA, USA). Differences were considered to be statistically significant at $p < 0.05$, or very significant at $p < 0.01$, by the Duncan test.

4. Conclusions

In conclusion, PUFA analysis in chia seed oil from Mexico was greater than 80% of fatty acids. This research indicates that α-linolenic acid was the most abundant fatty acid, followed by linoleic acid with an n-3/n-6 ratio of 3.21:1. Due to the large amount of easily-oxidized PUFAs in chia seed oil (induction period = 0.68 h), raw chia seeds might need special treatment during sample collection, processing, storage, and transportation, to prevent oxidation and maintain the quality of the oil. We also found that CSO presented some antioxidant effects in vitro, which are beneficial for human health. This is the first study to report the ability of CSO to reduce the levels of hepatic TG and LDL-C markedly, and increase the content of HDL-C in HepG2 cell model. Thus, this indicates that chia seed oil may help in reducing the risk for cardiovascular disease, and it can be widely used as cooking oil and in manufacture of healthy food supplements.

Supplementary Materials: The following are available online. Figure S1: HPLC chromatogram of fatty acids or chia seed oils, Figure S2: HPLC chromatography of sn-2 fatty acids from chia seed oil, Figure. S3: Analysis of triacylglycerols from chia seed oil by HPLC, Figure S4: HPLC chromatogram of tocopherols in chia seed oil. (A) standards of tocopherols, and (B) tocoperols in chia seed oil.

Author Contributions: Extraction, IR, and GC/MS analysis, Y.S. and L.Z.; Antioxidant and cytotoxicity assay, J.J. and M.W.; PAHs analysis, X.L. and J.F.; Writing improvement, M.W.; Project administration, supervision and writing, Y.G. and X.S. All authors read and approved the final manuscript.

Funding: This work was financially supported by China Postdoctoral Science Foundation funded project (Grant No. 2016M602605), Medical Scientific Research Foundation of Guangdong Province, China (Grant No. A2018037), National Natural Science Foundation of China (Grant No. 81503215 and 31670360), and the Research Start-up Funds for Institute of Marine Drugs in Guangxi University of Chinese Medicine (Grant No. 2018ZD005-A17).

Conflicts of Interest: There is no conflict of interest recorded for this work.

References

1. Simopoulos, A.P. The importance of the ratio of omega-6/omega-3 essential fatty acids. *Biomed. Pharmacother.* **2002**, *56*, 365–379. [CrossRef]
2. Singh, M. Essential fatty acids, DHA and human brain. *Indian J. Pediatr.* **2005**, *72*, 239–242. [CrossRef] [PubMed]
3. Watchareewan, J.T.; Phillips, R.; Geneve, J.; Cahill, D.H. Extending the range of an ancient crop, *Salvia hispanica L.*—A new x3 source. *Genet. Resour. Crop Evol.* **2012**, *59*, 171–178.

4. Timilsena, Y.P.; Wang, B.; Adhikari, R.; Adhikari, B. Preparation and characterization of chia seed protein isolate—Chia seed gum complex coacervates. *Food Hydrocoll.* **2016**, *52*, 554–563. [CrossRef]

5. Martinez-Cruz, O.; Paredes-Lopez, O. Phytochemical profile and nutraceutical potential of chia seeds (*Salvia hispanica L.*) by ultra high performance liquid chromatography. *J. Chromatogr. A* **2014**, *1346*, 43–48. [CrossRef] [PubMed]

6. Wu, H.; Sung, A.; Burns-Whitmore, B.; Jo, E.; Wien, M. Effect of chia seed (*Salvia hispanica, L.*) supplementation on body composition, weight, post-prandial glucose and satiety. *Faseb J.* **2016**, *30*, lb221.

7. Gazem, R.A.A.; Ramesh Puneeth, H.; Shivmadhu Madhu, C.; Chandrashekaraiah Sharada, A. In vitro anticancer and anti-lipoxygenase activities of chia seed oil and its blends with selected vegetable oils. *Asian J. Pharm. Clin. Res.* **2017**, *10*, 124–128. [CrossRef]

8. Authority, E.F.S. Opinion on the safety of 'Chia seeds (*Salvia hispanica L.*) and ground whole Chia seeds' as a food ingredient. *EFSA J.* **2009**, *996*, 1–26.

9. Ayerza, R. The Seed's protein and oil content, fatty acid aomposition, and growing cycle length of a single genotype of chia (*Salvia hispanica L.*) as affected by environmental factors. *J. Oleo Sci.* **2009**, *58*, 347–354. [CrossRef] [PubMed]

10. Dornbos, D.L.; Mullen, R.E. Soybean seed protein and oil contents and fatty-acid composition adjustments by drought and temperature. *J. Am. Oil Chem. Soc.* **1992**, *69*, 228–231. [CrossRef]

11. Silva, B.P.; Anunciacao, P.C.; Matyelka, J.; Della Lucia, C.M.; Martino, H.S.D.; Pinheiro-Sant'Ana, H.M. Chemical composition of Brazilian chia seeds grown in different places. *Food Chem.* **2017**, *221*, 1709–1716. [CrossRef] [PubMed]

12. Ayerza, R.; Coates, W. Composition of chia (*Salvia hispanica*) grown in six tropical and subtropical ecosystems of South America. *Trop. Sci.* **2004**, *44*, 131–135. [CrossRef]

13. Ratusz, K.; Popis, E.; Ciemniewska-Zytkiewicz, H.; Wroniak, M. Oxidative stability of camelina (*Camelina sativa L.*) oil using pressure differential scanning calorimetry and Rancimat method. *J. Therm. Anal. Calorim.* **2016**, *126*, 343–351. [CrossRef]

14. Farhoosh, R. The effect of operational parameters of the rancimat method on the determination of the oxidative stability measures and shelf-life prediction of soybean oil. *J. Am. Oil Chem. Soc.* **2007**, *84*, 205–209. [CrossRef]

15. Uzunova, G.; Nikolova, K.; Perifanova, M.; Gentscheva, G.; Marudova, M.; Antova, G. Physicochemical characterization of chia (*Salvia hispanica*) seed oil from Argentina. *Bulg. Chem. Commun.* **2016**, *48*, 131–135.

16. Hosseinian, F.S.; Rowland, G.G.; Bhirud, P.R.; Dyck, J.H.; Tyler, R.T. Chemical composition and physicochemical and hydrogenation characteristics of high-palmitic acid solin (low-linolenic acid flaxseed) oil. *J. Am. Oil Chem. Soc.* **2004**, *81*, 185–188. [CrossRef]

17. Patil, R.T.; Ali, N. Effect of pre-treatments on mechanical oil expression of soybean using a commercial oil expeller. *Int. J. Food Prop.* **2006**, *9*, 227–236. [CrossRef]

18. Coates, W.; Ayerza, R. Chia (*Salvia hispanica L.*) seed as an n-3 fatty acid source for finishing pigs: Effects on fatty acid composition and fat stability of the meat and internal fat, growth performance, and meat sensory characteristics. *J. Anim. Sci.* **2009**, *87*, 3798–3804. [CrossRef]

19. Ixtaina, V.Y.; Martínez, M.L.; Spotorno, V.; Mateo, C.M.; Maestri, D.M.; Diehl, B.W.K.; Nolasco, S.M.; Tomás, M.C. Characterization of chia seed oils obtained by pressing and solvent extraction. *J. Food Compos. Anal.* **2011**, *24*, 166–174. [CrossRef]

20. Boschin, G.; D'Agostina, A.; Annicchiarico, P.; Arnoldi, A. The fatty acid composition of the oil from *Lupinus albus cv. Luxe* as affected by environmental and agricultural factors. *Eur. Food Res. Technol.* **2007**, *225*, 769–776. [CrossRef]

21. Food and Nutrition Board. *Nutrient Recommendations: Dietary Reference Intakes (Dri), Dri Table: Recommended Dietary Allowances and Adequate Intakes, Total Water and Macronutrients*; National Academy of Sciences: Washington, DC, USA, 2017.

22. Reena, M.B.; Lokesh, B.R. Hypolipidemic effect of oils with balanced amounts of fatty acids obtained by blending and interesterification of coconut oil with rice bran oil or sesame oil. *J. Agric. Food Chem.* **2007**, *55*, 10461–10469. [CrossRef] [PubMed]

23. Mohd Ali, N.; Yeap, S.K.; Ho, W.Y.; Beh, B.K.; Tan, S.W.; Tan, S.G. The promising future of chia, *Salvia hispanica* L. *J. Biomed. Biotechnol.* **2012**, *2012*, 1–9. [CrossRef] [PubMed]

24. Matthaus, B.; Ozcan, M.M. Quantitation of fatty acids, sterols, and tocopherols in turpentine (*Pistacia terebinthus Chia*) growing wild in Turkey. *J. Agric. Food Chem.* **2006**, *54*, 7667–7671. [CrossRef] [PubMed]
25. Alvarez-Chavez, L.M.; Valdivia-Lopez, M.D.; Aburto-Juarez, M.D.; Tecante, A. Chemical characterization of the lipid fraction of Mexican chia seed (*Salvia hispanica L.*). *Int. J. Food Prop.* **2008**, *11*, 687–697. [CrossRef]
26. Bakowska-Barczak, A.M.; Schieber, A.; Kolodziejczyk, P. Characterization of saskatoon berry (*Amelanchier alnifolia Nutt.*) seed oil. *J. Agric. Food Chem.* **2009**, *57*, 5401–5406. [CrossRef]
27. Maguire, L.S.; O'Sullivan, S.M.; Galvin, K.; O'Connor, T.P.; O'Brien, N.M. Fatty acid profile, tocopherol, squalene and phytosterol content of walnuts, almonds, peanuts, hazelnuts and the macadamia nut. *Int. J. Food Sci. Nutr.* **2004**, *55*, 171–178. [CrossRef]
28. Karadjova, I.; Zachariadis, G.; Boskou, G.; Stratis, J. Electrothermal atomic absorption spectrometric determination of aluminium, cadmium, chromium, copper, iron, manganese, nickel and lead in olive oil. *J. Anal. Atom Spectrom.* **1998**, *13*, 201–204. [CrossRef]
29. Commission, C.A. Codex Standard for Edible Fats and Oils not Covered by Individual Standards (Codex stan 19–1981, Rev. 2-1999). Available online: http://www.fao.org/docrep/004/y2774e/y2774e03.htm (accessed on 30 October 2018).
30. Brevedan, M.I.V.; Carelli, A.A.; Crapiste, G.H. Changes in composition and quality of sunflower oils during extraction and degumming. *Grasas Aceites* **2000**, *51*, 417–423.
31. Ciecierska, M.; Obiedzinski, M.W. Polycyclic aromatic hydrocarbons in vegetable oils from unconventional sources. *Food Control* **2013**, *30*, 556–562. [CrossRef]
32. Mafra, I.; Amaral, J.S.; Oliveira, M.B.P.P. Polycyclic aromatic hydrocarbons (PAH) in olive oils and other vegetable oils; Potential for carcinogenesis. *Olives Olive Oil Health Dis. Prev.* **2010**, 489–498. [CrossRef]
33. Moreda, W.; Perez-Camino, M.C.; Cert, A. Gas and liquid chromatography of hydrocarbons in edible vegetable oils. *J. Chromatogr. A* **2001**, *936*, 159–171. [CrossRef]
34. Zelinkova, Z.; Wenzl, T. The occurrence of 16 EPA PAHs in food—A review. *Polycycl. Aromat. Comp.* **2015**, *35*, 248–284. [CrossRef] [PubMed]
35. Arrebola, F.J.; Frenich, A.G.; Rodriguez, M.J.G.; Bolanos, P.P.; Vidal, J.L.M. Determination of polycyclic aromatic hydrocarbons in olive oil by a completely automated headspace technique coupled to gas chromatography-mass spectrometry. *J. Mass Spectrom.* **2006**, *41*, 822–829. [CrossRef] [PubMed]
36. Guillen, M.D.; Cabo, N. Infrared spectroscopy in the study of edible oils and fats. *J. Sci. Food Agr.* **1997**, *75*, 1–11. [CrossRef]
37. Timilsena, Y.P.; Vongsvivut, J.; Adhikari, R.; Adhikari, B. Physicochemical and thermal characteristics of Australian chia seed oil. *Food Chem.* **2017**, *228*, 394–402. [CrossRef] [PubMed]
38. Xuan, T.D.; Gu, G.Q.; Minh, T.N.; Quy, T.N.; Khanh, T.D. An overview of chemical profiles, antioxidant and antimicrobial activities of aommercial vegetable edible oils marketed in Japan. *Foods* **2018**, *7*, 21. [CrossRef] [PubMed]
39. Scapin, G.; Schmidt, M.M.; Prestes, R.C.; Rosa, C.S. Phenolics compounds, flavonoids and antioxidant activity of chia seed extracts (*Salvia hispanica*) obtained by different extraction conditions. *Int. Food Res. J.* **2016**, *23*, 2341–2346.
40. Li, X.; Zhao, M.; Fan, L.; Cao, X.; Chen, L.; Chen, J.; Lo, Y.M.; Zhao, L. Chitobiose alleviates oleic acid-induced lipid accumulation by decreasing fatty acid uptake and triglyceride synthesis in HepG2 cells. *J. Funct. Foods* **2018**, *46*, 202–211. [CrossRef]
41. Ayerza, R.; Coates, W. Ground chia seed and chia oil effects on plasma lipids and fatty acids in the rat. *Nutr. Res.* **2005**, *25*, 995–1003. [CrossRef]
42. China National Standardization Management Committee. *Determination of Oil Content Valid GB/T 14488.2-2008*; China National Standardization Management Committee: Beijing, China, 2008.
43. Azadmard-Damirchi, S.; Habibi-Nodeh, F.; Hesari, J.; Nemati, M.; Achachlouei, B.F. Effect of pretreatment with microwaves on oxidative stability and nutraceuticals content of oil from rapeseed. *Food Chem.* **2010**, *121*, 1211–1215. [CrossRef]
44. Committee, A.C. Color of Fats and Oils, Lovibond (Wesson) Wesson Method Using Color Glasses Calibrated in Accordance with the Aocs-Tintometer Color Scale. AOCS Official Method Cc 13b-45. Available online: https://aocs.personifycloud.com/PersonifyEBusiness/Default.aspx?TabID=251&productId=111497 (accessed on 30 October 2018).

45. Jin, J.; Wang, Y.; Su, H.; Warda, P.; Xie, D.; Liu, Y.J.; Wang, X.S.; Huang, J.H.; Jin, Q.Z.; Wang, X.G. Oxidative stabilities of mango kernel fat fractions produced by three-stage fractionation. *Int. J. Food Prop.* **2017**, *20*, 2817–2829. [CrossRef]

46. Jin, J.; Warda, P.; Mu, H.Y.; Zhang, Y.F.; Jie, L.; Mao, J.H.; Xie, D.; Huang, J.H.; Jin, Q.Z.; Wang, X.G. Characteristics of mango kernel fats extracted from 11 China-specific varieties and their typically fractionated fractions. *J. Am. Oil Chem. Soc.* **2016**, *93*, 1115–1125. [CrossRef]

47. AOCS. Individual Triglycerides in Oils and Fats by HPLC. AOCS Official Method Ce 5c-93. Available online: https://www.aocs.org/attain-lab-services/methods/methods/method-detail?productId= 114619 (accessed on 30 October 2018).

48. Moreau, R.A.; Hicks, K.B. Reinvestigation of the effect of heat pretreatment of corn fiber and corn germ on the levels of extractable tocopherols and tocotrienols. *J. Agric. Food Chem.* **2006**, *54*, 8093–8102. [CrossRef] [PubMed]

49. Llorent-Martinez, E.J.; Ortega-Barrales, P.; Fernandez-de Cordova, M.L.; Dominguez-Vidal, A.; Ruiz-Medina, A. Investigation by ICP-MS of trace element levels in vegetable edible oils produced in Spain. *Food Chem.* **2011**, *127*, 1257–1262. [CrossRef] [PubMed]

50. Shen, Y.; Zhang, H.; Cheng, L.; Wang, L.; Qian, H.; Qi, X. In vitro and in vivo antioxidant activity of polyphenols extracted from black highland barley. *Food Chem.* **2016**, *194*, 1003–1012. [CrossRef]

51. Srivastava, M.; Nambiar, M.; Sharma, S.; Karki, S.S.; Goldsmith, G.; Hegde, M.; Kumar, S.; Pandey, M.; Singh, R.K.; Ray, P.; et al. An inhibitor of nonhomologous end-joining abrogates double-strand break repair and impedes cancer progression. *Cell* **2012**, *151*, 1474–1487. [CrossRef] [PubMed]

Sample Availability: Samples of the compounds are not available from the authors.

molecules

MDPI

Article

Development of New Analytical Microwave-Assisted Extraction Methods for Bioactive Compounds from Myrtle (*Myrtus communis* L.)

Ana V. González de Peredo [1], Mercedes Vázquez-Espinosa [1], Estrella Espada-Bellido [1], Ana Jiménez-Cantizano [2], Marta Ferreiro-González [1], Antonio Amores-Arrocha [2], Miguel Palma [1], Carmelo G. Barroso [1] and Gerardo F. Barbero [1,*]

[1] Department of Analytical Chemistry, Faculty of Sciences, University of Cadiz, Agrifood Campus of International Excellence (ceiA3), IVAGRO, Puerto Real, 11510 Cadiz, Spain; ana.velascogope@uca.es (A.V.G.d.P.); mercedes.vazquez@uca.es (M.V.-E.); estrella.espada@uca.es (E.E.-B.); marta.ferreiro@uca.es (M.F.-G.); miguel.palma@uca.es (M.P.); carmelo.garcia@uca.es (C.G.B.)

[2] Department of Chemical Engineering and Food Technology, Faculty of Sciences, University of Cadiz, Agrifood Campus of International Excellence (ceiA3), IVAGRO, Puerto Real, 11510 Cadiz, Spain; ana.jimenezcantizano@uca.es (A.J.-C.); antonio.amores@uca.es (A.A.-A.)

* Correspondence: gerardo.fernandez@uca.es; Tel.: +34-956-01-6355

Received: 16 October 2018; Accepted: 13 November 2018; Published: 16 November 2018

check for updates

Abstract: The phenolic compounds and anthocyanins present in myrtle berries are responsible for its beneficial health properties. In the present study, a new, microwave-assisted extraction for the analysis of both phenolic compounds and anthocyanins from myrtle pulp has been developed. Different extraction variables, including methanol composition, pH, temperature, and sample–solvent ratio were optimized by applying a Box–Behnken design and response surface methodology. Methanol composition and pH were the most influential variables for the total phenolic compounds (58.20% of the solvent in water at pH 2), and methanol composition and temperature for anthocyanins (50.4% of solvent at 50 °C). The methods developed showed high repeatability and intermediate precision (RSD < 5%). Both methods were applied to myrtle berries collected in two different areas of the province of Cadiz (Spain). Hierarchical clustering analysis results show that the concentration of bioactive compounds in myrtle is related to their geographical origin.

Keywords: anthocyanins; bioactive compounds; Box–Behnken design; microwave-assisted extraction; myrtle; *Myrtus communis*; phenolic compounds

---

1. Introduction

Myrtus communis L., the common myrtle, is an evergreen shrub that grows spontaneously in the Mediterranean area and in the Middle East. Myrtle berries have a maximum ripening period from October to February. These berries have multiple shapes and colors [1], but are mainly dark blue in color. The ancient Mediterranean populations already used myrtle mainly for ornamental and aromatic purposes [2]. Recent developments in the fields of health and food have markedly increased their interest in natural compounds with antioxidant potential [3]. The extraction of natural antioxidants in fruits is very useful when substituting synthetic antioxidants, which are being restricted because of their potential health risks and side-effects, and their safety has been questioned for a long time [4]. Therefore, nowadays, myrtle has gained greater recognition in the food and medicinal industries due to its potential beneficial effects [5]. Myrtle presents anti-diabetic, anti-inflammatory, anticancer, antioxidant, antihyperglycaemic, antimycotic, and antiseptic properties [6,7]. For example, myrtle oil is recommended for the treatment of respiratory diseases [8], and it is normally taken as an infusion. In addition, essential oils are used in the perfume and cosmetics industries [9]. Despite the aforementioned uses, myrtle is still mainly known for the production of an aromatic liqueur called "Mirto", which is obtained by alcoholic maceration of its leaves and fruit. This liquor is very popular and traditional in Sardinia, where it is usually served very cold after meals due to its digestive powers [10].

The phenolic compounds and anthocyanins present in myrtle berries are the main contributors to these beneficial health properties. The quantities of phenolic compounds and anthocyanins present in myrtle berries are extraordinarily high [11,12]. The major phenolic compounds identified in myrtle are quercetin 3-*O*-galactoside, quercetin 3-*O*-rhamnoside, myricetin 3-*O*-rhamnoside, quercetin 3-*O*-glucoside, ellagic acid, and myricetin [13,14]. The major anthocyanins identified are delphinidin 3,5-*O*-diglucoside, delphinidin 3-*O*-glucoside, cyanidin 3-*O*-galactoside, cyanidin 3-*O*-glucoside, cyanidin 3-*O*-arabinoside, petunidin 3-*O*-glucoside, delphinidin 3-*O*-arabinoside, peonidin 3-*O*-glucoside, malvidin 3-*O*-glucoside, petunidin 3-*O*-arabinoside, and malvidin 3-*O*-arabinoside [15,16].

For the extraction of bioactive compounds, specifically for the extraction of phenolic compounds and anthocyanins in vegetable matrices, solid–liquid extraction is usually carried out. Microwave-assisted extraction (MAE) is one of the more advanced extraction methods. MAE is widely used as it is a promising green-extraction method that has the advantage of reducing both extraction time and solvent consumption [17,18]. The microwave technique is based on the application of electromagnetic radiation, with a frequency from 0.3 to 300 GHz. This radiation, which leads to rapid and localized heating of the solvent and sample, is based on a direct effect on the molecules through ionic conduction and dipole rotation [19]. The localized heating leads to a pressure build-up within the cells of the sample, resulting in a rapid transfer of the compound of interest from the cells to the extraction solvent [20]. This extraction method has been widely employed for the extraction of phenolic compounds and anthocyanins from a wide variety of vegetable matrices, such as grapes [21], tomatoes [22], and blackberries [23]. In a recent study, MAE has been applied to extract polyphenols, tannins, and flavonoids from myrtle leaves [24]. Based on the previous results obtained from this study and the high content of bioactive compounds in myrtle leaves, the development of new methods for myrtle analyses to improve the quality of final products, such as liqueurs, is required. Besides, this extraction technique has not yet been completely developed and optimized for the extraction of phenolic compounds and anthocyanins from myrtle berries. As in myrtle leaves, its berries are expected to contain an important amount of the same bioactive compounds besides the anthocyanins due to its dark blue color. This approach is of great interest due to the importance of developing techniques for analysis of the raw material to improve the liqueur quality, which is the main use of myrtle.

The efficiency of MAE can be affected by several variables, such as the solvent (volume, composition, pH), the temperature, the time of application, and the power level [25]. For this reason, experimental designs are usually applied in order to study the effects of the different variables and their interactions, and to determine the optimal conditions. In the present study, a Box–Behnken design (BBD) with a response surface methodology (RSM) was chosen [26,27]. It was employed because the number of experiments necessary to provide sufficient information for statistically acceptable results in a BBD is lower than other statistical designs, and it also ensures that each experiment is in the region of interest, avoiding extreme conditions [28].

The aim of the present study was to develop and optimize MAE methods for the extraction of bioactive compounds (phenolic compounds and anthocyanins) in myrtle in order to evaluate the quality of myrtle berries and to study the possible effect of the geographical origin in the total amount of bioactive compounds.

2. Results and Discussion

2.1. Development of the MAE Methods

A Box–Behnken design was carried out for the development and optimization of the microwave-assisted extraction of both total phenolic compounds and total anthocyanins (as the sum of individual components) in the myrtle berries. Analysis of variance (ANOVA) was carried out to evaluate the effects of the factors and the possible interactions between them. The factors studied in this work were: solvent composition (% methanol in water), solvent pH, extraction temperature, and sample–solvent ratio. The results of this analysis are shown in Tables 1 and 2 for the total phenolic compounds and total anthocyanins, respectively. The coefficients for the different parameters of the quadratic polynomial equation and their significance (*p*-values) are presented. The factors and/or interactions that showed a *p*-value lower than 0.05 were considered to be significant factors that influenced the response at the selected level of significance (95%).

Table 1. Analysis of variance of the quadratic model adjusted to the extraction of total phenolic compounds. The studied ranges for each parameter were: methanol (50–100%), pH (2–7), temperature (50–100 °C), and sample–solvent ratio (0.5 g/10 mL–0.5 g/20 mL).

Variable	Source	Coefficient	Sum of Squares	Degrees of Freedom	Mean Square	F-Value	p-Value
	Model		1016.28	14	72.59	2.92	0.0352
Methanol	X_1	−4.39257	231.54	1	231.54	9.31	0.0101
pH	X_2	−3.62017	157.27	1	157.27	6.32	0.0272
Temperature	X_3	0.915742	10.06	1	10.06	0.4047	0.5366
Ratio	X_4	1.11842	15.01	1	15.01	0.6036	0.4522
Methanol-pH	X_1X_2	−5.57764	3.80	1	3.80	0.1528	0.7027
Methanol-Temperature	X_1X_3	0.974675	47.75	1	47.75	1.92	0.1911
Methanol-Ratio	X_1X_4	−3.45505	3.43	1	3.43	0.1380	0.7167
pH-Temperature	X_2X_3	0.926275	51.01	1	51.01	2.05	0.1776
pH-Ratio	X_2X_4	4.93415	25.11	1	25.11	1.01	0.3348
Temperature-Ratio	X_3X_4	−3.571	8.33	1	8.33	0.3350	0.5734
Methanol-Methanol	$X_1{}^2$	−2.50565	165.92	1	165.92	6.67	0.0240
pH-pH	$X_2{}^2$	1.45055	129.84	1	129.84	5.22	0.0413
Temperature-Temperature	$X_3{}^2$	−1.44313	11.22	1	11.22	0.4513	0.5145
Ratio-Ratio	$X_4{}^2$	0.889012	4.22	1	4.22	0.1695	0.6878
Residual		42.6799	298.41	12	24.87		
Lack of Fit			266.57	10	26.66	1.67	0.4311
Pure Error			31.83	2	15.92		
Total			1314.68	26			

With regard to the total phenolic compounds (Table 1), the *p*-values for solvent composition and pH were less than 0.05, meaning that these factors had significant effects. The quadratic interactions of solvent composition ($X_1{}^2$) and pH ($X_2{}^2$) had a significant effect on the extraction of phenolic compounds. The interactions between factors were not significant (*p*-value > 0.05). Among the linear terms, the most significant factor was the solvent composition (*p*-value < 0.01), and this had a negative effect ($b_1 = -4.39257$), which means that the phenolic compounds were extracted more efficiently when the solvent had a low methanol content in water in this range. The pH also had a negative effect ($b_2 = -3.62017$), which means that the extraction of phenolic compounds is more favorable at a low pH. Among the quadratic effects, the methanol effect was more significant than the pH effect, and methanol had a negative effect, whereas pH had a positive effect.

In the case of anthocyanins (Table 2), only the linear term temperature (X_3) had an influence on the response, with a *p*-value < 0.01. With regard to quadratic effects, $X_1{}^2$ (solvent composition) once again had a significant effect on the extraction. The interaction between the factors' pH and temperature (X_2X_3) had a significant effect. The temperature had a negative effect ($b_2 = -2.56032$), which indicates that a decrease in its value led to a higher recovery of anthocyanins. This should be due to the degradation of anthocyanins when high temperatures were used [29]. The quadratic effect for methanol and the interaction between pH-temperature also had positive coefficients.

Table 2. Analysis of variance of the quadratic model, adjusted to the extraction of total anthocyanins. The ranges studied for each parameter were: methanol (50–100%), pH (2–7), temperature (50–100 °C), and sample–solvent ratio (0.5 g/10 mL–0.5 g/20 mL).

Variable	Source	Coefficient	Sum of Squares	Degrees of Freedom	Mean Square	*F*-Value	*p*-Value
	Model		445.11	14	31.79	4.05	0.0100
Methanol	X_1	0.249033	0.7442	1	0.7442	0.0949	0.7633
pH	X_2	−0.207067	0.5145	1	0.5145	0.0656	0.8022
Temperature	X_3	−2.56032	78.66	1	78.66	10.03	0.0081
Ratio	X_4	−0.92215	10.20	1	10.20	1.30	0.2762
Methanol-pH	X_1X_2	5.98483	15.51	1	15.51	1.98	0.1850
Methanol-Temperature	X_1X_3	−1.9692	21.20	1	21.20	2.70	0.1260
Methanol-Ratio	X_1X_4	2.3022	0.6398	1	0.6398	0.0816	0.7800
pH-Temperature	X_2X_3	0.39995	41.29	1	41.29	5.27	0.0406
pH-Ratio	X_2X_4	−0.758496	0.5627	1	0.5627	0.0718	0.7933
Temperature-Ratio	X_3X_4	3.21277	9.76	1	9.76	1.24	0.2864
Methanol-Methanol	$X_1{}^2$	0.375075	191.03	1	191.03	24.36	0.0003
pH-pH	$X_2{}^2$	−0.830096	3.07	1	3.07	0.3913	0.5433
Temperature-Temperature	$X_3{}^2$	−1.56213	3.67	1	3.67	0.4687	0.5066
Ratio-Ratio	$X_4{}^2$	1.0868	6.30	1	6.30	0.8034	0.3877
Residual		18.6496	94.10	12	7.84		
Lack of Fit			87.43	10	8.74	2.62	0.3076
Pure Error			6.67	2	3.33		
Total			539.21	26			

The standardized Pareto chart, which allows for knowledge of the influencing variables and their order of influence from a graphical point of view, is presented in Figure 1. As mentioned earlier, for phenolic compounds (Figure 1a), it can be seen that the significant factors in decreasing order of influence on the response are: methanol percentage, the quadratic interaction of methanol percentage, and the quadratic interaction of pH. For anthocyanins (Figure 1b), the significant factors in the same order are: the quadratic interaction of methanol percentage, the temperature, and the interaction pH temperature.

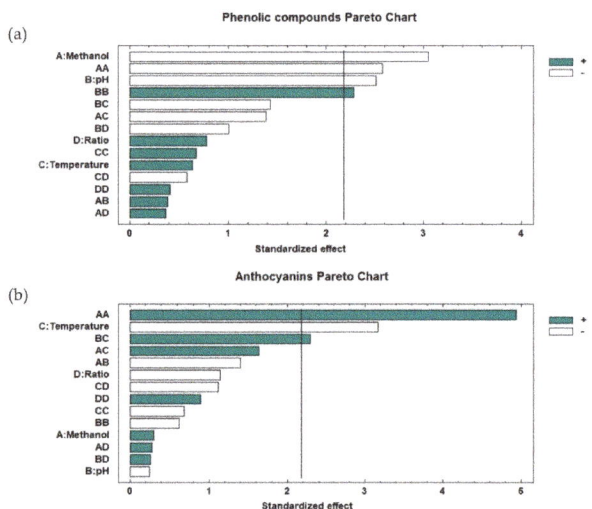

Figure 1. Standardized Pareto charts for: (**a**) total phenolic compounds; (**b**) anthocyanins.

The complete second-order polynomial model correlates the relationship between independent variables and responses. The correlation was evaluated using the squared correlation coefficients (R^2). The coefficients obtained for the total phenolic compounds, $R^2 = 77.30\%$, and the anthocyanins, $R^2 = 82.55\%$, indicate a statistically significant agreement between the measured and estimated responses. More specifically, the lack-of-fit test showed a *p*-value higher than 0.05 (not significant) for both phenolic compounds and anthocyanins, which means that the models fit well.

The reduced Equations (1) and (2), which show acceptable agreement between the experimental data and the estimated values, are expressed as follows:

$$Y_{TP}\ (\mu g\ g^{-1}) = 42.6799 - 4.3927X_1 - 3.62017X_2 - 5.57764X_{12} + 4.93415X_{22} \tag{1}$$

$$Y_{TA}\ (\mu g\ g^{-1}) = 18.6496 - 2.56032X_3 + 5.98483X_{12} + 2.3022X_1X_3 \tag{2}$$

The trends outlined above were recorded in three-dimensional surface plots obtained by using the polynomial equations. Solvent composition and pH were selected as the most significant factors for phenolic compounds (Figure 2a), and temperature and solvent composition as the most significant factors for anthocyanins (Figure 2b) according to the previous results mentioned above. The plots illustrate the combined effects of the most significant variables on: (**a**) total anthocyanins; and (**b**) total phenolic compound recovery, respectively.

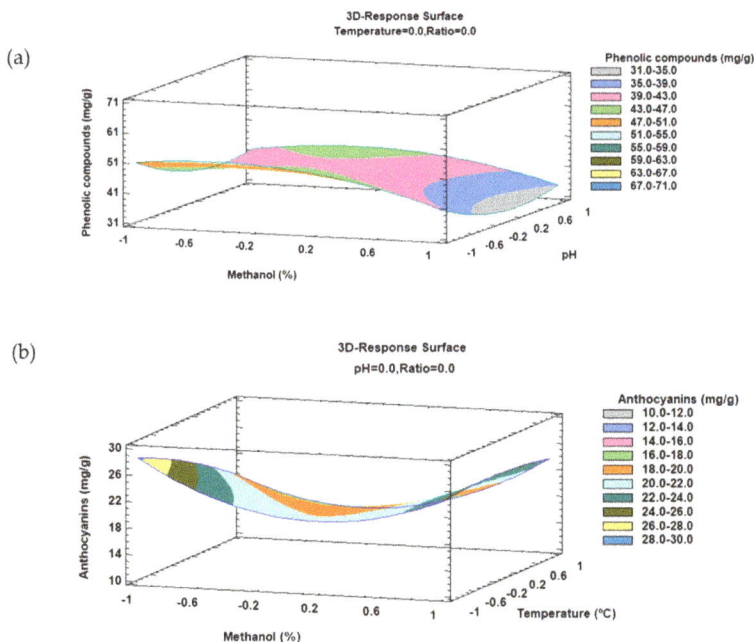

Figure 2. 3D surface plots of the Box–Behnken design using polynomial equations: (**a**) solvent composition and pH on the total phenolic compound extraction, and (**b**) temperature and solvent composition on the total anthocyanin extraction.

2.2. Optimal Conditions

From the Box–Behnken design, it is possible to extract information about the optimum values which show the maximum response for each factor. The optimum MAE conditions to extract the maximum amount of phenolic compounds are as follows: a solvent with 58.20% MeOH in water at pH 2, an extraction temperature of 100 °C, and a 0.5 g/20 mL sample–solvent ratio. The optimum

MAE conditions to extract the maximum amount of anthocyanins are as follows: a solvent with 50.4% MeOH in water at pH 3.33, an extraction temperature of 50 °C, and a 0.5 g/20 mL sample–solvent ratio. These results show that the optimal extraction of both phenolic compounds and anthocyanins occurs with values of methanol and pH closer to the lower end of the studied range (50% MeOH in water). With regard to temperature, numerous authors are in agreement that an increase in temperature favors extraction, but also that beyond a certain value, the compounds of interest can be denatured [30]. With respect to anthocyanins, high temperatures may reduce its recovery, since these compounds are thermally sensible and thus can be easily degraded [31]. With respect to phenolic compounds, although anthocyanins are also phenolic compounds, they are present at determinate levels in the overall mixture, so that the other compounds could be different phenolic compounds and less thermally sensible [32]. This possibility would explain why the optimal extraction temperature was high for the total phenolic compounds (100 °C), whereas for the anthocyanins, it was at the lower end of the range studied (50 °C). The phenolic compounds, which were less thermally sensible, increased the solubility in the solvent and the diffusion and mass transfer of the extracted molecules with high temperatures, favoring the extraction [33].

The results obtained at the optimum conditions using MAE were compared with those achieved by other extraction methods already developed from the same raw material (myrtle) [14,34–36]. Most of these studies employ traditional extraction techniques that imply long extraction times (in many cases, up to 24 h) without obtaining large recoveries. In comparison with traditional methods, such as maceration, MAE offers better extraction yields of the compounds of interest in a shorter time frame and with lower expense, regarding both solvents and costs. The higher extraction yield of the total phenolic compounds and total anthocyanins could be due to water dipole rotation and ionic conduction effects, which is the main mechanism of microwave heating.

2.3. Kinetics of the Extraction Process

Once the optimal values had been obtained, they were used to study the kinetics of the extraction process. Several extractions were carried out at different times, with fixed values for the factors already studied (percentage of methanol, temperature, sample–solvent ratio, and pH). The experiments were performed in triplicate, and the times employed were: 2, 5, 10, 15, 20, and 25 min. The results for the recovery of total anthocyanins and total phenolic compounds are represented in Figure 3.

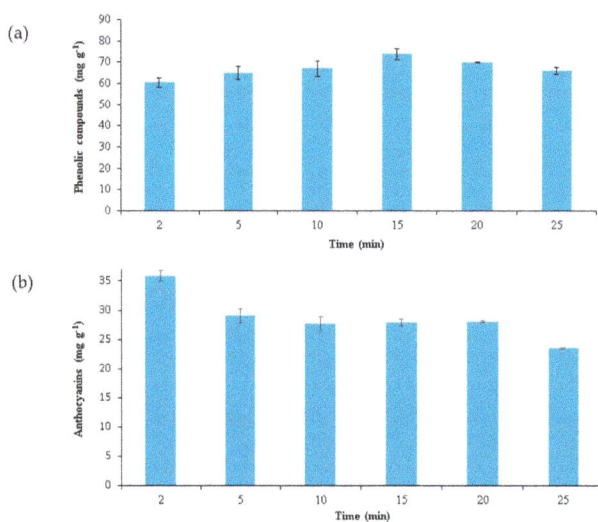

Figure 3. (**a**) Extraction kinetics of total phenolic compounds; (**b**) extraction kinetics of total anthocyanins.

It can be seen that for phenolic compounds (Figure 3a), the maximum recovery was achieved at 15 min, being lower when longer times were used. For anthocyanins (Figure 3b), it can be seen that the extracts that were subjected to microwave irradiation for 2 min gave better results—that is, the maximum quantity of anthocyanins was obtained. In addition, a longer extraction time of 5 min gave rise to worse results. This may be due to the degradation of anthocyanins when the extracts are subjected to prolonged microwave irradiation at that temperature [31]. Therefore, for anthocyanins, a shorter time of 2 min was selected as the optimum extraction time, and for phenolic compounds, a longer time of 15 min was chosen. In addition to these optimal extraction times, it is necessary to take into account the extra time required for the extracts to be tempered.

2.4. Precision of MAE Methods

The precision of the developed methods was evaluated in terms of repeatability and intermediate precision. Both parameters concern the precision of the MAE of myrtle samples under the same extraction conditions, but repeatability implies extractions carried out on the same day, whereas intermediate precision is related to different days. These terms were evaluated by following the methodology employed in several previous studies [32,37,38]. A total of 30 experiments were developed over three consecutive days by performing ten experiments each day. For repeatability, 10 extractions were performed on the first day of the study. For intermediate precision, 10 more extractions were carried out on each of the next two consecutive days. For phenolic compounds, the repeatability (RSD) was 3.98% and the intermediate precision was 4.54%. For anthocyanins, the repeatability (RSD) was 3.41% and the intermediate precision was 4.10%. Both methods were considered to have good repeatability and intermediate precision, since a maximum error of 5% is generally considered in this type of work [39].

2.5. Application to Real Sample

2.5.1. Study of Myrtle Berries from Different Locations

Both MAE methods for total phenolic compounds and for total and individual anthocyanins were applied to the entire myrtle ecotypes collected for this study. A total of 14 different ecotypes of myrtle were evaluated. 8 ecotypes were collected from local evergreen shrubs from the Puerto Real region (My-1, My-2, My-3, My-4, My-5, My-6, My-7, and My-8), and 6 ecotypes were collected from the San José del Valle region (My-9, My-10, My-11, My-12, My-13, and My-14). From each ecotype, the pulp was separated from the seed and processed in duplicate, using the optimum conditions for the developed MAE method for phenolic compounds and the optimum conditions for the extraction of anthocyanins. For the quantification of total phenolic compounds, the extracts were analyzed by the Folin–Ciocalteau (FC) spectrophotometric method. For the quantification of total and individual anthocyanins, the extracts were analyzed by UHPLC. The results of the analyses (total phenolic compounds, the eleven individual anthocyanins, and total anthocyanins) from the 14 myrtle ecotypes are shown in Table 3. First, it is noteworthy to highlight the high content of anthocyanins and total phenolic compounds found in this fruit, equaling or surpassing the contents of well-known superfruits [37,40,41]. The results indicate that myrtle ecotypes collected in the region of Puerto Real have a higher concentration of total phenolic compounds and total anthocyanins than the ecotypes collected in the region of San José del Valle. This information suggests, a priori, that the amount of bioactive compounds in myrtle is related to the location of the ecotype.

Table 3. Results of total phenolic compounds (mg g^{-1}) and total and individual anthocyanins (mg g^{-1}) for each myrtle ecotype ($n = 3$). Del-3,5-diGl: delphinidin 3,5-O-diglucoside; Del-3-Glu: delphinidin 3-O-glucoside; Cy-3-Ga: cyanidin 3-O-galactoside; Cy-3-Gl: cyanidin 3-O-glucoside; Cy-3-Ar: cyanidin 3-O-arabinoside; Pet-3-Gl: petunidin 3-O-glucoside; Del-3-Ara: delphinidin 3-O-arabinoside; Peo-3-Gl: peonidin 3-O-glucoside; Mal-3-Gl: malvidin 3-O-glucoside; Pet-3-Ar: petunidin 3-O-arabinoside; Mal-3-Ar: malvidin 3-O-arabinoside.

Compounds	Myrtle Ecotypes of Puerto Real									Myrtle Ecotypes of San José del Valle				
	MY-1	MY-2	MY-3	MY-4	MY-5	MY-6	MY-7	MY-8	MY-9	MY-10	MY-11	MY-12	MY-13	MY-14
Del-3,5-diGl	0.434 ± 0.012	0.440 ± 0.015	0.367 ± 0.013	0.352 ± 0.035	0.514 ± 0.0132	0.397 ± 0.018	0.210 ± 0.001	0.498 ± 0.023	0.156 ± 0.004	0.073 ± 0.003	0.1456 ± 0.004	0.181 ± 0.005	0.157 ± 0.004	0.180 ± 0.007
Del-3-Glu	9.405 ± 0.256	13.232 ± 0.369	9.555 ± 0.051	9.704 ± 0.159	15.110 ± 0.160	10.935 ± 0.001	10.164 ± 0.171	9.664 ± 0.497	8.049 ± 0.283	1.798 ± 0.072	7.897 ± 0.301	6.102 ± 0.234	5.432 ± 0.234	6.192 ± 0.236
Cy-3-Ga	0.159 ± 0.0006	0.288 ± 0.014	0.211 ± 0.010	0.309 ± 0.010	0.372 ± 0.010	0.326 ± 0.003	0.184 ± 0.004	0.426 ± 0.023	0.504 ± 0.023	0.047 ± 0.001	0.490 ± 0.023	0.133 ± 0.003	0.341 ± 0.013	0.154 ± 0.005
Cy-3-Gl	1.702 ± 0.063	2.326 ± 0.075	1.013 ± 0.019	1.067 ± 0.022	1.854 ± 0.048	1.142 ± 0.015	2.276 ± 0.016	1.626 ± 0.079	1.011 ± 0.039	0.321 ± 0.013	1.002 ± 0.035	1.191 ± 0.043	0.988 ± 0.038	1.235 ± 0.031
Cy-3-Ar	0.090 ± 0.002	0.098 ± 0.002	0.075 ± 0.002	0.145 ± 0.009	0.136 ± 0.0014	0.611 ± 0.656	2.124 ± 0.015	0.134 ± 0.005	0.943 ± 0.037	0.299 ± 0.012	0.898 ± 0.032	0.084 ± 0.002	0.765 ± 0.029	0.085 ± 0.038
Pet-3-Gl	4.738 ± 0.065	8.680 ± 0.239	6.512 ± 0.027	7.480 ± 0.159	9.958 ± 0.426	4.614 ± 0.196	2.036 ± 0.023	6.058 ± 0.330	0.094 ± 0.010	0.036 ± 0.001	0.091 ± 0.003	2.346 ± 0.087	1.247 ± 0.051	2.988 ± 0.002
Del-3-Ara	1.979 ± 0.073	1.838 ± 0.072	1.527 ± 0.069	1.601 ± 0.032	2.320 ± 0.062	4.974 ± 0.234	4.022 ± 0.052	1.627 ± 0.079	6.256 ± 0.222	1.634 ± 0.072	5.990 ± 0.189	0.900 ± 0.023	3.257 ± 0.138	0.912 ± 0.119
Peo-3-Gl	0.614 ± 0.021	0.856 ± 0.027	0.573 ± 0.023	0.608 ± 0.022	0.705 ± 0.019	0.7206 ± 0.056	0.869 ± 0.052	0.406 ± 0.022	0.627 ± 0.011	0.330 ± 0.011	0.599 ± 0.019	0.488 ± 0.023	0.178 ± 0.07	0.375 ± 0.029
Mal-3-Gl	6.893 ± 0.245	13.124 ± 0.417	14.346 ± 0.024	15.495 ± 0.331	0.750 ± 0.020	0.767 ± 0.060	0.925 ± 0.0004	6.227 ± 0.339	0.667 ± 0.024	0.352 ± 0.013	0.601 ± 0.023	9.832 ± 0.342	5.563 ± 0.190	6.877 ± 0.013
Pet-3-Ar	0.295 ± 0.0003	0.430 ± 0.010	0.411 ± 0.0174	0.504 ± 0.020	0.590 ± 0.013	0.656 ± 0.021	0.3891 ± 0.0004	0.378 ± 0.018	0.765 ± 0.023	0.203 ± 0.001	0.679 ± 0.021	0.184 ± 0.007	0.634 ± 0.022	0.199 ± 0.268
Mal-3-Ar	0.2451 ± 0.002	0.231 ± 0.007	0.356 ± 0.002	0.372 ± 0.002	0.328 ± 0.006	0.5402 ± 0.001	0.289 ± 0.002	0.154 ± 0.010	0.520 ± 0.024	0.263 ± 0.01	0.492 ± 0.012	0.299 ± 0.013	0.5757 ± 0.020	0.213 ± 0.007
Total anthocyanins	26.555 ± 0.395	41.544 ± 1.390	34.947 ± 0.183	37.638 ± 0.730	32.637 ± 0.854	25.682 ± 0.846	23.487 ± 0.234	35.846 ± 0.896	19.595 ± 0.577	5.355 ± 0.242	18.884 ± 0.762	21.740 ± 0.865	19.138 ± 0.645	19.409 ± 0.567
Total phenolic compounds	88.598 ± 2.983	124.684 ± 0.934	82.603 ± 0.343	63.457 ± 2.541	86.251 ± 2.934	81.225 ± 2.199	88.340 ± 3.899	90.682 ± 4.706	55.934 ± 2.743	69.550 ± 3.123	56.790 ± 2.065	60.654 ± 2.127	61.193 ± 2.356	59.898 ± 2.967

2.5.2. Multivariate Statistical Analysis

In order to assess whether the concentration of bioactive compounds in myrtle is related to the location of the ecotype, a non-supervised chemometric technique, namely, hierarchical cluster analysis (HCA), was carried out for the 14 myrtle ecotypes. HCA allows the trends in the myrtle ecotypes to be grouped according to the origin by using the concentrations of the studied bioactive compounds as independent variables for the formation of groups. Therefore, the variables employed in the differentiation were: the amount of total phenolic compounds (mg g^{-1}), the amount of each individual anthocyanins (mg g^{-1}), and the amount of total anthocyanins (mg g^{-1}). Ward´s method was used for the preparation of the clusters, and square Euclidean distance was employed to measure distances between clusters. The results of the HCA are graphically represented in the dendrogram in Figure 4, in which all of the ecotypes of myrtle are listed, along with the distance at which any of the two clusters are joined.

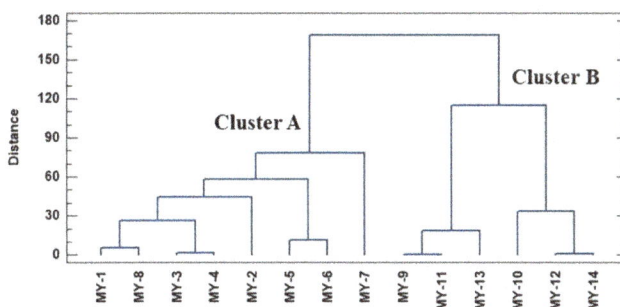

Figure 4. Dendrogram obtained by a hierarchical cluster analysis, based on the chemical parameters studied of the 14 samples by duplicated myrtle pulp extracts.

Based on these results, it can be observed that there are two main groups; Cluster A, which includes all of the myrtle ecotypes collected in Puerto Real, and Cluster B, which only includes the myrtle ecotypes collected in San José del Valle. Therefore, it can be concluded that the chemical information obtained—that is, the amounts of total phenolic compounds and anthocyanins, is related to the location of the ecotypes due to the tendency to be grouped according to their geographical origin. Specifically, samples from San José del Valle have a lower number of total phenolic compounds and total anthocyanins than samples collected in Puerto Real. This finding is consistent with bibliographic information, which highlights that myrtle is a shrub that prefers fertile and humid soils, and therefore, warm zones closer to sea level [42,43]. The wet climate of Puerto Real, due to its proximity to the sea, makes the myrtle shrubs of this region mature and grow more favorably than those located in inland areas, such as San José del Valle, where the climate is more variable and dry and the soil is less fertile.

3. Material and Methods

3.1. Myrtle Sample

Myrtle ecotypes were collected by the authors in December 2016 from local myrtle shrubs in their optimum ripeness stage from two areas (Puerto Real and San José del Valle) of the province of Cadiz, Andalusia, Spain. Specifically, 14 different myrtle ecotypes were collected: 8 ecotypes from the Puerto Real region, and 6 ecotypes from the San Jose del Valle region. Both regions are in the province of Cadiz but have different climatic characteristics, though they are fundamentally based in proximity to the sea. The area of Puerto Real is located on the coast, whereas the region of San José del Valle is located 50 km inland. Proximity to the coast results in very humid areas, with soils that have readily available water, and these provide more fertile conditions for the growth and maturation of many species. By contrast, San José del Valle, as an inland zone, experiences greater temperature changes and less water availability for plants, particularly in the summer. All this leads to the generation of myrtle ecotypes with different characteristics. In addition, a morphological characterization was made of both leaves and berries, applying the guidelines described by M., Mulas & M.R. Cani [43], to confirm that the samples collected came from different ecotypes.

The seeds of the myrtle berries were separated from the pulp. The pulp was lyophilized in a Virtis Benchtop K freeze-drier (SP Scientific, New York, NY, USA) and triturated in a spice grinder. The triturated and homogeneous sample was stored in a freezer at $-20\,^\circ$C prior to analysis.

3.2. Chemicals and Solvents

Methanol (HPLC grade) was purchased from Fischer Chemical (Loughborough, United Kingdom). Water was obtained from a Milli-Q water purification system from Millipore (Bedford, MA, USA). Hydrochloric acid and sodium hydroxide (both analytical grade) employed for the adjustment of pH were obtained from Panreac (Barcelona, Spain). The reagents necessary for the determination of total phenolic compounds were anhydrous sodium carbonate (Panreac, Barcelona, Spain), and Folin–Ciocalteu reagent (Merck Millipore, Darmstadt, Germany). The phenolic standard (gallic acid) and the anthocyanin standard (cyanidin chloride) were purchased from Sigma-Aldrich Chemical Co. (St. Louis, MO, USA).

3.3. Microwave-Assisted Extraction Procedure

The extraction of total phenolic compounds and total anthocyanins (as the sum of the individual components) from myrtle was performed by microwave-assisted extraction. The extracts were obtained using a MARS 6 One TouchTM Technology system (1800 W) (CEM Corporation, Matthews, NC, USA). Approximately 0.5 g of triturated myrtle was weighed into a MARSXpress vessel (CEM Corporation), and the appropriate volume of solvent was added. The vessel was closed securely and placed in the microwave with another eight tubes, which had the same solvent and volume. Each extraction was carried out under controlled MAE conditions. The parameters used were: solvent composition (50, 75, and 100% methanol in water), pH (2, 4.5, and 7), temperature (50, 75, and 100 $^\circ$C) and sample–solvent ratio (0.5 g/10 mL, 0.5 g/15 mL, and 0.5 g/20 mL). The initial extraction time was 5 min, and this was followed by a set time to temper the sample. Once the samples had been warmed, the extracts were centrifuged (7500 rpm, 5 min) and the supernatant was added to a 25 mL volumetric flask. The precipitates from the extraction were subsequently redissolved in 5 mL of the same extraction solvent. The extracts were centrifuged again (7500 rpm, 5 min) and the supernatant was placed in the volumetric flask (25 mL). The volume was completed with the same solvent. The extracts were stored at $-20\,^\circ$C prior to analysis.

3.4. Identification of Anthocyanins

A chromatographic method using ultra-high performance liquid chromatography (UHPLC) coupled to quadrupole-time-of-flight mass spectrometry (Q-ToF-MS) (Xevo G2 QToF, Waters Corp., Milford, MA, USA) was developed for the identification of anthocyanins in MAE extracts. The injection volume was set to 3 µL. The chromatographic separation was performed on a reverse-phase C18 analytical column (1.7 µm, 2.1 mm $\times$ 100 mm, Acquity UPLC BEH C18, Waters). A gradient method, using acidified water (2% formic acid, solvent A) and methanol (solvent B) at a flow rate of 0.4 mL min^{-1} was used. The gradient was as follows (time, % solvent B): 0.00 min, 15%; 3.30 min, 20%; 3.86 min, 30%; 5.05 min, 40%; 5.35 min, 55%; 5.64 min, 60%; 5.95 min, 95%; 7.50 min, 95%. The total run time was 12 min, including 4 min for re-equilibration. The analyses were carried out using an electrospray source operating in positive ionization mode under the following conditions: desolvation gas flow = 700 L h^{-1}, desolvation temperature = 500 $^\circ$C, cone gas flow =10 L h^{-1}, source temperature = 150 $^\circ$C, capillary voltage = 700 V, cone voltage = 30 V and collision energy = 20 eV. The full-scan mode was used (m/z 100–1200). The following eleven anthocyanins were identified in the samples: delphinidin 3,5-O-diglucoside (m/z 627.1561), delphinidin 3-O-glucoside (m/z 465.1033), cyanidin 3-O-galactoside

(m/z 449.1084), cyanidin 3-*O*-glucoside (m/z 449.1084), cyanidin 3-*O*-arabinoside (m/z 419.0978), petunidin 3-*O*-glucoside (m/z 479.1189), delphinidin 3-*O*-arabinoside (m/z 435.0927), peonidin 3-*O*-glucoside (m/z 463.1240), malvidin 3-*O*-glucoside (m/z 493.1346), petunidin 3-*O*-arabinoside (m/z 449.1084), and malvidin 3-*O*-arabinoside (m/z 463.1240). Prior to chromatographic analysis, the extracts were filtered through a 0.20 µm nylon syringe filter (Membrane Solutions, Dallas, TX, USA). Information regarding the mass spectra, theoretical and measured masses, as well as the structure of the compounds, are included as Supplementary Material (Table S1 and Figure S1).

3.5. Determination of Anthocyanins

Once the anthocyanins had been identified, they were separated and quantified by ultra-high performance liquid chromatography (UHPLC) (Elite LaChrom Ultra System, VWR Hitachi, Tokyo, Japan) available in our research group. The UHPLC chromatogram representing the eleven anthocyanins is shown in Figure 5.

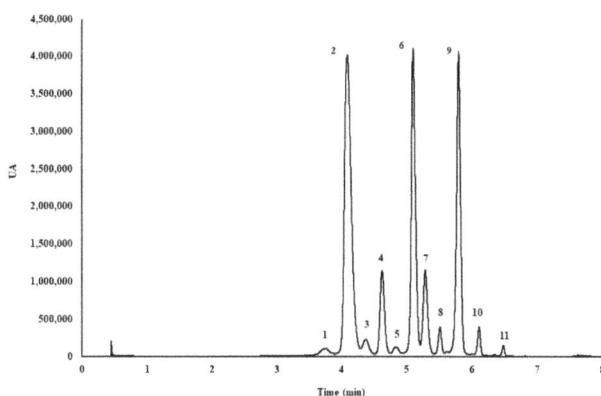

Figure 5. Ultra-high performance liquid chromatography (UHPLC) chromatogram of the eleven anthocyanins identified in the microwave-assisted extraction (MAE) extracts from myrtle berries. Peak assignment: (**1**) delphinidin 3,5-*O*-diglucoside; (**2**) delphinidin 3-*O*-glucoside; (**3**) cyanidin 3-*O*-galactoside; (**4**) cyanidin 3-*O*-glucoside; (**5**) cyanidin 3-*O*-arabinoside; (**6**) petunidin 3-*O*-glucoside; (**7**) delphinidin 3-*O*-arabinoside; (**8**) peonidin 3-*O*-glucoside; (**9**) malvidin 3-*O*-glucoside; (**10**) petunidin 3-*O*-arabinoside; (**11**) malvidin 3-*O*-arabinoside.

The UHPLC system was equipped with an autosampler (L-2200U), two pumps (L-2160U), a UV-vis detector (L-2420U) set at 520 nm for the analysis, and a column oven (L2300), set at 50 °C for the chromatographic analysis. Anthocyanins were analyzed on a "Fused Core" C18 column (2.6 µm, 2.1 × 100 nm, Phenomenex, Torrance, CA, USA). The separation and quantification were carried out using acidified water (5% formic acid, solvent A) and methanol (solvent B), working at a flow rate of 0.7 mL min^{-1}. After testing several methods and flows, amongst other parameters, the gradient employed was as follows (time, % solvent B): 0.00 min, 0%; 1.50 min, 5%; 3.30 min, 15%; 4.80 min, 25%; 5.40 min, 40%. The injection volume was set to 15 µL. This gradient provided optimum results in less than 7 min. Prior to chromatographic analysis, the extracts were filtered through a 0.20 µm nylon syringe filter (Membrane Solutions, Dallas, TX, USA). In order to quantify the eleven anthocyanins present in myrtle extracts, a calibration curve of cyanidin chloride ($y = 300568.8819x - 28462.4337$) was used as the anthocyanidin standard. The regression equation and the correlation coefficient ($R^2 = 0.9999$) were also calculated using Microsoft Office Excel 2013. The limit of detection (0.196 mg L^{-1}) was calculated as three times the standard deviation of the blank signal divided by the slope of the calibration curve. By analogy, the limit of quantification (0.653 mg L^{-1}) was calculated as ten times the standard deviation of the blank signal divided by the slope of the calibration curve.

The linear range studied was 0.06–35 mg L^{-1}. Each one of the eleven anthocyanins was quantified using the calibration curve for cyanidin chloride, and the molecular weight of the anthocyanins analyzed were taken into account. All analyses were carried out in duplicate, and the results were expressed as milligrams of anthocyanins per g of dry weight.

3.6. Determination of Total Phenolic Compounds

The Folin–Ciocalteu (FC) spectrophotometric method was used to determine the total phenolic compounds [44,45]. The FC method is based on the fact that phenolic compounds react at basic pH with the Folin–Ciocalteau reagent (a mixture of sodium tungstate and sodium molybdate). The products of the reduction have a blue color, and they have a broad absorption with a maximum of 765 nm. The FC assay was performed by transferring 0.25 mL of MAE extract, 1.25 mL of water, and 1.25 mL of the Folin–Ciocalteu reagent to a volumetric flask (25 mL). Then, 5 mL of a 20% aqueous sodium carbonate solution was also added, and the solution was made up to the mark with water. After 30 min, the absorbance of the solutions was measured at 765 nm. The range of absorbance obtained for the studied samples was 0.4–1.3. Prior to spectrophotometric analysis, the extracts were filtered through a 0.45 μm nylon filter (Membrane Solutions, Dallas, TX, USA). The absorbance was measured on a HeΛios-γ Unicam UV-Vis Spectrophotometer (Thermo Scientific, Waltham, MA, USA). In order to quantify the phenolic compounds present in myrtle extracts, a calibration curve was developed under the same conditions, using gallic acid as the reference standard. The results are expressed as milligrams of gallic acid equivalent per g of fresh weight. The regression equation ($y = 0.0010x + 0.0065$) and correlation coefficient ($R^2 = 0.9999$) were calculated using Microsoft Office Excel 2013. The linear range studied was 100–2600 mg L^{-1}. All analyses were carried out in duplicate.

3.7. Optimization Procedure and Data Analyses

The spherical response surface Box–Behnken method was employed for experimental design in the optimization procedure. In this approach, the treatment combinations are at the midpoints of the edges of the process space and at the centre, which ensures that all experiments are in the region of interest [46]. To obtain the significant factors and the optimal MAE conditions, a Box–Behnken design with four factors and three levels for each factor was used: a low level (-1), a medium level (0), and a high level (1).

The four factors (independent variables) used in this work were solvent composition (methanol in water) (X_1), solvent pH (X_2), extraction temperature (X_3), and sample–solvent ratio (X_4), while the response variables (dependent variables) used were the total amount of phenolic compounds (Y_{TP}, mg g^{-1}) on the one hand, and the total amount of anthocyanins (Y_{TA}, mg g^{-1}), as the sum of individual ones, on the other. The experimental design consisted of 27 trials performed in duplicate, with three repetitions at the center point to calculate the pure error sum of squares. The whole experimental design matrix and the results obtained are shown in Table 4. My-9 from the San Jose del Valle location was the myrtle sample used for the optimization procedure.

A quadratic model was used for model construction, and this gave a second-order polynomial equation that correlated the relationship between independent variables and responses (Equation (3)):

$$Y = \beta_0 + \sum_{i=1}^{k} \beta_i X_i + \beta_{ii} X_i^2 + \sum_{i}\sum_{i=1}^{k} \beta_{ij} X_i X_j + r \tag{3}$$

where Y represents the aforementioned responses (Y_{TP} and Y_{TA}) for the extractions of total phenolic compounds and anthocyanins, respectively; β_0 is the model constant; X_i are the independent variables; β_i are the linear coefficients; β_{ij} are the interactive coefficients; β_{ii} are the quadratic coefficients; and r is the pure error sum of squares.

The statistical significances of the model, lack of fit, and regression terms were evaluated from the analysis of variance (ANOVA). The fitting quality of the polynomial model was evaluated by the determination coefficient (R^2). All experimental data were compiled by Design Expert software 11 (Trial Version, Stat- Ease Inc., Minneapolis, MN, USA). This software was employed for experimental design, data analysis, and model building.

Hierarchical clustering analysis (HCA) was carried out using the Statgraphic Centurion XVII software (Statgraphics Technologies, Inc., The Plains, VA, USA). Pareto Charts were used to identify factors and combinations of factors that are statistically significant at the selected confidence level (95%) for total phenolic compounds and anthocyanins. Regarding the HCA analysis, the Ward method and the Euclidean square distance were used.

Table 4. Box–Behnken design matrix of four variables, and measured and predicted responses.

Run	Factors				Responses			
					Y_{TP} (mg g^{-1})		Y_{TA} (mg g^{-1})	
	Solvent	pH	Temp.	Ratio	Measured	Predicted	Measured	Predicted
1	−1	−1	0	0	50.66	51.02	26.04	21.86
2	1	−1	0	0	34.64	40.29	24.74	26.30
3	−1	1	0	0	43.86	41.83	26.73	25.39
4	1	1	0	0	31.74	35.00	17.54	21.95
5	0	0	−1	−1	44.30	41.54	20.05	20.83
6	0	0	1	−1	42.60	46.26	20.50	18.83
7	0	0	−1	1	46.71	46.67	20.21	22.11
8	0	0	1	1	39.23	45.61	14.41	13.86
9	0	0	0	0	38.17	42.68	16.56	18.65
10	−1	0	0	−1	38.51	42.19	25.34	26.79
11	1	0	0	−1	31.50	31.55	26.42	26.49
12	−1	0	0	1	40.14	42.58	25.04	24.15
13	1	0	0	1	36.83	35.64	27.71	25.45
14	0	−1	−1	0	44.07	48.20	22.79	23.04
15	0	1	−1	0	46.49	48.10	18.40	16.20
16	0	−1	1	0	56.28	57.17	10.11	11.50
17	0	1	1	0	44.42	42.79	18.57	17.51
18	0	0	0	0	44.14	42.68	19.45	18.65
19	0	−1	0	−1	53.28	48.50	20.59	20.48
20	0	1	0	−1	46.14	46.27	19.82	19.32
21	0	−1	0	1	62.00	55.75	16.79	17.89
22	0	1	0	1	44.84	43.50	17.52	18.22
23	−1	0	−1	0	39.22	38.57	26.60	28.42
24	1	0	−1	0	39.00	36.70	26.86	24.31
25	−1	0	1	0	51.14	47.32	15.60	18.69
26	1	0	1	0	37.10	31.62	25.03	23.80
27	0	0	0	0	45.74	42.68	19.94	18.65

4. Conclusions

To the best of our knowledge, this is the first study in which MAE has been optimized for the extraction of bioactive compounds, like phenolic compounds and anthocyanins, from myrtle berries. For total phenolic compounds, the solvent methanol-water (58.20% MeOH in water) and pH (pH 2) were the most influential variables. For the recovery of anthocyanins, methanol composition (50.4% MeOH in water) and temperature (50 °C) were found to be the most efficient variables. The optimum extraction times were 2 and 15 min for phenolic compounds and anthocyanins, respectively. Both of the developed methods showed high repeatability and intermediate precision (RSD < 5%). The methods were successfully applied to different myrtle ecotypes collected in two different geographical areas. HCA analysis showed a correlation between the bioactive compounds studied and the location of the ecotype. Due to these previous findings, myrtle berries from more geographical locations could be interesting for further studies.

Based on these results, MAE (under optimum conditions) can be considered as a suitable technique for the extraction of bioactive compounds in myrtle berries. In addition, this technique presents several advantages in comparison to other extraction techniques, since it is fast, economic, and eco-friendlier, as it does not require the use of a high amount of solvent.

Supplementary Materials: The following are available online, Figure S1: MS spectra and structure of the eleven anthocyanins identified in myrtle berries: (**a**) Delphinidin 3,5-*O*-diglucoside; (**b**) Delphinidin 3-*O*-glucoside; (**c**) Cyanidin 3-*O*-galactoside; (**d**) Cyanidin 3-*O*-glucoside; (**e**) Cyanidin 3-*O*-arabinoside; (**f**) Petunidin 3-*O*-glucoside; (**g**) Delphinidin 3-*O*-arabinoside; (**h**) Peonidin 3-*O*-glucoside; (**i**) Malvidin 3-*O*-glucoside; (**j**) Petunidin 3-*O*-arabinoside; (**k**) Malvidin 3-*O*-arabinoside, Table S1: Mass spectra information of the eleven anthocyanins present in myrtle berries.

Author Contributions: Conceptualization, E.E.-B. and G.F.B.; methodology, A.V.G.d.P. and M.V.-E.; software, M.F.-G.; formal analysis, A.V.G.d.P., M.V.-E and A.A.-A.; investigation, A.V.G.d.P. and M.V.-E.; resources, A.J.-C., C.G.B. and M.P.; data curation, E.E.-B., M.F.-G. and G.F.B.; writing—original draft preparation, A.V.G.d.P. and M.V.-E.; writing—review and editing, G.F.B. and E.E.-B.; supervision, E.E.-B. and G.F.B.; project administration, G.F.B. and E.E.-B.

Funding: This research was funded by University of Cadiz and V. la Andaluza (Project OT2017/032).

Acknowledgments: The authors express their acknowledgments to V. la Andaluza.

Conflicts of Interest: The authors declare no conflict of interest.

References

1. Bouzabata, A.; Casanova, J.; Bighelli, A.; Cavaleiro, C.; Salgueiro, L.; Tomi, F. The genus *Myrtus* L. in Algeria: Composition and biological aspects of essential oils from *M. communis* and *M. nivellei*: A review. *Chem. Biodivers.* **2016**, *13*, 672–680. [CrossRef] [PubMed]

2. Melito, S.; La Bella, S.; Martinelli, F.; Cammalleri, I.; Tuttolomondo, T.; Leto, C.; Fadda, A.; Molinu, M.G.; Mulas, M. Morphological, chemical, and genetic diversity of wild myrtle (*Myrtus communis* L.) populations in Sicily. *Turk. J. Agric. For.* **2016**, *40*, 249–261. [CrossRef]

3. Al-Yafeai, A.; Bellstedt, P.; Böhm, V. Bioactive compounds and antioxidant capacity of rosa rugosa depending on degree of ripeness. *Antioxidants* **2018**, *7*, 134. [CrossRef] [PubMed]

4. Yang, C.; Fu, L.; Du, B.; Chen, B.; Wang, F.; Wang, M. Isolation and characterization of new phenolic compounds with estrogen biosynthesis-inhibiting and antioxidation activities from *Broussonetia papyriferia* leaves. *PLoS ONE* **2014**, *9*. [CrossRef]

5. Wannes, A.; Mhamdi, B.; Sriti, J.; Ben Jemia, M.; Ouchikh, O.; Hamdaoui, G.; Kchouk, M.E.; Marzouk, B. Antioxidant activities of the essential oils and methanol extracts from myrtle (*Myrtus communis* var. italica L.) leaf, stem and flower. *Food Chem. Toxicol.* **2010**, *48*, 1362–1370. [CrossRef] [PubMed]

6. Aleksic, V.; Knezevic, P. Antimicrobial and antioxidative activity of extracts and essential oils of *Myrtus communis* L. *Microbiol. Res.* **2014**, *169*, 240–254. [CrossRef] [PubMed]

7. Wahid, N.; Chkichekh, A.; Bakry, M. Morphological traits and essential oil yield variation of three *Myrtus communis* L. populations: Implication for domestication. *J. Agric. Food Chem.* **2016**, *4*, 199–207.

8. Bouzabata, A.; Castola, V.; Bighelli, A.; Abed, L.; Casanova, J.; Tomi, F. Chemical variability of algerian *Myrtus communis* L. *Chem. Biodivers.* **2013**, *10*, 129–137. [CrossRef] [PubMed]

9. Fadda, A.; Mulas, M. Chemical changes during myrtle (*Myrtus communis* L.) fruit development and ripening. *Sci. Hortic.* **2010**, *125*, 477–485. [CrossRef]

10. Serreli, G.; Jerković, I.; Gil, K.A.; Marijanović, Z.; Pacini, V.; Tuberoso, C.I.G. Phenolic compounds, volatiles and antioxidant capacity of white myrtle berry liqueurs. *Plant Foods Hum. Nutr.* **2017**, *72*, 205–210. [CrossRef] [PubMed]

11. Messaoud, C.; Boussaid, M. *Myrtus communis* berry color morphs: A comparative analysis of essential oils, fatty acids, phenolic compounds, and antioxidant activities. *Chem. Biodivers.* **2011**, *8*, 300–310. [CrossRef] [PubMed]

12. De Oliveira, C.B.; Comunello, L.N.; Lunardelli, A.; Amaral, R.H.; Pires, M.G.S.; Da Silva, G.L.; Manfredini, V.; Vargas, C.R.; Gnoatto, S.C.B.; de Oliveira, J.R.; et al. Phenolic enriched extract of baccharis trimera presents anti-inflammatory and antioxidant activities. *Molecules* **2017**, *17*, 1113. [CrossRef] [PubMed]

13. Sarais, G.; D'Urso, G.; Lai, C.; Pirisi, FM.; Pizza, C.; Montoro, P. Targeted and untargeted mass spectrometric approaches in discrimination between *Myrtus communis* cultivars from Sardinia region. *J. Mass. Spectrom.* **2016**, *51*, 704–715. [CrossRef] [PubMed]

14. Tuberoso, C.I.G.; Rosa, A.; Bifulco, E.; Melis, M.P.; Atzeri, A.; Pirisi, F.M.; Dessì, M.A. Chemical composition and antioxidant activities of *Myrtus communis* L. berries extracts. *Food Chem.* **2010**, *123*, 1242–1251. [CrossRef]

15. Montoro, P.; Tuberoso, C.I.G.; Perrone, A.; Piacente, S.; Cabras, P.; Pizza, C. Characterisation by liquid chromatography-electrospray tandem mass spectrometry of anthocyanins in extracts of *Myrtus communis* L. berries used for the preparation of myrtle liqueur. *J. Chromatogr. A* **2006**, *1112*, 232–240. [CrossRef] [PubMed]

16. Tuberoso, C.I.G.; Melis, M.; Angioni, A.; Pala, M.; Cabras, P. Myrtle hydroalcoholic extracts obtained from different selections of *Myrtus communis* L. *Food Chem.* **2007**, *101*, 806–811. [CrossRef]

17. Spigno, G.; De Faveri, D.M. Microwave-assisted extraction of tea phenols: A phenomenological study. *J. Food Eng.* **2009**, *93*, 210–217. [CrossRef]

18. Zhao, C.-N.; Zhang, J.; Li, Y.; Meng, X.; Li, H. Microwave-assisted extraction of phenolic compounds from *Melastoma sanguineum* fruit: Optimization and identification. *Molecules* **2018**, *23*, 2498. [CrossRef] [PubMed]

19. Bouras, M.; Chadni, M.; Barba, F.J.; Grimi, N.; Bals, O.; Vorobiev, E. Optimization of microwave-assisted extraction of polyphenols from Quercus bark. *Ind. Crops. Prod.* **2015**, *77*, 590–601. [CrossRef]

20. Haddadi-Guemghar, H.; Janel, N.; Dairou, J.; Remini, H.; Madani, K. Optimisation of microwave-assisted extraction of prune (*Prunus domestica*) antioxidants by response surface methodology. *Int. J. Food Sci. Technol.* **2014**, *49*, 2158–2166. [CrossRef]

21. Liazid, A.; Guerrero, R.F.; Cantos, E.; Palma, M.; Barroso, C.G. Microwave assisted extraction of anthocyanins from grape skins. *Food Chem.* **2011**, *124*, 1238–1243. [CrossRef]

22. Li, H.; Deng, Z.; Wu, T.; Liu, R.; Loewen, S.; Tsao, R. Microwave-assisted extraction of phenolics with maximal antioxidant activities in tomatoes. *Food Chem.* **2012**, *130*, 928–936. [CrossRef]

23. Wen, Y.; Chen, H.; Zhou, X.; Deng, Q.; Zhao, Y.; Zhao, C.; Gong, X. Optimization of the microwave-assisted extraction and antioxidant activities of anthocyanins from blackberry using a response surface methodology. *RSC Adv.* **2015**, *5*, 19686–19695. [CrossRef]

24. Dahmoune, F.; Nayak, B.; Moussi, K.; Remini, H.; Madani, K. Optimization of microwave-assisted extraction of polyphenols from *Myrtus communis* L. leaves. *Food Chem.* **2015**, *166*, 585–595. [CrossRef] [PubMed]

25. Routray, W.; Orsat, V. Microwave-assisted extraction of flavonoids: A review. *Food Bioprocess Technol.* **2012**, *5*, 409–424. [CrossRef]

26. Dong, C.H.; Xie, X.Q.; Wang, X.L.; Zhan, Y.; Yao, Y.J. Application of Box–Behnken design in optimisation for polysaccharides extraction from cultured mycelium of Cordyceps sinensis. *Food Bioprod. Process* **2009**, *87*, 139–144. [CrossRef]

27. Hou, W.; Zhang, W.; Chen, G.; Luo, Y. Optimization of extraction conditions for maximal phenolic, flavonoid and antioxidant activity from melaleuca bracteata leaves using the response surface methodology. *PLoS ONE* **2016**, *11*. [CrossRef] [PubMed]

28. Razali, M.A.A.; Sanusi, N.; Ismail, H.; Othman, N.; Ariffin, A. Application of response surface methodology (RSM) for optimization of cassava starch grafted polyDADMAC synthesis for cationic properties. *Starch/Staerke* **2012**, *64*, 935–943. [CrossRef]

29. Castañeda-ovando, A.; Pacheco-hernández, M.D.L.; Páez-hernández, M.E.; Rodríguez, J.A.; Galán-vidal, C.A. Chemical studies of anthocyanins: A review. *Food Chem.* **2009**, *113*, 859–871. [CrossRef]

30. Spigno, G.; Tramelli, L.; De Faveri, D.M. Effects of extraction time, temperature and solvent on concentration and antioxidant activity of grape marc phenolics. *J. Food Eng.* **2007**, *81*, 200–208. [CrossRef]

31. Pereira, D.T.V.; Tarone, A.G.; Cazarin, C.B.B.; Barbero, G.F.; Martínez, J. Pressurized liquid extraction of bioactive compounds from grape marc. *J. Food Eng.* **2019**, *240*, 105–113. [CrossRef]

32. Espada-Bellido, E.; Ferreiro-González, M.; Barbero, G.F.; Carrera, C.; Palma, M.; Barroso, C.G. Alternative extraction method of bioactive compounds from Mulberry (*Morus nigra* L.) pulp using pressurized-liquid extraction. *Food Anal. Methods* **2018**, *11*, 2384–2395. [CrossRef]

33. D'Alessandro, L.G.; Kriaa, K.; Nikov, I.; Dimitrov, K. Ultrasound assited extraction of polyphenols from black chokeberry. *Sep. Purif. Technol.* **2012**, *93*, 42–47. [CrossRef]

34. Pereira, P.; Cebola, M.-J.; Oliveira, M.C.; Bernardo Gil, M.G. Antioxidant capacity and identification of bioactive compounds of *Myrtus communis* L. extract obtained by ultrasound-assisted extraction. *J. Food Sci. Technol.* **2017**, *54*, 4362–4369. [CrossRef] [PubMed]

35. Aidi Wannes, W.; Marzouk, B. Differences between myrtle fruit parts (*Myrtus communis* var. italica) in phenolics and antioxidant contents. *J. Food Biochem.* **2013**, *37*, 585–594. [CrossRef]

36. Scorrano, S.; Lazzoi, M.R.; Mergola, L.; Di Bello, M.P.; Del Sole, R.; Vasapollo, G. Anthocyanins profile by Q-TOF LC/MS in *Myrtus communis* berries from salento area. *Food Anal. Methods* **2017**, *10*, 2404–2411. [CrossRef]

37. Espada-Bellido, E.; Ferreiro-González, M.; Carrera, C.; Palma, M.; Barroso, C.G.; Barbero, G.F. Optimization of the ultrasound-assisted extraction of anthocyanins and total phenolic compounds in mulberry (*Morus nigra*) pulp. *Food Chem.* **2017**, *219*, 23–32. [CrossRef] [PubMed]

38. Stipcovich, T.; Barbero, G.F.; Ferreiro-González, M.; Palma, M.; Barroso, C.G. Fast analysis of capsaicinoids in Naga Jolokia extracts (*Capsicum chinense*) by high-performance liquid chromatography using fused core columns. *Food Chem.* **2018**, *239*, 217–224. [CrossRef] [PubMed]

39. AOAC Peer Verified Methods Advisory Committee. *AOAC Peer Verified Methods Program*; AOAC International: Gaithersburg, MD, USA, 1998; pp. 1–35.

40. Vázquez-Espinosa, M.; Espada-Bellido, E.; González de Peredo, A.V.; Ferreiro-González, M.; Carrera, C.; Palma, M.; Barroso, C.G.; Barbero, G.F. Optimization of microwave-assisted extraction for the recovery of bioactive compounds from the Chilean superfruit (*Aristotelia chilensis* (mol.) stuntz). *Agronomy* **2018**, *8*, 240. [CrossRef]

41. Garzón, G.A.; Narváez-Cuenca, C.E.; Vincken, J.P.; Gruppen, H. Polyphenolic composition and antioxidant activity of açai (*Euterpe oleracea* Mart.) from Colombia. *Food Chem.* **2017**, *217*, 364–372. [CrossRef] [PubMed]

42. González-Varo, J.P.; Albaladejo, R.G.; Aparicio, A. Mating patterns and spatial distribution of conspecific neighbours in the Mediterranean shrub *Myrtus communis* (Myrtaceae). *Plant Ecol.* **2009**, *203*, 207–215. [CrossRef]

43. Mulas, M.; Cani, M.R. Germplasm Evaluation of spontaneous Myrtle (*Myrtus communis* L.) for cultivar selection and crop development. *J. Herbs Spices Med. Plants* **1999**, *6*, 31–49. [CrossRef]

44. Singleton, V.L.; Orthofer, R.; Lamuela-Raventós, R.M. Analysis of total phenols and other oxidation substrates and antioxidants by means of folin-ciocalteu reagent. *Methods Enzymol.* **1999**, *299*, 152–178. [CrossRef]

45. Santos, U.P.; Campos, J.F.; Heron, F.V.; Paredes-Gamero, E.J.; Carollo, C.A.; Estevinho, L.M.; de Picol Souza, K.; Dos Santos, E.L. Antioxidant, antimicrobial and cytotoxic properties as well as the phenolic content of the extract from *Hancornia speciosa* Gomes. *PLoS ONE* **2016**, *11*. [CrossRef] [PubMed]

46. Ferreira, S.L.C.; Bruns, R.E.; Ferreira, H.S.; Matos, G.D.; David, J.M.; Brandão, G.C.; da Silva, E.G.P.; Portugal, L.A.; dos Reis, P.S.; Souza, A.S.; dos Santos, W.N.L. Box–Behnken design: An alternative for the optimization of analytical methods. *Anal. Chim. Acta* **2007**, *597*, 179–186. [CrossRef] [PubMed]

![molecules logo] *molecules*

MDPI

Article

Essential Oils from *Humulus Lupulus* scCO$_2$ Extract by Hydrodistillation and Microwave-Assisted Hydrodistillation

Katarzyna Tyśkiewicz [1,*] , **Roman Gieysztor** [1], **Marcin Konkol** [1], **Jan Szałas** [2] **and Edward Rój** [1]

[1] Supercritical Extraction Department, New Chemical Syntheses Institute, Al. Tysiąclecia Państwa Polskiego 13A, 24-110 Puławy, Poland; roman.gieysztor@gmail.com (R.G.); marcin.konkol@ins.pulawy.pl (M.K.); edward.roj@ins.pulawy.pl (E.R.)

[2] Import Export J.A. Szałas Company, Garbarska 125A, 26-600 Radom, Poland; jan.szalas@chmiel.com.pl

* Correspondence: katarzyna.tyskiewicz@ins.pulawy.pl; Tel.: (81)-473-1452

Received: 27 September 2018; Accepted: 31 October 2018; Published: 3 November 2018

check for updates

Abstract: Two different extraction methods were used for a comparative study of essential oils obtained from the *Humulus lupulus* scCO$_2$ (sc-supercritical) extract: microwave-assisted hydrodistillation (MAHD) and conventional hydrodistillation (HD). As a result, the best conditions for the maximum essential oil production were determined for the MAHD method at 335 W microwave power for 30 min at water to raw material ratio of 8:3. The obtained essential oils were enriched in β-myrcene in the amount of 74.13%–89.32% (wt%). Moreover, the first application for determination of the above-mentioned volatile compounds by supercritical fluid chromatography (SFC) with photo-diode array detection (PDA) is presented, which in comparison with gas chromatography coupled with mass spectrometry (GC-MS/MS) resulted in similar values for β-myrcene and α-humulene in obtained samples within less than 1 min.

Keywords: essential oil; extraction techniques; hops extracts; hydrodistillation; Marynka strain; microwave-assisted hydrodistillation

---

1. Introduction

Hops (*Humulus lupulus*) belong to a perennial plant, which is mainly used in the brewing industry due to its valuable properties [1,2]. These properties are attributed to alpha- and beta-acids (responsible for bitterness), essential oils, and polyphenols, which constitute 24, 1, and 5 wt%, respectively [3,4]. The above-mentioned groups of compounds are considered as secondary metabolites. The phenolic compounds present in hops have been investigated in terms of their anticarcinogenic, antioxidant, and anti-inflammatory properties [5,6]. Secondary metabolites' composition affects the assessment of the diversity of a given species. However, according to Biendl et al. [3] the exact functions of these metabolites are yet unknown. It is believed that they protect the plant against pests and pathogens. Due to their favorable physiological and pharmacological properties, secondary metabolites are used in the pharmaceutical industry as medicines and dietary supplements. The antibacterial properties were also evaluated for hops extract obtained by the supercritical fluid extraction [7].

The main methods used to obtain essential oils are, among other, hydrodistillation, steam distillation, maceration, absorption, and supercritical extraction. These methods are referred to as conventional methods [8]. In order to shorten the extraction time and improve the extraction efficiency, another form of hydrodistillation has gained a lot of interest. Microwave-assisted hydrodistillation (MAHD), which is an advanced distillation method using microwave heating, has an advantage over conventional hydrodistillation as was indicated by a number of researchers [8–11]. It is a technique widely used to extract essential oils from medicinal plants and herbs due to its economic and environmental advantages ('green' technology) [12]. It is used in the extraction of various chemical compounds from the group of aromatic compounds, phenolic compounds, pesticides and, especially, essential oils. It is proven that the 30-minute duration of the microwave hydrodistillation process corresponds to a duration of conventional hydrodistillation that is eight times longer [9]. Hydrodistillation and microwave-assisted hydrodistillation have been compared in terms of extraction time, extraction efficiency, chemical composition, and quality of the essential oil as well as energy consumption. Microwave action is based mainly on the weakening of cell membranes and facilitating the extraction of biologically active compounds [13].

Literature data confirm the superiority of the microwave assisted hydrodistillation over conventional hydrodistillation. In previous studies, microwave-assisted hydrodistillation was carried out on such materials as ginger [12], lemongrass [12], rosemary [14,15], cinnamon [8,16], mint [10], thyme [9], sandalwood [11], and orange [17]. However, there is no information available on the use of microwave hydrodistillation in order to obtain essential oils from hop extracts, and also to compare this method with hydrodistillation with the example of the mentioned materials. The mixture of hop extract and water required the addition of residues as a scattering element obtained after the scCO$_2$ extraction of *Humulus lupulus*.

β-myrcene and α-humulene, which belong to a group of terpenes, are characterized by their high oxidative sensitivity. During the brewing processes, thermal oxidation products of β-myrcene and α-humulene may affect both the quality of beer and its organoleptic properties [18]. Therefore, their use as beer additives is limited. However, if not in brewing, the oil rich in β-myrcene and α-humulene might be used in medicine [19]. For instance, β-myrcene is responsible for the anti-inflammatory effects in human osteoarthritis treatment [20]. The activity of α-humulene is described as similar to that of dexamethasone (anti-inflammatory, antiallergic, immunosuppressive) [21]. Figure 1 presents the chemical structures of β-myrcene and α-humulene.

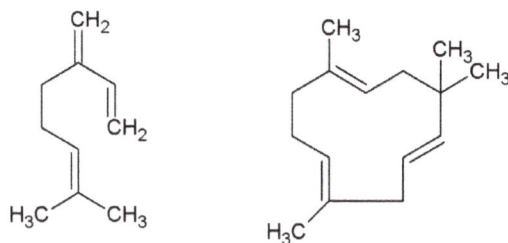

Figure 1. The chemical structures of β-myrcene and α-humulene.

2. Materials and Methods

2.1. Plant Material

Humulus lupulus scCO$_2$ extracts were used as experimental raw materials in this study. The extracts were obtained with the use of supercritical fluid extraction with carbon dioxide in a supercritical state from the Polish hops variety, Marynka. Hop cones from the 2017 harvest were used to produce the hop extract, "Marynka". The geographic origin of crops was the Lublin province, Poland. Cones with a moisture content of 10.6% were milled on a Retsch SM100 mill using a sieve with a mesh size of 1.5 mm. The average particle size was 0.8 mm. The produced material was subjected to the extraction with supercritical carbon dioxide using a pilot plant. A total of 4 kg of ground hop cones were loaded into a 20 dm^3 extractor. The extraction was carried out with the following parameters: Pressure -250 bar, extraction temperature -50 °C, and CO$_2$ consumption -50 kg/kg dm. The extraction yield was 11.4% by weight.

2.2. Sample Preparation

The samples for microwave-assisted hydrodistillation (MAHD) and hydrodistillation (HD) were prepared in the following way. The extract (ca. 150 g) was weighed separately in a 1 L flask and approx. 400 g of distilled water was added and the resulting emulsion was stirred for half an hour. Water to extract ratio was 8:3. For the experiments, 50 g of residues (residues obtained after the extraction of hops with scCO$_2$) were also added to the mixture. Such preparation of the material assures an optimal increase of the material contact with the solvent and enables satisfactory efficiency. Table 1 summarizes the exact compositions of the mixtures for microwave-assisted extraction and hydrodistillation for Marynka extract. For the sake of clarity and discussion, the oils were named according to the scheme: the name of the extract + power at which the extraction was carried out or the name of hydrodistillation apparatus (e.g., Marynka_275 W).

Table 1. Preparation of samples for Marynka extract.

Sample Name	Charge Mass (g)	Extract Mass (g)	Water Mass (g)	Residue Mass (g) *
Marynka_275W	603.34	150.25	402.51	50.58
Marynka _295W	605.99	150.99	404.99	50.01
Marynka_335W	602.71	150.47	402.16	50.08
Marynka_395W	608.65	156.00	401.94	50.71
Marynka_Deryng	602.61	151.83	400.78	50.00

* Residue mass is the mass of the residue obtained after the extraction of *Humulus lupulus* with scCO$_2$

2.3. Microwave-Assisted Hydrodistillation (MAHD)

The MAHD extractions of the essential oils from Marynka hops varieties were performed at atmospheric pressure using a NEOS-GR (Milestone Technologies) system, equipped with a glycol cooled condenser, lab-grade glycol chiller, collected specimens, as well as a video system for the visual control of the process. The extraction was continued at 100 °C until no more essential oil was obtained. The essential oils were dried over anhydrous sodium sulfate, and filtered and stored in dark vials at 4 °C. During the process, two parameters were tested, namely, microwave power as well as the influence of the extraction time (HD and MAHD) on the oils composition and obtained yields.

The efficiency of the processes was calculated from the amount of the extracts loaded into the extraction process (m_e) to the amount of obtained essential oils (m_{oil}) [4]. The yields (Y) of extractions were calculated according to the following equation (Equation (1)):

$$Y\ (\%) = \frac{m_e}{m_{oil}} \times 100\%$$ (1)

The total energy used for the extraction of essential oils by means of MAHD method was calculated according to the following equation (Equation (2)):

$$EC = \frac{P \times t}{1000}$$ (2)

in which, EC (kWh) is the energy consumed, P (W) is the power consumed during the extraction, and t (h) is the extraction time.

2.4. Hydrodistillation (HD)

The mixture was subjected to hydrodistillation at 150 °C until no more essential oils were obtained, which was approx. 276 min for Marynka using a Deryng apparatus. The heating was performed using a heater equipped with a magnetic stirrer (Heidolph, Schwabach, Germany) with 825 W and working temperature up to 300 °C. The extracted essential oils were dried over anhydrous sodium sulfate, and filtered, weighed and stored in dark sealed vials at 4 °C.

3. Analytical Procedure

3.1. Supercritical Fluid Chromatography (SFC)

The separation of β-myrcene and α-humulene was conducted on a Waters Acquity Ultra-Performance Convergence Chromatography (UPC^2) system (Waters, Milford, MA, USA) equipped with a PDA detector. For the data analysis, Empower 3 software was used. The column was Acquity UPC^2 HSS C18 SB (100 mm × 3.0 mm; 1.8 µm), packed with a chemically modified, highly strengthened silica with C18 groups. Carbon dioxide (99.998% purity), which was used as the mobile phase in SFC, was purchased from Air Liquide. The separation was performed with an isocratic elution with pure CO_2 and a flow rate of 2.3 mL/min as well as a column temperature of 35 °C and ABPR (automated back pressure regulator) set to 12.6 MPa. The compensation wavelength was set at 500−600 nm throughout the analyses. The wavelength for the compounds was set to 220 nm. The authentic analytical standards of β-myrcene and α-humulene were purchased from Sigma-Aldrich and were dissolved in hexane. Figure 2 presents the chromatogram of β-myrcene and α-humulene by UPC^2 method.

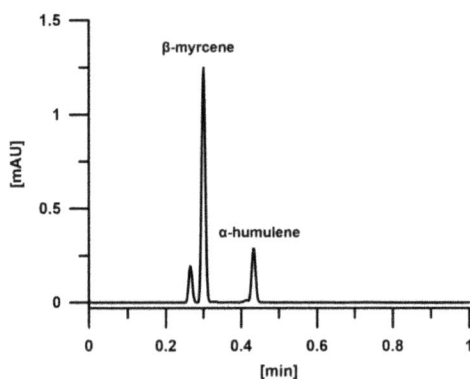

Figure 2. The chromatogram of β-myrcene and α-humulene by UPC^2 (ultra-performance convergence chromatography) on HSS C18 SB (100 mm × 3.0 mm; 1.8 µm) column; 35 °C, ABPR (automated back pressure regulator): 12.6 MPa; pure CO_2; isocratic elution.

3.1.1. Quantification

In order to quantify β-myrcene and α-humulene in hops oils, an external calibration method was applied. Individual stock solutions of authentic standards of β-myrcene and α-humulene were prepared in hexane. The appropriate amounts of β-myrcene and α-humulene were weighed and hexane was increased to the volume of 1 mL. The standard curves were obtained from five concentrations in the amount of 1000, 500, 250, 125, and 62.5 µg for β-myrcene and 500, 250, 125, 62.5, and 31.125 µg for α-humulene, adjusted to the real content of β-myrcene and α-humulene in the analyzed samples.

3.1.2. Partial Method Validation

Limits of detection (LOD), limits of quantification (LOQ), as well as repeatability and intermediate precision, were determined for the mixture of β-myrcene and α-humulene in order to evaluate the correctness of the provided method. Limits of detection (LOD) were 0.78 µg/mL for β-myrcene and 0.89 µg/mL for α-humulene, whereas limits of quantification were, respectively, 2.23 µg/mL and 1.95 µg/mL. The coefficients of determination (R^2) for calibration curves were 0.998 for both compounds. The repeatability was measured by 3 replicate analyses of a mixture of β-myrcene and α-humulene at three different concentrations at the same day, whereas the intermediate precision on three consecutive days. The RSD values, which are calculated according to the equation: RSD% = (standard deviation of the peak area/mean) × 100, show that both repeatability (RSD% < 3) and intermediate precision (RSD% < 6) are acceptable (Table 2).

Table 2. Calibration data as well as repeatability and intermediate precision for analyzing β-myrcene and α-humulene by UPC^2 system.

Compound	Linearity Range (μg/mL)	R2	LOD (μg/mL)	LOQ (μg/mL)	Concentration (μg/mL)	RSD % (*n* = 3)	
						Repeatability	Intermediate Precision
β-myrcene	62.5–1000	0.998	0.89	2.23	75	2.16	3.87
					100	1.02	2.91
					200	0.72	2.36
α-humulene	31.125–500	0.998	0.78	1.95	75	2.00	3.12
					100	1.27	2.74
					200	0.50	2.03

3.2. Gas Chromatography (GC-MS/MS)

The qualitative analyses of obtained hops oils were performed with a GC-MS instrument (Agilent 7890) equipped with mass spectrometry and NIST 2011 MS spectra library. As for the acquired results, Masshunter software (ver. C.01.03) was used. The following analyses parameters were used for *Humulus lupulus* essential oil samples: Capillary GC column DB-EUPAH (60 m × 250 μm; 0.25 μm) with helium as the carrier gas, and flow rate of 2 mL/min in a split injection mode (40:1). Oven temperature was set initially at 50 °C (held for 1 min) and increased to 260 °C at the rate of 6 °C/min (held for 6 min) and then finally reached 300 °C at the rate of 3 °C/min and held for 20 min. The system was operating in the EI mode (electron energy 70 eV) with the temperature of the source ion set to 230 °C and the scanning range from 40 to 650 amu.

3.2.1. Quantification

Concerning the quantitative analysis, internal standard calibration method was used with 4-androstene-3, and 17-dione as an internal standard. About 5 mg of the standard was weighed into a 5 mL volumetric flask. Then 5 mL of methanol/chloroform (9:1, *v/v*) mixture was added and 10 μL of such prepared standard was added to each oil sample in order to quantify β-myrcene and α-humulene. The standard curves were obtained from five concentration points: 200, 100, 50, 25, and 10 μg for β-myrcene and 50, 25, 10, 5, and 1 μg for α-humulene, adjusted to the real content of β-myrcene and α-humulene in the analyzed samples. The quantification was achieved in the MRM mode with different GC conditions as were used in the qualitative analyses. The quantitative method was shortened to 38 min with an oven temperature of 80 °C at the starting point, held for 1 min, and then ramped up to 300 °C with the rate of 10 °C/min and held for 15 min with nitrogen as a collision gas at the flow rate of 1.5 mL/min. The fragment ions for quantification and identification of each analyte were 136→121 (6) and 136→107 (6) for β-myrcene, 204→189 (10) and 204→161 (10) for α-humulene, as well as 286→148 (5), 124→109 (10) and 124→96 (10) for internal standard. The collision energies are provided in the brackets.

3.2.2. Partial Method Validation

Limits of detection (LOD) were 1.8 ng/mL and 2.5 ng/mL, whereas limits of quantification were 4.6 ng/mL and 8.0 ng/mL, respectively, for β-myrcene and α-humulene. The coefficients of determination (R^2) for the calibration curves were 0.999 for both compounds. The mixture of β-myrcene and α-humulene at three different concentrations was injected on the same day as well as on three consecutive days to evaluate the repeatability and intermediate precision of the developed method. The RSD values indicate that both repeatability (RSD% < 3) and intermediate precision (RSD% < 6) are acceptable (Table 3).

Table 3. Calibration data as well as repeatability and intermediate precision for analyzing β-myrcene and α-humulene by GC-MS/MS system.

Compound	Linearity Range [μg/mL]	R2	LOD [ng/mL]	LOQ [ng/mL]	Concentration [μg/mL]	RSD % (n = 3)	
						Repeatability	Intermediate Precision
β-myrcene	10–200	0.999	1.8	4.6	50	2.36	3.25
					100	1.54	2.84
					200	0.98	1.99
α-humulene	1–50	0.999	2.5	8.0	50	2.47	3.17
					100	1.69	2.99
					200	1.12	2.39

4. Results and Discussion

4.1. Material for Extraction

During our previous studies different forms of hops were tested before the final material was chosen for the extraction. One of the materials was post-extraction residue (residue after hops extraction using scCO$_2$) as a scattering element of the extract mixture with water. However, hop extract is characterized by its poor solubility in water [10,16]. The microwave hydrodistillation process was very rapid with the accompanying overheating of the raw material in an aqueous environment. During the optimization of the extraction process, attempts were made to replace the residues with an emulsifying substance. For this purpose, four different emulsifiers were used, including polysorbate 80 (E433), emulsifying biobase, soya lecithin, and cetyl alcohol. None of the used emulsifiers brought the expected results, meaning no adequate degree of mixing and homogenization was achieved. The local overheating of the material was also observed. Our studies were also performed on other forms of materials, such as raw hops and hops granules. In any case, no satisfactory effect and no amount of oils were obtained. It might have been caused by the low concentration of desirable compounds in the analyzed material.

4.2. Extraction Results

The microwave-assisted hydrodistillation (MAHD) with various microwave power and hydrodistillation (HD) were performed on extracts of Marynka hops variety. In case of the Marynka extract, the yield obtained by MAHD with the microwave power of 295 W was similar to the yield obtained by means of the HD method (1.80% for MAHD vs. 1.90% for HD). The extraction time of MAHD was significantly shorter than that of the conventional HD (40 vs. 276 min). The maximum yield of the essential oil obtained from the Marynka extract was using MAHD with microwave power of 335 W (3.03%). Table 4 summarizes the results of MAHD and HD methods for the Marynka extract.

Table 4. The results for Marynka extract essential oils microwave-assisted extraction.

Sample Name	Raw Material Mass (g)	Total Oil Amount (mg)	Yield (wt%)	Extraction Time (min)	Energy Consumption (kWh)
Marynka_275W	603.34	2248.14	1.50	46	0.215
Marynka_295W	605.99	2720.97	1.80	40	0.200
Marynka_335W	602.71	5676.44	3.77	30	0.170
Marynka_395W	608.65	4723.37	3.03	20	0.134
Marynka_Deryng	602.61	2807.31	1.90	276	1.897

According to Routray and Orsat [22], the higher the extraction temperature, the better the yield can be obtained. However, when the optimal temperature is achieved, the extraction efficiency decreases. During the studies, a relationship between the extraction power and the extraction time was observed. With increasing microwave power, the extraction time was shortened. When the extraction was performed at 395 W (3.03%) the extraction yield was relatively lower in a comparison with the process at 335 W (3.77%). Figure 3 presents the influence of the extraction power on the extraction time and yield.

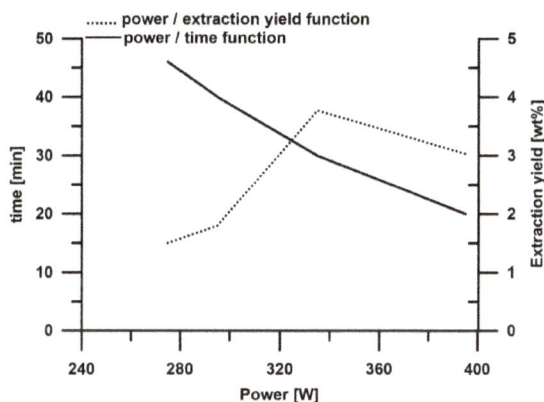

Figure 3. The influence of microwave power on the extraction time (———) and the extraction yield (. . . . .).

The function of the extraction time from the microwave power for the microwave-assisted hydrodistillation of the Marynka extract was determined. From the graph, the curve equation for the experimental data was read, from which the values for the extraction time with the applied microwave power values were determined. The obtained data showed that with the increase of the microwave power by 5 W, the extraction time is shorter by 1.07 min (Figure 4).

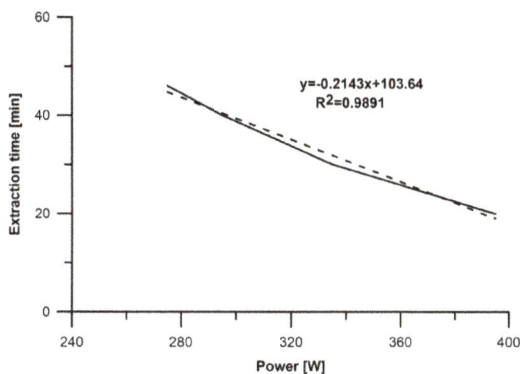

Figure 4. The function of the extraction time from the microwave power for the microwave-assisted hydrodistillation.

4.3. β-Myrcene and α-Humulezne Determination

Ligor et al. [23] obtained the essential oils from hops with a use of different techniques, such as supercritical fluid extraction, steam distillation, accelerated solvent extraction (ASE), and solid phase microextraction (SPME). The values for β-myrcene and α-humulene in obtained essential oils ranged from 15.7% to 21.1% and from 11.1% to 33.4%, respectively. Our study involves the use of microwave-assisted hydrodistillation, which is considered as a safe method [9]. The method resulted in obtaining the essential oil from *Humulus lupulus* scCO$_2$ extract at optimized parameters of 30 min and 335 W, which corresponds to one third of the maximal working power of the system.

The first application of the supercritical fluid chromatography with a photo-diode array detector to determine volatile compounds such as β-myrcene and α-humulene is provided in this study. This method was compared with a traditional method involving gas chromatography equipped with mass spectrometer (GC-MS/MS). As was suggested by Flament et al. [24], the use of scCO$_2$ for the separation of volatile compounds is inappropriate. However, in this study the obtained results for β-myrcene and α-humulene were comparable. The SFC method appeared to be significantly shorter than GC-MS/MS. Moreover, the use of supercritical carbon dioxide is understandable here due to the non-polar nature of volatile compounds being a subject of analysis. According to SFC quantitative analysis of β-myrcene and α-humulene, confirmed by GC-MS/MS analysis, the obtained oils were enriched in β-myrcene in the amount of approx. 74.13%–89.32%, which was several times higher in a comparison with the values obtained by Ligor et al. [23]. In the case of α-humulene, the concentration ranged from 7.36% to 10.14% in the essential oils obtained by microwave-assisted hydrodistillation, whereas only 1.59% in the essential oil obtained by hydrodistillation. Table 5 summarizes the concentrations of β-myrcene and α-humulene in obtained essential oils analyzed by UPC^2 and GC-MS/MS systems.

Table 5. The results of quantitative analyses of β-myrcene and α-humulene in essential oils obtained by microwave-assisted hydrodistillation (MAHD) and hydrodistillation (HD).

Sample Name	β-Myrcene		α-Humulene	
	UPC2 (wt%)	GC-MS/MS (wt%)	UPC2 (wt%)	GC-MS/MS (wt%)
Marynka-275W	89.32	88.89	10.14	10.56
Marynka-295W	74.13	75.02	8.23	8.01
Marynka-335W	77.24	77.36	9.33	9.47
Marynka-395W	75.12	74.66	7.36	7.96
Marynka-Deryng	84.73	84.99	1.59	1.87

5. Conclusions

Microwave-assisted hydrodistillation has been proven to offer advantages over hydrodistillation in determination of essential oils in *Humulus lupulus* scCO$_2$ extract. A similar extraction yield of the essential oil from *Humulus lupulus* scCO$_2$ extract is achieved in shorter extraction time (30 min for MAHD vs. 276 min for HD) and at a lower energy consumption (0.17 kWh/g of essential oil for MAHD vs. 1.89 kWh/g of essential oil for HD). Increasing the extraction time increases efficiency of the process as well as the amount of obtained compounds. Moreover, a significantly higher temperature and longer extraction time in the case of the hydrodistillation may have an influence on the compound's thermal degradation. Both chromatographic methods, such as supercritical fluid chromatography (SFC) and gas chromatography-mass spectrometer (GC-MS/MS), are comparable with the advantage of the SFC method in terms of shortened analysis time.

Author Contributions: Conceptualization, R.G., J.S. and E.R.; Methodology, K.T. and E.R.; Software, R.G. and J.S.; Validation, K.T. and M.K.; Formal Analysis, K.T., M.K. and R.G.; Investigation, K.T., R.G. and J.S; Resources, J.D. and E.R.; Data Curation, R.G.; Writing-Original Draft Preparation, K.T.; Writing, Reviewing and Editing, M.K.; Visualization, M.K.; Supervision, M.K. and E.R; Project Administration, E.R.; Funding Acquisition, E.R.

Funding: This research received no external funding.

Conflicts of Interest: The authors declare no conflict of interest.

References

1. Cermak, P.; Paleckova, V.; Houska, M.; Strohalm, J.; Novotna, P.; Mikyska, A.; Jurkova, M.; Sikorova, M. Inhibitory effects of fresh hops on *Helicobacter pylori* strains. *Czech J. Food Sci.* **2015**, *33*, 302–307. [CrossRef]
2. Skorek, U.; Hubicki, Z.; Rój, E. Intensification of the use of hop extract for beer production. *Chemik Sci.-Tech.-Market.* **2011**, *65*, 160–163.
3. Biendl, M.; Engelhard, B.; Forster, A.; Gahr, A.; Lutz, A.; Mitter, W.; Schmidt, R.; Schönberger, C. *Hops: Their Cultivation, Composition and Usage*; Fachverlag Hans Carl: Nuremberg, Germany, 2014.
4. Biendl, M.; Pinzl, C. Hops and health. *German Hop Museum Wolnzach* **2013**, 1–123. [CrossRef]

5. Luzak, B.; Kassassir, H.; Rój, E.; Stańczyk, L.; Watala, C.; Golanski, J. Xanthohumol from hop cones (*Humulus lupulus* L.) prevents ADP-induced platelet reactivity. *Arch. Physiol. Biochem.* **2017**, *123*, 54–60. [CrossRef] [PubMed]
6. Yamaguchi, N.; Yamaguchi, K.; Ono, M. In vitro evaluation of antibacterial anticollagenase and antioxidant activities of hop components (*Humulus lupulus*) addressing acne vulgaris. *Phytomedicine* **2009**, *16*, 369–376. [CrossRef] [PubMed]
7. Rój, E.; Tadić, V.M.; Mišić, D.; Žižović, I.; Arsić, I.; Dobrzyńska-Inger, A.; Kostrzewa, D. Supercritical carbon dioxide hops extract with antimicrobial properties. *Open Chem.* **2017**, *13*, 1157–1171. [CrossRef]
8. Jeyaratnam, N.; Abdurahman, H.N.; Ramesh, K.; Azhari, H.N.; Yuvaraj, A.R.; Akindoyo, J.O. Essential oil from *Cinnamomum cassia* bark through hydrodistillation and advanced microwave assisted hydrodistillation. *Ind. Crop. Prod.* **2016**, *92*, 57–66. [CrossRef]
9. Golmakani, M.T.; Rezaei, K. Comparison of microwave-assisted hydrodistillation with the traditional hydrodistillation method in the extraction of essential oils from *Thymus vulgaris* L. *Food Chem.* **2008**, *109*, 925–930. [CrossRef] [PubMed]
10. Gavahian, M.; Farahnaky, A.; Farhoosh, R.; Javidnia, K.; Shahidi, F. Extraction of essential oils from *Mentha piperita* using advanced techniques: Microwave versus ohmic assisted hydrodistillation. *Food Bioprod. Process.* **2015**, *94*, 50–58. [CrossRef]
11. Kusuma, H.S.; Mahfud, M. Kinetic studies on extraction of essential oil from sandalwood (*Santalum album*) by microwave air-hydrodistillation method. *Alex. Eng. J.* **2017**, *57*, 1163–1172. [CrossRef]
12. Abdurahman, H.N.; Ranitha, M.; Azhari, H.N. Extraction and characterization of essential oil from Ginger (*Zingiber Officinale* Roscoe) and Lemongrass (*Cymbopogon citratus*) by microwave-assisted hydrodistillation. *Int. J. Chem. Environ.* **2013**, *4*, 221–226.
13. Farzaneh, V.; Carvalho, I.S. Modelling of microwave assisted extraction (MAE) of anthocyanins (TMA). *J. Appl. Res. Med. Aromat. Plants* **2017**, *6*, 92–100. [CrossRef]
14. Karakaya, S.; El, S.N.; Karagozlu, N.; Sahin, S.; Sumnu, G.; Bayramoglu, B. Microwave-assisted hydrodistillation of essential oil from rosemary. *J. Food. Sci. Technol.* **2014**, *51*, 1056–1065. [CrossRef] [PubMed]
15. Fargat, A.; Benmoussa, H.; Bachoual, R.; Nasfi, Z.; Elfalleh, W.; Romdhane, M.; Bouajila, J. Efficiency of the optimized microwave assisted extractions on the yield, chemical composition and biological activities of *Tunisian Rosmarinus officinalis* L. essential oil. *Food Bioprod. Process.* **2017**, *105*, 224–233.
16. Wei, L.; Zhang, Y.; Jiang, B. Comparison of microwave-assisted hydrodistillation with the traditional hydrodistillation method in the extraction of essential oils from dwarfed *Cinnamomum camphora* var. *Linaolifera Fujita leaves and twigs. Adv. J. Food Sci. Technol.* **2013**, *5*, 1436–1442.
17. Franco-Vega, A.; Ramirez-Corona, N.; Palou, E.; Lopez-Malo, A. Estimation of mass transfer coefficients of the extraction process of essential oil from orange peel using microwave assisted extraction. *J. Food Eng.* **2016**, *170*, 136–143. [CrossRef]
18. Yang, X.; Lederer, C.; McDaniel, M.; Deinzer, M. Hydrolysis products of caryophyllene oxide in hops and beer. *J. Agric. Food Chem.* **1993**, *41*, 2082–2085. [CrossRef]
19. Hartsel, J.A.; Eades, J.; Hickory, B.; Makriyannis, A. Cannabis sativa and hemp. In *Nutraceuticals*; Elsevier: Amsterdam, The Netherlands, 2016.
20. Rufino, A.T.; Ribeiro, M.; Sousa, C.; Judas, F.; Salgueiro, L.; Cavaleiro, C.; Mendes, A.F. Evaluation of the anti-inflammatory, anti-catabolic and pro-anabolic effects of E.-caryophyllene, myrcene and limonene in a cell model of osteoarthritis. *Eur. J. Pharmacol.* **2015**, *750*, 141–150. [CrossRef] [PubMed]
21. Fernandes, E.S.; Passos, G.F.; Medeiros, R.; da Cunha, F.M.; Ferreira, J.; Campos, M.M.; Pianowski, L.F.; Calixto, J.B. Anti-inflammatory effects of compounds alpha-humulene and (−)-trans-caryophyllene isolated from the essential oil of Cordia verbenacea. *Eur. J. Pharmacol.* **2007**, *569*, 228–236. [CrossRef] [PubMed]
22. Routray, W.; Orsat, V. Microwave-assisted extraction of flavonoids: A review. *Food Bioprocess Technol.* **2011**, *5*, 409–424. [CrossRef]

23. Ligor, M.; Stankevičius, M.; Wenda-Piesik, A.; Obelevičius, K.; Ragažinskienė, O.; Stanius, Ž.; Maruška, A.; Buszewski, B. Comparative gas chromatographic-mass spectrometric evaluation of hop (*Humulus lupulus* L.) essential oils and extracts obtained using different sample preparation methods. *Food Anal. Method* **2014**, *7*, 1433–1442. [CrossRef]

24. Flament, I.; Chevalier, C.; Keller, U. Extraction and chromatography of food constituents with supercritical CO_2. In *Flavour Science and Technology*; Martens, M., Dalen, G.A., Russwurm, H., Chichester, H.R., Jr., Eds.; Wiley: New York, NY, USA, 1987; pp. 151–163.

Sample Availability: Samples of the compounds are not available from the authors.

molecules

MDPI

Article

Simultaneous Identification and Dynamic Analysis of Saccharides during Steam Processing of Rhizomes of *Polygonatum cyrtonema* by HPLC–QTOF–MS/MS

Jian Jin [1,2,3](ORCID), Jia Lao [4], Rongrong Zhou [5], Wei He [4], You Qin [1,2], Can Zhong [1,3], Jing Xie [1,3], Hao Liu [1,3], Dan Wan [1,3], Shuihan Zhang [1,3,*] and Yuhui Qin [1,2,*]

[1]   Institute of Chinese Materia Medica, Hunan Academy of Chinese Medicine, Changsha 410013, China;
      jinjian2016@163.com (J.J.); 20162010@stu.hnucm.edu.cn (Y.Q.); canzhong651@163.com (C.Z.);
      axxj2057@163.com (J.X.); 350013@hnucm.edu.cn (H.L.); 350017@hnucm.edu.cn (D.W.)
[2]   School of Chinese Medicine, Hunan University of Chinese Medicine, Changsha 410208, China
[3]   2011 Collaboration and Innovation Center for Digital Chinese Medicine in Hunan, Changsha 410208, China
[4]   Resgreen Group International Inc., Changsha 410329, China; laojia1973@163.com (J.L.);
      hewei3218@126.com (W.H.)
[5]   College of Pharmacy, Changchun University of Chinese Medicine, Changchun 130117, China;
      rz172@georgetown.edu
[*]   Correspondence: zhangshuihan0220@126.com (S.Z.); dlqyh@sohu.com (Y.Q.);
      Tel.: +86-0731-8888-1651 (S.Z.); +86-0731-8885-4257 (Y.Q.)

Received: 4 October 2018; Accepted: 31 October 2018; Published: 2 November 2018

check for updates

Abstract: The sweet rhizomes of *Polygonatum cyrtonema* are widely used as a tonic and functional food. A sensitive and rapid analytical method was developed for simultaneous identification and dynamic analysis of saccharides during steam processing in *P. cyrtonema* using HPLC–QTOF–MS/MS. Fructose, sorbitol, glucose, galactose, sucrose, and 1-kestose were identified, as well as a large number of oligosaccharides constituted of fructose units through β-(2→1) or β-(2→6). Polysaccharides and oligosaccharides were decomposed to monosaccharides during a steaming process, since the contents of glucose, galactose, and fructose were increased, while those of sucrose, 1-kestose, and polysaccharides were decreased. The high content of fructose was revealed to be the main determinant for increasing the level of sweetness after steaming. The samples of different repeated steaming times were shown to be well grouped and gradually shift along the PC1 (72.4%) axis by principal component analysis. The small-molecule saccharides, especially fructose, could be considered as markers for the steaming process of rhizomes of *P. cyrtonema*.

Keywords: *Polygonatum cyrtonema*; saccharides; oligosaccharides; fructose; HPLC–QTOF–MS/MS; steaming

1. Introduction

The rhizomes of *Polygonatum cyrtonema* are wildly used as a tonic and functional food in China. The use of this plant is documented in the well-known ancient Pharmacopoeia "*Shennong Bencao Jing*", and it is considered as a "Top grade" herb. Its main efficacy includes replenishing energy, strengthening immunity, and treating fatigue, weakness, loss of appetite, and so on. The genus *Polygonatum* has also been used in some other regions, such as India, Japan, Europe, and North America [1].

There are various ways to cook the rhizomes of *Polygonatum*. They are often cooked with meats or porridges, made into tea or medicated wine, or consumed as fruits or vegetables [2]. It is worth noting that raw rhizomes without processing treatment are rarely directly used. Raw rhizomes are often processed to enhance their tonic function by repeated steaming and drying. In particular, according to its traditional uses, the rhizomes of *Polygonatum* are processed by steaming nine times, which is recently combined with autoclave as an efficient approach. Traditional quality control is performed by qualitative index, described as changes such as rhizomes becoming black, soft, and especially, gaining a sweet taste. However, the changes of the chemical composition during the steaming process are not understood.

With respect to chemical composition, there could be steroidal saponins, triterpenoid saponins, homoisoflavanones, polysaccharides, and lectins in the *Polygonatum* plants [1]. Polysaccharides are the main chemical components, and possess various pharmacological actions, such as antioxidant protection and immune-regulation [3]. The quality of *P. cyrtonema* is evaluated depending on the content of polysaccharides, with a minimal content of 7% required by Chinese Pharmacopoeia. Polysaccharides from *Polygonatum* plants are composed of different ratios of monosaccharides, mainly including mannose, galactose, glucose, fructose, rhamnose, arabinose, and galacturonic acid [1]. A branched fructan was also isolated from *P. cyrtonema* [4]. A large number of monosaccharides, oligosaccharides, and polysaccharides were found in rhizomes of *Polygonatum* plants, making it worthwhile to try to evaluate the processed products of rhizomes of *P. cyrtonema* with the help of saccharides in order to clarify the reason why the processed rhizomes of *Polygonatum* gradually take on a sweet taste.

Commonly-used analytical modes of saccharides include nuclear magnetic resonance (NMR) [5], gas chromatography (GC) [6], and liquid chromatography (LC) [7,8]. LC is widely used thanks to its various detectors, while NMR cannot be used to detect trace amounts of saccharides, and GC requires tedious derivatization [9]. However, because the absence of chromophore and fluorophore groups, avoiding direct detection by ultraviolet and fluorescence or diode array detectors, pre-column derivatization is often employed in LC, which is time consuming and could introduce extra variations into the final data [10]. LC equipped with an evaporative light scattering detector (ELSD) could be used to detect the monosaccharides; for instance, glucose, fructose, and sucrose were examined in the raw materials of *agave* leaf [11]. Structural information can even be provided by well-developed analytical techniques; for example, mass spectrometry (MS) coupled with LC is currently a more powerful analytical tool for saccharide analysis [10]. Sucrose, fructose, and glucose in plant tissues can be unambiguously determined by LC–MS [10]. Three major tri–saccharides in wheat flour were also determined by LC–MS [12]. It is worth mentioning that most previous study focused on fructose, glucose, sucrose and other oligosaccharides, but seldom distinguished glucose and galactose, which are the most common isomeric hexoses present in physiology. LC coupled with atmospheric pressure chemical ionization (APCI)–MS was developed to identify most of the standard saccharides, but it is still hard to separate glucose and galactose [13]. It seems that no simple method is available for the differentiation of these two isomers, since current methods required either derivatization techniques based on the reaction of reducing sugars with 1-phenyl-3-methyl-5-pyrazolone, phenylhydrazones [14], or the help of a zinc diethylenetriamine (Zndien) metal-ligand system [15].

In *Polygonatum* plants, polyphenolic antioxidants have ever been characterized by high-performance liquid chromatography (HPLC)–quadrupole-time-of-flight (QTOF)–MS, a useful technology for structural identification, even when standard compounds are not available [16]. Nevertheless, few studies of simultaneous analysis of saccharides in rhizomes of *Polygonatum* were reported, let alone characterization of the changes of the saccharide composition during the nine times of repeated steaming process. In the present study, an efficient and sensitive method was developed by HPLC–QTOF–MS/MS to profile and identify saccharides, including glucose and galactose, in fresh and processed rhizomes of *P. cyrtonema* by nine-time steaming. The present study will provide not only a useful tool for the analysis of saccharides, but also information for the improvement of *Polygonatum* steam process.

2. Results and Discussion

2.1. Appearance Change

The rhizomes of *P. cyrtonema* were processed by repeated steaming for one to nine times, shown in Figure 1.

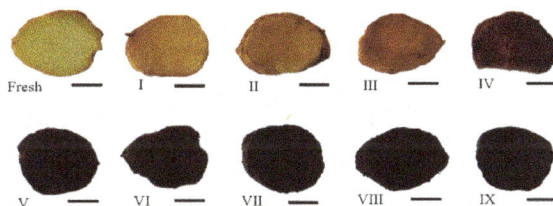

Figure 1. Samples of the rhizomes of *P. cyrtonema* are processed by different times of repeated steaming; "fresh" implies without steam process; roman numerals indicate the number of times of steaming; bar, 1 cm.

Steaming treatment is a usual food processing technology for the pretreatment of lignocellulosic materials [17]. In recent decades, an autoclaving steam process has emerged as an industrially-adopted method, not only to deliver a wide range of high quality products to meet the traditional Chinese requirements, but also to extend shelf life as a technique of food pasteurization to kill food pathogens and ensure food safety. Heavy metals in *Polygonatum* roots, such as Pb and Cd, could also be eliminated by exposure to high temperature, pressurized steaming conditions [18]. The well-known medicinal herb black ginseng is also usually treated with steaming; ginsenoside composition changes during the steaming process [19,20]. Traditionally, rhizomes of *Polygonatum* were used mainly as a crude drug, while in recent years, they are more popular as food when cooked with meats or porridges, made into tea or medicated wine, or consumed as fruits or vegetables [2]. The color and taste are two of the most crucial sensory properties of food. From fresh to processed samples with different numbers of repeated steaming, the samples changed to black, and became sweeter. But after steaming four times, the color appeared stable. The Maillard reaction, during which sugars react with amino acids under thermal conditions [21], would be largely responsible for the dark-colored appearance of processed rhizomes of *P. cyrtonema*. For the sweet taste, the present study was designed to systematically assess the saccharides, especially the small-molecular saccharides, between fresh and processed rhizomes of *P. cyrtonema*.

2.2. Optimization and Validation of the HPLC–QTOF–MS/MS Method

For simultaneous determination of saccharides and their derivatives, the optimal chromatographic conditions were investigated. Various mixtures of water, methanol, and acetonitrile at flow rate 1.2 mL/min, 1.0 mL/min, 0.8 mL/min, 0.5 mL/min and 0.3 mL/min were tested as the mobile phase. The detection temperature was tested at 15, 25, 35, and 45 °C. Eventually, the optimal chromatographic conditions were as follows: Prevail Carbohydrate column, with acetonitrile 75% and ultrapure water 25% as mobile phase at 1.0 mL/min at 25 °C. As shown in Figure 2A, the peaks in fresh and processed samples are separated by HPLC–QTOF–MS/MS. This condition was suitable for most of the small molecular standards (Figure 2B), except for glucose and galactose, which appeared at the same retention time 11.60 min. Surprisingly, when mobile phase flow rate decreased to 0.3 mL/min, only the standard glucose appeared at retention time 39.56 min, while no peak of the galactose appeared using sole standard solutions (Figure 2C). Then, the quantitative determination of glucose in samples was carried out with the mobile phase flow rate at 0.3 mL/min for the peak at retention time 39.56 min. The content of galactose was calculated by subtracting the glucose from the peak at retention time 11.60 min with mobile phase flow rate at 1.0 mL/min.

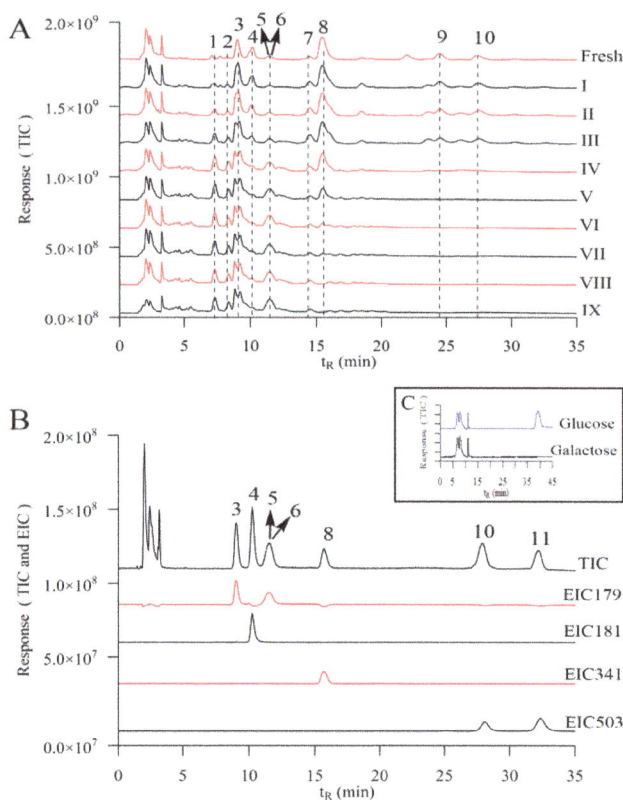

Figure 2. Chromatogram of the extract of the rhizomes of *P. cyrtonema* are processed by different times of repeated steaming (**A**), standards of small–molecular saccharides (**B**) by HPLC–QTOF–MS/MS at solvent flow rate 1.0 mL/min, glucose and galactose at solvent flow rate 0.3 mL/min (**C**); 1, fructose derivative; 2, glucose derivative; 3, fructose; 4, sorbitol; 5, glucose; 6, galactose; 7, hexose; 8, sucrose; 9, trisaccharide; 10, 1-kestose; 11, raffinose; TIC, total ion chromatogram; EIC, extracted ion chromatogram; "fresh" means without steaming; roman numerals indicate the number of times of steaming.

Noting that the peak of TIC and EIC will be saturated and even lead to bifurcation when the standard concentration is too high (Figure S1), the standard stock solutions containing fructose, glucose, galactose, sucrose and 1-kestose were prepared and diluted to a series of appropriate concentrations for the construction of calibration curves. These curves showed good linearity, and the correlation coefficients were found to be in the range with R^2 0.99 for all of the compounds in a certain concentration range (Table 1), consistent with saccharides determined by hydrophilic interaction liquid chromatography (HILIC)–TOF–MS technology [7]. The limits of detection (LOD) and limits of quantification (LOQ) were in the range of 0.41–3.63 µg/mL and 1.33–12.10 µg/mL, respectively. The intraday and interday variations of the analytes (RSDs) were within 0.93–1.91% and 1.82–3.51%, respectively. The RSDs for stability were lower than 3.01%. The average recoveries of the standard compounds ranged from 94.3% to 107.5%, and their RSDs were within 6.39% (Table 2). All these results demonstrated that the developed quantitative HPLC–QTOF–MS method was linear, precise, stable, sensitive, and accurate enough for the determination of small-molecular saccharides in rhizomes of *P. cyrtonema*. There are studies using HPLC system coupled with MS to investigate saccharides in plant tissues or products, but previous reports mainly investigated glucose, fructose and sucrose [10,22]. LC–APCI–MS was developed to distinguish saccharides with different types of columns, but it is still hard to separate glucose and galactose [13]. To the best of our knowledge, this is the first time that small-molecule saccharides, including glucose and galactose, were separated and determined by the highly sensitive and rapid HPLC–QTOF–MS/MS method using a common column without derivatization.

Table 1. Linear regression, LOD and LOQ of investigated compounds.

Compounds	Range (µg/mL)	Regression Equation [a]	R^2	LOD [b] (µg/mL)	LOQ [c] (µg/mL)
Fructose [d]	6.4–64	y = 71886x + 682691	0.996	0.48	1.60
Glucose [d]	15.2–152	y = 238502x + 2043001	0.991	0.90	2.99
Glucose [e]	15.2–152	y = 457396x + 2818970	0.992	0.87	2.89
Galactose [d]	15.7–157	y = 65662x + 479229	0.992	3.63	12.10
Sucrose [d]	7.1–71	y = 296456x + 1182954	0.994	0.41	1.33
1-Kestose [d]	73–366	y = 71886x + 922640	0.995	0.92	3.06

[a] y: peak area, x: concentration of the analyte (µg/mL); [b] LOD, limit of detection; [c] LOQ, limit of quantification; [d] solvent flow rate at 1.0 mL/min; [e] solvent flow rate at 0.3 mL/min.

Table 2. Precision, stability and recovery of investigated compounds.

Compounds	Precision (n = 6)		Stability (48 h) (RSD, %)	Recovery (n = 3)	
	Intra-Day (RSD, %)	Inter-Day (RSD, %)		Mean (%)	RSD (%)
Fructose [a]	1.37	1.82	2.29	96.7	2.37
Glucose [a]	1.24	2.56	3.01	94.3	2.46
Glucose [b]	1.62	2.29	1.48	98.3	2.13
Galactose [a]	1.91	3.51	2.12	96.7	3.13
Sucrose [a]	1.84	3.27	2.14	107.5	6.39
1-Kestose [a]	0.93	2.22	1.29	96.1	3.71

[a] solvent flow rate at 1.0 mL/min; [b] solvent flow rate at 0.3 mL/min.

2.3. Identification of Small-Molecule Saccharides

Since the chromatographic peaks could not be identified unambiguously by HPLC retention time alone, HPLC–QTOF–MS/MS was used to confirm identities by comparing the retention time and molecular ions for each peak. In this experiment, the mass spectral conditions were optimized in negative-ion mode, and the six compounds exhibiting distinct quasi-molecular ions $[M - H]^-$ and $[M + COOH]^-$ were optimized in this mode: fructose (peak 3), molecular weight (MW) 180 $m/z = 179$ $[M - H]^-$, sorbitol (peak 4) (MW 182) $m/z = 181$ $[M - H]^-$, glucose (peak 5) (MW 180) $m/z = 179$ $[M - H]^-$, galactose (peak 6) (MW 180) $m/z = 179$ $[M - H]^-$, sucrose (peak 8) (MW 342) $m/z = 341$ $[M - H]^-$ and $m/z = 387$ $[M + COOH]^-$, 1-kestose (peak 10) (MW 504) $m/z = 503$ $[M - H]^-$ and $m/z = 549$ $[M + COOH]^-$ and raffinose (peak 11) (MW 504) $m/z = 503$ $[M - H]^-$ and $m/z = 549$ $[M + COOH]^-$. For the monosaccharide, the fragment ions demonstrated the same m/z value at 179, 161, 143, 113 and 101. For the disaccharide sucrose, the fragment ions are at m/z 341, 179, 161, 143, 119 and 101. For the trisaccharides, i.e., 1-kestose and raffinose, the fragment ions are at m/z 503, 341, 323, 221, 179, 161, 143, 113 and 101 (Table 3).

Table 3. Identification of small–molecule saccharides by HPLC–QTOF–MS/MS technology

No	t_R (min)	$[M - H]^-$ (m/z) (Δppm)	Fragment Ions (m/z)	Molecular Formula	Compound
1	7.09	323.1003 (−4.98)	323.0990; 179.0564; 161.0461; 143.0364; 113.0258; 101.0251	$C_{12}H_{20}O_{10}$	Hexose derivative
2	8.16	323.1010 (−6.13)	323.1011; 179.0567; 161.0464; 143.0386; 113.0230; 101.0267	$C_{12}H_{20}O_{10}$	Hexose derivative
3	9.12	179.0581 (−0.58)	179.0533; 161.0458; 143.0354; 113.0242; 101.0245	$C_6H_{12}O_6$	Fructose
4	10.43	181.0739 (−0.36)	181.0715; 163.0629; 149.0436; 119.0338; 101.0247	$C_6H_{14}O_6$	Sorbitol
5 [a]	11.60	179.0566 (−2.94)	179.0583; 161.0433; 143.0350; 113.0248; 101.0248	$C_6H_{12}O_6$	Glucose
6 [a]	11.60	179.0585 (−0.08)	179.0574; 161.0480; 143.0336; 112.9853; 101.0246	$C_6H_{12}O_6$	Galactose
7	14.4	179.0587 (−0.65)	179.0583; 161.0451; 143.0378; 112.9846; 101.0241	$C_6H_{12}O_6$	Hexose
8	15.24	341.1125 (−1.14)	341.1090; 179.0555; 161.0465; 143.0338; 119.0350; 101.0243	$C_{12}H_{22}O_{11}$	Sucrose
9	23.71	503.1666 (−1.1)	503.1668; 341.1101; 323.0988; 221.0670; 179.0554; 161.0441; 143.0328; 113.0220; 101.0240	$C_{18}H_{32}O_{16}$	Trisaccharide
10	26.49	503.1670 (−4.19)	503.1668; 341.1054; 323.0986; 221.0655; 179.0553; 161.0467; 143.0368; 113.0255; 101.0244	$C_{18}H_{32}O_{16}$	1-Kestose
11	33.15	503.1653 (2.62)	503.1653; 341.1031; 323.0936; 221.0616; 179.0513; 161.0412; 143.0308; 113.0216; 101.0219	$C_{18}H_{32}O_{16}$	Raffinose

[a] Characterization with sole standard glucose or galactose.

In the HPLC–QTOF–MS/MS spectra, the molecular ions of each compound agreed well with the chemical structures. Basing on the above results, the standard compounds of fructose, sorbitol, glucose, galactose, sucrose, and 1-kestose were identified in the extract of the rhizomes of *P. cyrtonema*, and the identities of each peak were clearly confirmed. However, raffinose, a common oligosaccharide in food [23], was not found in fresh or processed samples.

It is worth noting that there were also obvious peaks 1, 2, 7, and 9 in the rhizomes of *P. cyrtonema*. Peak 1 and 2 were shown to be due to the same fragment ions to monosaccharide with m/z value at 179, 161, 143, 113, and 101, indicating a component of hexose. The fragment ion at m/z 323 could be due to $[341 - H_2O]^-$, where 341 is the characterized ion of disaccharide. Then, peak 1 and 2 could be hexose derivatives formed by dehydration from disaccharides. The m/z values of peak 7 fragment ions were at 179, 161, 143, 113, and 101, the same with that of monosaccharide, indicating a hexose structure. Peak 9 demonstrated the same fragment ions with those of 1-kestose and raffinose, with *m/z* value at 503, 341, 323, 221, 179, 161, 143, 113, and 101, also indicating a trisaccharide. 1-kestose is formed of two molecules of fructose linked by β-(2→1) D-fructosyl-fructose bonds, and terminated with one glucose unit [24]. Noting that for the retention time, the peak of fructose is 2.48 min before glucose, peak 9 could be a trisaccharide composed of three molecules of fructose, because the retention time of peak 9 was 2.78 min before that of 1-kestose. There could be two possibilities for the linkage of this identified trisaccharide. One is that the molecules of fructose are linked by β-(2→1) and β-(2→6) as mixed type F3 fructan in inulin [25], namely MF3 as is shown in Figure 3A peak 9-1. The other could be simpler that a molecule of fructose replaces the glucose in 1-kestose, keeping the straightforward structure, namely SF3 (Figure 3A peak 9-2).

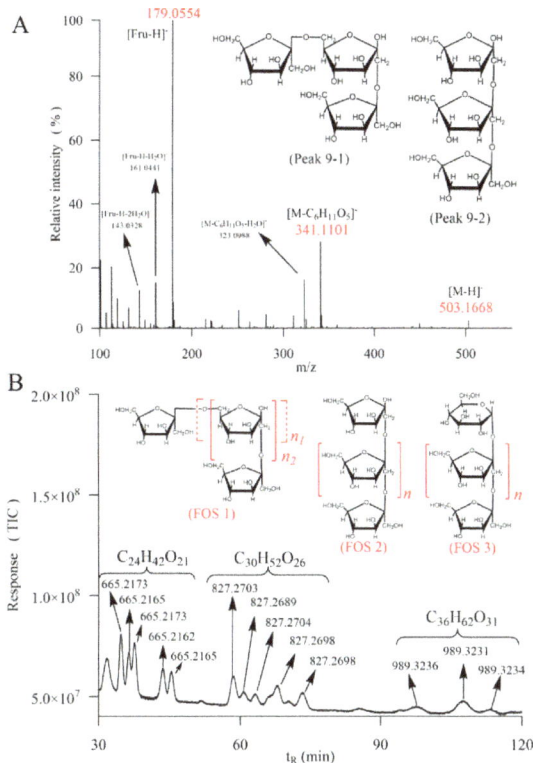

Figure 3. MS/MS spectrum, the two speculated structures of peak 9 (**A**) and LC spectrum with *m/z* values, speculated molecule formulas and structures in fresh rhizomes of *P. cyrtonema* by HPLC–QTOF–MS/MS with mobile phase flow rate at 1.2 mL/min (**B**). FOS, fructo-oligosaccharides.

In order to further characterize the fructo-oligosaccharides (FOS), the mobile phase increased to 1.2 mL/min for 120 min to overview the outcome. It could be speculated that the peaks appearing at *m/z* value 665, 827, and 989 could have molecular formulas $C_{24}H_{42}O_{21}$ (tetrasaccharide), $C_{30}H_{52}O_{26}$ (pentasaccharide) and $C_{36}H_{62}O_{31}$ (hexasaccharide) (Figure 3B). These FOS should be derived from the trisaccharides, i.e., 1-kestose, MF3 and SF3, by lengthening the chains with adding a variable number of fructose units through β-(2→1) or β-(2→6), as is shown in Figure 3B FOS 1, FOS 2 and FOS 3. Fibers of *P. odoratum* have been found to be composed of arabinose, xylose, sorbose, mannose and galactose [26] and lectin in *P. cyrtonema* could be binding to mannose [27]. Another monosaccharide analysis study showed that all the complete hydrolytic polysaccharides of *P. cyrtonema* are determined to be heteropolysaccharides containing rhamnose, arabinose, xylose, mannose, galactose and glucose [28].

However, the present investigation found that oligosaccharides in *P. cyrtonema* were mainly made up of fructose and glucose. The speculated structures of FOS in the present work are consistent with those of the previous study, in which the neutral polysaccharide in *P. cyrtonema* was found to be a branched fructan, composed of (2→6)-linked β-D-fructofuranosyl residues and (2→1)-linked β-D-fructofuranosyl residues in the backbone [4]. Since the intake of prebiotic fructans could sustain health and overall well-being [29], the observation of a large number of FOS in *P. cyrtonema* could be an evidence to explain its functional benefits.

2.4. Dynamic Change of Saccharides

The contents of fructose, glucose, galactose, sucrose, and 1-kestose in three samples treated with different times of individual processing steps were measured by the developed method, as well as the total small-molecule saccharides and polysaccharides. Even though the saccharides in different *Polygonatum* samples usually varied [30], the present results demonstrated that there was an obvious influence on the quantities of small-molecule saccharides between fresh and steam processed samples. The dynamic changes and proposed model are illustrated in Figure 4.

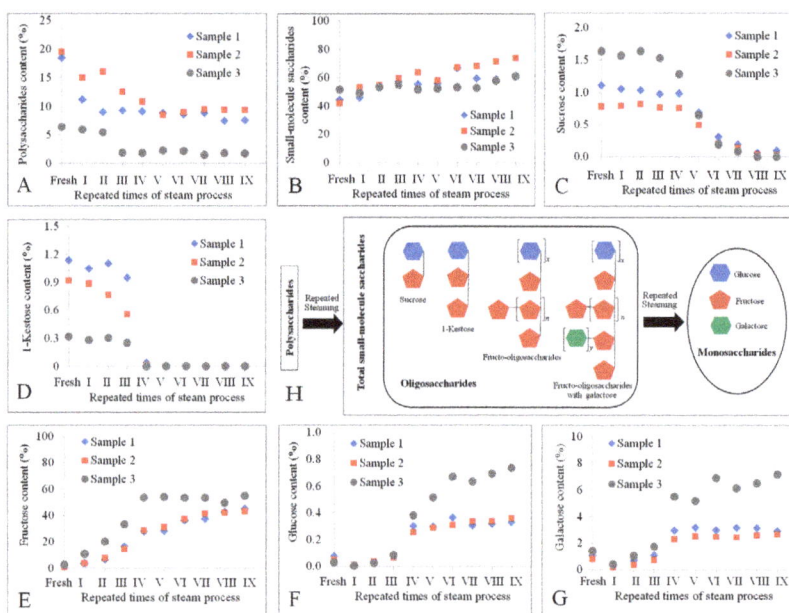

Figure 4. The content of polysaccharides (**A**), total small–molecule saccharides (**B**), sucrose (**C**), 1-kestose (**D**), fructose (**E**), glucose (**F**), galactose (**G**), in the rhizomes of *P. cyrtonema* processed by different times of repeated steaming and the proposed model of saccharides change during repeated steaming (**H**); "fresh" means without steam process; roman numerals indicate the number of times of steaming.

The content of polysaccharides decreased, while the total small-molecule saccharides increased, indicating the polysaccharides were decomposed to small-molecule saccharides with the repeated times of steam process (Figure 4A,B). The measured oligosaccharides sucrose and 1-kestose were broke down forming glucose and fructose, because sucrose and 1-kestose were decreased (Figure 4C,D), whereas glucose and fructose increased (Figure 4E,F). Unexpectedly, the content of galactose also increased (Figure 4G), indicating this monosaccharide could be a branch linked to FOS in *P. cyrtonema*, as proposed in the model Figure 4H. Interestingly, steaming four times seems to be a landmark for these small-molecule saccharides, since their content appeared to be relatively stable after that. All these dynamic changes of polysaccharides and small-molecule saccharides caused by the decomposition of the glycosidic linkages, which could be broken by steam processing [26], resulting in an increase in the contents of glucose, galactose, and fructose, and a decrease in oligosaccharide and polysaccharides. It is reported that the fructose in Radix Rehmanniae gradually decreased, and the glucose remained relatively stable after steaming [9]. These could be species dependent, since the glycosidic linkages might be different in different kinds of plants.

Surprisingly, the content of fructose gradually increased to an extremely high value of about 50% of dry sample weight, while that of most food is less than 10% [31]. Only in a few foods, for instance honey, fructose, and glucose together account for 85–95%, and fructose usually predominates, which is responsible for the sweetness [32]. The change of fructose could be the main reason why the taste of the rhizomes of *P. cyrtonema* became sweeter after the steaming processing, since fructose is a major determinant of sweetness in food [10]. The high content of fructose is also in accordance with our speculation that oligosaccharides were mainly composed of fructose. It is reported that fructans isolated from the rhizomes of *P. odoratum* could be composed of 29 units of fructose and 1 unit of glucose [33]. A neutral polysaccharide with an average degree of polymerization of 28 from *P. cyrtonema* Hua is also mainly made up of fructose [4]. Therefore, it could be concluded that the high concentration of fructose probably came from the decomposition of fructo-oligosaccharides and polysaccharides.

2.5. PCA Statistical Analysis

To overview the difference between fresh and processed rhizomes of *P. cyrtonema*, unsupervised PCA was performed. The contents of monosaccharides, oligosaccharides, identified saccharides derivatives, as well as polysaccharides, in all the samples of fresh and processed rhizomes were considered as the variables of PCA. The PCA biplot displayed the scores and loadings of the first two components (PC1 and PC2), revealing the projection of an observation on the subspace with score points (Figure 5). 72.4% and 15.8% of the variation in the pattern of concentrations of identified saccharides and derivatives were explained by PC1 and PC2, respectively. These two components together explained 88.2% of the variation.

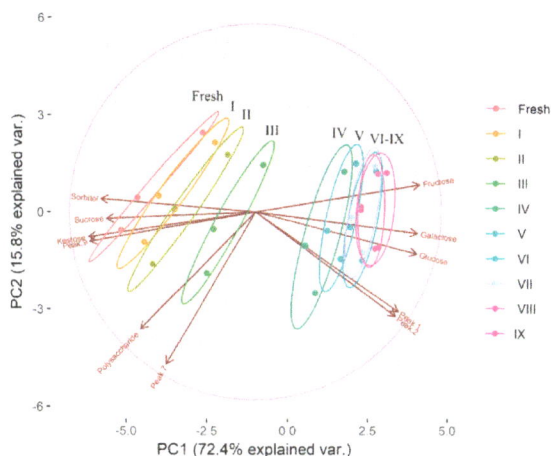

Figure 5. A biplot showing the samples of fresh and processed rhizomes of *P. cyrtonema* by different times of repeated steaming on a two-dimensional space derived from principal component analysis basing on identified saccharides and their derivatives; the arrows indicated the projections of the original features onto the principal components; "fresh" means without steam process; roman numerals indicate the number of times of steaming.

The variables of fructose, glucose, galactose, peak 1, and peak 2 shown in the same direction correlated along PC1, whereas sucrose, 1-kestose, sorbitol and peak 9 were oppositely correlated to PC1. Polysaccharides and peak 7 were primarily correlated with PC2. The samples gradually shifted along PC1 axis with the increasing times of repeated steaming. PCA results could substantially separate fresh and processed rhizomes by 1–4 times of repeated steaming, but failed to distinguish samples processed with 4–9 times steaming, as there were a large number of samples overlapping between these groups. This result was in consistent with the previous observation that the color and aforementioned compounds were relatively stable after steaming four times. To a certain degree, present PCA study was in accordance with previous report that the raw and steamed samples were separated into two groups by PCA in potato strips [34] and in Radix Rehmanniae samples [7,9]. To date, no compound has been used as a marker compound for processed *P. cyrtonema*. Then, these small–molecule saccharides might be considered as marker compounds. In particular, fructose, due to its high content, could be given special consideration as an important marker compound.

3. Materials and Methods

3.1. Chemicals and Reagents

Galactose (CAS 387116-33-2), 1-kestose (CAS 470-69-9) and D-raffinose (CAS 512-69-6) were obtained from National Institutes for Food and Drug Control (Beijing, China). D-fructose (CAS 57-48-7), D-glucose (CAS 50-99-7), D-sorbitol (CAS 50-70-4), and sucrose (CAS 57-50-1) were purchased from Hefei Bomei Biotechnology. Co., Ltd. (Hefei, China). Acetonitrile (HPLC–grade) was purchased from Merck (Darmstadt, Germany). Ultrapure water was purchased from Wahaha (Changsha, China). Other reagents were of analytical grade.

3.2. Preparation of Samples

Fresh rhizomes of *P. cyrtonema* were collected from Xinhua County, Hunan Province, China and authenticated by Professor Zhaoming Xie and Hao Liu from Hunan Academy of Chinese Medicine. The voucher specimens were kept in our department for future reference. Rhizomes of *P. cyrtonema* were cleaned, dried and then cut into thin slices (3 mm $\pm$ 1 mm) right after the harvest. The rhizomes in a glass bottle were processed by nine times repeated steaming with autoclave (121 °C, 0.12 MPa, 30 min); then, samples were kept in the autoclave until cooled to room temperature. Samples were collected after each cycle of steaming process.

3.3. Extraction of Saccharides

The steaming-processed rhizomes of *P. cyrtonema* were dried at 60 °C to a constant weight. Then, samples were ground by passing a 40-mesh screen and extracted by distilled water with ultrasonication (KM-500DB, 40 KHz, Kunshan Meimei Ultrasonic Instrument Co., Ltd., Jiangsu, China) for 30 min followed by maintenance at 90 °C for 60 min. After that, the extracts were centrifuged (3000$\times$ *g* for 15 min) to separate the supernatant from residues. Then, 10 mL of ethanol (95%) was added to 2 mL of the supernatant for precipitating the polysaccharides. The small-molecule saccharides, including monosaccharides and oligosaccharides, which were easier to dissolve in ethanol, were still dissolved in upper layer [35]. Then, the polysaccharides and small-molecule saccharides were separated by centrifugation (3000$\times$ *g*, 10 min). The precipitated polysaccharides were redissolved in distilled water for colorimetric quantitative measurement. The up layer was collected for colorimetric quantitative measurement of the total small-molecule saccharides and HPLC–QTOF–MS/MS analysis to identify monosaccharides and oligosaccharide compounds.

3.4. Spectrophotometric Quantitative Measurement

The concentration of total small-molecule saccharides and polysaccharides were determined with the phenol/sulfuric acid colorimetric method [36]. Briefly, 2 mL sample solution and 1 mL of 5%

aqueous solution of phenol were mixed in a test tube. Subsequently, 5 mL of concentrated sulfuric acid was added rapidly to the mixture. Test tubes were shaken in an ultrasonic bath for 10 min and then left at room temperature for 20 min for color development. The absorbance of the acquired solution was measured at 490 nm on a UV–Vis spectrophotometer (UV 2450, Shimadzu Corporation, Kyoto, Japan). A reference solution was prepared in an identical manner as that explained above, except that the 2 mL sample solution was replaced by deionized water. The quantification was done based on a calibration curve obtained with glucose (linear range 1.2–12 µg/mL, $R^2 = 0.998$).

3.5. HPLC–QTOF–MS/MS Analysis

The extracts of small-molecule saccharides were passed through 0.22 µm filter for HPLC–QTOF–MS/MS analysis. Chromatographic analysis was carried out on an Agilent 1200 liquid chromatography system coupled with QTOF–MS/MS, which was equipped with an electrospray interface (Agilent 6530, Agilent Technologies, Santa Clara, CA, USA), in accordance with our previous studies [37]. Various mixtures of water, methanol, and acetonitrile at flow rate 1.2 mL/min, 1.0 mL/min, 0.8 mL/min, 0.5 mL/min, and 0.3 mL/min were tested as the mobile phase. The detection temperature was tested at 15, 25, 35, and 45 °C. For most of the measurements, acetonitrile and ultrapure water were used as mobile phase A and B, respectively. Solvent flow rate was 1.0 mL/min and the column (Prevail Carbohydrate ES 5 µm, 250 mm × 4.6 mm, Alltech, New Westminster, Canada) was operated at 25 °C. The solvent gradient was used as follows: 75% A and 25% B. The condition of the QTOF was as follows: scan range 100–1500 *m/z*; drying gas (N_2) flow rate, 8.0 L/min; drying gas temperature, 320 °C; sheath gas temperature, 320 °C; capillary voltage, 3.5 kV; fragmentor, 110 V; collision energy at 10, 15 and 20 eV. The operation and acquisition of data were controlled by Agilent Mass Hunter LC/MS Acquisition console, while the analysis of data was done by Qualitative Analysis B.05.00. Monosaccharides and oligosaccharides were determined according to precursor ions, the fragment ions and retention time comparing to those of the standard compounds.

3.6. Validation of the Methods for Quantitative Characterization of Monosaccharides and Oligosaccharides

Monosaccharides and oligosaccharides were quantitatively determined as described previously, but with minor modifications [38]. Briefly, standard solutions of monosaccharides and oligosaccharides at a series of appropriate concentrations were prepared for the construction of calibration curves, which were constructed by plotting the extracted ion chromatograms (EIC) peak area versus the concentration. The LOD and LOQ were calculated at approximately three-fold and ten-fold of the signal-to-noise (S/N) ratios, respectively. The measurement of the interday and intraday variabilities was used to determine the precision of this method. The standard solution was analyzed for six replicates within the same day, and additionally, on three consecutive days for evaluating intraday and interday variabilities, respectively. For the stability assessment, the sample extract was analyzed at 0, 2, 4, 8, 12, 24, and 48 h at room temperature. Recovery was determined by adding an accurately-measured amount of the standard compounds to the sample. Relative standard deviation (RSD) was used to assess the results.

3.7. PCA Statistical Analysis

The experiments were done in triplicate with three samples, and statistical analyses were performed using the statistical software R (https://www.r--project.org). A 30 × 11 matrix was constructed for the multivariate data treatment. The determined peak area value of the EIC responses of monosaccharides, oligosaccharides, polysaccharides, and saccharides derivatives were defined as variables, and were therefore placed in the columns. The thirty extracts were defined as samples and placed in rows. The data were imported by R software and treated using ggbiplot package to perform Principal Component Analysis (PCA) [39].

4. Conclusions

A sensitive and rapid analytical method was developed for the simultaneous identification and dynamic analysis of saccharide compounds during steam processing of *P. cyrtonema* using HPLC–QTOF–MS/MS. Fructose, sorbitol, glucose, galactose, sucrose, and 1-kestose were identified in the extract of the rhizomes of *P. cyrtonema*. Additionally, a large number of oligosaccharides constituted of fructose units through β-(2→1) or β-(2→6). Polysaccharides and oligosaccharides were decomposed to monosaccharides during repeated steaming processes, since the contents of glucose, galactose, and fructose increased, while those of sucrose, 1-kestose and polysaccharides decreased. Fructose was revealed to be the main determinant for the increasing sweetness after steaming of *P. cyrtonema*. Principal component analysis using saccharides as variables could substantially separate fresh and processed rhizomes of *P. cyrtonema* by repeated steaming 1–4 times, while samples with 4–9 times repeated steaming were grouped together. The samples gradually shifted along the PC1 (72.4%) axis with increases in the number of repeated times of steaming. The small-molecule saccharides, especially fructose, could be considered as markers for the steam process of rhizomes of *P. cyrtonema*.

Supplementary Materials: The Supplementary Materials are available online. Figure S1: Total ion chromatogram of fructose at different concentration (A) and nine times steam-treated samples of P. cyrtonema diluted to 20-fold and 100-fold (B) by HPLC–QTOF–MS/MS at solvent flow rate 1.0 mL/min.

Author Contributions: J.J. performed the experiments, analyzed the data and wrote the manuscript. J.L., R.Z. and W.H. contributed to the design of experiments. Y.Q., C.Z. and J.X. contributed to the chemical experiments and data analysis. H.L. and D.W. helped on the collection of the samples and data analysis. Y.Q. and S.Z. conceived the project and directed the research.

Funding: This work is supported by China Postdoctoral Science Foundation (2018M630902), State Administration of Traditional Chinese Medicine ([2015] 21), Program of Survey and Monitoring of Chinese Medicines for National Drugs ([2017] 66), China Hunan Provincial Science & Technology Department (2016SK3056 and S2017SFXYZY0019) and Key project at central government level for the ability establishment of sustainable use for valuable Chinese medicine resources (2060302).

Acknowledgments: We thank Hunan Engineering Research Center of Anti-tumor TCM Creation & Technology for technical support.

Conflicts of Interest: The authors declare no conflict of interest.

References

1. Zhao, P.; Zhao, C.; Li, X.; Gao, Q.; Huang, L.; Xiao, P.; Gao, W. The genus *Polygonatum*: A review of ethnopharmacology, phytochemistry and pharmacology. *J. Ethnopharmacol.* **2018**, *214*, 274–291. [CrossRef] [PubMed]

2. Wujisguleng, W.; Liu, Y.; Long, C. Ethnobotanical review of food uses of *Polygonatum* (Convallariaceae) in China. *Acta Soc. Bot. Pol.* **2012**, *81*, 239–244. [CrossRef]

3. Liu, N.; Dong, Z.; Zhu, X.; Xu, H.; Zhao, Z. Characterization and protective effect of *Polygonatum sibiricum* polysaccharide against cyclophosphamide-induced immunosuppression in Balb/c mice. *Int. J. Biol. Macromol.* **2018**, *107*, 796–802. [CrossRef] [PubMed]

4. Liu, F.; Liu, Y.; Meng, Y.; Yang, M.; He, K. Structure of polysaccharide from *Polygonatum cyrtonema* Hua and the antiherpetic activity of its hydrolyzed fragments. *Antivir. Res.* **2004**, *63*, 183–189. [CrossRef] [PubMed]

5. Blunder, M.; Orthaber, A.; Bauer, R.; Bucar, F.; Kunert, O. Efficient identification of flavones, flavanones and their glycosides in routine analysis via off-line combination of sensitive NMR and HPLC experiments. *Food Chem.* **2017**, *218*, 600–609. [CrossRef] [PubMed]

6. Ruiz-Matute, A.I.; Brokl, M.; Soria, A.C.; Sanz, M.L.; Martínez-Castro, I. Gas chromatographic–mass spectrometric characterisation of tri- and tetrasaccharides in honey. *Food Chem.* **2010**, *120*, 637–642. [CrossRef]

7. Liu, Z.; Lou, Z.; Ding, X.; Li, X.; Qi, Y.; Zhu, Z.; Chai, Y. Global characterization of neutral saccharides in crude and processed *Radix Rehmanniae* by hydrophilic interaction liquid chromatography tandem electrospray ionization time-of-flight mass spectrometry. *Food Chem.* **2013**, *141*, 2833–2840. [CrossRef] [PubMed]

8. Wu, X.; Jiang, W.; Lu, J.; Yu, Y.; Wu, B. Analysis of the monosaccharide composition of water-soluble polysaccharides from *Sargassum fusiforme* by high performance liquid chromatography/electrospray ionisation mass spectrometry. *Food Chem.* **2014**, *145*, 976–983. [CrossRef] [PubMed]

9. Xue, S.; Wang, L.; Chen, S.; Cheng, Y. Simultaneous analysis of saccharides between fresh and processed Radix Rehmanniae by HPLC and UHPLC-LTQ-Orbitrap-MS with multivariate statistical analysis. *Molecules* **2018**, *23*, 541. [CrossRef] [PubMed]

10. Georgelis, N.; Fencil, K.; Richael, C.M. Validation of a rapid and sensitive HPLC/MS method for measuring sucrose, fructose and glucose in plant tissues. *Food Chem.* **2018**, *262*, 191–198. [CrossRef] [PubMed]

11. Corbin, K.R.; Byrt, C.S.; Bauer, S.; Debolt, S.; Chambers, D.; Holtum, J.A.; Karem, G.; Henderson, M.; Lahnstein, J.; Beahan, C.T. Prospecting for energy-rich renewable raw materials: *Agave* leaf case study. *PLoS ONE* **2015**, *10*, e0135382. [CrossRef] [PubMed]

12. Liu, Z.; Rochfort, S. Identification and quantitative analysis of oligosaccharides in wheat flour using LC–MS. *J. Cereal Sci.* **2015**, *63*, 128–133. [CrossRef]

13. Ricochon, G.; Paris, C.; Girardin, M.; Muniglia, L. Highly sensitive, quick and simple quantification method for mono and disaccharides in aqueous media using liquid chromatography–atmospheric pressure chemical ionization–mass spectrometry (LC–APCI–MS). *J. Chromatogr. B* **2011**, *879*, 1529–1536. [CrossRef] [PubMed]

14. Lattova, E.; Perreault, H. Labelling saccharides with phenylhydrazine for electrospray and matrix-assisted laser desorption–ionization mass spectrometry. *J. Chromatogr. B* **2003**, *793*, 167–179. [CrossRef]

15. Gaucher, S.P.; Leary, J.A. Stereochemical differentiation of mannose, glucose, galactose, and talose using zinc(II) diethylenetriamine and ESI-ion trap mass spectrometry. *Anal. Chem.* **1998**, *70*, 3009–3014. [CrossRef] [PubMed]

16. Hu, X.; Zhao, H.; Shi, S.; Li, H.; Zhou, X.; Jiao, F.; Jiang, X.; Peng, D.; Chen, X. Sensitive characterization of polyphenolic antioxidants in *Polygonatum odoratum* by selective solid phase extraction and high performance liquid chromatography-diode array detector-quadrupole time-of-flight tandem mass spectrometry. *J. Pharm. Biomed. Anal.* **2015**, *112*, 15–22. [CrossRef] [PubMed]

17. Wang, L.; Xu, H.; Yuan, F.; Fan, R.; Gao, Y. Preparation and physicochemical properties of soluble dietary fiber from orange peel assisted by steam explosion and dilute acid soaking. *Food Chem.* **2015**, *185*, 90–98. [CrossRef] [PubMed]

18. Kyung-Tae, K.; Jungeun, N.; Jungeun, L.; Jung-Ok, K.; Gee-Dong, L.; Joong-Ho, K. Elimination of heavy metals (Pb, Cd) by steaming and roasting conditions of *Polygonatum odoratum* Roots. *Korean J. Food Preserv.* **2005**, *12*, 209–215.

19. Kim, S.N.; Kang, S.-J. Effects of black Ginseng (9 times-steaming Ginseng) on hypoglycemic action and changes in the composition of ginsenosides on the steaming process. *Korean J. Food Sci. Technol.* **2009**, *41*, 77–81.

20. Wangyu, K.; Jongmoon, K.; Han, S.B.; Seungki, L.; Nakdoo, K.; Manki, P.; Chongkook, K.; Jeonghill, P. Steaming of ginseng at high temperature enhances biological activity. *J. Nat. Prod.* **2000**, *63*, 1702–1704.

21. Arena, S.; Renzone, G.; D'Ambrosio, C.; Salzano, A.M.; Scaloni, A. Dairy products and the Maillard reaction: A promising future for extensive food characterization by integrated proteomics studies. *Food Chem.* **2017**, *219*, 477–489. [CrossRef] [PubMed]

22. Zhang, T.-B.; Yue, R.-Q.; Xu, J.; Ho, H.-M.; Ma, D.-L.; Leung, C.-H.; Chau, S.-L.; Zhao, Z.-Z.; Chen, H.-B.; Han, Q.-B. Comprehensive quantitative analysis of Shuang-Huang-Lian oral liquid using UHPLC-Q-TOF-MS and HPLC-ELSD. *J. Pharm. Biomed. Anal.* **2015**, *102*, 1–8. [CrossRef] [PubMed]

23. Martínez-Villaluenga, C.; Frías, J.; Vidal-Valverde, C. Raffinose family oligosaccharides and sucrose contents in 13 Spanish lupin cultivars. *Food Chem.* **2005**, *91*, 645–649. [CrossRef]

24. Apolinário, A.C.; Damasceno, B.P.G.D.L.; Beltrão, N.E.D.M.; Pessoa, A.; Converti, A.; Silva, J.A.D. Inulin-type fructans: A review on different aspects of biochemical and pharmaceutical technology. *Carbohydr. Polym.* **2014**, *101*, 368–378. [CrossRef] [PubMed]

25. Laere, A.V.; Ende, W.V.D. Inulin metabolism in dicots: Chicory as a model system. *Plant Cell Environ.* **2002**, *25*, 803–813. [CrossRef]

26. Lan, G.; Chen, H.; Chen, S.; Tian, J. Chemical composition and physicochemical properties of dietary fiber from *Polygonatum odoratum* as affected by different processing methods. *Food Res. Int.* **2012**, *49*, 406–410. [CrossRef]

27. Liu, B.; Cheng, Y.; Bian, H.J.; Bao, J.K. Molecular mechanisms of *Polygonatum cyrtonema* lectin-induced apoptosis and autophagy in cancer cells. *Autophagy* **2009**, *5*, 253–255. [CrossRef] [PubMed]

28. Li, L.; Liao, B.Y.; Thakur, K.; Zhang, J.G.; Wei, Z.J. The rheological behavior of polysaccharides sequential extracted from *Polygonatum cyrtonema* Hua. *Int. J. Biol. Macromol.* **2018**, *109*, 76–771. [CrossRef] [PubMed]

29. Peshev, D.; Ende, W.V.D. Fructans: Prebiotics and immunomodulators. *J. Funct. Foods* **2014**, *8*, 348–357. [CrossRef]

30. Ervin, E.L.; Syperda, G. Seasonal effects on soluble sugars and cytological aspects of *Polygonatum canaliculatum* rhizomes. *Bull. Torrey Bot. Club* **1971**, *98*, 162–167. [CrossRef]

31. Chumpitazi, B.P.; Lim, J.; McMeans, A.R.; Shulman, R.J.; Hamaker, B.R. Evaluation of FODMAP carbohydrates content in selected foods in the United States. *J. Pediatr.* **2018**, *199*, 252–255. [CrossRef] [PubMed]

32. Cavia, M.M.; Fernández-Muiño, M.A.; Gömez-Alonso, E.; Montes-Pérez, M.J.; Huidobro, J.F.; Sancho, M.T. Evolution of fructose and glucose in honey over one year: Influence of induced granulation. *Food Chem.* **2002**, *78*, 157–161. [CrossRef]

33. Tomoda, M.; Satoh, N.; Sugiyama, A. Isolation and characterization of fructans from *Polygonatum odoratum* var.japonicum rhizomes. *Chem. Pharm. Bull. (Tokyo)* **1973**, *21*, 1806–1810. [CrossRef]

34. Aisala, H.; Sinkkonen, J.; Kalpio, M.; Sandell, M.; This, H.; Hopia, A. In situ quantitative (1)H nuclear magnetic resonance spectroscopy discriminates between raw and steam cooked potato strips based on their metabolites. *Talanta* **2016**, *161*, 245–252. [CrossRef] [PubMed]

35. Qi, X.; Cheng, L.; Li, X.; Zhang, D.; Wu, G.; Zhang, H.; Wang, L.; Qian, H.; Wang, Y.N. Effect of cooking methods on solubility and nutrition quality of brown rice powder. *Food Chem.* **2018**. [CrossRef] [PubMed]

36. Ravber, M.; Knez, Z.; Škerget, M. Simultaneous extraction of oil- and water-soluble phase from sunflower seeds with subcritical water. *Food Chem.* **2015**, *166*, 316–323. [CrossRef] [PubMed]

37. Jin, J.; Kang, W.; Zhong, C.; Qin, Y.; Zhou, R.; Liu, H.; Xie, J.; Chen, L.; Qin, Y.; Zhang, S. The pharmacological properties of *Ophiocordyceps xuefengensis* revealed by transcriptome analysis. *J. Ethnopharmacol.* **2018**, *219*, 195–201. [CrossRef] [PubMed]

38. Mohammat, A.; Yili, A.; Aisa, H.A. Rapid quantification and quantitation of alkaloids in Xinjiang *Fritillaria* by ultra performance liquid chromatography-quadrupole time-of-flight mass spectrometry. *Molecules* **2017**, *22*, 719. [CrossRef] [PubMed]

39. Wu, T.; Taylor, C.; Nebl, T.; Ng, K.; Bennett, L.E. Effects of chemical composition and baking on in vitro digestibility of proteins in breads made from selected gluten-containing and gluten-free flours. *Food Chem.* **2017**, *233*, 514–524. [CrossRef] [PubMed]

Sample Availability: Samples of the compounds are available from the authors.

molecules

MDPI

Article

Assessing Nutritional Traits and Phytochemical Composition of Artisan Jams Produced in Comoros Islands: Using Indigenous Fruits with High Health-Impact as an Example of Biodiversity Integration and Food Security in Rural Development

Dario Donno [1,*](ORCID), Maria Gabriella Mellano [1], Saandia Hassani [2], Marta De Biaggi [1], Isidoro Riondato [1], Giovanni Gamba [1], Cristina Giacoma [3] and Gabriele Loris Beccaro [1]

[1] Dipartimento di Scienze Agrarie, Forestali e Alimentari, Università degli Studi di Torino, 10095 Grugliasco, Italy; gabriella.mellano@unito.it (M.G.M.); marta.debiaggi@unito.it (M.D.B.); isidoro.riondato@unito.it (I.R.); giovanni.gamba@unito.it (G.G.); gabriele.beccaro@unito.it (G.L.B.)
[2] École National de Cuisine et d'Application-Codcom, 167 Moroni, Comoros; saandiacodcom@gmail.com
[3] Dipartimento di Scienze della Vita e Biologia dei Sistemi, Università degli Studi di Torino, 10123 Torino, Italy; cristina.giacoma@unito.it
*   Correspondence: dario.donno@unito.it; Tel.: +39-011-670-8751

Academic Editors: Alessandra Gentili and Chiara Fanali
Received: 19 September 2018; Accepted: 18 October 2018; Published: 20 October 2018

check for updates

Abstract: In the Comoros Islands, as in other developing countries, malnutrition and food insecurity affect a very large percentage of the population. Developing fruit-based products in order to make profit, reduce poverty and improve indigenous people diet could be very important for local population of countries as Comoros Islands. The aim of the present work was to study the chemical composition of jams and jellies produced from seven fruit species harvested in Grand Comore Island. The following parameters were studied sugars and organic acids, total phenolics, total anthocyanins and high-performance liquid chromatography (HPLC) fingerprint of the main phytochemicals. Antioxidant activity was also measured. A multivariate approach (Principal Component Analysis) was performed in order to better characterize the products and to set a potential analytical tool for jam characterisation. Results showed that the analysed products are a good source of polyphenolic constituents, as caffeic and gallic acids, catechin and quercetin and volatile compounds, as limonene and γ-terpinene: these molecules may be considered as suitable markers for these fruit-derived products as characterizing the chromatographic patterns. The characterisation of these products and their nutritional and nutraceutical traits is important as valorisation of local food production for poverty reduction and rural development. Further benefits of this approach include the maintenance of local agro-biodiversity as raw material for fruit-based products and the strengthening of food security practices.

Keywords: fruit jams; food security; phenolic acids; quercetin; agro-biodiversity; HPLC fingerprint

---

1. Introduction

The Comoros Islands are situated off the coast of East Africa (290 km), at the northern entrance of the channel of Mozambique between Madagascar and the southern-east African mainland. The archipelago is composed of four main islands: Grand Comore (Ngazidja), Anjouan (Ndzuani), Mohéli (Mwali) and Mayotte (Maore). Comoros Islands are characterized by high plant biodiversity but less than one-sixth of the land remains covered by forest due to a rapid deforestation mainly caused by domestic firewood consumption [1].

Comoros, one of the world's poorest countries, has an economy based on subsistence agriculture and fishing. In the Comoros Union, as in all the developing countries, malnutrition and food insecurity are the main challenges: in particular energy, malnutrition in children and micronutrient deficiencies (e.g., vitamin deficiency and nutritional anaemias), are important public health issues influencing productivity, maternal/infant health and intellectual development. The improvement of productivity and post-harvest techniques is pivotal to increase Comorian smallholder farmers' income in order to help fighting poverty and ensure a medium-high nutrition [2].

On the other hand, an abundance of tropical fruit species, often underexploited, grown in semi-natural conditions occurring in Comoros. Tropical fruits attract special attention as they usually have stronger antioxidant properties than common fruits, thanks to bioactive compounds as polyphenols (anthocyanins, flavonoids, phenolic acids and tannins), carotenoids, organic acids and vitamins (B2, B6, C, E, P, PP), as reported in several studies [3,4]. Health-promoting components occurring in these fruits, in particular polyphenols, also show anti-microbial, anti-carcinogenic and anti-viral effects [5]. Therefore, tropical fruits show a high considerable horticultural and nutritional importance for the diet of the rural population [6] in specific critical periods of the year providing sustenance to millions of people. Unfortunately, a significant amount (30–40%) of total fruit production in developing countries is wasted due to poor postharvest handling and inadequate marketing and/or storage facility [7]. Moreover, most of population in these countries lives in rural areas and depends on subsistence agriculture for their livelihoods. In these settlements, agriculture is dependent on rainfall. Inadequacy of food during the dry season and in early summer before the harvest period exposes people to inadequate intake of both macro- and micro-nutrients [6].

Fruit processing industries making preserves, jams, sauces and jellies play an important role in reducing these losses of fruit production, thriving largely as a domestic-product market [8]. In particular, some of the most popular postharvest-stable products made from fruit are jams and jellies, both at the household and commercial levels. Jams and jellies are defined as a mixture, brought to a suitable gelled consistency, of sugars, pulp of one or more fruits and water [9]. To manufacture these products, fruits and sugars are combined in a similar ratio, followed by cooking, to produce a tasty product of sufficiently high sugar content with satisfactory keeping qualities [10]. Jam processing is an important strategy to preserve perishable fruits and improve food security in developing countries [11].

Considering new potential agro-industrial and commercial activities related to jam and jelly production, including value-added product labelling, it is of essential importance to guarantee both high quality and compliance with the product specification. For these purposes, increasingly sophisticated analytical methodologies, based on chemical markers, have been developed. Chemical fingerprint methods include the analysis of organic acids, pigments (as carotenoids and anthocyanins), sugars, phenolics and other bioactive compounds [12]. In particular, secondary plant metabolites are very suitable as chemotaxonomic markers [13]. Quantitative differences may occur depending on fruit genotype (e.g., species and cultivar), maturity stages and environmental growth [14], storage conditions [15] and on presence of the skin in fruit-based products [16]. In previous research, they have been successfully used for the determination of the adulteration of some fruit jams and jellies as reported by Dragovic-Uzelac et al. [17].

The aim of the present work was to study chemical compositions of jams and jellies from seven tropical fruit species, harvested near Moroni in Comoros Islands. The following parameters were studied: sugar and organic acid contents, total phenolics (TPC), total anthocyanins (TAC) and fingerprint of the main phytochemicals with demonstrated health-promoting activity by high-performance liquid chromatography (HPLC). Furthermore, antioxidant activity was measured in these products. A multivariate approach (Principal Component Analysis—PCA) was performed in order to better characterize fruit-based products and set a potential analytical tool for analysis and characterisation of local jams and jellies. This study could contribute to the commercial valorisation of these fruit-derived products in rural communities in Comoros Islands thereby reducing post-harvest losses, promoting food security, enhancing small farmers' income and contributing to a sustainable rural development.

2. Results and Discussion

2.1. Nutraceutical Properties

Food processing plays an important role in the bioactive compound degradation, because several transformations of phenolics occur to produce yellowish or brownish pigments [18]: the final product outward appearance is a crucial in determination of consumers' choices and anthocyanins are the main food colorants responsible for intense colour (associated with raw material freshness and good quality) [19]. Moreover, physical and biological factors as temperature increase and enzymatic activity may be very important in degradation of polyphenolic compounds [10].

In this research, the used methods allowed a rapid measurement of TAC and TPC: the Folin-Ciocalteu method suffers many interferences but it can be a complementary technique applied to confirm and support chromatography results, as reported in this study. The TPC ranged from 10.98 ± 2.18 mg$_{GAE}$/100g$_{Pr}$ (mango jam, CM2) to 625.34 ± 67.86 mg$_{GAE}$/100g$_{Pr}$ (red guava jam, CM8), while TAC ranged from 0.59 ± 0.25 mg$_{C3G}$/100g$_{Pr}$ (mango jam, CM2) to 9.56 ± 0.46 mg$_{C3G}$/100g$_{Pr}$ (orange jam, CM7): values obtained from the analysed extracts (Table 1) were higher than values reported by Poiana et al. [20] and Rababah et al. [21]; the differences in phenolic and anthocyanin content could be due to the effects of several internal and external factors on plant material (genetic variability, climatic conditions and environmental factors) [22]. In particular, Comoros pedoclimatic conditions such as volcanic soil, high temperatures, well distributed rains, influence polyphenolic content in fresh fruits and related products.

Bioactive compound redox properties allow them to act as reducing agents, hydrogen donators and singlet oxygen quenchers [23]. In this research, the Ferric Reducing Antioxidant Power (FRAP) assay was used to evaluate antioxidant capacity of fruit jams and jellies, studying the ability of antioxidants to reduce Fe^{3+} ions to Fe^{2+} ions. Jams and jelly FRAP value ranged from 4.71 ± 2.07 mmol Fe^{2+} kg$_{Pr}{}^{-1}$ (mango jam, CM1) to 25.50 ± 0.28 mmol Fe^{2+} kg$_{Pr}{}^{-1}$ (red guava jam, CM8) as shown in Table 1, in accordance with other studies [24,25].

Table 1. Nutraceutical traits of the analysed fruit-derived products.

Sample	ID	Total Polyphenolic Content (mg$_{GAE}$/100 g$_{Pr}$)			Antioxidant Activity (mmol Fe^{2+}/kg$_{Pr}$)			Total Anthocyanin Content (mg$_{C3G}$/100 g$_{Pr}$)		
		Mean Value	SD	Tukey Test	Mean Value	SD	Tukey Test	Mean Value	SD	Tukey Test
mango jam	CM1	11.11	1.34	a	4.71	2.07	a	1.48	0.26	ab
mango jam	CM2	10.98	2.18	a	7.86	0.36	b	0.59	0.25	a
tamarind jam	CM3	484.95	81.89	d	23.97	0.28	d	0.87	0.43	ab
banana flower	CM4	13.06	1.85	a	19.23	1.29	c	4.49	1.62	c
guava jelly	CM5	437.13	2.44	d	25.40	0.35	d	8.35	1.13	d
lychee jam	CM6	266.96	7.95	c	23.01	0.07	d	3.15	0.66	bc
orange jam	CM7	124.50	27.49	b	16.62	0.89	c	9.56	0.46	d
red guava jam	CM8	625.34	67.86	e	25.50	0.28	d	4.99	0.67	c

Mean value and standard deviation (SD) of each sample is given ($N = 3$). Different letters for each class indicate the significant differences at $p < 0.05$. GAE = gallic acid equivalent; C3G = cyanidin 3-O-glucoside. Pr = product.

Antioxidant activity of fruit-derived products was determined by different bioactive molecules (e.g., polyphenols, as anthocyanins and vitamin C) [26]; for example, antioxidant activity of cyanidin is about 4 times higher compared to ascorbic acid [27]. In this study, the TPC/TAC/antioxidant activity correlation was positive: results showed a significant Pearson correlation coefficients (R = 0.6636 for TPC/antioxidant activity and R = 0.4347 for TAC/antioxidant activity).

2.2. Phytochemical Composition

Antioxidant (polyphenols and vitamin C) and anti-inflammatory (terpenes) compounds are the main biologically active substances in fresh fruits and derived products: synergistic or additive health-promoting effects of different phytochemicals (phytocomplex) contribute to biological activity better than a single molecule or a group of few compounds [28]. In this study 22 biologically active compounds together with 9 nutritional substances were selected as markers for HPLC fingerprinting due to their importance in humans [29]. In Supplementary material (Suppl. Figure S1) HPLC chromatograms of orange jams are reported as an example of analysed fruit-based products.

An important question in HPLC analysis is whether the peak comprises one or more components. In quality control and research analysis, impurities hidden behind the peak of interest can falsify results and an undetected component might lead to a loss of essential information. In this research, a peak purity check was assessed in order to control if peaks were pure or contained impurities comparing spectra recorded during the elution of each peak. No-coeluting peaks were detected. Moreover, HPLC-DAD does not allow a definitive identification of phytochemicals. Indeed, liquid chromatography (LC) coupled to mass (MS) or mass/mass spectrometry (MS^2) is one of the most effective technique for analysis on complex plant extract/fresh fruit/derived product providing a rapid and accurate identification of phytochemicals, as phenolics. For this reason, future developments are necessary but in this preliminary study HPLC-DAD was a simply, rapid and effective approach to describe considered samples in relation to the research aim. Additional markers with demonstrated biological activity could be also taken into consideration for a better identification of the chromatographic pattern of fruit-derived products, together with a mass spectrometry detection of unknown peaks.

The chemical fingerprint of analysed jams and jellies is reported in Tables 2–4 (phenolics, other health-promoting components and nutritional substances, respectively). The total bioactive compound content (TBCC) was calculated as the sum of the main molecules (polyphenols, monoterpenes and vitamin C) selected for their biological proved effects on humans and detected in the extracts: TBCC value ranged from 94.25 ± 4.13 mg/100 g_{Pr} (banana flower, CM4) to 357.50 ± 2.10 mg/100 g_{Pr} (orange jam, CM7). This is only a preliminary study on tropical fruit-based jams: in further fingerprint studies, other markers with health-promoting capacity or positive nutritional value should be added for a complete chromatographic pattern characterisation coupling a mass spectrometry (MS) detection of unknown peaks to a UV-visible determination.

Table 2. Phytochemical fingerprint of the polyphenolic compounds in analysed jams and jellies.

		Mango Jam, CM1		Mango Jam, CM2		Tamarind Jam, CM3		Banana Flower, CM4		Guava Jelly, CM5		Lychee Jam, CM6		Orange Jam, CM7		Red Guava Jam, CM8	
		Mean Value	SD	Mean Value	SD	Mean Value	SD	Mean Value	SD	Mean Value	SD	Mean Value	SD	Mean Value	SD	Mean Value	SD
Cinnamic acids	caffeic acid	0.678	0.170	0.761	0.101	0.498	0.134	0.749	0.154	0.687	0.034	0.764	0.116	0.571	0.110	0.744	0.204
	chlorogenic acid	n.d.	/	n.d.	/	11.454	0.731	n.d.	/	13.766	0.546	n.d.	/	12.971	0.185	n.d.	/
	coumaric acid	n.d.	/	n.d.	/	n.d.	/	0.556	0.509	n.d.	/	n.d.	/	n.d.	/	5.583	0.797
	ferulic acid	n.d.	/	n.d.	/	1.377	0.217	n.d.	/	n.d.	/	n.d.	/	1.001	0.110	n.d.	/
Flavonols	hyperoside	n.d.	/	n.d.	/	n.d.	/	0.356	0.033	n.d.	/	1.027	0.116	n.d.	/	n.d.	/
	isoquercitrin	n.d.	/	n.d.	/	0.491	0.006	n.d.	/	0.525	0.034	n.d.	/	1.140	0.110	0.541	0.155
	quercetin	n.d.	/	n.d.	/	11.441	0.963	n.d.	/	7.562	0.034	n.d.	/	15.694	1.107	7.621	0.204
	quercitrin	0.161	0.030	0.007	0.003	1.640	0.257	0.173	0.019	1.001	0.034	1.334	0.116	1.152	0.110	1.477	0.256
	rutin	0.936	0.036	0.858	0.034	n.d.	/	n.d.	/	n.d.	/	n.d.	/	n.d.	/	n.d.	/
Benzoic acids	ellagic acid	1.167	0.264	1.228	0.154	23.302	1.962	1.604	0.365	4.444	0.316	3.799	0.116	57.105	1.637	4.561	0.607
	gallic acid	0.668	0.105	1.535	0.232	8.571	0.314	0.611	0.070	42.023	0.582	30.650	0.886	87.724	1.812	40.955	1.727
Catechins	catechin	0.872	0.023	1.749	0.175	40.644	1.558	1.216	0.162	0.327	0.034	n.d.	/	0.436	0.110	4.287	0.204
	epicatechin	3.683	0.328	1.747	0.400	n.d.	/	1.348	0.184	2.523	0.034	n.d.	/	2.015	0.110	4.031	0.157
Tannins	castalagin	n.d.	/	n.d.	/	n.d.	/	2.595	0.425	4.384	0.034	n.d.	/	15.210	2.414	9.773	0.295
	vescalagin	3.639	0.174	0.359	0.008	n.d.	/	1.554	0.371	4.339	0.205	n.d.	/	3.031	0.110	10.012	0.204

Mean value and standard deviation (SD) of each sample is given (N = 3). Results are expressed as mg/100 g$_{Pr}$. Pr = product. n.d. = not detected.

Table 3. Phytochemical fingerprint of monoterpenes and vitamin C in analysed jams and jellies.

		Mango Jam, CM1		Mango Jam, CM2		Tamarind Jam, CM3		Banana Flower, CM4		Guava Jelly, CM5		Lychee Jam, CM6		Orange Jam, CM7		Red Guava Jam, CM8	
		Mean Value	SD	Mean Value	SD	Mean Value	SD	Mean Value	SD	Mean Value	SD	Mean Value	SD	Mean Value	SD	Mean Value	SD
Monoterpenes	limonene	103.529	3.116	31.201	0.588	n.d.	/	n.d.	/	102.750	0.582	33.602	0.687	18.818	0.472	51.319	3.204
	phellandrene	27.843	2.380	7.809	0.540	7.497	0.219	n.d.	/	4.671	0.034	4.668	0.116	n.d.	/	4.946	0.204
	sabinene	8.140	0.045	12.187	1.146	11.273	1.180	n.d.	/	12.847	1.020	12.811	0.687	n.d.	/	12.332	0.797
	γ-terpinene	24.552	2.464	48.067	2.301	n.d.	/	17.949	2.097	39.002	0.034	19.069	0.492	37.084	0.505	43.005	0.155
	terpinolene	n.d.	/	n.d.	/	26.944	2.718	8.228	0.213	7.731	0.034	7.473	0.886	7.926	0.110	7.715	0.564
Vitamin C	ascorbic acid	15.058	0.362	10.178	0.072	20.397	0.117	49.699	0.356	18.074	0.138	18.978	0.057	54.062	0.074	19.104	0.487
	dehydroascorbic acid	0.813	0.130	1.815	0.163	3.767	0.283	3.123	0.213	1.138	0.192	1.196	0.225	32.003	0.401	1.279	0.234

Mean value and standard deviation (SD) of each sample is given (N = 3). Results are expressed as mg/100g$_{Pr}$. Pr = product. n.d. = not detected.

Table 4. Phytochemical fingerprint of nutritional substances in analysed jams and jellies.

		Mango Jam, CM1		Mango Jam, CM2		Tamarind Jam, CM3		Banana Flower, CM4		Guava Jelly, CM5		Lychee Jam, CM6		Orange Jam, CM7		Red Guava Jam, CM8	
		Mean Value	SD	Mean Value	SD	Mean Value	SD	Mean Value	SD	Mean Value	SD	Mean Value	SD	Mean Value	SD	Mean Value	SD
Organic acids	citric acid	16.217	4.058	29.696	0.469	522.583	2.566	694.580	3.627	326.950	3.030	268.422	2.629	287.911	4.614	200.408	2.204
	malic acid	357.878	3.679	302.528	2.531	n.d.	/	21.926	1.696	n.d.	/	n.d.	/	n.d.	/	n.d.	/
	oxalic acid	4.741	0.198	4.573	0.416	21.444	0.530	8.441	0.102	8.194	0.117	n.d.	/	6.629	0.110	23.717	0.797
	quinic acid	n.d.	/	n.d.	/	n.d.	/	n.d.	/	n.d.	/	138.233	1.263	391.128	2.192	n.d.	/
	succinic acid	n.d.	/	n.d.	/	n.d.	/	n.d.	/	n.d.	/	n.d.	/	406.648	4.401	n.d.	/
	tartaric acid	n.d.	/	n.d.	/	n.d.	/	n.d.	/	76.724	1.513	105.358	4.157	193.497	3.661	134.637	1.729
Sugars	fructose	11.701	0.038	11.316	0.100	15.353	0.140	12.275	0.287	21.341	0.034	14.814	0.116	13.620	0.110	14.428	0.204
	glucose	14.505	0.321	13.620	0.401	27.637	2.326	12.342	0.111	12.615	0.034	11.044	0.116	13.090	0.110	13.413	0.155
	sucrose	13.444	0.339	14.831	0.680	13.509	0.231	11.131	1.084	11.801	0.867	14.573	0.492	14.935	0.110	14.904	0.157

Mean value and standard deviation (SD) of each sample is given ($N = 3$). Results are expressed as mg/100g$_{Pr}$. Pr = product. n.d. = not detected.

In Figure 1 identified health-promoting substances were grouped into bioactive classes for the evaluation of the single contribution of each class to total phytocomplex/TBCC (mean values were considered). The most important class in tamarind (CM3) and orange (CM6) jams was polyphenols (58.93% and 58.07%, respectively), expressed as the sum of anthocyanins, phenolic acids, flavanols, catechins and tannins, while monoterpenes were the first class in mango jam (83.79%), guava jelly (60.48%) and lychee jam (56.04%). Banana flower showed a high percentage of vitamin C (56.05%), while red guava jam presented similar content of polyphenols and monoterpenes (40.37% and 50.93%, respectively) with a positive percentage of vitamin C (8.70%).

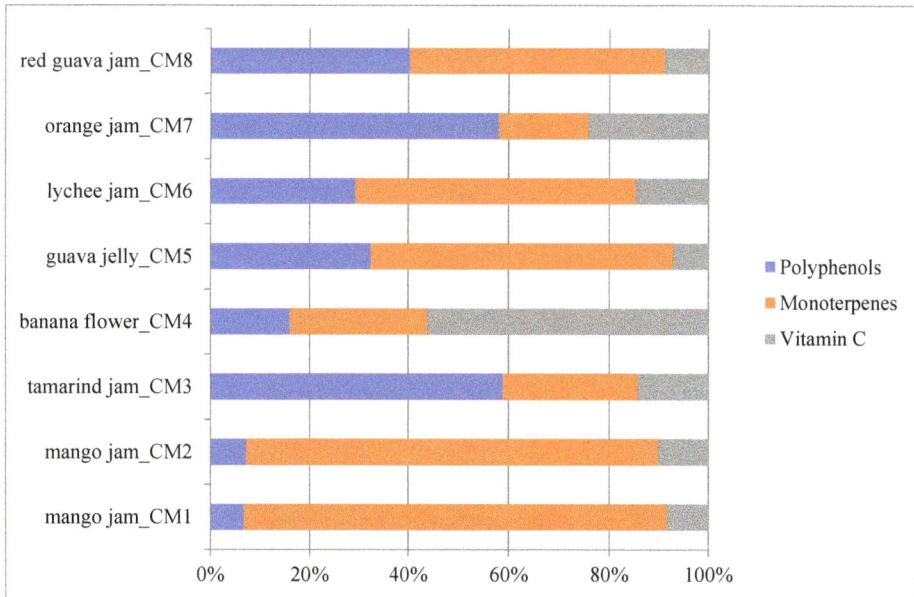

Figure 1. Phytocomplex representation of analysed fruit-derived products. The mean value of each analysed sample is given (*N* = 3).

Results showed that analysed fruit jams and jellies are a very good source of polyphenolic compounds. Figure 2 reports the single contribution of each polyphenolic class to total polyphenols detected by HPLC. Catechins were the most important class in mango and tamarind jams (36.93% and 40.53%, respectively), while phenolic acids (cinnamic acids plus benzoic acids) were the most important classes in guava jelly (67.75%) and in lychee (86.46%), orange (76.77%) and red guava (54.82%) jams. Banana flower showed a high percentage of tannins and anthocyanins (27.21% and 29.42%, respectively), followed by catechins (16.81%) and benzoic acids (14.53%).

Analysed jams and jellies presented interesting quali-quantitative polyphenolic profiles if compared to commercial products derived from common temperate fruits; in particular, they showed higher TPC values and relative antioxidant activity than strawberry (101.40 $mg_{GAE}/100\ g_{Pr}$) [25], apricot (51.49 $mg_{GAE}/100\ g_{Pr}$) [21], berry fruits (336.67 $mg_{GAE}/100\ g_{Pr}$) [24], peach (18.85 $mg_{GAE}/100\ g_{Pr}$) and apple (20.07 $mg_{GAE}/100\ g_{Pr}$) [16]. Moreover, jams from tropical fruits showed higher values of specific phenolic markers as coumaric, caffeic and ferulic acids (lychee, banana and guava), catechin (tamarind) and rutin (mango) than commercial products derived from berry fruits (the content values in berry fruits were 0.39 mg/100 g_{Pr} for coumaric acid, 1.38 mg/100 g_{Pr} for caffeic acid, 0.13 mg/100 g_{Pr} for ferulic acid, 3.93 mg/100 g_{Pr} for catechin and 0.26 mg/100 g_{Pr} for rutin) [19]. These results may contribute to better valorise the products derived from local biodiversity compared to imported commercial ones and improve food industry in the Comoros Islands with the potential exportation of these productions.

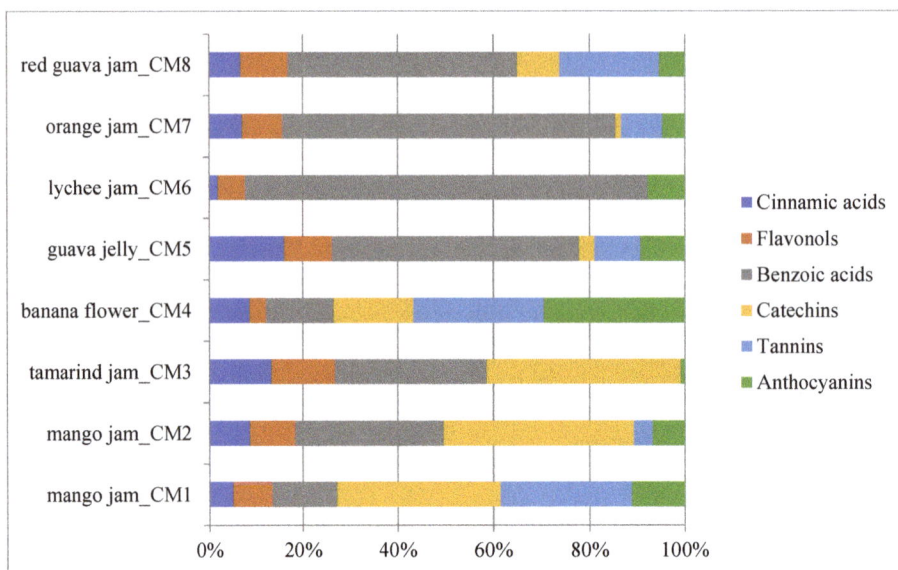

Figure 2. Polyphenolic phytocomplex representation of analysed fruit-derived products. The mean value of each analysed sample is given ($N = 3$).

In some studies, phenolic compound characterisation was mainly used for chemotaxonomic purposes; moreover, some researchers reported that fruit processing during jam/jelly production did not change much the qualitative polyphenolic profile [10,30]. In this research, each product showed a specific phenolic composition characterised by the presence of one or more specific markers. Chlorogenic acid proved to be characteristic of tamarind jam (11.45 $\pm$ 0.73 mg/100 g$_{Pr}$), guava jelly (13.77 $\pm$ 0.55 mg/100 g$_{Pr}$) and orange jam (12.97 $\pm$ 0.19 mg/100 g$_{Pr}$) and it could be used as a marker to prove the addition of these fruits to other jams. Moreover, chlorogenic acid is considered a preferential substrate for the catecholase activity of polyphenol oxidase and it may be important during fruit processing [31]. Caffeic acid was detected in all the fruit-derived products in a close range between 0.49 $\pm$ 0.13 mg/100 g$_{Pr}$ (tamarind jam, CM3) and 0.76 $\pm$ 0.12 mg/100 g$_{Pr}$ (lychee jam, CM6) similar to other studies [32,33]. Coumaric acid was the most important cinnamic acid in red guava jam-CM8 (5.58 $\pm$ 0.80 mg/100 g$_{Pr}$), while ferulic acid was present in tamarind jam-CM3 (1.38 $\pm$ 0.22 mg/100 g$_{Pr}$) and orange jam-CM7 (1.00 $\pm$ 0.11 mg/100 g$_{Pr}$) as reported by Jimohand Onabanjo [34] and Marquina et al. [35]. Caffeic, ferulic and coumaric acids could be involved in the oxidation processes and colour development during technological processing [36]. Flavonoids (flavanols and catechins) were also demonstrated to be important as markers for orange and tamarind jam quality control [37], because flavonoids are not affected by the manufacturing process. Flavonols quench active oxygen species [38] and inhibit in vitro oxidation of low-density lipoproteins reducing thrombotic tendency. In this research, the most effective flavonol selected as marker was quercetin in guava jelly (7.56 $\pm$ 0.03 mg/100 g$_{Pr}$) and jams of tamarind (11.44 $\pm$ 0.96 mg/100 g$_{Pr}$), orange (15.69 $\pm$ 1.11 mg/100 g$_{Pr}$) and red guava (7.62 $\pm$ 0.20 mg/100 g$_{Pr}$), while rutin proved to be characteristic of mango jams (0.94 $\pm$ 0.04 mg/100 g$_{Pr}$ for CM1 and 0.86 $\pm$ 0.03 mg/100 g$_{Pr}$ for CM2) according to other studies [39,40]. The identification of catechin (maximum value of 40.64 $\pm$ 1.56 mg/100 g$_{Pr}$ in tamarind jam) and epicatechin (maximum value of 4.03 $\pm$ 0.16 mg/100 g$_{Pr}$ in red guava jam) could be useful: indeed, they are involved in the lipid peroxidation inhibition and human cancer cell line proliferation as other similar compounds [41]. The presence of tannins in adequate amounts in orange and red guava jams (18.24 $\pm$ 2.52 mg/100 g$_{Pr}$ and 19.79 $\pm$ 0.50 mg/100 g$_{Pr}$, respectively) are positive as they are free radical quenchers [42]. High levels of ellagic acid in tamarind jam (23.30 ± 1.96 mg/100 g$_{Pr}$) and orange jam (57.11 ± 1.64 mg/100 g$_{Pr}$) as well as high content of gallic acid in guava jelly (42.02 ± 0.58 mg/100 g$_{Pr}$) and orange jam (87.72 ± 1.81 mg/100 g$_{Pr}$) were also detected: these molecules are endowed with numerous biological properties, as anticancer, anti-inflammatory and anti-HIV replication activities [43]. These preliminary results on phenolic composition demonstrate the need of identifying more bioactive substances for control of the authenticity of fruit-based products.

Anthocyanins have frequently been also considered for the cited purposes, because their specific patterns may allow the classification of fruit species and relative derived products and the characterization of their nutraceutical and nutritional traits (e.g., detection of admixtures of fruits with a more stable colour during jam processing). As previously discussed, anthocyanins are of prominent importance in guava jelly (8.35 ± 1.13 mg$_{C3G}$/100g$_{Pr}$), orange jam (9.56 ± 0.46 mg$_{C3G}$/100g$_{Pr}$) and red guava jam (4.99 ± 0.67 mg$_{C3G}$/100g$_{Pr}$) because i) they are important for quality traits, due to their levels directly related to the product colour and ii) they have been proved to show several health-promoting activities and a high potential phytochemical value [44]. As opposed to other polyphenolic compounds, composition in anthocyanins may be subject to modification during processing and storage steps as reported by Garzon and Wrolstad [45]: in particular, enzymation during jam processing may change anthocyanin patterns [46]. During jam and jelly storage new pyranoanthocyanins may also be formed by direct reaction of anthocyanins with cinnamic acids, as shown by Schwarz et al. [47]. For this reason, anthocyanins may be only used as quantitative markers in the quality control of jams and similar products.

Monoterpenes represent an important fraction of the TBCC in analysed fruit-based products, in particular limonene in mango jam (103.53 ± 3.12 mg/100 g$_{Pr}$) and guava jelly (102.75 ± 0.58 mg/100 g$_{Pr}$) and γ-terpinene in jams of mango (48.07 ± 2.30 mg/100 g$_{Pr}$) and red guava (43.01 ± 0.16 mg/100 g$_{Pr}$): the plant terpenoids are a large class of phytochemicals used for their aromatic qualities and antioxidant and anti-inflammatory activity [48]. Monoterpenes are non-nutritive dietary substances with antibacterial and antitumor activity found in the essential oils of several plants [49] and several studies reported their chemopreventive activity against several cancers [50].

Vitamin C value was obtained as the sum of ascorbic and dehydroascorbic acids due to their biological activity in humans [51]. In this research banana flower and orange jams showed a high vitamin C content (52.82 ± 0.57 mg/100 g$_{Pr}$ and 86.07 ± 0.48 mg/100 g$_{Pr}$, respectively). Other analysed products showed good values of vitamin C (about 10–25 mg/100 g$_{Pr}$) according to previous similar studies [52,53].

Organic acids in fruits are little influenced by changes during processing and storage and show a good stability if compared to pigments and flavour compounds. Accordingly, their identification may be suitable for the estimation of fruit amount and for the fruit quality control [54]. However, since organic acids (e.g., citric acid) are indispensable technological components of most derived-products, they are not applicable as quality control markers in fruit jams and jellies. Furthermore, organic acid composition is influenced by genetic factors (e.g., cultivar) and degree of ripeness, limiting their applicability as a quantitative marker in fruit-derived products [55]. In any case, they are important antioxidants with multi-purpose uses in pharmacology as reported by Eyduran et al. [56]. They were also utilised as food acidifiers by food companies [57]. In this research orange jam showed high variability in organic acid composition: in particular, succinic acid (406.65 ± 4.40 mg/100 g$_{Pr}$) and quinic acid (391.13 ± 2.19 mg/100 g$_{Pr}$) were found to be suitable markers for jam characterization because they were not detected in other analysed fruit-based products and represent specific molecules of orange jam chromatographic pattern as shown by Cejudo-Bastante et al. [58] and Flores et al. [59]. Similarly, malic acid could be a specific marker for mango jams and tartaric acid for guava jelly and jams of lychee and red guava. Figure 3 shows total organic acid content in analysed fruit-based products.

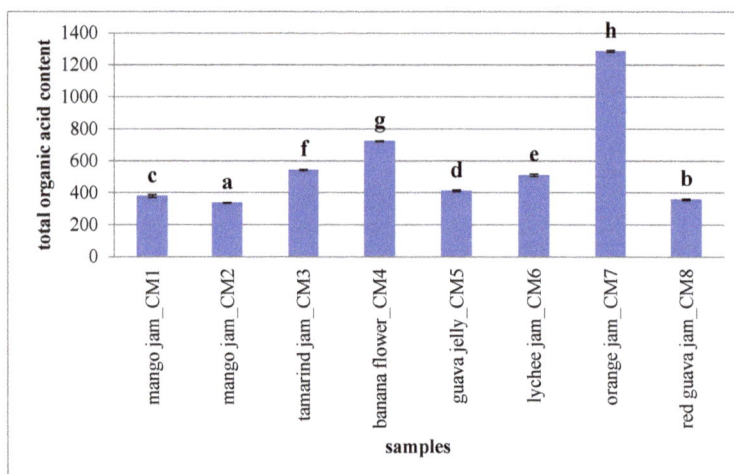

Figure 3. Total organic acid content of analysed fruit-derived products. The mean value of each analysed sample is given ($N = 3$). Different letters for each class indicate the significant differences at $p < 0.05$. Results are expressed as mg/100 g of product.

Apart from organic acids, the sugar pattern was also utilised for fruit species differentiation, while their use as quantifiers of fruit content is very limited. Identical sugar profiles, used often for the detection of illegal adulterations as the admixture of sugar solutions or fruit juices, were observed for fruits from different countries as well as for different genotype [55]. In this research sugar pattern (Figure 4) was studied in order to evaluate the nutritional potential of analysed fruit-derived products. Tamarind jam showed the highest sugar content (56.50 ± 2.70 g/100 g_{Pr}), expressed as sum of fructose (15.35 ± 0.14 g/100 g_{Pr}), glucose (27.64 ± 2.33 g/100 g_{Pr}) and sucrose (13.51 ± 0.23 g/100 g_{Pr}). Other analysed products presented a total sugar content of about 40 g/100 g_{Pr} as reported in other studies [60,61].

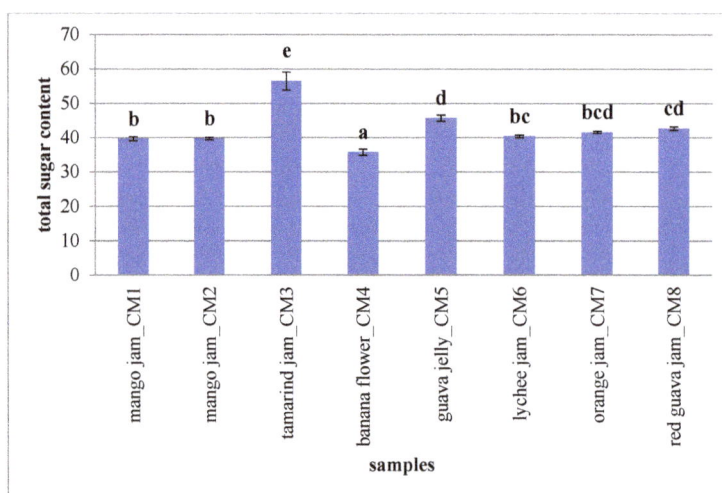

Figure 4. Total sugar content of analysed fruit-derived products. The mean value of each analysed sample is given ($N = 3$). Different letters for each class indicate the significant differences at $p < 0.05$. Results are expressed as g/100 g of product.

2.3. Multivariate Analysis

Single markers of the same phytochemical group were combined in bioactive classes for multivariate data handling. For the visualisation of potential differences in the products and easily characterize analysed jams and jellies, PCA was performed on all the data and it reduced the initial variables (TPC, antioxidant activity, TAC, content of 9 chemical classes) to three principal components (86.74% of total variance), placing the eight fruit-based products in the PCA score plot (Figure 5) in relation to phytochemical composition, nutritional properties and nutraceutical traits. PC1 and PC2 well represent the system information (70.20% of total variance); the PCA results showed five groups, highlighted in Figure 5 with circles, without statistical meaning, according to the phytochemical results; the groups were named α (guava jelly—CM5, red guava jam—CM8), β (tamarind jam—CM3), χ (lychee jam—CM6), δ (orange jam—CM7) and ε (mango jam, genotype 1—CM1, mango jam, genotype 2—CM2, banana flower—CM4).

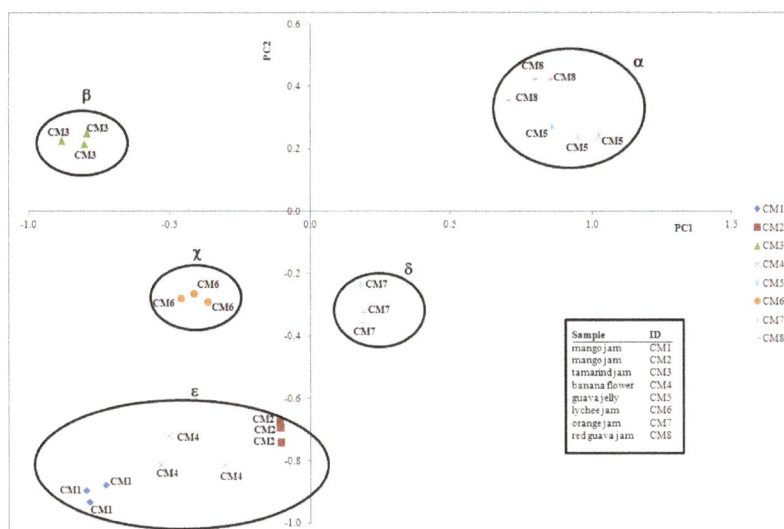

Figure 5. PCA score plot of fruit-derived products (eight samples and three replications per sample). The ellipses around each object group only indicate the position of a category in the plot without statistical meaning, based on the phytochemical results.

PCA loading plot showed a correlation between most of the polyphenolic classes (phenolic acids, tannins, anthocyanins and flavanols), vitamin C, organic acids and PC1 (43.15% of total variance) and a correlation between TPC, catechins, antioxidant activity, sugars and PC2 (27.05% of total variance). Monoterpenes showed an intermediate position between PC1 and PC2 (Figure 6). Polyphenolic compounds (in particular, benzoic acids, tannins and anthocyanins) and vitamin C presented a high discriminating power among different samples, as reported in other studies [37,55], as well as catechins.

These results showed that PCA classification obtained from main bioactive groups, nutritional traits and nutraceutical properties characterized the analysed products according to the different chemical pattern and provided information on the phytochemical markers with the most influence on the phytocomplex. A chemometric method was applied coupled to an HPLC fingerprint technique for a better recognition of analysed products as reported by Tzouros and Arvanitoyannis [62]. Different marker compounds were found to be the most discriminating variables, which could be applied to accurate composition control of fruit jams and jellies; in particular, the phytocomplex graphical view showed that genotypes included in the α and δ PCA group (guava jelly/red guava jam and orange jam, respectively) present the highest amount of antioxidant compound classes (polyphenols and vitamin C) together with the highest amount of volatile molecules (organic acids and monoterpenes), the most responsible compounds for the product aroma. The combination of chromatographic fingerprinting and chemometrics could be an effective potential tool for quality control of fruit-based products avoiding potential voluntary or involuntary adulterations and contaminations [63]. The analysis of other samples from different origin and nature are required to ensure that the proposed methodology is applicable to the authentication of jellies and jams from Comoros Islands or to the detection of their adulteration. In the present research, this methodology only showed that jams from Comoros Islands may be differentiated among them in relation to their fruit phytochemical composition. Therefore, further additional experiments are required for demonstrating that this methodology can be used for adulteration detection and/or authentication. These hyphenated techniques could also be positively used for the evaluation and differentiation of several products in local markets setting a potential tool to obtain label certifications for the valorisation of local productions.

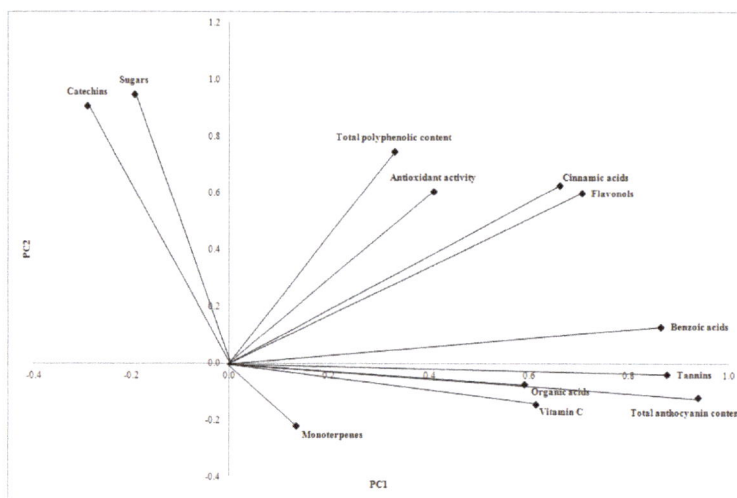

Figure 6. PCA loading plot of fruit-based products showing correlation among nutritional traits, nutraceutical properties and bioactive compound classes and PCs.

3. Materials and Methods

A detailed description of the used extraction protocols and analytical methods is reported in the Supplementary material.

3.1. Plant. Material

The investigated material consisted of fruit-derived products prepared from 7 species (*Mangifera indica* L.—two genotypes, *Tamarindus indica* L., *Musa* × *paradisiaca* L., *Psidium cattleyanum* Sabine, *Litchi chinensis* Sonn., *Citrus sinensis* (L.) Osbeck and *Psidium guajava* L.). Fruits were picked at commercial maturity stage during the 2017 from commercial orchards managed by the *Coopérative et Mutuelle des Comores pour le Développement* (Codcom) in Moroni, Comoros Islands. CODCOM is a non-governmental organization involved in poverty reduction in Comoros and in the countries of the Indian Ocean. Physiologically mature fruits were randomly selected from three plants for each biological replication (*n* = 3). Fruits were manually harvested from the plants based on selected qualitative parameters (e.g., colour, firmness and total soluble solids), considering also literature and experience of the local researchers and technicians. Fruits were then sorted, washed and stored in a 4 °C cold room for less than 5 days until jam and jelly preparation. Inedible parts were rejected.

3.2. Preparation of Fruit-Derived Products

A standard commercial procedure to manufacture jams and jellies was followed.

Fruit (1 kg) was reduced to fine particles with a commercial food processor for approximately 30 s. Jam formulation was 50% fruit, 48% sugar and 2% pectin mix (composed of dextrose, pectin and fumaric acid). pH was checked and adjusted if needed by addition of 50% citric acid solution (10–15 mL) to a target pH of 3.0. An essence of vanilla or cinnamon was added to improve the sensory properties of the final products. Sugar was added and the mixture was boiled for 30–40 min to a final concentration of 65°Brix (approximately 105 °C final boiling point). The jam was packed at 90 °C in 350 mL glass jars, immediately sealed with plastisol lined metal lids and inverted for 5 min to sterilize the lids. The jars were then returned to normal position for air-cooling.

For the preparation of 1 kg fruit jelly, 500 g filtered fruit juice, 550 g sugar, 5 g citric acid was used. Five g of pectin were added. The remaining sugar was mixed with fruit juice and heated until the total soluble solid (TSS) become near to 55°Brix. Then sugar mixed pectin was added and continued the heating until TSS becomes near to 58°Brix. The citric acid was added and continued the heating. When TSS of the jelly becomes 67°Brix, jelly was poured in a sterilized glass bottle and parafinning the cap.

3.3. Spectrophotometric Analysis

Total polyphenolic content (TPC) was determined according to the Folin-Ciocalteu colorimetric method [64,65]; results were expressed as mg of gallic acid equivalents (GAE) per 100 g of product (Pr).

Total anthocyanin content (TAC) was determined using the pH-differential method [66,67] and expressed as milligrams of cyanidin-3-O-glucoside (C3G) per 100 g of product ($mg_{C3G}/100\ g_{Pr}$).

Antioxidant activity was evaluated by Ferric Reducing Antioxidant Power (FRAP) assay [68], with the results expressed as millimoles of ferrous iron (Fe^{2+}) equivalents per kilogram (solid food) of Pr.

3.4. Chromatographic Analysis

Chromatographic analysis was carried out using an Agilent 1200 high-performance liquid chromatograph coupled to an Agilent UV-Vis diode array detector (Agilent Technologies, Santa Clara, CA, USA), based on HPLC methods previously validated for fresh fruits, herbal medicines and other food products [2,69].Composition of solvents, used gradient elution conditions and UV-Vis wavelengths were listed and described in Table 5, while calibration parameters for all the used analytical standards were reported in Table 6 [70].

Table 5. Chromatographic conditions of the used methods.

Method	Compounds of Interest	Stationary Phase	Mobile Phase	Flow (mL min^{-1})	Wavelength (nm)
A	cinnamic acids, flavanols	KINETEX-C18 column (4.6 × 150 mm, 5 µm)	A: 10 mM KH_2PO_4/H_3PO_4, pH = 2.8; B: CH_3CN	1.5	330
B	benzoic acids, catechins, tannins	KINETEX-C18 column (4.6 × 150 mm, 5 µm)	A: $H_2O/CH_3OH/HCOOH$ (5:95:0.1 $v/v/v$), pH = 2.5; B: $CH_3OH/HCOOH$ (100:0.1 v/v)	0.6	280
C	monoterpenes	KINETEX-C18 column (4.6 × 150 mm, 5 µm)	A: H_2O; B: CH_3CN	1.0	210, 220, 235, 250
D	organic acids	KINETEX-C18 column (4.6 × 150 mm, 5 µm)	A: 10 mM KH_2PO_4/H_3PO_4, pH = 2.8; B: CH_3CN	0.6	214
E	vitamins	KINETEX-C18 column (4.6 × 150 mm, 5 µm)	A: 5 mM $C_{16}H_{33}N(CH_3)_3Br/50$ mM KH_2PO_4, pH = 2.5; B: CH_3OH	0.9	261, 348
F	sugars	SphereClone-NH$_2$ column (4.6 × 250 mm, 5 µm)	A: H_2O; B: CH_3CN	0.5	200, 267, 286

Elution conditions. Method A: gradient analysis: 5%B to 21%B in 17 min + 21%B in 3 min (2 min conditioning time); Method B: gradient analysis: 3%B to 85%B in 22 min + 85%B in 1 min (2 min conditioning time); Method C: gradient analysis: 30%B to 56%B in 15 min + 56%B in 2 min (3 min conditioning); Method D: gradient analysis: 5%B to 14%B in 10 min + 14%B in 3 min (2 min conditioning time); Method E: isocratic analysis: ratio of phase A and B: 95:5 in 10 min (5 min conditioning time); Method F: isocratic analysis: ratio of phase A and B: 5:85 in 12 min (3 min conditioning time). Chromatographic separation was performed on a Kinetex-C18 column (Phenomenex, Torrance, CA, USA) and a SphereClone-NH2 column (Phenomenex, Torrance, CA, USA).

Table 6. Calibration parameters for all the used analytical standards.

Class	Standard	Calibration Curve Equation	R^2	Calibration Curve Range (mg L^{-1})
Cinnamic acids	caffeic acid	y = 59.046x + 200.6	0.996	111–500
	chlorogenic acid	y = 13.583x + 760.05	0.984	111–500
	coumaric acid	y = 8.9342x + 217.4	0.997	111–500
	ferulic acid	y = 3.3963x − 4.9524	1.000	111–500
Flavonols	hyperoside	y = 7.1322x − 4.583	0.999	111–500
	isoquercitrin	y = 8.3078x + 26.621	0.999	111–500
	quercetin	y = 3.4095x − 98.307	0.998	111–500
	quercitrin	y = 2.7413x + 5.6367	0.998	111–500
	rutin	y = 6.5808x + 30.831	0.999	111–500
Benzoic acids	ellagic acid	y = 29.954x + 184.52	0.998	62.5–250
	gallic acid	y = 44.996x + 261.86	0.999	62.5–250
Catechins	catechin	y = 8.9197x + 66.952	1.000	62.5–250
	epicatechin	y = 12.88x − 43.816	0.999	62.5–250
Tannins	castalagin	y = 4.236x − 8.535	1.000	62.5–250
	vescalagin	y = 4.939x − 1.232	1.000	62.5–250
Monoterpenes	limonene	y = 0.1894x − 5.420	0.999	125–1000
	phellandrene	y = 8.783x − 145.3	0.998	125–1000
	sabinene	y = 18.14x − 1004	0.998	125–1000
	γ-terpinene	y = 0.4886x − 23.02	0.999	125–1000
	terpinolene	y = 26.52x + 876.8	0.999	125–1000
Organic acids	citric acid	y = 1.0603x − 22.092	1.000	167–1000
	malic acid	y = 1.415x − 80.254	0.996	167–1000
	oxalic acid	y = 6.4502x + 6.1503	0.998	167–1000
	quinic acid	y = 0.8087x − 38.021	0.998	167–1000
	succinic acid	y = 0.9236x − 8.0823	0.995	167–1000
	tartaric acid	y = 1.8427x + 15.796	1.000	167–1000
Vitamins	ascorbic acid	y = 42.71x + 27.969	0.999	100–1000
	dehydroascorbic acid	y = 4.1628x + 140.01	0.999	30–300
Sugar	fructose	y = 1.8548x + 1.2324	0.999	125–1000
	glucose	y = 0.1269x − 0.1107	0.998	125–1000
	sucrose	y = 0.296x − 3.2202	1.000	125–1000

Single bioactive molecules were identified and quantified using selected biomarkers with a positive role in human health ("multi-marker approach") according to Mok and Chau [71]. All the results were expressed as mg/100 g of product (Pr), except sugars (expressed as g/100 g of Pr). Total bioactive compound content (TBCC) were determined as sum of selected compounds and expressed as mg/100 g of product.

3.5. Statistical Analysis

Uni- and multivariate analysis (MVA) were carried out on all of the samples. Data were treated by one-factor analysis of variance (ANOVA) and the averages were compared with Tukey's HSD post-hoc comparison test at significance level $p < 0.05$ ($N = 3$). Correlation between antioxidant activity and TPC/TAC was evaluated with Pearson's coefficient (R) at $p < 0.05$ ($N = 3$). For discrimination of the investigated samples, principal component analysis (PCA) was performed on the column-centred data. All calculations were performed with statistical software package IBM SPSS Statistics 22.0 (IBM, Armonk, NY, USA).

4. Conclusions

In this study, spectrophotometric and chromatographic methods coupled to chemometrics were used for phytochemical analysis in order to detect and quantify bioactive substances and characterize nutraceutical properties and nutritional traits in fruit-based products from Comoros Islands. HPLC profile of phenolic compounds may be used as "fingerprint" in the detection ofquali-quantitative differences in jellies and jams.

Results showed that analysed fruit jams and jellies could be a good source of polyphenolic constituents, as benzoic and cinnamic acids, catechins and flavanols and volatile compounds, as organic acids and monoterpenes: these molecules were found to be suitable markers for product characterization because they were specific compounds of the obtained chromatographic patterns. For this reason, even if each analytical approach has its limitations which restrict its applicability, the identification of these chemical markers could be a simple, rapid and generally available potential tool for quality control and labelling of local jams and jellies.

Moreover, the characterisation of these products and their nutritional and nutraceutical traits could be important to valorise local food production and to raise incomes for local population in the Comoros Islands, particularly women that work in agri-food industry. The obtained incomes could be useful to reduce poverty; indeed, the advances in local food production can be important in poverty reduction and deserves greater attention in rural development: it is necessary to link the evidence of poverty impact to simple policy recommendations in order to potentially integrate the promotion of local fruit-based products into national-level planning.

Finally, further benefits of this approach include the potential for a better nutrition, maintenance of biodiversity and environmentally sustainable food systems.

Supplementary Materials: The supplementary materials are available online.

Author Contributions: D.D., S.H., C.G. and G.L.B. conceived and designed the experiments; S.H. proceeded to collect plant material and prepare fruit-based products; D.D. performed the chemical experiments and instrumental HPLC analysis; D.D. and M.G.M. statistically analysed the data; M.D.B., I.R. and G.G. contributed to manuscript elaboration; D.D. wrote the paper. All authors have read and approved the final manuscript.

Funding: This research received no external funding.

Conflicts of Interest: Authors declare no conflict of interest.

References

1. Sewall, B.J.; Granek, E.F.; Trewhella, W.J. The endemic Comoros Islands fruit bat *Rousettus obliviosus*: Ecology, conservation, and red list status. *Oryx* **2003**, *37*, 344–352. [CrossRef]

2. Soifoini, T.; Donno, D.; Jeannoda, V.; Rakotoniaina, E.; Hamidou, S.; Achmet, S.M.; Solo, N.R.; Afraitane, K.; Giacoma, C.; Beccaro, G.L. Bioactive compounds, nutritional traits, and antioxidant properties of *Artocarpus altilis* (Parkinson) fruits: Exploiting a potential functional food for food security on the Comoros Islands. *J. Food Qual.* **2018**, in press. [CrossRef]

3. Clevidence, B. Tropical and subtropical fruits: Phytonutrients and anticipated health benefits. In Proceedings of the III International Symposium on Tropical and Subtropical Fruits, Fortaleza, Ceara, Brazil, 12–17 September 2004; pp. 485–498.

4. Lim, Y.Y.; Lim, T.T.; Tee, J.J. Antioxidant properties of several tropical fruits: A comparative study. *Food Chem.* **2007**, *103*, 1003–1008. [CrossRef]

5. Sajise, P.; Ahmad, I. Conservation and sustainable use of tropical and sub-tropical fruits: Current status and prospects. In Proceedings of the International Workshop on Tropical and Subtropical Fruits, Chiang Mai, Thailand, April 2007; pp. 97–112.

6. Chivandi, E.; Mukonowenzou, N.; Nyakudya, T.; Erlwanger, K.H. Potential of indigenous fruit-bearing trees to curb malnutrition, improve household food security, income and community health in sub-saharan africa: A review. *Food Res. Int.* **2015**, *76*, 980–985. [CrossRef]

7. Reardon, T.; Barrett, C.B.; Berdegué, J.A.; Swinnen, J.F.M. Agrifood industry transformation and small farmers in developing countries. *World Dev.* **2009**, *37*, 1717–1727. [CrossRef]

8. Singh, S.; Jain, S.; Singh, S.; Singh, D. Quality changes in fruit jams from combinations of different fruit pulps. *J. Food Process. Preserv.* **2009**, *33*, 41–57. [CrossRef]

9. Codex_Alimentarius. Codex Standard for Jams, Jellies and Marmalades (Codex Stan 296-2009). Available online: http://www.codexalimentarius.org (accessed on 27 March 2018).

10. Kim, D.O.; Padilla-Zakour, O. Jam processing effect on phenolics and antioxidant capacity in anthocyanin-Rich fruits: Cherry, plum, and raspberry. *J. Food Sci.* **2004**, *69*, S395–S400. [CrossRef]

11. Rosa, A.; Atzeri, A.; Deiana, M.; Scano, P.; Incani, A.; Piras, C.; Cesare Marincola, F. Comparative antioxidant activity and ^{1}H NMR profiling of Mediterranean fruit products. *Food Res. Int.* **2015**, *69*, 322–330. [CrossRef]

12. Puiggròs, F.; Solà, R.; Bladé, C.; Salvadó, M.-J.; Arola, L. Nutritional biomarkers and foodomic methodologies for qualitative and quantitative analysis of bioactive ingredients in dietary intervention studies. *J. Chromatogr. A* **2011**, *1218*, 7399–7414. [CrossRef] [PubMed]

13. Donno, D.; Cavanna, M.; Beccaro, G.L.; Mellano, M.G.; Torello-Marinoni, D.; Cerutti, A.K.; Bounous, G. Currants and strawberries as bioactive compound sources: Determination of antioxidant profiles with HPLC-DAD/MS. *J. App. Bot. Food Qual.* **2013**, *86*, 1–10.

14. Gonzalez-Molina, E.; Moreno, D.A.; Garcia-Viguera, C. Genotype and harvest time influence the phytochemical quality of fino lemon juice (*Citrus limon* (L.) Burm. F.) for industrial use. *J. Agric. Food Chem.* **2008**, *56*, 1669–1675. [CrossRef] [PubMed]

15. Tibaldi, G.; Fontana, E.; Nicola, S. Growing conditions and postharvest management can affect the essential oil of *Origanum vulgare* L. spp. *hirtum* (Link) ietswaart. *Ind. Crop. Prod.* **2011**, *34*, 1516–1522. [CrossRef]

16. Bengoechea, M.L.; Sancho, A.I.; Bartolomé, B.; Estrella, I.; Gómez-Cordovés, C.; Hernández, M.T. Phenolic composition of industrially manufactured purees and concentrates from peach and apple fruits. *J. Agric. Food Chem.* **1997**, *45*, 4071–4075. [CrossRef]

17. Dragovic-Uzelac, V.; Pospišil, J.; Levaj, B.; Delonga, K. The study of phenolic profiles of raw apricots and apples and their purees by HPLC for the evaluation of apricot nectars and jams authenticity. *Food Chem.* **2005**, *91*, 373–383. [CrossRef]

18. Clifford, M.N. Anthocyanins—Nature, occurrence and dietary burden. *J. Agric. Food Chem.* **2000**, *80*, 1063–1072. [CrossRef]

19. Banaś, A.; Korus, A.; Tabaszewska, M. Quality assessment of low-sugar jams enriched with plant raw materials exhibiting health-promoting properties. *J. Food Sci. Technol.* **2018**, *55*, 408–417. [CrossRef] [PubMed]

20. Poiana, M.-A.; Moigradean, D.; Dogaru, D.; Mateescu, C.; Raba, D.; Gergen, I. Processing and storage impact on the antioxidant properties and color quality of some low sugar fruit jams. *Rom. Biotechnol. Lett.* **2011**, *16*, 6504–6512.

21. Rababah, T.M.; Al-Mahasneh, M.A.; Kilani, I.; Yang, W.; Alhamad, M.N.; Ereifej, K.; Al-u'datt, M. Effect of jam processing and storage on total phenolics, antioxidant activity, and anthocyanins of different fruits. *J. Sci. Food Agric.* **2011**, *91*, 1096–1102. [CrossRef] [PubMed]

22. Heimler, D.; Romani, A.; Ieri, F. Plant polyphenol content, soil fertilization and agricultural management: A review. *Eur. Food Res. Technol.* **2017**, *243*, 1107–1115. [CrossRef]

23. Donno, D.; Randriamampionona, D.; Andriamaniraka, H.; Torti, V.; Mellano, M.G.; Giacoma, C.; Beccaro, G.L. Biodiversity and traditional medicinal plants from Madagascar: Phytochemical evaluation of *Brachylaena ramiflora* (DC.) Humbert decoctions and infusions. *J. Appl. Bot. Food Qual.* **2017**, *90*, 205–2013.

24. Šavikin, K.; Zdunić, G.; Janković, T.; Tasić, S.; Menković, N.; Stević, T.; Đorđević, B. Phenolic content and radical scavenging capacity of berries and related jams from certificated area in Serbia. *Plant. Food Hum. Nutr.* **2009**, *64*, 212–217. [CrossRef] [PubMed]

25. Da Silva Pinto, M.; Lajolo, F.M.; Genovese, M.I. Bioactive compounds and antioxidant capacity of strawberry jams. *Plant. Food Hum. Nutr.* **2007**, *62*, 127–131. [CrossRef] [PubMed]

26. Wang, H.; Cao, G.; Prior, R.L. Oxygen radical absorbing capacity of anthocyanins. *J. Agric. Food Chem.* **1997**, *45*, 304–309. [CrossRef]

27. Veberic, R.; Slatnar, A.; Bizjak, J.; Stampar, F.; Mikulic-Petkovsek, M. Anthocyanin composition of different wild and cultivated berry species. *LWT-Food Sci. Technol.* **2015**, *60*, 509–517. [CrossRef]

28. Jia, N.; Xiong, Y.L.; Kong, B.; Liu, Q.; Xia, X. Radical scavenging activity of black currant (*Ribes nigrum* L.) extract and its inhibitory effect on gastric cancer cell proliferation via induction of apoptosis. *J. Funct. Foods* **2012**, *4*, 382–390. [CrossRef]

29. Donno, D.; Beccaro, G.L.; Mellano, M.G.; Cerutti, A.K.; Bounous, G. Goji berry fruit (*Lycium* spp.): Antioxidant compound fingerprint and bioactivity evaluation. *J. Funct. Foods* **2015**, *18*, 1070–1085. [CrossRef]

30. Silva, B.M.; Andrade, P.B.; Gonçalves, A.C.; Seabra, R.M.; Oliveira, M.B.; Ferreira, M.A. Influence of jam processing upon the contents of phenolics, organic acids and free amino acids in quince fruit (*Cydonia oblonga* Miller). *Eur. Food Res. Technol.* **2004**, *218*, 385–389. [CrossRef]

31. Wojdyło, A.; Oszmiański, J.; Bielicki, P. Polyphenolic composition, antioxidant activity, and polyphenol oxidase (PPO) activity of quince (*Cydonia oblonga* Miller) varieties. *J. Agric. Food Chem.* **2013**, *61*, 2762–2772. [CrossRef] [PubMed]

32. Cabral, T.A.; de Morais Cardoso, L.; Pinheiro-Sant'Ana, H.M. Chemical composition, vitamins and minerals of a new cultivar of lychee (*Litchi chinensis* cv. Tailandes) grown in Brazil. *Fruits* **2014**, *69*, 425–434. [CrossRef]

33. Singh, D.; Wangchu, L.; Moond, S.K. Processed Products of Tamarind. *Nat. Prod. Radiance* **2007**, *6*, 315–321.

34. Jimoh, S.; Onabanjo, O. Potentials of *Tamarindus indica* (L.) in jam production. *J. Agric. Soc. Res.* **2012**, *12*, 29–43.

35. Marquina, V.; Araujo, L.; Ruíz, J.; Rodríguez-Malaver, A.; Vit, P. Composition and antioxidant capacity of the guava (*Psidium guajava* L.) fruit, pulp and jam. *Arch. Latinoam. Nutr.* **2008**, *58*, 98–102. [PubMed]

36. Sánchez-Salcedo, E.M.; Mena, P.; García-Viguera, C.; Martínez, J.J.; Hernández, F. Phytochemical evaluation of white (*Morus alba* L.) and black (*Morus nigra* L.) mulberry fruits, a starting point for the assessment of their beneficial properties. *J. Funct. Foods* **2015**, *12*, 399–408. [CrossRef]

37. Garcia-Viguera, C.; Tomás-Barberán, F.A.; Ferreres, F.; Artés, F.; Tomás-Lorente, F. Determination of *Citrus* jams genuineness by flavonoid analysis. *Eur. Food Res. Technol.* **1993**, *197*, 255–259.

38. Del Rio, D.; Rodriguez-Mateos, A.; Spencer, J.P.; Tognolini, M.; Borges, G.; Crozier, A. Dietary (poly) phenolics in human health: Structures, bioavailability, and evidence of protective effects against chronic diseases. *Antioxid. Redox Signal.* **2013**, *18*, 1818–1892. [CrossRef] [PubMed]

39. Kansci, G.; Koubala, B.B.; Mbome, I.L. Biochemical and physicochemical properties of four mango varieties and some quality characteristics of their jams. *J. Food Process. Preserv.* **2008**, *32*, 644–655. [CrossRef]

40. López, R.; Ramírez, A.; de Fariñas Graziani, L. Physicochemical and microbiological evaluation of 3 commercial guava jams (*Psidium guajava* L.). *Arch. Latinoam. Nutr.* **2000**, *50*, 291–295. [PubMed]

41. Seeram, N.P. Berry fruits: Compositional elements, biochemical activities, and the impact of their intake on human health, performance, and disease. *J. Agric. Food Chem.* **2008**, *56*, 627–629. [CrossRef] [PubMed]

42. Ammar, I.; Ennouri, M.; Bouaziz, M.; Ben Amira, A.; Attia, H. Phenolic profiles, phytchemicals and mineral content of decoction and infusion of *Opuntia ficus-indica* flowers. *Plant. Food Hum. Nutr.* **2015**, *70*, 388–394. [CrossRef] [PubMed]

43. Landete, J.M. Ellagitannins, ellagic acid and their derived metabolites: A review about source, metabolism, functions and health. *Food Res. Int.* **2011**, *44*, 1150–1160. [CrossRef]

44. Ştefănuţ, M.N.; Căta, A.; Pop, R.; Moşoarcă, C.; Zamfir, A.D. Anthocyanins HPLC-DAD and MS characterization, total phenolics, and antioxidant activity of some berries extracts. *Anal. Lett.* **2011**, *44*, 2843–2855. [CrossRef]

45. Garzon, G.; Wrolstad, R. Comparison of the stability of pelargonidin-based anthocyanins in strawberry juice and concentrate. *J. Food Sci.* **2002**, *67*, 1288–1299. [CrossRef]

46. Kammerer, D.; Carle, R.; Schieber, A. Detection of peonidin and pelargonidin glycosides in black carrots (*Daucus carota* ssp. *Sativus* var. *Atrorubens alef.*) by high-performance liquid chromatography/electrospray ionization mass spectrometry. *Rapid Commun. Mass Spectrom.* **2003**, *17*, 2407–2412. [CrossRef] [PubMed]

47. Schwarz, M.; Wray, V.; Winterhalter, P. Isolation and identification of novel pyranoanthocyanins from black carrot (*Daucus carota* L.) juice. *J. Agric. Food Chem.* **2004**, *52*, 5095–5101. [CrossRef] [PubMed]

48. Papaefthimiou, D.; Papanikolaou, A.; Falara, V.; Givanoudi, S.; Kostas, S.; Kanellis, A.K. Genus *Cistus*: A model for exploring labdane-type diterpenes' biosynthesis and a natural source of high value products with biological, aromatic, and pharmacological properties. *Front. Chem.* **2014**, *2*, 35. [CrossRef] [PubMed]

49. Trombetta, D.; Castelli, F.; Sarpietro, M.G.; Venuti, V.; Cristani, M.; Daniele, C.; Saija, A.; Mazzanti, G.; Bisignano, G. Mechanisms of antibacterial action of three monoterpenes. *Antimicrob. Agents Chemother.* **2005**, *49*, 2474–2478. [CrossRef] [PubMed]

50. Crowell, P.L. Prevention and therapy of cancer by dietary monoterpenes. *J. Nutr.* **1999**, *129*, S775–S778. [CrossRef] [PubMed]

51. Cazares-Franco, M.C.; Ramirez-Chimal, C.; Herrera-Hernandez, M.G.; Nunez-Colin, C.A.; Hernandez-Martinez, M.A.; Guzman-Maldonado, S.H. Physicochemical, nutritional and health-related component characterization of the underutilized mexican serviceberry fruit *Malacomeles denticulata* (Kunth) G. N. Jones. *Fruits* **2014**, *69*, 47–60. [CrossRef]

52. Mattila, P.H.; Hellström, J.; McDougall, G.; Dobson, G.; Pihlava, J.-M.; Tiirikka, T.; Stewart, D.; Karjalainen, R. Polyphenol and vitamin c contents in European commercial blackcurrant juice products. *Food Chem.* **2011**, *127*, 1216–1223. [CrossRef] [PubMed]

53. Georgé, S.; Brat, P.; Alter, P.; Amiot, M.J. Rapid determination of polyphenols and vitamin c in plant-derived products. *J. Agric. Food Chem.* **2005**, *53*, 1370–1373. [CrossRef] [PubMed]

54. Donno, D.; Cerutti, A.K.; Prgomet, I.; Mellano, M.G.; Beccaro, G.L. Foodomics for mulberry fruit (*Morus* spp.): Analytical fingerprint as antioxidants' and health properties' determination tool. *Food Res. Inter.* **2015**, *69*, 179–188. [CrossRef]

55. Fügel, R.; Carle, R.; Schieber, A. Quality and authenticity control of fruit purées, fruit preparations and jams—A review. *Trends Food Sci. Technol.* **2005**, *16*, 433–441. [CrossRef]

56. Eyduran, S.P.; Ercisli, S.; Akin, M.; Beyhan, O.; Gecer, M.K.; Eyduran, E.; Erturk, Y.E. Organic acids, sugars, vitamin c, antioxidant capacity and phenolic compounds in fruits of white (*Morus alba* L.) and black (*Morus nigra* L.) mulberry genotypes. *J. Appl. Bot. Food Qual.* **2015**, *88*, 134–138.

57. Soyer, Y.; Koca, N.; Karadeniz, F. Organic acid profile of turkish white grapes and grape juices. *J. Food Compos. Anal.* **2003**, *16*, 629–636. [CrossRef]

58. Cejudo-Bastante, C.; Castro-Mejías, R.; Natera-Marín, R.; García-Barroso, C.; Durán-Guerrero, E. Chemical and sensory characteristics of orange based vinegar. *J. Food Sci. Technol.* **2016**, *53*, 3147–3156. [CrossRef] [PubMed]

59. Flores, P.; Hellín, P.; Fenoll, J. Determination of organic acids in fruits and vegetables by liquid chromatography with tandem-mass spectrometry. *Food Chem.* **2012**, *132*, 1049–1054. [CrossRef]

60. Chinnici, F.; Spinabelli, U.; Riponi, C.; Amati, A. Optimization of the determination of organic acids and sugars in fruit juices by ion-exclusion liquid chromatography. *J. Food Compos. Anal.* **2005**, *18*, 121–130. [CrossRef]

61. Touati, N.; Tarazona-Díaz, M.P.; Aguayo, E.; Louaileche, H. Effect of storage time and temperature on the physicochemical and sensory characteristics of commercial apricot jam. *Food Chem.* **2014**, *145*, 23–27. [CrossRef] [PubMed]

62. Tzouros, N.; Arvanitoyannis, I. Agricultural produces: Synopsis of employed quality control methods for the authentication of foods and application of chemometrics for the classification of foods according to their variety or geographical origin. *Crit. Rev. Food Sci. Nutr.* **2001**, *41*, 287–319. [CrossRef] [PubMed]

63. Reid, L.M.; O'Donnell, C.P.; Downey, G. Potential of SPME-GC and chemometrics to detect adulteration of soft fruit purees. *J. Agric. Food Chem.* **2004**, *52*, 421–427. [CrossRef] [PubMed]

64. Slinkard, K.; Singleton, V.L. Total phenol analysis: Automation and comparison with manual methods. *Am. J. Enol. Vitic.* **1977**, *28*, 49–55.

65. Sánchez-Rangel, J.C.; Benavides, J.; Heredia, J.B.; Cisneros-Zevallos, L.; Jacobo-Velázquez, D.A. The Folin–Ciocalteu assay revisited: Improvement of its specificity for total phenolic content determination. *Anal. Methods* **2013**, *5*, 5990–5999. [CrossRef]

66. Lee, J.; Durst, R.W.; Wrolstad, R.E. Determination of total monomeric anthocyanin pigment content of fruit juices, beverages, natural colorants, and wines by the pH differential method: Collaborative study. *J. AOAC Int.* **2005**, *88*, 1269–1278. [PubMed]

67. Giusti, M.M.; Wrolstad, R.E. Characterization and measurement of anthocyanins by UV-visible spectroscopy. *Curr. Protoc. Food Anal. Chem.* **2001**, *00*, F1.2.1–F1.2.13. [CrossRef]

68. Benzie, I.F.; Strain, J.J. Ferric reducing/antioxidant power assay: Direct measure of total antioxidant activity of biological fluids and modified version for simultaneous measurement of total antioxidant power and ascorbic acid concentration. *Meth. Enzymol.* **1999**, *299*, 15–27. [PubMed]

69. Donno, D.; Mellano, M.G.; Prgomet, Z.; Beccaro, G.L. Advances in *Ribes* × *nidigrolaria* Rud. Bauer and A. Bauer fruits as potential source of natural molecules: A preliminary study on physico-chemical traits of an underutilized berry. *Sci. Hortic.* **2018**, *237*, 20–27. [CrossRef]

70. Donno, D.; Mellano, M.G.; De Biaggi, M.; Riondato, I.; Rakotoniaina, E.N.; Beccaro, G.L. New findings in *Prunus padus* L. fruits as a source of natural compounds: Characterization of metabolite profiles and preliminary evaluation of antioxidant activity. *Molecules* **2018**, *23*, 725. [CrossRef] [PubMed]

71. Mok, D.K.W.; Chau, F.T. Chemical information of chinese medicines: A challenge to chemist. *Chemom. Intell. Lab. Syst.* **2006**, *82*, 210–217. [CrossRef]

Sample Availability: Samples of the compounds are not available from the authors.

molecules

MDPI

Technical Note

Synthesis and Structural Identification of a Biaryl Ether-Linked Zearalenone Dimer

Julia Keller [1], Luisa Hantschke [1], Hajo Haase [2] and Matthias Koch [1,*]

[1] Department Analytical Chemistry, Reference Materials, Bundesanstalt für Materialforschung und -prüfung (BAM), Richard-Willstätter-Straße 11, 12489 Berlin, Germany; juliakeller19@yahoo.de (J.K.); luisa.hantschke@web.de (L.H.)
[2] Department of Food Chemistry and Toxicology, Technische Universität Berlin, Gustav-Meyer-Allee 25, 13355 Berlin, Germany; haase@tu-berlin.de
* Correspondence: matthias.koch@bam.de; Tel.: +49-30-8104-1170

check for updates

Received: 14 September 2018; Accepted: 11 October 2018; Published: 12 October 2018

Abstract: A new dimer of the food-relevant mycotoxin zearalenone was isolated after electrochemical and chemical oxidation. The structure was determined as a 16-O-15′-biaryl ether-linked dimer based on spectroscopic analyses (^{1}H- and ^{13}C-NMR, COSY, HMBC, and HSQCAD) and high-resolution mass spectrometry analysis (Q-TOF).

Keywords: mycotoxin; dimerization; HRMS; NMR

1. Introduction

The fungal secondary metabolite zearalenone (ZEN) is found worldwide and is primarily produced by *Fusarium* species [1–3]. Often found in common crops like corn, wheat, rice, soybeans, sorghum, spices, or walnuts, it poses a health risk to human and animals [4–6]. As mycoestrogen, it causes swelling of the uterus and vulva, infertility, and atrophy of ovaries reported in swine and cattle [7,8]. Several metabolites derived from plants, fungi, and mammalian metabolism are part of ongoing research due to unknown toxic effects and occurrence [9–12].

Oxidative reactions of ZEN lead to hydroxylated species obtained from in vitro assays with liver microsomes of rodent and non-rodent liver cells [13,14]. A recent study proposed the production of numerous hydroxylated as well as new dimeric species of ZEN by using electrochemistry coupled to mass spectrometry [15]. The production, isolation, and structural elucidation of the predominant dimeric species is now achieved.

2. Results and Discussion

Compound **1** was obtained as a pale orange solid substance after electrochemical and chemical oxidation of ZEN with a molecular formula of $C_{36}H_{42}O_{10}$, which was measured by high-resolution mass spectrometry with m/z 633.2658 [M − H]$^-$ in an ESI negative ionization mode (theoretical exact mass m/z 633.2702), as previously described [15]. The MS/MS measurements of compound 1 revealed a fragment with m/z 589.2756, which is due to the loss of CO_2, and a fragment with m/z 565.2746 because of a loss of C_3O_2. The signal with m/z 491.1658, corresponds to a loss of $C_8H_4O_2$, which led to the fragment m/z 447.1763 after the loss of CO_2 (Supplement Figure S1). The assumed fragments and their chemical formulas with theoretical exact masses are shown in Figure 1.

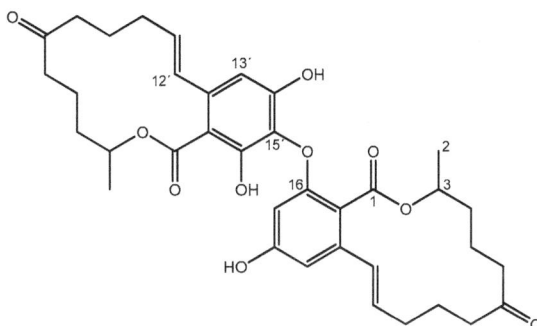

Figure 1. Structure of compound **1**.

The ^{1}H-NMR spectrum of compound 1 measured at 400 MHz in MeOH-d_4 gave the following information with 1.00–2.85 (m, H, J = 6.8 Hz), 1.34 (d, 3H, J = 6.2 Hz), 1.39 (d, 3H, J = 6.4 Hz), 5.06–5.16 (m, 1H), 5.20–5.36 (m, 1H), 5.71 (ddd, 1H, J = 4.0, 9.8, 15.6 Hz), 5.77 (d, 1H, J = 2.1 Hz), 6.03 (d, 1H, J = 4.4, 9.6, 14.6 Hz), 6.34–6.41 (m, 2H), 6.42 (s, 1H), and 6.58 (d, 1H, J = 2.1 Hz). The ^{13}C-NMR measurements with 100 MHz revealed the following chemical shifts with 20.2, 21.0, 22.1, 22.6, 23.4, 32.4, 35.9, 37.6, 38.3, 43.9, 44.7, 73.7, 74.3, 100.8, 104.3, 106.7, 126.4, 130.2, 134.4, 137.2, 139.4, 158.0, 161.0, 172.1, 213.8, and 214.1 ppm (Supplement Figures S2 and S3). The assignments of the carbons and protons are summarized in Table 1.

Table 1. ^{1}H-NMR and ^{13}C-NMR data of **1** (400/100 MHz, Methanol-d_4).

Position	^{1}H (ppm)	^{13}C (ppm)
1, 1'	-	213.8, 214.1
2	1.39 (d, J = 6.4 Hz, 3H)	35.9
3	5.20–5.36 (m, 1H)	73.7
10, 10'	-	32.4, 32.4
11	6.03 (ddd, J = 4.4, 9.6, 14.6 Hz, 1H)	134.4
12	6.38–6.41 (m, 1H)	130.2
13	6.58 (d, J = 2.1 Hz, 1H)	106.7
14	-	161.0
15	5.77 (d, J = 2.1 Hz, 1H)	100.8
2'	1.34 (d, J = 6.2 Hz, 3H)	35.9
3'	5.06–5.16 (m, 1H)	74.3
11'	5.71 (ddd, J = 4.0, 9.8, 15.6 Hz, 1H)	137.2
12'	6.34–6.37 (m, 1H)	126.4
13'	6.42 (s, 1H)	104.3

The ^{1}H-NMR and COSY spectra revealed two aromatic systems (Supplement Figures S2 and S4). One of these aromatic systems contained two protons (d 6.58 ppm, 1H, J = 2.1 Hz, and d 5.77 ppm, 1H, J = 2.1 Hz) while the second aromatic system had only one proton (s 6.42 ppm, 1H). From this observation, the structures connected over the C-C linkages 15, 15', 13, 13', and 13, 15' can be excluded since these dimers would have two aromatic systems with only one proton. Thus, only dimers having an C-O-C ether bridge between the monomers are possible. The COSY spectrum showed nearly identical chemical structures for the spectral part of the molecule (Supplement Figure S4). As a result, the dimerization of two ZEN molecules is only likely over an ether-link between 14-O-13', 14-O-15', 16-O-13', or 16-O-15'. The HMBC spectrum (8 Hz) indicates a common coupling partner of the single aromatic proton and 12'-H, which is located in the aromatic-olefinic region (Supplement Figure S5). Due to the spatial proximity, it should be at the singlet (6.42 ppm) and, as a result, act around the 13'-H position. Consequently, an O-15'-linkage is conceivable.

The observed significant difference of the chemical shifts of 3-H and 3′-H indicated a different chemical environment. For a C-14 link, the closer chemical environment of 3-H and 3′-H would be very similar. A C-16 link, on the other hand, would be a greater steric influence and inductive effects would occur. Thus, the compound 16-O-15′-biaryl ether bond is the most likely structure, which is shown in Figure 2. Whether this dimer can be found naturally in food or feed remains to be analyzed in detail. Especially in plants, lichen, bacteria, and fungi regio-selective and stereo-selective biaryl C-C and biaryl ether C-O-C linkages are often found and it is conceivable that dimers of ZEN might not be uncommon in nature [16].

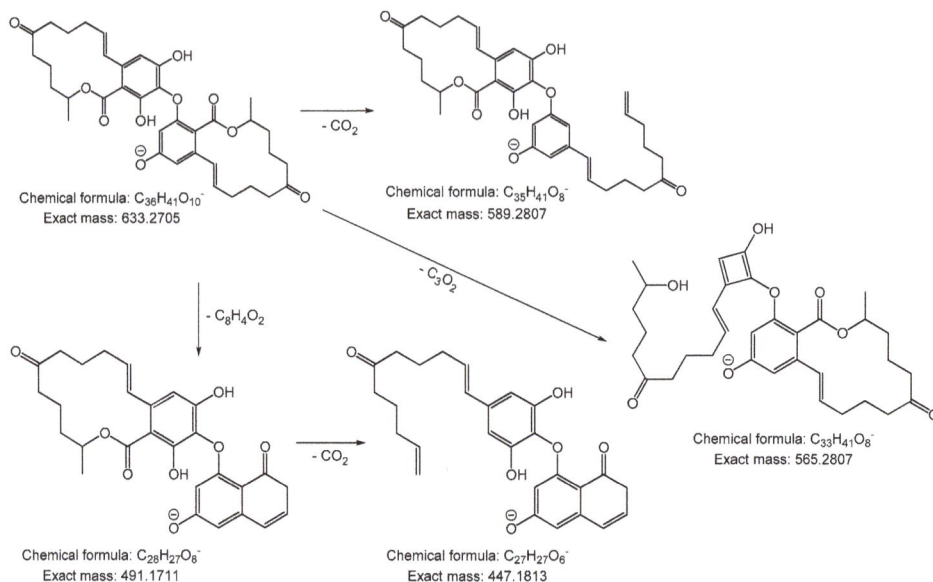

Chemical formula: $C_{36}H_{41}O_{10}^-$
Exact mass: 633.2705

$- CO_2$

Chemical formula: $C_{35}H_{41}O_8^-$
Exact mass: 589.2807

$- C_3O_2$

Chemical formula: $C_{33}H_{41}O_8^-$
Exact mass: 565.2807

$- C_8H_4O_2$

Chemical formula: $C_{28}H_{27}O_8^-$
Exact mass: 491.1711

$- CO_2$

Chemical formula: $C_{27}H_{27}O_6^-$
Exact mass: 447.1813

Figure 2. Postulated fragments of the ZEN-dimer, according to the MS/MS measurements with a molecular structure and formula along with the theoretical exact masses.

3. Materials and Methods

3.1. Chemicals and General Experimental Procedures

Zearalenone with a purity over 98% was obtained from Fermentek (Jerusalem, Israel) and Cerium(IV)sulfate was purchased from Sigma-Aldrich (Steinheim, Germany). Ultrapure water was generated by a Seralpur PRO 90 CN system (Ransbach-Baumbach, Germany). All standard chemicals were of p.a. grade and all solvents were of an HPLC grade. Electrochemical oxidation was achieved by using the Roxy® system synthesis cell (Antec, Zoeterwoude, The Netherlands) equipped with a platinum working electrode. The HPLC system used for fractionation consisted of an Agilent 1200 series autosampler, a 1260 series pump, a 1200 series diode array detector, and a column oven. A Macherey-Nagel Nucleosil C18 100-5 150 × 4.6 mm column (Düren, Germany) was used. The TripleTOF® 6600 Quadrupole Time-Of-Flight (QTOF) mass analyzer (Sciex, Darmstadt, Germany) was operated in negative ionization mode and 10 μM of the dimer sample were dissolved in methanol with 0.1% of formic acid. The used parameters were as follows:

Gas temperature 350 °C, ion source gas 1 (nitrogen) 20 L/min, ion source gas 2 (nitrogen) 15 L/min, curtain gas (nitrogen) 20 L/min, and ion spray voltage floating −4500 V. The MS/MS-spectrum was recorded in the targeted MS/MS mode with the following parameters: De-clustering potential (−15 V), collision energy (−40 V), and TOF Masses 100–640 Da. Confirmation of the 16-O-15′-biaryl ether-linked dimer structure was conducted by nuclear magnetic resonance spectroscopy (NMR).

The NMR spectra were recorded in methanol-d_4 on an Agilent 400-MR NMR spectrometer (Agilent Technologies, Waldbronn, Germany) at 30 °C. For the measurements, the ATB 5mm-probe head was operated at 399.8 MHz for ^{1}H and 100.5 MHz for ^{13}C data.

3.2. Electrochemical and Chemical Production of ZEN Dimer

Electrochemical: The optimal potential was tested by taking aliquots after different time points and using potentials of 0, 1.0, 1.4, and 1.8 V vs. Pd/H_2. For the electrochemical oxidation, 80 mL of 250 µM ZEN in acetonitrile/water (50/50, *v/v*) was stirred for 48 h using 1.4 V vs. Pd/H_2. The solution was subsequently evaporated to dryness by using a rotary evaporator and dissolved in water/acetonitrile (65/35, *v/v*) for HPLC fractionation.

Chemical: For the oxidative production of ZEN dimers, Ce(IV)sulfate was used. About 100 mg of ZEN and 350 mg of Ce(IV)sulfate were dissolved in 200 mL of acetonitrile/water (50/50, *v/v*) and stirred for two hours at 70 °C. Subsequently, the sample was stirred for 24 h at room temperature and a white precipitate was formed. The yellow solvent mixture was extracted three times with 20 mL of ethyl acetate. After the extraction, the ethyl acetate was colored yellow and the acetonitrile water mixture was colorless. After evaporation, a deep orange and highly viscous fluid was obtained. After freeze-drying, a pale orange solid was formed with a yield of 10%.

3.3. Purification of ZEN Dimer

For the separation of ZEN dimers, an already described HPLC method was adapted [15] by using a flowrate of 1.2 mL/min and an isocratic eluent consisting of water/acetonitrile, 65/35, *v/v* without modifiers. The ZEN dimer was isolated by collecting the fraction between 15.4 min and 16.2 min of the retention time by using a Foxy® R1 fraction collector (Teledyne ISCO, Nebraska, NE, USA). The purity of the dimer was determined to be 92% based on DAD spectra by using a wavelength at $\lambda = 254$ nm (Supplement Figure S7).

4. Conclusions

A new dimeric species of the food-relevant mycotoxin zearalenone was synthesized electrochemically and chemically with Ce(IV)sulfate and structurally identified. Among other possible dimers, the occurrence of the 16-*O*-15′-biaryl ether-linked dimer in food and feed is conceivable because the dimerization of phenolic compounds is often observed in plants, fungi, or lichen.

Supplementary Materials: The following are available online at http://www.mdpi.com/1420-3049/23/10/2624/s1, Figure S1. MS/MS fragmentation spectrum of the ZEN dimer obtained with Q-TOF in negative ionization mode. Figure S2. ^{1}H-NMR spectrum of the 16-*O*-15′-biaryl ether-linked zearalenone dimer. Figure S3. ^{13}C-NMR spectrum of the 16-*O*-15′-biaryl ether-linked zearalenone dimer. Figure S4. COSY spectrum of the 16-*O*-15′-biaryl ether-linked zearalenone dimer. Figure S5. HMBC spectrum of the 16-*O*-15′-biaryl ether-linked zearalenone dimer. Figure S6. HSQC spectrum of the 16-*O*-15′-biaryl ether-linked zearalenone dimer. Figure S7. HPLC-DAD chromatograms ($\lambda = 254$ nm) of the zearalenone dimer reaction mixture before fractionation (top) and after fractionation (bottom).

Author Contributions: Conceptualization, J.K. and L.H. Methodology, J.K., L.H., and M.K. Investigation, L.H. Data Curation, J.K. and M.K. Writing-Original Draft Preparation, J.K. Writing-Review & Editing, H.H. and M.K. Supervision, M.K and H.H.

Funding: This research received no external funding.

Acknowledgments: The authors would like to thank René Kudick (ASCA GmbH, Berlin, Germany) for the NMR measurements and analytical advice.

Conflicts of Interest: The authors declare no conflict of interest.

References

1. Caldwell, R.W.; Tuite, J.; Stob, M.; Baldwin, R. Zearalenone production by fusarium species. *Appl. Microbiol.* **1970**, *20*, 31–34. [PubMed]
2. Fraeyman, S.; Croubels, S.; Devreese, M.; Antonissen, G. Emerging fusarium and alternaria mycotoxins: Occurrence, toxicity and toxicokinetics. *Toxins* **2017**, *9*, 228. [CrossRef] [PubMed]
3. Placinta, C.M.; D'Mello, J.P.F.; Macdonald, A.M.C. A review of worldwide contamination of cereal grains and animal feed with fusarium mycotoxins. *Anim. Feed Sci. Technol.* **1999**, *78*, 21–37. [CrossRef]
4. Alexander, J.; Benford, D.; Boobis, A.; Ceccatelli, S.; Cottrill, B.; Cravedi, J.P.; Di Domenico, A.; Doerge, D.; Dogliotti, E.; Edler, L.; et al. Scientific opinion on the risks for public health related to the presence of zearalenone in food efsa panel on contaminants in the food chain. *EFSA J.* **2011**, *9*, 2197.
5. Hetmanski, M.T.; Scudamore, K.A. Detection of zearalenone in cereal extracts using high-performance liquid chromatography with post-column derivatization. *J. Chromatogr.* **1991**, *588*, 47–52. [CrossRef]
6. Koppen, R.; Riedel, J.; Proske, M.; Drzymala, S.; Rasenko, T.; Durmaz, V.; Weber, M.; Koch, M. Photochemical trans-/cis-isomerization and quantitation of zearalenone in edible oils. *J. Agric. Food Chem.* **2012**, *60*, 11733–11740. [CrossRef] [PubMed]
7. Drzymala, S.S.; Binder, J.; Brodehl, A.; Penkert, M.; Rosowski, M.; Garbe, L.-A.; Koch, M. Estrogenicity of novel phase i and phase ii metabolites of zearalenone and cis-zearalenone. *Toxicon* **2015**, *105*, 10–12. [CrossRef] [PubMed]
8. Pfohlleszkowicz, A.; Chekirghedira, L.; Bacha, H. Genotoxicity of zearalenone, an estrogenic mycotoxin—DNA adduct formation in female mouse-tissues. *Carcinogenesis* **1995**, *16*, 2315–2320. [CrossRef]
9. Binder, S.B.; Schwartz-Zimmermann, H.E.; Varga, E.; Bichl, G.; Michlmayr, H.; Adam, G.; Berthiller, F. Metabolism of zearalenone and its major modified forms in pigs. *Toxins* **2017**, *9*, 56. [CrossRef] [PubMed]
10. Borzekowski, A.; Drewitz, T.; Keller, J.; Pfeifer, D.; Kunte, H.-J.; Koch, M.; Rohn, S.; Maul, R. Biosynthesis and characterization of zearalenone-14-sulfate, zearalenone-14-glucoside and zearalenone-16-glucoside using common fungal strains. *Toxins* **2018**, *10*, 104. [CrossRef] [PubMed]
11. Brodehl, A.; Moller, A.; Kunte, H.J.; Koch, M.; Maul, R. Biotransformation of the mycotoxin zearalenone by fungi of the genera rhizopus and aspergillus. *FEMS Microbiol. Lett.* **2014**, *359*, 124–130. [CrossRef] [PubMed]
12. Kiessling, K.H.; Pettersson, H. Metabolism of zearalenone in rat liver. *Acta Pharmacol. Toxicol.* **1978**, *43*, 285–290. [CrossRef]
13. Bravin, F.; Duca, R.C.; Balaguer, P.; Delaforge, M. In vitro cytochrome p450 formation of a mono-hydroxylated metabolite of zearalenone exhibiting estrogenic activities: Possible occurrence of this metabolite in vivo. *Int. J. Mol. Sci.* **2009**, *10*, 1824–1837. [CrossRef] [PubMed]
14. Hildebrand, A.A.; Pfeiffer, E.; Rapp, A.; Metzler, M. Hydroxylation of the mycotoxin zearalenone at aliphatic positions: Novel mammalian metabolites. *Mycotoxin Res.* **2012**, *28*, 1–8. [CrossRef] [PubMed]
15. Keller, J.; Haase, H.; Koch, M. Hydroxylation and dimerization of zearalenone: Comparison of chemical, enzymatic and electrochemical oxidation methods. *World Mycotoxin J.* **2017**, *10*, 297–307. [CrossRef]
16. Wezeman, T.; Bräse, S.; Masters, K.-S. Xanthone dimers: A compound family which is both common and privileged. *Nat. Prod. Rep.* **2015**, *32*, 6–28. [CrossRef] [PubMed]

molecules

MDPI

Article

Ultrasensitive Electrochemical Sensor Based on Polyelectrolyte Composite Film Decorated Glassy Carbon Electrode for Detection of Nitrite in Curing Food at Sub-Micromolar Level

Jingheng Ning, Xin Luo, Min Wang, Jiaojiao Li, Donglin Liu, Hou Rong, Donger Chen and Jianhui Wang *

School of Chemistry and Biological Engineering, Changsha University of Science & Technology, Changsha 410110, China; jinghengning@csust.edu.cn (J.N.); luoxin_gl@126.com (X.L.); wang_min1993@126.com (M.W.); lijiaojiao_2013@126.com (J.L.); dong993@163.com (D.L.); hourong0406@163.com (H.R.); chendonger9@163.com (D.C.)
* Correspondence: wangjh0909@csust.edu.cn; Tel.: +86-731-8525-8309

Academic Editors: Alessandra Gentili and Chiara Fanali
Received: 3 September 2018; Accepted: 5 October 2018; Published: 9 October 2018

Abstract: To ensure food quality and safety, developing cost-effective, rapid and precision analytical techniques for quantitative detection of nitrite is highly desirable. Herein, a novel electrochemical sensor based on the sodium cellulose sulfate/poly (dimethyl diallyl ammonium chloride) (NaCS/PDMDAAC) composite film modified glass carbon electrode (NaCS/PDMDAAC/GCE) was proposed toward the detection of nitrite at sub-micromolar level, aiming to make full use of the inherent properties of individual component (biocompatible, low cost, good electrical conductivity for PDMDAAC; non-toxic, abundant raw materials, good film forming ability for NaCS) and synergistic enhancement effect. The NaCS/PDMDAAC/GCE was fabricated by a simple drop-casting method. Electrochemical behaviors of nitrite at NaCS/PDMDAAC/GCE were investigated by cyclic voltammetry (CV) and differential pulse voltammetry (DPV). Under optimum conditions, the NaCS/PDMDAAC/GCE exhibits a wide linear response region of 4.0×10^{-8} mol·L^{-1}~1.5×10^{-4} mol·L^{-1} and a low detection limit of 43 nmol·L^{-1}. The NaCS/PDMDAAC shows a synergetic enhancement effect toward the oxidation of nitrite, and the sensing performance is much better than the previous reports. Moreover, the NaCS/PDMDAAC also shows good stability and reproducibility. The NaCS/PDMDAAC/GCE was successfully applied to the determination of nitrite in ham sausage with satisfactory results.

Keywords: polyelectrolyte composite film; nitrite detection; differential pulse voltammetry; cyclic voltammetry

1. Introduction

Nitrite is widely used as a preservative in food industry, especially in the production and processing of cured meat and fishery products [1]. Additionally, it is also utilized as a fertilizing agent in agriculture and an inhibitor in corrosion science [2]. Furthermore, nitrite has become one of the widespread inorganic contaminants present in soil, food, ground water, and even physiological systems [3]. Nitrite has been proven to be of potential risk to human health and leads to many serious diseases, such as methemoglobinemia, esophageal cancer, spontaneous abortion, and birth defects of the central nervous system [4–7]. Therefore, the content of nitrite demands strict control. The World Health Organization (WHO) has set a fixed maximum limit in drinking water for nitrite as 65 µM [8]. In China, according to national standard GB 2760-2014, the maximum permitted level of nitrite in meat food is 0.15 g/kg [9]. To ensure food quality and safety, developing cost-effective, rapid, and precision analytical techniques for quantitative detection of nitrite is highly desirable.

To date, many techniques have been developed for the detection of nitrite, including chromatography [10,11], UV spectrophotometry [12,13], fluorometry [14,15], and flow injection analysis [16,17]. Although these analytical methods are well-recognized, they also suffered from limits, such as high cost, being time-consuming, cumbersome, and requiring complicated pretreatment. Recently, electrochemical analysis has become one of most preferred techniques to detection small biomolecules, food additives, and environmental contaminants, owing to its overwhelming merits including cost- and time-effective, facile operation, and high sensitivity and specificity [18–23]. In the last few years, electrochemical nitrite sensors have received increasing attention. Various modified materials have been proposed for nitrite sensors, including graphene [24,25], carbon nanotubes [26], metal or metal oxide nanoparticles [27–29], metal organic frameworks [30,31], and conducting polymers [32,33]. However, these modified electrodes have several drawbacks, such as low detection limits (almost at the mmol/L level, far lower than sub-micromolar level for HPLC and fluorometry), narrow linear range, and susceptibility to interference. Thus, it is urgent to develop novel electrochemical sensors for the detection of nitrite at sub-micromolar level.

Poly (dimethyldiallylammonium chloride) (PDMDAAC) is a ubiquitous linear cationic polymer that has unique properties, such as biocompatibility, low cost, high charge density, as well as good water solubility [34]. It has been reported that PDMDAAC film was successfully immobilized on electrochemical sensors [35–37]. These previous studies demonstrated that PDMDAAC film has excellent conductivity due to the presence of electron conductive networks, and good ion-exchange performances that can effectively increase the concentration of anions round the film as compared with the bulk solution. For these reasons, PDMDAAC film was chosen as a modified material to construct electrochemical sensors, aiming to facilitate preconcentration and enhance selectivity properties of nitrite sensors, and eventually enhance the electrochemical response. However, PDMDAAC was found to exhibit relatively poor performances in film formation in our preliminary experiments, which could influence the electrode stability. Additionally, the sensitivity of PDMDAAC modified electrode is very limited. To overcome these problems, PDMDAAC was combined with sodium cellulose sulfate (NaCS), which has good film-formation ability. NaCS/PDMDAAC is an emerging composite film that has unique properties such as excellent mechanical strength, transparency and small pore size [38]. NaCS/PDMDAAC has been widely used in the immobilization of bacteria, fungus and microalgae, drug delivery systems, and membrane systems [38–40]. However, to our best knowledge, NaCS/PDMDAAC has rarely been applied as a modified material in electrochemical sensors.

Herein, NaCS/PDMDAAC composite film as prepared to construct a novel electrochemical sensor for the detection of nitrite at the sub-micromolar level, aiming to make full use of the merits of individual components (biocompatible, low cost, good electrical conductivity for PDMDAAC; non-toxic, abundant raw materials, and the good film forming ability of NaCS) and synergistic enhancement effect (Scheme 1). NaCS/PDMDAAC/GCE was fabricated by coating the PDMDAAC and NaCS sequentially on the surface of GCE. The electrochemical performance were investigated by cyclic voltammetry (CV) and electrochemical impedance spectroscopy (EIS) using $[Fe(CN)_6]^{3-/4-}$ as redox probe solution. The analytical parameters (including pH) were systemically explored. Finally, the content of nitrite in ham sausage was determined by differential pulse voltammetry (DPV) with satisfactory results.

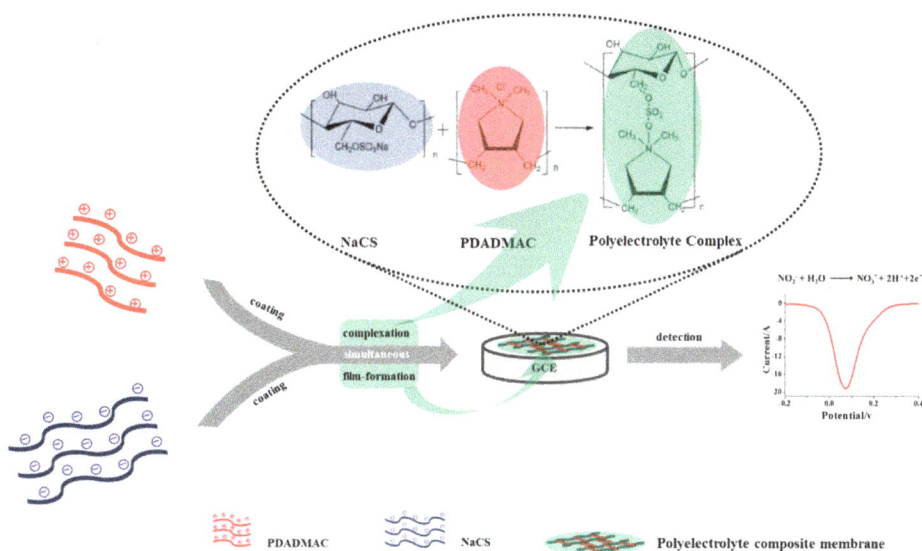

Scheme 1. Schematic of NaCS/PDMDAAC/GCE for the detection of nitrite detection.

2. Results and Discussion

2.1. Characterization of Surface Morphology

The surface morphologies of NaCS film and NaCS/PDMDAAC composite film were characterized by scanning electron microscope (SEM, Figure 1). As shown in Figure 1A, the NaCS film consists of acicular crystal structure. After compositing with NaCS/PDMDAAC, the surface is changed from acicular crystal to smooth film (Figure 2a), which is favorable for the electron transfer between the surface of modified electrode and analytes, and would eventually enhance electrocatalytic activity toward nitrite.

2.2. Electrochemical Characterization of NaCS/PDMDAAC/GCE

The electrochemical performance of bare GCE, NaCS/GCE, PDMDAAC/GCE, and NaCS/PDMDAAC/GCE were investigated using cyclic voltammetry (CV) recorded in 5×10^{-3} mol L^{-1} K$_3$[Fe(CN)$_6$] solution. The cyclic voltammograms recorded on various electrodes are shown in Figure 2. A pair of redox peaks appears on all the electrodes, suggesting the electrochemical process is quasi-reversible. A pair of weak and broad peaks was observed on the bare GCE (curve a), with redox peak separation (ΔE_p) of 137 mV. The anodic peak current (i_{pa}) and cathodic peak current (i_{pc}) is 43 μA, and 58 μA, respectively. On the NaCS modified electrode (curve b), the ΔE_p increases to 186 mV. Meanwhile, the peak currents also increase slightly, with i_{pa} of μA and i_{pc} of 69 μA, respectively. When modified with PDMDAAC alone (curve c), a pair of obvious and well-shaped redox peaks appears with a significant decrease on ΔE_p (101 mV). The corresponding redox peaks increase greatly, with i_{pa} of 81 μA and i_{pc} of 80 μA, respectively. When modified with NaCS/PDMDAAC composite film (curve d), the ΔE_p (143 mV) has a slight increase compared to that on the NaCS/GCE. The i_{pa} and i_{pc} is 72 μA and 89 μA, indicating that the NaCS/PDMDAAC film facilitates the electron transfer between the electrode and analyses, and eventually improves the electrochemical response.

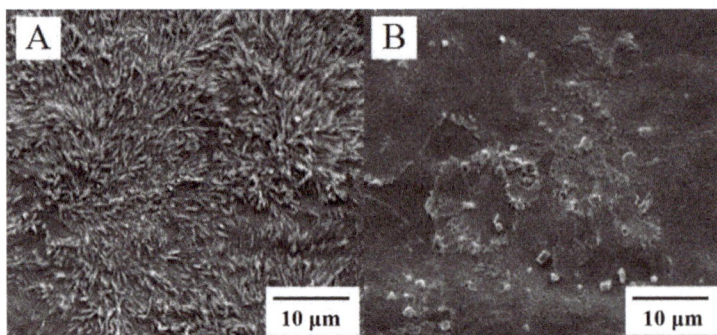

Figure 1. SEM images of NaCS film (**A**) and NaCS/PDMDAAC composite film (**B**).

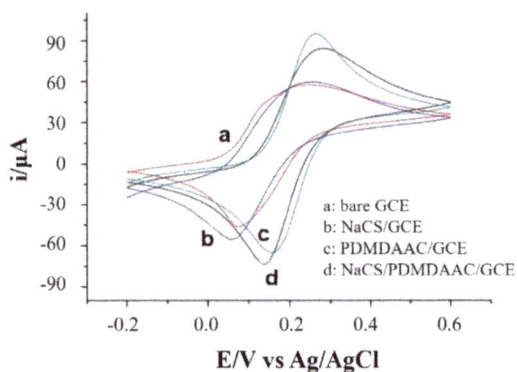

Figure 2. Cyclic voltammograms of bare GCE (a), NaCS/GCE (b), PDMDAAC/GCE (c), and NaCS/PDMDAAC/GCE (d) recorded in 5×10^{-3} mol·L^{-1} K$_3$[Fe(CN)$_6$] solution.

The influence of layers of NaCS/PDMDAAC composite film was also investigated in 5×10^{-3} mol L^{-1} K$_3$[Fe(CN)$_6$] solution by CV. As illustrated in Figure 3, the response peak currents decrease with the increase of layers of NaCS/PDMDAAC composite film. It is probably related to the charge transfer resistance that increases with the thickness increasing. Hence, monolayer NaCS/PDMDAAC composite film was employed for subsequent experiments.

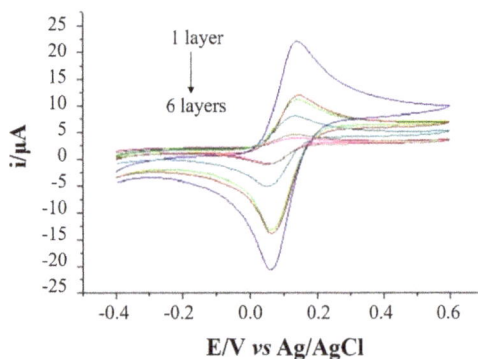

Figure 3. Cyclic voltammograms of different layer of PDMDAAC/NaCS modified GCE recorded in 5 $\times$ 10^{-3} mol·L^{-1} K$_3$[Fe(CN)$_6$] solution.

Electrochemical impedance spectroscopy (EIS) is a useful tool to acquire abundant information about electrode interface, which have been widely used in the electrocatalysis [41,42], chemosensors, and biosensors [43,44]. The EIS of bare GCE, NaCS/GCE, PDMDAAC/GCE, and NaCS/PDMDAAC/GCE were also measured in the 5×10^{-3} mol L^{-1} K$_3$[Fe(CN)$_6$] solution. The Nyquist plots of different electrodes were shown in Figure 4. The charge transfer resistance (R_{ct}) were estimated by EIS fitting with the equivalent circuit. The R_{ct} of bare GCE, NaCS/GCE, PDMDAAC/GCE, and NaCS/PDMDAAC/GCE are 37,830 Ω, 104,800 Ω, 36,810 Ω, and 52,190 Ω, respectively, i.e., the conductivity ($1/R_{ct}$) of them are 2.6434×10^{-5} S, 9.5420×10^{-6} S, 2.7167×10^{-5} S, and 1.9161×10^{-5} S, respectively. Apparently, the conductivity of the bare GCE (2.6434×10^{-5} S) is slightly lower than that of PDMDAAC/GCE (2.7167×10^{-5} S) but much higher than that of NaCS/GCE (9.5420×10^{-6} S), indicating that when each of them is independently modified onto the bare electrode, the polycation PDMDAAC can improve the conductivity while the polyanion NaCS is not favorable for electron transfer. As is well known, both PDMDAAC and NaCS are polyelectrolytes and, theoretically, should have good conductivities. However, these polyelectrolytes are quite different in charge density, solubility, and counter-ion effects in solution (ions will naturally attract the opposite-charged ions to form counter-ion layers), all of which are thought to be able to influence the conductivity, and the former two factors are thought to be favorable for conductivity while the last one is unfavorable. In fact, the molecular weight of PDMDAAC used in our experiment is less than 1.0×10^5, while NaCS has a much bigger average molecular weight more than 1.9×10^6, and so it is obvious that PDMDAAC contains much shorter molecular chains, which reinvests it with a higher charge density, a better solubility, as well as a weaker counter-ion effect due to narrower interfaces to attract less opposite-charged ions. Thus, although both being polyelectrolytes, PDMDAAC can improve the conductivity of the modified GCE while NaCS cannot, as the above results show. However, it is worth noting that when using the composite NaCS/PDMDAAC modifies the bare GCE, the R_{ct} of NaCS/PDMDAAC/GCE (52,190 Ω) is only about half of that on the NaCS/GCE (104,800 Ω), suggesting the unfavorable counter-ion effect of NaCS is significantly weakened after the introduction of PDMDAAC, which actually improves the charge transfer efficiency. Obviously, the conductivity of NaCS/PDMDAAC/GCE (1.9161×10^{-5} S) is better than that of NaCS/GCE (9.5420×10^{-6} S). Thus, by considering the outstanding conductivity of PDMDAAC as well as the good performances of NaCS for film formation, the composite NaCS/PDMDAAC is the most suitable material for the modification of GCE to construct the new and effective sensor for nitrite detection, herein with a moderate R_{ct}/conductivity value, which is consistent with the above research result conducted by the CV method.

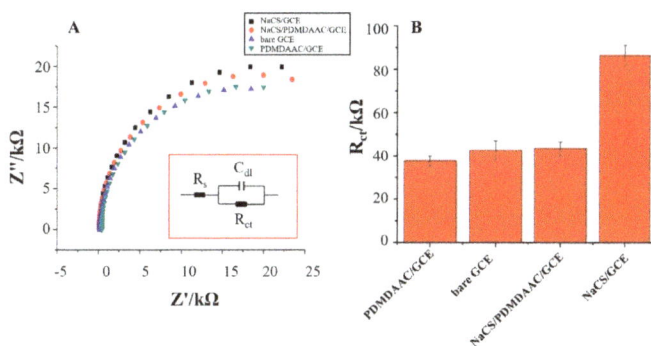

Figure 4. (**A**) Nyquist plots of different electrodes in 5×10^{-3} mol·L^{-1} K$_3$[Fe(CN)$_6$] solution obtained from electrochemical impedance spectroscopy (EIS). The frequency investigated was between 10,000 and 0.1 Hz with a pulse amplitude of 5 mV. The inset is the equivalent circuit of electrode/electrolyte interface. (**B**) The R_{ct} of different electrodes by EIS fitting.

2.3. Electrochemical Behavior of Nitrite on NaCS/PDMDAAC/GCE

The electrochemical behaviors of nitrite on bare GCE and NaCS/PDMDAAC/GCE were investigated by CV in 0.1 mol·L^{-1} PBS solution containing 4.0×10^{-3} mol·L^{-1} nitrite (Figure 5). The anodic peak current on the bare GCE is 59 μA. When modified with NaCS/PDMDAAC composite film, the anodic peak current increases to 97 μA, suggesting the NaCS/PDMDAAC composite film improve the electrochemical responses of nitrite.

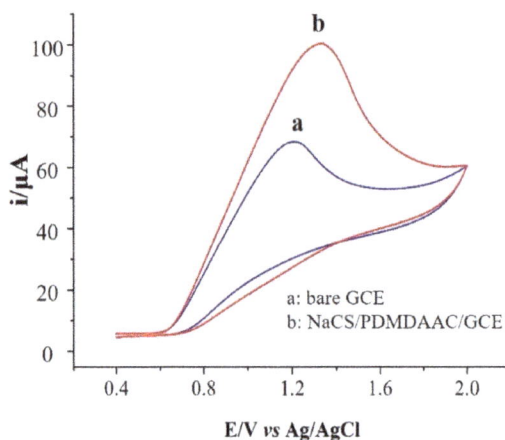

Figure 5. Cyclic voltammograms of bare GCE (a) and NaCS/PDMDAAC/GCE (b) in 0.1 mol·L^{-1} PBS solution (pH 7.0) containing 4.0×10^{-3} mol·L^{-1} nitrite.

2.4. Optimization of Analytical Conditions

2.4.1. Effect of the Modifier Loading Amount

The effect of the modifier loading amount is one of important factors that affect the electrochemical response of modified electrode. The bare GCEs were modified with various dosage of NaCS/PDMDAAC dispersion ($v/v = 1{:}1$) firstly. Then the response peak currents of 1.0×10^{-4} mol·L^{-1} nitrite on these electrodes were recorded and compared. The effect of modifier loading on the response peak current was shown in Figure 6A. The response peak currents of nitrite vary with the loading amount of NaCS/PDMDAAC composite. The response peak current increases gradually with the loading amount increasing from 2–6 μL, afterwards the response peak currents decrease with the further increase on loading amount. The maximum response peak current is obtained when the loading amount is 6 μL. Namely, the loading amounts of PDMDAAC and NaCS are 3 μL. Hence, the 6 μL loading amount was used in the subsequent experiments. The trend of response peak current to loading amount is directly related to the amount of reactive active sites and the thickness of modifier films. With loading amount increasing, the amount of available reactive active sites increases, which causes the increase on the response peak current. However, the thickness of the modified film increases when the loading amount beyond 6 μL, which hinders the electron transfer between the electrode and nitrite. As a result, the response peak current decreases with the further increase on the loading amount.

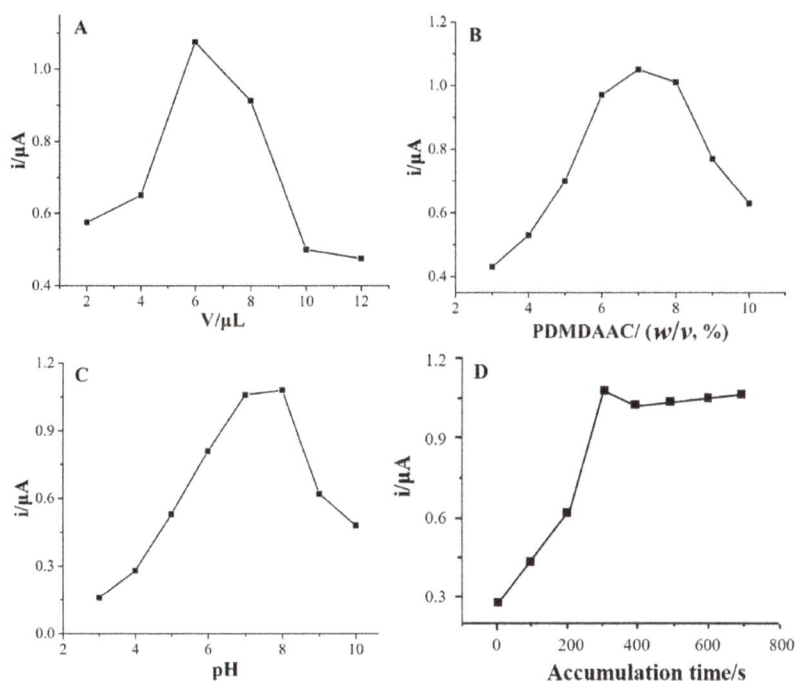

Figure 6. Effect of loading amount of NaCS/PDMDAAC (**A**), concentration of PDMDAAC(w/v) (**B**), pH (**C**) and accumulation time (**D**) on the response peak current of nitrite (1.0×10^{-4} mol·L^{-1}) at the NaCS/PDMDAAC/GCE.

2.4.2. Effect of PDMDAAC Concentration

The effect of the PDMDAAC concentration on the response peak current of nitrite was also explored. The results are shown in Figure 6B. It observed that the response peak current increases as the increase of PDMDAAC concentration. When the concentration (w/v) is 7%, the peak current reaches the maximum. Afterwards the response peak current decreases with further increase on PDMDAAC concentration. Therefore, 7% was selected to be the suitable PDMDAAC concentration for the following measurements.

2.4.3. Effect of pH

It is well known that pH plays a crucial role on the oxidation of nitrite, so it's well worth investigating the influence of pH. The response peak currents of 1.0×10^{-4} mol·L^{-1} nitrite were measured in various pH (3.0–10.0), and the result was presented in Figure 6C. When the electrolyte solution closes to neutral (pH 7.0 or 8.0), the response peak current is largest. Therefore, pH 7.0 was recommended as the optimum pH. Under acidic conditions, NO$_2^-$ is very unstable and easily reacted with H$^+$ to produce NO and NO$_2$ gas. As a result, the response peak current declines as the pH decreasing. In addition, most amount of the NO$_2^-$ is protonated in such acidic environment (pK_a(HNO$_2$) = 3.3). Under alkaline pH ranging of 8.0–10.0, the increase on pH leads to the decline of the response peak current due to the lack of sufficient H$^+$.

2.4.4. Effect of Accumulation Time

As we all know, accumulation time directly influences the adsorption of analyses at the surface of modified electrode. The NaCS/PDMDDA/GCE was accumulated for various times firstly, then corresponding response peak currents were record and compared (Figure 6D). During the first 300 s, the response peak currents increase with the time prolonging. It is probably due to the increasing amount of nitrite adsorbed on the reaction active sites. The highest response peak current was obtained for 300 s accumulation. Afterwards the response peak current remain stable, suggesting the nitrite adsorption has achieved saturation. Thus, 300 s was selected as the optimal accumulation time.

2.5. Standard Curves, Linear Ranges, and Detection Limit

Under the optimized analytical conditions, the response peak currents of various concentrations of nitrite were measured by NaCS/PDMDAAC/GCE using differential pulse voltammetry (DPV). The response DPV curves of various concentrations of nitrite are shown in Figure 7. There is a good relationship between the response peak current and the nitrite concentration ranging from 4.0×10^{-8} mol·L^{-1} to 1.5×10^{-4} mol·L^{-1}. The corresponding linear regression equation can expressed as i (µA) = 0.01027c (µmol·L^{-1}) + 0.02771 (r = 0.9984, n = 8). The limit of detection (LOD, S/N = 3) is estimated to 4.3×10^{-8} mol·L^{-1}. A comparison on sensing performances between reported electrodes and NaCS/PDMDAAC/GCE is summarized on the Table 1. The sensing performances (in terms of linear response range and limit of detection) of the NaCS/PDMDAAC/GCE are at least comparable to, and even better than, the most reported modified electrodes [45–50].

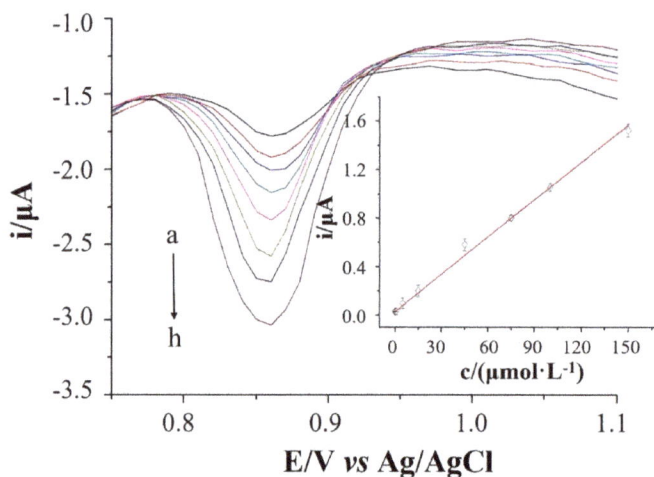

Figure 7. Differential pulse voltammograms of NaCS/PDMDAAC/GCE in the nitrite solution of different concentrations with 0.1 V pulse amplitude, 0.01 s pulse width and 0.1 V s^{-1} scan rate. (a) 1.5 $\times 10^{-4}$ mol·L^{-1}; (b) 1.0×10^{-4} mol·L^{-1}; (c) 7.5×10^{-5} mol·L^{-1}; (d) 4.5×10^{-5} mol·L^{-1}; (e) 1.5×10^{-5} mol·L^{-1}; (f) 5×10^{-6} mol·L^{-1}; (g) 5×10^{-7} mol·L^{-1}; (h) 4×10^{-8} mol·L^{-1}.

Table 1. A comparison on sensing performances between reported electrodes and NaCS/PDMDAAC/GCE.

Electrodes	Methods	Linear Ranges (µmol·L^{-1})	LOD (µmol·L^{-1})	References
POSS/rGO/GCE	CA	0.5–120	0.08	[45]
PNB/GCE	DPV	0.5–100	0.1	[46]
β-MnO$_2$ NRs/GCE	CA	0.29–26,090	0.29	[47]
Ag/HNT/MoS$_2$/CPE	CA	2–425	0.7	[48]
MWCNTs-TiN/GCE	CA	1–2000	0.0014	[49]
hemin/TNT/GCE	CA	0.6–130	0.084	[50]
NaCS/PDMDAAC/GCE	DPV	150–0.04	0.043	This work

2.6. Selectivity, Reproducibility, and Stability of NaCS/PDMDAAC/GCE

Before detection nitrite in real samples, the practicability (including selectivity, reproducibility, and stability) of the proposed NaCS/PDMDAAC/GCE were checked carefully. Firstly, the interference investigation was performed in 1×10^{-5} mol·L^{-1} sodium nitrite mixed with the equal concentration of potential interfering substances including sodium chloride, glucose, L-cysteine, and ascorbic acid. The response peak current is hardly changed with these interfering compounds (peak current variation within 5%), indicating the nitrite is bonded specifically to the NaCS/PDMDAAC composite film and interfering substances cannot interfere with the nitrite detection. In other words, the proposed sensor has good specificity and selectivity toward nitrite.

Then, the reproducibility of NaCS/PDMDAAC/GCE was validated by repeatedly measuring the response peak currents for 9 times in 8.0×10^{-5} mol·L^{-1} and 8.0×10^{-6} mol·L^{-1} nitrite solution alternately. The result is shown in Table 1. Obviously, the response peak current remains stable after repeated measurements for 9 times, with relative standard deviation (RSD) of 1.05% and 4.89%, respectively (Table 2). It demonstrates that the proposed NaCS/PDMDAAC exhibits good reproducibility for nitrite detection.

Table 2. The reproducibility of NaCS/PDMDAAC/GCE for nitrite determination.

No.	Current Response (µA)	
	$8.0 \times 10-5$ mol·L^{-1}	$8 \times 10-6$ mol·L^{-1}
1	0.896	0.115
2	0.874	0.123
3	0.887	0.112
4	0.888	0.121
5	0.878	0.114
6	0.890	0.117
7	0.898	0.112
8	0.894	0.129
9	0.903	0.122
Average	0.889	0.118
SD	0.009	0.006
RSD (%)	1.05%	4.89%

Finally, the stability of the proposed sensor was investigated by continuously monitoring the response peak current for 30 days. The NaCS/PDMDAAC/GCE was always immersed in 0.1 mol·L^{-1} PBS solution (pH 7.0) and stored at 4 °C for 30 days. The cyclic voltammetric response of nitrite at the NaCS/PDMDAAC/GCE was measured every 3 days. As shown in Figure 8, after 24 days, the current response of nitrite remains unchanged nearly. After 30 days, the response peak current value still retains 73.6% of the initial value. It can be inferred that the NaCS/PDMDAAC/GCE have good stability, which is very especially important for real sample detection.

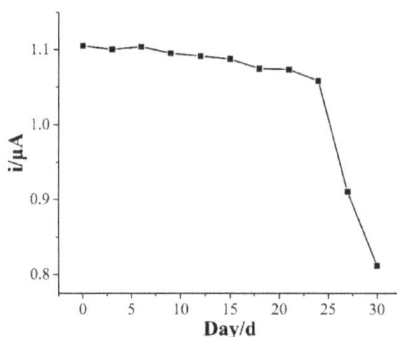

Figure 8. Stability for NaCS/PDMDAAC/GCE monitored in the 1.0×10^{-4} mol·L^{-1} nitrite for 30 days.

2.7. Determination of Actual Samples

Under the optimal analytical conditions, the pretreated ham sausage filtrate was used as the actual sample. The concentration of nitrite was detected by DPV method in 0.1 mol·L^{-1} PBS (pH 7.0) using the proposed NaCS/PDMDAAC/GCE, and determination results are listed in Table 3. The unspiked content of nitrite are 26.24−27.32 mg/kg, which does not exceed the maximum permitted content (0.15 g/kg) [9]. Then these samples were spiked with known concentration nitrite, the recovery is estimated to 98.2–104.0%, with relative standard deviation (RSD) lower than 3.3%, suggesting the satisfactory results are obtained. Together with outstanding advantages including low cost, sensitivity, and selectivity, the proposed NaCS/PDMDAAC/GCE shows great prospects on the detection of nitrite in diverse food samples.

Table 3. Determination of nitrite in ham sausage samples using the NaCS/PDMDAAC/GCE.

Samples	Content (mg/kg)	Added (mg/kg)	Founded (mg/kg)	Recovery (%)	RSD (%)
1	27.32	30	58.52	104.0	2.3
2	26.87	30	57.32	102.6	1.4
3	26.24	30	55.71	98.2	3.3

3. Conclusions

In this work, a novel electrochemical sensor based on NaCS/PDMDAAC composite film was proposed for the detection of nitrite at sub-micromolar level. The NaCS/PDMDAAC/GCE was fabricated by a facile drop-casting method. The surface morphology of NaCS film and NaCS/PDMDAAC composite was characterized by SEM. The electrochemical behavior of nitrite at the NaCS/PDMDAAC was investigated by CV and DPV. The NaCS/PDMDAAC not only shows the inherent properties from the individual component, but also exhibits synergetic enchantment effect toward sensing nitrite. Under optimal detection conditions, the proposed NaCS/PDMDAAC/GCE exhibits a wide linear response region of 4.0×10^{-8} mol·L^{-1}~1.5×10^{-4} mol·L^{-1}, and low detection limit of 43 nmol·L^{-1} (S/N = 3). Moreover, the proposed NaCS/PDMDAAC/GCE also has good stability and reproducibility. Finally, the NaCS/PDMDAAC/GCE was successfully used to detect the nitrite content in the ham sausage sample with satisfactory results. The proposed NaCS/PDMDAAC/GCE provided a promising sensing platform on the detection nitrite in diverse real samples at the sub-micromolar level.

4. Materials and Methods

4.1. Chemicals

Poly(dimethyldiallylammonium chloride) (PDMDAAC) was supplied by Aldrich Co., Ltd (Shanghai, China). Cellulose sulfate sodium salt was purchased from J&K SCIENTIFIC LTD (Shanghai, China). Borax and sodium nitrite were purchased from Hunan General Chemical Reagent Works (Changsha, China). Potassium ferricyanide, potassium hexacyanoferrate, sodium dihydrogen phosphate and disodium hydrogen phosphate were all bought from Sinopharm Chemical Reagent Co., Ltd (Shanghai, China). All the chemicals were of analytical reagent grade and were used as received without any further purification. Both PDMDAAC and NaCS were also synthesized in our experiments (details can be seen in the Supporting Information).

4.2. Fabrication of NaCS/PDMDAAC/GCE

The synthesis of NaCS and PDMDAAC has been described in the Supporting Information. The infrared spectra of cellulose sodium sulfate, ^{1}H NMR spectrum of DMDAAC, infrared spectra of (a) PDMDAAC & (b) DMDAAC have been also plotted in the Supporting Information (Figure S1,

Figure S2 & Figure S3). GCE was polished to mirror-like surface with 0.3 μm and 0.05 μm alumina powder, rinsed twice with ethanol and distilled water, and dried under natural condition. Firstly, 3 μL PDMDAAC (7% (w/v)) solution was drop-casted on the surface of GCE, dried naturally to obtain PDMDAAC/GCE. Secondly, 3 μL NaCS (3.5% (w/v)) solution was dropped-casted to the surface of PDMDAAC/GCE, dried naturally to obtain NaCS/PDMDAAC/GCE.

4.3. Electrochemical Measurements

All electrochemical experiments were performed on the CHI-760B electrochemical workstation (Shanghai Chenhua Instrument Co., Ltd., Shanghai, China). A standard three-electrode system is used for all electrochemical experiments, consisting of bare GCE or modified electrodes as the working electrode, platinum wire as the counter electrode, and an Ag/AgCl electrode (saturated potassium chloride) as the reference electrode. The electrochemical performance was measured by cyclic voltammetry (CV) and electrochemical impedance spectroscopy (EIS) in 5×10^{-3} mol·L^{-1} K$_3$[Fe(CN)$_6$] solution. The electrochemical behavior of nitrite at the NaCS/PDMDAAC/GCE was investigated by CV. The content of nitrite was detected by differential pulse voltammetry (DPV). The supporting electrolyte for all electrochemical measurements is 0.1 mol·L^{-1} Na$_2$HPO$_4$-NaH$_2$PO$_4$ buffer solution (PBS, pH 7.0, containing 0.1 M NaCl) unless otherwise stated.

4.4. Pretreatment of Ham Sausage Samples

The ham sausage samples were purchased from a local supermarket. 12.5 mL of borax saturation solution was added into 5 g ground ham sausage, and stirred to mix well. Then the ham sausage sample was washed into 50 mL volumetric flask by a small amount of hot water, and heated in boiling water bath for 15 min. Afterwards the volumetric flask was cooled to room temperature. Subsequently, 2.5 mL PBS solution was added into the volumetric flask, and then dilute with distilled water to 50 mL. After 30 min rest, the upper layer of fat or oil was removed, the spare liquor was extracted by suction filtration to obtain sample solution for quantitative analysis.

Supplementary Materials: The following are available online.

Author Contributions: J.N. and J.W. conceived and designed the experiments; J.N., X.L., M.W. and J.L. performed the experiments; D.L., R.H., D.C., J.N., and J.W. analyzed the data; J.N., and J.W. contributed reagents/materials/analysis tools; J.N. and J.W. wrote the paper; all authors read and approved the final manuscript.

Funding: This research was funded by the NSFC (no. 21505005), the National Key R&D Program of China (2017YFC1600306), Hunan Provincial Natural Science Foundation (no. 2018JJ2424), Open Project of National Engineering Laboratory of Hydrometallurgical Cleaner Production Technology of Chinese Academy of Sciences (No. 20120220CH1), Scientific Research Fund of Hunan Provincial Education Department (no. 17C0034), Open Research Program of Hunan Provincial Key Laboratory of Power and Transportation Materials (no. 2017CL07), and the Huxiang Youth Talent Support Program (2015RS4051).

Conflicts of Interest: The authors declare no conflict of interest.

References

1. Noor, N.S.; Tan, L.L.; Heng, L.Y.; Chong, K.F.; Tajuddin, S.N. Acrylic microspheres-based optosensor for visual detection of nitrite. *Food Chem.* **2016**, *207*, 132–138. [CrossRef] [PubMed]

2. Kumar, J.V.; Karthik, R.; Chen, S.M.; Balasubramanian, P.; Muthuraj, V.; Selvam, V. A novel cerium tungstate nanosheets modified electrode for the effective electrochemical detection of carcinogenic nitrite ions. *Electroanalysis* **2017**, *29*, 2385–2394. [CrossRef]

3. Rastogi, P.; Ganesan, V.; Krishnamoorthi, S. A promising electrochemical sensing platform based on a silver nanoparticles decorated copolymer for sensitive nitrite determination. *J. Mater. Chem. A* **2013**, *2*, 933–943. [CrossRef]

4. Manassaram, D.M.; Backer, L.C.; Moll, D.M. A review of nitrites in drinking water: Maternal exposure and adverse reproductive and developmental outcomes. *Environ. Health Perspect.* **2006**, *114*, 320–327. [CrossRef] [PubMed]

5. Bukowski, J.; Somers, G.; Bryanton, J. Agricultural contamination of groundwater as a possible risk factor for growth restriction or prematurity. *J. Occup. Environ. Med.* **2001**, *43*, 377–383. [CrossRef] [PubMed]

6. Brender, J.D.; Olive, J.M.; Felkner, M.; Suarez, L.; Marckwardt, W.; Hendricks, K.A. Dietary nitrites and nitrites, nitrosatable drugs, and neural tube defects. *Epidemiology* **2004**, *15*, 330–336. [CrossRef] [PubMed]

7. Kikura-Hanajiri, R.; Martin, R.S.; Lunte, S.M. Indirect measurement of nitric oxide production by monitoring nitrite and nitrite using microchip electrophoresis with electrochemical detection. *Anal. Chem.* **2002**, *74*, 6370–6377. [CrossRef] [PubMed]

8. WHO. *Guidelines for Drinking-Water Quality*, 3rd ed.; World Health Organization: Geneva, Switzerland, 2008; ISBN 9789241547611.

9. Mao, Y.; Bao, Y.; Han, D.; Zhao, B. Research progress on nitrite electrochemical sensor. *Chin. J. Anal. Chem.* **2018**, *46*, 147–156. [CrossRef]

10. Li, Y.; Whitaker, J.S.; McCarty, C.L. Reversed-phase liquid chromatography/electrospray ionization/mass spectrometry with isotope dilution for the analysis of nitrite and nitrite in water. *J. Chromatogr.* **2011**, *1218*, 476–483. [CrossRef] [PubMed]

11. He, L.; Zhang, K.; Wang, C.; Luo, X.; Zhang, S. Effective indirect enrichment and determination of nitrite ion in water and biological samples using ionic liquid-dispersive liquid-liquid microextraction combined with high-performance liquid chromatography. *J. Chromatogr.* **2011**, *1218*, 3595–3600. [CrossRef] [PubMed]

12. Manzoori, J.L.; Sorouraddin, M.H.; Hajishabani, A.M. Spectrophotometric determination of nitrite based on its catalytic effect on the oxidation of carminic acid by bromate. *Talanta* **1998**, *46*, 1379–1386. [CrossRef]

13. Pourreza, N.; Fat'Hi, M.R.; Hatami, A. Indirect cloud point extraction and spectrophotometric determination of nitrite in water and meat products. *Microchem. J.* **2012**, *104*, 22–25. [CrossRef]

14. Peng, Q.; Xie, M.G.; Koon-Gee, N.; Kang, E.T. A new nitrite-selective fluorescent sensor fabricated from surface-initiated atom-transfer radical polymerization. *Chem. Lett.* **2005**, *34*, 1628–1629. [CrossRef]

15. Jiao, C.X.; Niu, C.G.; Huan, S.Y.; Shen, Q.; Yang, Y.; Shen, G.L.; Yu, R.Q. A reversible chemosensor for nitrite based on the fluorescence quenching of a carbazole derivative. *Talanta* **2004**, *64*, 637–643. [CrossRef] [PubMed]

16. Lin, K.; Ma, J.; Yuan, D.; Feng, S.; Su, H.; Huang, Y.; Shangguan, Q. Sequential determination of multi-nutrient elements in natural water samples with a reverse flow injection system. *Talanta* **2017**, *167*, 166–171. [CrossRef] [PubMed]

17. Feng, S.; Zhang, M.; Huang, Y.; Yuan, D.; Zhu, Y. Simultaneous determination of nanomolar nitrite and nitrite in seawater using reverse flow injection analysis coupled with a long path length liquid waveguide capillary cell. *Talanta* **2013**, *117*, 456–462. [CrossRef] [PubMed]

18. Ning, J.; He, Q.; Luo, X.; Wang, M.; Liu, D.; Wang, J.; Liu, J.; Li, G. Rapid and sensitive determination of vanillin based on a glassy carbon electrode modified with Cu_2O-electrochemically reduced graphene oxide nanocomposite film. *Sensors* **2018**, *18*, 2762. [CrossRef] [PubMed]

19. He, Q.; Liu, J.; Liang, J.; Liu, X.; Li, W.; Liu, Z.; Ding, Z.; Tuo, D. Towards Improvements for Penetrating the Blood–Brain Barrier—Recent Progress from a Material and Pharmaceutical Perspective. *Cells* **2018**, *7*, 24. [CrossRef] [PubMed]

20. He, Q.; Liu, J.; Liu, X.; Li, G.; Deng, P.; Liang, J. Preparation of Cu_2O-reduced graphene nanocomposite modified electrodes towards ultrasensitive dopamine detection. *Sensors* **2018**, *18*, 199. [CrossRef] [PubMed]

21. He, Q.; Liu, J.; Liu, X.; Li, G.; Chen, D.; Deng, P.; Liang, J. Fabrication of amine-modified magnetite-electrochemically reduced graphene oxide nanocomposite modified glassy carbon electrode for sensitive dopamine determination. *Nanomaterials* **2018**, *8*, 194. [CrossRef] [PubMed]

22. He, Q.; Liu, J.; Liang, J.; Liu, X.; Tuo, D.; Li, W. Chemically Surface Tunable Solubility Parameter for Controllable Drug Delivery—An Example and Perspective from Hollow PAA-Coated Magnetite Nanoparticles with R6G Model Drug. *Materials* **2018**, *11*, 247. [CrossRef] [PubMed]

23. He, Q.; Liu, J.; Liu, X.; Li, G.; Deng, P.; Liang, J. Manganese dioxide Nanorods/electrochemically reduced graphene oxide nanocomposites modified electrodes for cost-effective and ultrasensitive detection of Amaranth. *Colloid Surf. B* **2018**, *172*, 565–572. [CrossRef] [PubMed]

24. Ma, Y.; Song, X.; Ge, X.; Zhang, H.; Wang, G.; Zhang, Y.; Zhao, H. In situ growth of α-Fe_2O_3 nanorod arrays on 3d carbon foam as an efficient binder-free electrode for highly sensitive and specific determination of nitrite. *J. Mater. Chem. A* **2017**, *5*, 4726–4736. [CrossRef]

25. Mehmeti, E.; Stanković, D.M.; Hajrizi, A.; Kalcher, K. The use of graphene nanoribbons as efficient electrochemical sensing material for nitrite determination. *Talanta* **2016**, *159*, 34–39. [CrossRef] [PubMed]

26. Rabti, A.; Aoun, S.B.; Raouafi, N. A sensitive nitrite sensor using an electrode consisting of reduced graphene oxide functionalized with ferrocene. *Microchim. Acta* **2016**, *183*, 3111–3117. [CrossRef]

27. Dai, J.; Deng, D.; Yuan, Y.; Zhang, J.; Deng, F.; He, S. Amperometric nitrite sensor based on a glassy carbon electrode modified with multi-walled carbon nanotubes and poly(toluidine blue). *Microchim. Acta* **2016**, *183*, 1553–1561. [CrossRef]

28. Shivakumar, M.; Nagashree, K.L.; Manjappa, S.; Dharmaprakash, M.S. Electrochemical detection of nitrite using glassy carbon electrode modified with silver nanospheres (AgNS) obtained by green synthesis using pre-hydrolysed liquor. *Electroanalysis* **2017**, *29*, 1434–1442. [CrossRef]

29. Seo, Y.; Manivannan, S.; Kang, I.; Lee, S.W.; Kim, K. Gold dendrites co-deposited with m13 virus as a biosensor platform for nitrite ions. *Biosens. Bioelectron.* **2017**, *94*, 87–93. [CrossRef] [PubMed]

30. Kung, C.W.; Li, Y.S.; Lee, M.H.; Wang, S.Y.; Chiang, W.H.; Ho, K.C. In-situ growth of porphyrinic metal-organic framework nanocrystals on graphene nanoribbons for electrocatalytic oxidation of nitrite. *J. Mater. Chem. A* **2016**, *4*, 10673–10682. [CrossRef]

31. Kung, C.W.; Chang, T.H.; Chou, L.Y.; Hupp, J.T.; Farha, O.K.; Ho, K.C. Porphyrin-based metal–organic framework thin films for electrochemical nitrite detection. *Electrochem. Commun.* **2015**, *58*, 51–56. [CrossRef]

32. Xu, G.; Liang, S.; Fan, J.; Sheng, G.; Luo, X. Amperometric sensing of nitrite using a glassy carbon electrode modified with a multilayer consisting of carboxylated nanocrystalline cellulose and poly(diallyldimethyl ammonium) ions in a pedot host. *Microchim. Acta* **2016**, *183*, 2031–2037. [CrossRef]

33. Chen, L.; Liu, X.; Wang, C.; Lv, S.; Chen, C. Amperometric nitrite sensor based on a glassy carbon electrode modified with electrodeposited poly(3,4-ethylenedioxythiophene) doped with a polyacene semiconductor. *Microchim. Acta* **2017**, *184*, 2073–2079. [CrossRef]

34. Masumi, T.; Matsushita, Y.; Dan, A.; Takama, R.; Saito, K.; Kuroda, K.; Fukushima, K. Adsorption behavior of poly(dimethyl-diallylammonium chloride) on pulp fiber studied by cryo-time-of-flight secondary ion mass spectrometry and cryo-scanning electron microscopy. *Appl. Surf. Sci.* **2014**, *289*, 155–159. [CrossRef]

35. Wang, B.; Okoth, O.K.; Yan, K.; Zhang, J. A highly selective electrochemical sensor for 4-chlorophenol determination based on molecularly imprinted polymer and PDDA-functionalized graphene. *Sens. Actuators B Chem.* **2016**, *236*, 294–303. [CrossRef]

36. Yang, J.; Lin, Q.; Yin, W.; Jiang, T.; Zhao, D.; Jiang, L. A novel nonenzymatic glucose sensor based on functionalized pdda-graphene/cuo nanocomposites. *Sens. Actuators B Chem.* **2017**, *253*, 1087–1095. [CrossRef]

37. Yang, T.; Zhang, W.; Du, M.; Jiao, K. A PDDA/poly(2,6-pyridinedicarboxylic acid)-CNTs composite film DNA electrochemical sensor and its application for the detection of specific sequences related to PAT gene and NOS gene. *Talanta* **2008**, *75*, 987–994. [CrossRef] [PubMed]

38. Chen, G.; Yao, S.J.; Guan, Y.X.; Lin, D.Q. Preparation and characterization of NaCS-cmc/PDMDAAC capsules. *Colloids Surf. B Biointerfaces* **2005**, *45*, 136–143. [CrossRef] [PubMed]

39. Zeng, X.H.; Danquah, M.K.; Zheng, C.; Potumarthi, R.; Chen, X.D.; Lu, Y.H. NaCS-PDMDAAC immobilized autotrophic cultivation of chlorella sp. For wastewater nitrogen and phosphate removal. *Chem. Eng. J.* **2004**, *99*, 185–192. [CrossRef]

40. Zhang, J.; Guan, Y.X.; Ji, Z.; Yao, S.J. Effects on membrane properties of NaCS–PDMDAAC capsules by adding inorganic salts. *J. Membr. Sci.* **2006**, *277*, 270–276. [CrossRef]

41. Wang, M.; Chen, K.; Liu, J.; He, Q.; Li, G.; Li, F. Efficiently enhancing electrocatalytic activity of α-MnO$_2$ nanorods/N-doped ketjenblack carbon for oxygen reduction reaction and oxygen evolution reaction using facile regulated hydrothermal treatment. *Catalysts* **2018**, *8*, 138. [CrossRef]

42. Chen, K.; Wang, M.; Li, G.; He, Q.; Liu, J.; Li, F. Spherical α-MnO$_2$ supported on N-KB as efficient electrocatalyst for oxygen reduction in al–air battery. *Materials* **2018**, *11*, 601. [CrossRef] [PubMed]

43. Li, G.; Wang, S.; Duan, Y.Y. Towards gel-free electrodes: A systematic study of electrode-skin impedance. *Sens. Actuators B Chem.* **2017**, *241*, 1244–1255. [CrossRef]

44. Li, G.; Zhang, D.; Wang, S.; Duan, Y.Y. Novel passive ceramic based semi-dry electrodes for recording electroencephalography signals from the hairy scalp. *Sens. Actuators B Chem.* **2016**, *237*, 167–178. [CrossRef]

45. Bai, W.; Sheng, Q.; Zheng, J. Hydrophobic interface controlled electrochemical sensing of nitrite based on one step synthesis of polyhedral oligomeric silsesquioxane/reduced graphene oxide nanocomposite. *Talanta* **2016**, *150*, 302–309. [CrossRef] [PubMed]

46. Chen, X.; Wang, F.; Chen, Z. An electropolymerized nile blue sensing film-based nitrite sensor and application in food analysis. *Anal. Chim. Acta* **2008**, *623*, 213–220. [CrossRef] [PubMed]

47. Feng, J.J.; Zhang, P.P.; Wang, A.J.; Zhang, Y.; Dong, W.J.; Chen, J.R. One-pot hydrothermal synthesis of uniform β-MnO_2 nanorods for nitrite sensing. *J. Colloid Interface Sci.* **2011**, *359*, 1–8. [CrossRef] [PubMed]

48. Ghanei-Motlagh, M.; Taher, M.A. A novel electrochemical sensor based on silver/halloysite nanotube/molybdenum disulfide nanocomposite for efficient nitrite sensing. *Biosens. Bioelectron.* **2018**, *109*, 279–285. [CrossRef] [PubMed]

49. Haldorai, Y.; Hwang, S.K.; Gopalan, A.I.; Huh, Y.S.; Han, Y.K.; Voit, W.; Saianand, G.; Lee, K.P. Direct electrochemistry of cytochrome c immobilized on titanium nitride/multi-walled carbon nanotube composite for amperometric nitrite biosensor. *Biosens. Bioelectron.* **2016**, *79*, 543–552. [CrossRef] [PubMed]

50. Ranjani, B.; Kalaiyarasi, J.; Pavithra, L.; Devasena, T.; Pandian, K.; Gopinath, S. Amperometric determination of nitrite using natural fibers as template for titanium dioxide nanotubes with immobilized hemin as electron transfer mediator. *Microchim. Acta* **2018**, *185*, 194. [CrossRef] [PubMed]

molecules

MDPI

Article

The Content of Biogenic Amines in Croatian Wines of Different Geographical Origins

Ivana Mitar [1], Ivica Ljubenkov [1,*], Nikolina Rohtek [2], Ante Prkić [3], Ivana Anđelić [1] and Nenad Vuletić [1]

[1] Department of Chemistry, Faculty of Science, University of Split, Ruđera Boškovića 33, 21000 Split, Croatia; imitar@pmfst.hr (I.M.); angel@pmfst.hr (I.A.); nenov@pmfst.hr (N.V.)
[2] University Department for Forensic Sciences, University of Split, Ruđera Boškovića 33, 21000 Split, Croatia; rohtekn@yahoo.com
[3] Department of Analytical Chemistry, Faculty of Chemistry and Technology, University of Split, Ruđera Boškovića 35, 21000 Split, Croatia; prkic@ktf-split.hr
*   Correspondence: iljubenk@pmfst.hr; Tel.: +385-21-619-291

Received: 6 September 2018; Accepted: 7 October 2018; Published: 9 October 2018

check for updates

Abstract: Samples of white and red wines produced in two different wine-growing regions, coastal (Dalmatia) and continental (Hrvatsko zagorje) of Croatia, were analysed for biogenic amines content. Biogenic amines content was determined, and its concentration levels were associated with the geographical origin of the wine. Due to its high sensitivity, HPLC method with ultraviolet detector was used, including the derivatisation step with dansyl chloride. The method was applied to detect and quantify 11 biogenic amines in 48 red and white wines. It was found that both Dalmatian red and white wines are characterised by tryptamine (0.23–1.22 mg L^{-1}), putrescine (0.41–7.5 mg L^{-1}) and ethanolamine (2.87–24.32 mg L^{-1}). White wines from the Hrvatsko zagorje region are characterised by content of isopentylamine (0.31–1.47 mg L^{-1}), putrescine (0.27–1.49 mg L^{-1}) and ethanolamine (3.80–17.96 mg L^{-1}). In contrast to white wines from the Hrvatsko zagorje region, in the red wines, all biogenic amines except ethylamine, were found and equally presented.

Keywords: Croatian wines; biogenic amines; HPLC; geographical origin

1. Introduction

Recently, there has been a great interest of scientists to find a way to control the global wine market. There are numerous parameters determining the quality of the wine. These parameters can be classified as chemical and sensory parameters. Wine is a beverage wherein the quality depends on many factors, among which grape variety, origin, vintage, grape growing conditions, winemaking practice and maturation process, physical conditions of production and way of storage, are the most significant, and they also influence sensory characteristics. Over the past century, chemists have played a significant role in the determination of wine chemical composition and its association with wine flavour and sensory attributes. In the global wine market, wine identity (brand, type, vintage and origin) is extremely important and all those characteristics are crucial for the determination of its price. In the past century, chemists had developed powerful tools for detecting adulteration of wine, such as the addition of water, glycerol, alcohol, dyes, sweeteners, flavour substances and a non-authorised addition of sugar or acidity adjustment [1]. Therefore, in the last few years, there has been a great interest from scientists as well as consumers and commercial wine producers, on the geographical origin and authenticity of wines in terms of quality and price determination. In the 19th and early 20th centuries, the focus was on detecting fraud, while more recently the emphasis has been on quantifying trace compounds, especially those that may be related to a grape variety [2]. There is a large number of studies dealing with the classification of wines according to geographical origin, e.g., assuming that the concentrations or ratios of some chemical parameters in wine depend on its geographical origin [1]. According to some studies, volatile components [3], polyphenols [3–7], as well as elemental composition [8–11], can be linked to the geographical origin or grape variety.

Biogenic amines (BAs) are in the focus of wine quality analysis, and generally of food quality control of many studies. BAs can be found in different fermented foods, such as milk, cheese or beer as well as in wine. They are low molecular weight organic compounds. In wine they can originate from grapes, or can be produced during fermentations (alcoholic and malolactic). In most wines, the level of BAs is low after alcoholic fermentation, while after malolactic fermentation their level increases [12]. Also, they can be formed during wine ageing or storage processes, especially if wine is exposed to the microorganisms' activity or free amino acids are present in it. The formations of BAs from free amino acids can be through reaction of decarboxylation (histamine, tyramine, putrescine, cadaverine), transamination, reductive amination or degradation of their amino acids precursors [12]. The conditions such as storage temperature, pH, and presence of oxygen, sulphur dioxide or sodium chloride content in wine are important factors that affect the concentration of BAs in the final product [13–18]. All mentioned conditions act synergistically. Obviously, with some combinations, BAs may be increased through progressive fermentation process, e.g., elevated storage temperature, pH and O_2 presence. On the other hand, elevated SO_2 and NaCl concentrations diminish fermentation process and decrease BAs content in wines.

Although BAs are considered essential for many physiological functions, such as: body temperature regulation, stomach pH value and brain activity, the frequent and prolonged intake of BAs through wine consummation causes various health problems, such as headaches, flushing, itching, skin irritation, hypertension, etc. [13,15]. BAs are found to be toxic in cases of the intake of foods or beverages that contain them in large amounts. BAs suspected of having toxicological effects are histamine, tryptamine and phenethylamine. Therefore, the level of BAs in wine can be a safety indicator as well as an important parameter for grading wine quality [15]. Polyamines are usually associated with deficient sanitary conditions, especially putrescine and cadaverine. There is a large variability in content and distribution of BAs in wine.

A large number of studies on BAs in commercially available wines have been carried out in the world's most important wine producing countries: Turkish red wines [19], Greek red wines [20,21], Chinese red wines [22], Chilean wines [23], Brazilian wines [16], Portuguese [24] and Spanish red and white wine samples [25–27], purchased from local stores with the aim of monitoring or determining the content of biogenic amines, the high content of which disrupts the quality or proves bad hygiene conditions of wine. More than 20 biogenic amines have been identified in the mentioned studies.

However, there are only a few studies on BAs content with the purpose of determining wine geographical origin or grape variety. BAs are naturally present in grapes (putrescine, spermidine, histamine and tyramine) [28] and in initial musts (ethanolamine, tyramine, putrescine, cadaverine, phenylethylamine and spermidine) [29]. Del Prete et al. detected ethanolamine, ethylamine and putrescine in grapes [30], while Ladente et al. found that differences in putrescine concentration may be attributed to certain grape varieties [31]. According to those studies, the concentrations of BAs that are naturally present in must are directly connected to a grape variety or a soil type, and finally, to the grape's geographical origin. Marques et al., explained a correlation between BAs and grape variety on the samples of red wines produced in three different Portuguese regions [32], while a group of Italian researchers in their study on red wines concluded that BAs composition is a feature of a particular geographic region [4]. Red wines are richer in the content of biogenic amines [15], which can be related to the fact that secondary fermentation (malolactic fermentation) is less usual in white wines [21,27]. Therefore, significantly smaller number of studies are conducted on white wine samples.

Croatia, like many other Mediterranean countries such as Spain, France, Italy, Greece and Turkey, has a long-standing tradition of wine production. In Croatia, there is a large number of small manufacturers who produce small quantities of wine, so the Croatian market is also known as a market of a large number of monovarietal wines [33]. As the authors know, there are only a few studies on BAs content in Croatian wines. Kovačević Ganić et al. investigated BAs content in red wine samples from Slavonia wine region and changes in BAs content during the winemaking and maturation processes. They reported the presence of 10 BAs: tryptamine, hydroxylamine, phenylethylamine, putrescine, cadaverine, histamine, tyramine, serotonin, spermine and spermidine in investigated samples [14]. Jeromel et al. investigated the concentration of BAs in red wines from northwest Croatia with the aim of comparison of BAs levels in wines produced by classic and cold maceration. In their study, the most abundant BAs were histamine, tryptamine and 2-phenylethylamine, while tyramine, putrescine, cadaverine, spermidine, spermine and serotonin were also detected in significantly lower concentrations [34]. There is also a lack of studies on BAs in Croatian white wine samples.

Although numerous analytical methods have been reported for the determination of the BAs in wines and other beverages or food samples [24,35,36], high-performance liquid chromatography (HPLC) is preferred by most researchers [27,30,37–39]. There are differences in the derivatisation procedures among studies, but the most commonly used reagents are o-phthalalhehyde (OPA), dabsyl chloride (DABS-Cl) and dansyl chloride (DNS-Cl).

Although, researchers are usually using OPA as a derivatisation agent, we used dansyl chloride because of its stability when exposed to UV-Vis spectra (in detection system). Also, OPA reacts only with primary amines, which prevents the determination of polyamines such as spermine and spermidine whose presence has been previously reported in Croatian wines.

In this study, 48 samples of Croatian white and red wines were analysed. The investigated wines were produced from native and introduced grape varieties, characteristic of two Croatian wine regions; Dalmatia (coastal wine region) and Hrvatsko zagorje (continental wine region). Also, this research was conducted with the aim of comparison of the BAs concentration in red and white wines from the same wine regions.

2. Results

2.1. Samples

A total of 48 samples of wines from two different Croatian wine regions are listed in Table 1 (samples from coastal wine region of Dalmatia) and Table 2 (samples from continental wine region of Hrvatsko zagorje) with corresponding grape variety and origin. All wine samples were obtained directly from small farmers of the corresponding region.

Table 1. The investigated red and white wine samples from Croatian coastal wine region of Dalmatia.

Number	Sample Mark	Grape Variety	Origin	Type of Wine
1	1DB	Debit	Drniš	white
2	2DB	Pošip	Bol, Brač	white
3	3DB	Pošip barrique	Bol, Brač	white
4	4DB	Kujundžuša	Rogoznica	white
5	5DB	Maraština	Kaštela	white
6	6DB	Maraština	Bol, Brač	white
7	7DB	Chardonnay	Kaštela	white
8	8DB	Pošip barrique	Kaštela	white
9	9DB	Kujundžuša + Graševina	Imotski	white
10	10DB	Debit	Sinj	white
11	11DB	Debit	Šibenik	white
12	1DC	Plavina + Lasin + Shiraz	Drniš	red
13	2DC	Bogondon	Bol, Brač	red
14	3DC	Tribidrag	Bol, Brač	red
15	4DC	Plavac	Rogoznica	red
16	5DC	Plavac	Kaštela	red
17	6DC	Plavac	Bol, Brač	red
18	7DC	Crljenak	Kaštela	red
19	8DC	Plavac	Kaštela	red
20	9DC	Crljenak	Bol, Brač	red
21	10DC	Crljenak	Imotski	red
22	11DC	Plavac	Sinj	red
23	12DC	Plavac	Šibenik	red
24	13DC	Plavac	Knin	red

Table 2. The investigated red and white wine samples from Croatian continental wine region of Hrvatsko zagorje.

Number	Sample Mark	Grape Variety	Origin	Type of Wine
1	1ZB	Graševina	Beretinec	white
2	2ZB	Pinot	Moslavec	white
3	3ZB	Rajnski rizling	Varaždin	white
4	4ZB	Graševina	Varaždin	white
5	5ZB	Muškat žuti	Varaždin	white
6	6ZB	Sivi pinot	Varaždin	white
7	7ZB	Manzoni	Beretinec	white
8	8ZB	Graševina	Varaždin	white
9	9ZB	Traminac	Beretinec	white
10	14ZB	Sauvignon	Cestica	white
11	1ZC	Frankovka	Beretinec	red
12	2ZC	Isabella + Farber	Kneginec	red
13	3ZC	Shiraz	Varaždin	red
14	4ZC	Frankovka + Plavac	Ludbreg	red
15	5ZC	Frankovka	Cestica	red
16	6ZC	Frankovka	Ledinec	red
17	7ZC	Merlot + Frankovka	Beretinec	red
18	8ZC	Portugizac	Ivanec	red
19	10ZC	Isabella	Ledinec	red
20	11ZC	Frankovka	Beretinec	red
21	12ZC	Frankovka + Merlot	Beretinec	red
22	13ZC	Isabella	Beretinec	red
23	14ZC	Isabella	Ledinec	red
24	15ZC	Frankovka	Beretinec	red

2.2. Biogenic Amines Determinations

The content of biogenic amines was determined by HPLC method, as described by Manetta et al. [40], with slight modifications.

The derivatisation was performed without pre-treatment of the samples as follows: 0.25 mL of BAs standard solution or wine sample was mixed with 70 µL of a saturated sodium hydrogen carbonate solution and 65 µL 0.1 M potassium hydroxide solution. Then, 1 mL of dansyl chloride (0.5% *w/v* in acetone) was added and the mixture was incubated for 45 min at 40 °C in thermoblock with occasional stirring. After that, 100 µL of ammonia solution (25% *w/w*) was added and after strong stirring by vortex, the reaction mixtures were left in dark for 30 min. The volume of samples was made up to 5 mL with acetonitrile, and after the shaking they were filtrated and ready for the analysis. The control samples (blank) was prepared by the same procedure, but instead of standards or wine samples, ultrapure water was used.

All samples were prepared and analysed in triplicate, and the data are presented as a mean value ± standard deviation.

Gradient elution was conducted using acetonitrile (solvent A), and ultrapure water (solvent B) according to the program presented in Table 3.

Table 3. The HPLC gradient elution program used for the analysis of biogenic amines.

Time (min)	Solvent A (%)	Solvent B (%)
0.5	40	60
25	80	20
30	95	5
34	95	5
35	40	60
43	40	60

The applied flow rate was 1 mL min^{-1}, column temperature was 25 °C, the sample injection volume was 10 µL and the detection wavelength was 254 nm. The identification of BAs was carried out by comparing their retention times, individually or in a mixture of standards as is shown at Figure 1. Peaks appearing in the chromatogram that are not assigned to any standard are secondary products of the derivatisation process. The quantification was done using linear calibration curves that were created for every standard compound of BAs.

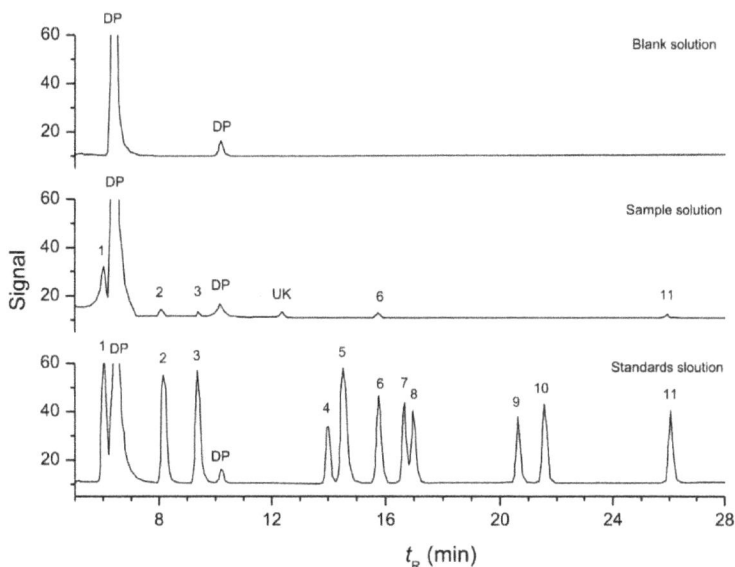

Figure 1. HPLC chromatogram of the blank sample, wine sample and standards' solution of biogenic amines. The numbers correspond to amines reported in Table 4, DP—derivation peak, UK—unknown peak.

As ethanolamine, ethylamine and methylamine were obtained as hydrochloride salts, their concentrations in standard solutions were corrected as for a free base.

The limits of detection (LODs) and the limits of quantification (LOQs) were determined using signal-to-noise-ratio (S/N) of 3 and 10 respectively for all standards. Table 4 is showing calculated analytical parameters for the applied method.

The detected concentrations of 11 investigated biogenic amines (tryptamine, putrescine, cadaverine, histamine, tyramine, spermidine, spermine, isopentylamine, ethanolamine, methylamine and ethylamine) in 48 samples of Croatian red and white wines from Hrvatsko zagorje and Dalmatia wine regions are given in Table 5. The samples can be divided into four groups: Dalmatian white wines (11 samples), Dalmatian red wines (13 samples), Hrvatsko zagorje white wines (10 samples) and Hrvatsko zagorje red wines (14 samples). The results are presented in Table 5 as the average value ± relative standard deviation (RSD) (%) of all samples.

Table 4. The analytical parameters of a chromatographic method.

Number	Amine Name (Short Name)	t_R (min) ± RSD (n = 3)	LOD (mg L^{-1})	LOQ (mg L^{-1})	R^2	Recovery White Wines (%)	Recovery Red Wines (%)	Linear Range (mg L^{-1})
1	Ethanolamine (ETHA)	5.79 ± 0.04	0.85	2.50	0.998	102	106	2.50–300.00
2	Methylamine (MA)	7.92 ± 0.05	0.14	0.41	0.999	123	88	0.41–100.00
3	Ethylamine (ETHYL)	9.07 ± 0.04	0.20	0.59	0.999	104	90	0.59–100.00
4	Tryptamine (TRP)	14.05 ± 0.04	0.03	0.06	0.998	103	82	0.06–215.00
5	Isopentylamine (IPA)	14.49 ± 0.04	0.22	0.66	0.999	107	93	0.66–200.00
6	Putrescine (PUT)	15.81 ± 0.05	0.03	0.09	0.998	97	88	0.09–96.70
7	Cadaverine (CAD)	16.71 ± 0.04	0.03	0.10	0.998	117	109	0.10–107.50
8	Histamine (HIS)	16.97 ± 0.04	0.03	0.10	0.998	99	91	0.10–107.50
9	Tyramine (TYR)	20.66 ± 0.04	0.04	0.13	0.999	107	92	0.13–107.50
10	Spermidine (SPD)	21.60 ± 0.03	0.03	0.09	0.998	107	101	0.09–96.70
11	Spermine (SPM)	26.12 ± 0.02	0.03	0.10	0.998	108	93	0.10–107.50

t_R—retention time, RSD—relative standard deviation, LOD—limit of detection, LOQ—limit of quantification, R—correlation coefficient.

Table 5. The content of biogenic amines (mg L^{-1}) in 48 investigated wine samples.

Sample	TRP	IPA	PUT	CAD	HIS	TYR	SPD	SPM	ETHA	MA	ETHYL
1DB	<0.03	<0.22	1.21 ± 0.04	<0.03	<0.03	<0.04	<0.03	<0.03	24.32 ± 8.05	<0.14	<0.59
2DB	<0.03	0.29 ± 0.21	1.19 ± 0.02	<0.03	<0.03	<0.04	<0.03	0.19 ± 0.09	10.42 ± 0.09	<0.14	<0.20
3DB	<0.03	<0.22	0.99 ± 0.04	<0.03	<0.10	0.39 ± 0.04	<0.03	0.39 ± 0.21	12.08 ± 0.83	1.40 ± 0.01	<0.20
4DB	0.49 ± 0.38	<0.22	0.41 ± 0.03	<0.03	<0.03	<0.04	<0.09	<0.10	9.09 ± 0.24	<0.14	<0.20
5DB	<0.03	<0.22	1.06 ± 0.00	<0.03	<0.03	<0.04	<0.09	<0.03	6.71 ± 0.00	<0.14	<0.20
6DB	0.42 ± 0.13	<0.22	0.86 ± 0.53	<0.03	<0.03	<0.04	<0.03	<0.03	2.87 ± 0.45	<0.14	<0.20
7DB	1.22 ± 1.12	<0.22	1.15 ± 0.01	<0.03	<0.03	<0.04	<0.03	0.11 ± 0.01	7.83 ± 0.33	<0.14	<0.20
8DB	0.68 ± 0.04	<0.22	1.25 ± 0.01	<0.03	0.56 ± 0.03	0.40 ± 0.01	<0.03	<0.10	4.22 ± 0.91	<0.14	<0.20
9DB	0.27 ± 0.18	<0.22	0.72 ± 0.02	<0.03	<0.03	<0.04	<0.03	<0.03	<2.5	<0.14	<0.20
10DB	0.38 ± 0.25	<0.22	2.10 ± 0.06	<0.10	<0.03	<0.04	<0.09	<0.03	<0.85	<0.14	<0.59
11DB	<0.03	<0.22	1.58 ± 0.02	<0.10	<0.03	<0.04	<0.03	<0.10	9.68 ± 0.04	<0.14	<0.59
1DC	<0.03	<0.22	1.58 ± 0.03	<0.03	<0.03	<0.04	<0.03	<0.03	4.85 ± 0.25	<0.14	<0.20
2DC	<0.03	1.13 ± 0.16	7.50 ± 0.17	<0.03	2.33 ± 0.08	2.00 ± 0.00	0.18 ± 0.04	0.31 ± 0.15	13.24 ± 0.13	<0.14	<0.20
3DC	0.54 ± 0.15	<0.22	1.76 ± 0.04	<0.10	0.89 ± 0.05	1.57 ± 0.01	<0.03	0.27 ± 0.13	14.00 ± 0.38	<0.14	<0.20
4DC	1.20 ± 0.01	<0.22	1.31 ± 0.00	<0.03	<0.03	<0.04	<0.03	<0.10	<0.85	<0.14	<0.20
5DC	0.43 ± 0.21	<0.22	1.35 ± 0.02	<0.03	<0.03	<0.04	0.09 ± 0.00	<0.03	<2.5	<0.14	<0.20
6DC	0.43 ± 0.18	<0.22	3.47 ± 0.09	<0.03	1.64 ± 0.01	1.88 ± 0.04	0.14 ± 0.05	0.24 ± 0.20	8.77 ± 0.50	<0.14	<0.20
7DC	0.23 ± 0.04	<0.22	0.78 ± 0.04	1.75 ± 0.03	0.89 ± 0.02	0.32 ± 0.03	<0.03	0.14 ± 0.03	5.11 ± 0.07	<0.14	<0.20
8DC	0.71 ± 0.04	<0.22	1.45 ± 0.02	<0.03	0.52 ± 0.01	<0.13	<0.03	<0.03	<2.5	<0.14	<0.20

Table 5. *Cont.*

Sample	TRP	IPA	PUT	CAD	HIS	TYR	SPD	SPM	ETHA	MA	ETHYL
9DC	0.34 ± 0.15	<0.22	0.78 ± 0.00	1.98 ± 0.09	1.17 ± 0.01	<0.13	0.46 ± 0.25	<0.10	9.22 ± 0.06	<0.14	<0.20
10DC	0.43 ± 0.35	<0.22	1.91 ± 0.11	<0.03	<0.03	<0.04	<0.03	0.17 ± 0.08	<0.85	<0.14	<0.20
11DC	<0.03	<0.22	1.77 ± 0.01	0.12 ± 0.06	<0.03	<0.04	0.12 ± 0.01	<0.03	10.75 ± 0.45	<0.14	<0.20
12DC	0.46 ± 0.30	<0.22	2.77 ± 0.07	<0.03	<0.03	<0.13	<0.09	<0.10	6.58 ± 0.17	<0.14	<0.20
13DC	0.66 ± 0.04	<0.22	0.89 ± 0.03	1.01 ± 0.14	0.36 ± 0.04	<0.04	<0.03	<0.03	6.36 ± 0.03	<0.14	<0.59
1ZB	0.68 ± 0.02	1.29 ± 0.04	0.27 ± 0.01	0.40 ± 0.00	<0.03	<0.04	<0.03	<0.10	4.07 ± 0.28	<0.14	<0.20
2ZB	<0.03	0.87 ± 0.49	1.49 ± 0.67	0.82 ± 0.31	<0.03	<0.04	<0.03	<0.03	17.96 ± 0.01	<0.14	<0.20
3ZB	<0.03	1.32 ± 0.00	0.62 ± 0.01	<0.03	<0.03	<0.04	<0.03	<0.03	4.80 ± 0.51	<0.14	<0.20
4ZB	<0.03	1.47 ± 0.32	1.19 ± 0.01	<0.03	<0.03	<0.04	<0.03	<0.10	8.61 ± 1.19	<0.14	<0.20
5ZB	<0.03	0.50 ± 0.09	1.08 ± 0.01	<0.03	<0.03	<0.04	<0.03	<0.03	<0.85	<0.14	<0.20
6ZB	<0.03	0.82 ± 0.21	<0.09	<0.03	<0.03	<0.04	<0.03	<0.03	3.80 ± 0.00	<0.14	<0.20
7ZB	<0.03	0.42 ± 0.46	1.42 ± 0.01	<0.03	<0.03	<0.04	<0.03	<0.03	4.61 ± 0.01	<0.14	<0.20
8ZB	<0.03	0.31 ± 0.22	0.83 ± 0.04	<0.03	<0.03	<0.04	0.12 ± 0.02	<0.10	4.16 ± 0.14	<0.14	<0.20
9ZB	0.33 ± 0.00	1.00 ± 0.80	1.16 ± 0.02	<0.03	<0.10	<0.04	<0.03	<0.10	7.19 ± 0.08	<0.14	<0.20
14ZB	<0.03	<0.22	0.94 ± 0.01	<0.10	<0.03	<0.04	<0.09	<0.03	<0.85	<0.14	<0.59
1ZC	0.75 ± 0.87	<0.22	0.17 ± 0.01	<0.03	<0.03	<0.13	0.21 ± 0.05	<0.03	<2.5	<0.14	<0.20
2ZC	1.82 ± 0.06	0.79 ± 0.32	2.06 ± 0.04	0.47 ± 0.05	1.06 ± 0.04	<0.13	0.75 ± 0.04	3.55 ± 0.04	15.41 ± 0.02	<0.14	<0.59
3ZC	3.31 ± 0.33	1.19 ± 0.01	3.75 ± 0.06	<0.10	9.63 ± 0.13	<0.13	0.26 ± 0.13	0.74 ± 0.04	4.10 ± 0.23	1.28 ± 0.01	<0.59
4ZC	4.89 ± 0.27	0.50 ± 0.06	0.39 ± 0.04	<0.10	0.35 ± 0.08	0.45 ± 0.01	0.15 ± 0.01	0.67 ± 0.07	8.04 ± 0.16	<0.14	<0.20
5ZC	3.61 ± 0.45	0.87 ± 0.04	3.17 ± 0.11	0.10 ± 0.00	0.50 ± 0.01	0.50 ± 0.04	0.22 ± 0.02	1.81 ± 0.01	16.80 ± 0.23	1.57 ± 0.04	<0.59
6ZC	0.88 ± 0.22	3.71 ± 0.14	<0.03	0.10	<0.03	0.31 ± 0.08	2.51 ± 0.21	<0.03	95.84 ± 0.20	0.99 ± 0.55	<0.20
7ZC	0.39 ± 0.31	2.44 ± 0.18	0.66 ± 0.78	0.21 ± 0.01	0.98 ± 0.02	2.10 ± 0.14	0.57 ± 0.07	<0.03	63.96 ± 0.20	<0.14	<0.20
8ZC	0.09 ± 0.01	2.97 ± 2.44	<0.09	<0.03	<0.10	0.78 ± 0.71	0.33 ± 0.00	<0.03	31.14 ± 0.17	0.65 ± 0.03	<0.20
10ZC	0.67 ± 0.37	1.81 ± 0.02	0.16 ± 0.09	0.44 ± 0.20	<0.03	2.31 ± 0.05	0.15 ± 0.00	<0.03	51.79 ± 3.21	<0.14	<0.20
11ZC	9.18 ± 0.07	2.14 ± 0.48	<0.09	<0.10	<0.03	1.45 ± 0.22	6.05 ± 1.22	<0.03	35.19 ± 10.47	0.56 ± 0.15	<0.59
12ZC	0.87 ± 0.61	3.27 ± 0.35	<0.09	0.27 ± 0.32	2.01 ± 0.14	2.97 ± 0.09	1.97 ± 0.20	<0.03	46.03 ± 12.33	<0.41	1.17 ± 0.21
13ZC	<0.03	1.55 ± 1.27	1.00 ± 0.04	<0.03	0.35 ± 0.06	0.13 ± 0.03	<0.03	<0.10	23.78 ± 1.50	<0.14	<0.20
14ZC	<0.03	0.33 ± 0.32	0.24 ± 0.02	<0.10	<0.03	<0.04	<0.09	<0.10	5.22 ± 0.28	<0.14	<0.20
15ZC	<0.03	0.70 ± 0.10	<0.03	<0.03	<0.03	<0.04	<0.03	<0.10	4.37 ± 0.20	<0.14	<0.20

The table contains abbreviations for biogenic amines, and the corresponding full names are reported in Table 4. DB = Dalmatian white samples, DC = Dalmatian red samples, ZB = Hrvatsko zagorje white samples and ZC = Hrvatsko zagorje red samples.

3. Discussion

In the case of polyamines (putrescine, cadaverine, spermidine and spermine), their high concentrations are usually associated with unsanitary conditions.

If we look at the detected concentrations of those polyamines in the investigated samples, in most white wines cadaverine, spermidine and spermine were not detected, while in red wines the highest concentrations detected were 1.98 mg L^{-1} of cadaverine, 2.51 mg L^{-1} of spermidine and 3.55 mg L^{-1} of spermine. Therefore, it can be concluded that the sanitary conditions of all the samples were satisfactory.

Histamine level also plays a special role as indicator amine. Histamine is the most toxic amine, although the toxicity is caused by histamine and the total content of amines, ethanol and acetaldehyde [41]. The allowed concentrations of histamine in wines are different across countries. According to available references, the highest histamine concentration of 10 mg L^{-1} is allowed in Switzerland [42]. According to the reported results, the highest concentration of histamine among tested samples was observed in the 3ZC sample (9.63 mg L^{-1}), while its content in other red wine samples was generally low (from 0.35 to 2.01 mg L^{-1}), and in white wines even below the detection limit, except for sample 8 DB, where it was found at concentration of 0.56 mg L^{-1}.

Putrescine and ethanolamine were the most prominent amines in all of the samples, regardless of the type of wine or its origin.

According to the Bover-Cid et al. [25], Glória et al. [43] and Kiss et al. [44] putrescine, spermine and spermidine are naturally present in grapes and their presence could be an indicator of the wine's geographical region or grape variety.

Putrescine concentrations were slightly higher in red wines, which can be explained by the fact that it can be formed during malolactic fermentation that usually occurs in the process of red wine production [25]. The highest concentration of putrescine in red wine samples was found in sample 2DC (7.5 mg L^{-1}). In comparison to results reported in other studies, these concentrations were relatively low [15,16,19,21–24,27,30,37,45–47]. The putrescine content in samples was investigated by Landete et al. ranged from 30 to 50 mg L^{-1} [31], while in our study they ranged from 0.16 to 3.75 mg L^{-1}. Spermidine was not detected in white wines, except in the sample 8ZB where its concentration was 0.12 mg L^{-1}. In red wine samples, spermidine was found in the range from 0.09 to 0.46 mg L^{-1} in Dalmatian red wines and from 0.15 to 6.05 mg L^{-1} in wine samples from Hrvatsko zagorje region.

Spermine was found in low concentrations in four samples of white wines from Dalmatian region (ranged from 0.11 to 0.39 mg L^{-1}), while in white wines from Hrvatsko zagorje region it was not detected. If we compare the results for spermine in red wines, slightly higher concentrations were detected in samples from Hrvatsko zagorje region than in Dalmatian wines but final concentrations correspond to those that are reported in the literature [20,43].

Del Prete et al. confirmed the presence of ethanolamine, ethylamine and putrescine in grapes [30].

Ethanolamine is an amine that was detected in almost all samples at significant concentrations and especially in red wines from Hrvatsko zagorje region. Its concentrations correspond to those reported in other studies of Mediterranean wines, such as in samples from Italy [4,30], Portugal [47], and Greece [20]. A large number of studies didn't research the ethanolamine content, but in our study, its concentration was found to be significant in almost all of the samples.

Ethylamine was detected only in sample 12ZC, and thus in a very low concentration of 1.17 mg L^{-1}, as well as methylamine, which was found in sample 3DB and in few red wines from Hrvatsko zagorje wine region at concentration range from 0.99 to 1.57 mg L^{-1}.

Jeromel et al. studied BAs in Croatian red wines, but in their study the content of ethylamine and methylamine were not investigated [34], while histamine and tryptamine were found to be the most abundant biogenic amines.

In this study, tryptamine was detected in the range from 0.2 to 1.2 mg L^{-1} in the wines from Dalmatia, while in the red samples from Hrvatsko zagorje region it was found in higher concentrations (from 0.09 to 9.18 mg L^{-1}).

It is interesting to see the distribution of isopentylamine in the samples. Its concentrations in the samples from Hrvatsko zagorje ranged from 0.3 to 1.5 mg L^{-1} in white wines and from 0.33 to 3.71 mg L^{-1} in red wines, while among Dalmatian samples, only samples 2DB and 2DC contained this compound in concentration from 0.29 and 1.13 mg L^{-1}, respectively.

Marques et al. in their research proved the connection between the content of tyramine and type of wine especially as its concentration is higher in red wines after malolactic fermentation [32].

Tyramine was found in very low concentrations in two white wines from Dalmatian region, while it was not detected at all in white wines from Hrvatsko zagorje region. As expected, red samples contained higher amounts of tyramine, but still significantly lower than those reported in the literature [32,45,46]. A very wide range of total BAs content has been reported, from not-detected to 130 mg L^{-1} with the main amines e.g., putrescine, histamine, tyramine and cadaverine [48].

A scree plot in Figure 2 suggests involving four principal components in the model. Those four PC (columns PC1–PC4) explain 70% of the total variance in the data, (Table 6). Since graphical presentation only allows for using two columns, the cut off point for loading values was >0.30 and it is marked throughout Table 6 in boldface type only for PC1 and PC2.

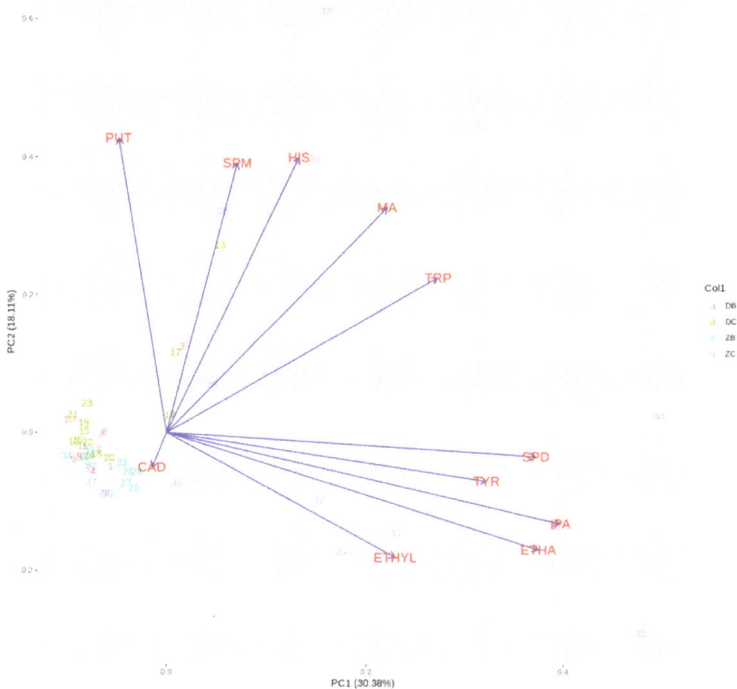

Figure 2. PCA plot. Table 4 contains abbreviations for biogenic amines. Numbers correspond to samples' in Tables 1 and 2. DB = Dalmatian white samples, DC = Dalmatian red samples, ZB = Hrvatsko zagorje white samples and ZC = Hrvatsko zagorje red samples.

Table 6. Loading values for PCA.

	PC1	PC2	PC3	PC4
TRP	**0.3160**	0.2615	−0.4331	0.1254
IPA	**0.4612**	−0.1534	0.0624	−0.1102
PUT	−0.0565	**0.4970**	0.3967	−0.1822
CAD	−0.0167	−0.0588	0.1155	0.9085
HIS	0.1533	**0.4641**	0.3395	0.0760
TYR	**0.3744**	−0.0824	0.4237	−0.0500
SPD	**0.4325**	−0.0409	−0.3308	0.0847
SPM	**0.3160**	0.2615	−0.4331	0.1254
ETHA	**0.4612**	−0.1534	0.0624	−0.1102
MA	−0.0565	**0.4970**	0.3967	−0.1822
ETHYL	−0.0167	−0.0588	0.1155	0.9085

Short amine names correspond to amines' name in Table 4.

The loading values express how well the new PCs correlate with old variables. The first PC, which explains 30.38% of the total variance correlates positively with Tryptamine (TRP), Isopentylamine (IPA), Tyramine (TYR), Spermidine (SPD) and Ethanolamine (ETHA). The second PC (18.11% of the total variance) correlates positively with Putrescine (PUT), Histamine (HIS), Spermine (SPM) and Methylamine (MA). On the other hand, it can be seen that CAD has small values. All data are given in Tables 6 and 7.

Table 7. Explanation of variance in statistical analysis.

	PC1	PC2	PC3	PC4
Standard deviation	1.8279	1.4114	1.1870	1.03047
Proportion of variance	0.3038	0.1811	0.1281	0.09653
Cumulative proportion	0.3038	0.4849	0.6130	0.70949

4. Materials and Methods

4.1. Chemical and Reagents

All used reagents were of analytical grade.

Dansyl chloride and amine standards (isopenthylamine, ethanolamine, methylamine, ethylamine, spermidine, spermine, putrescine, tyramine, histamine, cadaverine and tryptamine) were purchased from Sigma-Aldrich (Steinheim, Germany).

Hydrochloric acid (37%, *w/w*), ammonia solution (25%, *w/w*), sodium hydrogen carbonate, acetonitrile (HPLC grade) were also purchased from Sigma-Aldrich.

Ultrapure water was obtained from ELGA Purelab flex.

4.2. Apparatus and Software

A Perkin Elmer Series 200 HPLC system, equipped with an autosampler, binary pump and UV/Vis detector (all of Series 200), was applied with a TotalChrom Workstation software (PerkinElmer, Waltham, MA, USA).

Chromatographic separations were performed on a Restek Ultra IBD C18 column (5 μm particle size, 250 × 4.6 mm i.d.) with Ultra IBD guard column (5 μm particle size, 10 × 4 mm i.d.), Restek, Bellefonte, PA, USA.

4.3. Statistical Analysis

For exposing the underlying patterns in the data, principal component analysis (PCA) was used with the intention of showing which biogenic amines (wine samples) carry comparable information, and which of them are unique. The statistical analysis was carried out using the RStudio ver. 1.1.383 [49]

Molecules **2018**, *23*, 2570

while PCA analysis was done using 'prcomp' method by using singular value decomposition (SVD). Since SVD has slightly better numerical accuracy, therefore, 'prcomp' is the preferred function.

Biogenic amines were taken as variables (columns of the input matrix) and the various wines as cases (rows of the matrix). The underlying patterns, 'components' are represented by new variables called principal components.

5. Conclusions

This work shows that biogenic amines content can be a differentiation factor for a grape variety and geographical origin for red wines. It can be stated that Dalmatian white wines are characterised by tryptamine, putrescine and ethanolamine. Their content is in the following ranges: tryptamine from 0.23 to 1.22 mg L^{-1}; putrescine from 0.41 to 7.5 mg L^{-1} and ethanolamine from 2.87 to 24.32 mg L^{-1}. White wines from the Hrvatsko zagorje region are characterised by content of isopentylamine (from 0.31 to 1.47 mg L^{-1}), putrescine (from 0.27 to 1.49 mg L^{-1}) and ethanolamine (from 3.80 to 17.96 mg L^{-1}). On the other hand, in red wines from the Hrvatsko zagorje region all BAs, except ethylamine, were found. According to the PCA, the wines of the Hrvatsko zagorje red group samples are the most distinguished. Wines from the Hrvatsko zagorje red group marked as 36, 37 and 39 have a higher concentration of spermine, histamine, methylamine and putrescine, as well as lower concentration of spermidine, tyramine, isopentylamine, ethanolamine and ethylamine than wines from the same group in lines 41–45. All other wines are mostly concentrated around similar values, with the exceptions of Dalmatian red wines marked as 13 and 17, and Dalmatian white wine marked 3, respectively.

Author Contributions: Conceptualisation, I.M. and I.L.; methodology, I.L.; software, I.L. and A.P.; validation, I.M. and I.L.; formal analysis, A.P.; investigation, I.M. and N.R.; resources, N.R. and I.L.; data curation, I.M. and I.L.; writing—original draft preparation, I.M.; writing—review and editing; visualization, I.M.; supervision, I.L.; funding acquisition, I.A. and N.V.

Funding: This research received no external funding.

Acknowledgments: The authors would like to thank all companies that provided their samples to this research.

Conflicts of Interest: The authors declare no conflict of interest.

References

1. Schlesier, K.; Fauhl-Hassek, C.; Forina, M.; Cotea, V.; Kocsi, E.; Schoula, R.; van Jaarsveld, F.; Wittkowski, R. Characterisation and determination of the geographical origin of wines. Part I: Overview. *Eur. Food Res. Technol.* **2009**, *230*, 1–13. [CrossRef]

2. Ebeler, S.E.; Thorngate, J.H. Wine Chemistry and Flavor: Looking into the Crystal Glass. *J. Agric. Food Chem.* **2009**, *57*, 8098–8108. [CrossRef] [PubMed]

3. Marquez, R.; Castro, R.; Natera, R.; Garcia-Barroso, C. Characterisation of the volatile fraction of Andalusian sweet wines. *Eur. Food Res. Technol.* **2008**, *226*, 1479–1484. [CrossRef]

4. Galgano, F.; Caruso, M.; Perretti, G.; Favati, F. Authentication of Italian red wines on the basis of the polyphenols and biogenic amines. *Eur. Food Res. Technol.* **2011**, *232*, 889–897. [CrossRef]

5. Jaitz, L.; Siegl, K.; Eder, R.; Rak, G.; Abranko, L.; Koellensperger, G.; Hann, S. LC–MS/MS analysis of phenols for classification of red wine according to geographic origin, grape variety and vintage. *Food Chem.* **2010**, *122*, 366–372. [CrossRef]

6. Agatonovic-Kustrin, S.; Hettiarachchi, C.G.; Morton, D.W.; Razic, S. Analysis of phenolics in wine by high performance thin-layer chromatography with gradient elution and high resolution plate imaging. *J. Pharm. Biomed. Anal.* **2015**, *102*, 93–99. [CrossRef] [PubMed]

7. Marković, M.; Martinović Bevanda, A.; Talić, S. Antioxidant activity and total phenol content of white wine Žilavka. *Bull. Chem. Technol. Bosn. Herceg.* **2015**, *44*, 1–4.

8. Galgano, F.; Favati, F.; Camso, M.; Scarpa, T.; Palma, A. Analysis of trace elements in southern Italian wines and their classification according to provenance. *LWT-Food Sci. Technol.* **2008**, *41*, 1808–1815. [CrossRef]

9. Geana, I.; Iordache, A.; Ionete, R.; Marinescu, A.; Ranca, A.; Culea, M. Geographical origin identification of Romanian wines by ICP-MS elemental analysis. *Food Chem.* **2013**, *138*, 1125–1134. [CrossRef] [PubMed]

10. Ražić, S.; Čokeša, D.; Sremac, S. Multivariate data visualization methods based on elemental analysis of wines by atomic absorption spectrometry. *J. Serb. Chem. Soc.* **2007**, *72*, 1487–1492. [CrossRef]

11. Ražić, S.; Onjia, A. Trace Element Analysis and Pattern Recognition Techniques in Classification of Wine from Central Balkan Countries. *Am. J. Enol. Viticult.* **2010**, *61*, 506–511. [CrossRef]

12. Mo Dugo, G.; Vilasi, F.; La Torre, G.; Pellicano, T. Reverse phase HPLC/DAD determination of biogenic amines as dansyl derivatives in experimental red wines. *Food Chem.* **2006**, *95*, 672–676. [CrossRef]

13. Guo, Y.Y.; Yang, Y.P.; Peng, Q.; Han, Y. Biogenic amines in wine: A review. *Int. J. Food Sci. Technol.* **2015**, *50*, 1523–1532. [CrossRef]

14. Kovačević Ganić, K.; Gracin, L.; Komes, D.; Ćurko, N.; Lovrić, T. Changes of the content of biogenic amines during winemaking of Sauvignon wines. *Croat. J. Food Sci. Technol.* **2009**, *1*, 21–27.

15. Anlı, R.E.; Bayram, M. Biogenic Amines in Wines. *Food Rev. Int.* **2008**, *25*, 86–102. [CrossRef]

16. Souza, S.C.; Theodoro, K.H.; Souza, E.R.; da Motta, S.; Beatriz, M.; Gloria, A. Bioactive Amines in Brazilian Wines: Types, Levels and Correlation with Physico-Chemical Parameters. *Braz. Arch. Biol. Technol.* **2005**, *48*, 53–62. [CrossRef]

17. Košmerl, T.; Šućur, S.; Prosen, H. Biogenic amines in red wine: The impact of technological processing of grape and wine. *Acta Agric. Slov.* **2013**, *101*, 249–261. [CrossRef]

18. Cecchini, F.; Morassut, M. Effect of grape storage time on biogenic amines content in must. *Food Chem.* **2010**, *123*, 263–268. [CrossRef]

19. Anlı, R.E.; Vural, N.; Yılmaz, S.; Vural, Ỳ.H. The determination of biogenic amines in Turkish red wines. *J. Food Compos. Anal.* **2004**, *17*, 53–62. [CrossRef]

20. Soufleros, E.H.; Bouloumpasi, E.; Zotou, A.; Loukou, Z. Determination of biogenic amines in Greek wines by HPLC and ultraviolet detection after dansylation and examination of factors affecting their presence and concentration. *Food Chem.* **2007**, *101*, 704–716. [CrossRef]

21. Proestos, C.; Loukatos, P.; Komaitis, M. Determination of biogenic amines in wines by HPLC with precolumn dansylation and fluorimetric detection. *Food Chem.* **2008**, *106*, 1218–1224. [CrossRef]

22. Li, Z.; Wu, Y.; Zhang, G.; Zhao, Y.; Xue, C. A survey of biogenic amines in Chinese red wines. *Food Chem.* **2007**, *105*, 1530–1535. [CrossRef]

23. Pineda, A.; Carrasco, J.; Pena-Farfal, C.; Henriquez-Aedo, K.; Aranda, M. Preliminary evaluation of biogenic amines content in Chilean young varietal wines by HPLC. *Food Control* **2012**, *23*, 251–257. [CrossRef]

24. Fernandes, J.; Ferreira, M. Combined ion-pair extraction and gas chromatography–mass spectrometry for the simultaneous determination of diamines, polyamines and aromatic amines in Port wine and grape juice. *J. Chromatogr. A* **2000**, *886*, 183–195. [CrossRef]

25. Bover-Cid, S.; Iquierdo-Pulido, M.; Marine-Font, A.; Vidal-Carou, M.C. Biogenic mono-, di- and polyamine contents in Spanish wines and influence of a limited irrigation. *Food Chem.* **2006**, *96*, 43–47. [CrossRef]

26. Romero, R.; Gázquez, D.; Bagur, M.; Sánchez-Viñas, M. Optimization of chromatographic parameters for the determination of biogenic amines in wines by reversed-phase high-performance liquid chromatography. *J. Chromatogr. A* **2000**, *871*, 75–83. [CrossRef]

27. Romero, R.; Sanchez-Vinas, M.; Gazquez, D.; Bagur, M.G. Characterization of selected spanish table wine samples according to their biogenic amine content from liquid chromatographic determination. *J. Agric. Food Chem.* **2002**, *50*, 4713–4717. [CrossRef] [PubMed]

28. Moreno-Arribas, M.V.; Polo, M.C. *Wine Chemistry and Biochemistry*; Springer: New York, NY, USA, 2009; Volume 735.

29. Wang, Y.-Q.; Ye, D.-Q.; Zhu, B.-Q.; Wu, G.-F.; Duan, C.-Q. Rapid HPLC analysis of amino acids and biogenic amines in wines during fermentation and evaluation of matrix effect. *Food Chem.* **2014**, *163*, 6–15. [CrossRef] [PubMed]

30. Del Prete, V.; Costantini, A.; Cecchini, F.; Morassut, M.; Garcia-Moruno, E. Occurrence of biogenic amines in wine: The role of grapes. *Food Chem.* **2009**, *112*, 474–481. [CrossRef]

31. Landete, J.M.; Ferrer, S.; Polo, L.; Pardo, I. Biogenic amines in wines from three spanish regions. *J. Agric. Food Chem.* **2005**, *53*, 1119–1124. [CrossRef] [PubMed]

32. Marques, A.P.; Leitao, M.C.; Romao, M.V.S. Biogenic amines in wines: Influence of oenological factors. *Food Chem.* **2008**, *107*, 853–860. [CrossRef]

33. Alpeza, I.; Prša, I.; Mihaljević, B. Vinogradarstvo i vinarstvo Republike Hrvatske u okviru svijeta. *Glasnik Zaštite Bilja* **2014**, *37*, 6–13.

34. Jeromel, A.; Kovačević Ganić, K.; Herjavec, S.; Mihaljević, M.; Korenika Jagatić, A.M.; Rendulić, I.; Čolić, M. Concentration of Biogenic Amines in 'Pinot Noir' Wines Produced in Croatia. *Agric. Conspec. Sci.* **2012**, *77*, 37–40.

35. Almeida, C.; Fernandes, J.; Cunha, S. A novel dispersive liquid–liquid microextraction (DLLME) gas chromatography-mass spectrometry (GC–MS) method for the determination of eighteen biogenic amines in beer. *Food Control* **2012**, *25*, 380–388. [CrossRef]

36. Onal, A. A review: Current analytical methods for the determination of biogenic amines in foods. *Food Chem.* **2007**, *103*, 1475–1486. [CrossRef]

37. Busto, O.; Miracle, M.; Guasch, J.; Borrull, F. Determination of biogenic amines in wines by high-performance liquid chromatography with on-column fluorescence derivatization. *J. Chromatogr. A* **1997**, *757*, 311–318. [CrossRef]

38. Ganić, K.K.; Ćurko, N.; Kosić, U.; Komes, D.; Gracin, L. Determination of Biogenic Amines in Red Croatian Wines. In *32nd World Congress of Vine and Wine, Zagreb, Croatia 2009*; Kubanović, V., Ed.; Ministry of Agriculture, Fisheries and Rural Development: Zagreb, Croatia, 2009.

39. Galgano, F.; Caruso, M.; Favati, F.; Romano, P.; Caruso, M. HPLC determination on agmatine and other amines in wine. *J. Int. Sci. Vigne. Vin.* **2003**, *37*, 237–242. [CrossRef]

40. Manetta, A.C.; Di Giuseppe, L.; Tofalo, R.; Martuscelli, M.; Schirone, M.; Giammarco, M.; Suzzi, G. Evaluation of biogenic amines in wine: Determination by an improved HPLC-PDA method. *Food Control* **2016**, *62*, 351–356. [CrossRef]

41. Garcia-Marino, M.; Trigueros, A.; Escribano-Bailon, T. Influence of oenological practices on the formation of biogenic amines in quality red wines. *J. Food Compos. Anal.* **2010**, *23*, 455–462. [CrossRef]

42. Lehtonen, P. Determination of amines and amino acids in wine—A review. *Am. J. Enol. Vitic.* **1996**, *47*, 127–133.

43. Gloria, M.B.A.; Watson, B.T.; Simon-Sarkadi, L.; Daeschel, M.A. A survey of biogenic amines in Oregon Pinot noir and Cabernet Sauvignon wines. *Am. J. Enol. Viticult.* **1998**, *49*, 279–282.

44. Sass-Kiss, A.; Kiss, J.; Havadi, B.; Adanyi, N. Multivariate statistical analysis of botrytised wines of different origin. *Food Chem.* **2008**, *110*, 742–750. [CrossRef]

45. Lehtonen, P.; Saarinen, M.; Vesanto, M.; Riekkola, M.-L. Determination of wine amines by HPLC using automated precolumn derivatisation with o-phthalaldehyde and fluorescence detection. *Z. Lebensm.-Unters. Forsch.* **1992**, *194*, 434–437. [CrossRef]

46. Herbert, P.; Santos, L.; Alves, A. Simultaneous Quantification of Primary, Secondary Amino Acids, and Biogenic Amines in Musts and Wines Using OPA/3-MPA/FMOC-Cl Fluorescent Derivatives. *J. Food Sci.* **2001**, *66*, 1319–1325. [CrossRef]

47. Mafra, I.; Herbert, P.; Santos, L.; Barros, P.; Alves, A. Evaluation of Biogenic Amines in Some Portuguese Quality Wines by HPLC Fluorescence Detection of OPA Derivatives. *Am. J. Enol. Viticult.* **1999**, *50*, 128–132.

48. Restuccia, D.; Loizzio, M.R.; Spizzirri U., G. Accumulation of Biogenic Amines in Wine: Role of Alcoholic and Malolactic Fermentation. *Fermentation* **2018**, *4*, 6. [CrossRef]

49. R Development Core Team. *Integrated Development for R. Rstudio*; R Development Core Team: Vienna, Austria, 2016.

Sample Availability: Samples of the compounds are not available from the authors.

molecules

MDPI

Article

Quantification and Confirmation of Fifteen Carbamate Pesticide Residues by Multiple Reaction Monitoring and Enhanced Product Ion Scan Modes via LC-MS/MS QTRAP System

Ying Zhou, Jian Guan, Weiwei Gao, Shencong Lv and Miaohua Ge *

Jiaxing Center for Disease Control and Prevention, Zhejiang 314050, China; zhouyingand@sina.cn (Y.Z.); jguan001@sina.com (J.G.); dannyday@126.com (W.G.); thestorm2008@126.com (S.L.)
* Correspondence: gemiaohua0823@sina.com; Tel.: +86-0573-83683808

Academic Editor: Alessandra Gentili
Received: 29 August 2018; Accepted: 27 September 2018; Published: 29 September 2018

Abstract: In this research, fifteen carbamate pesticide residues were systematically analyzed by ultra-high performance liquid chromatography–quadrupole-linear ion trap mass spectrometry on a QTRAP 5500 system in both multiple reaction monitoring (MRM) and enhanced product ion (EPI) scan modes. The carbamate pesticide residues were extracted from a variety of samples by QuEChERS method and separated by a popular reverse phase column (Waters BEH C18). Except for the current conformation criteria including selected ion pairs, retention time and relative intensities from MRM scan mode, the presence of carbamate pesticide residues in diverse samples, especially some doubtful cases, could also be confirmed by the matching of carbamate pesticide spectra via EPI scan mode. Moreover, the fragmentation routes of fifteen carbamates were firstly explained based on the mass spectra obtained by a QTRAP system; the characteristic fragment ion from a neutral loss of CH_3NCO (-57 Da) could be observed. The limits of detection and quantification for fifteen carbamates were 0.2–2.0 $\mu g\ kg^{-1}$ and 0.5–5.0 $\mu g\ kg^{-1}$, respectively. For the intra- ($n = 3$) and inter-day ($n = 15$) precisions, the recoveries of fifteen carbamates from spiked samples ranged from 88.1% to 118.4%, and the coefficients of variation (CVs) were all below 10%. The method was applied to pesticide residues detection in fruit, vegetable and green tea samples taken from local markets, in which carbamates were extensively detected but all below the standard of maximum residue limit.

Keywords: carbamates; multiple reaction monitoring (MRM); enhanced product ion (EPI); mass fragmentation; confirmatory method; pesticide residues

1. Introduction

Carbamate pesticides, namely, N-methyl carbamate esters not only share with organophosphates the capacity to kill insects by inhibiting the enzyme acetylcholinesterase (AChE) but also show lower toxicity to human being [1–3]. Gradually, they have become one kind of most popular pesticides in agriculture to guarantee food production. However, the improper use of carbamate pesticides have adversely affected food safety g [4,5]. Carbamate pesticide residues are known to be frequently present in fruits, vegetables and green teas on the market at levels close to or below the standard of maximum residue limit [6]. The legitimate content has brought chronic and continuous intake of carbamate residues that could be relevant to multiple ailments and further pose a threat to public health [7–9]. Meeting the challenge of many samples in daily work, developing a best concise, highly specific and sensitive analytical method to detect the carbamate residues in diverse samples from the area of food safety is important.

The challenge to modify this analytical method could be ascribed to two main factors: the existence of a range of carbamate pesticides and samples with complex composition that depend on the instrumental analysis and the sample pretreatment. With the efforts of researchers for half a century, the sample pretreatment of pesticide residues has been developed from Soxhlet extraction, liquid–liquid partition chromatography and column chromatography to solid-phase extraction, solid-phase micro-extraction, molecularly imprinted solid-phase extraction, super critical fluid extraction, matrix solid phase dispersion and QuEChERS (Quick, Easy, Cheap, Effective, Rugged, and Safe) methods [10–12]. Developed in 2003 by Anastassiades, QuEChERS sample preparation has been readily accepted by many pesticide residue analysts over the years [13,14]. In the section of instrumental analysis, a literature survey (Table 1) shows several methods for the detection of carbamate pesticide residues have been developed at low concentration levels in various samples, such as electrochemical detection [15], liquid chromatography [16–24], high-performance thin-layer chromatography [25], liquid chromatography with post-column fluorescence derivatization [26,27] and liquid chromatography with different mass spectrometry [28–32]. Of all the instrumental methods, the LC-MS system that combines the separation ability of liquid chromatography along with the sensitivity and specificity of detection from mass spectrometry and abandoned the additional derivation procedure of whole analysis process has gone mainstream for pesticide residue analysis. Among the different mass spectrometers, the ion trap triple quadrupole mass spectrometer allows for multiple reaction monitoring (MRM) scan modes is the ascendant instrument to quantify the residues of pesticides at trace amount. In MRM mode, the selected precursor ion of relevant compound is pre-screened in the first quadrupole (Q1), dissociated in the second quadrupole (Q2) and identified and quantified in the third quadrupole (Q3). During the analysis procedure, both Q1 and Q3 are set at several specific masses, allowing only those distinct fragment ions from the certain precursor ions to be detected. The setting of certain precursor ions and specific masses results in increased sensitivity and the structural specificity of the analyte [33,34].Compared with traditional triple quadrupole mass spectrometry, the Triple Quadrupole Linear Ion Traps system (QTRAP, AB SCIEX) equips an extra linear ion trap (LIT) located in the third quadrupole set and represents the new generation of LC-MS/MS system with unique scan mode of multiple reaction monitoring–information dependent acquisition–enhanced product ion (MRM-IDA-EPI) scan mode, which could maximize the information of the analytes in a single run, including selected ion pairs, retention time, relative intensities and the mass spectra. In this mode, the ions of analytes enter Q3 and are stopped from leaving this region by the changing of voltages. Then, the trap is allowed to fill with target ions with a set amount of time. The voltages are changed to stop any more ions from entering. Finally, the packet of ions is scanned out from the trap in a controlled manner. The target analytes are quantified as ever, and further identified on the automatically acquired MS/MS spectra while the precursor ions intensity exceeds the fixed area threshold setting. In particular, the linear ion trap could keep on capturing the interested precursor ions and product ions for a few milliseconds. This accumulation signifies a 500-times increase in the scanning sensitivity and brings a high intensity of the mass spectra recorded in EPI mode. It double checks the doubtful samples by comparing the acquired mass spectra with the mass spectral library built from standards [35]. This study developed a new analytical strategy to combine QuEChERS sample preparation and QTRAP LC-MS/MS system to improve the efficiency of quantification and confirmation of carbamate residues while ensuring high degree of reproducibility. In addition, this method should be stable enough to avoid false negative or false positive results.

Table 1. Comparative study between the published analysis methods for carbamates.

Target	Object	Sample Pretreatment	Instrumental Analysis	Analysis Limits	Disadvantage	Ref.
Carbaryl	soil	ionic liquid–dispersive liquid–liquid microextraction	HPLC-FD (fluorescence detector)	0.63–4.0 ng g^{-1}	Ionic liquids are not commercially available No confirmation spectra	[11]
Seven carbamates	gram, wheat, lentil, soybean, fenugreek leaves apple	column chromatography	HPLC-UV	0.08–1.16 mg L^{-1}	Large amount of Solvents Lack of sensitivity No confirmation spectra	[17]
Seven carbamates	/	/	HPLC post-column derivatization and fluorescence detection	0.2–0.7 ng	No application No confirmation spectra	[26]
Eleven carbamates	/	/	HPLC post-column derivatization and fluorescence detection	0.5 ng mL^{-1}	The tempestuously hydrolyzation of standards in dilute hydrochloric acid solution No confirmation spectra	[27]
Fifteen carbamates	corn, cabbage, tomato	QuEChERS method	LC-MS (QDa)	1 µg mL^{-1}	Lack of sensitivity; expensive No confirmation spectra	[28]
Thirteen carbamates	orange, grape, onion, tomatoes	Matrix solid-phase dispersion (concentration)	LC-MS (ESI/APCI)	0.001–0.01 mg kg^{-1}	No confirmation spectra	[29]
Thirteen carbamates	Traditional Chinese Medicine	QuEChERS method	UPLC-MS (MRM)	5.0–10.0 µg kg^{-1}	No confirmation spectra	[30]

2. Results and Discussion

2.1. Optimization of Sample Preparation

As shown in Figure 1, the scheme of analysis method, extraction process and purification process are the two main components of the QuEChERS pretreatment. In this study, acetonitrile with different concentrations of formic acid supplement was compared for extraction efficiency. However, acetonitrile and acidified acetonitrile did not show any obvious distinction in the recovery experiment. Thus, it was more convenient to adopt acetonitrile in the extraction process. In the purification process, 1.0 g of sodium chloride and 2.0 g of anhydrous sodium sulfate were applied to separate the aqueous solution and acetonitrile part. In this study, the traditional anhydrous magnesium sulfate was replaced by anhydrous sodium sulfate because of the heat release in the moisture absorbing process. In the purification process, the commonly used cleaning agents are ethylenediamine-N-propylsilane (PSA), graphitized carbon black (GCB) and C18. PSA is pertinent to organic acids, some pigments, and some sugars and fats; GCB has a strong adsorption effect on pigments; and C18 aims to remove non-polar impurities such as fats. Different proportions of PSA, GCB, and C18 were combined to find the best recipe with the best average recoveries. It was proventhat the combination of 50 mg C18 and 150 mg PSA as cleaning agent was most suitable for determining quantitatively carbamate residues in multi-matrices by recovery tests and samples determination (Figure S1).

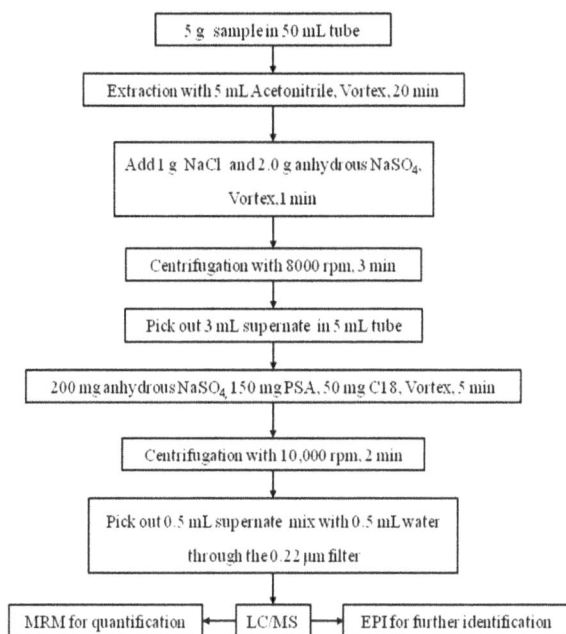

Figure 1. Scheme of analysis for the simultaneous determination and identification of fifteen carbamate pesticide residues in multi-matrices.

2.2. Optimization of Mass Spectrometry and Chromatographic Conditions

To confirm the ion pairs for quantification, the mixed standard solution of fifteen carbamate pesticides with a concentration of 100 μg L^{-1} was infused into the QTRAP mass spectrometer at a flow rate of 7 μL min^{-1} to obtain automatic analytes optimization by the ESI positive mode. In the optimal mass spectrometry conditions, all carbamates except tsumacide acquired two pairs of optimum ions for quantification and identification.

For the high sensitivity, symmetric shape of the ionic peaks and minimal retention time, this study examined the elution type, flow rate, gradient and the type of the chromatographic column. Once the most general chromatographic column C18 was decided, the separation and ionization of the analytes were mainly affected by the compositions of the elution type and the elution gradient. Therefore, several classical compositions of the mobile phase were performed including methanol, acetonitrile, water and water with ammonium acetates or formic acid. Finally, water and acetonitrile, both supplemented with 0.1% formic acid, were chosen as the optimal mobile phases. The final gradient elution with the total flow rate of 0.3 mL/min was as follows: 0–2.5 min, 15% A; 2.5–5 min, 15–50% A; 5–7 min, 50% A; 7–8 min, 50–15% A and 8–10 min, 15% A. The column oven was maintained at 40 °C and the injection volume was 5 μL. The representative LC-QTRAP-MS/MS chromatograms were merged (Figure 2). The retention time (RT) and MS information for each analyte including precursor and product ions, DP and CE are shown in Table 2.

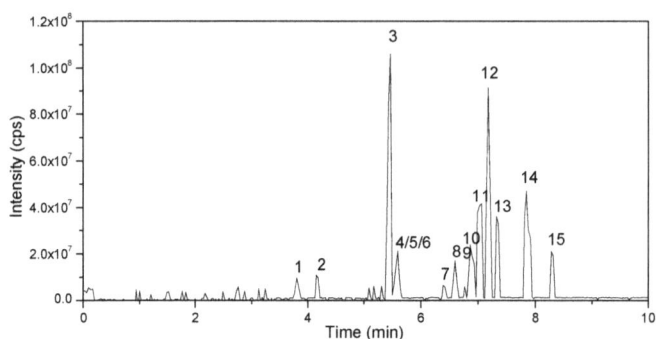

Figure 2. LC-MS/MS chromatograms of fifteen carbamate pesticides (50 ng mL^{-1}).

The EPI survey scans in IDA experiment should be triggered when the ionic intensity exceeded the threshold of 1000 cps. The total scan time (including pauses) was 1.08 s for all MRM transitions. The range of the mass spectra was set between 50 and 300 amu with a scan rate of 10,000 Da s^{-1}. The mass spectra of fifteen carbamates and their characteristic fragmentations are shown in the Supplementary Materials.

Table 2. Retention time and MS parameters of the fifteen carbamates pesticides.

No.	Compound	Retention Time (min)	[1] CAS No.	Precursor Ion (*m/z*)	Product Ion (*m/z*)	Declustering Potential (V)	Collision Energy (eV)
1	Aldicarb-sulfoxide	3.81	1646-87-3	206.9	132.0	130	8
					89.0 *	130	17
2	Aldicarb-sulfone	4.17	1646-88-3	222.8	166.1 *	140	11
					148.0	140	18
3	Pirimicarb	5.45	23103-98-2	239.1	182.1 *	80	20
					72.0	80	30
4	Carbofuran-3-hydroxy	5.57	16655-82-6	237.8	181.0*	150	15
					163.0	150	19
5	Methomyl	5.58	16752-77-5	162.8	135.0	180	26
					106.0 *	180	28
6	Oxamyl	5.59	23135-22-0	220.0	72.0	40	18
					163.1 *	40	10
7	Aldicarb	6.40	116-06-3	212.7	89.0 *	150	18
					156.0	150	19
8	Tsumacide	6.60	1129-41-5	166.0	109.0 *	50	13
9	Propoxur	6.87	114-26-1	210.0	168.0 *	60	18
					153.0	60	10
10	Carbofuran	6.98	1563-66-2	221.7	165.0 *	120	15
					123.0	120	27
11	Carbaryl	7.04	63-25-2	202.0	145.0 *	60	12
					117.0	80	35
12	Isoprocarb	7.18	2631-40-5	194.0	95.0	100	19
					137.0 *	100	11
13	Methiocarb	7.33	2032-65-7	226.0	169.0 *	55	12
					121.0	55	24
14	Fenobucarb	7.84	3766-81-2	208.0	95.0 *	70	19
					152.0	70	11
15	Banol	8.29	671-04-5	214.1	157.1 *	60	12
					121.2	60	16

[1] CAS: chemical abstracts service; *: quantitative ion.

2.3. Fragmentation Manner of Carbamate Pesticides

Based on the structure, the fifteen carbamates can be divided into three kinds: N-methyl amino formic acid aromatic ester, N-methyl amino formic acid oxime ester and heterocyclic N-methyl amino formic acid ester. The former two include carbaryl, tsumacide, methiocarb and methomyl; and aldicarb sulfone, aldicarb sulfoxide, etc., respectively, while the latter includes pirimicarb, isolan (highly regulated and unavailable), etc. Generally speaking, in the positive mode of electrospray ionization mass spectrometry, the proton first attaches to the protonation site, and then triggers the cleavages by migrating to the reactive center [36]. With the common structure of amino nitrogen and carbonyl oxygen, the protons could gravitate to these preferred protonation sites, and then trigger the dissociation reaction by migrating to the reactive center. Therefore, the oxygen of ester near the carbonyl carbon is a suitable active site to induce this fragmentation of carbamates. For example, 3-hydroxyl carbofuran belonging to N-methyl amino formic acid aromatic ester, could generate a stable fragment ion *m/z* 181.1 by a neutral loss of CH_3NCO (−57 Da). The resulting spectrum is shown in Figure 3. According to this fragment mechanism, all fifteen carbamate pesticides were investigated and the vast majority of carbamate pesticides, except pirimicarb, could dissolve at the same active site and obtain the same characteristic loss of 57 Da (Figures 4 and 5).

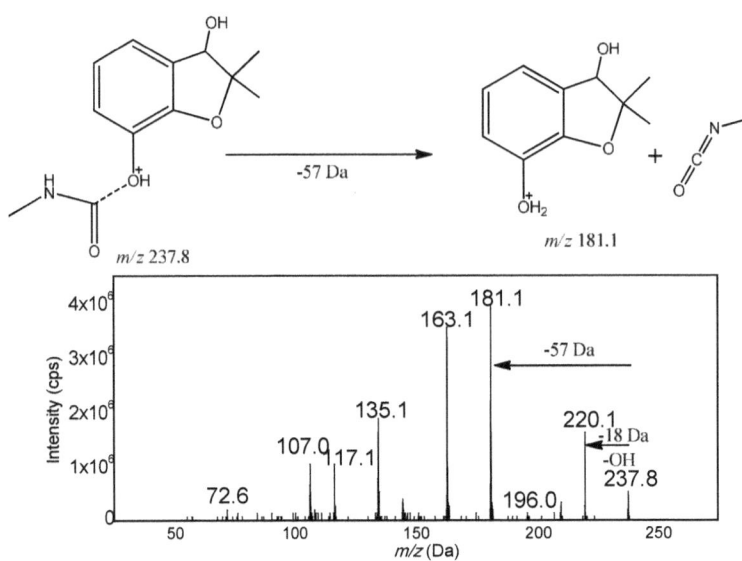

Figure 3. Product ion spectrum and probable fragmentation routes of Carbofuran-3-hydroxy.

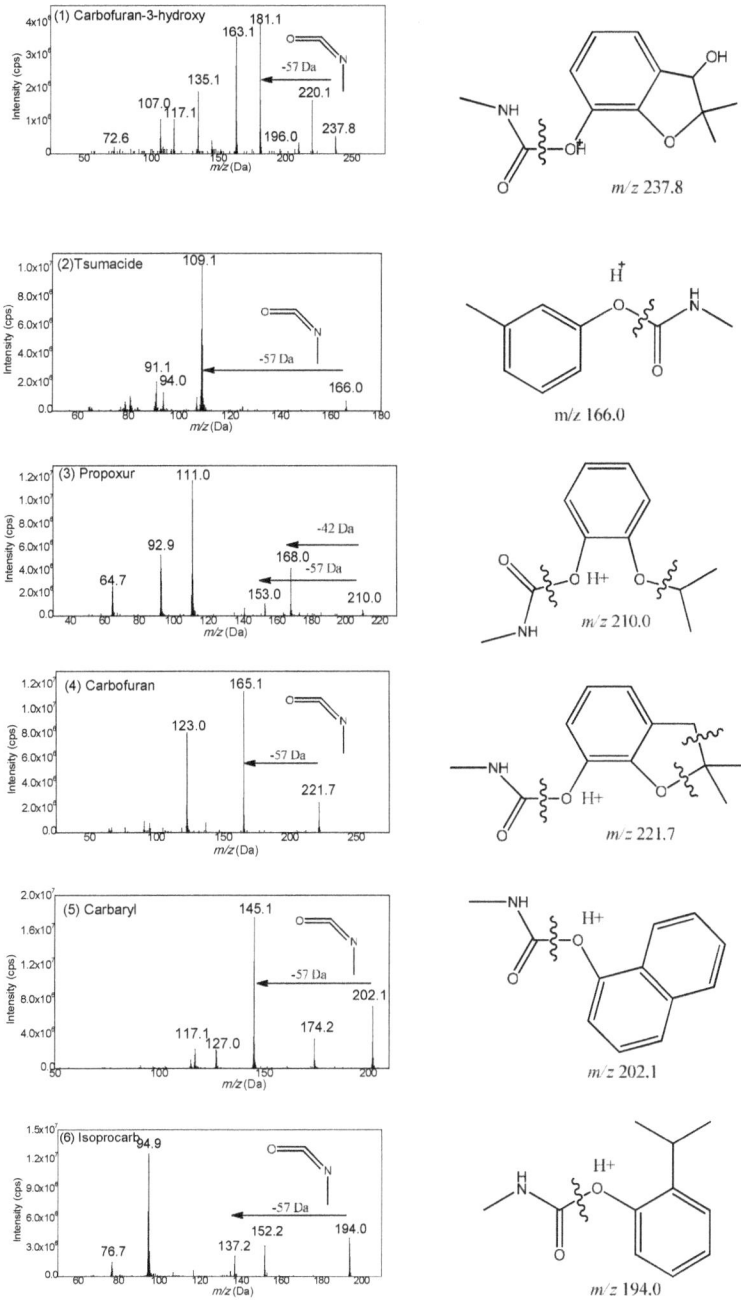

Figure 4. *Cont.*

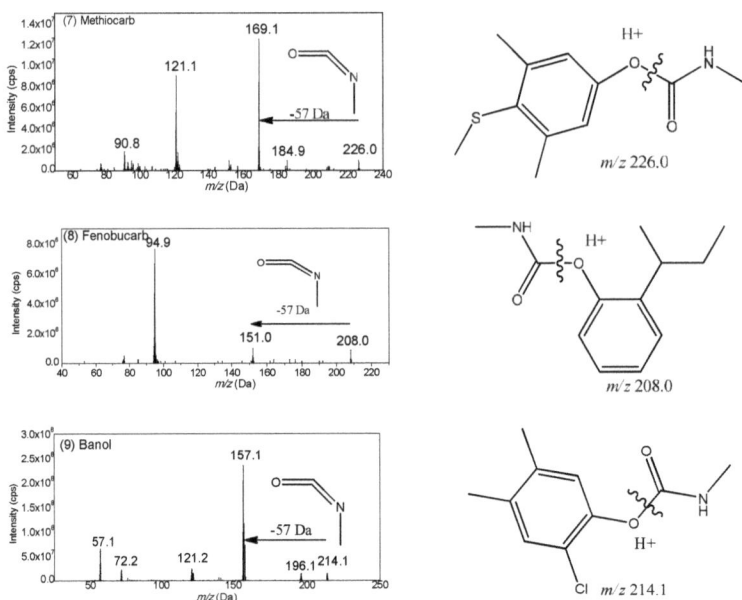

Figure 4. Product ion spectra and proposed fragmentation pathway of nine N-methyl amino formic acid aromatic ester: (**1**) Carbofuran-3-hydroxy; (**2**) Tsumacide; (**3**) Propoxur; (**4**) Carbofuran; (**5**) Carbaryl; (**6**) Isoprocarb; (**7**) Methiocarb; (**8**) Fenobucarb; and (**9**) Banol.

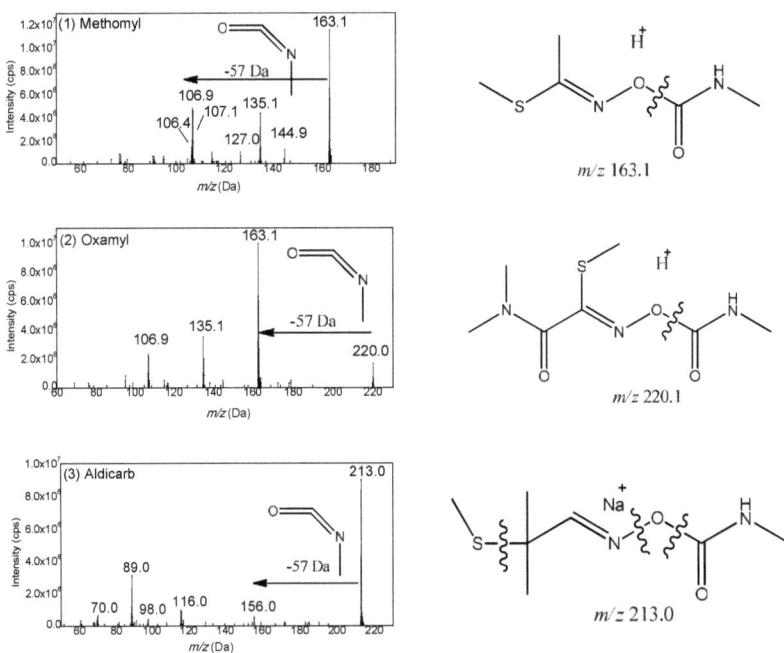

Figure 5. *Cont.*

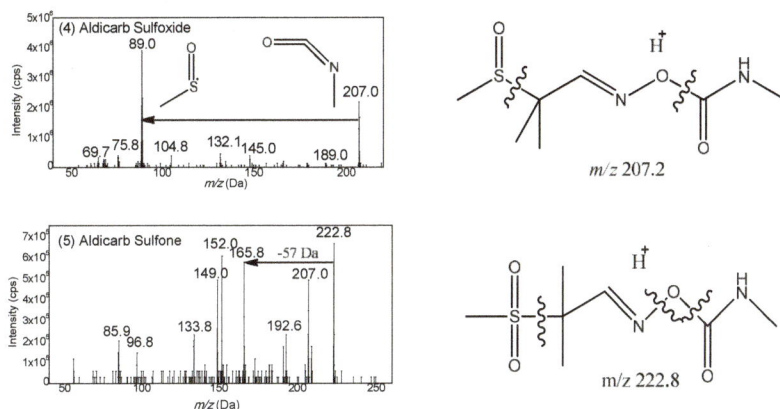

Figure 5. Product ion spectra and proposed fragmentation pathway of three N-methyl amino formic acid oxime ester: (**1**) Methomyl; (**2**) Oxamyl; (**3**) Aldicarb; (**4**) Aldicarb-sulfoxide; and (**5**) Aldicarb-sulfone.

Unexpectedly, 57 Da was lost in pirimicarb. Structurally speaking, it has two methyl groups linked to the amino group and it should produce a loss of 72 Da. In fact, the mass spectrum of pirimicarb indicated an unexpected loss of 57 Da rather than 72 Da, as shown in Figure 6. This is because of the nitrogen-containing pyrimidine heterocyclic ring adjoined the ester oxygen that could induce a methyl transfer reaction. According to a literature survey [37–39], the proton is localized at the pyrimidine nitrogen first; and then the charge-remote become the reaction center that forming an ion-neutral complex and induce an alkyl cation transfer.

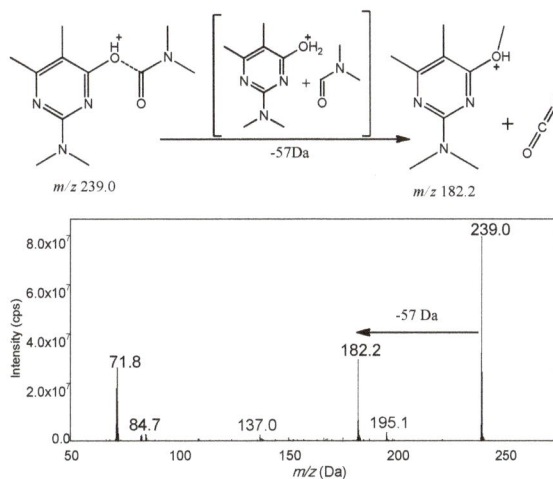

Figure 6. Product ion spectra and probable fragmentation routes of Pirimicarb.

2.4. Method Validation

2.4.1. Matrix Effect

As shown in Figure 7, the complex composition of diverse matrices might have different effects on the accuracy of identification and quantification. It was necessary to correct the matrix effects in LC-MS detection. Eight distinct types of organic vegetable samples or pesticide-free samples were used as matrix-matched blank to evaluate the matrix effects on the ionization of fifteen carbamate residues. The evaluation of matrix effects was calculated using the following equation [40]:

$$ME = \frac{A_{Matrix}}{A_S} \times 100\%$$

where A_{Matrix} is the peak area of the standard solution with the matrix-matched blank and A_S is the peak area of the standard solution with initial mobile phase. As shown in Table 3, the percentages of the matrix effects of 15 carbamate pesticides at three different concentrations (2, 20, and 200 ng mL^{-1}) ranged 95.4–111.2% (see Table 2 for details). Since there was no obvious ion suppression or ion enhancement at the chosen levels of quantification for 15 carbamate pesticides in eight distinct types of samples, the matrix effect can be ignored.

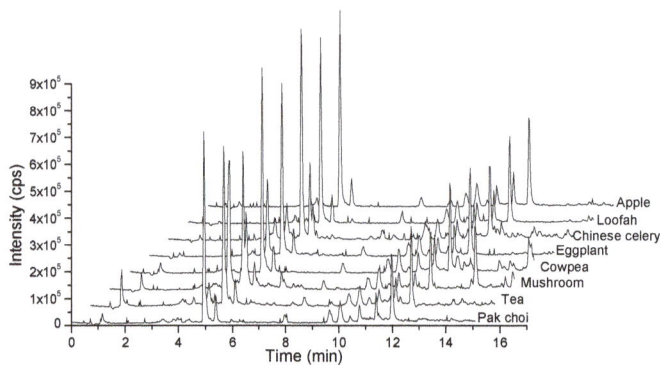

Figure 7. Typical total ion chromatograms of fifteen carbamates at 2 μg/kg spiked in eight matrices (cleaning by 50 mg C18 and 150 mg PSA).

Table 3. Matrix effects of the fifteen carbamates pesticides in distinct samples.

Compound	Matrix Effects/% (n = 3; 2, 20, 200 ng mL^{-1})							
	Pak Choi	Chinese Celery	Loofah	Eggplant	Cowpea	Apple	Mushroom	Tea
Aldicarb-sulfoxide	98.6	98.5	96.3	98.8	96.0	95.4	96.6	97.6
Aldicarb-sulfone	97.8	96.5	98.3	99.1	95.7	96.2	98.4	96.9
Pirimicarb	105.1	102.0	98.0	98.7	96.3	98.9	101.2	99.2
Carbofuran-3-hydroxy	102.3	101.3	103.5	107.5	101.2	102.5	101.0	105.4
Methomyl	110.2	108.6	107.2	108.1	105.4	111.2	105.7	109.2
Oxamyl	102.5	98.6	105.2	106.3	102.0	102.1	105.2	106.2
Aldicarb	107.5	102.1	100.2	101.3	100.8	103.2	100.5	102.3
Tsumacide	105.0	100.9	102.3	98.6	97.9	99.0	102.6	102.7
Propoxur	102.3	105.4	102.7	104.9	101.8	102.4	103.1	101.5
Carbofuran	102.5	105.9	106.2	103.1	98.7	99.2	97.4	98.5
Carbaryl	102.3	104.1	106.5	104.5	101.7	102.4	106.9	107.4
Isoprocarb	102.3	104.6	102.5	104.8	101.2	107.5	108.4	101.8
Methiocarb	105.9	104.1	102.4	106.2	107.1	105.1	106.3	105.1
Fenobucarb	102.1	105.2	102.6	107.8	110.0	108.4	107.6	106.8
Banol	105.6	104.2	102.9	104.1	103.2	102.5	102.3	101.1

2.4.2. Linearity and Analytical Limits

Satisfactory correlation coefficients (R > 0.996) were obtained for fifteen carbamate residues in matrix-matched blank over the concentration range of 0.05–200.0 ng mL^{-1}. The analytical limits of the proposed method were measured by the limit of detection (LOD) and the limit of quantification (LOQ) using the following equations [41]:

$$LOD = C_S \frac{3}{S/N}$$

$$LOQ = C_S \frac{10}{S/N}$$

where S/N is the average signal to noise ratio and C_S is the concentration of the specific pesticide. The estimated values were tested with a suitable number and kind of samples containing the 15 carbamate pesticides at the corresponding concentrations. All related parameters of the proposed method are summarized in Table 4. The LOD and LOQ were 0.2–2.0 μg kg^{-1} and 0.5–5.0 μg kg^{-1}, respectively, which demonstrated a good analytical limit of this method for carbamate residues.

Table 4. Linear ranges, linear equations, correlation coefficients (R) limits of detection (LODs) and limits of quantification (LOQs) of the fifteen carbamate pesticides.

Compound	Linear Range (ng mL^{-1})	Linear Equation	R (1/X^2)	LOD (μg kg^{-1})	LOQ (μg kg^{-1})
Aldicarb-sulfoxide	0.05–200	Y = 2.42 × 10^{4}X + 812.9	0.9963	0.2	0.5
Aldicarb-sulfone	0.50–200	Y = 1.44 × 10^{4}X + 10,476	0.9972	2.0	5.0
Pirimicarb	0.05–200	Y = 5.49 × 10^{5}X + 15,110.0	0.9968	0.2	0.5
Carbofuran-3-hydroxy	0.10–200	Y = 1.75 × 10^{4}X + 457.2	0.9964	0.3	1.0
Methomyl	0.05–200	Y = 2.91 × 10^{4}X + 102,771	0.9962	0.2	0.5
Oxamyl	0.05–200	Y = 8.38 × 10^{4}X + 2287.1	0.9965	0.2	0.5
Aldicarb	0.05–200	Y = 1.96 × 10^{4}X + 7283.2	0.9973	0.2	0.5
Tsumacide	0.05–200	Y = 6.64 × 10^{4}X + 3330.4	0.9967	0.2	0.5
Propoxur	0.05–200	Y = 1.27 × 10^{5}X + 3441.9	0.9963	0.2	0.5
Carbofuran	0.20–200	Y = 6.31 × 10^{4}X + 543.2	0.9982	0.3	1.0
Carbaryl	0.05–200	Y = 6.01 × 10^{4}X + 4445.6	0.9968	0.2	0.5
Isoprocarb	0.05–200	Y = 9.33 × 10^{4}X + 15,304	0.9957	0.2	0.5
Methiocarb	0.05–200	Y = 9.86 × 10^{4}X + 6565.0	0.9965	0.2	0.5
Fenobucarb	0.05–200	Y = 1.24 × 10^{5}X + 7436.7	0.9962	0.2	0.5
Banol	0.05–200	Y = 1.01 × 10^{5}X + 2436.7	0.9971	0.2	0.5

Y, peak area; X, mass concentration, ng mL^{-1}; 1/X^2, least square method.

2.4.3. Accuracy and Precision

The accuracy and precision of the method were measured by the recoveries and coefficients of variation (CVs) for intra- and inter-day. Therefore, the standard mixtures solution of fifteen carbamate residues was spiked into distinct types of samples including greengrocery, Chinese celery, loofah, eggplant, cowpea, apple, mushroom and tea leaves, and 24 spiked samples (eight types at three concentrations of 2, 20, and 200 μg kg^{-1}) were obtained. All spiked samples were detected three times a day on five different days following the process of Figure 1. The results are shown in the Supplementary Materials. The recovery of fifteen carbamate residues (88.1–118.4%) fell within the recommended Eurachem guidelines of 80–120% [42]. The analysis precision, measured as the coefficient of variation percentage (% CV) of the recovery (3.2–9.8%), was well under the criteria of 10% [42].

2.5. Sample Analyses

After validation of the analyti cal methodology through above experimentation, it was used to detect numerous types of samples including apple, carrot, chili pepper, Chinese celery, coriander, cowpea, crown daisy, eggplant, garlic sprout, grape, ginger, leek, lettuce, loofah, mushroom, needle mushroom, orange, pakchoi, radish, shiitake, spinach and tea leaves. Carbofuran and its metabolite Carbofuran-3-hydroxy were the predominant detected residues of the fifteen carbamates. Meanwhile, of the 26 samples (Table S3), eggplant contributed the detected residues of maximum frequency, with concentrations of carbofuran at 85.5, 77.4, 68.4, 35.6, 25.6, 17.3, 16.5 and 11.8 μg kg^{-1}, respectively. Pokchoi, spinach, and Chinese celery all had considerable detection rate and concentration of carbofuran. Among all samples, although green tea had the second highest detection rate, the concentration of carbamate residues were localized at relative low concentration of 6.4, 1.5, 1.2, 0.8, 0.6, 0.6, 0.4, 0.3, 0.2 and 0.2 μg kg^{-1}, respectively. This was mainly due to the further degradation of carbamate residues in the drying process of tea manufacturing. Cowpea occasionally had the highest concentration of carbofuran at 180.2 μg kg^{-1}.

3. Materials and Methods

3.1. Chemicals and Reagents

Acetonitrile and Methanol were of HPLC grade and obtained from Merck (Darmstadt, Germany). Formic acid (≥98%) was of HPLC grade and purchased from Aladdin (Shanghai, China). High purity water was obtained using a Milli-Q water purification system from Millipore (Bedford, MA, USA). Cleanert® Pesticarb, Cleanert® PSA and Cleanert® C18 for QuEChERS were purchased from Agela Technologies (Beijing, China). Millipore filters of polytetrafluoroethylene (0.22 μm) were purchased from ANPEL Lab (Shanghai, China).

Standards of fifteen carbamates were purchased from Dr. Ehrenstorfer GmbH (Augsburg, Germany): Aldicarb-sulfoxide, Aldicarb-sulfone, Pirimicarb, Carbofuran-3-hydroxy, Methomyl, Oxamyl, Banol, Aldicarb, Metolcarb, Propoxur, Carbofuran, Carbaryl, Isoprocarb, Mercaptodimethur and Fenobucarb. Individual standard solutions were prepared in HPLC-grade methanol at a concentration of 1.0 mg/mL and stored at −28 °C temporarily.

Samples were collected from local supermarkets, including vegetables (leeks, pakchoi, crowndaisy, coriander, spinach, lettuce, Chinese celery, garlic sprout, loofah, chili pepper, eggplants, cowpea, radish, and turnip), fruits (apples, kiwis, grapes, and oranges), tea (green tea, black tea, and flower tea), and edible fungus (shiitake, mushroom, and needle mushroom). Each fresh sample was crushed and mixed into homogenate, and stored in plastic bottles at −20 °C until analysis.

3.2. Instrumentation

HPLC analysis was performed on a Shimadzu LC-30AD system (Shimadzu, Kyoto, Japan) which consist of two interconnected pump units: one with an integrated degasser and the other with a mixer, and is comprised of a UHPLC gradient system, a refrigerated autosampler, and a column oven compartment. Mass spectrometric detection was performed on an AB SCIEX QTRAP® 5500 (AB SCIEX instruments, Foster, CA, USA) in MRM mode and EPI mode. A Turbo V™ Ion Source (ESI) interface in positive ionization mode was used. Both the UHPLC and mass spectrometer were controlled remotely using Analyst® software v. 1.6.2 (AB SCIEX instruments, Foster City, CA, USA). A Waters BEH C18 column (1.7 μm 2.1 mm × 100 mm, Waters, Milford, MA, USA) was applied for analysis. The mass spectrometer was equipped with an electrospray ionization source and spectra were acquired in the positive ion multiple reaction monitoring (MRM) mode and enhanced product ion (EPI) scan modes. MS was optimized using a capillary voltage of 5.50 kV and desolvation temperature of 500 °C. The cone gas pressure and desolvation gas pressure were 50 psi. Nitrogen was used as the cone and collision gasses, respectively. The raw data were analyzed using an Analyst 1.6.2 workstation (AB SCIEX, Foster, CA, USA).

3.3. Sample Extraction

Five grams of the homogenized sample was weighed in 50 mL Corning extraction tubes, and 5 mL of acetonitrile was added. The tubes were vibrated vigorously for 20 min in batches by a Multi Reax Vortexer (Heidolph, Schwabach, Germany). After the vortex, 1.0 g of sodium chloride and 2.0 g of sodium sulfate were separately added to each tube for the elimination of moisture, and the tubes were immediately manually shaken vigorously for 1 min to prevent sodium sulfate agglomeration. The process was followed by centrifugation at 8000 rpm for 3 min at 4 °C in a Sigma 2–16 K. centrifuge (Sartorius AG, Göttingen, Germany). The supernatant was collected in 10 mL extraction tubes and subjected to QuEChERS dispersive SPE cleaning. The dSPE agent contained 100 mg of primary secondary amine (PSA), 50 mg of C18EC, and 200 mg of magnesium sulfate. The tubes were vibrated for 30 s and centrifuged in the Sigma 2–16 K centrifuge at 13,000 rpm for 2 min at 20 °C. The supernatants were collected and filtered through 0.22 µm Anpel Syringe Hydrophilic PTFE filters (Anpel, Shanghai, China) before LC-MS/MS analysis.

Organic vegetables were cultivated by Ying Zhou's mother from her garden. The samples were homogenized with an Ultra-Turrax T 25 homogeniser (IKA® Werke, Staufen, Germany) and extracted using the QuEChERS method.

4. Conclusions

In this paper, a quick and credible method to quantify and confirm fifteen carbamate pesticide residues by modified QuEChERS sample pretreatment combined ultra-high performance liquid chromatography–quadrupole-linear ion trap mass spectrometry was systematically developed. For modified QuEChERS part, the composition of the cleaning agents was optimized to adapt the general samples. For the instrumental analysis, based on traditional conformation criteria including selected ion pairs, retention time and relative intensities between transitions from MRM scan mode, the presence of carbamate pesticides in doubtful samples was further confirmed by the matching of carbamate pesticide spectra via EPI scan mode. Then, the fragmentation routes of all the carbamates were preliminarily explained based on the mass spectra that the carbamate could generate a stable characteristic fragment ion by a neutral loss of CH_3NCO (-57 Da). The linearity, matrix effect, analysis limits, accuracy and repeatability were validated in diverse samples. All fifteen carbamates have good correlation coefficients above 0.996 at the range of 0.05–200 ng mL^{-1}. The recoveries of intra- and inter-day experiments were in the range of 80–120% at three concentrations with coefficients of variation all better than 10%. Moreover, the whole process of one dozen samples from pretreatment to final report did not exceed 2 h, which is shorter and exacter than traditional methods. Finally, the method was applied to carbamate residues detection in 26 kinds of samples from local markets, in which carbamates were extensively detected but all below the standard of maximum residue limit.

Supplementary Materials: The following are available online, Figure S1: Recovery tests of different proportions of PSA, GCB, and C18 in multi-matrices, Table S1: Intra-day accuracy and precision ($n = 3$), Table S2: Inter-day accuracy and precision ($n = 15$), Table S3: The maximum residues of 15 carbamate pesticide residues in 26 different samples ($n = 28$).

Author Contributions: Y.Z. conceived and designed the study; M.G., J.G., S.L. and W.G. performed the experiments; and Y.Z. analyzed the data and wrote the paper.

Funding: This research received no external funding and the APC was funded by [Jiaxing Center for Disease Control and Prevention].

Acknowledgments: This work was supported by all colleagues of Jiaxing Center for Disease Control and Prevention. Special thanks are given to Ying Zhou's mother, Yafen Wang, for the organic vegetables and fruits from her garden.

Conflicts of Interest: The authors declare no conflict of interest.

References

1. Carbamate. Available online: https://en.wikipedia.org/wiki/Carbamate#Carbamate_insecticides (accessed on 15 August 2018).
2. Fukuto, T.R. Mechanism of action of organophosphorus and carbamate insecticides. *Environ. Health Persp.* **1990**, *87*, 245–254. [CrossRef] [PubMed]
3. Richard, D. Insecticides. Action and Metabolism. *Bull. Entomol. Soc. Ameri.* **1968**, *14*, 258–259. [CrossRef]
4. Geoffrey, M.C.; Wayne, T.S.; Margot, B.; Jerome, M.B.; Louise, N.M. Surveillance of pesticide-related illness and injury in humans. In *Handbook of Pesticide Toxicology*, 2nd ed.; Robert, K., Ed.; Academic Press: Cambridge, MA, USA, 2001; Volume 1, pp. 603–641.
5. Michal, R.; Shlomo, A. Evaluation of the decarbamylation process of cholinesterase during assay of enzyme activity. *Clin. Chim. Acta* **1995**, *240*, 107–116. [CrossRef]
6. Announcement of Food Safety Inspection. Qinhai Food and Drug Administration. No. 68, 2018. Available online: http://www.sdaqh.gov.cn/html/201896/n140026114.html (accessed on 10 September 2018).

7. Tsiplakou, E.; Anagnostopoulos, C.J.; Liapis, K.; Haroutounian, S.A.; Zervas, G. Pesticides residues in milks and feedstuff of farm animals drawn from Greece. *Chemosphere* **2010**, *80*, 504–512. [CrossRef] [PubMed]

8. Daphne, B.M. Public health impacts of organophosphate and carbamates. In *Toxicology of Organophosphate & Carbamate Compounds*; Gupta, R.C., Ed.; Academic Press: London, UK, 2005; pp. 599–601.

9. Marina, P.G.; Margarita, L.D.; Rajnarayanan, R.V. Carbamate Insecticides Target Human Melatonin Receptors. *Chem. Res. Toxicol.* **2017**, *30*, 574–582. [CrossRef]

10. Pinto, M.I.; Gerhard, S.; Bernardino, R.J.; Noronha, J.P. Pesticides in water and the performance of the liquid-phase microextraction based techniques. A review. *Microchem. J.* **2010**, *96*, 225–237. [CrossRef]

11. María, A.R.; Javier, H.B.; Teresa, M.B.M.; Miguel, Á.R.D. Ionic liquid-dispersive liquid-liquid microextraction for the simultaneous determination of pesticides and metabolites in soils using high-performance liquid chromatography and fluorescence detection. *J. Chromatogr. A* **2011**, *1218*, 4808–4816. [CrossRef]

12. Wang, J.; Zhang, K.; Tech, K.; Brown, D. Multiresidue Pesticide Analysis of Ginseng Powders Using Acetonitrile- or Acetone-Based Extraction, Solid-Phase Extraction Cleanup, and Gas Chromatography-Mass Spectrometry/Selective Ion Monitoring (GC-MS/SIM) or -Tandem Mass Spectrometry (GC-MS/MS). *J. Agric. Food Chem.* **2010**, *58*, 5884–5896. [CrossRef] [PubMed]

13. Anastassiades, M.; Scherbaum, E.; Tasdelen, B.; Stajnbaher, D. Recent Developments in QuEChERS Methodology for Pesticide Multiresidue Analysis. In *Pesticide Chemistry: Crop Protection, Public Health, Environmental Safety*; Ohkawa, H., Miyagawa, H., Lee, P.W., Eds.; Wiley-VCH: Weinheim, Germany, 2007; pp. 439–458.

14. Wilkowska, A.; Biziuk, M. Determination of pesticides residues in food matrices using the QuEChERS methology. *Food Chem.* **2011**, *125*, 803–812. [CrossRef]

15. Rao, T.N.; Loo, B.H.; Sarada, B.V.; Terashima, C.; Fujishima, A. Electrochemical Detection of Carbamate Pesticides at Conductive Diamond Electrodes. *Anal. Chem.* **2002**, *74*, 1578–1583. [CrossRef] [PubMed]

16. Tao, L.; Su, H.; Qu, X.; Ju, P.; Lin, C.; Ai, S. Acetylcholinesterase biosensor based on 3-carboxyphenylboronic acid/reduced graphene oxide-gold nanocomposites modified electrode for amperometric detection of organophosphorus and carbamate pesticides. *Sens. Actuators B Chem.* **2011**, *160*, 1255–1261. [CrossRef]

17. Rajendra, P.; Niraj, U.; Vijay, K. Simultaneous determination of seven carbamate pesticide residues in gram, wheat, lentil, soybean, fenugreek leaves and apple matrices. *Microchem. J.* **2013**, *111*, 91–96. [CrossRef]

18. Sanchez, B.C.; Albero, B.; Tadeo, J.L. High-performance liquid chromatography multiresidue method for the determination of N-methyl carbamates in fruit and vegetable juices. *J. Food Prot.* **2004**, *67*, 2565–2569. [CrossRef]

19. Tsumura, Y.; Ujita, K.; Tonogai, Y.; Ito, Y. Simultaneous determination of aldicarb, ethiofencarb, methiocarb and their oxidized metabolites in grains, fruits and vegetables by high performance liquid chromatography. *J. Food Prot.* **1995**, *58*, 217–222. [CrossRef]

20. Bernal, J.L.; Nozal, M.J.; Torlblo, L.; Jlmenez, J.J.; Atienza, J. High-performance liquid chromatographic determination of benomyl and carbendazim residues in apiarian samples. *J. Chromatogr. A* **1997**, *787*, 129–136. [CrossRef]

21. Sandahl, M.; Mathiasson, L.; Jonsson, J.A. Determination of thiophanate-methyl and its metabolites at trace level in spiked natural water using the supported liquid membrane extraction and the microporous membrane liquid-liquid extraction techniques combined on-line with high-performance liquid chromatography. *J. Chromatogr. A* **2000**, *893*, 123–131. [CrossRef] [PubMed]

22. Vandecasteele, K.; Gaus, I.; Debreuck, W.; Walraevens, K. Identification and quantification of 77 pesticides in groundwater using solid phase coupled to liquid–liquid microextraction and reversed-phase liquid chromatography. *Anal. Chem.* **2000**, *72*, 3093–3101. [CrossRef] [PubMed]

23. Sanchez, B.C.; Albero, B.; Tadeo, J.L. Multiresidue analysis of carbamate pesticides in soil by sonication-assisted extraction in small columns and liquid chromatography. *J. Chromatogr. A* **2003**, *1007*, 85–91. [CrossRef]

24. Gou, Y.; Eisert, R.; Pawliszyn, J. In-tube solid-phase microextraction coupled to capillary LC for carbamate analysis in water samples. *Anal. Chem.* **2000**, *72*, 2774–2779. [CrossRef] [PubMed]

25. Rami, A.; Wolfgang, S. Effect of bromine oxidation on high-performance thin-layer chromatography multi-enzyme inhibition assay detection of organophosphates and carbamate insecticides. *J. Chromatogr. A* **2011**, *1218*, 2775–2784. [CrossRef]

26. Sabala, A.; Portillo, J.L.; Broto-Puig, F.; Comellas, L. Development of a new high-performance liquid chromatography method to analyse N-methylcarbamate insecticides by a simple post-column derivatization system and fluorescence detection. *J. Chromatogr. A* **1997**, *778*, 103–110. [CrossRef]

27. Waters Technologies Ltd. *Waters ACQUITY UPLC H-Class FLR System for Carbamate Analysis Method Guide*; Waters Technologies Ltd.: Shanghai, China, 2010.

28. Huang, D.F. *Quantitative Analysis of 15 Carbamates in Vegetables Using DisQuE Cleanup and UHPLC with Mass Detection*; Waters Technologies Ltd.: Shanghai, China, 2008.

29. Fernández, M.; Picó, Y.; Mañes, J. Determination of carbamate residues in fruits and vegetables by matrix solid-phase dispersion and liquid chromatography-mass spectrometry. *J. Chromatogr. A* **2000**, *871*, 43–56. [CrossRef]

30. Chen, L.; Song, F.; Liu, Z.; Zheng, Z.; Xing, J.; Liu, S. Multi-residue method for fast determination of pesticide residues in plants used in traditional Chinese medicine by ultra-high-performance liquid chromatography coupled to tandem mass spectrometry. *J. Chromatogr. A* **2012**, *1225*, 132–140. [CrossRef] [PubMed]

31. Romero, R.G.; Frenich, A.G.; Vidal, J.L.M. Multiresidue method for fast determination of pesticides in fruit juices by ultra performance liquid chromatography coupled to tandem mass spectrometry. *Talanta* **2008**, *76*, 211–225. [CrossRef] [PubMed]

32. Didier, O.; Patrick, E.; Claude, C. Multiresidue analysis of 74 pesticides in fruits and vegetables by liquid chromatography-electrospray-tandem mass spectrometry. *Anal. Chim. Acta* **2004**, *520*, 33–45. [CrossRef]

33. Anderson, N.L.; Hunter, C.L. Quantitative Mass Spectrometric Multiple Reaction Monitoring Assays for Major Plasma Proteins. *Mol. Cell Proteom.* **2006**, *5*, 573–588. [CrossRef] [PubMed]

34. Multiple Quadrupoles, Hybrids and Variations. Available online: https://en.wikipedia.org/wiki/Quadrupole_mass_analyzer#Multiple_quadrupoles,_hybrids_and_variations (accessed on 28 August 2018).

35. Iwona, M.Z.; Barbara, W.; Andrzej, P. Comparison of the Multiple Reaction Monitoring and Enhanced Product Ion Scan Modes for Confirmation of Stilbenes in Bovine Urine Samples Using LC–MS/MS QTRAP® System. *Chromatographia* **2016**, *79*, 1003–1012. [CrossRef]

36. Sun, H.; Chai, Y.; Pan, Y. Dissociative Benzyl Cation Transfer versus Proton Transfer: Loss of Benzene from Protonated N-Benzylaniline. *J. Org. Chem.* **2012**, *77*, 7098–7102. [CrossRef] [PubMed]

37. Guo, C.; Yue, L.; Guo, M.; Pan, Y. Elimination of Benzene from Protonated N-Benzylindoline: Benzyl Cation/Proton Transfer or Direct Proton Transfer? *J. Am. Soc. Mass Spectrom* **2013**, *24*, 381–387. [CrossRef] [PubMed]

38. Li, F.; Zhang, X.; Zhang, H.; Jiang, K. Gas-phase fragmentation of the protonated benzyl ester of proline: Intramolecular electrophilic substitution versus hydride transfer. *J. Mass Spectrom.* **2013**, *48*, 423–429. [CrossRef] [PubMed]

39. Yue, L.; Guo, C.; Chai, Y.; Yin, X.; Pan, Y. Gas-phase reaction: Alkyl cation transfer in the dissociation of protonated pyridyl carbamates in mass spectrometry. *Tetrahedron* **2014**, *70*, 9500–9505. [CrossRef]

40. Bo, H.B.; Wang, J.H.; Guo, C.H. Determination of Strobilurin Fungicide Residues in Food by Gas Chromatography-Mass Spectrometry. *Chin. J. Anal. Chem.* **2008**, *36*, 1471–1475.

41. Chen, F.; Huber, C.; May, R.; Schröder, P. Metabolism of oxybenzone in a hairy root culture: Perspectives for phytoremediation of a widely used sunscreen agent. *J. Hazard. Mater.* **2016**, *306*, 230–236. [CrossRef] [PubMed]

42. Commission of the European Communities. *Commission Decision 2002/657/EC Implementing Council Directive 96/23/EC Concerning the Performance of Analytical Methods and the Interpretation of Results*; Commission of the European: Brussels, Belgium, 2002.

Sample Availability: Samples of the compounds are not available from the authors.

molecules

MDPI

Article

Selection and Identification of Novel Aptamers Specific for Clenbuterol Based on ssDNA Library Immobilized SELEX and Gold Nanoparticles Biosensor

Xixia Liu [1], Qi Lu [1], Sirui Chen [1], Fang Wang [1], Jianjun Hou [1,*], Zhenlin Xu [2,*], Chen Meng [1], Tianyuan Hu [1] and Yaoyao Hou [1]

[1] Hubei Key Laboratory of Edible Wild Plants Conservation and Utilization, Hubei Normal University, Cihu Road, Huangshigang District, Huangshi 435002, China; liuxixia1@163.com (X.L.); 18271691835@163.com (Q.L.); 17371958850@163.com (S.C.); WF_0129@163.com (F.W.); 13329967633@163.com (C.M.); 18771808527@sina.cn (T.H.); dxhe@163.com (Y.H.)
[2] Guangdong Provincial Key Laboratory of Food Quality and Safety, South China Agricultural University, Wushan Road, Tianhe District, Guangzhou 510642, China
* Correspondence: jjhou@mail.hzau.edu.cn (J.H.); jallent@163.com (Z.X.); Tel.: +86-714-651-1613 (J.H.); +86-020-8528-3448 (Z.X.)

Received: 21 August 2018; Accepted: 7 September 2018; Published: 13 September 2018

Abstract: We describe a multiple combined strategy to discover novel aptamers specific for clenbuterol (CBL). An immobilized ssDNA library was used for the selection of specific aptamers using the systematic evolution of ligands by exponential enrichment (SELEX). Progress was monitored using real-time quantitative PCR (Q-PCR), and the enriched library was sequenced by high-throughput sequencing. Candidate aptamers were picked and preliminarily identified using a gold nanoparticles (AuNPs) biosensor. Bioactive aptamers were characterized for affinity, circular dichroism (CD), specificity and sensitivity. The Q-PCR amplification curve increased and the retention rate was about 1% at the eighth round. Use of the AuNPs biosensor and CD analyses determined that six aptamers had binding activity. Affinity analysis showed that aptamer 47 had the highest affinity (Kd = 42.17 ± 8.98 nM) with no cross reactivity to CBL analogs. Indirect competitive enzyme linked aptamer assay (IC-ELAA) based on a 5′-biotin aptamer 47 indicated the limit of detection (LOD) was 0.18 ± 0.02 ng/L (n = 3), and it was used to detect pork samples with a mean recovery of 83.33–97.03%. This is the first report of a universal strategy including library fixation, Q-PCR monitoring, high-throughput sequencing, and AuNPs biosensor identification to select aptamers specific for small molecules.

Keywords: clenbuterol; systematic evolution of ligands by exponential enrichment; real-time quantitative PCR; high-throughput sequencing technology; aptamers; gold nanoparticles biosensor

---

1. Introduction

Clenbuterol (CBL) is a beta-adrenergic receptor agonist originally used to treat asthma [1]. After CBL enters the body of livestock and poultry, it promotes the growth of animals and increases the lean meat rate [2]. However, food poisoning can result from the human consumption of the edible parts of animals containing CBL. China's Ministry of Agriculture banned the use of adrenaline in animal production in March 1997. However, driven by economic interests, excess CBL in livestock and poultry products frequently causes human poisoning. Therefore, it is necessary to monitor the residue of CBL in livestock and poultry products over a long period. CBL residues are mainly found in livestock,

Molecules **2018**, *23*, 2337; doi:10.3390/molecules23092337   444   www.mdpi.com/journal/molecules

and there is a high demand for domestic animal foods; thus, the detection of CBL residue-related products needs to be rapid. Currently, gold immunochromatography strips [3] and enzyme linked immunosorbent assay [4] are commonly used for this purpose. The key reagents of these two domain detection methods are antibodies. However, the production of antibodies is extremely complex and batch-to-batch variation occurs. Therefore, it is necessary to identify molecules that function similar to antibodies for the development of rapid detection products.

Aptamers are single-stranded DNA (ssDNA) or RNA selected and prepared by Systematic Evolution of Ligands by Exponential Enrichment (SELEX) [5]. Compared with antibodies, they have many advantages including different classes of target, easy modification, timesaving, strong specificity, and high affinity [6]. In the field of food detection, they have an extremely broad usage for the detection of small molecules, and they rival antibodies to some degree. Aptamers have been used to detect small molecules including patulin [7], ochratoxin [8], ractopamine [9] chloramphenicol [10], clenbuterol [11], aflatoxin B1 [12], and tetracycline [13]. The ochratoxin aptamer [14–16] is the most commonly used aptamer, also known as the star aptamer. Currently, there are insufficient numbers of aptamers to detect small molecules residual in food, because of their small molecular weight. How to quickly and efficiently select and identify novel aptamers specific for small molecules is an important question that needs to be answered urgently.

Many SELEX techniques are used to select aptamers, such as magnetic bead SELEX [17], graphene oxide (GO) SELEX [18], capillary electrophoresis SELEX [19], and affinity chromatography SELEX [20]. These methods mainly fix target molecules on the surface of carriers, and then enrich the specific aptamers in the library. Compared with biological macromolecules, the fixation of small molecules is difficult. Before the fixation, small molecules must be derivated, and a carboxyl group or amino active group generated. Then, they are coupled with a solid phase carrier. However, the selected aptamers often recognize the derivatives rather than the small molecules themselves, and therefore cannot be used for the selection of specific aptamers. To solve this problem, a previous study fixed the library on the surface of a solid phase carrier and small molecules were used to capture the aptamers on the solid phase carrier surface [21,22]. This method allowed the simultaneous selection of multiple targets and was also used to select aptamers against small molecules. The current method is to fix the library on the surface of magnetic beads [23] or graphene oxide (GO) [24]. Compared with GO, the magnetic bead method is based on the binding principle of streptavidin-biotin affinity. The library is more stable for fixation and elution compared with GO, and the number of selection rounds is reduced, which saves the selection cost and time. Duan [23] used self-made magnetic beads to select an aptamer specific for CBL. However, there were too many selection rounds (16 rounds). In addition, conventional cloning and sequencing were used to pick out candidate aptamers, resulting in few aptamer candidate sequences. Furthermore, fluorescence labeled aptamers and GO were used for affinity identification, although the cost of the fluorescence labeled aptamer was high. We propose that unmodified gold nanoparticles (AuNPs) based colorimetric biosensors could be used to identify active aptamers from mass enriched sequences. The principle is that AuNPs aggregate after an aptamer specifically combines with its target and the color changes from red to purple (blue). This method is simple, cheap, and quick, and it is suitable for picking out active aptamers from candidate aptamers. Although this method has been used to detect different kinds of small molecules [25,26], no report has identified active aptamers specific for small molecules from an enriched library using this method.

This study reports the first use of a multiple combination selection and identifying strategy, including ssDNA library fixation, Q-PCR monitoring, high-throughput sequencing and AuNPs biosensor identification. This universal technology allows the rapid selection and identification of novel aptamers specific for small molecules. This multiple combination selection and identifying strategy was used to select aptamers specific for CBL, and the specific aptamer successfully detected residual CBL in a pork sample.

2. Results

2.1. Magnetic Bead SELEX Selection of an Aptamer Specific for Clenbuterol

Q-PCR was used to monitor the selection process according to the Ct values and amplification curve for qualitative and quantitative analysis. The quantitative standard curve is shown in Figure 1 and the standard curve correlation coefficient R^2 was 0.9972, showing that the linear correlation between Ct value and logarithm DNA concentration of the template by Q-PCR was good. Therefore, this method can be used for the quantitative calculation of the enriched library amount for each round. The amplification curve and retention rate for all eight rounds of SELEX selection are shown in Figure 2. Figure 2A shows that the amplification curve changed with an increase in enriched rounds. The amplification curve was used to determine when to stop selection, and was related to the initial concentration of the eluent library, the counter selection conditions and specific or non-specific binding. For these reasons, the amplification curves were disordered, but this did not prevent us determining when to stop the selection. According to a report by Luo [27], the diversity of DNA libraries decreased during the SELEX selection, and this reduced mismatching of the DNA sequences. Consequently, the drop-in fluorescence disappeared and the change in fluorescence reached a plateau. Changes in the amplification curves directly represent the convergence of the aptamer species during the SELEX procedure. Therefore, in our study, by the end of the eighth round of selection, the amplification curve showed that the fluorescence had increased, which indicated that the enriched aptamer library was specific for CBL; thus, the selection was stopped. Figure 2B shows that the retention rate increased gradually from the first to fourth round of selection because the library was enriched for specific and non-specific binders. After adding counter selection, the retention rate decreased gradually from the fifth to seventh round of selection compared with the fourth round of selection because most of the non-specific binders were selected out. For the eighth round of selection, the retention rate was increased because this enriched library contained the most specific binders although counter selection was added. In summary, the amplification curve and retention rate indicated that the aptamer library was enriched specifically after eight rounds of selection.

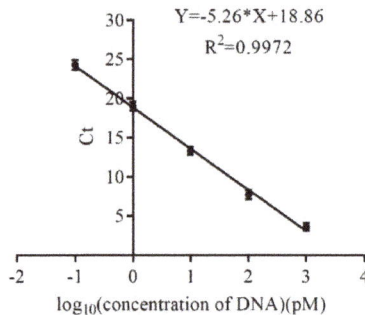

Figure 1. The quantitative standard curve of Q-PCR.

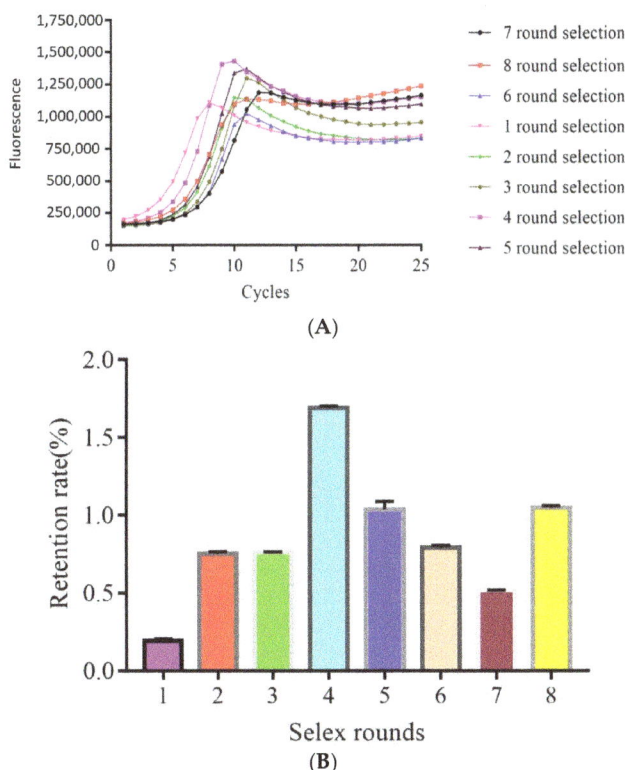

Figure 2. Q-PCR monitoring of the selection process: (**A**) amplification curve of Q-PCR for each round of selection; and (**B**) retention rate for each round of selection.

2.2. High-Throughput Sequencing and Sequence Analysis of the Enriched Library

High-throughput sequencing technology was used to sequence the enriched library of the eighth round. Overall, 105 sequences were obtained, and 172 sequences with a greater frequency of occurrence were selected, as shown in Figure S1. The 40 bp random area sequences underwent homologous alignment using clustal software, and the sequence homology rate was the highest when two sequences were nearest. Phylogenetic tree analysis of these sequences is shown in Figure S2. After removing the sequences that could not be clustered with other sequences, 104 sequences were selected. The mfold online analysis platform showed that 63 aptamers contained one ring region structure or more, and these were selected to identify binding activity. To save the cost of sequence synthesis, 11 bp were removed from the 5′ and 3′ constant regions of the aptamers. The truncated aptamers with 58 bp remaining had unchanged secondary structures, and these were used for binding activity analysis.

2.3. Establishment of a Gold Nanoparticles Biosensor to Identify Active Aptamers

The preliminarily selection results of the AuNPs biosensor are shown in Figure 3. The ratio between the experimental group and the blank group was close to 1 for most of the aptamers, indicating no binding activity between CBL and aptamers, and that the aptamer had no affinity. Interestingly, six aptamers had a ratio of >1.1 between the experimental group and the blank group. The results indicated that the $A_{620\,nm}/A_{520\,nm}$ value of the experimental group was greater than that of the blank group, which indicated that CBL and the aptamers produced a large number of combined complexes. Thus, these six aptamers had affinity with CBL. This method of identifying the binding activity of aptamers specific for a small molecule is the first to identify aptamers from an enriched library. The method is simple, rapid, and low-cost, and can be used to identify other active aptamers from the enriched library.

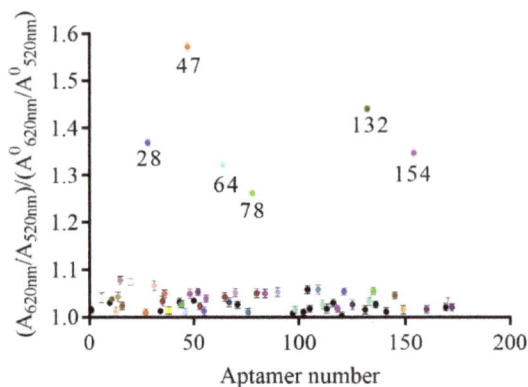

Figure 3. Preliminarily selection of active aptamers using the AuNPs biosensor.

2.4. Characterization of Active Aptamers

The sequence information of the six active aptamers are shown in Table 1. The secondary structures of these six aptamers were analyzed on the mfold platform, which showed that they contained two ring regions and Gibbs free energy (Table 1). The affinities of these six aptamers were analyzed, and the affinity constants were different. The affinity of aptamer 47 was the highest, with a Kd = 42.17 ± 8.98 nM. The secondary structure and affinity curve of aptamer 47 are shown in Figure 4.

Table 1. Sequence ($5'-3'$), dissociation constant (Kd) and dG values of aptamer candidates.

No.	Sequences	Kd (nM)	dG
28	5′ACGCATAGGATGCAACAACGCGAATGCCACCAATTCGTGCTTGTCGTGTCCTATGCGT3′	321 ± 104.5	−15.09
47	5′ACGCATAGGTCATGCCAAGCTGTACACCGTCCTGGCCTGGTTGGGATGTCCTATGCGT3′	42.17 ± 8.98	−13.76
64	5′ACGCATAGGTACACCACATCGGATCGTAATGCGTATGTGACTCAGTGTGCCTATGCGT3′	123.3 ± 14.28	−16.26
78	5′ACGCATAGGGTACGCCACAGATCGACCTCTGGTTGGTCCTGTTGTGTGGCCTATGCGT3′	201.1 ± 42.42	−16.00
132	5′ACGCATAGGGACCATGCGCAGTGAACTTGGTTCTTTGTGCTCATGTGTGCCTATGCGT3′	221.6 ± 41.87	−14.77
154	5′ACGCATAGGGCATCACACATGTCCTTGCCATTGCTGACTTGTTTGGTGTCCTATGCGT3′	268.8 ± 40.43	−15.40

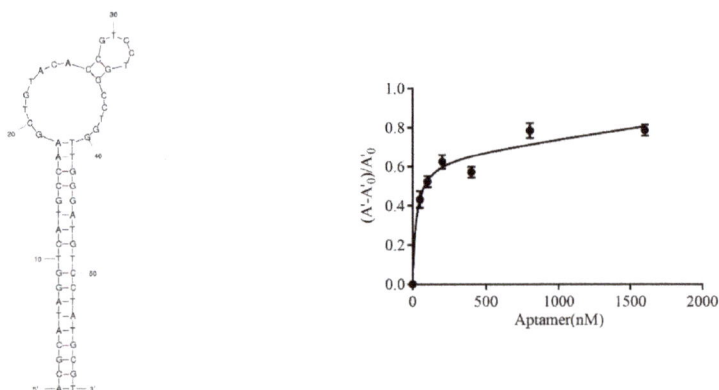

Figure 4. The secondary structure predicted by mfold and corresponding saturation curve of aptamer 47.

To confirm the binding activity between CBL and aptamers, conformation changes of the aptamer before and after CBL binding were studied by circular dichroism (CD) spectrum, a powerful and sensitive tool for studying the interaction between DNA and small molecules [28]. Figure 5 shows that CBL has an extremely weak CD signal, while the aptamer presents a positive and negative peak at about 275 nm and 250 nm, respectively, which was consistent with the reported CD spectrum of a single chain oligonucleotide [29]. When the ssDNA-CBL complex was formed by the binding of free CBL with the aptamer, the characteristic peak at 275 nm was significantly increased, while the characteristic peak at 250 nm was only slightly changed. This suggests that the binding of CBL to the aptamer embedded form led to the reduction of DNA base accumulation. Therefore, the aptamer conformation changed after binding with CBL, which confirmed the interaction between the six aptamers and CBL.

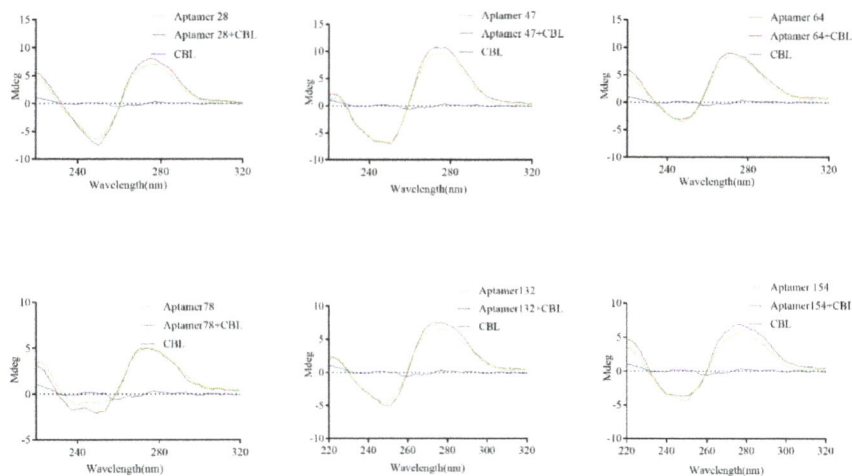

Figure 5. CD spectrum identification of binding activity between six aptamers and CBL.

The specific analysis results are shown in Figure 6. The response signals of the six aptamers binding with CBL exceeded that observed with the CBL analogs, which indicated that there was no cross reaction with the CBL analogs and that the six aptamers had high specificity.

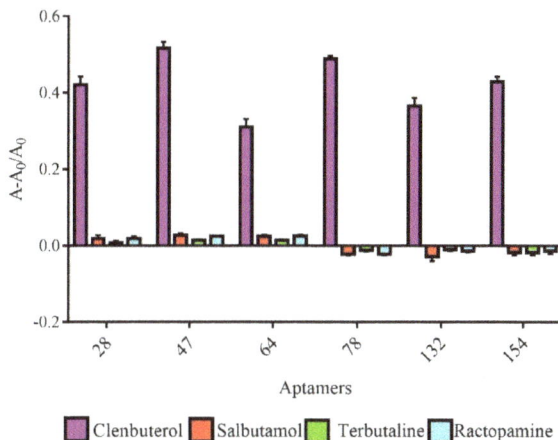

Figure 6. Identification of the specificity of six active aptamers against CBL.

2.5. Detection of CBL Using a Gold Nanoparticles Biosensor and Indirect Competitive Enzyme Linked Aptamer Assay

The sensitivity analysis of aptamer 47 and biotin-aptamer 47 are shown in Figure 7. The AuNPs biosensor results are shown in Figure 7A. The limit of detection (LOD) was 1.2 ± 0.005 ng/mL (n = 3), as estimated by the equation LOD = 3SD/slope, where SD represents the standard deviation of the blank, and the slope was obtained from the calibration curve. The linear detection range was 0.01–4 μg/mL. However, this method was affected by the salt concentration, and therefore was not suitable for the detection of real samples. The complex sample matrix in food may cause AuNPs to change color; therefore, it is necessary to establish another method for real sample detection.

The results of the indirect competitive enzyme-linked aptamer assay (ic-ELAA) are shown in Figure 7B. The mean concentration of CBL required for 50% inhibition of binding (IC_{50}) and LOD (LOD for 10% inhibition of binding) were 5.44 ± 0.01 ng/mL (n = 3) and 0.18 ± 0.02 ng/L (n = 3), respectively, and the linear response range extended from 0.61 to 92.20 ng/mL, showing that the 5′-biotin aptamer 47 could be used to develop ic-ELAA. We used the 5′-biotin aptamer 47 for pork analysis (Table 2). The mean recovery was 83.33–97.03%, indicating that the 5′-biotin aptamer 47 was suitable for the detection of residual CBL in real samples.

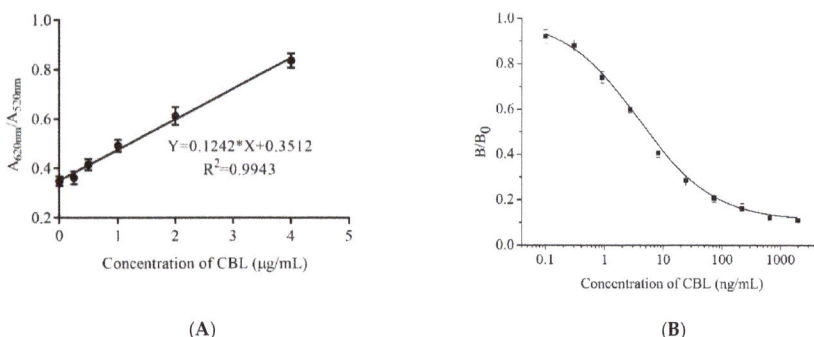

(A)   (B)

Figure 7. Standard curve for CBL detection: (**A**) standard curve for CBL detection with aptamer 47 using AuNPs biosensor; and (**B**) standard curve for CBL detection with 5′-biotin aptamer 47 using ic-ELAA.

Table 2. Recovery test results with pork samples (n = 3).

Samples	Spiked Concentration (ng/g)	Detection Concentration (ng/g)	Recovery (%)	RSD (%)
	1	0.83 ± 0.02	83.33 ± 1.53	1.83
Pork	5	4.66 ± 0.09	93.27 ± 1.70	1.82
	10	9.70 ± 0.05	97.03 ± 0.45	0.46

3. Discussion

SELEX selection generally has to undergo multiple rounds of library enrichment. It is very important to monitor whether the specific aptamer is enriched in the selection process. At present, the most widely used method is Q-PCR, and studies by Mencin [30] and Michiels [31] monitored the SELEX progress according to the melt curve of Q-PCR. However, Luo [27] found that a single melting curve analysis was insufficient to monitor a pool especially during small molecule aptamer selection, because the melting curve did not change during small molecule aptamer selection. They monitored the selection process using the Ct value and Q-PCR amplification curve. When the library presented with specific enrichment, the Ct value was significantly reduced and the amplification curve increased. This method was suitable to monitor the selection progress of aptamers specific for different classes of targets. However, it has not been used to monitor the selection of aptamers specific for CBL. In our study, we used this method to monitor SELEX progress, and after only eight rounds of selection,

SELEX was stopped. The number of selection rounds was lower than reported by a previous report for small molecule selection, which used 10–16 rounds. Therefore, this method might save time and cost, and aptamer selection was simple.

The preparation of a secondary ssDNA library is a key step of SELEX technology, and existing methods include asymmetric PCR [32], lambda exonuclease digestion [33], and unequal length primer PCR [34]. Among them, the asymmetric PCR method is often accompanied by the generation of double-stranded DNA during the preparation of ssDNA. The purity of the secondary library is not high and it is difficult to accurately determine concentrations. The lambda DNA enzyme method has high efficiency, but the enzyme cutting system is large, the cost of enzymes is high, and thus the selection cost is greatly increased. In the current study, we used unequal length primer PCR combined with fluorescence labeling technology to add a fluorescence label to short primers. Through PAGE degenerative electrophoresis separation, fluorescence bands were cut and prepared as for a secondary ssDNA library. This allows the secondary ssDNA library to be separated rapidly and at low cost. The main disadvantage of conventional PCR amplification during unequal length primer PCR is that overamplification increases nonspecific hybridization among different products and by-products, which may cause the loss of potential specific aptamers, inefficient selection, and even selection failure. It was reported [35] that emulsion PCR could overcome the shortcomings of conventional PCR. During emulsion PCR, different templates are separated by emulsion particles, allowing single-molecule PCR, and avoiding nonspecific hybridization. In our study, emulsion PCR was used for the preparation of a secondary ssDNA library.

After multiple rounds of SELEX selection, the active candidate aptamers were picked out. Currently, the conventional method for this is T cloning [36], which randomly picks a certain number of clones for sequencing and activity identification. The disadvantages of this method are that a limited number of clones were picked and some clones were duplicated. Therefore, in our study, high-throughput sequencing technology was used to analyze sequence information in the enriched library containing a large amount of sequence information, to avoid the multiple identification of repeated clones.

A method to identify the binding active aptamers from the enriched library is critical for aptamer selection. The AuNPs biosensor based on aptamers had the advantages of being simple, rapid, sensitive, and visible, which was convenient for the development of portable and on-site rapid detection [37,38]. Of note, the free-labeled aptamers could be identified, which saved cost and time. Therefore, this method was used to batch identify aptamers from the enriched library in our study. By using this convenient method, six novel aptamers were discovered, and aptamer 47 had the highest affinity. For the AuNPs biosensor based on aptamer 47, the LOD was lower than that of Han's patent [39] with a LOD = 4 μg/mL, and which used the ELAA method. Therefore, aptamer 47 has a greater potential for the development of a rapid detection method than the aptamer reported by Han. However, the AuNPs biosensor is not suitable for real sample detection. First, the sample matrix causes AuNPs to aggregate and change color, although this can be prevented by blocking the nanoparticle surface including adding DNA Spacer and changing aptamer graft density on AuNPs surface [40] to ensure greater stability in the conjugated colloids. However, this modification process is complicated. Second, the LOD of the AuNPs biosensor in our study was high and could not be used to detect samples with a low concentration of CBL residues. To solve this issue, we established an ic-ELAA based on the 5′-biotin aptamer 47. In the ic-ELAA analysis, the LOD was slightly higher than a previously reported aptamer with a LOD = 0.07 ng/mL [11] for CBL detection. The sequence of aptamer 47 is novel and the affinity constant is lower than a previously reported aptamer. The 5′-biotin aptamer 47 was also suitable for the development of a rapid detection kit to monitor low levels of residual CBL in pork samples. The sensitivity of the ic-ELAA was higher compared with the AuNPs biosensor, possibly because the reaction system and analysis principles were different. To avoid the interference of salt when using the AuNPs biosensor, aptamer 47 was reacted with CBL in water. For the ic-ELAA, the 5′-biotin aptamer 47 was reacted with CBL in binding buffer. Therefore, the sensitive ic-ELAA method is suitable for real sample detection.

4. Materials and Methods

4.1. Chemicals and Reagents

The clenbuterol and ractopamine standards were purchased from Shanghai Yuanye Biotechnology Ltd. (Shanghai, China). Terbutaline and salbutamol standards were purchased from Aladdin Co., Ltd. (Shanghai, China). Taq polymerase, dNTPs and 2× TBE-urea buffer were purchased from Sangon Biotech Ltd. (Shanghai, China). Streptavidin labeled magnetic beads were purchased from Thermo Fisher Scientific Ltd. (Shanghai, China). $HAuCl_4 \cdot 4H_2O$ was purchased from Sinopharm Chemical Reagent Co., Ltd. (Shanghai, China). Evagreen was purchased from Shanghai Open Biotechnology Ltd. (Shanghai, China). All other analytical pure chemical reagents were purchased from Sinopharm Chemical Reagents Ltd. (Shanghai, China). streptavidin-horseradish peroxidase was purchased from Boster Biological Technology Co. Itd. (Wuhan, China). Other solutions used in the experiments were treated with sterilization ultrapure water. The aptamer library was synthesized by Sangon Biotech Ltd. (Shanghai, China). The primers (Biotin-P, FAM-Forward, polyA-Reverse, Q-Forward, Q-Reverse, Table 3) were synthesized by Nanjing Genscript Biotechnology Ltd. (Nanjing, China). The aptamers and 5′-biotin aptamer 47 were synthesized by Suzhou Hongxun Biotechnology Ltd. (Suzhou, China). High-throughput sequencing was performed by Anhui Angputuomai Biotechnology Ltd. (Hefei, China). Secondary structures were calculated with mfold online bioinformatics platforms (http://unafold.rna.albany.edu/?q=mfold/DNA-Folding-Form).

Table 3. Primer sequences.

Name	Sequences (5′-3′)
Biotin-P	5′-CCTATGCGTGGAGTGCCAAT-3′-biotin
FAM-Forward	6-FAM-5′-ATTGGCACTCCACGCATAGG-3′
polyA-Reverse	5′-AAAAAAAAAAAAAAAAAAAAAspacer18TTCACGGTAGCACGCATAGG-3′
Q-Forward	5′-ATTGGCACTCCACGCATAGG-3′
Q-Reverse	5′-TTCACGGTAGCACGCATAGG-3′

4.2. Selection of Aptamers Specific for CBL Based on ssDNA Library Immobilized SELEX

The procedure of ssDNA Library Immobilized SELEX for aptamers against CBL is shown in Figure 8. The synthetic ssDNA library (5′-ATTGGCACTCCACGCATAGG(N)$_{40}$CCTATGCGTGCTACCGTGAA-3′) was dissolved in binding buffer (0.1 g $CaCl_2$, 0.2 g KCl, 0.2 g KH_2PO_4, 0.1 g $MgCl_2.6H_2O$, 8 g NaCl, and 1.15 g Na_2HPO_4, 1 L). Biotin-P was mixed with the library at a ratio of 2:1, and slowly denatured and renatured (95 °C for 10 min, 60 °C for 1 min, 25 °C for 1 min). The above mixture was added to the streptavidin magnetic beads and incubated (the magnetic beads were washed 4 times with the binding buffer before use, and 700 μL magnetic beads was used in the first round of selection) for 45 min. The magnetic beads were washed 6 times, and incubated for 90 min with CBL in binding buffer (200 μL) to a final concentration of 100 μM at room temperature. The supernatant was collected by magnetic separation, the magnetic beads were washed with 200 μL binding buffer containing the CBL at a final concentration of 100 μM, and the supernatant was collected using magnetic separation. The supernatant from two elutions were pooled as the eluent library with a volume of 400 μL.

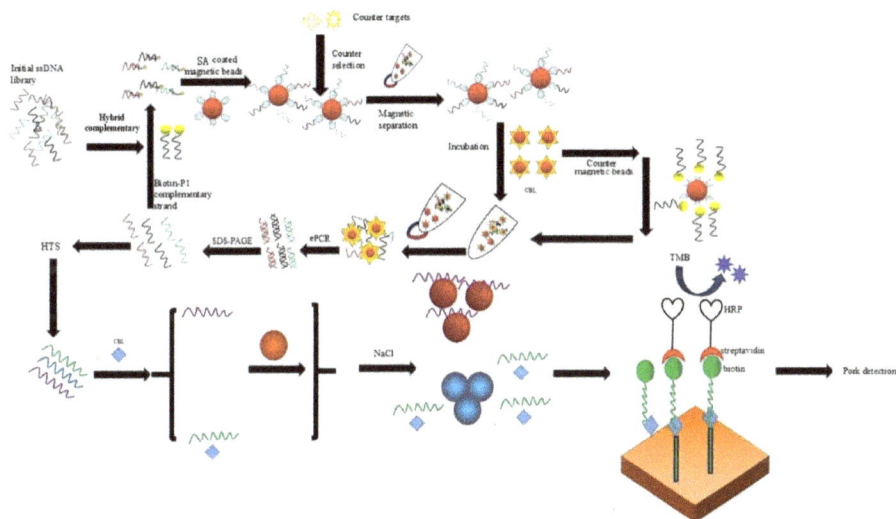

Figure 8. Graphical abstract.

4.3. Establishment of a Real-Time Quantitative PCR Method for the Monitoring Selection Process

A standard curve was prepared as follows: the initial ssDNA library was diluted to a concentration of 16,000, 1600, 160, 16, and 1.6 pM as a real-time quantitative PCR template. Then, 2 μL of template was mixed with 30 μL Q-PCR mix (1 μL Q-Forward at a concentration of 10 μM, 1 μL Q-Reverse at a concentration of 10 μM, 0.5 μL dNTP mix at a concentration of 10 mM, 3 μL 10× PCR buffer, 1 μL Taq DNA polymerase, 1 μL Evagreen, 15.5 μL sterile water, total volume of 30 μL), and the following Q-PCR program (StepOnePlus, ABI, Carlsbad, CA, USA) was used: 95 °C for 2 min, 95 °C for 0.5 min, 60 min at 0.5 °C, and 72 °C for 0.5 min with 25 cycles. The enriched libraries from each round were monitored using Q-PCR amplification according to the above method, and the template was the eluent library with a volume of 2 μL.

4.4. Preparation of Secondary Libraries

After each round of selection, all of the eluent library was added into the emulsion PCR mix (398 μL template DNA, 10 μL FAM-Forward at a concentration of 100 μM, 10 μL polyA-Reverse at a concentration of 100 μM, 40 μL dNTP mix at a concentration of 10 mM, 200 μL 10× PCR buffer, 8 μL Taq DNA polymerase, 1732 μL sterile water, total volume of 2 mL). After adding 8 mL emulsifier (1 mL EM 90, 25 μL triton X-100, 49 mL mineral oil), the emulsion was prepared by vortexing for 2 min, and left to stand for 5 min. The PCR program was as follows: (95 °C 2 min, 95 °C 1 min, 60 °C 1 min, 72 °C 1 min, 25 cycles). The PCR product was concentrated with n-butyl alcohol, then it was mixed with 2× TBE-urea buffer and boiled for 10 min. The sample was separated by denatured SDS-PAGE (400 V, 15 min). The fluorescent strip was cut off and boiled to separate the secondary ssDNA library, which was dialyzed overnight in binding buffer. After measuring of the secondary library concentration, the next round of selection was conducted. The first, second, third and fourth rounds of the selection procedures were consistent. The fifth, sixth and seventh rounds of selection were counter selection with analogs (salbutamol, ractopamine and terbutaline) and positive selection with CBL. The eighth round of selection was positive selection with CBL, and then biotin-P magnetic beads were added for counter selection. The amount of ssDNA library in the first round of selection was 1.3 nm, and in the later rounds it was 100 pM.

4.5. High-Throughput Sequencing and Sequence Analysis of the Enrichment Library

In the eighth round, the library was enriched and sent to Anhui Angputuomai Biotechnology Co., Ltd. for high-throughput sequencing by an Illumina high-throughput sequencing platform (HiSeq/MiSeq). Homologous comparison and evolutionary tree analysis were carried out by clustalX2 software (downloaded from http://www.clustal.org/download/current/) and then visualized using TreeView tool version 1.6.6 for 172 sequences, which were selected from high-throughput sequencing results according to the frequency of occurrence. Then, 104 sequences with a high homology rate were selected for secondary structure prediction using mfold online bioinformatics platforms (http://unafold.rna.albany.edu/?q=mfold/DNA-Folding-Form). Sixty-three sequences with several ring regions ≥ 2 were selected, and 11 bases were removed from the 5′ and 3′ constant regions, respectively. Each aptamer contained 58 bp and was synthesized.

4.6. Determination of Binding Activity between Aptamers and CBL Using a Gold Nanoparticles Biosensor

The binding activity of candidate aptamers were preliminarily determined using an AuNPs biosensor (Figure 8). AuNPs were prepared as previous report [41] by reducing chloroauric acid with sodium citrate. This was centrifuged at 12,000 r/min, 4 °C for 25 min, the supernatant was discarded, and the precipitate was dissolved in ultrapure water, as the concentration of AuNPs in ultrapure water was determined to be 8.76 nM using UV-vis spectroscopy according to the previously reported method [42]. First, 50 μL/well of aptamers were diluted with ultrapure water until the final concentration was 0.4 μM, and then they were incubated with 1 μg/mL CBL (50 μL/well) for 30 min at room temperature. AuNPs (50 μL/well, 8.76 nM) were added and incubated for 30 min at room temperature, and then 0.9 M NaCl (10 μL/well) was added. The absorbance values at 620 nm ($A_{620\,nm}$) and 520 nm ($A_{520\,nm}$) were measured on an automatic microplate reader (I3X, Molecular Devices, San Jose, CA, USA), and aptamers with binding activity were determined by the ratio of $A_{620\,nm}/A_{520\,nm}$ between the experimental group and the blank group.

4.7. Affinity Analysis of Active Aptamers

Aptamers with different concentrations (0, 50, 100, 200, 400, 800 and 1600 nM) were incubated with 1 μg/mL CBL for 30 min at room temperature. AuNPs were added and incubated for 30 min at room temperature, and then 0.9 M NaCl was added. The A_{520nm} was measured on an automatic microplate reader, $(A'-A'_0)/A'_0$ was used as the ordinate and the aptamer concentration was used as the abscissa. The affinity constant of aptamers was calculated by Graphpad Prism 7.0 software (GraphPad Software, La Jolla, CA, USA). A′ represents the A_{520nm} value at the concentration of each aptamer and A'_0 represents the value of A_{520nm} when the concentration of the aptamer is zero.

4.8. Circular Dichroism Spectrum Analysis of Active Aptamers

The aptamers were diluted to 1 μM with binding buffer, and incubated with 10 μM CBL for 30 min at room temperature. The circular dichroism of the mixture was scanned at a scanning wavelength of 220–320 nm. At the same time, the CD spectrum (J-810, Jasco, Tokyo, Japan) analysis of CBL or aptamers were performed under the same conditions.

4.9. Specificity Analysis of Active Aptamers

Clenbuterol, terbutaline, salbutamol, and ractopamine (1 μg/mL) were incubated with 0.4 μM of aptamers for 30 min at room temperature. AuNPs were added and incubated for 30 min at room temperature, and then 0.9 M NaCl was added. The $A_{620\,nm}$ and $A_{520\,nm}$ were measured on an automatic microplate reader. A: represents the value of ($A_{620\,nm}/A_{520\,nm}$) when the concentration of Clenbuterol, terbutaline, salbutamol, and ractopamine is 1 μg/mL and A_0: represents the value of (A_{620nm}/A_{520nm}) when the concentration of clenbuterol, terbutaline, salbutamol, and ractopamine is 0 μg/mL.

4.10. Detection of CBL Using a Gold Nanoparticles Biosensor and ic-ELAA

Aptamer 47 was diluted with ultrapure water until the final concentration was 0.4 µM, and it was then incubated with a series of CBL concentrations (0.01, 0.25, 0.50, 1.0, 2.0, and 4.0 µg/mL) for 30 min at room temperature. AuNPs were added and incubated for 30 min at room temperature, and then 0.9 M NaCl was added. The $A_{620\,nm}$ and $A_{520\,nm}$ was measured on an automatic microplate reader. ($A_{620\,nm}/A_{520\,nm}$) was used as the ordinate and the CBL concentration was used as the abscissa.

The sensitivity of the 5'-biotin aptamer 47 was also determined by ic-ELAA (Figure 8). A microtiter plate was coated with 5 µg/mL CBL-Bovine serum albumin (BSA) in coating buffer (15 mM Na_2CO_3 and 35 mM $NaHCO_3$, pH 9.6) at 37 °C overnight. The CBL-BSA was prepared as previously reported [43]. The plate was washed twice with washing buffer (6 g $Na_2HPO_4 \cdot 12H_2O$, 16 g NaCl, 1.2 mL Tween 20, pH 7.4, 2 L) and then blocked with 150 µL 3% (*w/v*) skim milk powder in PBS for 2 h. Subsequently, a series concentration of CBL standard (0, 0.1, 0.3, 0.9, 2.7, 8.1, 24.3, 72.9, 218.7, 656.1, and 1968.3 ng/mL, 50 µL) and 5'-biotin aptamer 47 (1 µM, 50 µL) were respectively added into each well, and the resulting solution was incubated for 1 h at 37 °C. After washing six times as described above, 100 µL of streptavidin-horseradish peroxidase (HRP) diluted 1:5000 was added, and the plate was incubated at 37 °C for 1 h. After washing the plate six times, 3,3',5,5'-tetramethylbenzidine (TMB) was added to the reaction for 10 min. The reaction was stopped with 10% H_2SO_4, and A_{450nm} was measured on an automatic microplate reader.

The detection of CBL in real samples was also studied using ic-ELAA. Pork was purchased from a supermarket and pretreated according the following method. One hundred grams of pork was homogenized for 10 min. Next, 10 µL of CLB standard solutions at different concentrations (1, 5, and 10 ng/µL) were individually mixed with 10 g of pork paste and homogenized for 10 min. Then, 1 g of the above homogenized sample was mixed with 1 mL of 0.01 M HCl, vortexed for 2 min and centrifuged at 12,000 r/min for 10 min. The supernatant was adjusted to pH 7.4 and centrifuged at 12,000 r/min for 10 min. The supernatant was filtered through a 0.45 µm filtration membrane and collected for the next assay.

5. Conclusions

Aptamers are molecules that complement the function of antibodies for the development of rapid detection kits, where the biggest bottleneck is the selection of novel specific aptamers. For small molecules, it is difficult to select aptamers with specific binding activity. This is the first study to report the use of multiple combination selection and identification strategy including library fixation, Q-PCR monitoring, high-throughput sequencing, and AuNPs biosensor identification for the selection of novel specific aptamers. The selection, monitoring, and identification process was simple, intuitive, reliable, and high-throughput. After eight rounds of selection, six novel aptamers were discovered from the ssDNA aptamer library. Aptamer 47 had the highest affinity with a binding constant of 42.17 ± 8.98 nM. The ic-ELAA based on 5'-biotin aptamer 47 was used for pork sample detection with a mean recovery of 83.33–97.03%. Most importantly, the multiple combination selection and identification strategy proposed in this study is a universal technical method, which can be used for the selection and identification of other novel aptamers specific for small molecules.

6. Patents

Liu, X.X.; Hou, J.J.; Lu, Q.; Wang, F.; Hou, Y.Y.; Meng, C.; Hu, T.Y.; Chen, S.Y. An aptamer used to detect CBL of clenbuterol and its screening method and application. Patent 2018, Application number: 201810874140.X.

Supplementary Materials: Figure S1: The 172 sequences with a greater frequency of occurrence from high-throughput sequencing sequences. Figure S2: Phylogenetic tree analysis of 172 sequences.

Molecules **2018**, *23*, 2337

Author Contributions: Formal analysis, C.M., T.H. and Y.H.; Funding acquisition, X.L. and J.H.; Investigation, Z.X.; Methodology, Q.L., S.C. and F.W.; Project administration, X.L.; Supervision, X.L., J.H., and Z.X.; and Writing—original draft, X.L.

Funding: This research was funded by grants from the Natural Science Foundation of Hubei Province (2016CFB218), National Natural Science Foundation of China (31601548), Open Foundation of Guangdong Provincial Key Laboratory of Food Quality and Safety (GPKLFQS201702), and the Research Fund for Science and Technology Innovation Team of Hubei Normal University (T201504).

Conflicts of Interest: The authors declare no conflicts of interest.

References

1. Torneke, K.; Larsson, C.I.; Appelgren, L.E. A comparison between clenbuterol, salbutamol and terbutaline in relation to receptor binding and in vitro relaxation of equine tracheal muscle. *J. Vet. Pharmacol. Ther.* **1998**, *21*, 388–392. [CrossRef] [PubMed]

2. Li, Y.; Qi, P.; Ma, X.; Zhong, J.G. Quick detection technique for clenbuterol hydrochloride by using surface plasmon resonance biosensor. *Eur. Food Res. Technol.* **2014**, *239*, 195–201. [CrossRef]

3. Wang, J.Y.; Zhang, L.; Huang, Y.J.; Dandapat, A.; Dai, L.W.; Zhang, G.G.; Lu, X.F.; Zhang, J.W.; Lai, W.H.; Chen, T. Hollow Au-Ag Nanoparticles labeled immunochromatography strip for highly sensitive detection of clenbuterol. *Sci. Rep.-UK* **2017**, *7*, 41419. [CrossRef] [PubMed]

4. Ma, L.Y.; Nilghaz, A.; Choi, J.R.; Liu, X.Q.; Lu, X.N. Rapid detection of clenbuterol in milk using microfluidic paper-based ELISA. *Food Chem.* **2018**, *246*, 437–441. [CrossRef] [PubMed]

5. Tian, R.Y.; Lin, C.; Yu, S.Y.; Gong, S.; Hu, P.; Li, Y.S.; Wu, Z.C.; Gao, Y.; Zhou, Y.; Liu, Z.S.; et al. Preparation of a specific ssDNA aptamer for brevetoxin-2 using SELEX. *J. Anal. Methods Chem.* **2016**, *2016*, 9241860. [CrossRef] [PubMed]

6. Wang, Y.; Li, J.; Qiao, P.; Jing, L.; Song, Y.Z.; Zhang, J.Y.; Chen, Q.; Han, Q.Q. Screening and application of a new aptamer for the rapid detection of Sudan Dye III. *Eur. J. Lipid Sci. Tech.* **2018**, *120*, 1700112. [CrossRef]

7. Wu, Z.Z.; Xu, E.; Jin, Z.Y.; Irudayaraj, J. An ultrasensitive aptasensor based on fluorescent resonant energy transfer and exonuclease-assisted target recycling for patulin detection. *Food Chem.* **2018**, *249*, 136–142. [CrossRef] [PubMed]

8. Zhu, X.; Kou, F.X.; Xu, H.F.; Han, Y.J.; Yang, G.D.; Huang, X.J.; Chen, W.; Chi, Y.W.; Lin, Z.Y. Label-free ochratoxin A electrochemical aptasensor based on target-induced noncovalent assembly of peroxidase-like graphitic carbon nitride nanosheet. *Sensor. Actuat. B-Chem.* **2018**, *270*, 263–269. [CrossRef]

9. Zhu, C.; Zhang, G.L.; Huang, Y.F.; Yan, J.; Chen, A.L. Aptamer based ultrasensitive determination of the beta-adrenergic agonist ractopamine using PicoGreen as a fluorescent DNA probe. *Microchim. Acta* **2017**, *184*, 439–444. [CrossRef]

10. Duan, Y.; Wang, L.H.; Gao, Z.Q.; Wang, H.S.; Zhang, H.X.; Li, H. An aptamer-based effective method for highly sensitive detection of chloramphenicol residues in animal-sourced food using real-time fluorescent quantitative PCR. *Talanta* **2017**, *165*, 671–676. [CrossRef] [PubMed]

11. Chen, D.; Yang, M.; Zheng, N.J.; Xie, N.; Liu, D.L.; Xie, C.F.; Yao, D.S. A novel aptasensor for electrochemical detection of ractopamine, clenbuterol, salbutamol, phenylethanolamine and procaterol. *Biosens.Bioelectron.* **2016**, *80*, 525–531. [CrossRef] [PubMed]

12. Peng, G.; Li, X.Y.; Cui, F.; Qiu, Q.Y.; Chen, X.J.; Huang, H. Aflatoxin B1 electrochemical aptasensor based on tetrahedral DNA nanostructures functionalized three dimensionally ordered macroporous MoS2-AuNPs film. *ACS Appl. Mater. Int.* **2018**, *10*, 17551–17559. [CrossRef] [PubMed]

13. Qi, M.Y.; Tu, C.Y.; Dai, Y.Y.; Wang, W.P.; Wang, A.J.; Chen, J.R. A simple colorimetric analytical assay using gold nanoparticles for specific detection of tetracycline in environmental water samples. *Anal. Methods-UK* **2018**, *10*, 3402–3407. [CrossRef]

14. Chen, Y.Q.; Chen, M.L.; Chi, J.X.; Yu, X.; Chen, Y.X.; Lin, X.C.; Xie, Z.H. Aptamer-based polyhedral oligomeric silsesquioxane (POSS)-containing hybrid affinity monolith prepared via a "one-pot" process for selective extraction of ochratoxin A. *J. Chromatogr. A* **2018**, *1563*, 37–46. [CrossRef] [PubMed]

15. Shen, P.; Li, W.; Ding, Z.; Deng, Y.; Liu, Y.; Zhu, X.R.; Cai, T.T.; Li, J.L.; Zheng, T.S. A competitive aptamer chemiluminescence assay for ochratoxin A using a single silica photonic crystal microsphere. *Anal. Biochem.* **2018**, *554*, 28–33. [CrossRef] [PubMed]

16. Al Rubaye, A.; Nabok, A.; Catanante, G.; Marty, J.L.; Takacs, E.; Szekacs, A. Detection of ochratoxin A in aptamer assay using total internal reflection ellipsometry. *Sensor. Actuat. B-Chem.* **2018**, *263*, 248–251. [CrossRef]

17. Lin, C.; Liu, Z.S.; Wang, D.X.; Li, L.; Hu, P.; Gong, S.; Li, Y.S.; Cui, C.; Wu, Z.C.; Gao, Y.; et al. Generation of internal-image functional aptamers of okadaic acid via magnetic-bead SELEX. *Mar. Drugs* **2015**, *13*, 7433–7445. [CrossRef] [PubMed]

18. Nguyen, V.T.; Kwon, Y.S.; Kim, J.H.; Gu, M.B. Multiple GO-SELEX for efficient screening of flexible aptamers. *Chem. Commun.* **2014**, *50*, 10513–10516. [CrossRef] [PubMed]

19. Luo, Z.F.; Zhou, H.M.; Jiang, H.; Ou, H.C.; Li, X.; Zhang, L.Y. Development of a fraction collection approach in capillary electrophoresis SELEX for aptamer selection. *Analyst* **2015**, *140*, 2664–2670. [CrossRef] [PubMed]

20. Setlem, K.; Monde, B.; Ramlal, S.; Kingston, J. Immuno affinity SELEX for simple, rapid, and cost-effective aptamer enrichment and identification against aflatoxin B_1. *Front. Microbiol.* **2016**, *7*, 1909. [CrossRef] [PubMed]

21. Wu, Y.G.; Zhan, S.S.; Wang, L.M.; Zhou, P. Selection of a DNA aptamer for cadmium detection based on cationic polymer mediated aggregation of gold nanoparticles. *Analyst* **2014**, *139*, 1550–1561. [CrossRef] [PubMed]

22. He, J.A.; Liu, Y.A.; Fan, M.T.; Liu, X.J. Isolation and identification of the DNA aptamer target to acetamiprid. *J. Agr. Food Chem.* **2011**, *59*, 1582–1586. [CrossRef] [PubMed]

23. Duan, N.; Gong, W.H.; Wu, S.J.; Wang, Z.P. Selection and application of ssDNA aptamers against clenbuterol hydrochloride based on ssDNA library immobilized SELEX. *J. Agric. Food Chem.* **2017**, *65*, 1771–1777. [CrossRef] [PubMed]

24. Duan, N.; Gong, W.H.; Wu, S.J.; Wang, Z.P. An ssDNA library immobilized SELEX technique for selection of an aptamer against ractopamine. *Anal. Chim. Acta* **2017**, *961*, 100–105. [CrossRef] [PubMed]

25. Zhang, D.W.; Yang, J.Y.; Ye, J.; Xu, L.R.; Xu, H.C.; Zhan, S.S. Colorimetric detection of bisphenol A based on unmodified aptamer and cationic polymer aggregated gold nanoparticles. *Anal. Biochem.* **2016**, *499*, 51–56. [CrossRef] [PubMed]

26. Yang, C.; Wang, Y.; Martyc, J.L.; Yang, X.R. Aptamer-based colorimetric biosensing of Ochratoxin A using unmodified gold nanoparticles indicator. *Biosens. Bioelectron.* **2011**, *26*, 2724–2727. [CrossRef] [PubMed]

27. Luo, Z.F.; He, L.; Wang, J.J.; Fang, X.N.; Zhang, L.Y. Developing a combined strategy for monitoring the progress of aptamer selection. *Analyst* **2017**, *142*, 3136–3139. [CrossRef] [PubMed]

28. Ramana, M.M.V.; Betkar, R.; Nimkar, A.; Ranade, P.; Mundhe, B.; Pardeshi, S. In vitro DNA binding studies of antiretroviral drug nelfinavir using ethidium bromide as fluorescence probe. *J. Photoch. Photobio. B* **2015**, *151*, 194–200. [CrossRef] [PubMed]

29. Zhang, A.M.; Jia, L.P.; Zhang, J.J. Study on the interaction between amikacin and DNA by molecular spectroscopy. *J. Liaocheng Univer.* **2008**, *21*, 67–70.

30. Mencin, N.; Smuc, T.; Vranicar, M.; Mavri, J.; Hren, M.; Galesa, K.; Krkoc, P.; Ulrich, H.; Solar, B. Optimization of SELEX: Comparison of different methods for monitoring the progress of in vitro selection of aptamers. *J. Pharmaceyt Biomed.* **2014**, *91*, 151–159. [CrossRef] [PubMed]

31. Vanbrabant, J.; Leirs, K.; Vanschoenbeek, K.; Lammertyn, J.; Michiels, L. Remelting curve analysis as a tool for enrichment monitoring in the SELEX process. *Analyst* **2014**, *139*, 589–595. [CrossRef] [PubMed]

32. Heiat, M.; Ranjbar, R.; Latifi, A.M.; Rasaee, M.J.; Farnoosh, G. Essential strategies to optimize asymmetric PCR conditions as a reliable method to generate large amount of ssDNA aptamers. *Biotechnol. Appl. Bioc.* **2017**, *64*, 541–548. [CrossRef] [PubMed]

33. Citartan, M.; Tang, T.H.; Tan, S.C.; Gopinath, S.C.B. Conditions optimized for the preparation of single-stranded DNA (ssDNA) employing lambda exonuclease digestion in generating DNA aptamer. *World J. Microb. Biot.* **2011**, *27*, 1167–1173. [CrossRef]

34. Williams, K.P.; Bartel, D.P. PCR product with strands of unequal length. *Nucleic. Acids Res.* **1995**, *23*, 4220–4221. [CrossRef] [PubMed]

35. Shao, K.K.; Shi, X.H.; Zhu, X.J.; Cui, L.L.; Shao, Q.X.; Ma, D. Construction and optimization of an efficient amplification method of a random ssDNA library by asymmetric emulsion PCR. *Biotechnol. Appl. Bioc.* **2017**, *64*, 239–243. [CrossRef] [PubMed]

36. Wu, S.J.; Li, Q.; Duan, N.; Ma, H.L.; Wang, Z.P. DNA aptamer selection and aptamer-based fluorometric displacement assay for the hepatotoxinmicrocystin-RR. *Microchim. Acta* **2016**, *183*, 2555–2562. [CrossRef]

37. Zhou, X.T.; Wang, L.M.; Shen, G.Q.; Zhang, D.W.; Xie, J.L.; Mamut, A.; Huang, W.W.; Zhou, S.S. Colorimetric determination of ofloxacin using unmodified aptamers and the aggregation of goldnanoparticles. *Microchim. Acta* **2018**, *185*, 355. [CrossRef] [PubMed]

38. Shi, Q.A.; Shi, Y.P.; Pan, Y.; Yue, Z.F.; Zhang, H.; Yi, C.Q. Colorimetric and bare eye determination of urinary methylamphetamine based on the use of aptamers and the salt-induced aggregation of unmodified gold nanoparticles. *Microchim. Acta* **2015**, *182*, 505–511. [CrossRef]

39. Han, Q.Q.; Qiao, P.; Song, Y.Z.; Zhang, J.Y.; Chen, Q. A Aptamer Specific CBL and Its Application. Patent 201510485760.0, 23 December 2015.

40. Zhao, W.A.; Chiuman, W.; Lam, J.C.F.; McManus, S.A.; Chen, W.; Cui, Y.G.; Pelton, R.; Brook, M.A.; Li, Y.F. DNA aptamer folding on gold nanoparticles: From colloid chemistry to biosensors. *J. Am. Chem. Soc.* **2008**, *130*, 3610–3618. [CrossRef] [PubMed]

41. Yao, X.; Ma, X.D.; Ding, C.; Jia, L. Colorimetric determination of lysozyme based on the aggregation of gold nanoparticlescontrolled by a cationic polymer and an aptamer. *Microchim. Acta* **2016**, *183*, 2353–2359. [CrossRef]

42. Haiss, W.; Thanh, N.T.K.; Aveyard, J.; Fernig, D.G. Determination of size and concentration of nanoparticles from UV-Vis spectra. *Anal. Chem.* **2007**, *79*, 4215–4221. [CrossRef] [PubMed]

43. Kim, Y.H.; Kim, Y.S. Effects of active immunization against clenbuterol on the growth-promoting effect of clenbuterolin rats. *J. Anim. Sci.* **1997**, *75*, 446–453. [CrossRef] [PubMed]

Sample Availability: Samples of the compounds are not available from the authors.

molecules

MDPI

Article

Novel Electrochemical Sensors Based on Cuprous Oxide-Electrochemically Reduced Graphene Oxide Nanocomposites Modified Electrode toward Sensitive Detection of Sunset Yellow

Quanguo He [1,2,†], Jun Liu [1,2,†], Xiaopeng Liu [2], Yonghui Xia [3], Guangli Li [1,2,*], Peihong Deng [4] and Dongchu Chen [1,*]

[1]   School of Materials Science and Energy Engineering, Foshan University, Foshan 528000, China; hequanguo@126.com (Q.H.); liu.jun.1015@163.com (J.L.)
[2]   College of Life Sciences and Chemistry, Hunan University of Technology, Zhuzhou 412007, China; amituo321@163.com
[3]   Zhuzhou Institute for Food and Drug Control, Zhuzhou 412000, China; sunnyxia0710@163.com
[4]   Department of Chemistry and Material Science, Hengyang Normal University, Hengyang 421008, China; dph1975@163.com
*   Correspondence: guangli010@hut.edu.cn (G.L.); chendc@fosu.edu.cn (D.C.);
    Tel./Fax: +86-731-22183382 (G.L. & D.C.)
†   These authors contributed equally to this work.

Received: 30 July 2018; Accepted: 21 August 2018; Published: 24 August 2018

check for updates

Abstract: Control and detection of sunset yellow is an utmost demanding issue, due to the presence of potential risks for human health if excessively consumed or added. Herein, cuprous oxide-electrochemically reduced graphene nanocomposite modified glassy carbon electrode (Cu_2O-ErGO/GCE) was developed for the determination of sunset yellow. The Cu_2O-ErGO/GCE was fabricated by drop-casting Cu_2O-GO dispersion on the GCE surface following a potentiostatic reduction of graphene oxide (GO). Scanning electron microscope and X-ray powder diffractometer was used to characterize the morphology and microstructure of the modification materials, such as Cu_2O nanoparticles and Cu_2O-ErGO nanocomposites. The electrochemical behavior of sunset yellow on the bare GCE, ErGO/GCE, and Cu_2O-ErGO/GCE were investigated by cyclic voltammetry and second-derivative linear sweep voltammetry, respectively. The analytical parameters (including pH value, sweep rate, and accumulation parameters) were explored systematically. The results show that the anodic peak currents of Cu_2O-ErGO /GCE are 25-fold higher than that of the bare GCE, due to the synergistic enhancement effect between Cu_2O nanoparticles and ErGO sheets. Under the optimum detection conditions, the anodic peak currents are well linear to the concentrations of sunset yellow, ranging from 2.0×10^{-8} mol/L to 2.0×10^{-5} mol/L and from 2.0×10^{-5} mol/L to 1.0×10^{-4} mol/L with a low limit of detection (S/N = 3, 6.0×10^{-9} mol/L). Moreover, Cu_2O-ErGO/GCE was successfully used for the determination of sunset yellow in beverages and food with good recovery. This proposed Cu_2O-ErGO/GCE has an attractive prospect applications on the determination of sunset yellow in diverse real samples.

Keywords: cuprous oxide nanoparticles; reduced graphene oxide; modified electrode; sunset yellow; second-derivative linear sweep voltammetry

1. Introduction

Sunset yellow, as a common azo colorant, has been widely added in several beverages (such as carbonated beverage, orange juice, and Fanta drink) and food (i.e. candies, cakes, cheese) to improve the appearance and appetite [1,2]. Sunset yellow contains an azo group (-N=N-) and aromatic ring structure, which may cause mutagenic and carcinogenic risk for human [3,4]. As a result, it can bring many serious health problems, such as hepatocellular damage, kidney failures, headache, cancers, and attention deficit hyperactivity disorder (ADHD), when it is excessively consumed [5]. Hence, the additive amount of sunset yellow in beverages and food demands strict control and regulation. European Food Safety Authority (EFSA) recommends an acceptable daily intake (ADI) for sunset yellow of 1.0 mg kg^{-1} body weight [6]. In some countries, it was explicitly stated that the permitted maximum content of sunset yellow in nonalcoholic beverages is 100 µg mL^{-1} [7]. Moreover, Finland and Norway even has already banned the use of sunset yellow in foods [6]. While considering the food quality and safety, it is urgent to develop reliable analytical techniques for the quick detection of this food dye.

At present, various techniques have been developed for sunset yellow, including high performance liquid chromatography (HPLC) [8], thin layer chromatography [9], spectrophotometry [10], fluorometry [11], and capillary electrophoresis [12]. Although these methods have been proved to be reliable, they still have some disadvantages, such as time-consuming, complicated preprocessing, and expensive equipment. Recently, electrochemical analysis has been widely used for the detection of bioactive molecules, nutrients, food additives, as well as contaminants, due to its considerable merits such as low cost, simple operation, rapid response, high sensitivity and sensitivity. As we all know, the key issue for electrochemical detection toward sunset yellow is to develop ultrasensitive modified electrodes.

Nanostructure precious metals or alloys have become the preferred electrode modification materials for the sensitive detection of sunset yellow, due to their superior electrocatalytic activity [7,13–17]. For example, Wang and coworkers developed a promising electrochemical sensor based on Au-Pd and reduced graphene oxide nanocomposite decorated electrode (Au-Pd-RGO/ GCE) [17]. The Au-Pd-RGO/GCE exhibited good stability, superior electrocatalytic performance, low detection limit (1.5 nmol/L), and wide response range (0.686–331.686 µmol/L). Pd-Ru nanoparticles incorporated carbon aerogel nanocomposites (Pd-Ru/CA) have been successfully used for electrochemical detection and catalytic degradation of sunset yellow [16]. The Pd-Ru/CA nanocomposite decorated screen printed carbon electrode (Pd-Ru/CA/SPCE) showed a low detection limit (7.1 nmol/L) and high sensitivity (3.571 µA/(µmol/L cm^2)). Electrochemical sensing platform based on Au NPs and reduced graphene modified GCE (Au NPs/RGO/GCE) was constructed for the quantitative analysis of sunset yellow, and the Au NPs/RGO/GCE had excellent catalytic activity toward the oxidation of sunset yellow [15]. The proposed electrode showed wide linear response range of 0.002–109.14 µmol/L and low detection limit of 2 nmo/L (S/N = 3). Au NPs decorated carbon-paste electrode (Au NPs/CPE) was also fabricated for the simultaneous detection sunset yellow and Tartrazine [7]. The Au NPs/CPE displayed low detection limits of 3.0×10^{-8} and 2.0×10^{-9} mol/L for sunset yellow and Tartrazine, respectively. These precious metals-based modified electrodes can detect sunset yellow at the nanomole level; however, the scarcity and high cost of precious metals or alloys have seriously hindered the broad practical applications.

When compared to precious metals or alloys, transition metals and metal oxides have outstanding advantages in terms of abundance and cost. What is more, transition metals and metal oxides also have excellent catalytic activity. In our previous work, Cu$_2$O-RGO nanocomposite [18], NH$_2$-Fe$_3$O$_4$-RGO nanocomposite [19], MnO$_2$-RGO nanocomposite [20], TiO$_2$-RGO nanocomposite [21], and α-MnO$_2$/N-doped ketjenblack carbon composite [22,23] demonstrated excellent electrochemical sensing performance toward the detection of dopamine and Tartrazine and superior electrocatalytic activity in oxygen reduction reaction (ORR) & oxygen evolution reaction (OER). Recently, few researchers have devoted to developing transition metal oxide modified electrodes for the determination of sunset yellow. For example, Dorraji et al. [24] developed ZnO/cysteic acid nanocomposite modified electrode (ZnO/CA/GCE), and successfully used for simultaneous determination of sunset yellow and Tartrazine. The ZnO/CA/GCE exhibited two linear response ranges in the concentration ranges of 0.1–3.0 μmol/L, and 0.07–1.86 μmol/L, and detection limits of 0.03 μmol/L and 0.01 μmol/L for sunset yellow and Tartrazine, respectively. However, the linear dynamic response range is limited for trace detection of sunset yellow. The detection capacity of sunset yellow has been improved with graphene and mesoporous TiO_2 composite [25] and ZnO/RGO/ZnO@Zn [26] modified electrodes, and they showed superior sensing performance (i.e., linear ranges, detection of limit) that was comparable to precious metal modified electrode. Although some progress has been made, there are only few related reports concerning the transition metal oxide modified electrodes. Therefore, it is still worthwhile to develop novel transition metal oxide modified electrodes for sensitive detection of sunset yellow.

Among transition oxides, cuprous oxide (Cu_2O) is an environmentally friendly p-type semiconductor material, which has been widely used in solar cells and photo catalysis [27,28], due to its unique electronic structure and excellent catalytic performances. However, its electrical conductivity is poor due to the nature of semiconductor. To resolve this problem, Cu_2O nanoparticles are often composited or hybrided with conductive materials [29–33], to decrease the charge transfer resistance and eventually enhance the electrochemical performance. Graphene, as an emerging two-dimensional (2D) carbon material, has been usually used as conductive materials in modified electrodes, owing to its high specific surface area, excellent electrical conductivity, superior electrochemical performance, and fast heterogeneous electron transfer rate. It has been reported that graphene-based modified electrodes have been widely employed for the determination of azo dyes, such as sunset yellow, Tartrazine, and Amaranth [34–36]. However, to our best knowledge, Cu_2O/reduced graphene oxide nanocomposite modified electrode toward sensitive detection of sunset yellow has not been reported.

Graphene usually prepared from graphene oxide (GO) by electrochemical reduction method in the field of electrochemical analysis. The chemically reduced graphene oxide is hydrophobic, due to the removal of most oxygen-containing functional groups. As a result, the chemically graphene oxide tends to agglomerate resulting in a degradation on sensing performance. The agglomeration issue can be overcomed by introducing the surfactants [37], which can effectively improve the dispensability. However, the electrical conductivity also declined due to the use of surfactants. Electrochemically reduction method is a green and efficient method to obtain reduced graphene oxide that not require any reductants. Moreover, the residual oxygen-containing functional groups can be tuned by facial adjusting the electrochemical parameter, such as reduction potential, reduction time, and scanning cycles [18,19,21]. In other words, the property of reduced graphene oxide can be tailored by electrochemical parameters. For these reasons, the electrochemically reduced graphene oxide (ErGO) have been widely used for constructing diverse sensors.

Inspired by the foregoing reports, herein ErGO was composited with low cost and excellent electrocatalytic activity Cu_2O nanoparticles, aiming to develop a cost-effective, high sensitive, and good selective modification materials to substitute the precious metal-based materials. Meanwhile, the Cu_2O-ErGO nanocomposites are expected to exert their synergistic sensitizing effects to improve the sensing performance. Then, Cu_2O-ErGO nanocomposites was modified on the surface of the glassy carbon electrode (GCE) to construct a novel sensor toward sunset yellow. The Cu_2O-ErGO modified glassy carbon electrode (Cu_2O-ErGO/GCE) was prepared while using a facile drop-casting technique in combination with electrochemical reduction. The electrochemical behavior of sunset yellow on the Cu_2O-ErGO/GCE were investigated by cyclic voltammetry and second-derivative linear sweep voltammetry. The effect of detection conditions (such as pH value, sweep rate, and accumulation parameters) on the electrochemical response were also explored. Finally, the proposed Cu_2O-ErGO/GCE was used to detect the content of sunset yellow in soda drinks, orange juice, and candies samples while using second-derivative linear sweep voltammetry.

2. Results and Discussion

2.1. Morphology and Microstructural Characterization

The surface morphologies of Cu_2O nanoparticles (Cu_2O NPs) and Cu_2O-ErGO nanocomposites were characterized by scanning electron microscope (SEM, Hitachi S-3000N, Tokyo, Japan). The SEM images are shown in Figure 1A,B, respectively. The Cu_2O NPs exhibit cubic-like structure with uniform size, and the particle size is estimated to about 150 nm. Obviously, the thin layer ErGO sheets were successfully coated on the surface of Cu_2O nanoparticles. Moreover, the particle size of Cu_2O-ErGO nanocomposite increases slightly, which facilitates the adsorption of sunset yellow. The Cu_2O NPs were further characterized by X-ray diffraction (XRD, JEOL JEM-2010 (HT, Tokyo, Japan) and the XRD pattern of Cu_2O NPs is plotted in Figure 1C. The diffraction peaks of Cu_2O nanoparticles are clearly indexed into the pure cubic phase of Cu_2O (JSPDS78-2076), suggesting that the cubic phase of Cu_2O nanoparticles was prepared.

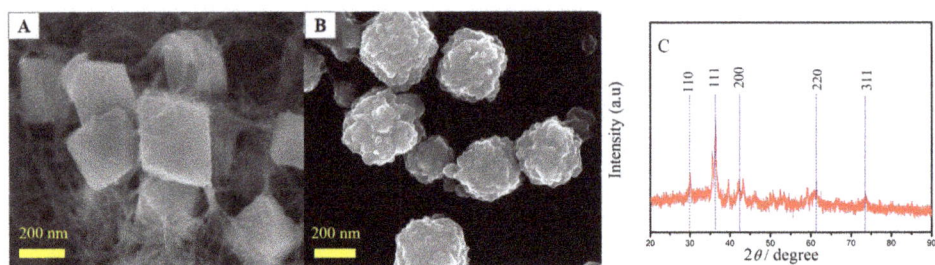

Figure 1. The scanning electron microscope (SEM) photos of cuprous oxide (Cu_2O) nanoparticles (**A**) and cuprous oxide-electrochemically reduced graphene oxide (Cu_2O-ErGO) nanocomposites (**B**); and, (**C**) The X-ray diffraction (XRD) pattern of Cu_2O nanoparticles.

2.2. Electrochemical Behavior of Sunset Yellow on Modified Electrodes

Second-derivative linear sweep voltammetric responses of 1.0×10^{-5} mol/L sunset yellow on different electrodes are presented in Figure 2. On the bare GCE, a weak anodic peak of sunset yellow appears at 798 mV with the anodic peak current (i_{pa}) of 0.725 µA. On the GO/GCE, the oxidation peak current of sunset yellow decreases to 0.497 µA, which is mainly due to the present of the poor electrical conductivity of GO. On the Cu_2O-GO/GCE, an apparent anodic peak of occurred at 770 mV, and the i_{pa} increases to 0.925 µA, probably owing to the electrocatalytic activity of Cu_2O nanoparticles. When the GO was electrochemically reduced to ErGO, the anodic peak appears at 792 mV and the i_{pa} increases to 16.93 µA. This phenomenon may be related to the high electrical conductivity, large specific surface area, and rapid heterogeneous electron transfer rate of ErGO. Moreover, the adsorption capacity of sunset yellow on the electrode surface is improved greatly by the π-π interaction, because the conductive carbon-conjugated networks are restored after the reduction process. When the GCE was modified with Cu_2O-ErGO nanocomposites, the i_{pa} is the largest (18.08 µA), which is about 25 fold greater than that of bare GCE. It mainly due to the synergistic enhancement effect between Cu_2O nanoparticles and ErGO sheets, which significantly improves the sensitivity of sunset yellow detection.

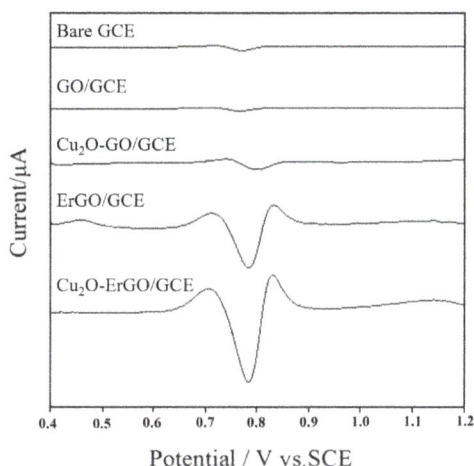

Figure 2. Second-derivative linear sweep voltammograms of 1.0×10^{-5} mol/L sunset yellow on the different electrodes.

The electrochemical behavior of 1.0×10^{-5} mol/L sunset yellow on the GCE (a), GO/GCE (b), Cu_2O-GO/GCE (c), ErGO/GCE (d), and Cu_2O-ErGO/GCE (e) were further investigated by cyclic voltammetry (Figure 3). All of the electrodes appear a pair of redox peaks, meaning that sunset yellow undergoes a quasi-reversible process. Obviously, a pair of sharp redox peaks occurs on the ErGO/GCE and Cu_2O-ErGO/GCE. Furthermore, the order of anodic peak currents obtained from cyclic voltammograms is consistent with the second-derivative linear sweep voltammograms, which further confirms that Cu_2O-ErGO nanocomposites can significantly enhance the electrochemical response toward sunset yellow.

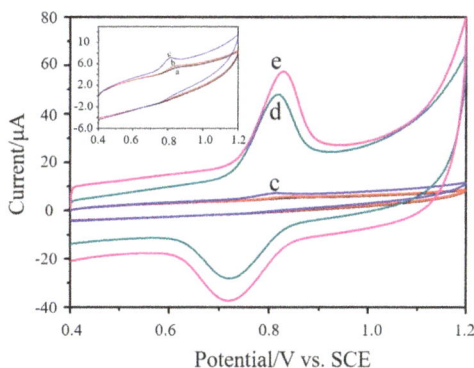

Figure 3. Cyclic voltammograms of 1.0×10^{-5} mol/L sunset yellow recorded on the glassy carbon electrode (GCE) (a), graphene oxide/glassy carbon electrode (GO/GCE) (b), Cu_2O-graphene oxide nanocomposite modified glass carbon electrode (Cu_2O-GO/GCE) (c), electrochemically reduced graphene oxide modified glass carbon electrode (ErGO/GCE) (d) and Cu_2O-ErGO modified glassy carbon electrode (Cu_2O-ErGO/GCE) (e). The inset is the magnification of the cyclic voltammograms recorded on the GCE (a), GO/GCE (b) and Cu_2O-GO/GCE (c).

2.3. Effect of pH Value

Since proton (H$^+$) plays an important role on the redox of sunset yellow, so it is worthwhile investigating the influence of pH value on the response peak current of sunset yellow. The i_{pa} of sunset yellow recorded in various pH PBS solution are depicted in Figure 4A. The i_{pa} of sunset yellow increases gradually as the pH value increases. When the pH value increases to 3.8, the largest i_{pa} is obtained. Afterwards the i_{pa} decreases slowly with the pH value further increasing. Hence, the pH 3.8 PBS solution was employed as supporting electrolytes on the subsequent experiments. Furthermore, the anodic peak potential (E_{pa}) of sunset yellow is negatively shifted with the increase of pH value. As plotted in Figure 4B, there is a good linear relationship between E_{pa} and pH value, confirming that protons are involved in the oxidation of sunset yellow. The linear regression equation can expressed as E_{pa} (V) = −0.0570 pH + 1.0167 (R^2 = 0.998). According to Nernst equation, its slope (−0.0570 V/pH) approaches to the theoretical value (−0.0590 V/pH), suggesting that the equal amounts of proton and electron involve in the electrochemical oxidation of sunset yellow [38].

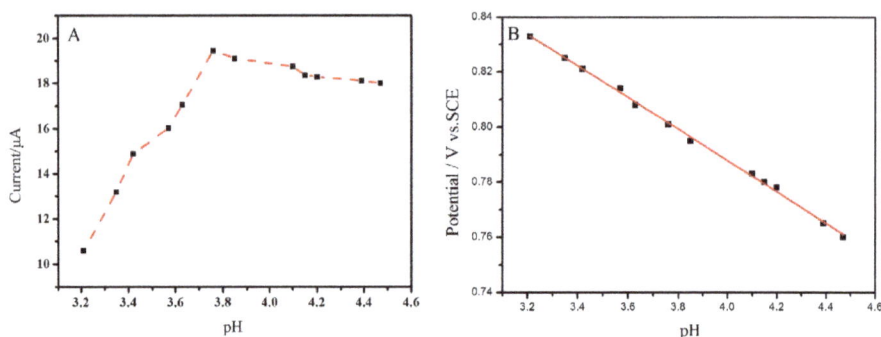

Figure 4. (**A**) Effects of pH value on the anodic peak currents of 1.0 × 10^{-5} mol/L sunset yellow on the Cu$_2$O-ErGO/GCE; and, (**B**) The plot of the anodic peak potential of 1.0 × 10^{-5} mol/L sunset yellow against pH value.

2.4. Effect of Sweep Rates

Sweep rate is a crucial parameter that directly affects the electrochemical response of analysts on the modified electrodes. Moreover, it is a powerful tool to reveal the electrochemical reaction mechanism. Cyclic voltammograms at various sweep rates were recorded at 0.1 mol/L PBS solution containing 1 × 10^{-5} mol/L sunset yellow while using the Cu$_2$O-ErGO/GCE, and their corresponding cyclic voltammograms are shown in the Figure 5A. As expected, with sweep rates increasing, the anodic peaks shift toward positive direction, while the cathodic peaks shift negatively, indicating that the oxidation of sunset yellow on the Cu$_2$O-ErGO/GCE is quasi-reversible. Both the anodic peak currents (i_{pa}) and cathodic peak currents (i_{pc}) increase with the sweep speeding up, however, the background currents also increase. To purist high signal-to-noise (S/N), a suitable sweep rate is recommended as 100 mV/s. It can be clearly seen from Figure 5B that both the anodic peak currents (i_{pa}) and cathodic peak currents (i_{pc}) of sunset yellow is nearly linear with the sweep rates (v). Their linear equations are expressed as: i_{pa} (μA) = 0.2968 v (mV/s) − 4.791 (R^2 = 0.998) and i_{pc} (μA) = −0.1471 v (mV/s) + 1.659 (R^2 = 0.997), suggesting that the oxidation of sunset yellow on the Cu$_2$O-ErGO/GCE is controlled by the adsorption process [39].

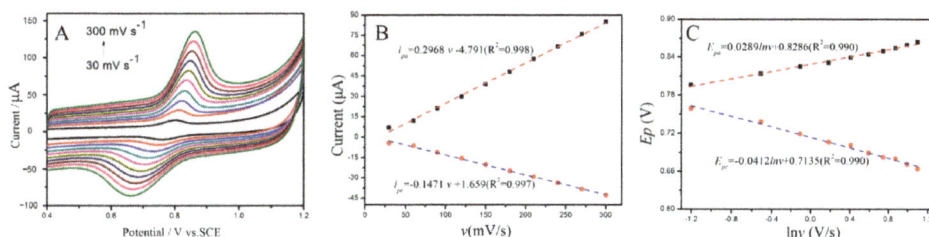

Figure 5. (**A**) Cyclic voltammograms of 1.0×10^{-5} mol/L sunset yellow on the Cu_2O-ErGO/GCE recorded at various sweep rates (30 mV/s–300 mV/s); (**B**) Linear plots of anodic and cathodic peak currents (i_{pa} and i_{pc}) of sunset yellow against sweep rates (v); and (**C**) Linear plots of andic and cathodic peak potential (E_{pa} and E_{pc}) of sunset yellow against Napierian Logarithm of sweep rates ($\ln v$). Supporting electrolytes: 0.1 mol/L PBS.

It is observed that both the anodic peak potential (E_{pa}) and the cathodic peak potential (E_{pc}) is well linear to the Napierian Logarithm of sweep rates ($\ln v$). Their corresponding linear equations are E_{pa} (V) = $0.0289\ln v$ (V/s) + 0.8286 (R^2 = 0.990) and E_{pc} (V) = $-0.0412\ln v$ (V/s) + 0.7135 (R^2 = 0.990). As for an adsorption-controlled and quasi-reversible process, according to the Lavrion equation [40], the peak potential and the sweep rate follows the following relationship:

$$E_{pa} = E^{0'} + \frac{RT}{(1-\alpha)nF}[0.78 + \ln(\frac{D^{1/2}}{K_s}) - \frac{1}{2}\ln\frac{RT}{(1-\alpha)nF}] + \frac{RT}{2(1-\alpha)nF}\ln v \quad (1)$$

$$E_{pc} = E^{0'} + \frac{RT}{\alpha nF}[0.78 + \ln(\frac{D^{1/2}}{K_s}) - \frac{1}{2}\ln\frac{RT}{\alpha nF}] - \frac{RT}{2\alpha nF}\ln v \quad (2)$$

where E_{pa} (V) and E_{pc} (V) represents the anodic peak potential and the cathodic peak potential, respectively; v (V/s) denotes the sweep rate; α is the charge transfer coefficient; k_s is the heterogeneous electron transfer rate; n is the electron transferred number; T is Kelvin temperature; F is Faraday constant (96,480 C/mol); and, R is molar gas constant (8.314 J/(mol·K)). Combining the slopes of Equations (1) and (2) with the E_{pa}/E_{pc} vs. $\ln v$ equations, the charge transfer coefficient α is estimated to be 0.45 and the electron transferred number n is around 1. Since the equal amount of proton and electron participates in the oxidation process, the electrochemical oxidation of sunset yellow is 1 electron and 1 proton process, which is in accordance with the previous studies [39,41]. Hence, the electrochemical oxidation mechanism of sunset yellow on the Cu_2O-ErGO/GCE can be inferred in Figure 6.

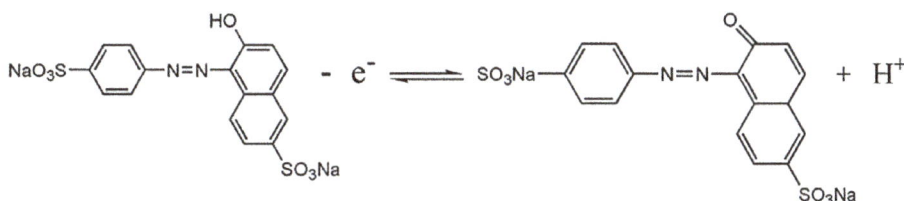

Figure 6. The mechanism for the electrochemical processes of sunset yellow on the Cu_2O-ErGO/GCE.

2.5. Effect of Acumualtion Parameters

Accumulation is a simple and effective technique to improve the electrochemical response. Since the electrochemical oxidation of sunset yellow is an adsorption-controlled process, so accumulation were performed before second-derivative linear sweep voltammetry. As we all know, accumulation potential as well as time are two important parameters that affect the response peak current greatly, so it is a worthwhile optimization. The Cu_2O-ErGO/GCE was accumulated at different accumulation potential for 240 s firstly. Then, their anodic peak currents (i_{pa}) of sunset yellow were recorded in 0.1 mol/L PBS solution (pH 3.8) while using second-derivative linear sweep voltammetry. The effect of the accumulation potential on the i_{pa} of sunset yellow is presented in Figure 7A. The i_{pa} increases gradually with the rising of accumulation potential. When the accumulation potential reaches 0.4 V, the strongest i_{pa} is obtained. Afterwards, the i_{pa} decreases with the accumulation potential further increasing. Therefore, 0.4 V was selected in the subsequent experiments. Furthermore, the influence of accumulation time was also explored. Similarly, the Cu_2O-ErGO/GCE was accumulated at an optimized accumulation potential for various time. Then, their i_{pa} of sunset yellow were recorded and compared. As shown in Figure 7B, the i_{pa} increases with the prolong of the accumulation during the first 180 s; then i_{pa} keep stable with the accumulation time further prolonging, demonstrating that the adsorption of sunset yellow achieved saturated. Hence, 180 s is recommended as the optimum accumulation time.

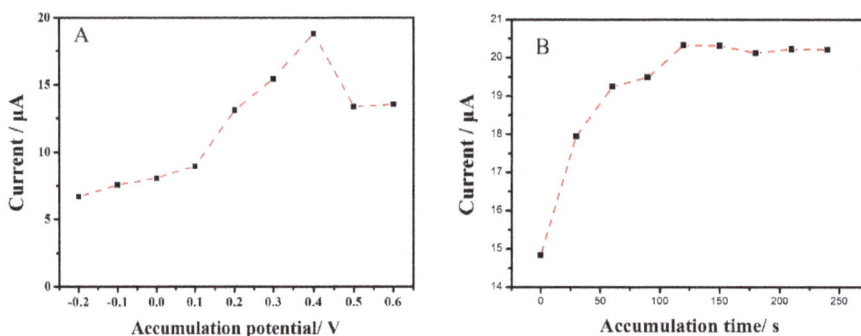

Figure 7. Influences of accumulation potential (**A**), and accumulation time (**B**) on the anodic peak currents of 1.0×10^{-5} mol/L sunset yellow on the Cu_2O-ErGO/GCE.

2.6. Standard Curves, Linear Range and Limit of Detection

With the optimal analytical parameters, the i_{pa} of different concentrations of sunset yellow standard solution was determined by second-derivative linear sweep voltammetry. Figure 8A shows the second-derivative linear sweep voltammograms of various concentrations of sunset yellow. There are two linear response ranges for the detection of sunset yellow, namely 2.0×10^{-8} ~2.0×10^{-5} mol/L (Figure 8B) and 2.0×10^{-5} mol/L ~1.0×10^{-4} mol/L (Figure 8C). Their corresponding linear equations are i_{pa} (μA) = 1.597c (μmol/L) + 2.628 (R^2 = 0.973) and i_{pa} (μA) = 0.0775c (μmol/L) + 30.36 (R^2 = 0.992), respectively. The limit of detection (LOD, S/N = 3) is estimated to be 6.0×10^{-9} mol/L. The linear response range is lower than the permitted maximum content of sunset yellow (2.2×10^{-4} mol/L), so that the Cu_2O-ErGO/GCE can be applied to detection sunset yellow by a direct or the dilution method. A comparison on sensing performances toward sunset yellow between the existing modified electrodes and Cu_2O-ErGO/GCE is summarized on Table 1.

Table 1. Comparison the sensing performances toward the detection of sunset yellow between the existing modified electrodes and the proposed Cu_2O-ErGO/GCE.

Modified Electrodes	Method	Metrological Parameters	Linear Range (μmol/L)	LOD (μmol/L)	Reference
PLPA/GCE	DPV	0.1 M phosphate-citrate buffer solution (pH 7.0); accumulation for 60 s	0.04–14	0.040	[42]
MIP/f-MWCNTs/GCE	DPV	0.1 M CBS solution (pH 5.0); accumulation for 30 min; scanned at 10 mV/s	0.05–100	0.005	[43]
Au-Pd-RGO/GCE	DPV	0.1 M PBS (pH 4.0); scanned at 50 mV/s	0.69–332	0.0015	[17]
CTAB-Gr-Pt/GCE	DPV	0.1 M PBS (pH 3.0); accumulation for 3 min	0.0085–1.0; 1.0–30	0.0042	[44]
GO/AgNPs-MIPs/GCE	LSV	0.1 M PBS (pH 5.5); accumulation for 7 min; scanned at 50 mV/s	0.1–0.6; 0.6–12	0.02	[45]
ILRGO-Au/GCE	SWV	0.1 M BR buffer solution (pH 7.0); accumulation for 300 s	0.004–1.0	0.00052	[13]
Au NPs/CPE	DPV	0.1 M PBS (pH 4.0); accumulation for 1 min; modulation amplitude = 60 mV and scan rate = 60 mV/s	0.1–2.0	0.03	[7]
MWCNT/GCE	DPV	0.1 M PBS (pH 7.0); accumulation at open circuit potential for 2 min; potential increment of 0.004 V, pulse amplitude of 0.05 V, and pulse period of 0.2 s	0.55–7.0	0.12	[46]
PDDA-Gr-Pd/GCE	DPV	0.1 M PBS (pH 3.0); accumulation for 5 min	0.01–10	0.002	[47]
Fe_3O_4@SiO_2-NPs@MIP/Gr/GCE	DPV	0.1 M PBS (pH 8.0); pulse amplitude = 0.05 V; pulse interval time = 0.05 s, and scan rate = 0.02 V/s for differential pulse voltammetry	0.02–20	0.0055	[48]
Cu_2O-ErGO/GCE	SDLSV	0.1 M PBS (pH 3.8); accumulation at 0.4 V for 180 s; scanned at 100 mV/s	0.02–20; 20–100	0.006	This work

The linear ranges and LOD of the proposed Cu_2O-ErGO/GCE are at least comparable to and even better than most of the previous reports. Moreover, Cu_2O-ErGO/GCE have outstanding advantages over noble metal modified electrodes (such as Au NPs/CPE [7], Au-Pd-RGO/GCE [17], CTAB-Gr-Pt/GCE [44], GO/AgNPs-MIPs/GCE [45] and PDDA-Gr-Pd/GCE [47]) in terms of the cost and electrode fabrication.

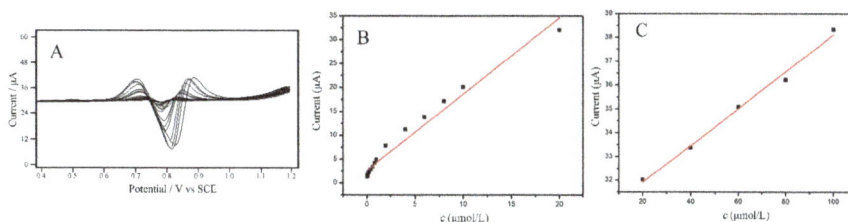

Figure 8. (**A**) Second-derivative linear sweep voltammograms of sunset yellow with various concentrations on the Cu_2O-ErGO/GCE; Calibration curves between the anodic peak current and the concentrations of sunset yellow ranging from 2.0×10^{-8} to 2.0×10^{-5} mol/L (**B**) and from 2.0×10^{-5} to 1.0×10^{-4} mol/L(**C**).

2.7. Interference and Reproducibility Investigation

Prior to the detection of real samples, the anti-interference and reproducibility was also investigated to validate the practicability of the proposed Cu_2O-ErGO/GCE. The response peak current of pure sunset yellow solution and potential interfering compounds mixture solution were recorded and compared. It is observed that the change of i_{pa} of 10 μmol/L sunset yellow is less than 5% in the presence of a 100-fold concentration of glucose, benzoic acid, citric acid, Na^+, K^+, Fe^{3+}, and 10-fold concentration of Tartrazine, quinoline yellow (Figure 9). It is demonstrating that our proposed Cu_2O-ErGO/GCE exhibits good selectivity toward sunset yellow. The reproducibility of Cu_2O-ErGO/GCEs were examined by continuous measurement the response peak currents of 10 μmol/L sunset yellow seven times. The result shows that the i_{pa} remain stable with relative standard deviation (RSD) of 2.78% (Table 2), indicating that the Cu_2O-ErGO/GCE has good reproducibility.

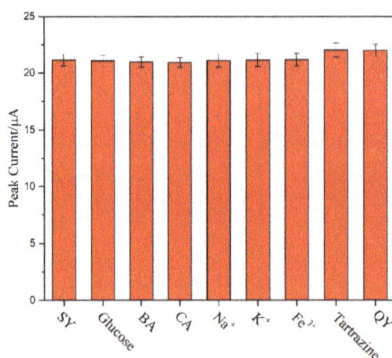

Figure 9. The response peak current of Cu_2O-ErGO/GCE for 1×10^{-5} mol/L sunset yellow (SY) in the presence of interfering substances ($n = 3$). The concentration of glucose, benzoic acid (BA), citric acid (CA), Na^+, K^+, Fe^{3+}, were 1×10^{-3} mol/L. The concentrations of Tartrazine, quinoline yellow (QY) were 1×10^{-4} mol/L.

Table 2. The reproducibility of Cu_2O-ErGO/GCE for detection of sunset yellow.

No.	1	2	3	4	5	6	7
i_{pa} (µA)	21.07	22.34	21.36	22.04	22.29	22.85	22.24
Average value (µA)				22.02			
RSD (%)				2.78			

2.8. Detection Sunset Yellow in Real Samples

Finally, Cu_2O-ErGO/GCE was applied to determinate sunset yellow in real samples, including carbonated drinks, orange juice, and candy samples. The detect results are listed in Table 3. No anodic peak current is presented in the carbonated beverage sample near 798 mV, indicating that the concentration of sunset yellow is low than the limit of detection. The concentrations of sunset yellow in orange juice and candy samples are detected by 0.085 µmol/L and 0.162 µmol/L, respectively. The content of sunset yellow in these samples is lower than the permitted maximum content that is recommended by national standard (100 µg/mL, namely 2.2×10^{-4} mol/L). Then, an appropriate amount of sunset yellow standard solution was added to the above three samples. Standard addition test results suggests that the recoveries of sunset yellow were 98.75~102.0%, with the relative standard deviation being less than 2.85%, indicating that satisfactory results were obtained using the proposed Cu_2O-ErGO/GCE. Together with cost-effective, quick response, high sensitive and good selectivity, the Cu_2O-ErGO/GCE exhibits great prospects on the determination of sunset yellow in different real samples, including but not limiting to beverages, food, and nutrients.

Table 3. The results of determination of sunset yellow in soft drink and candies ($n = 3$).

Samples	Original (µmol/L)	Added (µmol/L)	RSD (%)	Found (µmol/L)	Recovery (%)	RSD (%)
Sodas	ND [1]	4	1.28	4.06	102.0	1.50
Orange juice	0.085	0.080	2.46	0.082	102.5	2.31
Candies	0.162	0.160	3.25	0.158	98.75	2.85

[1] ND: Not detected.

3. Materials and Methods

3.1. Chemical and Solution

Graphite powder, $NaNO_3$, concentrated H_2SO_4, NaOH, $KMnO_4$, H_2O_2, $CuSO_4 \cdot 5H_2O$, Na_2HPO_4, NaH_2PO_4, hydrochloric acid (HCl), polyvinylpyrrolidone (PVP), hydrate hydrazine ($N_2H_4 \cdot H_2O$), and ethyl alcohol were analytic grade and supplied by Sinopharm Chemical Reagent Co., Ltd. (Shanghai, China). Sunset yellow was purchased from Aladdin (http://www.aladdin-e.com). All of these chemicals were directly used without further purification. A series of sunset yellow standard solutions were prepared by diluting appropriately the stock sunset yellow (1×10^{-3} mol/L). These standard solutions were kept in 4 °C refrigerator when not in use. Deionized water (18.2 MΩ) was used throughout the experiments.

3.2. Synthesis of Cu₂O Nanoparticles

Cu₂O nanoparticles were synthesized by the hydrothermal method referred to our previous work [18]. Specifically, 50 mg of CuSO₄·5H₂O and 24 mg of PVP were added into 10 mL deionized water and then stirred with ultrasonication for 30 min. Afterwards, 2 mL of 0.2 mol/L NaOH solution was added and stirred for 30 min at room temperature to obtain blue Cu(OH)₂ precipitates. Subsequently, 6 μL of N₂H₄·H₂O was added as reductant and then stirred for 20 min at room temperature to form a brick red suspension. The precipitate was separated by centrifugation at 5000 rpm, and washed repeatedly with deionized water and ethanol for three times, and vacuum-dried at 60 °C to obtain Cu₂O nanoparticles.

3.3. Preparation Cu₂O-GO Nanocomposite Dispersion

Graphene oxide (GO) is prepared from cheap graphite powder by a modified Hummers method according to our previous reports [18,19,21]. The as-prepared GO was dispersed in 100 mL of deionized water under ultrasonication for 2 h, and then centrifuged twice to obtain a golden yellow GO solution (1 mg/mL). 2 mg of Cu₂O nanoparticles were added to 5 mL of the above GO solution, and then ultrasonically dispersed for 2 h to obtain a Cu₂O-GO nanocomposite dispersion.

3.4. Preparation of Cu₂O-ErGO/GCE

The glassy carbon electrode (GCE, φ = 3 mm) was polished to a form mirror-like surface with 0.05 μm alumina slurry. Then, the electrode was rinsed by deionized water and ethanol alternately (each for 1 min), and then dried by ultrapure N₂ gas. Firstly, 5 μL of Cu₂O-GO dispersion were transferred and coated on the surface of the GCE, and then dried under an infrared lamp to obtain Cu₂O-graphene oxide nanocomposite modified glass carbon electrode (Cu₂O-GO/GCE). Then, the GO component was electrochemically reduced by potentiostatic method. Specially, the Cu₂O-GO/GCE was immersed into 0.1 mol/L PBS solution, and electrochemically reduced at −1.2 V for 120 s. For comparison, GO modified glass carbon electrode (GO/GCE), electrochemically reduced graphene oxide modified glass carbon electrode (ErGO/GCE) and Cu₂O-GO nanocomposite modified glass carbon electrode (Cu₂O-GO/GCE) were also fabricated by a similar method.

3.5. Electrochemical Measurements

All of the electrochemical measurements were carried out using a standard three-electrode assemble, comprising of bare or modified electrodes as working electrode, saturated calomel electrode (SCE) as reference electrode, and platinum wire electrode as counter electrode. 0.1 M PBS solution was used as supporting electrolytes in all electrochemical experiments. The electrochemical performances of 1×10^{-5} mol/L sunset yellow on various modified electrodes were investigated by cyclic voltammetry (CV) and second-derivative linear sweep voltammetry (SDLSV), with the potential scanning range of 0.4 ~1.2 V. Prior to all electrochemical measurements, an accumulation was performed under stirring at 500 rpm to improve the sensitivity. After 5 s rest, the CV or SDLSV were recorded at a scan rate of 100 mV/s, except where stated otherwise. All of the electrochemical tests were carried on the electrochemical workstation (CHI 760E, Shanghai Chenhua Inc., Shanghai, China).

3.6. Analysis of Real Samples

Carbonated drinks, orange juice, and candy samples were purchased from a local supermarket. The same amount of candies was taken out from five packages and carefully grounded into fine powder. Then, the candies powder (about 1.0 g) was accurately weighed and dispersed in 10 mL deionized water under sonication for 1 h. Subsequently, the mixture was centrifuged at 4000 rpm for 10 min to remove insoluble substances. The 0.5 mL as-obtained supernatant was diluted to 10 mL with 0.1 mol/L PBS solution. The liquid samples (carbonated drinks and orange juice) were added into beaker and ultrasonicated for degasification. Then, 1.0 mL of the liquid samples was diluted to 10 mL with 0.1 mol/L PBS solution. Prior to electrochemical detection, accumulation step was performed in the sample solution to enhance the electrochemical response. Then, these sample solutions were detected by second-derivative linear sweep voltammetry while using our proposed Cu_2O-ErGO/GCE. After each test, Cu_2O-ErGO/GCE was scanned by cyclic voltammetry a 0.1 mol/L blank PBS solution for serval cycles to remove any adsorbents. The modified electrode can reused only when the response peak disappeared in the blank PBS solution.

4. Conclusions

In this study, a promising electrochemical sensor based on the Cu_2O-ErGO/GCE was developed toward sensing sunset yellow. The Cu_2O-ErGO nanocomposites not only possess the advantages from individual component materials, but also exhibit obvious synergistic enhancement effects toward sunset yellow. The anodic peak current of sunset yellow on the Cu_2O-ErGO/GCE increases by about 25 times as compared to that on the bare GCE. The proposed Cu_2O-ErGO/GCE exhibits two linear regions, namely 2.0×10^{-8} mol/L–2.0×10^{-5} mol/L and 2.0×10^{-5} mol/L–1.0×10^{-4} mol/L, and the limit of detection is 6.0×10^{-9} mol/L (S/N = 3). The sensing performances in terms of linear response ranges and detection limit are comparable to, and even exceed the most reported modified electrodes, such as precious metal-based modified electrodes. Obviously, the Cu_2O-ErGO/GCE have outstanding advantages over precious metal-based modified electrodes in term of the cost. Moreover, the response current is basically not affected by potential interfering compounds, suggesting the Cu_2O-ErGO shows good selectivity. Besides, the good reproducibility was also obtained on the Cu_2O-ErGO/GCE. Finally, Cu_2O-ErGO/GCE have been successfully used for the quantitative detection of sunset yellow in real samples (i.e., carbonated beverage, orange juice, and candies) while using second-derivative linear sweep voltammetry. The satisfactory results are obtained with recovery rate is 98.75 ~102.5% and RSD is less than 2.85%. When compared with conventional analytical techniques (Table 4), the proposed method does not require expensive equipment, and time-consuming and complicated pretreatment procedures. Considering the considerable merits including low cost, rapid response, high sensitivity, as well as good selectivity and good reproducibility, the Cu_2O-ErGO/GCE will have broad application prospects in the detection of sunset yellow in diverse beverages, foods, and nutrients.

Table 4. Advantages and drawbacks of previous reported protocols and our proposed protocol.

Protocols	Advantages	Drawbacks
Liquid chromatography	Reliable; good repeatability; high sensitivity; low LOD	Limited separation ability; Time-consuming; expensive equipment
Thin layer chromatography	Low cost apparatus	Organic solvents are often used; they have intensive disagreeable smell and cancerogenic activity
Spectrophotometry	Simultaneous identification and quantification; simple technique	Low sensitivity; extraction separation is needed for detection of dyes in complex product composition
Capillary electrophoresis	High column efficiency, short analysis time and minimal amounts of samples	Limited sensitivity & selectivity; severe matrix interferences
Electrochemical analysis (This work)	Low cost, rapid response, facile operate, high sensitivity and good selectivity	Portability needs to be improved; not disposable

Molecules **2018**, *23*, 2130

Author Contributions: Q.H., G.L. and D.C. conceived and designed the experiments; J.L., X.L., and Y.X. performed the experiments; G.L., P.D., and X.L. analyzed the data; Q.H., J.L. and G.L. wrote the manuscript; Q.H. and D.C. contributed reagents and materials; All authors read and approved the final manuscript.

Funding: This research was funded by the NSFC (61703152), Hunan Provincial Natural Science Foundation (2016JJ4010, 2018JJ3134), Project of Science and Technology Department of Hunan Province (GD16K02), Project of Science and Technology Plan in Zhuzhou (201706-201806), Opening Project of Key Discipline of Materials Science in Guangdong (CNXY2017001, CNXY2017002 and CNXY2017003) and the key Project of Department of Education of Guangdong Province (2016GCZX008).

Conflicts of Interest: The authors declare no conflict of interest.

References

1. Ni, Y.; Wang, Y.; Kokot, S. Simultaneous kinetic spectrophotometric analysis of five synthetic food colorants with the aid of chemometrics. *Talanta* **2009**, *78*, 432–441. [CrossRef] [PubMed]

2. Al-Degs, Y.S. Determination of three dyes in commercial soft drinks using HLA/GO and liquid chromatography. *Food Chem.* **2009**, *117*, 485–490. [CrossRef]

3. Mao, Y.; Fan, Q.; Li, J.; Yu, L.; Qu, L. A novel and green ctab-functionalized graphene nanosheets electrochemical sensor for Sudan I determination. *Sens. Actuators B Chem.* **2014**, *203*, 759–765. [CrossRef]

4. Chung, K.-T.; Cerniglia, C.E. Mutagenicity of azo dyes: Structure-activity relationships. *Mutat. Res.* **1992**, *277*, 201–220. [CrossRef]

5. Aguilar, F.; Charrondiere, U.R.; Dusemund, B.; Galtier, P.; Gilbert, J.; Gott, D.M.; Grilli, S.; Guertler, R.; Koenig, J.; Lambre, C.; et al. Scientific Opinion on the re-evaluation of Sunset Yellow FCF (E 110) as a food additive on request from the European Commission. *EFSA J.* **2009**, *7*, 1–44.

6. Ye, X.; Du, Y.; Lu, D.; Wang, C. Fabrication of β-cyclodextrin-coated poly (diallyldimethylammonium chloride)-functionalized graphene composite film modified glassy carbon-rotating disk electrode and its application for simultaneous electrochemical determination colorants of sunset yellow and tartrazine. *Anal. Chim. Acta* **2013**, *779*, 22–34. [PubMed]

7. Ghoreishi, S.M.; Behpour, M.; Golestaneh, M. Simultaneous determination of Sunset yellow and Tartrazine in soft drinks using gold nanoparticles carbon paste electrode. *Food Chem.* **2012**, *132*, 637–641. [CrossRef] [PubMed]

8. Minioti, K.S.; Sakellariou, C.F.; Thomaidis, N.S. Determination of 13 synthetic food colorants in water-soluble foods by reversed-phase high-performance liquid chromatography coupled with diode-array detector. *Anal. Chim. Acta* **2007**, *583*, 103–110. [CrossRef] [PubMed]

9. Soponar, F.; Cătălin Moţ, A.; Sârbu, C. Quantitative determination of some food dyes using digital processing of images obtained by thin-layer chromatography. *J. Chromatogr.* **2008**, *1188*, 295–300. [CrossRef] [PubMed]

10. De León-Rodríguez, L.M.; Basuil-Tobias, D.A. Testing the possibility of using UV–vis spectrophotometric techniques to determine non-absorbing analytes by inclusion complex competition in cyclodextrins. *Anal. Chim. Acta* **2005**, *543*, 282–290. [CrossRef]

11. Yuan, Y.; Zhao, X.; Qiao, M.; Zhu, J.; Liu, S.; Yang, J.; Hu, X. Determination of sunset yellow in soft drinks based on fluorescence quenching of carbon dots. *Spectrochim. Acta A* **2016**, *167*, 106–110. [CrossRef] [PubMed]

12. Ryvolová, M.; Táborský, P.; Vrábel, P.; Krásenský, P.; Preisler, J. Sensitive determination of erythrosine and other red food colorants using capillary electrophoresis with laser-induced fluorescence detection. *J. Chromatogr.* **2007**, *1141*, 206–211. [CrossRef] [PubMed]

13. Wang, M.; Zhao, J. Facile synthesis of Au supported on ionic liquid functionalized reduced graphene oxide for simultaneous determination of Sunset yellow and Tartrazine in drinks. *Sens. Actuators B Chem.* **2015**, *216*, 578–585. [CrossRef]

14. Rovina, K.; Siddiquee, S.; Shaarani, S.M. Highly sensitive electrochemical determination of sunset yellow in commercial food products based on CHIT/GO/MWCNTs/AuNPs/GCE. *Food Control* **2017**, *82*, 66–73. [CrossRef]

15. Wang, J.; Yang, B.; Wang, H.; Yang, P.; Du, Y. Highly sensitive electrochemical determination of Sunset Yellow based on gold nanoparticles/graphene electrode. *Anal. Chim. Acta* **2015**, *893*, 41–48. [CrossRef] [PubMed]

16. Thirumalraj, B.; Rajkumar, C.; Chen, S.M.; Veerakumar, P.; Perumal, P.; Liu, S.B. Carbon aerogel supported palladium-ruthenium nanoparticles for electrochemical sensing and catalytic reduction of food dye. *Sens. Actuators B Chem.* **2018**, *257*, 48–59. [CrossRef]

17. Wang, J.; Yang, B.; Zhang, K.; Bin, D.; Shiraishi, Y.; Yang, P.; Du, Y. Highly sensitive electrochemical determination of Sunset Yellow based on the ultrafine Au-Pd and reduced graphene oxide nanocomposites. *J. Colloid Interface Sci.* **2016**, *481*, 229–235. [CrossRef] [PubMed]

18. He, Q.; Liu, J.; Liu, X.; Li, G.; Deng, P.; Liang, J. Preparation of Cu_2O-reduced graphene nanocomposite modified electrodes towards ultrasensitive dopamine detection. *Sensors* **2018**, *18*, 199. [CrossRef] [PubMed]

19. He, Q.; Liu, J.; Liu, X.; Li, G.; Chen, D.; Deng, P.; Liang, J. Fabrication of amine-modified magnetite-electrochemically reduced graphene oxide nanocomposite modified glassy carbon electrode for sensitive dopamine determination. *Nanomaterials* **2018**, *8*, 194. [CrossRef] [PubMed]

20. He, Q.; Li, G.; Liu, X.; Liu, J.; Deng, P.; Chen, D. Morphologically tunable MnO_2 nanoparticles fabrication, modelling and their influences on electrochemical sensing performance toward dopamine. *Catalysts* **2018**, *8*, 323. [CrossRef]

21. He, Q.; Liu, J.; Liu, X.; Li, G.; Deng, P.; Liang, J.; Chen, D. Sensitive and selective detection of Tartrazine based on TiO_2-electrochemically reduced graphene oxide composite-modified electrodes. *Sensors* **2018**, *18*, 1911. [CrossRef] [PubMed]

22. Chen, K.; Wang, M.; Li, G.; He, Q.; Liu, J.; Li, F. Spherical α-MnO_2 supported on N-KB as efficient electrocatalyst for oxygen reduction in Al–Air battery. *Materials* **2018**, *11*, 601. [CrossRef] [PubMed]

23. Wang, M.; Chen, K.; Liu, J.; He, Q.; Li, G.; Li, F. Efficiently enhancing electrocatalytic activity of α-MnO_2 nanorods/N-doped ketjenblack carbon for oxygen reduction reaction and oxygen evolution reaction using facile regulated hydrothermal treatment. *Catalysts* **2018**, *8*, 138. [CrossRef]

24. Dorraji, P.S.; Jalali, F. Electrochemical fabrication of a novel ZnO/cysteic acid nanocomposite modified electrode and its application to simultaneous determination of sunset yellow and tartrazine. *Food Chem.* **2017**, *227*, 73–77. [CrossRef] [PubMed]

25. Gan, T.; Sun, J.; Meng, W.; Song, L.; Zhang, Y. Electrochemical sensor based on graphene and mesoporous TiO_2 for the simultaneous determination of trace colourants in food. *Food Chem.* **2013**, *141*, 3731–3737. [CrossRef] [PubMed]

26. Wu, X.; Zhang, X.; Zhao, C.; Qian, X. One-pot hydrothermal synthesis of ZnO/RGO/ZnO@Zn sensor for sunset yellow in soft drinks. *Talanta* **2018**, *179*, 836–844. [CrossRef] [PubMed]

27. Hara, M.; Kondo, T.; Komoda, M.; Ikeda, S.; Kondo, J.N.; Domen, K.; Hara, M.; Shinohara, K.; Tanaka, A. Cu_2O as a photocatalyst for overall water splitting under visible light irradiation. *Chem. Commun.* **1998**, *3*, 357–358. [CrossRef]

28. Shao, F.; Sun, J.; Gao, L.; Luo, J.; Liu, Y.; Yang, S. High Efficiency Semiconductor-liquid junction solar cells based on Cu/Cu_2O. *Adv. Funct. Mater.* **2012**, *22*, 3907–3913. [CrossRef]

29. Meng, H.; Yang, W.; Ding, K.; Feng, L.; Guan, Y. Cu_2O nanorods modified by reduced graphene oxide for NH3 sensing at room temperature. *J. Mater. Chem. A* **2014**, *3*, 1174–1181. [CrossRef]

30. Wang, Y.; Ji, Z.; Shen, X.; Zhu, G.; Wang, J.; Yue, X. Facile growth of Cu_2O hollow cubes on reduced graphene oxide with remarkable electrocatalytic performance for non-enzymatic glucose detection. *New J. Chem.* **2017**, *41*, 9223–9229. [CrossRef]

31. Wang, A.; Li, X.; Zhao, Y.; Wei, W.; Chen, J.; Hong, M. Preparation and characterizations of Cu_2O/reduced graphene oxide nanocomposites with high photo-catalytic performances. *Powder Technol.* **2014**, *261*, 42–48. [CrossRef]

32. Tran, P.D.; Batabyal, S.K.; Pramana, S.S.; Barber, J.; Wong, L.H.; Loo, S.C. A cuprous oxide-reduced graphene oxide (Cu_2O-rGO) composite photocatalyst for hydrogen generation: employing rGO as an electron acceptor to enhance the photocatalytic activity and stability of Cu_2O. *Nanoscale* **2012**, *4*, 3875–3878. [CrossRef] [PubMed]

33. Xie, H.; Duan, K.; Xue, M.; Du, Y.; Wang, C. Photoelectrocatalytic analysis and electrocatalytic determination of hydroquinone by using a Cu_2O-reduced graphene oxide nanocomposite modified rotating ring-disk electrode. *Analyst* **2016**, *141*, 4772–4781. [CrossRef] [PubMed]

34. Li, J.; Wang, X.; Duan, H.; Wang, Y.; Bu, Y.; Luo, C. Based on magnetic graphene oxide highly sensitive and selective imprinted sensor for determination of sunset yellow. *Talanta* **2016**, *147*, 169–176. [CrossRef] [PubMed]

35. Deng, K.; Li, C.; Li, X.; Huang, H. Simultaneous detection of sunset yellow and tartrazine using the nanohybrid of gold nanorods decorated graphene oxide. *J. Electroanal. Chem.* **2016**, *780*, 296–302. [CrossRef]

36. Jampasa, S.; Siangproh, W.; Duangmal, K.; Chailapakul, O. Electrochemically reduced graphene oxide-modified screen-printed carbon electrodes for a simple and highly sensitive electrochemical detection of synthetic colorants in beverages. *Talanta* **2016**, *160*, 113–124. [CrossRef] [PubMed]

37. Tkalya, E.E.; Ghislandi, M.; With, G.D.; Koning, C.E. The use of surfactants for dispersing carbon nanotubes and graphene to make conductive nanocomposites. *Curr. Opin. Colloid Interface Sci.* **2012**, *17*, 225–232. [CrossRef]

38. Songyang, Y.; Yang, X.; Xie, S.; Hao, H.; Song, J. Highly-sensitive and rapid determination of sunset yellow using functionalized montmorillonite-modified electrode. *Food Chem.* **2015**, *173*, 640–644. [CrossRef] [PubMed]

39. Qiu, X.; Lu, L.; Leng, J.; Yu, Y.; Wang, W.; Jiang, M.; Bai, L. An enhanced electrochemical platform based on graphene oxide and multi-walled carbon nanotubes nanocomposite for sensitive determination of Sunset Yellow and Tartrazine. *Food Chem.* **2016**, *190*, 889–895. [CrossRef] [PubMed]

40. Laviron, E. General expression of the linear potential sweep voltammogram in the case of diffusionless electrochemical systems. *J. Electroanal. Chem.* **1979**, *101*, 19–28. [CrossRef]

41. Vladislavić, N.; Buzuk, M.; Rončević, I.Š.; Brinić, S. Electroanalytical Methods for Sunset Yellow Determination—A Review. *Int. J. Electrochem. Sci.* **2018**, *13*, 7008–7019. [CrossRef]

42. Chao, M.; Ma, X. Convenient electrochemical determination of sunset yellow and tartrazine in food samples using a poly(L-phenylalanine)-modified glassy carbon electrode. *Food Anal. Method* **2015**, *8*, 130–138. [CrossRef]

43. Arvand, M.; Zamani, M.; Sayyar Ardaki, M. Rapid electrochemical synthesis of molecularly imprinted polymers on functionalized multi-walled carbon nanotubes for selective recognition of sunset yellow in food samples. *Sens. Actuators B Chem.* **2017**, *243*, 927–939. [CrossRef]

44. Yu, L.; Shi, M.; Yue, X.; Qu, L. A novel and sensitive hexadecyltrimethyl ammonium bromide functionalized graphene supported platinum nanoparticles composite modified glassy carbon electrode for determination of sunset yellow in soft drinks. *Sens. Actuators B Chem.* **2015**, *209*, 1–8. [CrossRef]

45. Qin, C.; Guo, W.; Liu, Y.; Liu, Z.; Qiu, J.; Peng, J. A novel electrochemical sensor based on graphene oxide decorated with silver nanoparticles–molecular imprinted polymers for determination of sunset yellow in soft drinks. *Food Anal. Method* **2017**, *10*, 1–9. [CrossRef]

46. Sierra-Rosales, P.; Toledo-Neira, C.; Squella, J.A. Electrochemical determination of food colorants in soft drinks using MWCNT-modified GCEs. *Sens. Actuators B Chem.* **2017**, *240*, 1257–1264. [CrossRef]

47. Yu, L.; Zheng, H.; Shi, M.; Jing, S.; Qu, L. A novel electrochemical sensor based on poly (diallyldimethylammonium chloride)-dispersed graphene supported palladium nanoparticles for simultaneous determination of sunset yellow and tartrazine in soft drinks. *Food Anal. Method* **2016**, *10*, 1–10. [CrossRef]

48. Arvand, M.; Erfanifar, Z.; Ardaki, M.S. A new core@shell silica-coated magnetic molecular imprinted nanoparticles for selective detection of sunset yellow in food samples. *Food Anal. Method* **2017**, *10*, 1–14. [CrossRef]

Sample Availability: Samples of the compounds are not available from the authors.